# 13th International Conference on Muon Spin Rotation, Relaxation and Resonance 2014



Journal of Physics: Conference Series Volume 551

## Grindelwald, Switzerland
## 1 – 6 June 2014

**Editors:**

Z. Salman
A. Amato
H. Luetkens
E. Morenzoni

ISBN: 978-1-63439-749-0
ISSN: 1742-6588

# 13th International Conference on Muon Spin Rotation, Relaxation and Resonance 2014

**Journal of Physics: Conference Series Volume 551**

**Grindelwald, Switzerland**
**1 – 6 June 2014**

**ISBN: 978-1-63439-749-0**
**ISSN: 1742-6588**

**IOP** Institute of Physics

**Printed from e-media with permission by:**

Curran Associates, Inc.
57 Morehouse Lane
Red Hook, NY 12571

**Some format issues inherent in the e-media version may also appear in this print version.**



Printed by Curran Associates, Inc. (2015)

For permission requests, please contact the Institute of Physics
at the address below.

Institute of Physics
Dirac House, Temple Back
Bristol BS1 6BE UK

Phone: 44 1 17 929 7481
Fax: 44 1 17 920 0979

techtracking@iop.org

**Additional copies of this publication are available from:**

Curran Associates, Inc.
57 Morehouse Lane
Red Hook, NY 12571 USA
Phone: 845-758-0400
Fax: 845-758-2634
Email: curran@proceedings.com
Web: www.proceedings.com

# TABLE OF CONTENTS

# Proceedings of the Thirteenth International Conference on Muon Spin Rotation, Relaxation and Resonance

# $\mu$SR 2014

## held in Grindelwald, Switzerland

## 1–6 June 2014

## Editors:

Z. Salman, A. Amato, H. Luetkens and E. Morenzoni

Journal of Physics: Conference Series **551** (2014) 011001 doi:10.1088/1742-6596/551/1/011001

# Preface

The 13th International Conference on Muon Spin Rotation, Relaxation and Resonance ($\mu$SR2014) organized by the Laboratory for Muon Spin Spectroscopy, Paul Scherrer Institute in collaboration with the University of Zurich and the University of Fribourg, was held in Grindelwald, Switzerland from $1^{st}$ to $6^{th}$ June 2014. The conference provided a forum for researchers from around the world with interests in the applications of $\mu$SR to study a wide range of topics including condensed matter physics, materials and molecular sciences, chemistry and biology. Polarized muons provide a unique and versatile probe of matter, enabling studies at the atomic level of electronic structure and dynamics in a wide range of systems. The conference was the thirteenth in a series, which began in Rorschach in 1978 and took place for the third time in Switzerland. The previous conferences were held in Cancun, Mexico (2011), Tsukuba, Japan (2008), Oxford, UK (2005), Williamsburg, USA (2002), Les Diablerets, Switzerland (1999), Nikko, Japan (1996), Maui, USA (1993), Oxford, UK (1990), Uppsala, Sweden (1986), Shimoda, Japan (1983), Vancouver, Canada (1980), and Rorschach, Switzerland (1978).

These conference proceedings contain 67 refereed publications from presentations covering magnetism, superconductivity, chemistry, semiconductors, biophysics and techniques. The conference logo, displayed on the front page of these proceedings, represents both the location of $\mu$SR2014 in the Alps and the muon-spin rotation technique. The silhouette is that of the famous local mountains Eiger, Mönch and Jungfrau as drawn by the Swiss painter Ferdinand Hodler and the apple with arrow is at the same time a citation of the Wilhelm Tell legend and a remembrance of the key role played by the muon spin and the asymmetric muon decay (which for the highest positron energy has an apple like shape). More than 160 participants (including 32 registered as students and 13 as accompanying persons) from 19 countries submitted 227 contributions, which were intensively discussed during day and evening sessions. The scientific program was centered around invited talks from speakers outside the $\mu$SR community, who presented lectures on topics where $\mu$SR is giving or expected to give significant contributions. The invited speakers, covering various fields of interest, included Radu Coldea (Oxford, Quantum Magnetism), Claude-Henri Delmas (Bordeaux, Electro- and Solid State Chemistry), Dirk Johrendt (Munich, Iron Based Materials), Marc-Henri Julien (Grenoble, Cuprate Superconductors), Manfred Fiebig (Zurich, Multiferroics), Allan MacDonald (Austin, Topological Electronic States), Hidenori Takagi (Stuttgart and Tokyo, Transition Metal Oxides), and Jean-Marc Triscone (Geneva, Oxide Heterostructures). In addition to an overview about status and progress of the existing facilities in Europe, Canada and Japan, future projects and new ideas for $\mu$SR facilities in South Korea, China and the USA were presented. A special evening session was held to discuss about muon site and muon states calculations by DFT and other techniques. Several talks and posters can be found on the conference web page www.psi.ch/muSR2014.

In a ceremony at the beginning of the conference, Roberto De Renzi from the University of Parma was awarded the 2014 Yamazaki Prize for muon science by the International Society for Muon Spin Spectroscopy (ISMS) for his sustained and exceptional contributions to the development of the muon spin relaxation technique to investigate magnetism and superconductivity and for promoting synergies between $\mu$SR and NMR. In the closing session Rob Kiefl (UBC Vancouver and TRIUMF) very effectively summarized the five days of meeting, while giving an enlightening personal impression. In the same session five best poster prizes were awarded and ISMS gave two prizes to young researchers presenting outstanding work at the conference. The conference organizers also on behalf of the entire $\mu$SR community are extremely grateful to the conference sponsors: PSI, University of Zurich, University of Fribourg and the Swiss National Foundation as well as to the various industrial sponsors, which are listed in these proceedings. Special mention should also be made of the local organizing committee who has been active during a long period first preparing and then running the conference. Finally we would like to thank the participants whose stimulating research presentations and lively discussion made $\mu$SR2014 such a success.

Elvezio Morenzoni
Paul Scherrer Institute

Journal of Physics: Conference Series **551** (2014) 011001 doi:10.1088/1742-6596/551/1/011001

## Conference Chair

Elvezio Morenzoni (PSI, Switzerland)
Christian Bernhard (University of Fribourg, Switzerland)

## Honorary Conference Chair

Hugo Keller (University of Zurich, Switzerland)

## Local Organizing Committee

Isolde Fuchs, conference secretary (PSI, Switzerland)
Alex Amato (PSI, Switzerland)
Christopher Baines (PSI, Switzerland)
Rustem Khasanov (PSI, Switzerland)
Hubertus Luetkens (PSI, Switzerland)
Thomas Prokscha (PSI, Switzerland)
Zaher Salman (PSI, Switzerland)
Robert Scheuermann (PSI, Switzerland)
Andreas Suter (PSI, Switzerland)

## Program Committee

Stephen J. Blundell (University of Oxford, UK)
João Campos Gil (Universidade de Coimbra, Portugal)
Roberto De Renzi (University of Parma, Italy)
Ryosuke Kadono (KEK/J-PARC, Japan)
Robert F. Kiefl (University of British Columbia, Canada)
Hans-Henning Klauss (TU Dresden, Germany)
Phillippe Mendels (Université Paris-Sud, France)
Yasutomo J. Uemura (Columbia University, USA)

## International Advisory Committee

Helena V. Alberto (Coimbra University, Portugal)
Daniel Andreica (Babes-Bolyai University, Romania)
Fabrice Bert (Université Paris-Sud, France)
Alexander V. Boris (MPI Stuttgart, Germany)
Pierre Dalmas de Réotier (CEA Grenoble, France)
Antheunis de Visser (University of Amsterdam, Netherlands)
Alan J. Drew (Queen Mary, University of London, UK)
Sarah Dunsiger (Techn. University Munich, Germany)
Sean Giblin (Cardiff University, UK)
Michel Gingras (University of Waterloo, Canada)
Michael Graf (Boston College, USA)
Wataru Higemoto (JAEA/J-PARC, Japan)
Adrian Hillier (ISIS/RAL, UK)
Amit Keren (Technion, Israel)
Michael K. Kubo (International Christian University, Japan)
Gianrico Lamura (CNR Napoli, Italy)

## Industrial sponsors

CAEN - Costruzioni Apparecchiature Elettroniche Nucleari S.p.A., Italy

X-TRONIX AG, Switzerland

BlueFors Cryogenics Oy Ltd, Finland

TECO René Koch, Switzerland

Teledyne LeCroy SA, Switzerland

# Peer review statement

All papers published in this volume of *Journal of Physics: Conference Series* have been peer reviewed through processes administered by the proceedings Editors. Reviews were conducted by expert referees to the professional and scientific standards expected of a proceedings journal published by IOP Publishing.

# $\mu$SR Measurements of the Magnetism of $Sr_3Fe_2O_7$

**W A MacFarlane[1], D C Peets[2,7], T Buck[3], K H Chow[4], D L Cortie[1,3], A Jain[2], C Pomerantz[5], C B L Tschense[5,6], D Wang[3] and B Keimer[2]**

[1] Chemistry Department, University of British Columbia, Vancouver, BC, V6T 1Z1, Canada
[2] Max-Planck-Institut für Festkörperforschung, D-70569 Stuttgart, Germany
[3] Department of Physics, University of British Columbia, Vancouver, BC, V6T 1Z1, Canada
[4] Department of Physics, University of Alberta, Edmonton, AB, T6G 2E1, Canada
[5] TRIUMF, Vancouver, B.C. Canada, V6T 2A3
[6] Department of Inorganic Chemistry, University of Bayreuth, D-95440 Bayreuth, Germany

E-mail: wam@chem.ubc.ca

**Abstract.** We present a preliminary report of zero and transverse field $\mu$SR measurements in a high quality single crystal of fully oxygenated $Sr_3Fe_2O_7$[1], the insulating $3d$ magnetic analogue of the well studied $Sr_3Ru_2O_7$, in its charge-disproportionated state, through the magnetic transition to the incommensurate phase at $T_N \sim 115$ K. A very broad distribution of static fields below $T_N$ yields a large (2/3) missing fraction. Rapidly damped zero field oscillations indicate a large internal field at the muon, estimated to be 0.46 T at 0 K. In the paramagnetic state, we find two muon precession frequencies at 0.25 T applied field, reminiscent of the Mössbauer spectra.

## 1. Introduction

The Ruddlesden-Popper series of mixed alkaline-earth transition-metal oxides, with general formula $A_{n+1}M_nO_{3n+1}$, exhibit a wealth of interesting properties, such as superconductivity, colossal magnetoresistance, and charge, orbital, and spin order. Among these, the fully oxygenated $n = 2$ compound $Sr_3Fe_2O_7$ offers unique insight into the differences between the magnetic properties of the manganates, ruthenates, and ferrates. $Sr_3Fe_2O_7$ crystallizes in a tetragonal layered perovskite structure (space group $I4/mmm$) and its crystal structure contains double sheets of $FeO_6$ octahedra. Mössbauer spectroscopy indicates charge disproportionation of Fe (nominally $2Fe^{4+} \rightarrow Fe^{3+} + Fe^{5+}$, but realistically the charge difference is likely substantially less) at $T_{charge} = 340$ K[2]. The layered crystal structure of this compound gives rise to interesting quasi-two-dimensional magnetic properties. The analogous (metallic) ruthenate compound $Sr_3Ru_2O_7$ has been studied extensively, both in relation to the well-known related $Sr_2RuO_4$ superconducting phase, as well as for its metamagnetism. This has been interpreted as a magnetically tuned quantum critical transition[3], with a high field state that may be an electronic nematic analogous to quantum Hall systems[4]. In contrast, $3d$ Fe has stronger electron interactions (Hubbard $U$) and smaller bandwidths, making $Sr_3Fe_2O_7$ nonmetallic at ambient pressure below $T_{charge}$[8], but exhibits interesting low dimensional magnetism that may bear some relation to the analogous $4d$ ruthenate. The two dimensionality is expected to result

[7] Present address: IBS-CCES, Seoul National University Gwanak-ro 1, Gwanak-gu, Seoul 151-747, Korea
[8] It conducts above $\sim 15$ GPa and, at ambient pressure, above $T_{charge}$.

in a strongly 2D correlated paramagnetic state, with long range order brought about by small 3D couplings. Here we present preliminary $\mu$SR results on the low field magnetic state of fully oxygenated $Sr_3Fe_2O_7$ in the range from 300 K to 2 K.

## 2. Experiment

We have synthesized and oxygen-annealed large ($cm^3$ sized) single crystals of $Sr_3Fe_2O_7$. The oxygen content was confirmed by thermogravimetric analysis and refinement of neutron diffraction[1]. The DC magnetization indicates an antiferromagnetic transition at Néel temperature $T_N \sim 115$ K, e.g. see Fig. 1. The single crystal boule was mounted on a low background sample frame with thin aluminized mylar tape. A $c$-axis facet on the sample was aligned within a few degrees of the beam direction in the LAMPF spectrometer on the M15 beamline at TRIUMF. Two measurements were performed: 1) a zero applied field (ZF) experiment with the muon spin polarized along the beam direction parallel to the crystalline $c$-axis and 2) a precession experiment in an applied transverse field (along $c$) with the muon spin rotated (TF). Temperature was varied using a helium gas flow cryostat. Calibration measurements were made on a high purity Ag blank sample before the measurement.

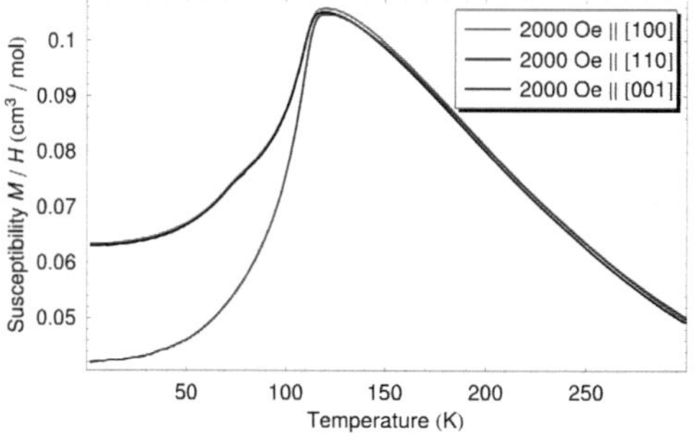


**Figure 1.** The molar magnetic susceptibility of $Sr_3Fe_2O_7$ as a function of temperature in a magnetic field of 0.2 Tesla applied along three different high symmetry directions of the tetragonal single crystal. The two in-plane directions are nearly identical.

## 3. Results and Discussion

The ZF data shows an increasing relaxation rate as the sample is cooled towards $T_N$, consistent with a slowing down of spin fluctuations, see Fig. 2. Below $T_N$ there is a very fast relaxing component, that exhibits measurable, but extremely rapidly damped oscillations over a narrow range of temperature (Fig. 3). Below 75 K, this signal is lost as its relaxation becomes too fast. There is also a non-oscillating slower relaxing component that, at low temperature, accounts for about 1/3 of the total asymmetry in the paramagnetic state. This signal, likely corresponding to muons stopping in a local longitudinal field, can be fit to a single exponential in the vicinity of $T_N$, but below 75 K is better described by a stretched exponential, $\exp(-\sqrt{t/T_1})$. The relaxation rates from such fits are shown in Fig. 4a). Interestingly, the relaxation rate does not peak close to $T_N$ but rather continues to increase well below the transition, eventually slowing below 50 K.

We fit the oscillating signal with a rapidly damped exponential and could extract the frequency reliably at only two temperatures, Fig. 4b). This signal corresponds to muons in large [4.0(2) kG at 75 K] local transverse fields, in the range seen in related Fe oxides, such as

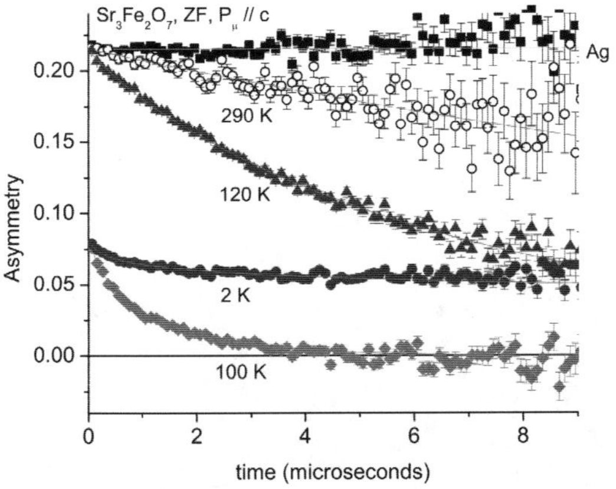


**Figure 2.** Temperature dependence of the zero field $\mu$SR spectra. Below $T_N$ there is a large missing fraction. The nonrelaxing calibration spectrum in Ag at 290 K is shown for comparison.

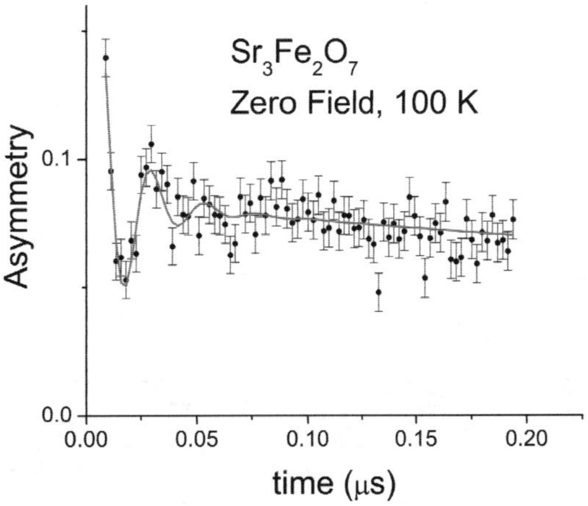


**Figure 3.** Rapidly damped zero field oscillations at 100 K are visible in the early times of the same data shown in Fig. 2.

$YFeO_3$[5] and Co-doped $SrFeO_3$[6]. We note it is not the frequency that limits our ability to follow this signal (with a time resolution of a few ns), rather it is the very fast damping that eliminates it within the first 50 ns or less. The rapid damping is consistent with a magnetic state with a broad distribution of local fields, likely a consequence of the incommensurate magnetic order[1]. In order to extrapolate the field at the $\mu^+$ to low temperature where the oscillations are lost, we fit the published temperature dependence of the Mössbauer splitting[2] to a smooth function and scaled it to the ZF frequency, see Fig. 4b), yielding a $T = 0$ frequency of 59(3) MHz for the $\mu^+$. Once the details of the magnetic order are firmly established, we will be able to use the precession signal to infer the muon site in this, the simplest fully oxygenated end-member compound. This will be important for studying the full field, temperature and doping phase diagram.

Measurements in transverse field in the paramagnetic charge disproportionated state reveal a long-lived precession above $T_N$ characterized by two distinct, but not very well-resolved, frequencies, see Fig. 5. Below $T_N$ the precession is rapidly damped by the broad distribution of static internal fields. In preliminary analysis, the splitting appears to increase with decreasing temperature (though the linewidth is similarly increasing). Here one might expect a muon shift to scale with the susceptibility (Fig.1). It is remarkable that there are two frequencies. The obvious interpretation is that there are 2 distinct muon sites with different couplings to the surrounding Fe moments. This seems possible in the disproportionated state, where there are already 2 distinct Fe sites, but it is not clear how to obtain *only* two muon sites. The $\mu^+$ sites are likely adjacent to an $O^{2-}$, of which 2 of the 7 are away from the Fe layer and 5 bridge adjacent Fe ions. In the simplest charge alternating structure, analogous to $CaFeO_3$, these would be equivalent. It would be interesting to see if this difference collapses above $T_{charge}$. One may expect a better resolved spectrum at higher applied field, that might be more easily tracked with temperature. The bulk $\chi$ above $T_N$ appears unaffected by field, so we do not anticipate field induced changes, but the TF relaxation rate might also scale with field. Note that the Mössbauer shows Fe sites that differ in hyperfine field by about 50% in the magnetic state and

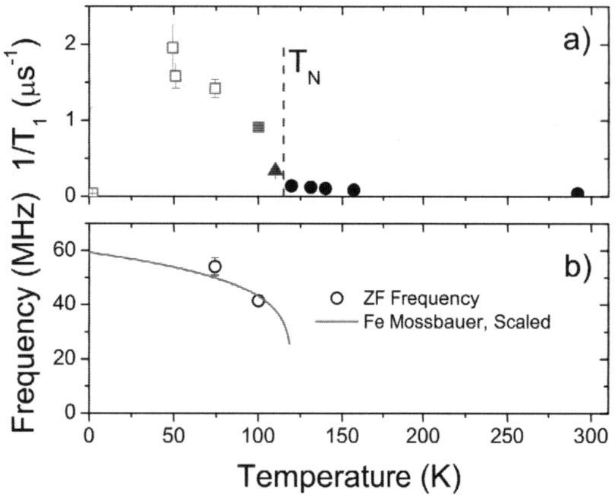


**Figure 4.** a) Temperature dependence of the zero field spin relaxation rate. From single (solid symbols) and stretched (open symbols) exponential fits to the relaxation with the full (circles), 1/3 (squares), and intermediate (triangle) asymmetry. b) The ZF oscillation frequency with the Mössbauer hyperfine field for the two Fe sites from Ref. [2] fit to a smooth function and scaled to match the observed frequencies.

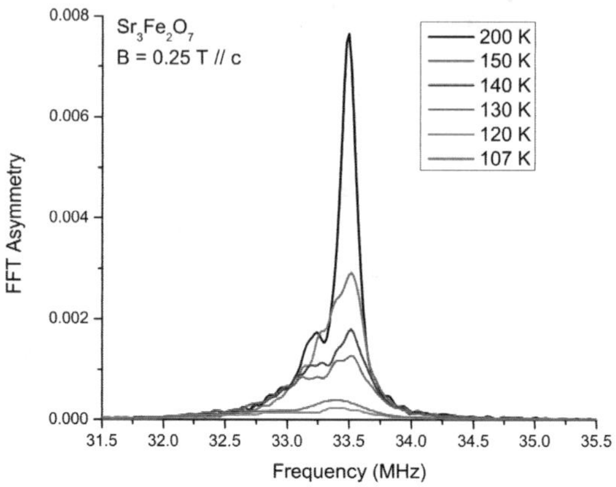


**Figure 5.** Temperature dependence of the fast Fourier transform of the transverse spin precession signal in an applied field of 0.25 T. The line is clearly two component, with a low frequency shoulder. The precession signal is rapidly damped as $T$ approaches $T_N$.

with a very different chemical shift in the paramagnetic state. In contrast, if the structure in the TF-$\mu$SR precession is due to two muon sites, they are rather similar.

## 4. Summary

In summary, we have studied the magnetism of $Sr_3Fe_2O_7$ using ZF and TF $\mu$SR, with the objective of studying the evolution of the magnetic phase with applied field. The preliminary data presented here establishes the $\mu$SR signals *in the low field regime.* Below $T_N = 115$ K, we find very rapidly damped oscillations over a narrow range of temperatures. Using the temperature dependence of the Mössbauer hyperfine field as a guide [Fig. 4b)], we estimate the average static field at the muon is on the order of 0.46 T at $T = 0$. The non-oscillating ZF signal yields a relaxation, likely dynamic in origin, that increases on cooling towards $T_N$ but then continues to increase below this until eventually slowing at low temperature. A better understanding of the ZF spectra will be aided by neutron data on the long range magnetism[7]. Structure in the TF precession signal in the paramagnetic phase may reflect the subtle charge disproportionation that remains unobserved by diffraction but is clear in the Mössbauer data. It would be interesting to study the precession through $T_{charge}$ to test this.

**Acknowledgments**
We thank B. Hitti for technical support, and K. Mogare and M. Jansen of Max-Planck-Institut für Chemische Physik fester Stoffe, Dresden for assistance with the high pressure oxygen annealing.

Journal of Physics: Conference Series **551** (2014) 012001 doi:10.1088/1742-6596/551/1/012001

## References

[1] Peets D C, Kim J H, Dosanjh P, Reehuis M, Maljuk A, Aliouane N, Ulrich C and Keimer B 2013 *Phys. Rev.* B **87** 214410

[2] Kuzushita K, Morimoto S, Nasu S and Nakamura S 2000 *J. Phys. Soc. Jpn.* **69** 2767–2770

[3] Grigera S A, Perry R S, Schofield A J, Chiao M, Julian S R, Lonzarich G G, Ikeda S I, Maeno Y, Millis A J and Mackenzie A P 2001 *Science* **294** 329–332

[4] Borzi R A, Grigera S A, Farrell J, Perry R S, Lister S J S, Lee S L, Tennant D A, Maeno Y and Mackenzie A P 2007 *Science* **315** 214–217

[5] Holzschuh E, Denison A B, Kündig W, Meier P F and Patterson B D 1983 *Phys. Rev.* B **27** 5294–5307

[6] Marshall I, Blundell S, Husmann A, Jestädt T, Lovett B, Pratt F, Lago J, Battle P and Rosseinsky M 2000 *Physica* B **289–290** 89–93

[7] Kim J H, Jain A, Reehuis M, Khaliullin G, Peets D C, Ulrich C, Park J T, Faulhaber E, Hoser A, Walker H C, Adroja D T, Walters A C, Inosov D S, Maljuk A and Keimer B 2014 *Phys. Rev. Lett.* **113**(14) 147206

Journal of Physics: Conference Series **551** (2014) 012002 doi:10.1088/1742-6596/551/1/012002

# Muon Knight shift in *d*-electron heavy fermion compound $Y_{0.95}Sc_{0.05}Mn_2$

**I Yamauchi[1], M Miyazaki[1], M Hiraishi[1], A Koda[1], K M Kojima[1], R Kadono[1] and H Nakamura[2]**

[1] Muon Science Laboratory and Condensed Matter Research Center, Institute for Materials Structure Science, High Energy Accelerator Research Organization (KEK), Tsukuba, Ibaraki 305-0801, Japan

[2] Department of Materials Science and Engineering, Kyoto University, Kyoto 606-8501, Japan

E-mail: ichihiro@post.kek.jp

**Abstract.** We report on the muon Knight shift in a polycrystalline sample of $Y_{0.95}Sc_{0.05}Mn_2$ that is known as one of the *d*-electron heavy fermion compounds. Since the muon site is presumed to have a trigonal symmetry, it is expected that the Fourier-transformed (FT) spectrum line shape would exhibit a uni-axial powder pattern which has two edges determined by the parallel and perpendicular components of the Knight shift, $K_\parallel$ and $K_\perp$, where $K_\parallel$ ($K_\perp$) is proportional to the parallel (perpendicular) component of the spin susceptibility, $\chi_\parallel$ ($\chi_\perp$). The observed FT spectrum at 2 K largely disagrees with the calculated line shape in isotropic ($\chi_\parallel = \chi_\perp$) and anisotropic ($\chi_\parallel \neq \chi_\perp$) cases, suggesting that there must be field-induced staggered magnetization due to strong antiferromagnetic spin correlations.

## 1. Introduction

The origin of heavy fermion behavior in *d*-electron systems such as $LiV_2O_4$ [1] and $Y_{1-x}Sc_xMn_2$ [2, 3, 4] remains as an enigmatic issue in condensed matter physics. $Y_{1-x}Sc_xMn_2$ is a series of inter-metallic compound with the cubic C15 Laves phase structure (space group $Fd\bar{3}m$), where the Mn atoms form a three-dimensional (3D) network of corner-sharing tetrahedra called pyrochlore lattice. $YMn_2$ ($x = 0$) shows a first-order magnetic transition at ~100 K accompanied by a long-wavelength (~400 Å) helical spin structure [5]. A large volume change ($\sim 5\%$) and small tetragonal distortion were also observed at the transition temperature [6]. These peculiar physical properties are presumed to be related with the quasi-one-dimensional antiferromagnetic (AF) correlation that is characteristic to the frustrated AF spin system on pyrochlore lattice as a manifestation of reduced dimensionality in the frustrated 3D systems [7, 8].

The magnetic ordering is suppressed by relatively low hydrostatic [9, 10] or chemical pressure due to partial substitution of Y by Sc [11]. Antiferromagnetic spin fluctuation [12, 13, 14] and large electronic contribution to the specific heat $\gamma$ ($\simeq 150$ mJ·mol$^{-1}$K$^{-2}$ [3, 15]) were found in the paramagnetic state in place of the magnetic ordering. As is widely observed in heavy fermion systems, the ratio $A/\gamma^2 \sim 1.0\times10^{-5}$ Ω·cm (mJ/mol K)$^{-2}$ satisfies the Kadowaki-Woods relation, where $A$ is the coefficient of the $T^2$ term in the temperature ($T$) dependence of resistivity[4].

Previously, we have conducted a systematic study of $Y_{1-x}Sc_xMn_2$ (with a series of specimens with $x$ =0.03, 0.05, 0.07, and 0.08) by muon spin rotation/relaxation ($\mu$SR) under a longitudinal-field up to 5 T to elucidate the spin fluctuation and its relevance to the heavy fermion behavior[16]. The observed $\mu$SR time spectra exhibit a stretched exponential-damping, and the depolarization rate ($\lambda$) is mostly independent of the magnitude of the external field for $x \geq 0.05$, while it shows a slight variation for $x = 0.03$ due to slowing down of spin fluctuation towards spin glass state at lower temperatures. Based on the conventional model of metals, $\lambda$ is given by an equation,

$$\lambda(\nu, H_0) = \frac{2(\gamma_\mu A_\mu)^2 \nu}{\nu^2 + (\gamma_\mu H_0)^2} k_B T \frac{\chi}{N_A \mu_B^2}, \tag{1}$$

where $\gamma_\mu$ ($= 2\pi \times 135.539$ MHz/T) is the muon gyromagnetic ratio, $A_\mu$ is the hyperfine coupling constant, $\nu$ is the fluctuation rate of $A_\mu$, and $H_0$ is the external field. We have found that $\nu$ deduced from $\lambda$ using eq. (1) exhibits a power low, $\nu \propto T^\alpha$, at lower temperatures (below 20$\sim$40 K). More interestingly, $\alpha$ approaches unity as the compound goes away from instability to spin glass state with increasing $x$. Such $T$-linear dependence of the fluctuation rate is also reported in $LiV_2O_4$ [17, 18] and discussed in terms of the hybridized one-dimensional (1D) Hubbard chain model[19, 20].

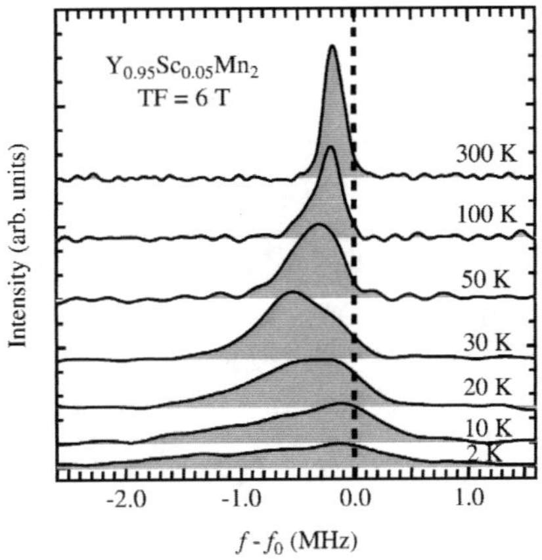


**Figure 1.** Fourier-transformed $\mu$SR spectra observed in $Y_{0.95}Sc_{0.05}Mn_2$ under a transverse-field of 6 T. The spectra are plotted against frequency shift from the reference, $f - f_0$.

However, there remains an issue in the above mentioned analysis that we rely on the bulk magnetic susceptibility for evaluating $\chi$ in eq. (1). Since the bulk susceptibility is a sum of many different contributions averaged over the specimen, possibility of extrinsic contributions such as magnetic impurities and defects cannot be ruled out. It is often difficult to separate all these different contributions in the bulk property measurements. The muon Knight shift ($K$) provides useful information to extract intrinsic electronic property out of those extrinsic contributions because $K$ monitors local magnetic susceptibility in the atomic scale.

## 2. Experiments

Polycrystalline sample of $Y_{0.95}Sc_{0.05}Mn_2$ was prepared by arc melting in an argon atmosphere. The prepared compound was annealed for a week in a evacuated quartz tube at 800°C. $\mu$SR measurements under a transverse-field (TF) of 6 T were conducted at M15 beamline of TRIUMF, Canada. The shift was determined by the spin rotation frequency observed in the sample ($f$) and that in a reference silver plate ($f_0$) as

$$K = \frac{f - f_0}{f_0}, \tag{2}$$

where a care was taken to keep the muon stopping condition unchanged in replacing the sample to the silver plate.

## 3. Results and discussion

Typical examples of Fourier-transformed (FT) $\mu$SR spectra at various temperatures are shown in Fig. 1, where the spectra are plotted against the frequency shift from the reference, $f - f_0$. observed a sharp and symmetric single line above ~125 K. $YMn_2$ is known as a hydrogen-absorbing alloy, and the hydrogen site is identified to be 16$c$ site by $^1$H NMR experiments [21]. It is then naturally expected that muon also occupies the 16$c$ site in $Y_{1-x}Sc_xMn_2$ as a hydrogen isotope (see Fig.2). Since the muon site has a trigonal $\bar{3}m$ point symmetry for cubic

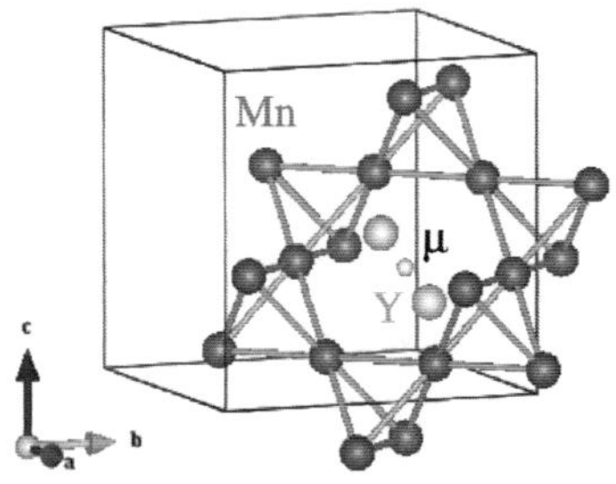


**Figure 2.** Local atomic structure of $YMn_2$ with muon shown at the 16$c$ site.

$Fd\bar{3}m$ structure, it is predicted that a uni-axial powder pattern will be observed in the FT $\mu$SR spectrum measured in polycrystalline samples. However, as shown in Fig. 1, no such fine structure is observed in the spectra above 125 K. This indicates that the anisotropy of $A_\mu$ is small and that the detailed structure of the spectra is masked by the natural line width determined by the muon decay lifetime.

Below ~100 K, the observed FT spectra gradually become broad and asymmetric with decreasing $T$. The uni-axial powder pattern has two edges determined by the parallel and perpendicular components of the Knight shift, $K_\parallel$ and $K_\perp$, where $K_\parallel$ ($K_\perp$) is defined as the Knight shift with an external field parallel (perpendicular) to the principal axis. They are generally related with the local magnetic susceptibility by the equations

$$K_\parallel = \frac{A_\mu^\parallel}{N_A \mu_B}\chi_\parallel, \quad K_\perp = \frac{A_\mu^\perp}{N_A \mu_B}\chi_\perp, \tag{3}$$

where $A_\mu^{\parallel,\perp}$ and $\chi_{\parallel,\perp}$ are the corresponding hyperfine parameters and the local magnetic susceptibility per 1 mol Mn atoms, respectively. Since the sign of $A_\perp$ is opposite to $A_\parallel$, $A_\perp = -\frac{1}{2}A_\parallel$, the line width increases with an enhancement of $\chi$. The FT spectrum at 2 K and calculated powder patterns are displayed in Fig. 3, where the red broken line represents an ideal uni-axial powder pattern at 2 K without magnetic anisotropy (i.e., $\chi_\parallel = \chi_\perp$). The hyperfine fields are obtained by taking a summation of the classical dipole fields from 1$\mu_B$ Mn moments at the nearest-neighboring six Mn atoms. Generally, FT $\mu$SR spectrum has unavoidable line broadening due to the limitation of time window and finite muon lifetime. The line broadening is estimated to be less than ~0.2 MHz from the FT spectra at 300 K. Thus the observed spectra at 2 K is significantly broader than the calculated ones.

One possibility is that the broadening comes from local magnetic anisotropy. In general, bulk average of spin susceptibility ($\chi_{\rm spin}$) is given as the average of the three principal values of

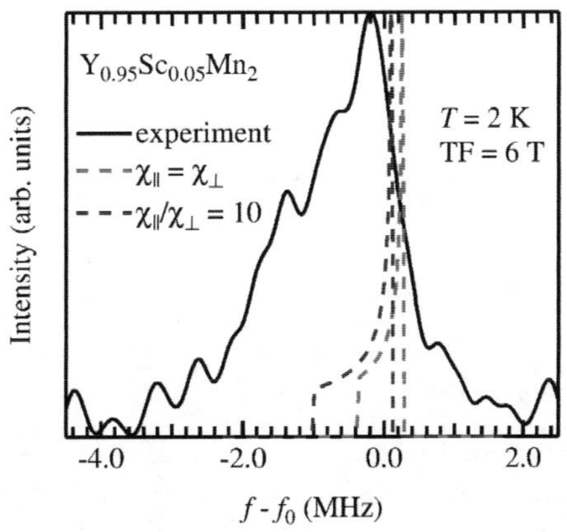


**Figure 3.** Fourier-transformed $\mu$SR spectrum at 2 K under a transverse-field of 6 T. The red and blue broken lines represent calculated spectra for the expected muon site (16$c$) assuming magnetically isotropic ($\chi_{\parallel} = \chi_{\perp}$) and anisotropic ($\chi_{\parallel} \neq \chi_{\perp}$) cases.

local magnetic susceptibility, i.e., $\chi_{\rm spin} = \frac{1}{3}\chi_{\parallel} + \frac{2}{3}\chi_{\perp}$ in the case of uni-axial anisotropy. Then, we can simulate the influence of anisotropy by changing $\chi_{\parallel}$ and $\chi_{\perp}$ with the above mentioned constraints. For instance, the blue broken line in Fig. 3 shows a calculated spectrum assuming $\chi_{\parallel}/\chi_{\perp} = 10$. It is clear that the observed FT spectrum is not reproduced by the calculated FT spectrum. Although we can calculate more anisotropic FT spectra with $\chi_{\parallel}/\chi_{\perp} > 10$, the rotation frequency for lower edge of the spectrum is limited to $f - f_0 \sim -1$ MHz because the $\chi_{\parallel}$ can not exceed the $\chi_{\rm spin}$ value. Then, the observed spectra is not explained only by taking into the magnetic anisotropy.

These facts suggest the appearance of staggered moments due to strong AF correlation that does not contribute to the uniform magnetic susceptibility even though the local magnetizations are enhanced. We made curve-fit analysis of time dependent asymmetry assuming two components,

$$A_0 G(t) = A_0 e^{-i\phi_0 t} \sum_{i=1}^{2} a_i \exp\left(2\pi i f_i t - \lambda_i t\right), \quad (4)$$

where $A_0$ is the initial muon-positron decay asymmetry, $\phi_0$ is the initial phase, $a_i$ is the fractional yield, $f_i$ is the precession frequency, and $\lambda_i$ is the relaxation rate. Fig. 4 shows the temperature dependence of the observed Knight shift $K_i$ ($i = 1, 2$) determined from $f_i$ as $K_i = (f_i - f_0)/f_0$. Below $\sim$100 K, $K_1$ decreases with $T$ whereas $K_2$ shows increasing behavior, indicating that the sign of the internal field for $K_1$ and $K_2$ are mutually opposite (i.e., negative for $K_1$ and positive for $K_2$).

Here, we briefly discuss the origin of unusually broad line shape shown in Fig. 3. It is expected that a characteristic quasi-one-dimensional AF correlation would develop in $Y_{1-x}Sc_xMn_2$[7, 8]. In this situation, two types of internal field associated with the staggered moments may be generated at the presumed muon site; it accompanies six nearest-neighboring Mn atoms, where a half of the implanted muons are surrounded by four-up and two-down spins (4U-2D) while those for another half are next to two-up and four-down spins (2U-4D). It is naturally expected that the sign of the internal fields is opposite between these two sites. While no quasistatic field is expected from those moments, the spin fluctuation may become slow enough at lower temperatures in comparison with the time window of the present $\mu$SR experiments so that the line shape might be described as a sum of the two types of powder patters. This assumption is also supported by the shoulder-like structure of the FT spectra in the $T$ range between 50 K and 20 K. In such a situation, the $K_{\parallel}$ and $K_{\perp}$ are given as,

$$K_{\parallel} = \frac{2A_{\mu}^{\parallel}m_{\uparrow} + A_{\mu}^{\parallel}m_{\downarrow}}{3H_0}, \quad K_{\perp} = -\frac{7A_{\mu}^{\parallel}m_{\uparrow} - A_{\mu}^{\parallel}m_{\downarrow}}{12H_0}, \tag{5}$$

for the 4U-2D site, and

$$K_{\parallel} = \frac{A_{\mu}^{\parallel}m_{\uparrow} + 2A_{\mu}^{\parallel}m_{\downarrow}}{3H_0}, \quad K_{\perp} = \frac{A_{\mu}^{\parallel}m_{\uparrow} - 7A_{\mu}^{\parallel}m_{\downarrow}}{12H_0}, \tag{6}$$

for the 2U-4D site, where $m_{\uparrow}$ and $m_{\downarrow}$ are the magnitude of the Mn moments with spin up and down, respectively. Then, we can simulate the line shape by varying $m_{\uparrow}$ and $m_{\downarrow}$ as parameters. We note that the model exhibits reasonable agreement with the FT spectrum in Fig. 3 for $m_{\uparrow} = 0.02 \sim 0.03\mu_B$ and $m_{\downarrow} = -0.005 \sim -0.006\mu_B$. It is confirmed that the uniform magnetization per Mn atom, which is deduced from the present results as $(m_{\uparrow} + m_{\downarrow})/2 = 0.007 \sim 0.009\mu_B$, is consistent with the bulk magnetization.

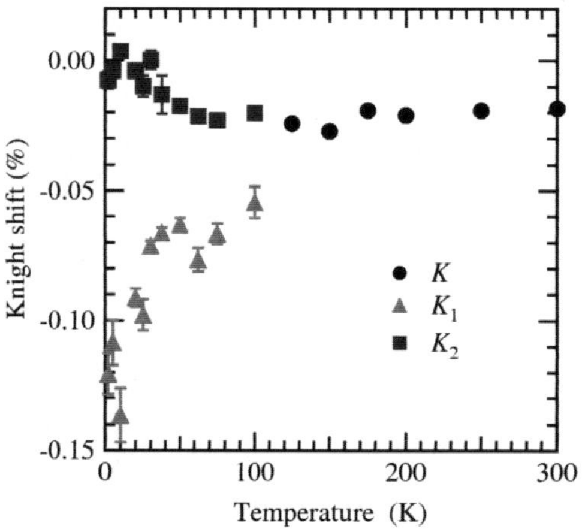


**Figure 4.** Temperature dependence of the muon Knight shift for $Y_{0.95}Sc_{0.05}Mn_2$.

Finally, we stress that a similar broadening of TF-$\mu$SR line shape has been observed in our previous study on $LiV_2O_4$ [22]. These features led us to speculate that the same microscopic magnetism may be present in $LiV_2O_4$. Further experimental investigation such as $H_0$ and $x$ dependence of the line broadening would be useful to clarify the local magnetism of these compounds.

## 4. Summary

We measured the muon Knight shift in the $d$-electron heavy fermion system $Y_{0.95}Sc_{0.05}Mn_2$ under a TF of 6 T. The FT spectra gradually became broad and asymmetric with decreasing temperature, where those at lower temperatures were not reproduced by the powder pattern even for an extreme case of magnetic anisotropy as large as $\chi_{\parallel}/\chi_{\perp} = 10$. This result suggests occurrence of short-range ordered staggered moments due to a strong antiferromagnetic correlation.

## References

[1] Kondo S, Johnston D C, Swenson C A, Borsa F, Mahajan A V, Miller L L, Gu T, Goldman A I, Maple M B, Gajewski D A, Freeman E J, Dilley N R, Dickey R P, Merrin J, Kojima K, Luke G M, Uemura Y J, Chmaissem O and Jorgensen J D 1997 *Phys. Rev. Lett.* **78** 3729
[2] Shiga M 1988 *Physica* B **149** 293
[3] Shiga M, Fujiwara K and Wada H 1993 *J. Phys. Soc. Jpn.* **62** 1329
[4] Hauser R, Bauer E, Gratz E, Dubenko I S and Markosyan A S 1994 *Physica* B **199&200** 662
[5] Cywinski R, Kilcoyne S H and Scott C A 1991 *J. Phys.: Condens. Matter.* **3** 6473
[6] Shiga M, Wada H and Nakamura Y 1983 *J. Mag. Mag. Matt.* **31-34** 119
[7] Nakamura H, Metoki N, Suzuki S, Takayanagi F and Shiga M 2001 *J. Phys.: Condens. Matter* **13** 475
[8] Nakamura H and Shiga M 2001 *J. Alloys Comp.* **326** 157
[9] Fisher R A, Ballou R, Emerson J P, Leievre-Berna E and Phillips N E 1993 *Int. J. Mod. Phys.* B **7** 830
[10] Zheng G-q, Nishikido K, Ohnishi K, Kitaoka Y and Asayama K 1999 *Phys. Rev.* B **59** 13973
[11] Nakamura H, Wada H, Yoshimura K, Shiga M, Nakamura Y, Sakurai J and Komura Y 1988 *J. Phys. F: Met. Phys.* **18** 981
[12] Shiga M, Wada H, Nakamura Y, Deportes J, Ouladdiaf B and Ziebeck K R A, 1988 *J. Phys. Soc. Jpn.* **57** 3141
[13] Ballou R, Leliévre-Berm E and Fåk B 1996 *Phys. Rev. Lett.* **76** 2125
[14] Nakamura H, Takayanagi F, Shiga M, Nishi M and Kakurai K 1996 *J. Phys. Soc. Jpn.* **65** 2779
[15] Wada H, Shiga M and Nakamura Y 1989 *Physica* B **161** 197
[16] Miyazaki M, Kadono R, Hiraishi M, Masuda T, Koda A, Kojima K M, Yamazaki T, Tabata Y and Nakamura H 2011 *J. Phys. Soc. Jpn.* **80** 063707
[17] Lee S H, Qiu Y, Broholm C, Ueda Y and Rush J J 2001 *Phys. Rev. Lett.* **86** 5554
[18] Kadono R, Koda A, Higemoto W, Ohishi K, Ueda H, Urano C, Kondo S, Nohara M and Takagi H 2012 *J. Phys. Soc. Jpn.* **81** 014709
[19] Fujimoto S 2002 *Phys. Rev.* B **65** 155108
[20] Lee J D 2003 *Phys. Rev.* B **67** 153108
[21] Fujiwara K, Ichinose K, Nagai H and Tsujimura A 1987 *J. Mag. Mag. Matter.* **70** 184
[22] Koda A, Kadono R, Ohishi K, Saha S R, Higemoto W, Matsushita Y and Ueda Y 2005 *J. Phys.: Condens. Matter.* **17** L257

Journal of Physics: Conference Series **551** (2014) 012003 doi:10.1088/1742-6596/551/1/012003

# Probing the magnetic phases in the Ni-V alloy close to the disordered ferromagnetic quantum critical point with $\mu$SR

**A Schroeder[1], R Wang[1], P J Baker[2], F L Pratt[2], S J Blundell[3], T Lancaster[3,4], I Franke[3] and J S Möller[3]**

[1] Department of Physics, Kent State University, Kent OH 44242, USA
[2] ISIS Facility, STFC Rutherford Appleton Laboratory, Harwell Oxford, OX11 0QX, UK
[3] Oxford University Department of Physics, Clarendon Laboratory, Parks Road, Oxford, OX1 3PU, UK

E-mail: aschroe2@kent.edu

**Abstract.** Zero (ZF) and longitudinal field (LF) muon spin relaxation data of the $d$-metal alloy $Ni_{1-x}V_x$ are presented at several vanadium concentrations $x$ below and above the critical $x_c \approx 11$ % where long-range ferromagnetic (FM) order is suppressed. The clear single precession frequency observed for Ni, as expected for a homogeneous FM, changes to rather damped osciallation with small V substitution at $x = 4$ %, confirming magnetic inhomogeneities caused by the less magnetic V environments in the magnetic Ni matrix. Furthermore, local fields and spatial field distributions can be estimated to characterize different inhomogeneous regimes developing with $x$ in the FM phase of $Ni_{1-x}V_x$. In the regime of $x = 7 - 10$ % a Kubo Toyabe function well describes the low temperature ZF and LF asymmetry data supporting a static Gaussian field distribution. Closer to the quantum critical concentration a single scale static Kubo Toyabe function with one field distribution is not sufficient to describe the muon relaxation. These data indicate that further changes in spatial distributions and dynamics are evolving as expected within the critical regime of a disordered quantum critical point.

## 1. Introduction

The study of quantum phase transitions (QPT) is a promising route to understand the origin of unconventional properties and in finding new phases in correlated many body systems. Introduction of "disorder", through e.g. impurities or sample imperfection can significantly change the nature of the QPT. In itinerant system with Heisenberg symmetry a new quantum critical point is predicted with exotic properties such as observable quantum Griffiths effects [1]. Examples for disordered systems are not rare, but the degree and significance of the "disorder" is often not clear. The binary alloy $Ni_{1-x}V_x$ shows the obvious signs of such a disordered QPT from bulk investigations [2, 3] and is therefore an ideal example for further microscopic studies. muSR is ideally suited as spectroscopic and local probe [4] to reveal the first insights into these disordered magnetic phases as it has been successfully employed for spin glasses [5], reduced moment correlated electron systems like heavy fermions [6, 7, 8] and phase separated systems close to first order transitions like transition metal compounds [9].

[4] Present address: Durham University, Centre for Materials Physics, South Road, Durham DH1 3LE, UK

The ferromagnetic (FM) order of Ni is rapidly suppressed replacing Ni by V. The critical temperature $T_c$ decreases approximately linearly and vanishes towards a V-concentration of $x_c \approx 11.4\%$ [2] as shown in Fig. 1(a). Magnetization measurements show the typical weak itinerant FM response in the FM phase for $x \leq 11\%$, the "non FM ordered", paramagnetic (PM) phase beyond $x_c$ is not compatible with independent magnetic moments. Instead, the magnetization displays power laws in temperature (T) and magnetic field with non universal exponents which vary with $x$ in a wide regime $x \geq x_c$ [2, 3]. This marks the quantum Griffiths phase (GP) (see e.g. in [10]), characterized by a distribution of magnetic clusters with different fluctuation rates. This phase is distinctly different than in clean QPT, where universal critical exponents are ideally observed only at $x_c$. We suspect that $Ni_{1-x}V_x$ is a disordered magnet but direct evidence is lacking. Friedel [11] already proposed that the drastic average magnetic moment reduction (see e.g. $m_s$ in Fig 2(a)) in Ni introducing V is not homogeneous, as V (with 5 d-electron less than Ni) might reduce severely the spin of all the Ni-neighbors creating large voids in the magnetization density. Early neutron measurements detected a large magnetic moment disturbance in Ni-V alloys [12]. Our bulk magnetization measurements only show (a mean field behavior of) the order parameter of a FM and no clear signs of the disorder. Only close to $x_c$ towards very low temperatures deviations from a long range FM become apparent, cluster glass freezing is observed [2] which does not show on this linear T-scale and is not discussed here. Here we use $\mu$SR to display the "disorder" in $Ni_{1-x}V_x$, probing the local fields in the FM phase. We present how to characterize the inhomogeneities in the FM and how they evolve upon dilution approaching a disordered QCP.

## 2. Experimentals

Polycrystalline samples of the alloy $Ni_{1-x}V_x$ were prepared as spherical pellets in the concentration range $x = 0$ to $x = 12.3$ % as before [2]. The fcc structure remains unchanged with $x$, the lattice constant increases by 1% for $x = 15\%$ [2]. The muon data were collected using the Dolly instrument at PSI, using several pellets of each composition wrapped in silver foil, at low temperatures, in zero (ZF) and longitudinal magnetic fields (LF).

## 3. Evolution of disorder in the ferromagnetic state in $Ni_{1-x}V_x$

Fig. 2 shows an overview of the muon asymmetry for several $Ni_{1-x}V_x$ samples with different V concentration $x$ at low temperatures $T$ in zero field. In pure Ni a nearly undisturbed muon precession is observed as expected for a clean FM with a single frequency of $f = 20.2$ MHz or $\omega = 127$ rad/$\mu$s corresponding to a local field of $B_0 = \omega/\gamma = 0.15$ T as seen before [13] (using the known gyromagnetic ratio of the muon as $\gamma = 2\pi \times 135.5$ MHz/T). Upon doping with V these oscillations appear more "damped", shift to larger time scales and finally disappear beyond the critical concentration in the PM phase. Already for $x \geq 4$ % only the first minimum remains visible due to a distribution of fields revealing an inhomogeneous FM. Some subtle qualitative changes can already be seen in Fig. 2: Note that $x = 4$ % still shows a second minor maximum or "nose", and that the minimum for $x = 11\%$ is rather shallow.

In order to model the total asymmetry $A(t)$ for all temperatures we use different polarization functions $P(t)$ for the FM component, a simple exponential function for the PM component (approximating effectively electronic spins of clusters and the diluted V nuclear moments) and a small constant background $A_{BG}$ due to the Ag.

$$A(t) = A_{FM}P(t) + A_{PM}\exp(-\lambda t) + A_{BG} \tag{1}$$

$$P_{osci}(t;\omega,\Gamma,\lambda_L) = \frac{1}{3}\exp(-\lambda_L t) + \frac{2}{3}\cos(\omega t + \phi)\exp(-\Gamma t). \tag{2}$$

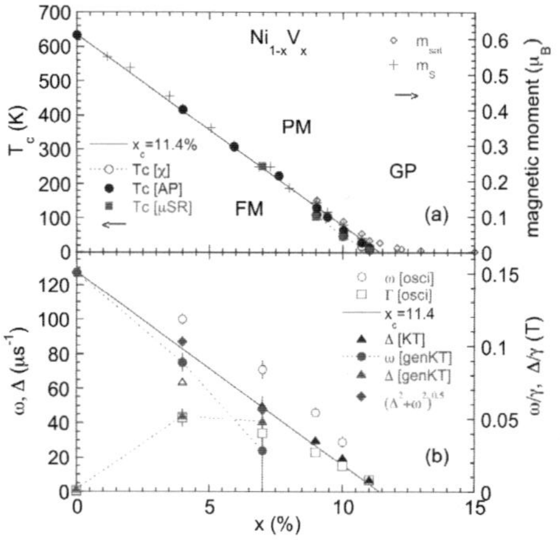


**Figure 1.** (a) Phase diagram for $Ni_{1-x}V_x$: critical temperature $T_c$ vs. V-concentration $x$, separating ferromagnetic (FM), paramagnetic (PM) and quantum Griffiths phases (GP). $T_c$ from magnetization (AP, $\chi$) [2] and $\mu$SR. Magnetic moment determined at high ($m_{sat}$) and low fields ($m_s$). Straight line indicates $x_c$ = 11.4 %. (b) Angular frequency or local field scales characterizing the center ($\omega$) and distribution width ($\Delta$) from various $P(t)$ models vs. $x$.

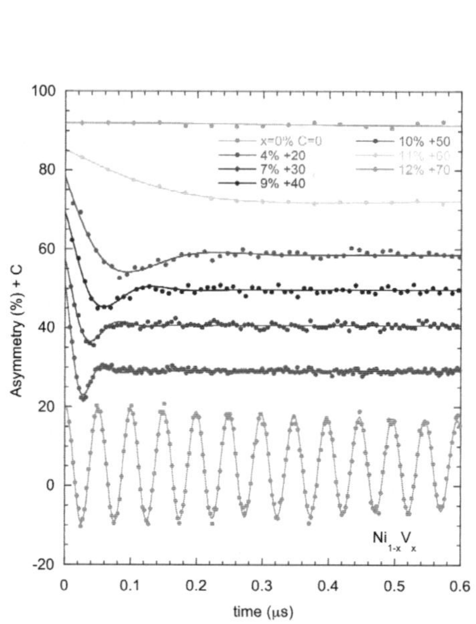


**Figure 2.** Asymmetry of $Ni_{1-x}V_x$ vs time for different V-concentrations $x$, ranging from $x$ = 0 to $x$ = 12.3 % at low temperatures (1.5K $< T < T_c/4$). Each is shifted by a constant $C$ for clarity. The line represents a fit to the oscillator model with $P_{osci}$.

First, we apply an oscillator model $P_{osci}$ (see lines in Fig. 2), a simplified "damped" expansion of a single precession frequency as expected in a perfect clean FM polycrystal like Ni with one preferred stopping site. This form is often used and also provides here a parametrization for all $x \leq 11$ %. For low $T$, $A_{PM}$ and $\lambda$ are essentially zero, later we need to set $\lambda_L = \lambda$ for all $T$. Also we keep $\phi = 0$ for higher $x$ to produce consistent parameters. The main frequency $\omega$ decreases with $x$ as shown in Fig. 1(b). The damping constant $\Gamma$ or transverse field depolarization rate is increased for $x$ = 4 % then decreases further with $x$. $\Gamma$ can be seen as a measure of field distribution, as $\omega$ is indicating some local field, but comparison to other models like $P_{genKT}$ (see eq.(4) below) show that in the "damped" cases $\omega$ is not exactly the mean field nor the field distribution. $\Gamma$ has half the value of $\omega$ at lowest T for all samples with 7% $< x <$ 10 %, which marks the regime with a Gaussian field distribution as discussed below.

More revealing simple models can be applied, but are restricted to certain $x$-regimes. In effect, for $x = 7 - 10$ % a static Kubo-Toyabe (KT) function $P_{KT}$ describes the data well for low T, indicating a Gaussian distribution of local fields [14] in the FM state.

$$P_{KT}(t; \Delta, \nu = 0) = \frac{1}{3} + \frac{2}{3}(1 - (\Delta t)^2) \exp(-\frac{1}{2}(\Delta t)^2), \tag{3}$$

The single parameter, the width $\Delta$ of the distribution decreases with $x$ towards $x_c$ as shown in Fig. 1(b). This field distribution follows the magnetic moment $m_s$ on the same straight line in Fig.1(a,b). It is remarkable that inside a FM a "random" magnetic field distribution is found as originally shown for randomly oriented frozen moments [14]. We suspect here that the distribution of different local fields indicates a spatial variation of magnetic moments in a long range ordered FM state. These muon data provide evidence that this system is indeed inhomogeneous or "disordered" as required to show this disordered QCP with distinct properties.

A Kubo Toyabe fit $P_{KT}$ is possible for $x = 4$ % but can be improved by a generalized form $P_{genKT}$ [15] allowing a finite center field $H_0 = \omega/\gamma \neq 0$ from large domains or crystallites which is the center of a Gaussian distribution with width $\Delta H = \Delta/\gamma$.

$$P_{genKT}(t; \Delta, \omega, \nu = 0) = \frac{1}{3} + \frac{2}{3}\exp(-\frac{1}{2}(\Delta t)^2)[\cos(\omega t) - \frac{1}{\omega}(\Delta t)^2 \sin(\omega t)]. \quad (4)$$

The two parameters $\omega$ and $\Delta$ are shown in Fig. 1(b). $\Delta$ is similar for both $x = 4$ % and $x = 7$ %, overall the maximum field distribution appears at about $\frac{1}{2}x_c$. The characteristic field $H_0$ in $x = 4$ % is still high due to intact magnetic domains, while $H_0$ is rather close to zero in $x = 7$ %, the field distribution $\Delta H$ dominates. The transition from dominating FM domains to random distribution occurs in between these concentrations at about $\frac{1}{2}x_c$. That matches the concentration $x = 5.6$ % where the number of reduced moment Ni-atoms (due to one or more V neighbors) exceeds the number of undisturbed "magnetic" Ni-atoms (with only Ni neighbors) in a fcc lattice. The local muon probe distinguishes the different degrees of inhomogeneities in this FM introduced by V substitution. The model $P_{genKT}$ provides even a direct quantitative assessment of an internal field distribution width $\Delta/\gamma$ and a center local field $\omega/\gamma$. An effective field $\sqrt{\Delta^2 + \omega^2}/\gamma$ finally scales with the mean magnetic moment $m_s(x)$.

The KT function $P_{KT}$ does not work as well towards the more diluted regime approaching $x_c$, the minimum becomes very shallow, for e.g. 11 %, questioning the long range order of this compound. Beyond $x_c$, for $x = 12.3$ %, a small depolarization rate ($\lambda = 0.07\mu s^{-1}$ for 2K) describes the data in ZF. Extended time, more temperature and magnetic field data and analysis are required to reveal more details about the fluctuation rates beyond the FM regime.

## 4. Contrasting the field distributions in $Ni_{1-x}V_x$ with $x = 9$ % and $x = 11$ %

The asymmetry spectra for $x = 9$ % can be modeled well by a KT function. For higher temperature we used a dynamic KT approximation with finite fluctuation rates $\nu$ [14, 16] for the FM contribution (as presented in Fig. 3(a)). Application of a longitudinal field (LF) at low T confirms the main parameter choices of a KT description (see Fig.3(b)). The straight lines shown are a KT fit with unchanged parameters from ZF with internal field = external LF, better fit results are achieved with reduced internal fields in these FM spherical samples. Fig.4 (a-d) presents the T-dependence of the fit-results of the exploratory oscillator model and the KT model. Panel (a) shows the main field distribution $\Delta$ together with $\omega$ and $\Gamma$. $\Delta(T)$ follows nearly a mean field T-dependence like the magnetic moment $m_s(T)$ vanishing at $T_c$. The fixed ratio $\omega/\Gamma$ of 2 (observed for several $x$ at low T) changes towards higher T towards 1, before $\omega$ vanishes. The ferromagnetic amplitude ratio, the ferromagnetic contribution divided through the total contribution, $A_{FM}/(A_{FM} + A_{PM})$, increases rapidly below $T_c$ and then saturates to about 1 independent of the model (see Fig.4(b)). The depolarization rate $\lambda$ is quite small in the PM regime and also very small, close to the detection limit, in the FM for $T < 0.8T_c$ (as $\lambda_L$) in $P_{osci}$. This rate peaks towards $T_c$, indicating some critical dynamics. The dynamic KT model reveals directly critical fluctuations increasing rapidly at $T_c$ (see Fig.4(d)). We see that below about $0.8T_c$ magnetic fluctuations become irrelevant. All quantities as $A_{AF}$, $\Delta$, $\nu$ indicate a critical temperature of $T_c = 108$ K consistent with low field susceptibility $\chi$ measurements [3], a bit lower than the usual high field Arrott plot (AP) determination of $T_c = 130$ K.

Magnetization measurements indicate that $x = 11$ % is still a ferromagnet displaying Arrott plots and no indication of a cluster glass transition is found in magnetic susceptibility. A strong ZF depolarization is observed, but the minimum is too shallow for the static KT form even at lowest T (see Fig. 3(d)). The ZF data can be modeled quite well applying a *dynamic* KT function ($P(t, \Delta, \nu > \Delta)$), using the strongly fluctuating approximation of a Keren function

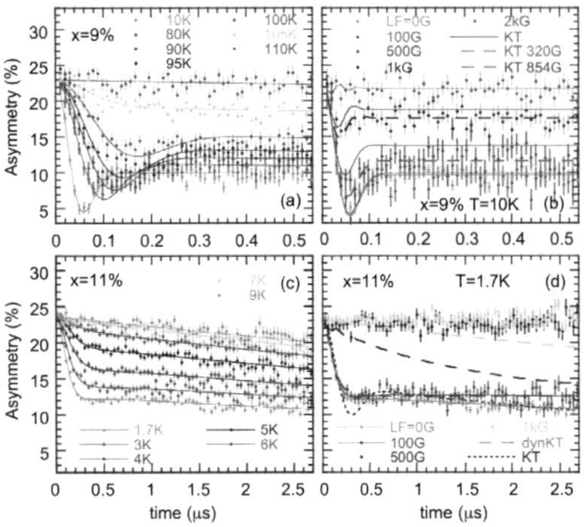

**Figure 3.** (a) ZF spectra for $Ni_{1-x}V_x$ with $x = 9$ %. Lines from $P_{KT}$. (b) LF spectra for $x = 9$ % at $T = 10$ K. Lines from $P_{KT}$: field = external LF (solid), adjusted internal field (dashed). (c) ZF spectra for $x = 11$ %. Lines from $P_{GBG}$. (d) LF spectra for $x = 11$ %. Dotted line: static $P_{KT}$. Dashed lines: dynamic $P_{KT}$. Solid lines: $P_{GBG}$.

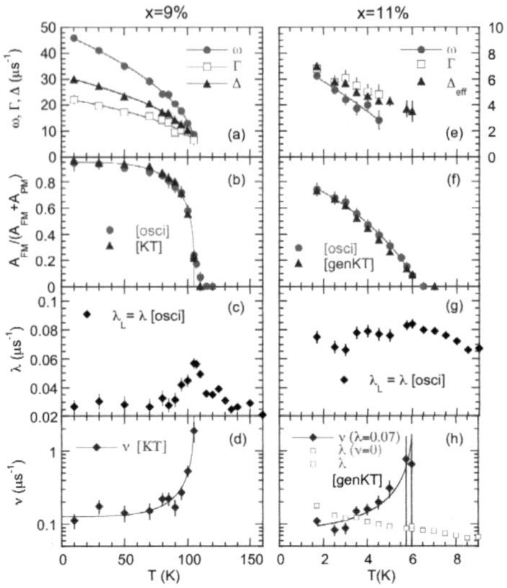

**Figure 4.** T-dependence of various fit parameters using $P_{osci}$, dynamic $P_{KT}$ or $P_{GBG}$ to model $A(t)$ as shown in Fig. 3 for $x = 9$ % (a-d) and for $x = 11$ % (e-h). Lines are guide to the eye.

[17, 16]. But the present LF data (e.g. in 500G) cannot be reproduced keeping the same fluctuation rate as in ZF (see dashed line in Fig. 3(d)). More intermediate fields are needed to study field dependent changes. We alternatively applied a *static* distribution function which describes multiple field distributions, narrow and wide ones to increase the weight towards very small field regions. Taking the response of many Gaussian distribution of fields with different $\Delta's$ following a Gaussian distribution with width $\Delta_W$ and center $\Delta_0$ gives the so called simple Gaussian broadened Gaussian function $P_{GBG}$ [18, 19].

$$P_{GBG}(t;\Delta_{eff},R,\nu) = \frac{1}{3}\exp(-\frac{2}{3}\nu t) + \frac{2}{3}((1+R^2)/N)^{3/2}(1-(\Delta t)^2/N)\exp(-\frac{1}{2N}(\Delta t)^2), \quad (5)$$

with $N = 1 + R^2 + R^2\Delta^2 t^2$, where $R = \Delta_w/\Delta_0$ and $\Delta^2 = \Delta_{eff}^2 = \Delta_0^2 + \Delta_w^2$. This form describes the data well as shown in Fig. 3(a) (with R=0.7). Note that an additional PM contribution with a small depolarization rate is always present even for low T. The effective field distribution width $\Delta_{eff}$ is shown in Fig. 4(e) together with the $P_{osci}$ parameters. They decrease towards higher $T$, a sharp critical temperature $T_c$ cannot be determined. Also, the ratio of these parameters at $x = 11$ % is different than those at $x = 9$ % at low T, more similar to those at $x = 9$ % close to $T_c$. The FM amplitude ratio increases only gradually towards low T (Fig. 4(f)), not as sharp as in $x = 9$ %. At the lowest T, only 75 % of the amplitude can be described by the FM component $P(t)$, while an extra (rest of 25 %) exponential "paramagnetic" contribution remains, independent of the model. The depolarization rate $\lambda$ (=$\lambda_L$) shows barely a maximum at $T_c$ and does not decrease towards low values as for $x = 9$ % in the oscillator model (Fig.4(g)). Keeping $P_{GBG}$ static ($\nu = 0$) the depolarization rate $\lambda$ of the PM contribution increases further below $T_c$ (Fig.4(h)). A small LF=100G suppresses these fluctuations (see Fig.3(d)), while the steep FM depolarization seems unaffected. Alternatively, allowing small fluctuations ($\nu > 0$) and keeping $\lambda = 0.07\mu s^{-1}$ constant produces some critical dynamic towards $T_c$. Estimates of

$T_c \approx 6$ K from these muon data is in agreement with low field susceptibility $\chi$ measurements [3], much lower than Arrott plot (AP) extrapolations of $T_c = 14$ K [2].

While the inhomogeneous magnetism in $x = 9$ % could be well characterized by one scale, the KT distribution width $\Delta$, multiple field distributions seemed better in $x = 11$ %. Furthermore, a PM contribution due to fluctuating clusters never vanishes at low T. These models are far too simple but indicate that fluctuating moments and the spatial inhomogeneities further evolve towards $x_c$. 11 % is not just a scaled down version of 9 % with a smaller field range.

## 5. Conclusion

$Ni_{1-x}V_x$ offers the opportunity to study a FM with different degrees of "spatial disorder" or inhomogeneities, introduced through V substitution from a pristine FM to a disordered QCP. Analyzing the muon asymmetry with a generalized KT model, $P_{genKT}$, the mean local field and the local field distribution could be determined for a large $x$ regime. These parameters measure the degree of "disorder" and allow a distinction between different regimes within the FM state. For lower $x$ intact domains with finite fields dominate, while for larger $x$ the field distribution dominates. This distribution width is the dominating energy scale in a wide region $7\,\% \leq x \leq 10\,\%$, it follows the order parameter, $m_s$. Approaching the critical concentration, $x_c$, simple single scale fit functions cease to work. Multiple distributions with multiple time scales evolve at the edge of the Griffiths phase, dominated by clusters with different time scales. So far we only used simple models to probe local field distributions in the FM phase. To relate this information to the spatial variation of inhomogeneities and correlation length a microscopic model is needed. The simple fcc lattice of Ni-V looks promising as a good model system. The study of the complex scenario in the quantum critical regime and towards the quantum Griffiths phase is a further challenge in this *d*-metal alloy.

## Acknowledgments

We thank S. Ubaid-Kassis for the original samples and W. Hayes for helpful discussions. This work was supported in part by OBR Research Challenge from KSU and part of this work was carried out at the Swiss Muon Source, Paul Scherrer Institute, CH.

## References

[1] Vojta T and Schmalian J 2005 *Phys. Rev.* B **72** 045438
[2] Ubaid-Kassis S, Vojta T and Schroeder A 2010 *Phys. Rev. Lett.* **104** 066402 and refs therein
[3] Schroeder A, Ubaid-Kassis S, and Vojta T 2011 *J. Phys.: Condens. Matter* **23** 094205
[4] Yaouanc A and Dalman de Réotier P 2011 *Muon Spin Rotation, Relaxation, and Resonance* (Oxford: Oxford University Press) and refs therein
[5] Uemura Y J, Yamazaki T, Harshman D R, Senba M and Ansaldo E J 1985 *Phys. Rev.* B **31** 546–563
[6] Amato A 1997 *Rev. Mod. Phys.* **69** 1119–1180
[7] MacLaughlin D E, Heffner R H, Bernal O O, Ishida K, Sonier J E, Nieuwenhuys G J, Maple M B and Stewart G R 2004 *Journal of Physics: Condensed Matter* **16** S4479
[8] Adroja D T, Hillier A D, Park J G, Kockelmann W, McEwen K A, Rainford B D, Jang K H, Geibel C and Takabatake T 2008 *Phys. Rev.* B **78** 014412
[9] Uemura Y J, Goko T, Gat-Malureanu I M et al 2007 *Nature Phys* **3** 29 – 35
[10] Vojta T 2006 *J. Phys. A.: Math Gen.* **39** R143–R205
[11] Friedel J 1958 *Nuovo Cimento* 7, 287-311
[12] Collins M F and Low G G 1965 *Proceedings of the Physical Society* **86** 535
[13] Foy M L G, Heiman N, Kossler W J and Stronach C E 1973 *Phys. Rev. Lett.* **30** 1064–1067
[14] Hayano R S, Uemura Y J, Imazato J, Nishida N, Yamazaki T and Kubo R 1979 *Phys. Rev.* B **20** 850–859
[15] Kornilov E I, Pomjakushin V Yu 1991 *Physics Letters* A **153** 364–367
[16] Pratt F L 2000 *Physica* B **289-290** 710–714
[17] Keren A 1994 *Phys. Rev. B* **50** 10039–10042
[18] Noakes D R and Kalvius G M 1997 *Phys. Rev.* B **56** 2352–2355
[19] Noakes D R 1999 *Journal of Physics: Condensed Matter* **11** 1589

Journal of Physics: Conference Series **551** (2014) 012004 doi:10.1088/1742-6596/551/1/012004

# $\mu$SR study of a quantum spin liquid candidate: the $S$=1/2 vanadium oxyfluoride kagome antiferromagnet

**J C Orain[1], L Clark[2], F Bert[1], P Mendels[1], P Attfield[2], F H Aidoudi[3], R E Morris[3], P Lightfoot[3], A Amato[4] and C Baines[4]**

[1] Laboratoire de Physique des Solides, UMR 8502 CNRS, Université Paris-Sud, 91400 Orsay, France

[2] CSEC and School of Chemistry, The University of Edimburgh, Edimburgh EH9 3JZ, United Kingdom

[3] School of Chemistry and EaSTChem, University of ST. Andrews, St. Andrews KY16 9ST, United Kingdom

[4] Laboratory for Muon Spin Spectroscopy, Paul Scherrer Institute, 5232 Villigen PSI, Switzerland

E-mail: jean-christophe.orain@u-psud.fr

**Abstract.** We present a detailed $\mu$SR study of the recently synthesized compound, $[NH_4]_2[C_7H_{14}N][V_7O_6F_{18}]$ (DQVOF), a geometrically frustrated magnetic material, both in longitudinal and transverse configurations. The $\mu$SR measurements in zero and longitudinal field show that there is no spin freezing down to 20 mK which is the key requirement for a quantum spin liquid state. Further experiments in transverse field single out two contributions with different shift and broadening which shed a new light on the location of the muons stopping sites.

## 1. Introduction

Understanding quantum spin liquid states (QSL) such as the resonating valence bond state (RVB) formed by macroscopic resonance between the various spin singlet coverings of the lattice, has been a long standing issue in condensed matter research [1, 2, 3, 4]. In two dimensions, the search for such states has focused on the $S$=1/2 quantum kagome antiferromagnet (QKAFM), which features a highly frustrated lattice of corner sharing equilateral triangles. Among the rare experimental candidates for a QKAFM, the recently synthesized $[NH_4]_2[C_7H_{14}N][V_7O_6F_{18}]$ (DQVOF) compound [5] is the first one based on $V^{4+}$ ($d^1$) ions rather than the more usual $Cu^{2+}$ ($d^9$) ions [6, 7, 8, 9], thus allowing the investigation of the effects of different perturbations to the ideal Heisenberg Hamiltonian.

The DQVOF material exhibits a rather complex bi-kagome layer structure (see Fig. 1 c)). It contains two different types of vanadium ions; $V^{4+}$ $S$=1/2 ions forming the kagome layers and additional $V^{3+}$ $S$=1 ions forming a triangular layer in between the kagome ones. These kagome bi-layers are well separated from each other by large organic molecules [5]. Previous low temperature heat capacity, susceptibility and magnetization measurements showed that the $V^{3+}$ ions are very weakly coupled to the $V^{4+}$ kagome layers so that the material is a good candidate to investigate the physics of the QKAFM [10]. From heat capacity analysis, the system shows a gapless behavior down to at least 350 mK with an original $T$-linear behavior

Journal of Physics: Conference Series **551** (2014) 012004 doi:10.1088/1742-6596/551/1/012004

suggesting fermionic excitations. $\mu$SR experiments in zero and low longitudinal field [10] have clearly demonstrated the absence of any spin freezing down to at least 40 mK. Further, these measurements have evidenced a slowing down of the spin dynamics below about 10 K and down to 1 K where the muon relaxation rate reaches a dynamical plateau as often observed in highly frustrated magnets. In such a complex system featuring strongly correlated quantum spins together with nearly free $S$=1 spins, asserting the origin of the spin dynamics remains a challenging problem. In this proceeding we present further $\mu$SR experiments on this compound: zero field measurements which demonstrates the absence of spin freezing down to 20 mK and longitudinal and transverse field experiments which allow to characterize the evolution of the spin dynamics at low temperature and clarify the role of both types of vanadium spin on the muon relaxation.

## 2. Experimental

The experiments were performed on a DQVOF powder sample synthesized as explained in Ref. [5]. The measurements were performed on the GPS and the LTF spectrometers at the Laboratory for Muon Spin Spectroscopy at the Paul Scherrer Institute. For the GPS experiment, we used 300 mg of sample wrapped in a thin aluminium foil inserted in a He-flow cryostat to vary the temperature in the range 1.6 K-200 K. The sample was first measured in a background free configuration and then mounted on a standard silver sample holder. Within our accuracy, we could not detect any fraction of signal arising from muons falling in the silver sample holder. Therefore we used the latter setup for all the measurements in longitudinal and transverse configurations. For the LTF experiment, where the temperature varies from 2.3 K down to 20 mK, about 200 mg of sample powder was deposited on a silver plate and then fixed using diluted varnish. By comparison to the GPS experiment we estimated the background asymmetry corresponding to muons falling in the LTF sample holder to be $\sim$ 0.021, i.e. less than 10 % of the total asymmetry.

## 3. Results

From the analysis performed in Ref. [10], the relaxation of the muon polarization in zero field consists of three different contributions; about 10% of the muons strongly couple to the fluorine ions to form a F-$\mu$-F complex, another minor fraction of muons ($\sim$ 10 %) stop far from the kagome planes, likely in the organic molecule in between the magnetic layers, and most of the muons ($\sim$ 80 %) stop near the oxygen ions, close to the $V^{4+}$ kagome planes. The two minor fractions can be decoupled easily by applying a longitudinal field of 0.02 T. As a consequence, following Ref.[10], we analyze the data obtained by applying longitudinal fields larger than 0.02 T to track the dynamical electronic contribution in isolation. The sizable relaxation observed at 20 mK even under the strongest 0.6 T applied field, see Fig. 1 a), leaves no doubt on the dynamical character of the relaxation. This extends down to 20 mK, the conclusion of Ref. [10] that DQVOF possesses a dynamical ground state.

We fitted the relaxation measured with an applied longitudinal field larger than 0.02 T to a stretched exponential

$$P(t) = exp(-\lambda t)^{\beta} + B \tag{1}$$

where $\lambda$ is the muon depolarization rate, $\beta$ is the stretched exponent and $B$ accounts for the fraction of muons fully decoupled in fields stronger than 0.02 T. The two latter parameters were fixed to the values obtained by fitting the 0.02 T field data, namely $\beta$=0.48 and $B$=0.35 at $T$=20 mK, and $\beta$=0.51 and $B$=0.35 at $T$=2.3 K in the crossover regime where the spin dynamics slows down drastically. The variation as a function of the applied field of the only free parameter $\lambda$ is very similar at the two studied temperatures, see Fig. 1 b). Coming from low fields, the

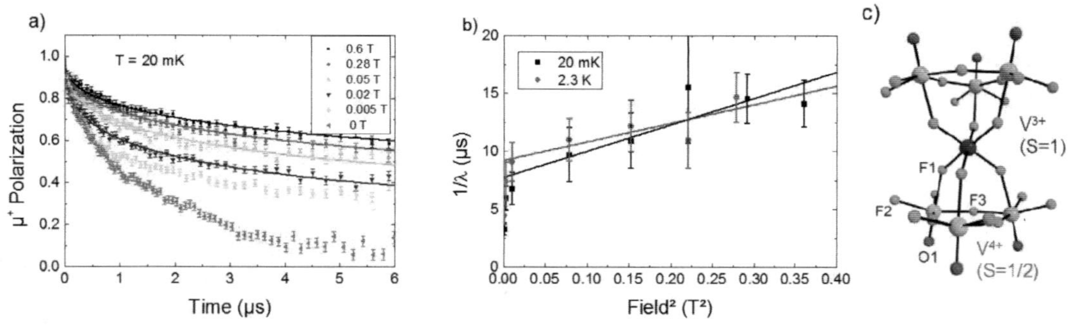


**Figure 1.** a) Time dependent muon spin polarization in DQVOF at 20 mK with different applied longitudinal fields. Solid lines are fits to the data (see text). b) Inverse of the relaxation rate versus field square. The solid lines are fits to the data. c) Local structure of the compound.

relaxation time $1/\lambda$ first increases rather rapidly with the applied field up to $H \sim 0.1$ T, before showing a weaker field dependence which can be approximated by a $H^2$ variation. Although the initial increase is poorly understood, the higher field regime suggests an analysis within the Bloembergen Purcell and Pound (BPP) theory which predicts for a simple spin dynamics $\langle S(t)S(0)\rangle \propto \exp(-\nu t)$

$$\lambda = \frac{2\gamma_\mu^2 H_{Fluc}^2 \nu}{\nu^2 + \gamma_\mu^2 H_{LF}^2} \tag{2}$$

where $\gamma_\mu$ is the muon gyromagnetic ratio, $H_{Fluc}$ is the internal fluctuating field, $\nu$ is the fluctuating frequency and $H_{LF}$ is the longitudinal applied field. Within this model, one can calculate the fluctuating field and frequency for each temperature. We find $H_{Fluc}$=0.007(1) T and $\nu$=470(80) MHz, and $H_{Fluc}$=0.007(1) T and $\nu$=620(125) MHz respectively at $T$=20 mK and $T$=2.3 K. This represents a drastic slowing down as compared to both $V^{3+}$ and $V^{4+}$ estimated fluctuation frequencies in the paramagnetic limit, respectively $\nu^{3+}$=523 GHz and $\nu^{4+}$=13.337 THz.

To gain a better understanding of the local magnetic properties, we performed transverse field experiments at different temperatures. For all the experiments we used a transverse field of 0.5 T. On the GPS experiment, the field is created by a resistive magnet powered by a stable power supply. In order to determine accurately the shift reference, we measured the signal from a bare silver plate at 60 K and assumed that the field remained constant during the whole experiment. On the LTF spectrometer, the field is produced by a superconducting magnet used in persistent mode. The sizable fraction of muons falling in the silver sample holder used in this experiment allowed us to determine the shift reference. For the analysis, the relaxations were transformed into a rotating reference frame at 70 MHz (RRF), about 2.5 MHz higher than the free muons rotation frequency at 0.5 T.

Although the relaxations were fitted in the time domain, the general evolution versus temperature is clearer on the Fourier transforms shown in Fig. 2 a) and 3 b). In the GPS experiment, the signal at high temperature consists of two different peaks, one narrow which remains clearly visible down to 2.3 K and one additional peak which broadens drastically upon cooling and eventually becomes too broad below 10 K to be detected. At low temperatures, in the LTF experiment the Fourier transforms consist of two peaks, one narrow and one broader (see Fig. 3 b)). The narrow peak corresponds to the silver background while the broad peak is

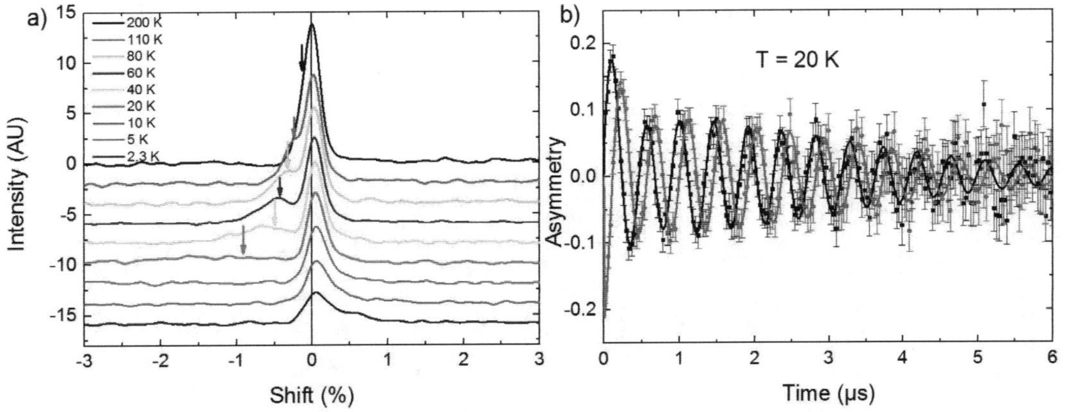


**Figure 2.** a) Fourier transforms of the asymmetries at different temperature with an applied transverse field of 0.5 T versus shift. The arrows are showing the broad peak center. The datas were taken on the GPS spectrometer. b) Time dependent signal at 20 K. The solid lines are fits to the data (see text).

the sample signal, inline with the GPS low temperature results.

In order to extract the shift and the depolarization rate of each components, we fitted the time dependent relaxations at each temperature. From 200 K to 10 K, given that two components, a gaussian and a lorentzian, are clearly seen in the Fourier transform spectra, we fitted the relaxations to two oscillations with different broadenings:

$$A(t) = A_1 cos(2\pi\nu_1 t + \phi_1)e^{(-\lambda_1 t)^2} + A_2 cos(2\pi\nu_2 t + \phi_2)e^{-\lambda_2 t} \quad (3)$$

where $\nu_i$ are the oscillation frequencies, $\phi_i$ their phases, $A_i$ their initial asymmetries and $\lambda_i$ their depolarization rates. We fixed the different $A_i$ to the values found at 20 K, $A_1$=0.099 and $A_2$=0.129. Below 10 K we fitted the data only after 0.1 $\mu$s keeping only the first oscillation in equation (3).

In the LTF experiment, we still used Eq. (3) to fit the time domain relaxation in the RRF, ignoring the initial fast relaxation as below 10 K in GPS. Here the second oscillation accounts for the hardly relaxing silver background and hence we fixed $\lambda_2$=0. The remaining parameters $A_1$=0.040, $A_2$=0.016 and $\nu_2$=2.43 MHz (in the rotating reference frame) were kept constant over the investigated 20 mK-2.3 K temperature range.

## 4. Discussion

The results of all fits are reported in Fig. 4. We defined the shift as $K_i = (\nu_i - \nu_s)/\nu_s$ where $\nu_s$ is the rotation frequency in silver. As one can see in Fig. 4 a) $K_1$ is increasing below about 5 K and saturates at a value of $\sim$ 0.27 % below 100 mK. The lines in the main panel of Fig. 4 a) are fits of the shift to a pure $S$=1 Brillouin function where the only free parameter is the amplitude. Similarly, as shown in Fig. 4 b) the depolarization rate related to the muons referenced as 1 also saturates below 100 mK. This leads us to believe that they are mainly influenced by the vanadium $V^{3+}$ ions ($S$=1) which are in between the kagome layers and hence track their susceptibility. Due to the low value of the saturating shift, the muons are likely stopping in the material far from the $V^{3+}$, i.e. near the kagome planes. The shift at the second muon site can be fitted to a paramagnetic Curie-Weiss law with a small $\theta$=8(6) K, which reminds also the magnetic behavior

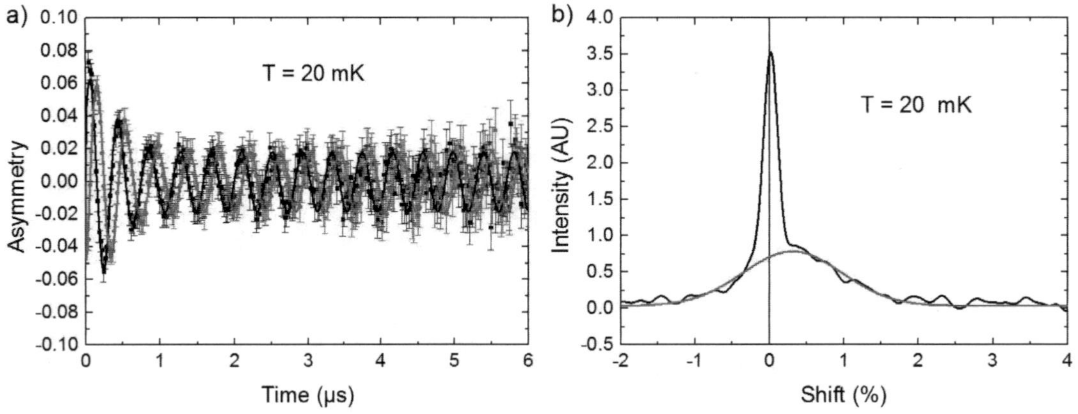


**Figure 3.** a) Time dependent signal at 20 mK with an applied transverse field of 0.5 T. The solid lines are fits to the data (see text). b) Fourier transform of the signal. The red line is a fit of the broad peak hidden under the silver narrow peak. The datas were taken on the LTF spectrometer

of the interlayer $V^{3+}$ ions. Therefore, the muons referenced as 2 seem also mainly influenced by the nearly free $V^{3+}$. The small coupling $\theta$ between the $V^{3+}$ interlayer ions is consistent with the coupling (~1 K) found in [10] by analyzing the specific heat and magnetization data. From the amplitude of the Curie like law in the inset of Fig. 4 a) we estimate the average distance from the muons corresponding to the second signal to the vanadium $V^{3+}$ to be $r_{max}$=2.4(1) Å. This locates them quite near the interlayer vanadium. Finally, those transverse field experiments

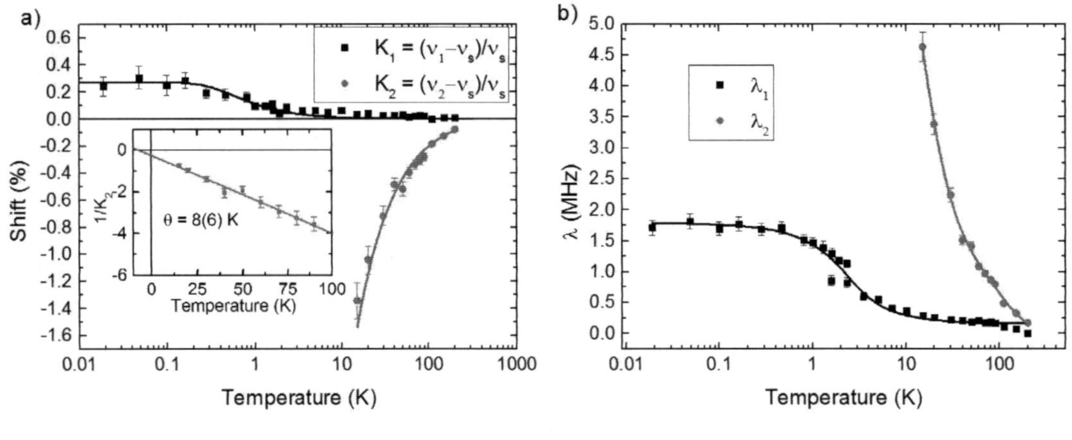


**Figure 4.** a) Shift versus temperature. $K_1$ relates to the narrow peak at high temperature and $K_2$ to the broad one. Inset : $1/K_2$ versus temperature. The solid lines are fits to the data (see text). b) Depolarization rate versus temperature. $\lambda_1$ is relative to the narrow peak at high temperature and $\lambda_2$ to the broad one. Solid lines are guides to the eyes.

suggest that the muons are experiencing a predominant influence from the interlayer vanadium ($S$=1) and allow the identification of two different muons sites that were not separated in the longitudinal field relaxation. Given the nearly similar values of $A_1$ and $A_2$, it is tempting to assign the corresponding signals to muons stopping near the two different triangles which constitute the kagome planes of DQVOF [5]. One kind of triangles is located above and below $V^{3+}$ ions as shown in Fig. 1 c). Muons stopping close to those triangles strongly feel the $V^{3+}$ dipolar field. The other kind of triangles made of $V^{4+}$ coupled by the $F2$ fluorine anions are not bridged by the $V^{3+}$ ions and the muons stopping nearby should feel a weaker dipolar field.

## 5. Conclusions

In summary, this $\mu$SR study shows that there is no spin freezing down to 20 mK in DQVOF and that the muons are preferentially implanted in two different sites in the material. First the longitudinal field experiments leave no doubt about the dynamical nature of the relaxation down to 20 mK. Furthermore, the low values of the fluctuating fields and frequencies at 20 mK and 2.3 K tend to prove that the dynamical behavior of the compound can not be simply understood in the paramagnetic limit of neither $V^{3+}$ nor $V^{4+}$ ions. It should arise from the existence of strong correlations in the low temperature phase of this highly frustrated compound. Second, the transverse field experiments clearly show that the signal consists of two main contributions. The analysis of the shifts suggests that they are related to the two different kind of triangles forming the kagome planes of DQVOF.

## Acknowledgments

This work was supported by the ANR-Grant No. ANR-SPINLIQ-86998. J C Orain acknowledges support from DIM OxyMore.

## References

[1] Anderson P W 1973 *Mater. Res. Bull.* **8** 153
[2] Balents L 2010 *Nature* **464** 199
[3] *Introduction to Frustrated Magnetism* ed Lacroix C, Mendels P and Mila F 2011 (Berlin: Springer-Verlag)
[4] Yan S, House D A and White S R 2011 *Science* **332** 1173
[5] Aidoudi F H, Aldous D W, Goff R J, Slawin A M Z, Attfield J P, Morris R E and Lightfoot P 2011 *Nat. Chem.* **3** 801
[6] Olariu A, Mendels P, Bert F, Trombe J C, de Vries M A and Harrison A 2008 *Phys. Rev. Lett.* **100** 087202
[7] Han T H, Helton J S, Chu S, Nocera D G, Rodriguez-Rivera J A, Broholm C and Lee Y S 2012 *Nature* **492** 406
[8] Fåk B *et al.* 2012 *Phys. Rev. Lett.* **109** 037208
[9] Zorko A, Bert F, Ozarowski A, van Tol J, Boldrin D, Wills A S and Mendels P 2013 *Phys. Rev.* B **88** 144419
[10] Clark L *et al.* 2013 *Phys. Rev. Lett.* **110** 207208

Journal of Physics: Conference Series **551** (2014) 012005 doi:10.1088/1742-6596/551/1/012005

# New insights for the description of magnetic correlations inferred from $\mu$SR

**P Dalmas de Réotier[1,2], A Yaouanc[1,2] and A Maisuradze[3,4,5]**

[1] Univ. Grenoble Alpes, INAC-SPSMS, F-38000 Grenoble, France
[2] CEA, INAC-SPSMS, F-38000 Grenoble, France
[3] Department of Physics, Tbilisi State University, Chavchavadze 3, GE-0128 Tbilisi, Georgia
[4] Physik-Institut der Universität Zürich, CH-8057 Zürich, Switzerland
[5] Laboratory for Muon-Spin Spectroscopy, Paul Scherrer Institute, CH-5232 Villigen-PSI, Switzerland

E-mail: pierre.dalmas-de-reotier@cea.fr

**Abstract.** Whenever a compound exhibits a spontaneous $\mu$SR oscillation, long-range magnetic ordering is usually inferred. Here we show that some caution is required. The coherence length needs not to be large for a spontaneous muon spin precession to be observed. Establishing the incommensurate nature of a magnetic structure, solely based on $\mu$SR measurements, may not be reliable. The absence of a spontaneous muon precession at low temperature does not mean that the system under investigation does not display long-range magnetic ordering.

The relaxation measured in zero and longitudinal field in the quasi-static limit is usually analyzed in the framework of the strong-collision model, the static polarization function being taken to be the famous Kubo-Toyabe function. This might not be satisfactory if short-range correlation effects are strong. Here we propose a method based on the maximum entropy concept and reverse Monte Carlo technique which gives results consistent with those obtained in 2013 by analytical means for the considered example.

## 1. Introduction

In this report we consider problems relevant for the analysis of experimental positive muon spectroscopy ($\mu$SR) data recorded for a magnetic compound. First, we discuss the meaning of an observed spontaneous muon precession in relation to the possible long-range nature of the magnetic order. Using published examples, we point out the difficulty of reliably establishing solely with $\mu$SR the incommensurate nature of the magnetic structure. In Sec. 4 we present a numerical method for the determination of the field distribution at the muon site for a compound characterized by quasi-static spin dynamics. As an example, we apply the method for the quantum spin-ice system $Yb_2Ti_2O_7$. A discussion of our results is presented in Sec. 5. An appendix details the reverse Monte Carlo algorithm.

## 2. Long- versus short-range magnetic order

A large fraction of the $\mu$SR studies are performed with the purpose of determining whether a magnetic material exhibits a magnetic transition. If a spontaneous $\mu$SR oscillation is detected, the material is usually proposed to be characterized by a long-range magnetic order. This may

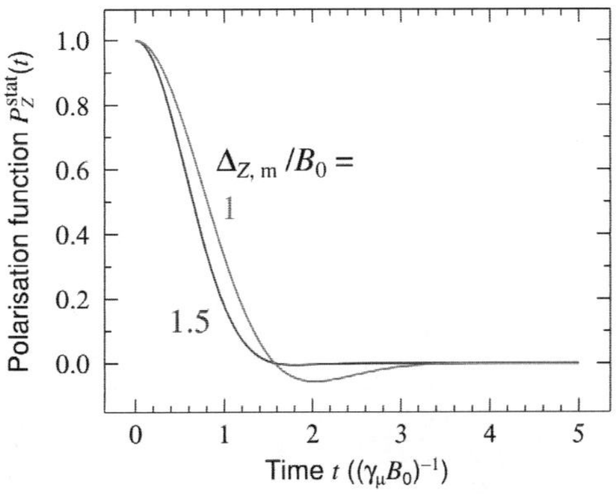


**Figure 1.** The polarization function for two values of the ratio of the standard deviation of the field at the muon site over the intensity of the spontaneous field; from Ref. [1] by permission of Oxford University Press.

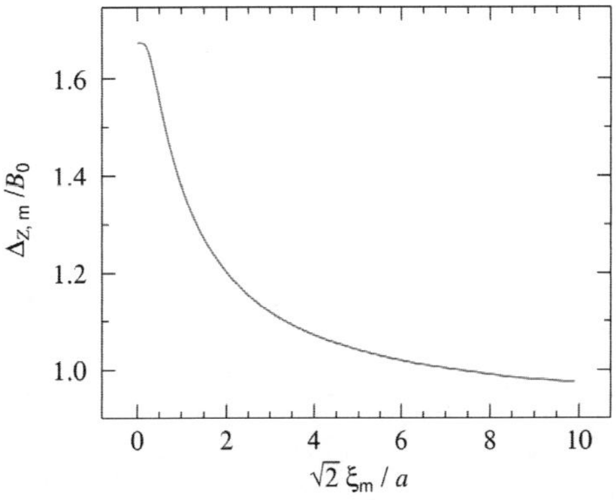


**Figure 2.** Ratio of the standard deviation of the field at the muon site over the intensity of the spontaneous field versus the correlation length; from Ref. [1] by permission of Oxford University Press.

not always be justified as it was shown theoretically in Ref. [1]. Let us consider the polarization function $P_Z^{\rm stat}(t) = \exp(-\gamma_\mu^2 \Delta_{Z,\rm m}^2 t^2/2)\cos(\gamma_\mu B_0 t)$. It describes a conventional Gaussian damped spontaneous oscillation. In Fig. 1 is plotted $P_Z^{\rm stat}(t)$ for two values of the ratio $\Delta_{Z,\rm m}/B_0$. From this plot we deduce that this ratio should be smaller than $\approx 1$ to conclude a spontaneous precession is detected. It is theoretically possible to express the ratio in terms of the correlation length of the magnetic structure $\xi_{\rm m}$. Assuming a ferromagnetic order in a face-centered-cubic compound with lattice parameter $a$ and the muon in the octahedral site, we compute $\Delta_{Z,\rm m}/B_0$ versus $\sqrt{2}\xi_{\rm m}/a$. The result is displayed in Fig. 2. A relatively short $\xi_{\rm m}$ of about ten interatomic distances is enough to give rise to a detectable spontaneous field. This result is qualitatively independent of the type of magnetic structure and the muon site. Are ten atomic distances long enough to call the magnetic structure long-range? In Fig. 3 we display a zero-field $\mu$SR spectrum recorded for the triangular system $NiGa_2S_4$. A spontaneous oscillation is clearly observed at 2.3 K. Yet, as shown in Fig. 4 the system only exhibits short-range correlations at low temperature. In fact, the lack of long-range magnetic order is the main interest of $NiGa_2S_4$.

To round up our discussion of the meaning of a damped oscillation for a zero-field $\mu$SR asymmetry, we point out that we have recently shown that such a signal may in fact reflect short-range correlations [4]. We recall that the Gaussian component field distributions as assumed by Kubo and Toyabe explicitly neglect them. A simple analytical method to account for them is to extend the component field distribution beyond the Gaussian function. This method successfully describes the spectrum recorded for the spin-ice system $Yb_2Ti_2O_7$ at low temperature [4].

### 3. Commensurate versus incommensurate magnetic order

An increasing number of observed spontaneous oscillations are found to be well accounted for by a zeroth-order Bessel function of the first kind, rather than a simple cosine. Usually from this observation it is inferred that the magnetic structure of the compound under study is incommensurate. However, the observation of such oscillations does not guarantee the inference to be correct. To support our argument, we cite two counter examples. A relatively complex yet commensurate structure was found in 1992 to yield a field distribution at a nuclear probe reminiscent of the one expected for an incommensurate modulated structure [5]. Recently, the triangular system $Ag_2NiO_2$ previously suggested to be incommensurate using zero-field $\mu$SR

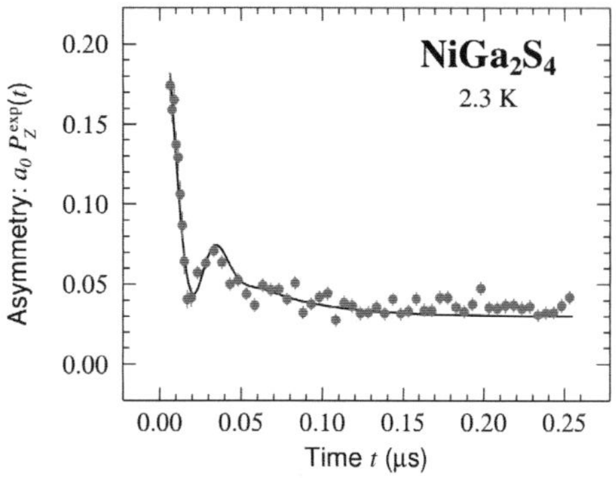


**Figure 3.** A zero-field $\mu$SR spectrum for a powder of $NiGa_2S_4$ taken at 2.3 K; adapted from Ref. [2]. The solid line results from a fit.

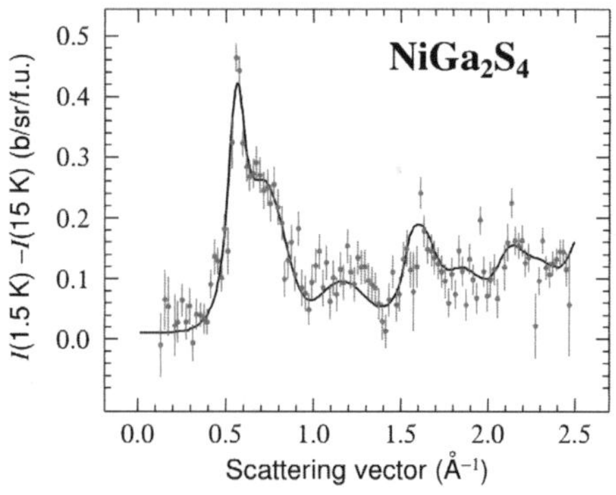


**Figure 4.** A magnetic neutron powder diffraction pattern for $NiGa_2S_4$ resulting from the difference between the patterns recorded at 1.5 and 15 K. The magnetic reflections are quite broad. They correspond to a correlation length $\xi_{\rm m} = 2.5\,(3)$ nm. The solid line results from a fit. Adapted from Ref. [3].

data was discovered to be commensurate by neutron diffraction [6, 7]. While the determination of a magnetic structure solely from $\mu$SR is questionable, it is pertinent to test the reality of a neutron-determined structure by zero-field $\mu$SR measurements. This is so because of the difficulty sometimes encountered to determine a magnetic structure from neutron diffraction data. In fact, a good practice is to record zero-field $\mu$SR and neutron diffraction data for the same sample and to extract the magnetic structure from a combined analysis of the data. We stress that information on $\xi_{\rm m}$ can be derived from neutron diffraction and that the field distribution at the muon site depends on it. This is further discussed in the next section.

## 4. Numerical determination of the field distribution at the muon site

Here we further consider the effect of short-range correlations on the longitudinal polarization function $P_Z(t)$. There is *a priori* no reason for limiting the description of the field distribution to the kurtosis as done recently in Ref. [4]. Higher terms could matter. Here we introduce a numerical method which does not have this limitation. At the end of this section it is applied to the analysis of the aforementioned $Yb_2Ti_2O_7$ spectrum.

Taking the vector field distribution at the muon sites, $D_{\rm v}(\mathbf{B}_{\rm loc})$, to be the product of the three Cartesian component field distributions,

$$D_{\rm v}(\mathbf{B}_{\rm loc}){\rm d}^3\mathbf{B}_{\rm loc} = D_{\rm c}(B^X_{\rm loc})D_{\rm c}(B^Y_{\rm loc})D_{\rm c}(B^Z_{\rm loc}){\rm d}B^X_{\rm loc}{\rm d}B^Y_{\rm loc}{\rm d}B^Z_{\rm loc}, \quad (1)$$

where $D_{\rm c}(B^\alpha_{\rm loc})$ is the component field distribution along the $\alpha$ direction, the zero-field static longitudinal polarization function can be written as follows:

$$\begin{aligned} P_Z^{\rm stat}(t) &= \int_{-\infty}^{\infty}\int_{-\infty}^{\infty}\int_{-\infty}^{\infty}\left\{\left(\frac{B^Z_{\rm loc}}{B_{\rm loc}}\right)^2 + \left[1-\left(\frac{B^Z_{\rm loc}}{B_{\rm loc}}\right)^2\right]\cos(\omega_\mu t)\right\} \\ &\times\ D_{\rm c}(B^X_{\rm loc})D_{\rm c}(B^Y_{\rm loc})D_{\rm c}(B^Z_{\rm loc}){\rm d}B^X_{\rm loc}{\rm d}B^Y_{\rm loc}{\rm d}B^Z_{\rm loc}, \end{aligned} \quad (2)$$

with $B^2_{\rm loc} = (B^X_{\rm loc})^2 + (B^Y_{\rm loc})^2 + (B^Z_{\rm loc})^2$ and $\omega_\mu = \gamma_\mu B$ ($\gamma_\mu = 2\pi \times 135.54$ MHz/T).

It is interesting to examine the $P_Z^{\rm stat}(t)$ expression. Let us replace $B_{\rm loc}^X$ by $-B_{\rm loc}^X$ in Eq. 2. The function $P_Z^{\rm stat}(t)$ does not change, but now it is expressed in terms of $D_{\rm c}(-B^X)$ rather than $D_{\rm c}(B^X)$. The same analysis applies for the $B_{\rm loc}^Y$ and $B_{\rm loc}^Z$ variables. Consequently, $P_Z^{\rm stat}(t)$ is conveniently written in terms of the symmetrized component field distribution $D_{\rm c}^{\rm sym}(B_{\rm loc}^\alpha) = \frac{1}{2}\left[D_{\rm c}(B_{\rm loc}^\alpha) + D_{\rm c}(-B_{\rm loc}^\alpha)\right]$.

It is easy to extend our scope to the case of a longitudinal-field measurement of the relaxation. By definition, an external field $\mathbf{B}_{\rm ext}$ is applied along the $Z$ direction. To account for it we only need to substitute $D_{\rm c}(B_{\rm loc}^Z)$ by $D_{\rm c}(B_{\rm loc}^Z - B_{\rm ext})$ in Eq. 2. Thus, we implicitly assume that the system is not modified by the external field. Then it must be noted that, contrary to the zero-field case, it is in principle possible to resolve an asymmetric component field distribution.

In fact, there is always some magnetic dynamics. Assuming the strong-collision model to be valid, $P_Z(t)$ can be expressed with an integral equation in terms of the fluctuation rate $\nu_{\rm c}$ and $P_Z^{\rm stat}(t)$:

$$P_Z(t) = P_Z^{\rm stat}(t)\exp(-\nu_{\rm c}t) + \nu_{\rm c}\int_0^t P_Z(t-t')P_Z^{\rm stat}(t')\exp(-\nu_{\rm c}t'){\rm d}t'. \tag{3}$$

To proceed further, as for the analytical analysis of $P_Z(t)$ given in Ref. [4], we shall assume the three Cartesian component field distributions to be identical. Thus we can write $P_Z(t) = P_Z(t, \nu_{\rm c}, \{D_{\rm c}(B_{\rm loc}^\alpha)\})$. Experimentally, the measured asymmetry is described as the sum of two components:

$$A(t) = a_0 P_Z^{\rm exp}(t) = a_{\rm s}P_Z(t) + a_{\rm bg}, \tag{4}$$

where $a_{\rm s}$ and $a_{\rm bg}$ are the initial asymmetry for the sample and background, respectively. We stress that experimentally there is always some background, i.e. some implanted muons miss the sample. Our experience shows that it should always be taken into account for a reliable analysis. For the numerical analysis we shall consider $N_B$ discrete field values $B_{{\rm loc},i}^\alpha$ in the finite field range $[B_{\rm min}, B_{\rm max}]$. To determine $D_{\rm c}(B_{\rm loc}^\alpha)$ we need to minimize the following chi-square:

$$\chi^2 = \sum_{i=1}^{N_{\rm t}} \frac{[d(t_i) - A(t_i)]^2}{\sigma_i^2}. \tag{5}$$

Here $N_{\rm t}$ is the number of discrete time channels of the measured asymmetry data, $d(t_i)$ the measured asymmetry at time $t_i$, and $\sigma_i$ the related standard deviation. For a reliable fit $\chi^2$ should be roughly equal to the number of degrees of freedom, i.e. $\chi^2 \simeq N_{\rm dof} = N_{\rm t} - N_{\rm p}$, where $N_{\rm p}$ is the number of free parameters. If we were to solely minimize $\chi^2$ the noise for $D_{\rm c}(B_{\rm loc}^\alpha)$ would be substantial. To proceed we shall use the maximum entropy (ME) concept, and therefore we shall minimize

$$F = \chi^2 - \lambda\mathcal{E}, \tag{6}$$

where $\lambda$ is a Lagrange parameter and $\mathcal{E}$ the entropy, i.e.

$$\mathcal{E} = -\sum_{i=1}^{N_{\rm B}} \left[D_{\rm c}(B_{{\rm loc},i}^\alpha)\delta B_{{\rm loc},i}^\alpha\right] \ln\left[D_{\rm c}(B_{{\rm loc},i}^\alpha)\delta B_{{\rm loc},i}^\alpha\right], \tag{7}$$

where $\delta B_{{\rm loc},i}^\alpha$ is the field distribution step which is taken uniform here. The basic idea is to minimize $\chi^2$ while maximizing $\mathcal{E}$.

As a first example of the use of the technique described above, we analyzed the $Yb_2Ti_2O_2$ spectrum already considered with the analytical method in Ref. [4]. There it was shown that

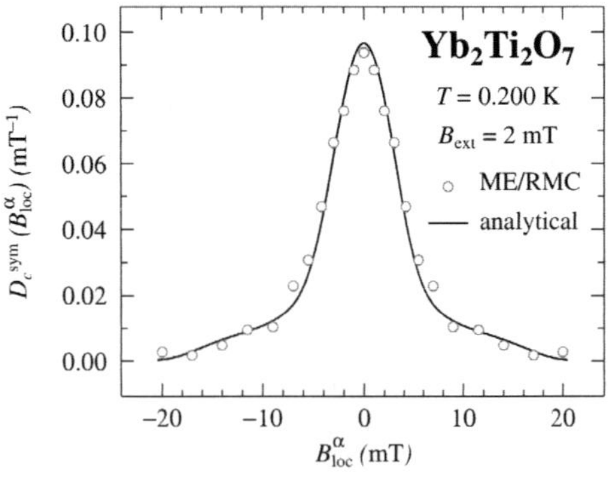


**Figure 5.** Comparison of the results from two methods for the determination of the symmetrized component field distribution at the muon site derived from measurements of the longitudinal asymmetry for a powder sample of $Yb_2Ti_2O_7$ [8] below the temperature at which the specific heat displays a sharp peak. While the result from the maximum entropy concept is obtained in this work, the analytical analysis was previously presented [4].

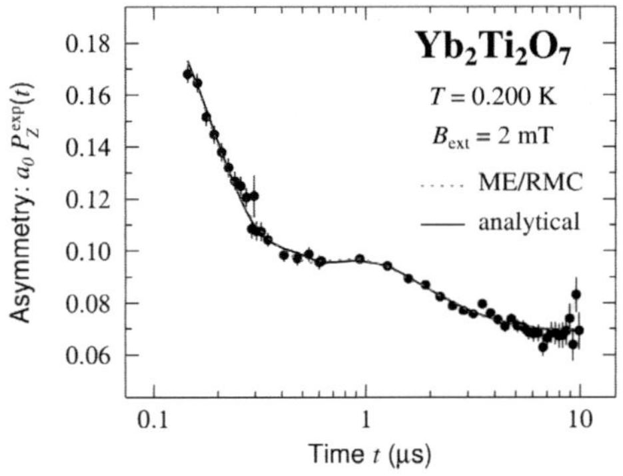


**Figure 6.** The measured longitudinal asymmetry $a_0 P_Z^{\rm exp}(t)$ displayed in a semi-logarithmic plot and results from its analysis in terms of the two symmetrized field distributions displayed in Fig. 5.

the small applied longitudinal field did not have any effect on the spectrum. So it will be analyzed as a zero-field spectrum. We take $B_{\rm max} = -B_{\rm min} = 20$ mT and $N_{\rm B} = 23$ points to describe the field distribution. Since a symmetrized distribution is considered, it counts only to 12 field points besides the $N_{\rm p} = 3$ parameters $a_{\rm s}$, $a_{\rm bg}$, and $\nu_{\rm c}$. Two methods were tested for the $F$ minimization: the reverse Monte Carlo (RMC) and the conventional Gauss-Newton (GN) methods. We found the first method better in terms of speed of convergence. In addition, it has the ability to converge to the global minimum, i.e. local minima are avoided. The RMC method is described in the appendix to this report. In Fig. 5 we display the symmetrized field distributions extracted from the spectrum shown in Fig. 6. It is rewarding that the previous analytical result [4] is consistent with the ME/RMC data. This provides a justification to limit the description of the field distribution to the kurtosis in this case. Obviously, the analytical analysis is performed much faster than the numerical one.

## 5. Discussion

The $\mu$SR method is recognized for its ability to easily detect magnetic transitions. However, one should not overinterpret the experimental data, such as proposing a long-range magnetic order when the features of the asymmetry does not allow to do it. Inferring a magnetic structure to be incommensurate with only $\mu$SR data at hand does not seem reasonable. To further complicate the state of affairs, the observation of the expected initial asymmetry without a spontaneous oscillation is not a proof of the absence of a magnetic order. The first recognized counter example was published for the ordered spin-ice $Tb_2Sn_2O_7$ in Ref. [9].

The $\mu$SR technique is well adapted for the study of quasi-static magnetic dynamics, i.e. with dynamics in the micro- and nanosecond time scale. Such dynamics is seen in exotic magnetic materials such as some heavy fermion and frustrated systems. The longitudinal relaxation is usually found to strongly deviate from the conventional exponential relaxation. The number of analytical formulas available for modeling the data is limited. We have introduced a numerical method to extract from the measured relaxation the field distribution at its origin. It supplements a recently proposed analytical method. The numerical method combines the best of two worlds: the maximum entropy concept first introduced by Rainford and Daniell in the $\mu$SR community [10] and the reverse Monte Carlo algorithm of common use for analyzing neutron diffraction data; see Ref. [11] and references therein. The numerical method proposed here can probably be extended to the analysis of spontaneous oscillations, at least in the case of only one muon magnetic site. The computation discussed here involves three-dimensional field integrations; see Eq. 2. If $B_0$ is sufficiently large relative to $\Delta_{\rm Z,m}$, the field integrations get restricted to one dimension [1]. Then the number $N_{\rm B}$ of discrete field values $B^{\alpha}_{{\rm loc},i}$ can be increased relative to the case considered here such that the asymmetry spectrum in a rotating frame is properly described, while keeping the computing time reasonable.

**Acknowledgments**
AY gratefully acknowledges Dr J. Sugiyama for a useful communication.

## Appendix A. The reverse Monte Carlo method

We write $F = F\left(a_0, a_{\rm bg}, \nu_{\rm c}, \left\{D_{\rm c}(B^{\alpha}_{\rm loc,i})\right\}\right) \equiv F(\mathbf{r})$, i.e. the set of parameters for $F$ — $a_0$, $a_{\rm bg}$, $\nu_{\rm c}$, $\left\{D_{\rm c}(B^{\alpha}_{\rm loc,i})\right\}$ — is denoted with the single vector $\mathbf{r}$ of dimension $N_r = N_{\rm p} + N_{\rm B}$.

Each iteration $j$ of the analysis consists of the following steps (see e.g. [11, 12]):
(1) A component $r_i$ of $\mathbf{r}$ is randomly chosen ($i = 1, 2, ..., N_r$).
(2) We change $r_i$ by a factor of $(1 + \mathcal{S}\epsilon)$, where, in our case, $\epsilon = 0.03$ (i.e. 3% of change) with the sign $\mathcal{S} = 1$ or $\mathcal{S} = -1$ randomly chosen. Thus, we obtain the new vector $\mathbf{r}_{\rm new}$.
(3) $F$ at iteration $j$, that is $F_j$, is computed using Eq. 6 for $\mathbf{r} = \mathbf{r}_{\rm new}$.
(4) If $\Delta F_j \equiv F_j - F_{j-1} < 0$ the iteration is successful and $\mathbf{r}_{\rm new}$ is accepted and stored ($\mathbf{r}_{\rm new} \to \mathbf{r}$). Otherwise $\mathbf{r}_{\rm new}$ is rejected and previous $\mathbf{r}$ is kept. The next iteration is repeated from step (1). There is an exception at this stage: for $\Delta F_j$ positive there is still a possibility to accept the iteration, and therefore $\mathbf{r}_{\rm new}$, with the probability $\mathcal{P} = \exp(-\Delta F_j/p)$. The parameter $p$ is chosen such that about half of the iterations are accepted [11]. We found $p = 0.003$ to be a proper choice. The $p$ value depends on $\Delta F_j$ and consequently on $\epsilon$. This mechanism of accepting the iteration with a positive $\Delta F_j$ turned out to be important for escaping local minima of $F(\mathbf{r})$. The larger the number of independent parameters $N_r$ the higher the probability to converge to a local minimum.

This algorithm depends on several parameters: $\epsilon$, $p$, and $\lambda$. The larger $\epsilon$ the faster the convergence at the early stage of iterations. However, one should consider that $\epsilon$ determines also the accuracy of the fitted parameters. As a variation of this algorithm one can gradually reduce $\epsilon$ with increasing iteration number $j$. The parameter $\epsilon$ is related to $\Delta F_j$ and consequently to the parameter $p$, determining the probability to accept an iteration with a positive $\Delta F_j$. Obviously the larger $\epsilon$ the larger the absolute value of $\Delta F_j$. In our implementation of the algorithm we analyze the average number of accepted and rejected iterations within the last 10 iterations and we increase or reduce parameter $p$ to reach the 50% goal for accepted iterations. The value of $\lambda$ determines the weight of entropy in the function $F(\mathbf{r})$. Entropy determines the probability of a distribution, while $\chi^2$ determines the fit quality. $\chi^2$ is a statistical quantity with an expectation value of $N_{\rm dof}$ and standard deviation $\sqrt{N_{\rm dof}}$. Consequently, for a good fit $N_{\rm dof} - \sqrt{N_{\rm dof}} \lesssim \chi^2 \lesssim N_{\rm dof} + \sqrt{N_{\rm dof}}$. Note that the number of free parameters used for the evaluation of $N_{\rm dof}$ is $N_{\rm p}$ i.e. the parameters entering the entropy calculation are excluded. If $\chi^2$ is too small one should increase $\lambda$ and conversely for a large $\chi^2$, $\lambda$ should be reduced. The result does not change significantly with a reasonable variation of $\lambda$ for $\chi^2 \simeq N_{\rm dof}$. The particular value of $\lambda$ must be adapted according to the number of field points $N_B$. To ensure the reliability of the result it is important to check that the condition $\chi^2 \simeq N_{\rm dof}$ is satisfied. The value of $N_B$ can be arbitrarily chosen, however one should consider that the computation time increases as $N_B^3$. A minimum number of points in the distribution is obviously required.

In the case of our spectrum, a stable solution was reached after about 500 successful RMC iterations. Its stability was checked by extending the computation up to 2000 successful RMC iterations. The computation was started with an unbiased distribution. i.e. assuming a flat distribution. Note that despite the large number of iterations required to converge, $F(\mathbf{r})$ is evaluated only once per iteration. Thus the convergence rate is quite fast considering the large number of parameters $r_i$.

Obviously, other than RMC methods may be used for a nonlinear minimization such as that of $F$. Note that the linear optimization method suggested for transverse field $\mu$SR analysis in Ref. [10] does not apply in present case. The relation between time and field domains given by Eq. 2 is highly nonlinear and consequently iterative nonlinear minimization algorithms should be used. The obvious merits of RMC are its simplicity, a reasonable computation efficiency, and most importantly, possibility to avoid local $\chi^2$ minima in parameter space due to the mechanism of accepting also the iterations with reduced fit quality. The presented method can be very useful in analysis of $\mu$SR data of correlated and magnetically ordered systems in zero or longitudinal field.

## References

[1] Yaouanc A and Dalmas de Réotier P 2011 *Muon Spin Rotation, Relaxation, and Resonance: Applications to Condensed Matter* International Series of Monographs on Physics 147 (Oxford: Oxford University Press)
[2] Yaouanc A, Dalmas de Réotier P, Chapuis Y, Marin C, Lapertot G, Cervellino A and Amato A 2008 *Phys. Rev.* B **77** 092403
[3] Nakatsuji S, Nambu Y, Tonomura H, Sakai O, Jonas S, Broholm C, Tsunetsugu H, Qiu Y and Maeno Y 2005 *Science* **309** 1697–1700
[4] Yaouanc A, Maisuradze A and Dalmas de Réotier P 2013 *Phys. Rev.* B **87** 134405
[5] Chiba M, Yoshitami A, Kikuchi H, Maegawa S and Morimoto T 1992 *J. Phys. Soc. Jpn.* **61** 1758
[6] Sugiyama J, Ikedo Y, Mukai K, Brewer J H, Ansaldo E J, Morris G D, Chow K H, Yoshida H and Hiroi Z 2006 *Phys. Rev.* B **73** 224437
[7] Nozaki H, Roessli M M B, Pomjakushin V, Kamazawa K, Ikedo Y, Fischer H E, Hansen T C, Yoshida H, Hiroi Z and Sugiyama J 2013 *J. Phys.: Condens. Matter* **25** 286005
[8] Hodges J A, Bonville P, Forget A, Yaouanc A, Dalmas de Réotier P, André G, Rams M, Królas K, Ritter C, Gubbens P C M, Kaiser C T, King P J C and Baines C 2002 *Phys. Rev. Lett.* **88** 077204
[9] Dalmas de Réotier P, Yaouanc A, Keller L, Cervellino A, Roessli B, Baines C, Forget A, Vaju C, Gubbens P C M, Amato A and King P J C 2006 *Phys. Rev. Lett.* **96** 127202
[10] Rainford B D and Daniell G J 1994 *Hyperfine Interactions* **87** 1129–1134
[11] Laver M, Forgan E M, Abrahansen A B, Bowell C, Geue T and Cubitt R 2008 *Phys. Rev. Lett.* **100** 107001
[12] McGreevy R L 2001 *J. Phys.: Condens. Matter* **13** R877

# Spin Depolarization of Muonium in Mesoporous Silica

**M H Dehn[1,†], D J Arseneau[2], M D Bridges[3], T Buck[4], D L Cortie[2,4], S P Cottrell[5], D G Fleming[2,6], J A Kelly[6], W A MacFarlane[2,6], M J MacLachlan[6], G D Morris[2], I McKenzie[2], J Xiao[6] and R F Kiefl[2,4,‡]**

[1]Physik-Department, Technische Universität München, James-Franck-Strasse 1, 85748 Garching bei München, Germany
[2]TRIUMF, 4004 Wesbrook Mall, Vancouver, BC V6T 2A3, Canada
[3]Department of Chemistry and Biochemistry, California State University, Fullerton, CA 92831, USA
[4] Department of Physics and Astronomy, University of British Columbia, Vancouver, BC, V6T 1Z1, Canada
[5] STFC Rutherford Appleton Laboratory, Harwell Oxford, Didcot OX11 0QX, UK
[6] Department of Chemistry, University of British Columbia, Vancouver, BC, V6T 1Z1, Canada

E-mail: [†]martin.dehn@tum.de [‡]kiefl@triumf.ca

**Abstract.** We report muon spin rotation/relaxation measurements of muonium in mesoporous silica (SBA-15) with a high specific surface area of 600 $m^2/g$. Up to 70 percent of the incoming muons form muonium and escape efficiently into the open pores at all temperatures between 3 and 300K. We present evidence that the interaction with the silica surfaces involves both spin exchange and a transition to a diamagnetic state, possibly due to dangling bonds on the surface. At very low temperatures, below 20K, the interaction between muonium and the silica surfaces is suppressed due to a He film coating the surfaces. These results indicate that it should be possible to use muonium to probe the surfaces of uncapped nanoparticles supported in silica.

## 1. Introduction

Finite size can have a profound effect on the properties of a material. On general grounds one expects that any surface will have different structural, electronic and magnetic properties compared to the bulk as a result of the broken translational symmetry and in some special cases topology of the electronic structure (e.g. topological insulators). One important experimental consideration, in order to study the intrinsic finite size effects, is to isolate nanoparticles from each other. One possibility is to disperse the nanoparticles on the surface of a highly porous inert medium such as silica. Materials like silica are well known for their performance as high surface area supports for catalysts, such as Au nanoparticles, which can be decorated uniformly inside the porous silica [1]. This approach has the advantage that it can easily be extended to study uncapped nanoparticles of other materials. Although the muon is primarily known as a bulk probe of matter the high specific surface area of nanoparticles means that the muon can in special circumstances be used to probe surfaces. However it is important to first understand the behavior of $\mu^+$ and muonium in pure silica, before studies of embedded nanoparticles can begin.

In this paper we report measurements of the transverse and longitudinal spin relaxation rate of muonium in mesoporous silica (SBA-15). We find that a large fraction of up to 70% of the muons form muonium and reach the open pore region [2, 3, 4, 5, 6, 7, 8]. This is comparable with previous experiments on Mu in $SiO_2$ both in bulk [3] and thin films [8]. Surprisingly this holds true even at low temperatures, indicating that the process for muonium to reach the pores is epithermal. We also observe slow depolarization of muonium attributed to muonium scattering off the silica surfaces. There is evidence that the collisions involve both spin exchange and a transition to a diamagnetic state. We propose the active sites are dangling bonds on the surface (e.g. Si-O- ) which have an unpaired spin but can also bind the muonium in a diamagnetic hydroxyl-like group Si-O-Mu. Finally at the lowest temperatures we observe that the interaction of muonium with the silica surfaces is suppressed in a He atmosphere due to coating of the surfaces with He. These results indicate that it may be possible to use muonium as a probe of the surfaces of nanoparticles supported in silica.

## 2. Experimental

The mesoporous silica (SBA-15) was chosen for this experiment because of its relatively stable channel structure compared to other mesoporous silicas. The sample was made at the University of British Columbia using a hydro-thermal method developed by Somorjai *et al* [9]. More details are given in a separate publication presented at this meeting (see J. Xiao et al). The end product is a white fluffy powder. The specific surface area measured with the Brunauer-Emmett-Teller (BET) method is 600 $m^2/g$ and the average pore diameter is about $\approx$ 10nm . Non-magnetic Al sample cells with a Ag backing plate were loaded with the powder and sealed with a Kapton window using a gold O-ring. The cell was heated and pumped for 24 hours to remove moisture and oxygen. The pumping outlet tube was then crimped to close the cell, which was subsequently stored in an evacuated desiccator until the experiment. At the beginning of the experiment the cell was then loaded into a He horizontal gas flow cryostat mounted on the SFUMU spectrometer on the M15 beamline at TRIUMF.

All measurements were taken in a spin rotated mode in which the initial muon spin polarization was oriented close to the vertical direction ($\hat{x}$) and perpendicular to the beam direction ($\hat{z}$). Two types of measurements were made as a function of temperature (1) $1/T_2$ spin depolarization rate of the muonium precession in small transverse field (TF) of 7G applied along $\hat{z}$ and (2) $1/T_1$ spin relaxation of the muonium in longitudinal field of 28G applied along $\hat{x}$. Typical asymmetry plots of raw data are shown in Figs. 1 and 3 respectively. Additional TF measurements were also made for detector calibration purposes. After measurements were taken on the closed cell it was removed from the cryostat and the outlet tube was cut to open the cell to the outside. Precautions were taken to minimize any exposure to moisture in the air. The open sample cell was reinserted into the cryostat and the sample space was flushed with warm He gas and pumped down to $\approx$ 0.015 mbar to remove any residual oxygen that may have leaked into the cell after it was crimped. A significant decrease in the muonium $1/T_2$ was observed at room temperature as may be seen by comparing Figs. 1a and 1b. This is strong indication that the closed cell had a small partial pressure of $O_2$. We estimate that 1.5 mbar of $O_2$ [3] at 300K would be sufficient to cause the additional $1/T_2$ relaxation in Fig. 1a compared to 1b. A second set of measurements of $1/T_1$ and $1/T_2$ were made on the open/pumped cell for comparison. In this case the powder was in direct contact with He gas of the flow cryostat, except at 300K where the sample space was evacuated. The He pressure is very low near room temperature but increases gradually at lower temperatures and reaches a few hundred mbar at 3K.

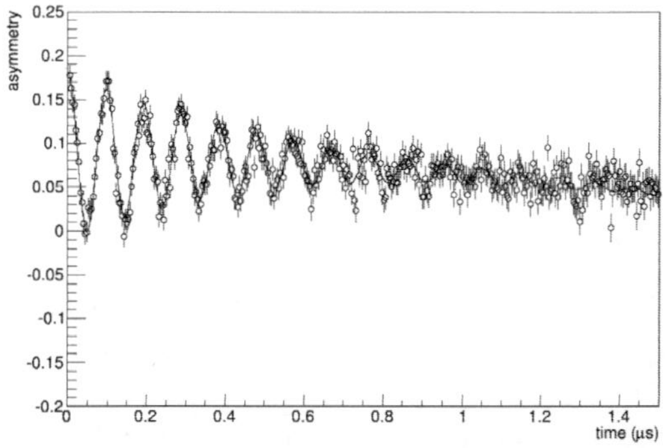


(a) $SiO_2$ closed cell at 293K

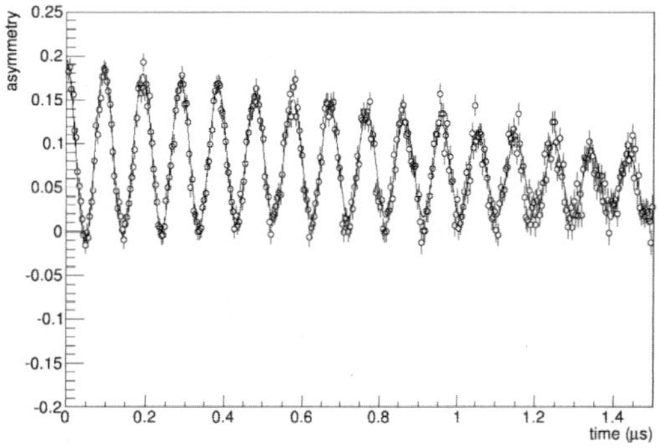


(b) $SiO_2$ open and pumped cell at 293K

**Figure 1.** (a) Muonium precession signal in a weak transverse field of 7G at room temperature in porous silica (SBA-15)in a closed cell (b) same conditions as (a) except the cell has been evacuated. Note the decrease in relaxation rate which is attributed to residual amounts of $O_2$ trapped in the closed cell.

## 3. Results and Discussion

### *3.1.* $1/T_2$ *Spin Depolarization of Muonium*

Figure 1 shows the TF muonium signal in SBA-15 silica at room temperature both in a closed cell (a) and and open cell environments (b). Note the $1/T_2$ relaxation rate in the closed cell is much larger than in the open cell. From this we conclude there was a small partial pressure of $O_2$ in the closed cell which causes spin exchange with muonium. This also demonstrates that the muonium reaches the open pores since that is the only place the oxygen gas can be. It would be interesting to compare our results with the rates measured by Himmer et al. for molecular O2 in the gas phase [10]. This will require further measurements with a controlled amount of O2 admitted into the cell. This will be subject of future studies in a controlled environment. The temperature dependence of $1/T_2$ in both the open and closed cells is shown in Fig. 2 (a). Note the difference between the open and closed cells is largest at room temperature where the oxygen in the closed cell is all in the gas phase. As the temperature is reduced the difference becomes smaller, indicating the effect of the oxygen is less pronounced, so that at 50K there is almost no difference. This is evidence that oxygen on the surface of the silica is much less effective in causing spin exchange. It is likely that the oxygen adsorbed on the surface dimerizes and is not paramagnetic. The two curves separate again below 20K. In particular there is a sudden drop in $1/T_2$ seen in the open cell not seen in the closed cell. We attribute this to the He atmosphere in the open cell since one expects He to form a film on the silica surfaces which shields muonium from the depolarizing collisions with the silica. A similar effect was observed in fine oxide powders [2]. In some silica powders the Mu formation probability decreases when a film of He or other gas is coating the surface [7]. However, no such decrease was observed in SBA-15 at least for He.

### *3.2.* $1/T_1$ *Spin Relaxation of Muonium*

Typical asymmetry plots showing the $1/T_1$ spin relaxation rate of muonium in a longitudinal field of 28 G are shown in Fig 3. Note there is much faster relaxation near room temperature in the closed cell. This is consistent with spin exchange from $O_2$ in the closed cell. However there is still observable $1/T_1$ spin relaxation in the open cell where all the oxygen has been removed. This is attributed to spin exchange with unpaired electronic spins on the silica surface, most likely Si-O- dangling bonds.

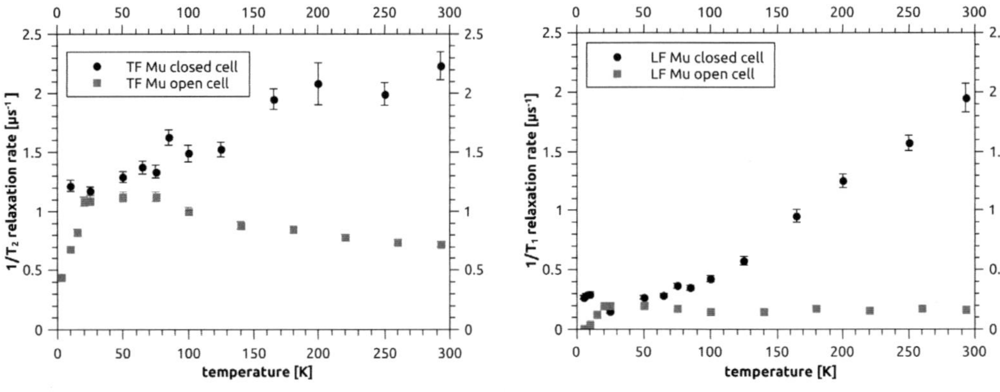


(a)

(b)

**Figure 2.** (a) $1/\mathrm{T}_2$ spin relaxation rate of Mu in a small transverse field of 7G versus temperature.Black points are for the closed cell and red points are the open cell. (b) $1/\mathrm{T}_1$ spin relaxation rate of Mu in a longitudinal field of 28G versus temperature. Note both $1/\mathrm{T}_1$ and $1/\mathrm{T}_2$ decrease sharply below 20K in the open cell due to the coating of the surfaces with a thin film of He.

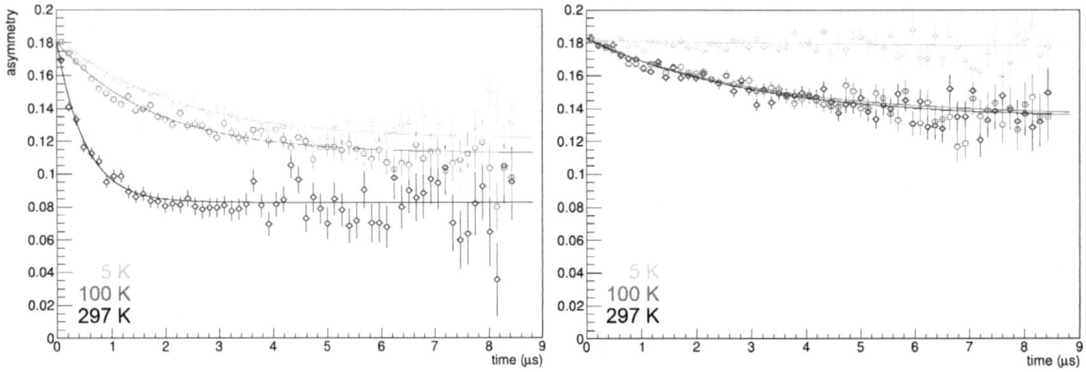


(a) $SiO_2$ closed cell

(b) $SiO_2$ open cell

**Figure 3.** LF spectra in a field of 28G at various temperatures for both samples

The data fit well to a two component transition model over the whole temperature range in which there is a relaxing term and non relaxing term. As mentioned above the relaxation is due to spin exchange from oxygen and unpaired spins on the silica surface, whereas the non relaxing component is due to a transition to a diamagnetic muon state. A simple transition model yields the following form:

$$P_z(t) = A_{\mu^+} + A_{Mu} \cdot \left\{ \frac{\nu}{\lambda+\nu} + \left(1 - \frac{\nu}{\lambda+\nu}\right) \exp(-(\lambda+\nu)\cdot t) \right\}, \quad (1)$$

where $\lambda = 1/T_1$ is the relaxation rate from spin exchange and $\nu$ is the reaction rate describing the transformation from Mu to $\mu^+$. $A_{Mu}$ describes the muons that form Mu when implanted into the sample. $A_{\mu^+}$, the fraction of the muons that stay diamagnetic, was held constant in the fit based on TF results. From the fitted values at 7G TF (Fig. 1) we obtain an average relative

Mu fraction $F_{Mu} = 2A_{Mu}/(2A_{Mu} + A_{\mu^+})$ of 0.67±0.02% for the closed cell and 0.69±0.02% for the open cell. The fitted value for the reaction rate ranges between 0.1 and 0.3 $\mu s^{-1}$ and varies weakly with temperature. The fitted values for $1/T_1$ are shown in Fig. 2. Note at 293K $1/T_1$ is much larger in the closed cell (with oxygen) than in the open cell (without oxygen). This is consistent with a large spin exchange rate from $O_2$ in the pores at room temperature in the closed cell. Also the difference between the open and closed cell diminishes with decreasing temperature as in the case of $1/T_2$. This is confirmation that the spin exchange rate for oxygen on the silica surface is much less than in the gas phase. Finally we note there is a dramatic drop in $1/T_1$ in the open cell below 20K which is not seen in the closed cell. This also confirms our interpretation that a He film shields the muonium from spin depolarizing interactions with the silica surface.

## 4. Summary and Conclusions

We have measured the spin depolarization of muonium in mesoporous silica (SBA-15) with a high specific surface area in the presence of oxygen and helium. A large fraction of the incoming muons form muonium and escape efficiently into the open pores at all temperatures down to 3K. We have identified two distinct processes for the way muonium interacts with the silica surface. The muonium may spin exchange with unpaired spins on the surface but there is also evidence for a transition to a diamagnetic state. The active sites are most likely dangling bonds which have an unpaired electronic spin but may also bind the muonium in a diamagnetic state. At very low temperatures, below 20K, the interaction between muonium and the silica surfaces can be suppressed in a He atmosphere due to the formation of a thin He film. These results suggest it may be possible to land polarized muons efficiently to the surface of nanoparticles supported in silica.

## Acknowledgments

We would like to thank Bassam Hitti and Deepak Vyas for their excellent technical support. All data was analyzed using the MUSRFIT package [11].

## References

[1] Kantcheva M and Sayan S 1999 *Catalysis Letters* **60** 27–38
[2] Kiefl R F, Warren J B, Oram C J, Marshall G M, Brewer J H, Harshman D R and Clawson C W 1982 *Phys. Rev.* **B26** 2432–2441
[3] Marshall G M, Warren J B, Garner D M, Clark G S, Brewer J H and Fleming D G 1978 *Phys, Lett.* A **65** 351
[4] Harshman D R, Keitel R, Kiefl R F, Senba M, Ansaldo E J, and Brewer J H 1984 *Phys. Lett.* A **104** 472–476
[5] Marzke R F , Glaunsinger W.S. , Harshman D R, Ansaldo E J, Keitel R, Senba M, Noakes D R, Spencer D P, Brewer J H 1985 *Chem. Phys. Lett.* **120** 6
[6] Harshman D R, (1986) *Hyperfine Interact.* **32** 847
[7] van der Werf D P, Donnelly P A, Charlton M, Cottrell S P and Cox S F J 2006 *Physica B* **374–375** 355–358
[8] Antognini A, Crivelli P, Prokscha T, Khaw K S, Barbiellini B, Liszkay L, Kirch K, Kwuida K, Morenzoni E, Piegsa F M, Salman Z and Suter A 2012 *Phys. Rev. Lett.* **108** 142401
[9] Song H, Rioux R M, Hoefelmeyer J D, Komor R, Niesz K, Grass M, Yang P and Somorjai G A 2006 *J. Am. Chem. Soc.* **128** 3027–37
[10] Himmer U, Dilger H, Roduner E, Pan, J J, Arseneau D J, Fleming D G and Senba M 1991 *J. Phys. Chem. A* **103** 2076
[11] Suter A and Wojek B M (2012) *Physics Procedia* **30** 69

# $\mu$SR study on quantum spin system $NH_4CuCl_3$

**K Matsui[1], A Oosawa[1], K Yoshizawa[1], T Goto[1], I Watanabe[2], T Suzuki[3], M Fujisawa[4], H Tanaka[4], P K Biswas[5] and A Amato[5]**

[1] Physics Division, Sophia University, Chiyodaku, Tokyo 102-8554, Japan
[2] Advanced Meson Science Laboratory, RIKEN Nishina Center for Accelerator-Based Science, 2-1 Hirosawa, Wako, Saitama 351-0198, Japan
[3] Faculty of Engineering, Shibaura Institute of Technology, Saitama 337-8570, Japan
[4] Department of Physics, Tokyo Institute of Technology, Meguro-ku, Tokyo 152-8551, Japan
[5] Laboratory for Muon Spin Spectroscopy, Paul Scherrer Institut, CH-5232 Villigen, Switzerland

E-mail: `k703861@sophia.ac.jp`

**Abstract.** The quantum spin dimer system $NH_4CuCl_3$ has a gapless ground state with $T_N = 1.3$ K, and shows plateaus in the high-field magnetization at 1/4 and 3/4 of the saturation magnetization, the mechanism of which has not yet been resolved until now. The three dimer model recently proposed by Matsumoto, which seems well accounting for the emergence of plateaus, predicts the existence of three magnetically-inequivalent dimers. In order to test the model, we have measured transverse-field (TF) muon depolarization curves. In fourier transform of TF muon depolarization curves, we observed homogeneous single peak at $T > 100$ K. Below 70 K, the peak showed rapid broadening, and split into multiple peaks in still lower temperatures, supporting the idea of the three dimer subsystems.

## 1. Introduction

Quantum spin systems have been attracting much interests in last two decades both experimentally and theoretically, because these systems exhibit numerous peculiar magnetic features, as represented by the triplon Bose-Einstein Condensation (BEC) in spin gap systems[1, 2] and the magnetization plateaus[3], which cannot be interpreted by conventional classical spin vector models. The former is observed as a field-induced magnetic ordering in the spin gap systems such as $KCuCl_3$ and $TlCuCl_3$[4, 5, 6], which are isomorphous of the title compound $NH_4CuCl_3$[7, 5, 8]. The field-induced magnetic ordering was captured as the Bose-Einstein condensation of triplons by mapping the spin gap system to the system consisting of boson with magnetic moment, triplon[2]. Namely the ordering can be interpreted as the superfluid-insulator transition induced by varying the chemical potential $\mu$ corresponding to the magnetic field $H$ in the original spin system. The superfluid-insulator transition occurs when the hopping of triplon $t$ is more dominant than the repulsive interaction of triplons $U$, and hence the triplons are mobile[9].

On the other hand, when $U$ is more dominant than $t$, the created triplons prefer to distance from one another so that a superlattice of triplons, which can be interpreted as the formation of the Wigner crystal of triplons, may appear at a certain fractional number of triplons. In the superlattice phase, a finite energy is needed to create an additional triplon, *i.e.* $dn/d\mu = 0$, where $n$ is the total triplon density corresponding to the magnetization $M$ in the original spin system so that the magnetization plateau emerges in the superlattice phase. Actually, such superlattice of triplons has been observed in the orthogonal dimer system $SrCu_2(BO_3)_2$[10] at the 1/8-plateau phase[11, 12]. However, the case is not directly applied to $NH_4CuCl_3$ system, which basically has the same crystal structure as those BEC-systems of $KCuCl_3$ and $TlCuCl_3$; its only major difference is the existence of non-centrosymmetric molecule $NH_4$[7]. At room temperatures, where $NH_4$ molecules are rapidly spinning, the space group belongs to $P2_1/c$, the same as that for $KCuCl_3$ and $TlCuCl_3$. With lowering temperature as reported recently by the elastic neutron experiment, the $NH_4$ system exhibits the two structural phase transitions at $T_1 \approx 156$ K and at $T_2 \approx 70$ K. At $T = T_1$, crystal structure changes into $P\bar{1}$. At $T < T_2$, the unit cell size along the $b$ direction is doubled[13]. These phase transitions are believed to be concerned with a freezing of rotational motions of $NH_4$ molecules.

Based on these experimental results, Matsumoto[14] has recently proposed a completely different theoretical model, which assumed three distinct magnetic dimer sublattices and successfully reproduced the magnetization plateaus as well as the field dependence of the resonance frequencies of ESR[15, 16]. He also proposed a possible geometrical configuration of field-induced triplons, which however was not in accordance with the recent NMR results[17]. The motivation of this paper is to test the model by investigating whether or not there exist magnetically-inequivalent dimers in the paramagnetic state by $\mu$SR technique, which is a powerful tool for the detection of microscopic inhomogeneity, both static or dynamic[18, 19, 20].

Before showing experimental procedures, we briefly describe the structural and magnetic properties of $NH_4CuCl_3$. The crystal structure shown schematically in Fig. 1 is composed of planar dimers of $Cu_2Cl_6$, in which $Cu^{2+}$ ions have spin-1/2. The dimers are stacked on top of one another to form infinite double chains parallel to the crystallographic $a$-axis. These double chains are located at the corners and centre of the unit cell in the $b$-$c$ plane, and are separated by $NH_4^+$ molecules. Under zero field, $NH_4CuCl_3$ is a gapless antiferromagnet with $T_N = 1.3$ K[13]. It shows in the magnetization process at one-quarter and three-quarters of the saturation magnetization[8], *i.e.*, for $H||a$, these plateaus are observed in 5.0 T $< H <$ 12.8 T, and 17.9 T $< H <$ 24.7 T, respectively, and the magnetization saturates at 29.1 T. The magnetization plateaus are observed irrespective of the field direction, suggesting that the origin of the plateau can be attributed to quantum effect.

## 2. Experimental details

Single crystals of $NH_4CuCl_3$ were made by evaporation method[6]. They were distorted hexagonal rod-shaped with a typical size of 100 mm$^3$. Transverse-field (TF) muon spin depolarization was measured on the DOLLY instrument at Paul Scherrer Institute (PSI) utilizing positive muons with an energy of 4.1 MeV implanted into the samples. The angle between muon spin polarization and muon-beam direction is set as 45 degrees by the spin rotator. About 10 crystals were mounted on the silver plate so as that $b$-axis was parallel to the muon-beam direction. An external magnetic field of 0.49 T was applied parallel to the sample $b$-axis. All TF experiments were carried out after a field-cooling procedure. A conventional $^4$He flowing cryostat was used for temperature control for the temperature range between 2.3 to 140 K. Each relaxation curve contains approximately 10000 data bin with a width of 0.977 ns. Fast fourier transform was applied to 16384 data extended by zeros-filling, which enabled us to obtain the frequency resolution of 0.0625 MHz. The typical statistical average for each curve was 50 Mev.

## 3. Results and discussion

Observed TF muon depolarization curves are shown in Fig. 2, where one can clearly see the muon spin rotation with a frequency approximately 66.4 MHz in the TF of $H_0 = 0.49$ T. The amplitude of oscillation decays monotonically with time. The decay becomes faster with lowering temperatures. In fourier transform of TF muon depolarization curves shown in Fig. 3,

Journal of Physics: Conference Series **551** (2014) 012007 doi:10.1088/1742-6596/551/1/012007

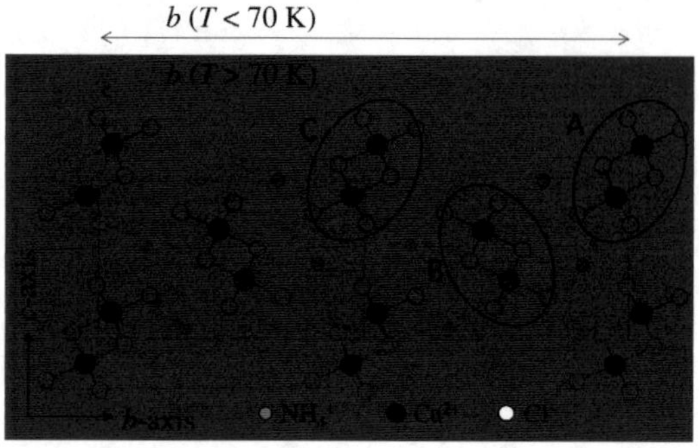


**Figure 1.** The unit cell of $NH_4CuCl_3$ in the *bc* plane below 70 K. A-, B- and C-dimers denote the magnetic dimer structure of $Cu^{2+}$ ions proposed by Matsumoto et al[14]. The temperature dependent *b*-axis is shown by horizontal arrows[13].

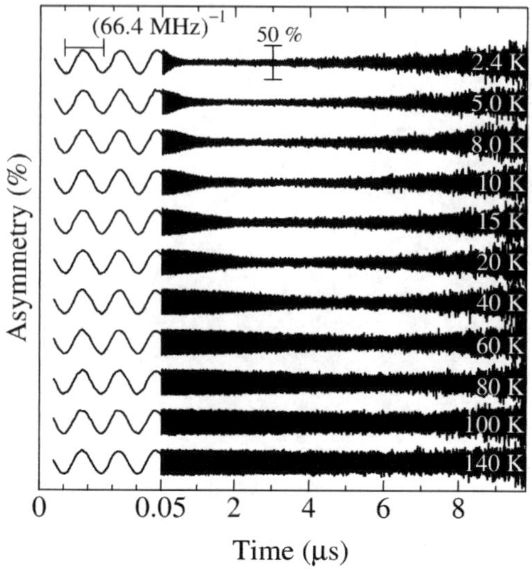


**Figure 2.** Transverse-field (TF) $\mu$SR time spectra obtained under $H = 0.49$ T at various temperatures.

we observed homogeneous single peak with a narrow width of nearly the resolution limit in high temperature region $T > 100$ K. With decreasing temperature, it shows a gradual broadening below 70 K and bears three peaks overlapping with each other as can be seen by the Gaussian deconvolution shown in Fig. 3. With further decreasing temperature, the distance between these three peaks increases, and at around 10 K, still other peaks emerge. At 2.4 K, the entire width reaches 4 MHz, which corresponds to 0.029 T of the hyperfine field. We plotted the peak positions against the temperature in Fig. 4.

The temperature where the peak starts to split is around 70 K just corresponds to the structural phase transition reported by the elastic neutron experiments[13]. Below this temperature, and above 20 K, muons sense three different hyperfine fields, the fact of which indicates two possiblities. One is that there are three muons' stopping sites in the elongated unit cell show in Fig. 1, and that each site has a different hyperfine field. In this case, observed peak split shows a direct evidence for the existence of three magnetically-inequivalent dimers in the unit cell. These dimers are expected to show a magnetic order with many sublattices. Actually, the recent zero-field (ZF) $\mu$SR in the $^3$He temperature range reports a muon rotation spectrum with multiple components below $T_N = 1.3$ K[13, 21].

On the other hand, however, if there is only one muons stopping site in the unit cell, the observed split should indicate the macroscopic phase separation of the sample in the paramagnetic state. Since no anomaly is reported sofar in the paramagnetic susceptibility [8], the latter possibility seems to be unlikely. However, in order to confirm the existence of multiple muon stopping sites, the potential calculation by the density functional theory (DFT) is indispensable, and is now in the progress. The knowledge of the hyperfine fields obtained by the present work must be of help for the calculation.

Finally we refer the emergence of a wide spectrum bearing multiple peaks below 10 K. This can be due to the development of antiferromagnetic instability toward the phase transition at $T_N = 1.3$ K[13]. Generally, in the vicinity of the critical temperature, the slow fluctuation dominated with a long wavelength component grows in samples. However, considering the hyperfine coupling constant in other usual quantum spin magnets [18], the magnitude observed hyperfine fields in spectra seems to be a little bit too large to be considered as broaden by this instability. So we also consider the other possibility that the number of the muon stopping site may be larger than three.

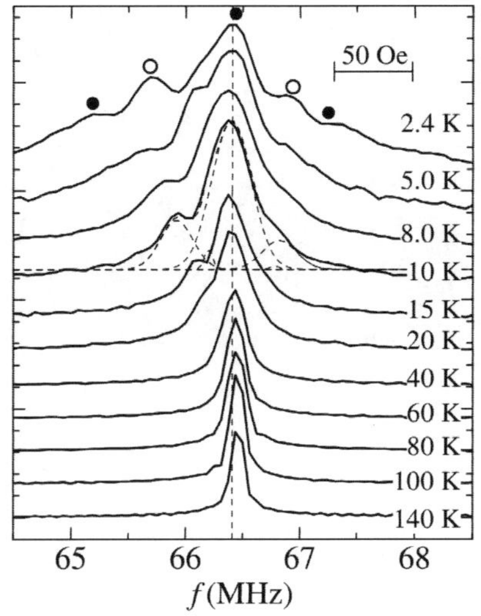


**Figure 3.** Fourier transform of TF muon spin depolarization curves. The vertical line shows the zero shift position. Dots at $T$ = 2.4 K correspond to Fig. 4. Dashed curves show Gaussian deconvolution.

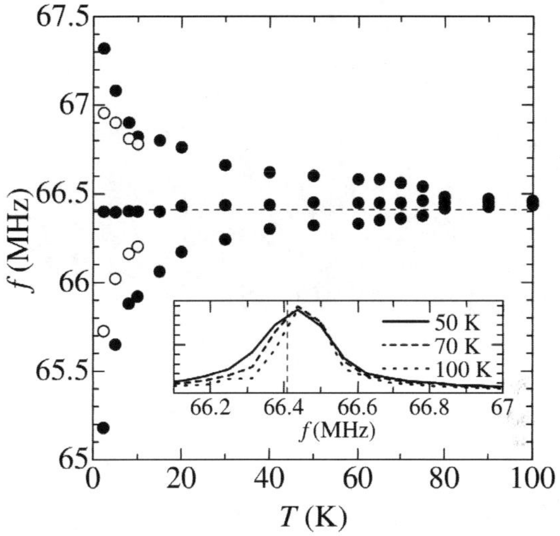


**Figure 4.** The temperature dependence of the peak positions in the FT spectra. Horizontal line shows the zero shift position. Inset shows the enlarged spectra around 70 K.

## 4. Conclusion

We have measured TF-$\mu$SR spectra for $NH_4CuCl_3$, and observed that a peak splits into three below 70 K, indicating the possibility of the existence of three inequivalent dimers in unit cell.

## Acknowledgments

This work was partly supported by JSPS KAKENHI Grant Number 26108713.

## References

[1] Totsuka K 2001 *Phys. Rev.* B **64** 134420
[2] Nikuni T, Oshikawa M, Oosawa A and Tanaka H 2000 *Phys. Rev. Lett.* **84** 5868
[3] Rice T M 2002 *Science* **298** 760 and references therein
[4] Tanaka H, Takatsu K, Shiramura W and Ono T 1996 *J. Phys. Soc. Jpn.* **65** 1945
[5] Takatsu K, Shiramura W and Tanaka H 1997 *J. Phys. Soc. Jpn.* **66** 1611
[6] Shiramura W, Takatsu K, Tanaka H, Kamishima K, Takahashi M, Mitamura H and Goto T 1997 *J. Phys. Soc. Jpn.* **66** 1900
[7] Willett R D, Dwiggins C, Kruh R F and Rundle R E 1963 *J. Chem. Phys.* **38** 2429
[8] Shiramura W, Takatsu K, Kurniawan B, Tanaka H, Uekusa H, Ohashi Y, Takizawa K, Mitamura H and Goto T 1998 *J. Phys. Soc. Jpn.* **67** 1548
[9] Cavadini N, Heigold G, Henggeler W, Furrer A, Güdel H -U, Krämer K and Mutka H 2001 *Phys. Rev.* B **63** 172414
[10] Kodama K, Takigawa M, Horvatić M, Berthier C, Kageyama H, Ueda Y, Miyahara S, Becca F and Mila F 2002 *Science* **298** 395

[11] Kageyama H, Yoshimura K, Stern R, Mushnikov N V, Onizuka K, Kato M, Kosuge K, Slichter C P, Goto T and Ueda Y 1999 *Phys. Rev. Lett.* **82** 3168
[12] Onizuka K, Kageyama H, Narumi Y, Kindo K, Ueda Y and Goto T 2000 *J. Phys. Soc. Jpn.* **69** 1016
[13] Rüegg Ch, Oettli M, Zaharko O, Furrer A, Tanaka H, Krämer K W, Güdel H -U, Vorderwisch P, Habicht K, Polinski T and Meissner M 2004 *Phys. Rev. Lett.* **93** 037207
[14] Matsumoto M 2003 *Phys. Rev.* B **68** 180403(R)
[15] Kurniawan B, Tanaka H, Takatsu K, Shiramura W, Fukuda T, Nojiri H and Motokawa M 1999 *Phys. Rev. Lett.* **82** 1281
[16] Nojiri H, Tanaka H, Kurniawan B and Motokawa M 2002 *Proceedings of French-Japanese Symposium on Quantum Properties of Low-Dimensional Antiferromagnets, Fukuoka, Japan, Nov. 20-22, 2001* ed Ajiro Y and Boucher J P (Fukuoka: Kyushu University Press) p 19
[17] Inoue H, Tani S, Hosoya S, Inokuchi K, Fujiwara T, Saito T, Suzuki T, Oosawa A, Goto T, Fujisawa M, Tanaka H, Sasaki T, Awaji S, Watanabe K and Kobayashi N 2009 *Phys. Rev.* B **79** 174418
[18] Saito T, Oosawa A, Goto T, Suzuki T and Watanabe I 2006 *Phys. Rev.* B **74** 134423
[19] Goto T, Suzuki T, Kaneda K, Saito T, Oosawa A, Watanabe I and Manaka H 2008 *Phys. Rev.* B **78** 054422
[20] Suzuki T, Yamada M, Ishii Y, Watanabe I, Goto T, Tanaka H and Kubo K 2011 *Phys. Rev.* B **83** 174436
[21] Matsui K, Guo H, Watanabe I, Suzuki T, Fujisawa M, Tanaka H and Goto T *in preparation*

# Spin fluctuations above 100 K in stoichiometric $LiCoO_2$

**K Mukai[1], Y Aoki[1], D Andreica[2], A Amato[3], I Watanabe[4], S R Giblin[5] and J Sugiyama[1]**

[1] Toyota Central Research & Development Laboratories, Inc., 41-1 Yokomichi, Nagakute, Aichi 480-1192, Japan
[2] Faculty of Physics, Babes-Bolyai University, 400084 Cluj-Napoca, Romania
[3] Laboratory for Muon Spin Spectroscopy, Paul Scherrer Institut, CH-5232 Villigen PSI, Switzerland
[4] Advanced Meson Science Laboratory, The Institute of Physical and Chemical Research (RIKEN), 2-1 Hirosawa, Wako, Saitama 351-0198, Japan
[5] School of Physics and Astronomy, Cardiff University, Queens Building, The Parade, Cardiff, CF24 3AA, UK

E-mail: e1089@mosk.tytlabs.co.jp

**Abstract.** Although stoichiometric lithium cobalt dioxide $LiCoO_2$ (ST-LCO) is the most common positive electrode material for Li-ion batteries, the magnetic nature of ST-LCO is still not fully understood. Therefore, we measured susceptibility ($\chi$), electron paramagnetic resonance (EPR), and $\mu$SR for ST-LCO, particularly above 100 K. The temperature dependence of $\chi$ shows a Pauli paramagnetic behavior, supporting the previous conclusion that $Co^{3+}$ ions are in a low-spin state with $S = 0$ ($t_{2g}^6$). However, the EPR and $\mu$SR measurements reveal a "dynamical" magnetic phase in ST-LCO above 100 K. Because the volume fraction of this magnetic phase reaches about 50% at 300 K, the appearance of the magnetic phase is not caused by impurities and/or muonium formation but is an intrinsic characteristic of ST-LCO. By considering the time windows of the three measurement techniques used in this study, we conclude that the origin of the dynamical magnetism is most likely spin fluctuations of the Co ions.

## 1. Introduction

Since the discovery that $Li^+$ ions are reversibly extracted and inserted by an electrochemical reaction [1], lithium cobalt oxide $LiCoO_2$ (LCO) has been widely used as a positive electrode material for lithium-ion batteries (LIBs). Stoichiometric LCO (ST-LCO) has a layered structure with the $R\bar{3}m$ space group. The $Li^+$ and $Co^{3+}$ ions occupy the octahedral $3b$ and $3a$ sites, respectively, in the cubic close-packed array of $O^{2-}$ ions (Fig. 1). In addition to LIB applications, ST-LCO has received much attention in the field of condensed matter physics because the Co ions form a two-dimensional triangular lattice, on which geometrical frustration often induces interesting magnetic behavior.

ST-LCO exhibits complicated magnetic properties as described by Chernova *et al* [2]. For instance, although susceptibility ($\chi$) measurements indicate that the $Co^{3+}$ ions are in a nonmagnetic low-spin (LS) state with $S = 0$ ($t_{2g}^6$) [3], $\mu$SR measurements clarify the complex magnetic nature below 60 K. That is, as temperature ($T$) decreases from 100 K, localized magnetic moments appear below ~60 K, and then, an antiferromagnetic (AF) ordered phase appears below 30 K. But, the volume fraction of such AF phase is ~10% [4, 5]. Since ST-LCO is very stable in air, such magnetic properties are most unlikely to be caused by the degradation of a sample, but are an intrinsic behavior of ST-LCO. Furthermore, it was reported that, even for a large single crystal sample, $\chi$ rapidly increases with decreasing $T$ particularly below 50 K [6]. This also supports that the appearance of localized moments at low $T$ is an intrinsic feature for $LiCoO_2$. This is quite different from $NaCoO_2$ [7], which is an isostructural compound of ST-LCO. In this paper, we report another unique magnetic feature of ST-LCO, i.e., "dynamical magnetism" at 100 K $\leq T \leq$ 500 K clarified by electron paramagnetic resonance (EPR) and $\mu$SR measurements up to 500 K.

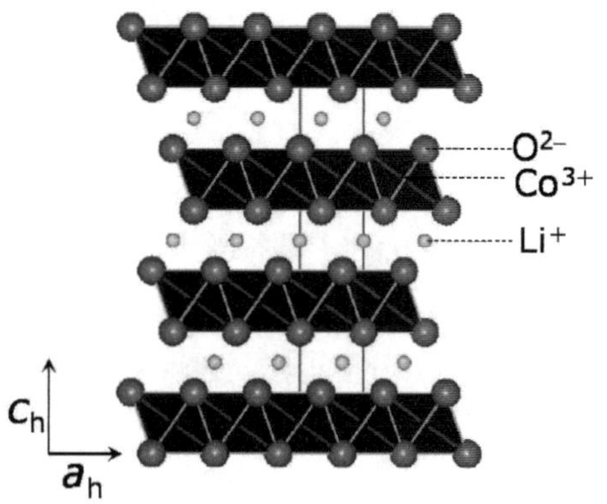


Figure 1. Crystal structure of stoichiometric $LiCoO_2$.

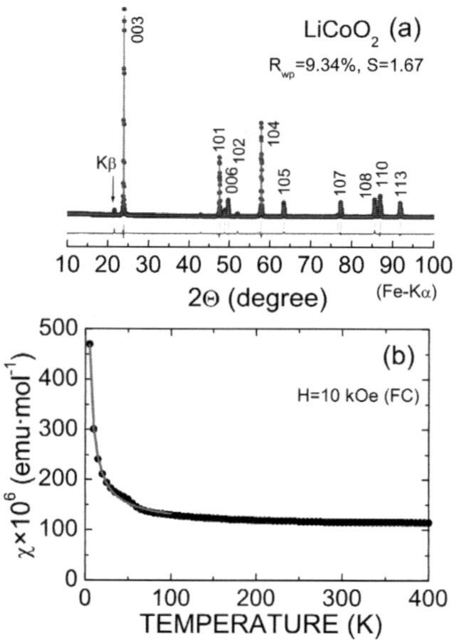


**Figure 2.** (a) Rietveld analysis and (b) $T$ dependence of $\chi$ for stoichiometric $LiCoO_2$ sample.

## 2. Synthesis and Characterization

A polycrystalline sample of ST-LCO was prepared by the solid-state reaction technique, as reported previously [8, 9]. The reaction mixture of $LiOH{\cdot}H_2O$ and $Co_3O_4$ was heated at 1173 K for 12 h in air. A Rietveld analysis of the powder x-ray diffraction (XRD) pattern of the sample indicated a single-phase of the $R\bar{3}m$ space group with the lattice parameters of $a_h$ = 2.8135(4) Å and $c_h$ = 14.042(1) Å [Fig. 2(a)]. Impurities, such as CoO and $Co_3O_4$, were not detected in the XRD pattern.

Susceptibility ($\chi$) was measured in the $T$ range between 5 and 400 K using a field-cooling (FC) mode with $H$ = 10 kOe [Fig. 2(b)]. As $T$ decreases from 400 K, $\chi$ is almost $T$-independent down to ~100 K, and then rapidly increases with further decreasing $T$. By fitting the $\chi(T)$ curve below 100 K with a Curie-Weiss formula [Fig. 2(b)], we obtain that $\chi_0 = 112(2)\times10^{-6}$ emu·mol$^{-1}$, where $\chi_0$ is the $T$-independent $\chi$ and consists of a diamagnetic ($\chi_{dia}$) and an orbital ($\chi_{orb}$) contribution. Since $\chi_0$ for the ST-LCO sample is comparable to that for $NaCoO_2$ ($\chi_{orb}$

$= 150\times10^{-6}$ emu·mol$^{-1}$) [7], $Co^{3+}$ ions in ST-LCO are assigned in a nonmagnetic LS state with $S = 0$ ($t_{2g}^6$), as reported previously [3].

EPR measurements were made in the $T$ range from 100 to 400 K by using an ESP300E spectrometer (Bruker BioSpin K. K.). The $\mu$SR spectra was measured in the $T$ range from 2 to 500 K at both S$\mu$S of the Paul Scherrer Institut and ISIS of the Rutherford Appleton Laboratory. Details of the experimental setup for the EPR and $\mu$SR measurements are described elsewhere [8].

## 3. Results and discussion

Figure 3 shows the EPR spectra for the ST-LCO sample at 100 and 300 K. According to the $\chi$ measurements, the ST-LCO sample was predicted to be EPR-inactive because of the lack of unpaired electrons. However, an EPR signal is clearly observed at 100 K. The $g$ value and the

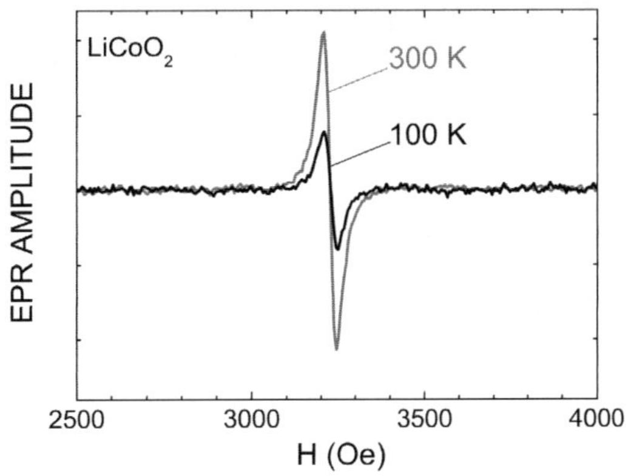


**Figure 3.** EPR spectra for stoichiometric $LiCoO_2$ sample at 100 and 300 K. The number of spins in $LiCoO_2$ has not been clarified.

EPR line width between the peaks ($\Delta H_{pp}$) is 2.12(1) and 42(1) Oe, respectively. The EPR signal is more notable at 300 K, although the values of $g$ and $\Delta H_{pp}$ are $T$ independent. Note that $Co_3O_4$ typically shows a broad EPR signal with $g \sim 5$ at room $T$ [10], whereas CoO has no EPR signal at 300 K [8]. Thus, the EPR signal observed at 100 and 300 K is not caused by the CoO or $Co_3O_4$ phases.

To further investigate the magnetism of ST-LCO, $\mu$SR was measured. Figure 4(a) shows the $T$ dependence of the normalized weak-transverse-field asymmetry ($N_{ATF}$) for the ST-LCO sample. The slight decrease in $N_{ATF}$ below ~60 K is attributed to the appearance of the localized moments before the formation of AF order, which is characteristic of ST-LCO [4, 5]. As $T$ increases from~100 K, $N_{ATF}$ decreases almost linearly up to 300 K, then rapidly increases with changing slope ($dN_{ATF}/dT$), and finally reaches ~1 at 500 K. Because $N_{ATF} = 0.55$ at 300 K, 45 vol% of the sample is found to be in a magnetic phase. Such a large volume fraction is unlikely to be due to impurities such as CoO and $Co_3O_4$, because the Rietveld analysis clarified that the ST-LCO sample is single-phase with $R\bar{3}m$ [Fig. 2(a)]. In addition, the $T$ dependence of $N_{ATF}$ clearly excludes the possibility of muonium formation.

To clarify whether the sample contains a static magnetic ordered phase, Figure 4(b) shows the zero-field (ZF) $\mu$SR spectrum at 300 K. Although the asymmetry rapidly decreases in the time domain below ~1 $\mu$s, there is no oscillatory signal, which indicates a lack of static magnetic order in the magnetic phase. Thus, we suggest that this phase is a "dynamical" magnetic phase. The ZF-$\mu$SR spectrum is fit by a combination of a fast-relaxing signal from the dynamical magnetic

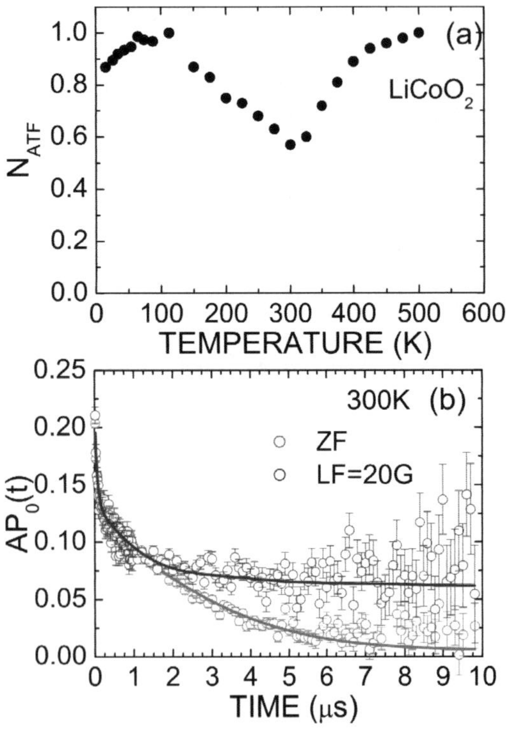


**Figure 4.** (a) Temperature dependence of normalized weak-transverse-field asymmetry ($N_{\rm ATF}$) for ST-LCO sample. (b) Zero-field $\mu$SR spectrum for the ST-LCO sample at 300 K. The longitudinal-field $\mu$SR spectrum with a magnetic field of 20 G is also shown for comparison.

phase, a dynamic Gaussian Kubo-Toyabe [11] (DGKT) signal from the nuclear moments of Li and Co in a nonmagnetic phase, and a background (BG) signal mainly from an Ag or Ti in the sample holder:

$$A_0P_{\rm ZF}(t) = A_{\rm fast}\exp(-\lambda_{\rm fast}t) + A_{\rm KT}G_{\rm DGKT}(t,\Delta,\nu) + A_{\rm BG}, \quad (1)$$

where $A_{\rm fast}$, $A_{\rm KT}$, and $A_{\rm BG}$ are the asymmetries associated with the three signals, $\lambda_{\rm fast}$ is the relaxation rate of the dynamical magnetic phase, and $\Delta$ and $\nu$ are the field distribution width and the field fluctuation rate, respectively. The values of $\Delta$ and $\nu$ were determined by a global fit with ZF and longitudinal field (LF) $\mu$SR spectra [Fig. 4(b)]. The $A_{\rm fast}(T)$ curve shows a maximum (~0.5) at 300 K, and if it is flipped upwards down, such curve is almost equivalent to the $N_{\rm ATF}(T)$ curve [Fig. 4(a)]. Furthermore, the $\nu(T)$ curve exhibits a maximum at around 300 K (not shown), which isprobably due to a slight displacement of the Li ions, as for $LiMn_2O_4$ [12].

As described briefly, not only the AF phase below 30 K, but also the dynamical magnetism at 100 K $\leq T \leq$ 500 K is found to be an intrinsic character of ST-LCO. It should be emphasized again that both phenomena are neither due to impurities, degradation of the sample, nor muonium formation. The discrepancy of magnetic properties obtained with $\chi$, EPR, and $\mu$SR are most likely caused by different time windows among the three techniques. That is, $\chi$ reflects the static internal magnetic fields ($H_{\rm int}$), whereas both EPR and $\mu$SR reflect the static and dynamic characteristics of $H_{\rm int}$ from $\sim 10^{10}$ Hz and $10^4$–$10^{12}$ Hz, respectively. Therefore, the hypothesis that the dynamical magnetism in ST-LCO is caused by the spin fluctuation of Co ions is very reasonable.

There are two possible origins for such spin fluctuations. One is a spin-state transition of the Co ions, as for $LaCoO_3$ [13]. As $T$ increases from low $T$, the spin state of the Co ions in

$LaCoO_3$ changes from an LS state with $S = 0(t_{2g}^6)$ to a high-spin (HS) state with $S = 2(t_{2g}^4 e_g^2)$ via an intermediate-spin (IS) state with $S = 1(t_{2g}^5 e_g^1)$ [13]. Thus, a thermally induced spin-state transition from LS $\to$ IS $\to$ HS is a possible origin of the spin fluctuation. The other possibility is charge disproportionation of the $Co^{3+}$ ions:

$$2Co^{3+}(LS, S = 0) \leftrightarrow Co^{2+}(LS, S = 1/2) + Co^{4+}(LS, S = 1/2), \tag{2}$$

or

$$2Co^{3+} \leftrightarrow Co^{3+\delta} + Co^{3-\delta}. \tag{3}$$

Because of thermal activation, both equilibrium reactions progress to the right in the $T$ range from 100 to 300 K, resulting in the formation of magnetic Co ions. However, the $N_{ATF}(T)$ curve above 300 K suggests the presence of the opposite contribution that shifts the equilibrium to the left side. On the basis of the $\nu(T)$ curve for ST-LCO and delithiated $Li_xCoO_2$ [14], Li diffusion could be the origin of this opposite contribution. In other words, in the $T$ range from 100 to 300 K, the $CoO_6$ octahedron would be most likely distorted because of the unusual spin states or charge states. However, at $T \geq 300$ K, the Li ions start to diffuse between the regular $3b$ site and the interstitial site, which tends to suppress the local distortions of the $CoO_6$ octahedron. As a result, as $T$ increases from 300 K, the nonmagnetic $Co^{3+}$ ions stabilize with $T$, and the volume fraction of the dynamical magnetic phase decreases with $T$.

It would be interesting to compare the LCO result with those for $NaCoO_2$ (NCO). Note that stoichiometric-NCO is very unstable in air, and a Na-deficient phase ($Na_{\sim 0.7}CoO_2$) is usually obtained by a conventional solid state reaction technique [7]. Since the exact 1:1 ratio between Li and Co is essential for the appearance of the dynamical magnetism in LCO [8], it is not surprising that the Na-deficient NCO does not exhibit a similar magnetic behavior.

In conclusion, we find a thermally induced dynamical magnetic phase in ST-LCO above 100 K, which we attribute to the spin fluctuations of Co ions. To better understand the nature of such spin fluctuations, neutron-scattering experiments with ST-LCO are planned.

## Acknowledgments

We thank the staff of PSI and RAL for their help with the $\mu$SR measurements. K.M. was supported in part by a Grant-in-Aid for Scientific Research (C), 25410207, from MEXT, Japan and Grant-in-Aid for Scientific Research (B), 26286084, JSPS, Japan. D.A. acknowledges financial support from the Romanian UEFISCDI Project No. PN-II-ID-PCE-2011-3-0583 (85/2011). J.S. was supported by a Grant-in-Aid for Scientific Research on Innovative Areas, 23108003, MEXT, Japan.

## References

[1] Mizushima K, Jones P C, Wiseman P J and Goodenough J B 1980 *Mater. Res. Bull.* **15** 783
[2] Chernova N A, Nolis G M, Omenya F O, Zhou H, Li Z and Whittingham M S 2011 *J. Mater. Chem.* **21** 9865
[3] Kikkawa S, Miyazaki, S and Kozumi M 1986 *J. Solid State Chem.* **62** 35
[4] Sugiyama J, Nozaki K, Brewer J H, Ansaldo, E J, Morris G D, and Delmas C 2005 *Phys. Rev.* B **72** 144424
[5] Mukai K, Ikedo Y, Nozaki H, Sugiyama J, Nishiyama K, Andreica D, Amato A, Russo, P L, Ansaldo, E J, Brewer J H, Chow K H, Ariyoshi K and Ohzuku T 2007 *Phys. Rev. Lett.* **99** 087601
[6] Pinsard-Gaudart L, Ciomaga V.-C, Dragos O, Guillot R and Dragoe N, 2011 *J. Crystal Growth* **334** 165
[7] Lang G, Bobroff J, Alloul H, Mendels P, Blanchard N and Collin G 2005 *Phys. Rev.* B **72** 094404
[8] Mukai K, Aoki Y, Andreica D, Amato A, Watanabe I, Giblin S R and Sugiyama J 2014 *Phys. Rev.* B **89** 094406
[9] Mukai K, Kishida Y, Nozaki H and Dohmae K 2013 *Chem. Mater.* **25** 2828
[10] Artemenko A, Ménétrier, Pollet, M and Delmas C 2005 *J. Appl. Phys.* **106** 064914
[11] Hayano R S, Uemura Y J, Imazato J, Nishida N, Yamazaki T and Kubo R 1979 *Phys. Rev.* B **20** 850

[12] Sugiyama J, Mukai K, Ikedo Y, Russo P L, Suzuki T, Watanabe I, Brewer J H, Ansaldo, E J, Chow K H, Ariyoshi K and Ohzuku T 2007 *Phys. Rev.* B **75** 174424
[13] Yamaguchi S, Okimoto Y, Taniguchi H, Tokura Y, 1996 *Phys. Rev.* B **53** R2926
[14] Sugiyama J, Mukai K, Ikedo Y, Nozaki H, Månsson M and Watanabe I 2009 *Phys. Rev. Lett.* **103** 147601

# $^8$Li $\beta$-NMR study of epitaxial $Li_xCoO_2$ films

**J Sugiyama[1], M Harada[1], H Oki[2], S Shiraki[3], T Hitosugi[3] O Ofer[4], Z Salman[5], Q Song[6], D Wang[7], H Saadaoui[5], G D Morris[4], K H Chow[8], W A MacFarlane[6] and R F Kiefl[4,7]**

[1] Toyota Central Research & Development Laboratories, Inc., 41-1 Yokomichi, Nagakute, Aichi 480-1192, Japan
[2] Higashifuji Technical Center, Toyota Motor Co., Susono, Shizuoka 410-1193 Japan
[3] Advanced Institute for Materials Research, Tohoku University, Sendai, Miyagi 980-8577 Japan
[4] TRIUMF, 4004 Wesbrook Mall, Vancouver, BC, V6T 2A3 Canada
[5] Laboratory for Muon-Spin Spectroscopy, Paul Scherrer Institut, CH-5232 Villigen PSI, Switzerland
[6] Department of Chemistry, University of British Columbia, Vancouver, BC, V6T 1Z1 Canada
[7] Department of Physics & Astronomy, University of British Columbia, Vancouver, BC, V6T 1Z1 Canada
[8] Department of Physics, University of Alberta, Edmonton, AB, T6G 2G7 Canada

E-mail: e0589@mosk.tytlabs.co.jp

**Abstract.** In order to investigate the diffusive motion of $Li^+$ in a thin film electrode material for Li-ion batteries, we have measured $\beta$-NMR spectra of $^8Li^+$ ions implanted into epitaxial films of $Li_{0.7}CoO_2$ and $LiCoO_2$ in the temperature range between 10 and 310 K. Below 100 K, the spin-lattice relaxation rate ($1/T_1$) in the $Li_{0.7}CoO_2$ film increased with decreasing temperature, indicating the appearance and evolution of localized magnetic moments, as observed with $\mu^+$SR. As temperature is increased from 100 K, $1/T_1$ starts to increase above $\sim$ 200 K, where both Li-NMR and $\mu^+$SR also sensed an increase in $1/T_1$ due to Li-diffusion. Interestingly, such diffusive behavior was found to depend on the implantation energy, possibly because the surface of the film is decomposed due to chemical instability of the $Li_{0.7}CoO_2$ phase in air. Such diffusive behavior was not observed for the $LiCoO_2$ film up to 310 K.

## 1. Introduction

In order to improve the safety and increase the volumetric energy density of Li-ion batteries, the present solid-liquid battery will be replaced by an all-solid-state battery in the near future (Fig. 1). For such solid-state batteries, the overall performance, particularly the charge/discharge rate, is mainly governed by Li diffusion in the electrodes and the interface between electrode and electrolyte. There is, thus, a pressing need to measure the Li diffusive behavior along the direction from cathode to anode, through the electrolyte of a solid-state layered Li-ion battery (vertical direction in Fig. 1). Therefore, we have initiated a program utilizing the "$\beta$-NMR" technique as a probe of the diffusion coefficient of $Li^+$ ions ($D_{Li}$) [1, 2] in battery materials and hopefully as a tool to obtain the depth profile of $D_{Li}$ in solid-state battery materials, and hence to elucidate the role of the interface. Since mobile Li ions always pass through such an interface, the change in $D_{Li}$ at the interface is crucial for developing solid-state batteries. It is worth noting that such information is detectable by $\beta$-NMR and/or low-energy

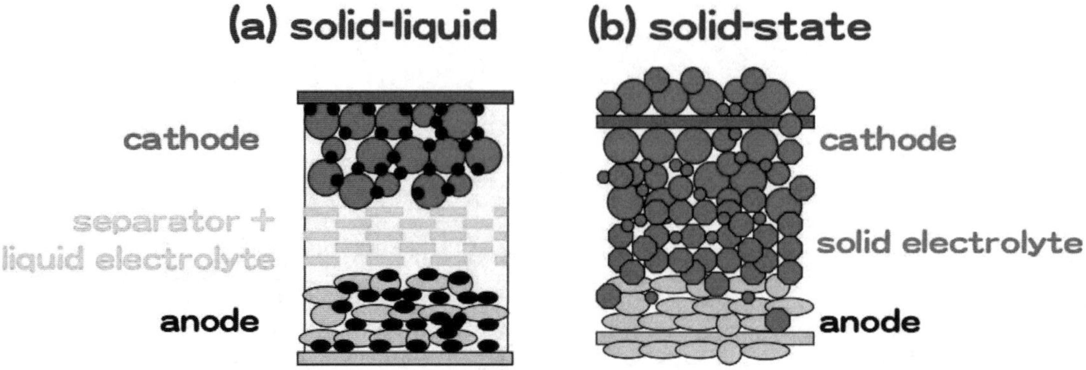


**Figure 1.** (a) The present solid-liquid battery design and (b) the near-future solid-state battery. In the present battery, a typical cathode material is $LiCoO_2$, electrolyte an ethylene carbonate-diethyl carbonate solution of $LiPF_6$, and anode material graphite.

$\mu^+$SR, but is very difficult, if not impossible, to obtain by electrochemical, NMR or neutron scattering measurements.

For NMR and other spin relaxation techniques, one principle mechanism for spin-lattice relaxation (the only one for pure magnetic species, like the spin 1/2 muon) is the existence of a magnetic field transverse to the spin polarization that fluctuates at the Larmor frequency of the spin in the static magnetic field. The pioneering work of Bloembergen, Purcell, and Pound [3] showed how diffusive motion that varies the local magnetic environment of a nucleus can produce such fluctuations and cause spin relaxation, and hence how spin relaxation (in the absence of other relaxation pathways) could be used to measure diffusion rates [4]. In contrast to conventional Li-NMR [5], the effect of localized moments[1] in the paramagnetic (PM) state on the $\beta$-NMR signal is very limited at high $T$, because the PM fluctuations are usually too fast to cause a measurable spin relaxation, as in the case for $\mu^+$SR. In fact, making comparison with the NMR results, $\mu^+$SR provides relatively reliable $D_{\rm Li}$ for $Li_xCoO_2$ [6]. However, due to the mass difference between $\mu^+$ and $Li^+$, the $D_{\rm Li}$ estimated from $\mu^+$SR might be affected by muon diffusion.

Here we use $^8Li^+$ that is, unlike the $\mu^+$, unambiguously an isotope of Li. Therefore, we could, in principle, estimate $D_{\rm Li}$ directly from the dynamic behavior of $^8Li^+$ detected by $\beta$-NMR. In addition, since the $\beta$-NMR timescale is much longer than $\mu^+$SR, as the radioactive lifetime $\tau$ is 1.21 s, rather than the $2.2 \times 10^{-6}$ s for $\mu^+$, it is expected to provide unique information on a different timescale. This is our primary motivation to perform these $\beta$-NMR experiments.

## 2. Experimental

The $LiCoO_2$ (110) epitaxial film of $\alpha$-$NaFeO_2$ structure with space group $R\bar{3}m$ was grown on an Au(110) single crystal substrate using a pulsed laser deposition technique (Fig 2) [7]. The thickness of the film was about 200 nm. The Li-deficient $Li_{0.7}CoO_2$ film was prepared by an electrochemical reaction between the $LiCoO_2$ film and Li metal sheet using a Li|$LiPF_6$-ethylene carbonate-diethyl carbonate|$LiCoO_2$ cell.

The $\beta$-NMR spectra were measured using the $^8Li$ beam produced at the Isotope Separator and Accelerator (ISAC) at TRIUMF in Canada. The nuclear spin is polarized using a collinear optical pumping method, producing a spin polarized $^8Li^+$ beam with about 70% polarization.

[1] In this compound the valence (and magnetic moment) of Co ions changes with the degree of Lithiation, $x$.

**Figure 2.** The $LiCoO_2$ epitaxial film on an Au substrate; that is, a dark blue film on a gold plate. For scale, the Au is affixed to an 8 mm × 10 mm sapphire plate that is clamped (by the visible screws) to the Al sample holder that mounts into the cold finger of an ultrahigh vacuum cryostat.

Finally, the beam is implanted into the $Li_xCoO_2$ sample. The implanted beam energy ($E_{\rm Li}^{\rm im}$) was 28 keV, for which the $^8$Li stops at an average depth of about 120 nm.

In the $\beta$-decay of $^8$Li, an electron is emitted preferentially opposite to the direction of the nuclear polarization and is detected by scintillation counters, as in $\mu^+$SR. Therefore, we can measure the change in asymmetry as a function of time for a pulsed beam or as a function of frequency of an applied RF magnetic field. The details of setup and experimental procedure of $\beta$-NMR are described elsewhere [8, 9, 10].

## 3. Results and discussion

### *3.1. Spin-lattice relaxation*

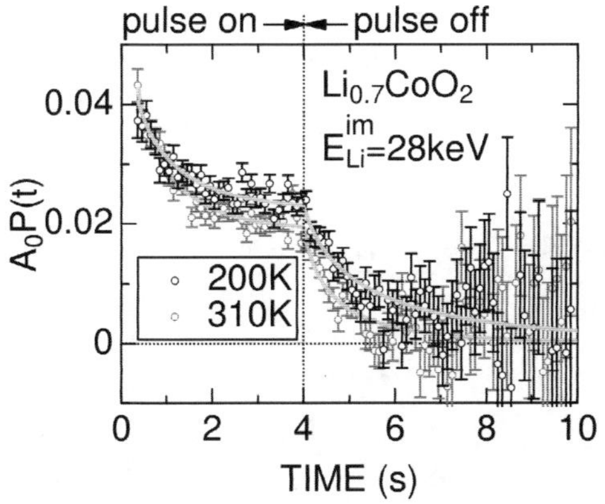


**Figure 3.** The asymmetry time spectrum of $^8$Li in the $Li_{0.7}CoO_2$ film measured at 310 and 200 K. Solid lines represent the best fit using eq. (1).

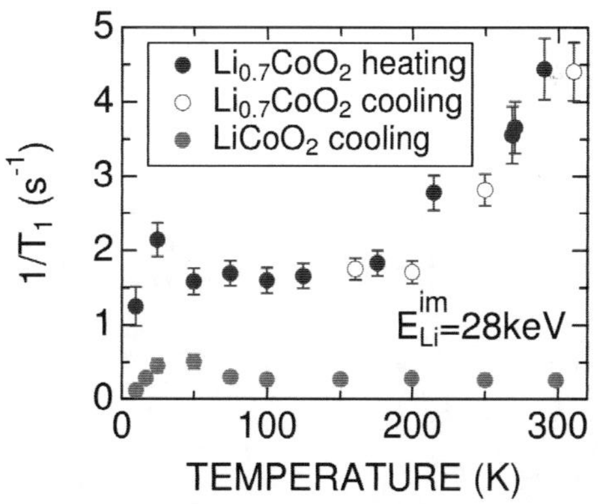


**Figure 4.** The temperature dependence of $1/T_1$ for the $Li_{0.7}CoO_2$ and $LiCoO_2$ film. The data were obtained by fitting the $\beta$-NMR time spectrum using eq. (1).

Figure 3 shows the $\beta$-NMR time spectrum for the $Li_{0.7}CoO_2$ film obtained at 310 and 200 K in an applied field of 6.55 Tesla. At this field, the $^8$Li is effectively decoupled from the stable $^6$Li spins that have nearly the same gyromagnetic ratio [11]. The $^8$Li pulse starts at $t = 0$ and continues for $\delta = 4$  s, resulting in the pronounced kink in the time spectrum at $t = \delta$. The $\beta$-decay asymmetry is measured both during and after the beam pulse [10]. As can be seen in Fig. 3, the relaxation at 310 K is clearly faster than at 200 K, indicating an increase in the spin-lattice relaxation rate ($1/T_1$) with temperature. The time spectrum was best fit by a power exponential relaxation function for the fraction of $^8$Li implanted in the sample at $t_p$ [10];

$$\begin{aligned} A_0\,P_z(t) &= A\frac{\int_0^t \exp[-\frac{t-t_p}{\tau}]\exp[-(\frac{t-t_p}{T_1})^\beta]dt_p}{\int_0^t \exp(-\frac{t}{\tau})dt} & t \le \delta, \\ A_0\,P_z(t) &= A\frac{\int_0^\delta \exp[-\frac{\delta-t_p}{\tau}]\exp[-(\frac{\delta-t_p}{T_1})^\beta]dt_p}{\int_0^\delta \exp(-\frac{t}{\tau})dt} & t > \delta, \end{aligned} \tag{1}$$

where $A_0$ is the initial asymmetry, $P_z(t)$ is the $^8$Li spin polarization function, $\beta$ is the power (fixed at 0.5), and $A$ is the asymmetry for the power exponential relaxation signal. Since we assumed a single component decay, $A_0 = A$.

Figure 4 shows the temperature dependence of $1/T_1$ for the $Li_{0.7}CoO_2$ and $LiCoO_2$ films. For $Li_{0.7}CoO_2$, as temperature increases from 10 K, $1/T_1$ shows a local maximum around 30 K due to localized Co moments [12], then, levels off to a constant value ($\sim$ 1.7 s $^{-1}$) up to 200 K, and then, increases monotonically with further increasing temperature. Since Li-NMR [13] and $\mu^+$SR [6] measurements on bulk $Li_{\sim 0.7}CoO_2$ also sensed an increase in $1/T_1$ above $\sim$ 200 K, we attribute this increase to Li-diffusion. On the other hand, for $LiCoO_2$, $1/T_1$ is almost temperature independent up to 310 K, besides the anomaly below 75 K due to antiferromagnetic order in $LiCoO_2$ with $T_N$ = 30 K [14, 12]. Considering the absence of Li vacancies in stoichiometric $LiCoO_2$, Li-diffusion in $LiCoO_2$ is predicted to be strongly suppressed compared to $Li_{0.7}CoO_2$ [15]. The present $\beta$-NMR result is very consistent with this.

Interestingly, when $E_{\rm Li}^{\rm im}$ is decreased from 28 keV to 20 and 10 keV for the $Li_{0.7}CoO_2$ film, such diffusive behavior was not observed up to the maximum available temperature, 310 K. This is probably due to a partial decomposition of the surface of the $Li_{0.7}CoO_2$ film, because the $Li_{0.7}CoO_2$ phase is very hygroscopic, easily reacting with water in the air. Currently, mounting the sample into the spectrometer requires exposure to ambient air for a period up to several minutes.

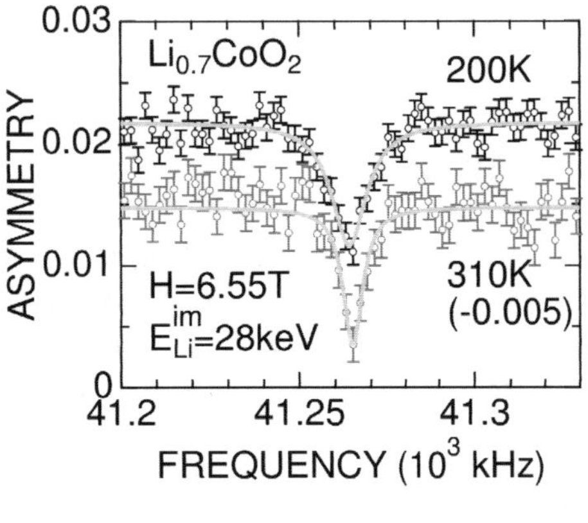


**Figure 5.** The nuclear resonance spectra of $^8$Li for the $Li_{0.7}CoO_2$ film measured at 310 and 200 K. The data at 310 K is offset by -0.005 for clarity. Solid lines represent the fit using a Lorentz function.

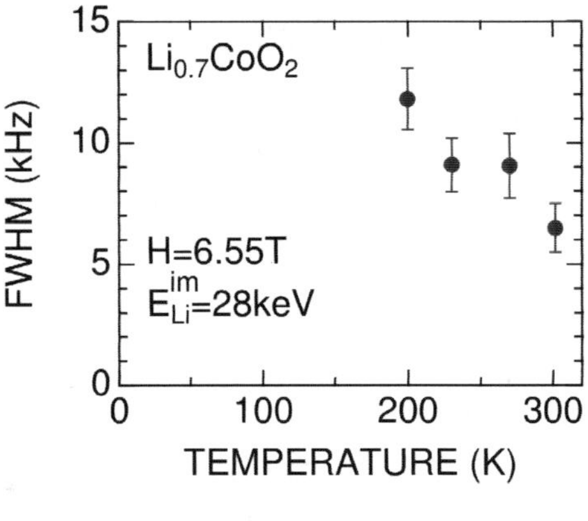


**Figure 6.** The temperature dependence of the full width at half maximum (FWHM) of the resonance for the $Li_{0.7}CoO_2$ film. The data were obtained by fitting the $\beta$-NMR resonance spectrum using a Lorentz function.

### *3.2. Resonance line*

In a noncubic material like $Li_xCoO_2$, the $I = 2$ nuclear spin of $^8$Li couples through its nuclear quadrupole moment $Q$ to the electric field gradient (EFG) tensor created by its charge environment. This usually yields a quadrupole split resonance [16, 17, 18]. However, we observed only a single broad line, even at 200 K, see Fig. 5. We note that the $^7$Li-NMR exhibits a quadrupole splitting on the order of 30-40 kHz in bulk $LiCoO_2$ [19]. If the implanted $^8Li^+$ stops at the same site, we would expect a quadrupole splitting scaled down by the ratio of the quadrupole moments, $|Q(^8\mathrm{Li})/Q(^7\mathrm{Li}|)$, i.e. about 0.75, which is larger than the observed linewidth by about a factor of 2. Static crystalline disorder resulting in a random distribution of EFGs, may account for the lack of a resolved quadrupole splitting. Alternatively, diffusive motion of Li may dynamically average the quadrupole splitting, yielding a narrowed line. But, this would imply that $^8$Li moves rapidly even at 200 K, which seems unreasonable. We note that a single broad line is sometimes observed for the materials without Li-diffusion[16], likely due to the effects of disorder. At 310 K, the linewidth is found to decrease, suggesting motional narrowing (dynamical averaging). In fact, the temperature dependence of the full width at half maximum (FWHM) of the resonance line (Fig. 6) shows a clear decrease with temperature above 200 K. However, more measurements over a wider range of temperature are required to confirm the relation between the resonance width and the diffusion of Li in this system.

## 4. Summary

We have measured $\beta$-NMR spectra for $Li_{0.7}CoO_2$ and $LiCoO_2$ film samples below 310 K. This is, to our knowledge, the first attempt to use $\beta$-NMR to measure Li-diffusion in a film sample containing paramagnetic ions. For $Li_{0.7}CoO_2$, possible motional narrowing behavior due to Li-diffusion was observed above 200 K both in the spin-lattice relaxation and resonance line measurements. For $LiCoO_2$, such behavior was not observed up to 300 K due to the absence of Li vacancies.

## Acknowledgments

We thank the staff of TRIUMF for help with the $\beta$-NMR experiments. This work was supported by MEXT KAKENHI Grant No. 23108003 and JSPS KAKENHI Grant No. 26286084.

## References

[1] Heitjans P, Faber W and Schirmer A 1981 *J. Non-crystalline Solids* **131-133** 1053
[2] Körblein A, Heitjans P, Stöckmann H-J, Fujara F, Ackermann H, Buttler W, Dörr K and Grupp H 1985 *J. Phys. F: Met. Phys.* **15** 561
[3] Bloembergen N, Purcell E M and Pound R V 1948 *Phys. Rev.* **73** 679
[4] Abragam A 1961 *The Principles of Nuclear Magnetism* (Oxford: Oxford Univ.)
[5] Grey C P and Dupré N 2004 *Chem. Rev.* **104** 4493
[6] Sugiyama J, Mukai K, Ikedo Y, Nozaki H, Månsson M and Watanabe I 2009 *Phys. Rev. Lett.* **103** 147601
[7] Tsuruhama T, Hitosugi T, Oki H, Hirose Y and Hasegawa T 2009 *Appl. Phys. Express.* **2** 085502
[8] Salman Z, Reynard E P, MacFarlane W A, Chow K H, Chakhalian J, Kreitzman S R, Daviel S, Levy C D P, Poutissou R and Kiefl R F 2004 *Phys. Rev.* B **70** 104404
[9] Morris G D, MacFarlane W A, Chow K H, Salman Z, Arseneau D J, Daviel S, Hatakeyama A, Kreitzman S R, Levy C D P, Poutissou R, Heffner R H, Elenewski J E, Greene L H and Kiefl R F 2004 *Phys. Rev. Lett.* **93** 157601
[10] Salman Z, Kiefl R F, Chow K H, Hossain M D, Keeler T A, Kreitzman S R, Levy C D P, Miller R I, Parolin T J, Pearson M R, Saadaoui H, Schultz J D, Smadella M, Wang D and MacFarlane W A 2006 *Phys. Rev. Lett.* **96** 147601
[11] Abov Y G, Gulko A D and Dzheparov F S 2006 *Phys. Atomic Nuclei* **69** 1701
[12] Mukai K, Ikedo Y, Nozaki H, Sugiyama J, Russo P L, Nishiyama K, Andreica D, Amato A, Brewer J H, Ansaldo E J, Chow K H, Ariyoshi K and Ohzuku T 2007 *Phys. Rev. Lett.* **99** 087601
[13] Nakamura K, Ohno H, Okamura K, Michihiro Y, Moriga T, Nakabayashi I and Kanashiro T 2006 *Solid State Ionics* **177** 821

[14] Sugiyama J, Nozaki H, Brewer J H, Ansaldo E J, Morris G D and Delmas C 2005 *Phys. Rev.* B **72** 144424
[15] Van der Ven A and Ceder G 2000 *Electrochem. Solid-State Lett.* **3** 301
[16] MacFarlane W A, Morris G D, Chow K H, Baartman R A, Daviel S, Dunsiger S R, Hatakeyama A, Kreitzman S R, Levy C D P, Miller R I, Nichol K M, Poutissou R, Dumont E, Greene L H and Kiefl R F 2003 *Physica* B **326** 209
[17] Salman Z, Kiefl R F, Chow K H, MacFarlane W A, Kreitzman S R, Arseneau D J, Daviel S, Levy C D P, Maeno Y and Poutissou R 2006 *Physica* B **374-375** 468
[18] Parolin T J, Shi J, Salman Z, Chow K H, Dosanjh P, Saadaoui H, Song Q, Hossain M D, Kiefl R F, Levy C D P, Pearson M R and MacFarlane W A 2009 *Phys. Rev.* B **80** 174109
[19] Siegel R, Hirschinger J, Carlier D, Matar S, Ménétrier M and Delmas C 2001 *J. Phys. Chem.* B **105** 4166

Journal of Physics: Conference Series **551** (2014) 012010 doi:10.1088/1742-6596/551/1/012010

# Magnetic phase diagram of $Sr_{1-x}Ca_xCo_2P_2$

**J Sugiyama[1], H Nozaki[1], I Umegaki[1], M Harada[1], Y Higuchi[1], E J Ansaldo[2], J H Brewer[2,3], M Imai[4], C Michioka[4], K Yoshimura[4] and M Månsson[5,6]**

[1] Toyota Central Research & Development Laboratories, Inc., 41-1 Yokomichi, Nagakute, Aichi 480-1192, Japan

[2] TRIUMF, 4004 Wesbrook Mall, Vancouver, BC, V6T 2A3 Canada

[3] Department of Physics & Astronomy, University of British Columbia, Vancouver, BC, V6T 1Z1 Canada

[4] Department of Chemistry, Graduate School of Science, Kyoto Univ., Kyoto, 606-8502 Japan

[5] Laboratory for Quantum Magnetism, École Polytechnique Fédérale de Lausanne, CH-1015, Switzerland

[6] Laboratory for Neutron Scattering, Paul Scherrer Institut, CH-5232 Villigen PSI, Switzerland

E-mail: e0589@mosk.tytlabs.co.jp

**Abstract.** In order to study the phase diagram from a microscopic viewpoint, we have measured wTF- and ZF-$\mu^+$SR spectra for the $Sr_{1-x}Ca_xCo_2P_2$ powder samples with $x = 0$, 0.2, 0.4, 0.5, 0.6, 0.8, and 1. Due to a characteristic time window and spatial resolution of $\mu^+$SR, the obtained phase diagram was found to be rather different from that determined by magnetization measurements. That is, as $x$ increases from 0, a Pauli-paramagnetic phase is observed even at the lowest $T$ measured (1.8 K) until $x = 0.4$, then, a spin-glass like phase appears at $0.5 \leq x \leq 0.6$, and then, a phase with wide field distribution probably due to incommensurate AF order is detected for $x = 0.8$, and finally, a commensurate $A$-type AF ordered phase (for $x = 1$) is stabilized below $T_N \sim 80$ K. Such change is most likely reasonable and connected to the shrink of the $c$-axis length with $x$, which naturally enhances the magnetic interaction between the two adjacent Co planes.

## 1. Introduction

The appearance of unconventional superconductivity in iron pnictides with the $ThCr_2Si_2$-type (122) structure, i.e. $Ba_{1-x}K_xFe_2As_2$ reconfirmed the competition between antiferromagnetism and superconductivity [1, 2]. Similar to the K-doping effect, the magnetic state of $CaFe_2As_2$ varies with pressure from antiferromagnetic (AF) to superconducting, and finally to nonmagnetic [3], while its structure changes from an "uncollapsed tetragonal" (uc$T$) phase, an orthorhombic phase, and a "collapsed tetragonal" (c$T$) phase [4]. Here, in the collapsed phase, the ratio of stacking to in-plane lattice parameters ($c/a$) is significantly smaller than that expected from simple atomic size considerations [5].

In contrast to $CaFe_2As_2$ and $BaFe_2As_2$, the related compounds, $AM_2P_2$ with $A =$ Ca, Sr, and Ba, and $M =$ Fe, Co, and Ni do not show unconventional superconductivity, but exhibit an interesting magnetic transition with the structural change. For the present target system, $Sr_{1-x}Ca_xCo_2P_2$, the crystal structure changes from uc$T$ for $x = 0$ to c$T$ for $x = 1$ [6]. According to the electronic and magnetic phase diagram determined from the macroscopic measurements [6], the system evolves from a nonmagnetic metallic ground state to an AF metallic ground state through a crossover composition regime at $x \sim 0.5$. Then, in the $x$ range between 0.8 and 0.9, the system manifests a FM-like ground state within which the magnetic ordering temperature is highest at $x \sim 0.9$. For the sample with $x \geq 0.9$, an AF ground state reappears [7]. A similar behavior was also reported for $Ca(Fe_{1-x}Co_x)_2P_2$ [8], $Ca(Ni_{1-x}Co_x)_2P_2$ [8], and $SrCo_2(Ge_{1-x}P_x)_2$ [9].

Note that $\chi$ measurements usually give us a significant insight into the ground state of magnetically ordered solids. However, such measurements are sometimes not suitable, particularly for the materials exhibiting order with a broad field distribution, *i.e.* when short-range order, random, nearly-random order, or incommensurate order appears in a material, due to the absence of periodic structure and/or the presence of rapid fluctuations. We have therefore performed a $\mu^+$SR experiment on $Sr_{1-x}Ca_xCo_2P_2$ in order to investigate the variation of the magnetic nature with the Ca content ($x$).

## 2. Experimental

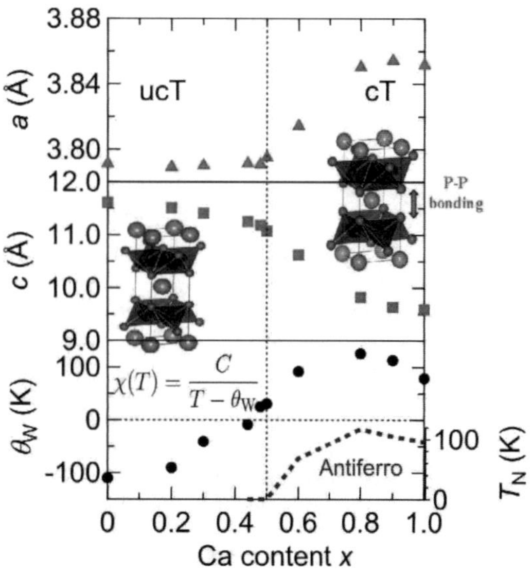


**Figure 1.** The change in the lattice constants and magnetic properties of $Sr_{1-x}Ca_xCo_2P_2$ with $x$ determined by XRD analyses and $\chi$ measurements [11].

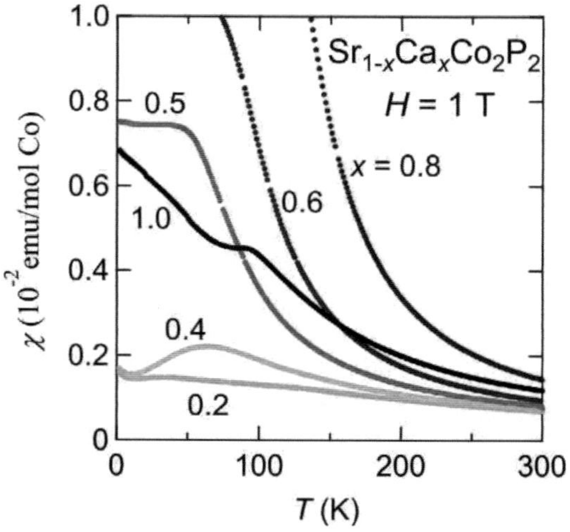


**Figure 2.** $T$ dependence of $\chi$ for $Sr_{1-x}Ca_xCo_2P_2$ with $x \geq 0.2$ [11].

Polycrystalline samples of $Sr_{1-x}Ca_xCo_2P_2$ with $x$ = 0, 0.2, 0.4, 0.5, 0.6, 0.8, and 1 were prepared from elemental P, Sr, Ca, and Co using two step reaction. For the first step, SrP, CaP, and $Co_2P$ were synthesized by a solid state reaction between Sr (Ca, Co) and P in an evacuated quartz tube at 800°C (700°C for $Co_2P$). Then, for the second step, $Sr_{1-x}Ca_xCo_2P_2$ were synthesized by a solid state reaction between SrP, CaP and $Co_2P$ at 1000°C for 20 hours in an Ar atmosphere. After grinding, the obtained powder was fired two times at 1000°C for 40 hours in an Ar atmosphere [10, 11].

According to powder x-ray diffraction (XRD) analyses, all the samples were single phase of tetragonal symmetry with space group $I4/mmm$. Also, as $x$ increases from 0, the $c$-axis length monotonically decreases with $x$ until $\sim 0.5$, then rapidly decreases until 0.9, and finally, levels off to a constant value above $x = 0.9$. This behavior is consistent with the literature [6]. Susceptibility ($\chi$) measurements suggested the presence of magnetic transition above $\sim$ 60 K for the samples with $x \geq 0.6$.

The $\mu^+$SR spectra were measured at surface muon beam lines using the **LAMPF** spectrometer of M20/TRIUMF in Canada. An approximately 500 mg powder sample was placed in an envelope with $1 \times 1$ cm$^2$ area, made with Al-coated Mylar tape with 0.05 mm thickness in order to minimize the signal from the envelope. Then, the envelope was attached to a low-back ground sample holder in a liquid-He flow-type cryostat for measurements in the $T$ range between 1.8 and 150 K.

## 3. Results and discussion

Figure 3 shows the ZF-$\mu^+$SR spectrum for $Sr_{1-x}Ca_xCo_2P_2$ with $x$ = 0.5, 0.6, 0.8, and 1 at the lowest $T$ measured. The ZF-spectrum for $CaCo_2P_2$ exhibits a clear oscillation due to the formation of static AF order below $T_N \sim 80$ K, as reported by neutron diffraction measurements [7]. However, according to the detailed analysis, it was found that the ZF-spectrum consists of two muon-spin precession signals with different frequencies [$f_{AF1}$ = 34.6(1) MHz and $f_{AF2}$ = 9.5(3) MHz at 2 K]. As $T$ increases from 2 K, the two signals disappeared at $T_N$. This suggests that there are two magnetically different muon sites in the $CaCo_2P_2$ lattice.

Although a clear oscillation is also observed in the ZF-spectrum for $Sr_{0.2}Ca_{0.8}Co_2P_2$, the initial phase ($\phi$) was found to delay by $12 - 60°$, when we attempted to fit the oscillatory signal by an exponential relaxing cosine function: $A_0P_{ZF}(t) = A_{AF}\cos(2\pi f_{AF}t + \phi)$. This implies the wide distribution of an internal field at the muon site(s). The most probable reason is that the magnetic order is incommensurate (IC) to the lattice [12]. This is because, due to the mismatch of the period between the muon site(s) and magnetic order, IC order always provides an oscillatory signal with wide field distribution in the ZF-$\mu^+$SR spectrum. But, the

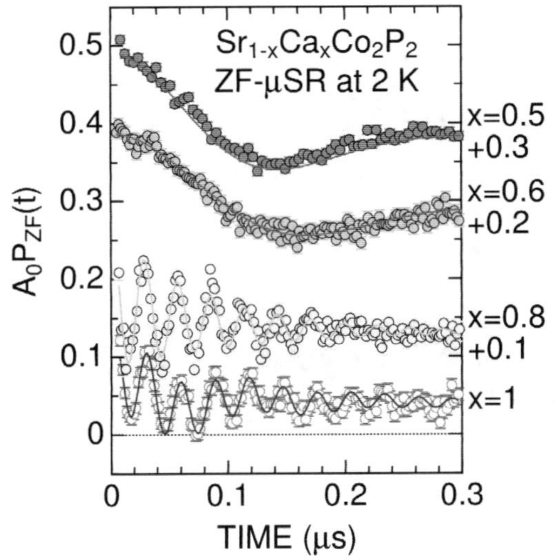


**Figure 3.** The ZF-$\mu^+$SR spectra at 2 K for the $Sr_{1-x}Ca_xCo_2P_2$ sample with $x$ = 0, 0.2, 0.4, and 0.5.

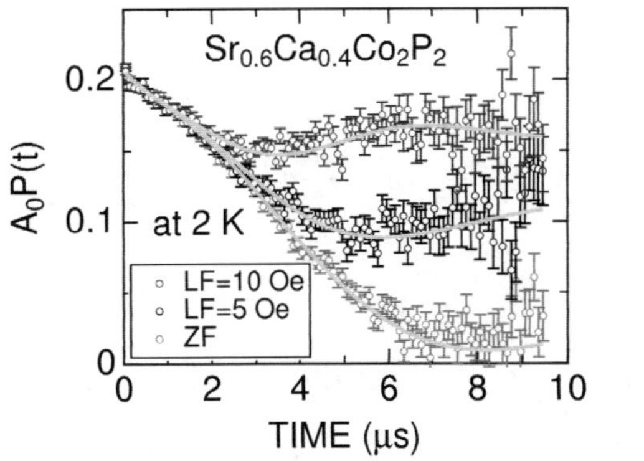


**Figure 4.** The ZF- and LF-$\mu^+$SR spectra at 2 K for the $Sr_{0.6}Ca_{0.4}Co_2P_2$ sample. The applied LF was 5 and 10 Oe. Solid lines represent the fit result using a static Gaussian Kubo-Toyabe function. By fitting both ZF- and LF-spectra using a common field distribution width ($\Delta$), we can obtain reliable $\Delta$.

other explanation is also available, as in the case for $Ag_2NiO_2$ [13] and $LiCrO_2$ [14]. Therefore, neutron diffraction measurements is required for further clarifying the magnetic nature of the $x = 0.8$ sample.

For $Sr_{0.5}Ca_{0.5}Co_2P_2$, the ZF-spectrum shows a static Kubo-Toyabe-type relaxation. Since the field distribution width ($\Delta$) was estimated as $12 \times 10^6$ $s^{-1}$($\sim$ 141 Oe), such KT behavior is naturally caused by disordered Co moments. Furthermore, weal transverse field $\mu^+$SR measurements revealed the appearance of a large internal magnetic field below $\sim$ 50 K. Therefore, the ground state of $Sr_{0.5}Ca_{0.5}Co_2P_2$ is most likely a spin-glass like state, although it is more preferable to measure AC-$\chi$ to confirm the spin-glass state. The situation for $Sr_{0.4}Ca_{0.6}Co_2P_2$ was the same to that for $Sr_{0.5}Ca_{0.5}Co_2P_2$.

On the other hand, for the samples with $x \leq 0.4$, the ZF-spectrum exhibits an almost static KT behavior even at 2 K (Fig. 4), but $\Delta = 0.22 \times 10^6$ $s^{-1}$($\sim$ 2.6 Oe). This is a typical value for nuclear magnetic fields at the muon site(s), and for this case, the origin is mainly due to the nuclear magnetic moments of $^{59}$Co and $^{31}$P. Hence, the ground state for $Sr_{1-x}Ca_xCo_2P_2$ with $x \leq 0.4$ is assigned to be a Pauli-paramagnet.

Based on these results, the magnetic phase diagram for $Sr_{1-x}Ca_xCo_2P_2$ was determined (Fig. 5). That is, thanks to a unique power of $\mu^+$SR, the phase with $x \sim 0.8$ was found to be not an FM phase [6] but an incommensurate (IC) AF phase. Then, next to the IC-AF phase, a spin-glass (SG) like phase appears in the $x$ range between $\sim$ 0.7 and $\sim$ 0.4. The samples with $x \leq 0.4$ was found to be nonmagnetic down to 2 K. However, since the phase boundaries between C and IC, between IC and SG, and SG and PM are still not clarified. we need additional measurements on several $Sr_{1-x}Ca_xCo_2P_2$ samples.

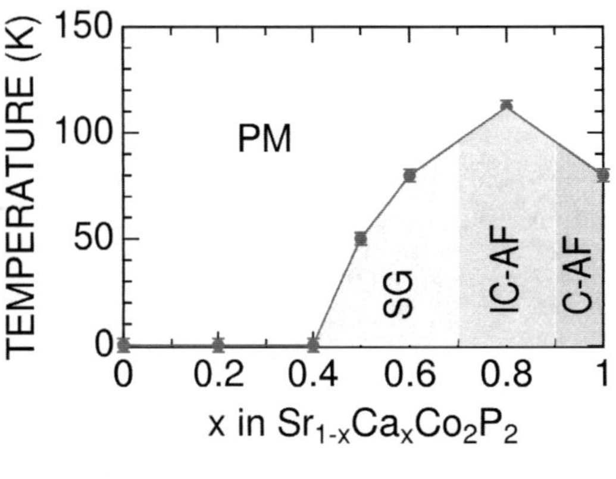


**Figure 5.** The tentative phase diagram for $Sr_{1-x}Ca_xCo_2P_2$ determined by $\mu^+$SR, where PM is a paramagnetic phase, SG is a spin-glass like phase, IC-AF is a phase with wide field distribution probably due to incommensurate AF order, and C-AF is a commensurate AF phase.

**Acknowledgments**
We thank the staff of TRIUMF for help with the $\mu^+$SR experiments. This work was supported by MEXT KAKENHI Grant No. 23108003 and JSPS KAKENHI Grant No. 26286084.

**References**

[1] Rotter M, Tegel M, Johrendt D, Schellenberg I, Hermes W and Pöttgen R 2008 *Phys. Rev.* B **78** 020503
[2] Rotter M, Tegel M and Johrendt D 2008 *Phys. Rev. Lett.* **101** 107006

[3] Torikachvili M S, Bud'ko S L, Ni N and Canfield P C 2008 *Phys. Rev. Lett.* **101** 057006
[4] Kreyssig A, Green M A, Lee Y, Samolyuk G D, Zajde P, Lynn J W, Bud'ko S L, Torikachvili M S, Ni N, Nandi S, Leão J B, Poulton S J, Argyriou D N, Harmon B N, McQueeney R J, Canfield P C and Goldman A I 2008 *Phys. Rev.* B **78** 184517
[5] Reehuis M and Jeitschko W 1990 *J. Phys. Chem. Solids* **51** 961
[6] Jia S, Williams A J, Stephens P W and Cava R J 209 *Phys. Rev.* B **80** 165107
[7] Reehuis M, Jeitschko W, Kotzyba G, Zimmer B and Hu X 1998 *J. Alloys Compounds* **266** 54
[8] Jia S, Chi S, Lynn J W and Cava R J 2010 *Phys. Rev.* B **81** 214446
[9] Jia S, Jiramongkolchai P, Suchomel M R, Toby B H, Checkelsky J G, Ong N P and Cava R J 2011 *Nature Phys.* **7** 207
[10] Imai M 2012 *Syntheses and itinerant-electron magnetic properties of the layered cobalt phosphide* $ACo_2P_2$ *(A = Ca, Sr, Ba, La)* (Kyoto: Kyoto University)
[11] Imai M, Michioka C, Ohta H, Matsuo A, Kindo K, Ueda H and Yoshimura K 2014 *Phys. Rev.* B **90** 014407
[12] Kalvius G M, Noakes D R and Hartmann O 2001 *Handbook on the Physics and Chemistry of Rare Earths* (Amsterdam: North-Holland) vol 32 chapter 206 pp 55–451
[13] Sugiyama J, Ikedo Y, Mukai K, Brewer J H, Ansaldo E J, Chow K H, Yoshida H and Hiroi Z 2006 *Phys. Rev.* B **73** 224437
[14] Sugiyama J, Månsson M, Ikedo Y, Goko T, Mukai K, Andreica D, Amato A, Ariyoshi K and Ohzuku T 2009 *Phys. Rev.* B **79** 184411

Journal of Physics: Conference Series **551** (2014) 012011 doi:10.1088/1742-6596/551/1/012011

# Microscopic magnetic nature of $K_2NiF_4$-type 3*d* transition metal oxides

**J Sugiyama[1,2], H Nozaki[1], I Umegaki[1], W Higemoto[2], E J Ansaldo[3], J H Brewer[3,4], H Sakurai[5], T-H Kao[5,6], H-D Yang[6] and M Månsson[7,8]**

[1] Toyota Central Research & Development Laboratories, Inc., 41-1 Yokomichi, Nagakute, Aichi 480-1192, Japan
[2] Advanced Science Research Center, Japan Atomic Energy Agency, Tokai, Ibaraki 319-1195, Japan
[3] TRIUMF, 4004 Wesbrook Mall, Vancouver, BC, V6T 2A3 Canada
[4] Department of Physics & Astronomy, University of British Columbia, Vancouver, BC, V6T 1Z1 Canada
[5] National Institute for Materials Science, Tsukuba, Ibaraki 305-0044, Japan
[6] Department of Physics, National Sun Yat-Sen University, Kaohsiung 804, Taiwan
[7] Laboratory for Quantum Magnetism, Ecole Polytechnique Fédérale de Lausanne, CH-1015, Switzerland
[8] Laboratory for Neutron Scattering and Imaging, Paul Scherrer Institut, CH-5232 Villigen PSI, Switzerland

E-mail: e0589@mosk.tytlabs.co.jp

**Abstract.** In order to elucidate the magnetic nature of $K_2NiF_4$-type 3*d* transition metal oxides, we have measured $\mu^+$SR spectra for $Sr_2VO_4$, $LaSrVO_4$, and $Sr_2CrO_4$ using powder samples. ZF- and wTF-$\mu^+$SR measurements propose that $Sr_2VO_4$ enters into the static antiferromagnetic (AF) order phase below 8 K. In addition, TF-$\mu^+$SR measurements evidence that the transition at 105 K is not magnetic but structural and/or electronic in origin. For $LaSrVO_4$, static long-range order has not been observed down to 20 K, while, as $T$ decreases from 145 K, wTF asymmetry starts to decrease below 60 K, suggesting the appearance and evolution of localized magnetic moments below 60 K. For $Sr_2CrO_4$, by contrast, both ZF- and wTF-$\mu^+$SR have confirmed the presence of antiferromagnetic order below 117 K, as predicted in the $\chi(T)$ curve.

## 1. Introduction

The discovery of high-$T_c$ superconductivity in Ba-doped $La_2CuO_4$ [1] led to extensive studies of the structural, electrical and magnetic properties of layered perovskites $A_2MO_4$ with a $K_2NiF_4$-type structure, particularly for the compounds with $M$ = Co, Ni, and Cu. However, due to the difficulty of preparing high-quality $A_2MO_4$ samples with $M$ = V, Cr, and Mn, the electronic and magnetic properties of these compounds are still not fully clarified. Very recently, the following high-quality and/or novel $A_2MO_4$ compounds were successfully synthesized under high pressures [2, 3];

- $3d^1$ system: $Sr_2VO_4$
- $3d^2$ system: $LaSrVO_4$, $Sr_2CrO_4$, and $Ca_2CrO_4$

- $3d^3$ system: $LaSrCrO_4$, $YSrCrO_4$, $YCaCrO_4$, $Ca_2MnO_4$ and $Sr_2MnO_4$

Based on resistivity ($\rho$) and susceptibility ($\chi$) measurements, all these compounds were found to be insulators showing magnetic transitions below ambient temperature [2]. This opens the opportunity for a systematic study on the relationship between the number of $d$ electrons and magnetic order and/or magnetic nature in the $K_2NiF_4$-type lattice, particularly with $\mu^+$SR. This is important because it is very difficult (though not impossible) to synthesize large enough samples for neutron scattering measurements using the high-pressure technique.

We have therefore initiated $\mu^+$SR experiments on these materials and a preliminary result on $Sr_2VO_4$ was already published elsewhere [4]. Here we report the $\mu^+$SR results on the $3d^1$ and $3d^2$ system compounds, namely, $Sr_2VO_4$, $LaSrVO_4$ and $Sr_2CrO_4$.

## 2. Experimental

Powder samples of $Sr_2VO_4$, $LaSrVO_4$, and $Sr_2CrO_4$ were prepared by a solid-state reaction technique at high pressures [2, 3]. The $\mu^+$SR spectra were measured at the surface muon beam lines using the **LAMPF** and **HiTime** spectrometer of TRIUMF in Canada.

## 3. Results and discussion

### 3.1. $3d^1$ system: $Sr_2VO_4$

Figure 1 shows the temperature variation of the ZF-$\mu^+$SR spectrum for $Sr_2VO_4$. The ZF-spectrum obtained at the lowest $T$ measured (1.8 K) looks to be fitted by a static Kubo-Toyabe function. However, for such fit, we need two additional exponential relaxing signals in order to reproduce a rapid decay in an early time domain (below 0.1 $\mu$s) and the behavior that $A_0P_{\rm ZF}(t) \sim 0$ at $t \geq 5$ $\mu$s, meaning three different muon sites in the lattice This also indicates that the muon-spin responsible for the KT signal is static, while those for the exponential relaxing signals are dynamic. This is an inconsistent situation, even though the three muon sites are crystallographically equivalent, but magnetically nonequivalent. We therefore fitted the data by a combination of an exponentially relaxing cosine signal for a static internal field and an exponentially relaxing non-oscillatory signal for the 1/3 tail signal;

$$A_0\,P_{\rm ZF}(t) = A_{\rm AF}\exp(-\lambda_{\rm AF}t)\cos(\omega^{\mu}_{\rm AF}t) + A_{\rm tail}\exp(-\lambda_{\rm tail}t). \tag{1}$$

Figure 2 shows the temperature dependences of the $\mu^+$SR parameters in Eq. (1). Combining with wTF-$\mu^+$SR measurements, it is proposed that $Sr_2VO_4$ enters into an antiferromagnetic ordered phase below $T_{\rm N} = 8$ K. However, even at the lowest temperature measured (1.8 K), the distribution of the internal magnetic field ($H_{\rm int}$) is very broad, and the exponential relaxation rate ($\lambda_{\rm AF}$) of the oscillatory signal is eventually comparable to the muon-spin precession frequency [$f_{\rm AF} \equiv \omega^{\mu}_{\rm AF}/(2\pi)$] [Fig. 2(b)]. This supports orbital-stripe order and collinear AF spin order [5, 6] for the ground state of $Sr_2VO_4$ rather than alternating spin-orbital order [7]. This is because, even for the single muon site in the lattice, dipole field calculations showed that there are several $H_{\rm int}$s in the former case, while only one well defined $H_{\rm int}$ would be expected in the latter case (see Fig. 3).

Concerning the 1/3 tail signal, although $A_{\rm tail}$ is only $\sim$ 22% of $A_0$, the magnitude and temperature dependence of $\lambda_{\rm tail}$ are reasonable for the 1/3 tail signal (Fig. 2). The smaller $A_{\rm tail}$ is probably due to the coexistence of a spin-glass like phase, which provides no 1/3 tail signal.

Then, we have measured muonic Knight shift measurements above $T_{\rm N}$ with $H_{\rm TF} = 50$ kOe for further studying the nature of the transition around $T_{\rm c} \sim 105$ K, which is observed by $\chi$ measurements but was not clearly detected by wTF and ZF-$\mu^+$SR. Figure 4 shows the variation of the Fourier transform spectrum of the TF-$\mu^+$SR time spectrum with $T$. The Fourier transform spectrum exhibits a Gauss distribution at 300 K, while it changes to a Lorentz distribution at

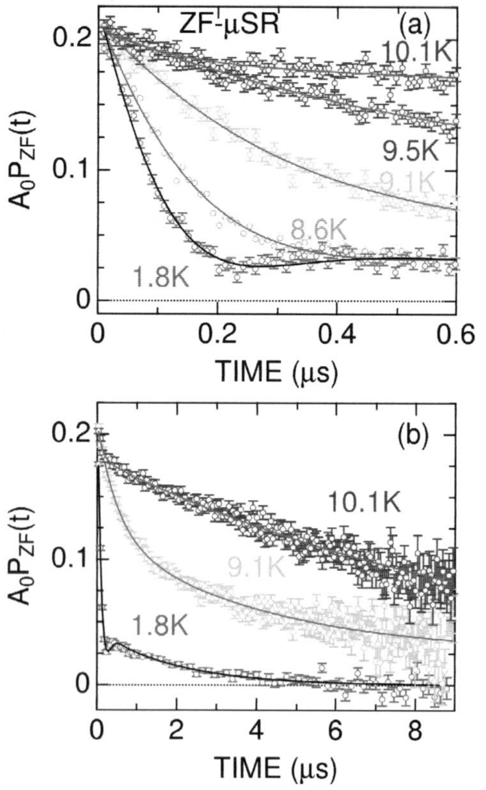


**Figure 1.** $T$ dependence of the ZF-$\mu^+$SR spectrum for $Sr_2VO_4$ below 10 K (a) in an early time domain, $t \leq 0.6$ $\mu$s and (b) in a full time domain measured. Solid lines represent the fit result using Eq. (1).

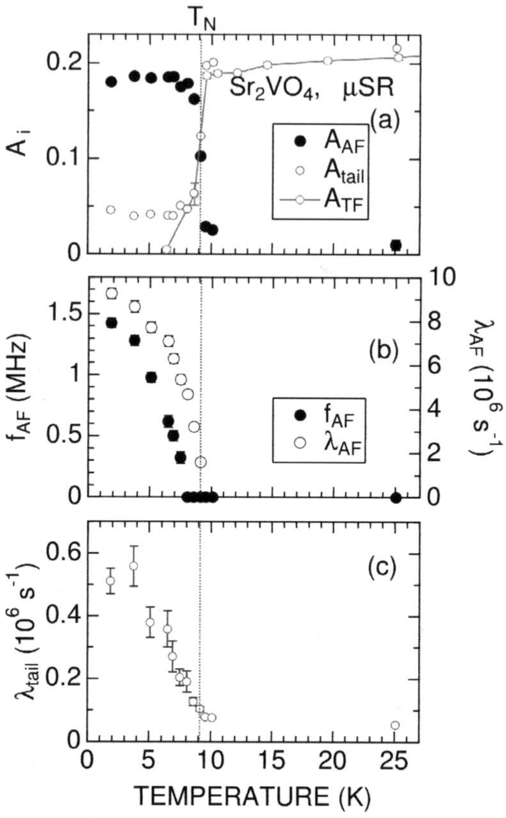


**Figure 2.** $T$ dependences of (a) the asymmetries ($A_{\rm AF}$, $A_{\rm tail}$, and $A_{\rm TF}$), (b) the muon spin precession frequency for the $A_{\rm AF}$ signal ($f_{\rm AF} \equiv \omega_{\rm AF}/2\pi$) and its exponential relaxation rate $\lambda_{\rm AF}$, and (c) the exponential relaxation rate of the tail signal ($\lambda_{\rm tail}$).

10 K. In fact, the TF-$\mu^+$SR spectrum was well fitted with a power exponential relaxed cosine oscillation (using a rotating frame analysis [8]) in the whole $T$ range measured;

$$A_0\,P_{\rm TF}(t) \;=\; A\exp[-(\lambda t)^{\beta}]\cos(\omega^{\mu}t+\phi). \tag{2}$$

Figure 5 shows the temperature dependences of the TF-$\mu^+$SR parameters together with $\chi$ and $1/\chi$. As $T$ decreases from 300 K, $\beta$ is almost $T$-independent (=2) down to $T_{\rm c}$, then decreases gradually with decreasing $T$ and reaches around 1 at $\sim$ 80 K, and then, $\beta$ is $T$-independent again until 30 K, and finally, decreases slightly with further decreasing $T$. Roughly speaking, $\beta = 2$ above $T_{\rm c}$, whereas $\beta = 1$ below $T_{\rm c}$. This means that the phase above $T_{\rm c}$ is paramagnetic and the implanted muons *see* only the nuclear magnetic moments, since the V moments are fluctuating very rapidly. However, below $T_{\rm c}$, thermal fluctuations of V moments become so slow that the muons sense the fluctuating V moments. Such picture is consistent with the $T$ dependences of $\lambda$ and $\chi$.

On the other hand, above $T_{\rm c}$, $\lambda$ increases slightly with decreasing $T$, but below $T_{\rm c}$, $\lambda$

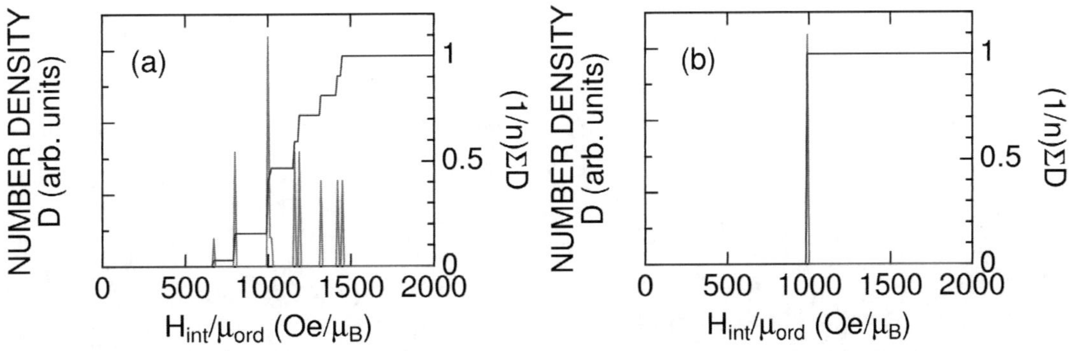


**Figure 3.** Predicted distribution of the internal magnetic field for (a) the orbital-stripe order and collinear AF spin order [5, 6] and (b) alternating spin-orbital order [7] obtained by dipole field calculations. Here, we assumed that the implanted muons locate at the vicinity of the apical oxygen anion 1 Å away; namely, crystallographically one muon site in the lattice.

increases rapidly with increasing a slope ($d\chi/dT$). The muonic Knight shift ($K$) is defined as $K \equiv (\omega^{\mu}_{\rm sample} - \omega^{\mu}_{\rm ref.})/\omega^{\mu}_{\rm ref.}$, where $\omega^{\mu}_{\rm ref.}$ is the angular frequency of a reference ($CaCO_3$). $K$ is found to be negative below 300 K, and decreases with decreasing $T$ down to $T_c$, then, below $T_c$ $K$ increases with further lowering $T$. From the $K-\chi$ plot, the hyperfine coupling constant ($A_{\rm hf} \equiv K/\chi$) above $T_c$ is estimated as -480 Oe/$\mu_B$. However, below $T_c$, $A_{\rm hf}$ increases with decreasing $T$ and approaches 0, when $T \to 0$.

From the $1/\chi(T)$ curve [Fig. 5(c)], the effective magnetic moment ($\mu_{\rm eff}$) of V ions below $T_c$ is found to be smaller by 33% than that above $T_c$. This implies either the formation of a spin-singlet like state below $T_c$ or a ferrimagnetic coupling between the nearest neighboring V ions. In addition, since the $\lambda(T)$ curve rapidly increases with decreasing $T$ below 50 K and the $\beta(T)$ curve starts to decrease from 1 also below 50 K, short-range order would be formed below 50 K. Therefore, it is highly desirable to perform careful inelastic neutron scattering studies in this $T$ range.

### *3.2. 3d$^2$ system: LaSrVO$_4$ and Sr$_2$CrO$_4$*

For $LaSrVO_4$, the $\chi(T)$ curve shows a sharp maximum around $T_N = 10$ K, implying the presence of an AF transition. However, since the past work on $LaSrVO_4$ reported the absence of such maximum [9], it is very important to confirm whether the AF transition exists at low $T$ by $\mu^+$SR. The weak transverse field asymmetry ($A_{\rm TF}$) starts to decrease below $\sim$ 70 K, and decreases linearly with lowering $T$ and reaches 0 at around $T_N$. This means the appearance of localized moments below 70 K and the evolution of them with decreasing $T$, and finally the whole volume of the sample enters into a magnetic ordered or disordered state. Moreover, below around $T_N$, ZF-$\mu^+$SR detected a damped oscillatory signal, which demonstrates the appearance of static magnetic order below $T_N$.

For $Sr_2CrO_4$, the $\chi(T)$ curve shows a broad maximum around $T_N \sim 117$ K, indicating the presence of an AF transition. The $A_{\rm TF}(T)$ curve also shows a step-like change from a full value ($\sim$ 0.23) to zero at $T_N$. This means that the whole volume of the sample enters into an AF ordered state at $T_N$. Below $T_N$, ZF-$\mu^+$SR spectrum includes only one oscillatory signal down to 45 K, but two oscillatory signals below 45 K. This also shows the appearance of static magnetic order below $T_N$ and a slight change in the magnetic structure at 45 K. Although $\chi$ rapidly

Journal of Physics: Conference Series **551** (2014) 012011 doi:10.1088/1742-6596/551/1/012011

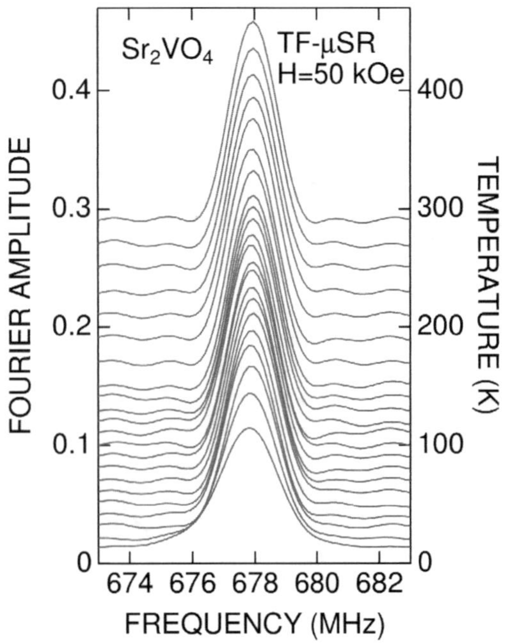


**Figure 4.** $T$ dependence of the Fourier transform spectrum of the TF-$\mu^+$SR time spectrum for $Sr_2VO_4$. The TF spectrum was obtained with $H_{TF} = 50$ kOe.

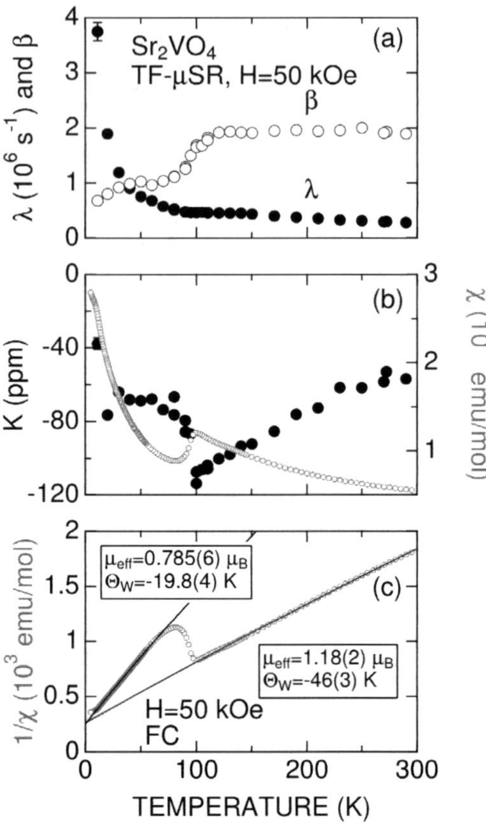


**Figure 5.** $T$ dependences of (a) the power ($\beta$) and relaxation rate ($\lambda$) and (b) the Knight shift ($K$) and $\chi$, and (c) $1/\chi$. The data were obtained by fitting the TF-$\mu^+$SR spectrum with eq. (2). $\chi$ was measured with $H = 50$ kOe in a field cooling mode. In (c), the Curie-Weiss fit [$\chi = \chi_0 + C/(T - \Theta_W)$] in the temperature range between 110 and 320 K provides that $\chi_0 = 4.1(9) \times 10^{-5}$ emu/mol, $\mu_{eff} = 1.18(2)$ $\mu_B$, and $\Theta_W = -46(3)$ K, while the fit in the temperature range between 10 and 70 K provides hat $\chi_0 = 6(1) \times 10^{-5}$ emu/mol, $\mu_{eff} = 0.785(6)$ $\mu_B$, and $\Theta_W = -19.8(4)$ K.

increases with decreasing $T$ below 15 K, there is no critical change in the $f_{AF}(T)$ curve at this $T$ range. Therefore, the reason of the increase in $\chi$ at low $T$ is most likely due to magnetic impurities.

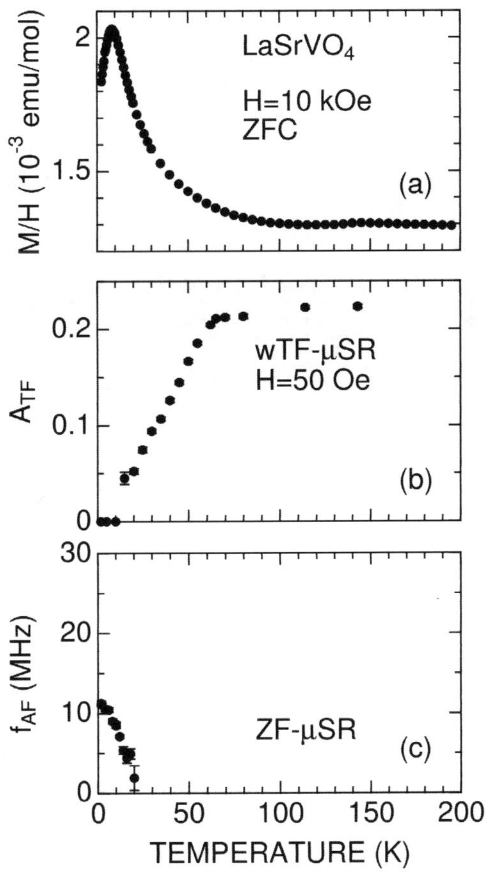


**Figure 6.** $T$ dependences of (a) susceptibility ($\chi$), (b) the weak transverse field asymmetry ($A_{TF}$), and (c) the muon-spin precession frequency ($f_{AF}$) below $T_N$ for $LaSrVO_4$.

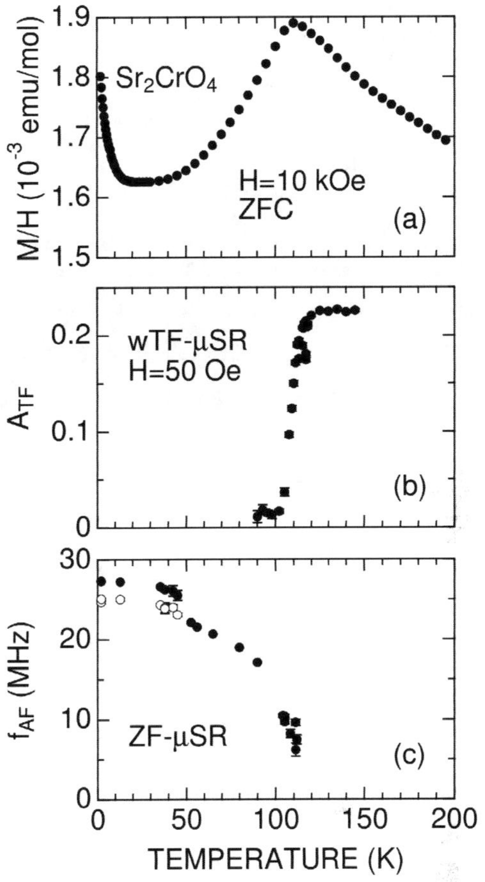


**Figure 7.** $T$ dependences of (a) susceptibility ($\chi$), (b) the weak transverse field asymmetry ($A_{TF}$), and (c) the muon-spin precession frequencies ($f_{AF}$s) below $T_N$ for $Sr_2CrO_4$.

## 4. Summary

We have measured $\mu^+$SR spectra for $Sr_2VO_4$, $LaSrVO_4$, and $Sr_2CrO_4$ using powder samples. For the two complex vanadium oxides, the microscopic magnetic nature detected with $\mu^+$SR is very different from that obtained by susceptibility ($\chi$) measurements. Such discrepancy is probably due to the difference of time-windows between $\mu^+$SR and $\chi$ measurements, together with the different spatial resolution. That is, $\mu^+$SR is a local probe, while $\chi$ provides average information. For $Sr_2CrO_4$, on the other hand, static magnetic order is found to appear below $T_N = 107$ K, at which the $\chi(T)$ curve exhibits a broad maximum.

## Acknowledgments

We thank the staff of TRIUMF and J-PARC for help with the $\mu^+$SR experiments. This work was supported by MEXT KAKENHI Grant No. 23108003 and JSPS KAKENHI Grant No. 26286084.

Journal of Physics: Conference Series **551** (2014) 012011 doi:10.1088/1742-6596/551/1/012011

## References

[1] Bednorz J G and Mueller K A 1986 *Zeitschrift für Physik* B **64** 189
[2] Sakurai H 2013 *in Meeting Abstracts of Phys. Soc. Jpn.* **68**(1)
[3] Sakurai H, Kao T-H, and Yang H-D 2014 *in Meeting Abstracts of Phys. Soc. Jpn.* **69**(1) 565
[4] Sugiyama J, Nozaki H, Umegaki I, Higemoto W, Brewer J H, Ansaldo E J, Sakurai H, Kao T-H, Yang H-D, and Månsson M 2014 *Phys. Rev.* B **89** 020402(R)
[5] Imai Y, Solovyev I, and Imada M 2005 *Phys. Rev. Lett.* **95** 176405
[6] Imai Y and Imada M 2006 *J. Phys. Soc. Jpn.* **75** 094713
[7] Eremin M V, Deisenhofer J, Eremina R M, Teyssier J, van der Marel D, and Loidl A 2011 *Phys. Rev.* B **84** 212407
[8] Riseman T M and Brewer J H 1991 *Hyp. Int.* **65** 1107
[9] Greedan J E and Gong W 1992 *J. Alloys and Comp.* **180** 281

Journal of Physics: Conference Series **551** (2014) 012012 doi:10.1088/1742-6596/551/1/012012

# Structural, magnetic, and diffusive nature of olivine-type $Na_xFePO_4$

**J Sugiyama[1], H Nozaki[1], I Umegaki[1], M Harada[1], Y Higuchi[1], E J Ansaldo[2], J H Brewer[2,3], Y Miyake[4], G Kobayashi[5] and R Kanno[6]**

[1] Toyota Central Research & Development Laboratories, Inc., 41-1 Yokomichi, Nagakute, Aichi 480-1192, Japan

[2] TRIUMF, 4004 Wesbrook Mall, Vancouver, BC, V6T 2A3 Canada

[3] Department of Physics & Astronomy, University of British Columbia, Vancouver, BC, V6T 1Z1 Canada

[4] Muon Science Laboratory, Institute of Materials Structure Science, KEK, Tsukuba, Ibaraki 305-0801, Japan

[5] Research Center of Integrative Molecular Systems, Institute for Molecular Science, Okazaki, Aichi 444-8585, Japan

[6] Department of Electronic Chemistry, Interdisciplinary Graduate School of Science and Engineering, Tokyo Institute of Technology, Yokohama 226-8503, Japan

E-mail: e0589@mosk.tytlabs.co.jp

**Abstract.** In order to investigate the microscopic magnetic nature of a sodium phospho-olivine, we have measured $\mu^+$SR spectra for $Na_xFePO_4$ using a powder sample in the temperature range between 2 and 500 K. ZF-$\mu^+$SR measurements below 200 K at TRIUMF showed the appearance of static magnetic order below $T_N \sim 61$ K, where the susceptibility-vs.-temperature curve exhibits a sharp maximum due to an antiferromagnetic (AF) transition. The wide distribution found for the internal magnetic field ($H_{int}$) was explained by the formation of AF order with the spread in $H_{int}$ due to random distribution of Na vacancies. At higher temperatures, the ZF-spectrum for $Na_{0.7}FePO_4$, obtained at J-PARC, is found to change from a low-$T$ static behavior to a high-$T$ dynamic behavior above $\sim 300$ K. This is consistent with the fact that $Na^+$ ions are reversibly extracted from and intercalated into the $Na_xFePO_4$ lattice, while there is no NMR study on Na-diffusion in $Na_xFePO_4$ due to large Fe moments.

## 1. Introduction

Recently, Na-ion battery materials attracted great attention [1, 2, 3], since sodium is more abundant than Li, resulting in lower material costs. For instance, $Na_xCoO_2$, which is an analogue to $LiCoO_2$, was originally studied as a positive electrode material for Na-ion batteries [4], before the huge amount of work on its thermoelectric [5], magnetic [6, 7], and superconducting properties [8]. In 2010, phospho-olivines with Na, i.e. $NaFePO_4$ and $Na_{0.7}FePO_4$ were reported to show reversible insertion and extraction of Na into olivine $FePO_4$ [9]. In the initial study, only the above two compositions were obtained as single phase. More detailed study, however, suggests that $Na_xFePO_4$ with $2/3 \le x \le 1$ is a single phase region below ambient temperature [10]. Since then, the electrochemical properties of $Na_xFePO_4$ have been naturally eagerly investigated [11, 12, 13, 14, 15, 16], while the physical properties are still not fully clarified, partially due to difficulty of sample preparation.

Very recent specific heat and neutron diffraction (ND) measurements showed that $NaFePO_4$ is also an antiferromagnet with $T_N \sim 50$ K and a collinear spin structure along the $b$-axis [17], as for $LiFePO_4$ with $T_N = 53$ K. But, since $Li_xFePO_4$ is known to be a mixture between $x$ $LiFePO_4$ and $(1-x)$ $FePO_4$ [18, 19], a single phase sample of $Li_xFePO_4$ with $0 < x < 1$ can not be obtained by usual chemical and electrochemical reactions. Fortunately, that is not the case for $Na_xFePO_4$, for which the change in physical properties can be investigated in single-phase samples with $x$ down to 2/3. Particularly, due to Na deficiency together with low-dimensionality, a complex magnetic behavior, such as short-range order or incommensurate order could be expected to appear at low temperatures, as for $Li_xCoO_2$ [20] and $Li_xNiO_2$ [21]. Here, we report initial $\mu^+$SR studies of the local magnetic environment and diffusion properties of $Na_{0.7}FePO_4$

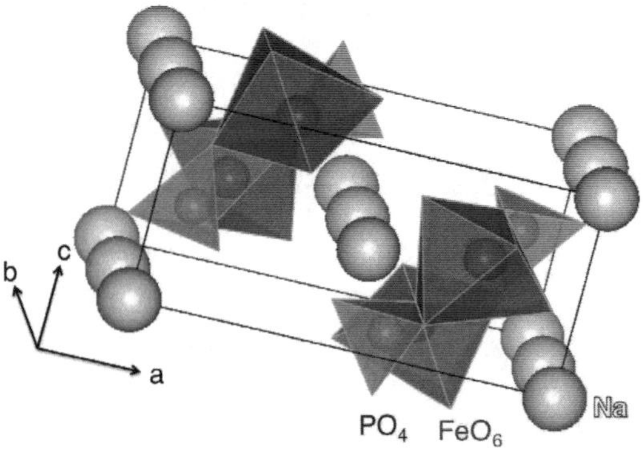


**Figure 1.** Crystal structure of $Na_{0.7}FePO_4$. The Rietvelt refinement of a power XRD pattern obtained at 295 K indicates as follows; space group = $Pnma$, $a$ = 10.2841(2) Å, $b$ = 6.0800(2) Å, and $c$ = 4.9373(2) Å. The atomic positions are; Na: (0, 0, 0), Fe: [0.28392(6), 3/4, 0.9860(2)], P: [0.1065(2), 3/4, 0.4391(3)], O1: [-0.0391(4), 3/4, 0.3502(6)], O2: [0.6200(3), 3/4, -0.2460(6)], and O3: [0.1766(3), 0.5520(4), 0.3035(4)]. The occupancy of the Na site is 0.7, indicating random distribution of Na.

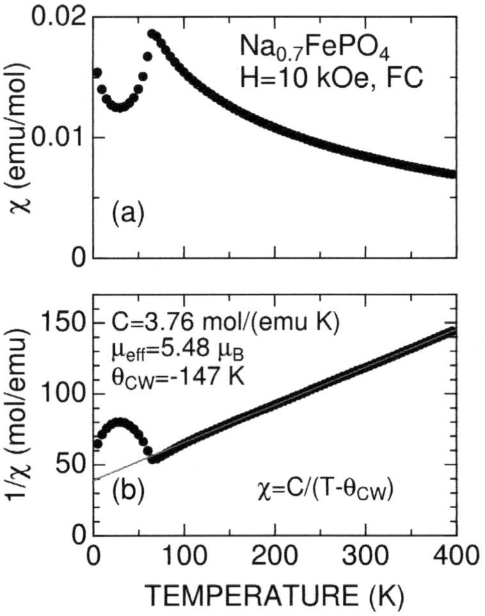


**Figure 2.** $T$ dependences of (a) $\chi$, and (b) $1/\chi$ for $Na_{0.7}FePO_4$. In (b), solid lines represent a linear fit in the $T$ range between 100 and 400 K using a Curie-Weiss formula.

## 2. Experimental

As starting material a powder sample of $LiFePO_4$ was prepared by a solid-state reaction technique using reagent grade $Li_2CO_3$, $Fe(II)C_2O_4{*}2H_2O$, and $(NH_4)_2HPO_4$. A stoichiometric mixture of the raw materials was thoroughly mixed by a conventional planetary milling apparatus, and then sintered at 700°C for 6 h in a purified argon gas flow. Then, in order to obtain $FePO_4$, the $LiFePO_4$ powder was immersed into an acetonitrile solution of $NO_2BF_4$ and was refluxed at room temperature for a day. Finally, in order to obtain $Na_xFePO_4$, the $FePO_4$ powder was immersed into an acetonitrile solution of NaI and was refluxed at room temperature for a day.

According to powder x-ray diffraction (XRD) analysis, the $Na_xFePO_4$ sample consists of a major $Na_{0.7}FePO_4$ phase (see Fig. 1) and a minor $FePO_4$ phase. The volume fraction of the $FePO_4$ phase was estimated below 3%. The XRD result is very consistent with that of the composition analyses by an inductively coupled plasma atomic emission spectral (ICP-AES) technique. In order to know the macroscopic magnetic properties, susceptibility ($\chi$) was measured below 400 K under a $H \leq 10$ kOe field with a SQUID magnetometer (MPMS, Quantum Design). The $\chi(T)$ curve clearly indicates the presence of the AF transition at $T_N = 66$ K (Fig. 2).

The $\mu^+$SR spectra were measured at the surface muon beam lines using the **LAMPF** spectrometer of TRIUMF in Canada and the **D-OMEGA1** spectrometer of MUSE/MLF/J-PARC in Japan. In TRIUMF, the approximately 500 mg powder sample was placed in an envelope with 1 cm×1 cm area, which is made of very thin Al-coated Mylar tape, and then the envelope was attached to a low-background sample holder in a liquid-He flow-type cryostat for the $T$ range between 1.8 and 250 K. In J-PARC, a $\sim$ 2 g powder sample was pressed into a disc with 23 mm diameter and 1 mm thickness, and packed into an Au O-ring sealed titanium cell. The window of the cell was made of a titanium film with 25 $\mu$m thickness. The cell was mounted onto the Cu plate of a liquid-He flow-type cryostat for measurements between 10 and 500 K.

## 3. Results and discussion

### *3.1. low temperature magnetism*

We have measured the wTF-$\mu^+$SR spectrum with $H_{TF} = 50$ Oe to study the volume fraction of non-magnetic phases in the sample. Figure 3 shows the $T$ dependences of the normalized $A_{TF}$ ($N_{ATF} \equiv A_{TF}/A_0$) and $\lambda_{TF}$. The $A_{TF}(T)$ curve shows a step-like change at $T_N^{TF} = 64.5$ K, at which $N_{ATF} = 0.5$, indicating the sample is almost single phase.

Figure 4 shows the ZF-$\mu^+$SR time spectrum at the lowest $T$ measured (2 K). One can clearly

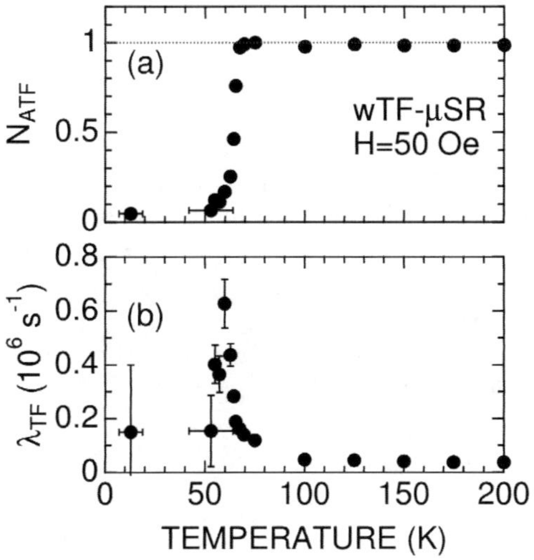


**Figure 3.** $T$ dependences of (a) the normalized wTF asymmetry and (b) its exponential relaxation rate ($\lambda_{TF}$) for $Na_{0.7}FePO_4$. The data were obtained by fitting the wTF-$\mu^+$SR spectra measured with $H_{TF} = 50$ Oe.

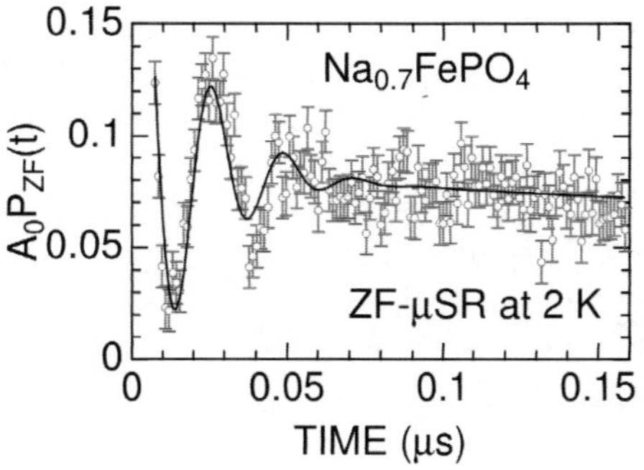


**Figure 4.** The ZF-$\mu^+$SR time-spectrum for $Na_{0.7}FePO_4$ at 2.0 K.

see a rapidly damping oscillation. Indeed, it is hard to find a third maximum even at 2 K. Such situation is easily understood from the Fourier Transform of the time spectrum. That is, the full width at half maximum (FWHM) of the peak is comparable to the oscillation frequency, reflecting the wide distribution of $H_{\rm int}$. The ZF-spectrum was fitted by a combination of a Gaussian relaxing cosine oscillation due to the ordered field, and two exponentially relaxing non-oscillatory signals; one is for a fast relaxing signal apparent even in the ordered state, and the other, labeled slow signal, for the typical ZF "1/3 tail" due to the field component parallel to the initial muon-spin polarization.

$$\begin{aligned} A_0\, P_{\rm ZF}(t) &= A_{\rm AF}\cos(2\pi f_{\rm AF}t+\phi_{\rm AF})\exp(-\frac{\sigma_{\rm AF}^2 t^2}{2}) \\ &+ A_{\rm f}\exp(-\lambda_{\rm f}t)+A_{\rm s}\exp(-\lambda_{\rm s}t), \end{aligned} \quad (1)$$

where $P_{\rm ZF}(t)$ is the muon spin polarization function under ZF, $A_{\rm AF}$, $A_{\rm f}$, and $A_{\rm s}$ are the asymmetries associated with the three signals, and $\sigma_{\rm AF}$, $\lambda_{\rm f}$, and $\lambda_{\rm s}$ ($\lambda_{\rm f} \gg \lambda_{\rm s}$) are their Gaussian and exponential relaxation rates. The fit using Eq. (1) at each $T$ showed that $A_{\rm AF}$, $A_f$, $A_s$, and $\lambda_{\rm f}$ are almost $T$ independent. At 2 K, $\phi_{\rm AF} = -17$ degree and, as $T$ increases from 2 K to 45 K, $\phi_{\rm AF}$ decreases slowly down to -30 degree, then decreases rapidly down to -70 deg with further increasing $T$. Therefore, we finally fitted all the ZF-spectra using common $A_{\rm AF}$, $A_{\rm f}$, $\lambda_{\rm f}$, and $A_{\rm s}$. Such a "global fit" provides that $A_{\rm AF} = 0.090(2)$, $A_{\rm f} = 0.0250(9)$, $\lambda_{\rm f} = 1.1(1)\times 10^6$ s$^{-1}$, and $A_{\rm s} = 0.0577(12)$.

Figure 5 show the $T$ dependences of the four independent $\mu^+$SR parameters, i.e. $f_{\rm AF}$, $\phi_{\rm AF}$, $\sigma_{\rm AF}$ and $\lambda_{\rm s}$. The $f_{\rm AF}(T)$ curve exhibits an order-parameter-like $T$-dependence. Namely, as $T$ decreases from $T_{\rm N}$, $f_{\rm AF}$ increases monotonically down to the lowest $T$ measured with decreasing slope ($df_i/dT$). $\phi_{\rm AF}$ shows a remarkable $T$-variation below the vicinity of $T_{\rm N}$ probably due to the change in the field distribution, while it looks almost $T$-independent below 45 K. The relaxation rate $\sigma_{\rm AF}$ exhibits a maximum at $T_{\rm N}$, then decreases with decreasing $T$, and levels off to a constant value below $\sim$ 50 K [Fig. 5(b)]. On the contrary, $\lambda_{\rm s}$ approximately decreases with decreasing $T$, and finally approaches 0 with $T \to 0$. The fact that $\lambda_{\rm s} \ll \sigma_{\rm AF}$ is also very reasonable for a 1/3 tail signal.

Since there are four Na ions in the $NaFePO_4$ lattice, we attempted to calculate dipole fields in $Na_3^+\square Fe_3^{2+}Fe^{3+}(PO_4)_4^{3-}$ (= $Na_{3/4}FePO_4$) using the same structural data for $NaFePO_4$ and the assumption that the implanted muons locate at the vicinity of $O^{2-}$ ions 1 Å away. We further assumed that the valence of the Fe ion close to the vacant Na site (□) is not 2+ but 3+ in order to keep charge neutrality. Such a simple model drastically altered the electrostatic potential around each oxygen, resulting in 16 different muon sites in the lattice. In addition, if we assume a collinear AF spin structure along the $b$-axis with full Fe moments for $Na_{3/4}FePO_4$ as proposed for $NaFePO_4$ [17], dipole field calculations provided an internal field $[(\gamma_\mu/2\pi)H_{{\rm int},i} = f_i, (i = 1-16)]$ for each muon site, where $\gamma_\mu/2\pi(= 13.554$ kHz/Oe) is the muon gyromagnetic ratio. The predicted distribution of $H_{\rm int}$ reasonably reproduces the measured $H_{\rm int}$ (see Fig. 6). Here, we define the number density ($D_i = 1$) for each $f_i$ and $A_{\rm real}$ for the real amplitude of the Fourier transform of the ZF-$\mu^+$SR time spectrum. Assuming that $f_{\rm calc}$ at which $\frac{1}{16}\sum_{i=1}^{16} D_i = 0.5$ is equivalent to $f$ at which $\int_0^f A_{\rm real}df = 0.5\int_0^{100{\rm MHz}} A_{\rm real}df$, $\mu_{\rm ord}$ is estimated as 48% for the full Fe moments at 2 K; that is, 2.3 $\mu_{\rm B}$ for $Fe^{3+}$ and 2.8 $\mu_{\rm B}$ for $Fe^{2+}$. This is consistent with the presence of the fast relaxing signal in the ZF-spectrum even at 2 K. Note that the employed model naturally leads to the order of Na vacancies and $Fe^{3+}$. For the disordered state, which is more suitable for $Na_{0.7}FePO_4$, the field distribution is expected to be broader and, therefore, this model is the more reasonable approach for interpreting the data.

In the above discussion, we simply reduced both $Fe^3$ and $Fe^{2+}$ moments with the same factor (48%) to the full moments. This is because $\chi$ measurements with $H = 10$ kOe (Fig. 2) show a typical AF nature and luck an indication for ferrimagnetism, which might change the sum

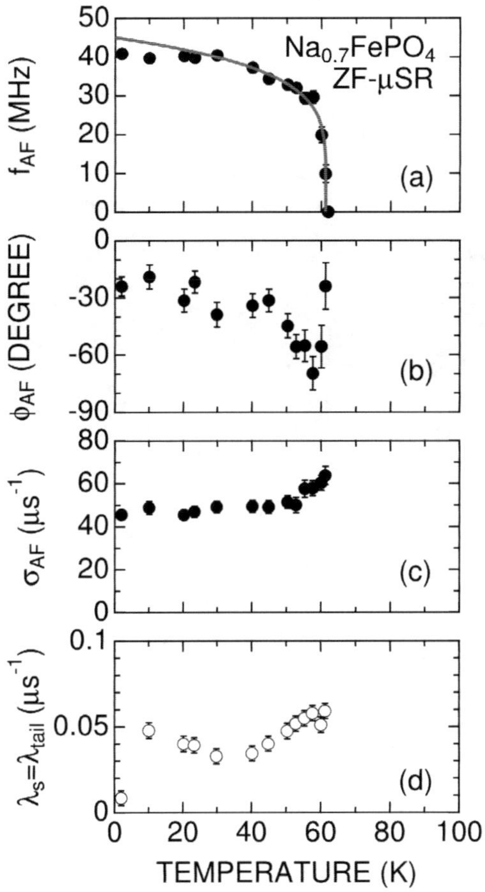


**Figure 5.** $T$ dependences of (a) the muon-spin precession frequency ($f_{\rm AF}$), (b) its initial phase ($\phi_{\rm AF}$), (c) its Gaussian relaxation rate ($\sigma_{\rm AM}$), and (d) the exponential relaxation rate ($\lambda_{\rm tail}$) for the 1/3 tail signal. In (a), the red solid line on the $f_{\rm AF}(T)$ curve corresponds to $f(T)/f(0\ {\rm K}) = [(T_{\rm N} - T)/T_{\rm N}]^{\beta}$ with $f(0\ {\rm K}) = 45(2)$ MHz, $T_{\rm N} = 61.22(2)$ K, and $\beta = 0.18(3)$.

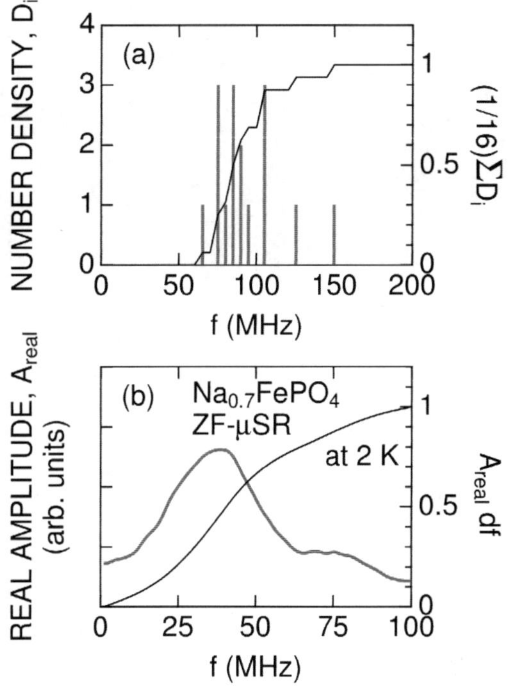


**Figure 6.** (a) Dipolar field distribution for $Na_{3/4}FePO_4$ and (b) the Fourier transformed frequency spectrum of the ZF-$\mu^+$SR time spectrum at 2 K. In (a) and (b), black solid lines represent the integral of the number density or amplitude.

of the moments. In order to further clarify the magnetic structure, we definitely need neutron diffraction measurements for $Na_{0.7}FePO_4$.

Back to the fit results, the fact that $\phi_{\rm AF} \sim -30$ degree below 45 K implies the wide distribution of an internal field at the muon site(s). Because of the mismatch of the period between the muon site(s) and magnetic order, incommensurate (IC) order always provides an oscillatory signal with wide field distribution in the ZF-$\mu^+$SR spectrum. Thus, the delay of $\phi_{\rm AF}$ is thought to be a signature of IC magnetic order [22]. However, if there are multiple muon sites with different $H_{\rm int}$s, the ZF-$\mu^+$SR spectrum also shows a damped oscillation due to wide field distribution, as in the case for $Ag_2NiO_2$ [23] and $LiCrO_2$ [24]. Based on the above discussion, the latter is most likely for $Na_{0.7}FePO_4$.

*3.2. high temperature diffusion*
In the paramagnetic state above $T_N$, the ZF-$\mu^+$SR spectrum is well explained by an exponentially relaxing dynamic Gaussian Kubo-Toyabe function $[\exp(-\lambda t)G^{DGKT}(\Delta, \nu, t, H_{LF})]$ caused by a fluctuating nuclear magnetic field due to $^{23}$Na and $^{31}$P and an offset background (BG) signal from the fraction of muons stopped mainly in the sample cell, which is made of high-purity titanium (Fig. 7). The multiplied exponential relaxation $[\exp(-\lambda t)]$ is caused by paramagnetic Fe moments [25].

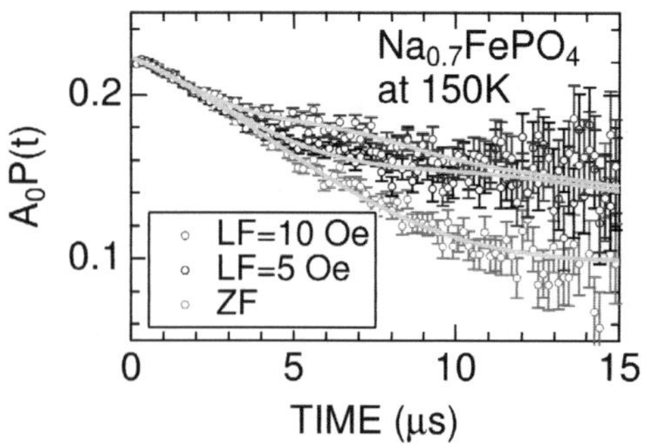


**Figure 7.** ZF- and LF-$\mu^+$SR spectra for $Na_{0.7}FePO_4$ obtained at 150 K. Applied LF were 5 and 10 Oe.

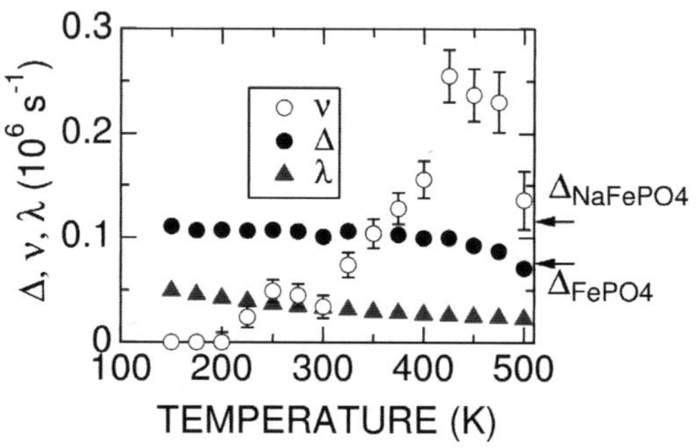


**Figure 8.** $T$-dependences of the field distribution width ($\Delta$), field fluctuation rate ($\nu$), and exponential relaxation rate ($\lambda$). Two arrows on the right vertical axis show the predicted $\Delta$ at the most stable muon site in $NaFePO_4$ and $FePO_4$.

Figure 8 shows the $T$ dependences of $\mu^+$SR parameters. Here, we assumed that the $T$ dependence of $\lambda$ is similar to that of $\chi$. As $T$ increases from 150 K, $\Delta$ is almost $T$-independent until 400 K, and decreases slightly with $T$. On the other hand, $\nu$ starts to increase above around 200 K, reaches a maximum at 425 K, and then decreases with further increasing $T$. Making comparison with the result of $LiFePO_4$ [26, 27], the reason for the increase in $\nu$ is most likely due to Na-diffusion in the lattice.

## Acknowledgments

We thank the staff of TRIUMF and J-PARC for help with the $\mu^+$SR experiments. This work was supported by MEXT KAKENHI Grant No. 23108003 and JSPS KAKENHI Grant No. 26286084.

## References

[1] Palomares V, Serras P, Villaluenga I, Hueso K B, Carretero-Gonzalez J and Rojo T 2012 *Energy Environ. Sci.* **5** 5884
[2] Pan H, Hu Y-S and Chen L 2013 *Energy Environ. Sci.* **6** 2338
[3] Månsson M and Sugiyama J 2013 *Physica Scripta* **88** 068509
[4] Molenda J, Delmas C and Dordor P 1984 *Solid State Ionics* **12** 473
[5] Terasaki I, Sasago Y and Uchinokura K 1997 *Phys. Rev.* B **56** R12685

[6] Sugiyama J, Itahara H, Brewer J H, Ansaldo E J, Motohashi T, Karppinen M and Yamauchi H 2003 *Phys. Rev.* B **67** 214420
[7] Sugiyama J, Brewer J H, Ansaldo E J, Itahara H, Tani T, Mikami M, Mori Y, Sasaki T, Hébert S and Maignan A 2004 *Phys. Rev. Lett.* **92** 017602
[8] Takada K, Sakurai H, Takayama-Muromachi E, Izumi F, Dilanian R A and Sasaki T 2003 *Nature* **422** 53
[9] Moreau P, Guyomard D, Gaubicher J and Boucher F 2010 *Chem. Mater.* **22** 4126
[10] Lu J, Chung S C, Nishimura S-, and Yamada A 2013 *Chem. Mater.* **25** 4557
[11] Zaghib K, Trottier J, Hovington P, Brochu F, Guerfi A, Mauger A and Julien C 2011 *J. Power Sources* **196** 9612
[12] Oh S-M, Myung S-T, Hassoun J, Scrosati B and Sun Y-K 2012 *Electrochem. Commun.* **22** 149
[13] Casas-Cabanas M, Roddatis V V, Saurel D, Kubiak P, Carretero-Gonzalez J, Palomares V, Serras P and Rojo T 2012 *J. Mater. Chem.* **22** 17421
[14] Zhu Y, Xu Y, Liu Y, Luo C and Wang C 2013 *Nanoscale* **5** 780
[15] Wongittharom N, Lee T-C, Wang C-H, Wang Y-C and Chang J-K 2014 *J. Mater. Chem.* A **2** 5655
[16] Galceran M, Saurel D, Acebedo B, Roddatis V V, Martin E, Rojo T and Casas-Cabanas M 2014 *Phys. Chem. Chem. Phys.* **16** 8837
[17] Avdeev M, Mohamed Z, Ling C D, Lu J, Tamaru M, Yamada A and Barpanda P 2013 *Inorg. Chem.* **52** 8685
[18] Padhi A K, Nanjundaswamy K S, and Goodenough J B 1997 *J. Electrochem. Soc.* **144** 1188
[19] Zhang W-J 2011 *J. Power Sources* **196** 2962
[20] Mukai K, Ikedo Y, Nozaki H, Sugiyama J, Nishiyama K, Andreica D, Amato A, Russo P L, Ansaldo E J, Brewer J H, Chow K H, Ariyoshi K and Ohzuku T, 2007 *Phys. Rev. Lett.* **99** 087601
[21] Sugiyama J, Mukai K, Ikedo Y, Russo P L, Nozaki H, Andreica D, Amato A, Ariyoshi K and Ohzuku T 2008 *Phys. Rev.* B **78** 144412
[22] Kalvius G M, Noakes D R, and Hartmann O 2001 *Handbook on the Physics and Chemistry of Rare Earths* (Amsterdam: North-Holland) vol 32 chapter 206 pp 55–451
[23] Sugiyama J, Ikedo Y, Mukai K, Brewer J H, Ansaldo E J, Chow K H, Yoshida H and Hiroi Z 2006 *Phys. Rev.* B **73** 224437
[24] Sugiyama J, Månsson M, Ikedo Y, Goko T, Mukai K, Andreica D, Amato A, Ariyoshi K and Ohzuku T 2009 *Phys. Rev.* B **79** 184411
[25] Matsuzaki T, Nishiyama K, Nagamine K, Yamazaki T, Senba M, Bailey J M and Brewer J H 1989 *Phys. Lett.* A **123** 91
[26] Sugiyama J, Nozaki H, Harada M, Kamazawa K, Ofer O, Månsson M, Brewer J H, Ansaldo E J, Chow K H, Ikedo Y, Miyake Y, Ohishi K, Watanabe I, Kobayashi G and Kanno R 2011 *Phys. Rev.* B **84** 054430
[27] Baker P J, Franke I, Pratt F L, Lancaster T, Prabhakaran D, Hayes W and Blundell S J 2011 *Phys. Rev.* B **84** 174403

Journal of Physics: Conference Series **551** (2014) 012013 doi:10.1088/1742-6596/551/1/012013

# Internal magnetic field in the zigzag-chain family $(Na,Ca)Cr_2O_4$

**H Nozaki[1], H Sakurai[2], M Harada[1], Y Higuchi[1], J H Brewer[3,4], E J Ansaldo[3] and J Sugiyama[1]**

[1] Toyota Central Research & Development Laboratories, Inc., 41-1 Yokomichi, Nagakute, Aichi 480-1192, Japan
[2] National Institute for Materials Science, Namiki 1-1, Tsukuba, Ibaraki 305-0044, Japan
[3] TRIUMF, 4004 Wesbrook Mall, Vancouver, BC, V6T 2A3 Canada
[4] Department of Physics & Astronomy, University of British Columbia, Vancouver, BC, V6T 1Z1 Canada

E-mail: h-nozaki@mosk.tytlabs.co.jp

**Abstract.** In order to elucidate the magnetic nature for a novel one-dimensional zigzag chain compound, $NaCr_2O_4$, we have measured $\mu^+$SR spectra using a powder sample in the temperature range between 2 and 200 K. Weak transverse field (wTF-) $\mu^+$SR measurements indicated that the whole volume of the sample enters into an antiferromagnetic (AF) phase below $T_N$ = 125 K. The zero field (ZF-) $\mu^+$SR spectrum obtained below $T_N$ exhibits a clear oscillation with a single muon-spin precession frequency ($f_\mu$). This suggests that static AF order is formed below $T_N$ and that all the implanted muons sense the same internal magnetic field. The temperature dependence of $f_\mu$ was found to be very similar to that for the intensity of the magnetic Bragg peak in neutron diffraction (ND) measurements. On the other hand, the ZF-$\mu^+$SR spectrum for the isostructural compound, $\beta$-$CaCr_2O_4$, showed a rapidly damped oscillation below $T_N$ = 21 K, supporting the formation of incommensurate AF order, as proposed by ND.

## 1. Introduction

The magnetic interaction on triangular crystal lattices is often frustrated in the sense that not all pair-wise interactions are satisfied. Such frustration leads to unconventional ground states, such as glassy, spin-ice and spin-liquid-like states. Among such frustrated systems, one-dimensional systems offer a typical showcase that relatively simple systems display glamorous magnetic phenomena [1, 2] and complex phases [3, 4, 5, 6, 7]. This is mainly due to the competition between nearest-neighbor (intra-chain) and next-nearest-neighbor (inter-chain) interactions. Following progress in the development of sample synthesis techniques, particularly synthesis under high-pressures, many novel geometrically frustrated materials have been found, resulting in a new dawn for the research of this field.

In order to identify the magnetic nature of several one-dimensional (1D) systems, we have previously studied with $\mu^+$SR the $Na_xCa_{1-x}V_2O_4$ family [8, 9, 10], the $AMn_2O_4$ with $A$ =Li, Na, [11] the $EuL_2O_4$ where $L$=Eu, Gd, Yb, Lu [12], and the $A_2Cr_8O_{16}$ with $A$ =K, Rb [13]. Following upon these experiments, we have started a $\mu^+$SR experiment on the new 1D *zigzag*-chain family $Na_xCa_{1-x}Cr_2O_4$ [14], which is a solid solution system between $\beta$-$CaCr_2O_4$ [15, 16]

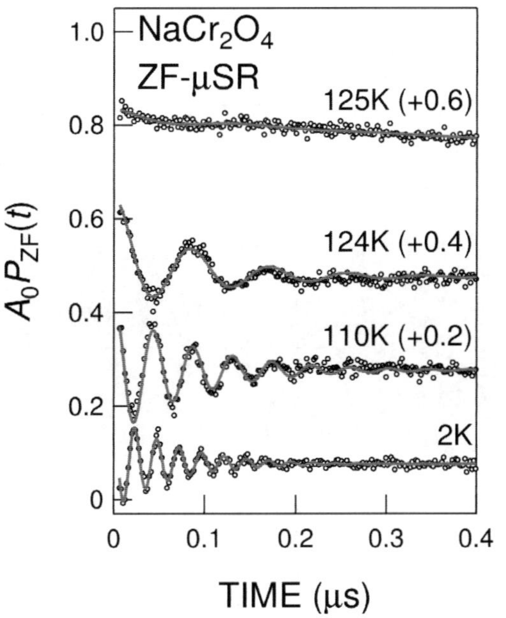


**Figure 1.** $T$ dependence of the ZF-$\mu^+$SR time-spectrum for $NaCr_2O_4$. Besides the spectrum at 2 K, each spectrum is shifted by 0.2 for clarity of display.

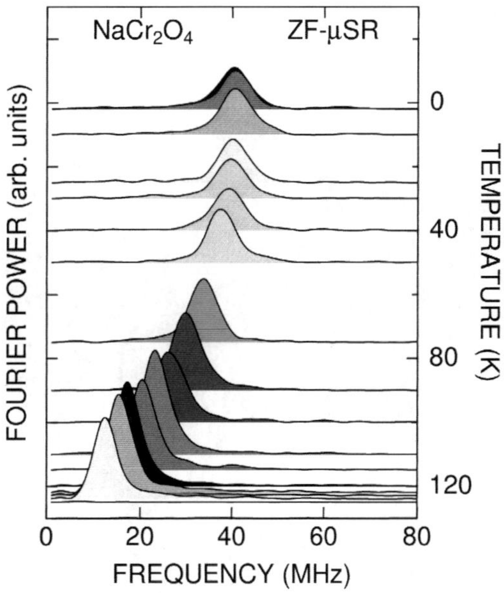


**Figure 2.** $T$ dependence of the Fourier power spectrum for the ZF-$\mu^+$SR time-spectrum.

and novel $NaCr_2O_4$ [17]. Both compounds possess a $CaFe_2O_4$ (CFO)-type *Pnma* structure and the goal in this experiment is to clarify the magnetic ground state of $NaCr_2O_4$ and to determine the magnetic phase diagram as a function of Ca content. Here, we report the $\mu^+$SR result on $NaCr_2O_4$ and $\beta$-$CaCr_2O_4$, and compare it with the result obtained by neutron diffraction (ND) measurements.

## 2. Experimental

A polycrystalline sample of $NaCr_2O_4$ was synthesized under 7 GPa at 1573 K for 12 h. A stoichiometric mixture of $NaCrO_2$, $Cr_2O_3$, and $CrO_3$ was mixed and sealed in a gold capsule in a glove box filled with Ar gas. Then, the mixture was pressed in a belt-type press and then heated for 12 h. $NaCrO_2$ was prepared by firing a stoichioemetric mixture of $Na_2CO_3$ and $Cr_2O_3$ at 1123 K for 15 h in a flowing Ar gas.

A polycrystalline sample of $\beta$-$CaCr_2O_4$ was synthesized by a conventional solid-state reaction under ambient pressure. The stoichiometric mixture of $CaCO_3$ and $Cr_2O_3$ was fired twice in an Ar gas flow at 1573 K for ~12 h with an intermediate regrinding.

The details of the synthesis were described elsewhere [14]. A powder x-ray diffraction (XRD) analysis showed that the sample was almost single phase with an orthorhombic system of space group *Pnma* at ambient $T$, but contains a small amount of $Cr_2O_3$ (less than 2 wt%). Magnetization measurements using a SQUID magnetometer indicated the presence of antiferromagnetic transition at 125 K ($= T_N$). The $\mu^+$SR measurements were performed on the M20 surface muon beam line at TRIUMF.

Journal of Physics: Conference Series **551** (2014) 012013 doi:10.1088/1742-6596/551/1/012013

## 3. Results and discussion

### *3.1. $NaCr_2O_4$*

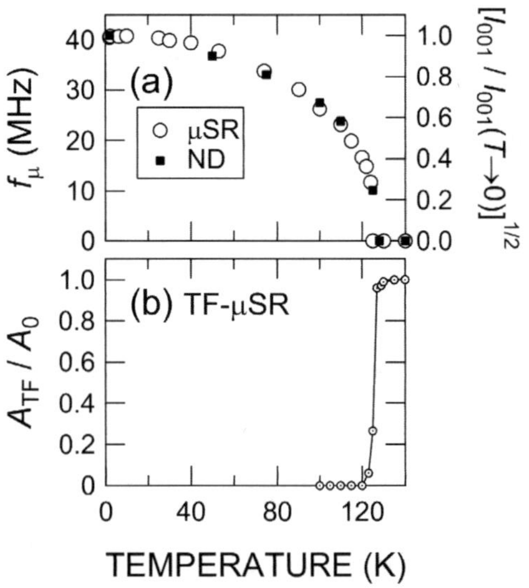


**Figure 3.** $T$ dependences of (a) the muon-spin precession frequency ($f_\mu$) and (b) the weak transverse field asymmetry ($A_{\rm TF}$) for $NaCr_2O_4$. The data in (a) were obtained by fitting the ZF-$\mu^+$SR spectrum with eq. (1), while the data in (b) were estimated by fitting the wTF-$\mu^+$SR spectrum with $A_0P_{\rm TF}(t) = A_{\rm TF}\cos(2\pi f_{\rm TF}+\phi_{\rm TF})\exp(-\lambda_{\rm TF}t)$. In (a), the normalized structure factor for the magnetic Bragg peak 001 extracted from the neutron diffraction data [19] is also plotted for comparison.

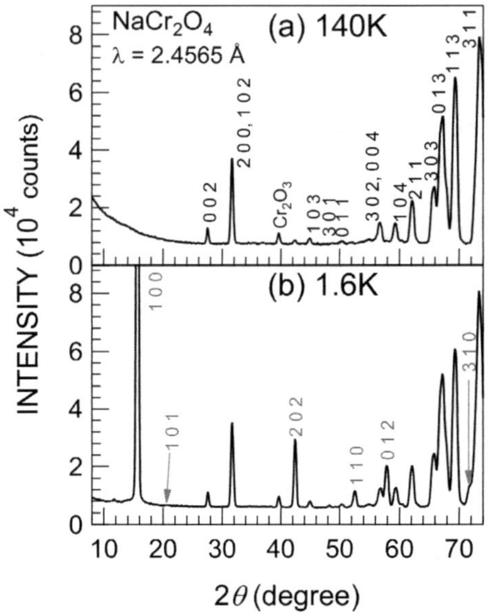


**Figure 4.** Powder neutron diffraction patters obtained at (a) 140 K and (b) 1.6 K for $NaCr_2O_4$ [19]. In (b), only the magnetic Bragg peaks are indexed, while the nuclear Bragg peaks are indexed in (a).

Figure 1 shows the variation of the ZF-$\mu^+$SR time-spectrum with $T$ for $NaCr_2O_4$. The spectra obtained below $T_{\rm N}$ exhibit a damped oscillation, which demonstrates the formation of static magnetic order. Note that there are four oscillatory signals in the ZF-spectrum below 120 K for the isostructural compound $NaV_2O_4$ , due to the presence of four different muon sites in the lattice [8, 10]. In order to confirm the existence or absence of multiple oscillatory signals in the ZF-spectrum, Fig. 2 shows the $T$ dependence of the Fourier transform frequency-spectra of the ZF-$\mu^+$SR time-spectra. It is clearly seen that there is only one signal with a wide frequency distribution. This is consistent with the damped oscillatory signal in the time-spectrum (Fig. 1). Furthermore, the full width at half maximum of the peak appears roughly $T$-independent, while the frequency of the oscillatory signal depends on $T$. This provides a clue to understand the origin of the wide distribution of the internal magnetic field ($H_{\rm int}$) in the $NaCr_2O_4$ lattice. The amplitude of the oscillatory signal of the ZF-$\mu^+$SR spectrum at 2 K is smaller by $\sim$ 10% than that at 110 K (see Fig. 1). In fact, as $T$ decreases from $T_{\rm N}$, the peak

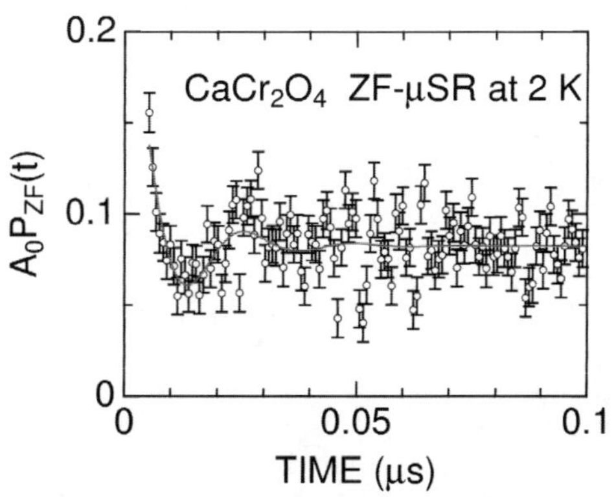


**Figure 5.** The ZF-$\mu^+$SR time-spectrum for $\beta$-$CaCr_2O_4$ obtained at 2 K. A solid line represents the best fit using an exponentially relaxing Bessel function $[J(2\pi f_{AF} t)]$.

height of the Fourier power spectrum decreases, while the peak width increases (see Fig. 2). This suggests that there are additional oscillatory signals at low $T$ due to the existence of the multiple muon sites. However, since it is difficult to fit the ZF-$\mu^+$SR spectra with multiple oscillatory signals, we tentatively fitted the spectrum by a combination of an exponentially relaxing cosine signal and a slowly relaxing non-oscillatory signal based on the above Fourier transform result. The former corresponds to the muon-spin precession signal due to $H_{\rm int}$, while the latter is the "1/3 tail" signal for a powder sample in an ordered state.

$$A_0 P_{\rm ZF}(t) = A_{\rm AF} \cos(2\pi f_\mu t + \phi) \exp(-\lambda_{\rm AF} t) + A_{\rm tail} \exp(-\lambda_{\rm tail} t), \tag{1}$$

where $A_0$ is the initial asymmetry at $t = 0$, $f_\mu$ is the muon-spin precession frequency, $\phi$ is the initial phase of the precession, $A_{\rm AF}$ and $A_{\rm tail}$ are the asymmetries for the two signals, and $\lambda_{\rm AF}$ and $\lambda_{\rm tail}$ are their exponential relaxation rates.

Figure 3 shows the $T$ dependences of (a) $f_\mu$ and (b) the normalized wTF asymmetry $[A_{\rm TF}(T)/A_0]$ obtained with $H_{\rm TF} = 50$ Oe in order to check the volume fraction of nonmagnetic phase(s) in the sample. The $f_\mu(T)$ curve shows an order-parameter-like $T$ dependence. In (a), the normalized magnetic moment $[\sqrt{I^{\rm ND}(T)/I^{\rm ND}(T \to 0\ {\rm K})}]$ derived from the intensity of the magnetic Bragg peak $[I^{\rm ND}(T)]$ is also shown as a function of $T$. The two $T$ dependences are almost identical each other, as expected. The $A_{\rm TF}(T)/A_0$ curve shows a step-like change between 0 and 1 at $T_{\rm N}$, suggesting the whole volume of the sample enters into a magnetic ordered state within the wTF-$\mu^+$SR resolution (around a few %). This is also consistent with the result of the XRD analysis that suggests the volume of impurity phase is below 2%.

Prior to the $\mu^+$SR result of $CaCr_2O_4$, we briefly explain the result of powder ND measurements on $NaCr_2O_4$ [19]. Figure 4 shows the powder ND pattern obtained at 1.6 and 140 K. Several magnetic Bragg peaks are clearly seen at 1.7 K and they are well indexed by the magnetic structure with $\vec{k} = (1, 0, 1)$, where $\vec{k}$ is the propagation vector. Our preliminary Rietveld analysis on the ND pattern suggested that the magnetic moments of the Cr ions aligns ferromagnetically along the $b$-axis, but antiferromagnetically along the $a$-axis. Although the Cr moments are proposed to be canted in the $ac$-plane [20], it is very difficult to determine the canted AF structure based only on the powder ND pattern.

### *3.2. β-CaCr$_2$O$_4$*

Although the $T$ dependence of wTF-$\mu^+$SR parameters for $\beta$-$CaCr_2O_4$ with $T_N$ = 21 K were studied with a pulsed muon beam together with ND [16], the microscopic magnetic nature was not fully clarified due to the limited time resolution of the pulsed muon beam.

Figure 5 shows the ZF-$\mu^+$SR spectrum for $\beta$-$CaCr_2O_4$ obtained at the lowest $T$ measured. One can clearly see a rapidly damped oscillation due to an inhomogeneous distribution of $H_{\rm int}$, although static magnetic order is formed. When we fitted the ZF-spectrum with eq. (1) as for $NaCr_2O_4$, such fit provided that $\phi = -38.55°$. This suggests the formation of incommensurate (IC) magnetic order [18], which is consistent with the ND result [16]. In fact, the ZF-spectrum was well fitted by a zeroth-order Bessel function of the first kind $[J(2\pi f_{\rm AF}t)]$, which is usually used for IC magnetic order [18, 21, 22].

Therefore, for the solid solution system, $Na_xCa_{1-x}Cr_2O_4$, it is expected that IC order changes into commensurate order with $x$. This means the existence of a magnetic lock-in transition at a certain $x$. In order to clarify the magnetic phase diagram of $Na_xCa_{1-x}Cr_2O_4$, further $\mu^+$SR study is in progress.

**Acknowledgments**

We thank the staff of TRIUMF for help with the $\mu^+$SR experiments. This work was supported by MEXT KAKENHI Grant No. 23108003 and JSPS KAKENHI Grant No. 26286084.

**References**

[1] Mao Z Q, He T, Rosario M M, Nelson K D, Okuno D, Ueland B, Deac I G, Schiffer P, Liu Y and Cava R J 2003 *Phys. Rev. Lett.* **90** 186601
[2] Kimura S, Takeuchi T, Okunishi K, Hagiwara M, He Z, Kindo K , Taniyama T and Itoh M 2008 *Phys. Rev. Lett.* **100** 057202
[3] Sugiyama J, Nozaki H, Brewer J H, Ansaldo E J, Takami T, Ikuta H and Mizutani U 2005 *Phys. Rev.* B **72** 064418
[4] Sugiyama J, Nozaki H, Ikedo Y, Mukai K, Andreica D, Amato A, Brewer J H, Ansaldo E J, Morris G D, Takami T and Ikuta H 2006 *Phys. Rev. Lett.* **96** 197206
[5] Sugiyama J, Nozaki H, Ikedo Y, Russo P L, Mukai K, Andreica D, Amato A, Takami T and Ikuta H 2008 *Phys. Rev.* B **77** 092409
[6] Yamauchi T, Ueda Y and Môri N 2002 *Phys. Rev. Lett.* **89** 057002
[7] Hikihara T, Kecke L, Momoi T and Furusaki A 2008 *Phys. Rev.* B **78** 144404
[8] Sugiyama J, Ikedo Y, Goko T, Ansaldo E J, Brewer J H, Russo P L, Chow K H and Sakurai H 2008 *Phys. Rev.* B **78** 224406
[9] Nozaki H, Sugiyama J, Månsson M, Harada M, Pomjakushin V, Sikolenko V, Cervellino A, Roessli B and Sakurai H 2010 *Phys. Rev.* B **81** 100410(R)
[10] Ofer O, Ikedo Y, Goko T, Månsson M, Sugiyama J, Ansaldo E J, Brewer J H, Chow K H and Sakurai H 2010 *Phys. Rev.* B **82** 094410
[11] Sugiyama J, Ikedo Y, Ofer O, Månsson M, Ansaldo E J, Brewer J H, Chow K H, Sakurai H and Takayama-Muromachi E 2009 *J. Phys. Soc. Jpn* **78** 084715
[12] Ofer O, Sugiyama J, Brewer J H, Ansaldo E J, Månsson M, Chow K H, Kamazawa K, Doi Y and Hinatsu Y 2011 *Phys. Rev.* B **84** 054428
[13] Sugiyama J, Nozaki H, Månsson M, Prša K, Andreica D, Amato A, Isobe M and Ueda Y 2012 *Phys. Rev.* B **85** 214407
[14] Sakurai H 2014 *Phys. Rev.* B **89** 024416
[15] Hill P M, Peiser H S and Rait J R 1956 *Acta Cryst.* **9** 981
[16] Damay F, Martin C, Hardy V, Maignan A, André G, Knight K, Giblin S R and Chapon L C 2010 *Phys. Rev.* B **81** 214405
[17] Sakurai H, Kolodiazhnyi T, Michiue Y, Takayama-Muromachi E, Tanabe Y and Kikuchi H 2012 *Angew. Chem. Int. Ed.* **51** 6653
[18] Kalvius G M, Noakes D R and Hartmann O 2001 *Handbook on the Physics and Chemistry of Rare Earths* (Amsterdam: North-Holland) vol 32 chapter 206 pp 55–451
[19] Nozaki H 2012 unpublished work obtained in PSI
[20] Takeda H, Shimizu Y, Itoh M, Sakurai H and Takayama-Muromachi E 2013 *J. Korean Phys. Soc.* **62** 1914

[21] Andreica D 2001 *PhD Thesis* IPP/ETH-Zurich
[22] Sugiyama J, Ikedo Y, Mukai K, Brewer J H, Ansaldo E J, Chow K H, Yoshida H and Hiroi Z 2006 *Phys. Rev.* B **73** 224437

Journal of Physics: Conference Series **551** (2014) 012014 doi:10.1088/1742-6596/551/1/012014

# Magnetic properties of the multiferroic compounds $Eu_{1-x}Y_xMnO_3$ ($x =$ 0.2 and 0.3)

**G M Kalvius[1], F J Litterst[2], O Hartmann[3], R Wäppling[3], A Krimmel[4], A A Mukhin[5], A M Balbashov[6] and A Loidl[4]**

[1] Physics Department, Technical University Munich, 85747 Garching, Germany
[2] IPCM, Technical University Braunschweig, 38106 Braunschweig, Germany
[3] Department of Physics and Astronomy, Uppsala University, 75120 Uppsala, Sweden
[4] Experimental Physics V, University of Augsburg, 86159 Augsburg, Germany
[5] General Physics Institute, Russian Academy of Sciences, 119991 Moscow, Russia
[6] Moscow Power Engineering Institute, 105835 Moscow, Russia

E-mail: kalvius@ph.tum.de

**Abstract.** Magnetic properties of multiferroic $Eu_{1-x}Y_xMnO_3$ with $x = 0.2, 0.3$ were studied by $\mu$SR and Mössbauer spectroscopy. Both compounds are known to become antiferromagnetic with $Mn^{3+}$ being the only magnetic ion. We find $T_N = 47 \pm 0.5$ K for $x = 0.2$ and $45 \pm 0.5$ K for $x = 0.3$. Below $T_N$ three different magnetic states (AFM-1, AFM-2, AFM-3) are formed. The $\mu$SR parameters of the uppermost magnetic state (AFM-1) are alike in both compounds, and compatible with a commensurate modulated collinear spin structure. The magnetic ground state in $x = 0.2$ (AFM-3) is shown to be single phase. Its spectral parameters support the proposal of a cone-like spin structure, yet with incommensurate modulation. The ground state for $x = 0.3$ is found to have an incommensurate spiral spin structure. $^{151}$Eu Mössbauer spectroscopy measured the Eu hyperfine field induced by the $Mn^{3+}$ ions. Its low value ($\sim$4 T) means that mixing with higher Eu electronic states is small and that the Mn-Eu exchange coupling is weak ($\sim$0.5 K). The various magnetic transitions appear as small irregularities in the temperature dependence of the hyperfine fields. The present results are discussed in terms of different published magnetic phase diagrams of $Eu_{1-x}Y_xMnO_3$.

## 1. Introduction

Multiferroics are materials where long-range magnetic and polar order coexist with both order parameters being strongly coupled [1]. The perovskites $Eu_{1-x}Y_xMnO_3$ show multiferroicity without rare earth magnetism since Eu is in the non-magnetic 3+ state. The only magnetic ion is $Mn^{3+}$ which occupies the B-site. Replacing Eu in part by Y changes the A-site volume (see Fig. 1-a) and with it the multiferroic properties, free from the additional influence of a magnetic rare earth moment. Standard $\mu$SR is not suited to explore internal electric fields since the muon has no quadrupole moment. For a small number of cases the magneto-electric coupling has been probed with muons in electric field measurements. The present $\mu$SR studies investigate the pure volume effect on the magnetic properties of $Mn^{3+}$ in $Eu_{1-x}Y_xMnO_3$.

A $(x, T)$ phase diagram establishing the different ordered magnetic and polar states of $Eu_{1-x}Y_xMnO_3$, obtained from structural, magnetic, dielectric and thermodynamic data by [2] and [3], is shown in Fig. 1-b. Details of the sequence of the magnetic states for the compounds studied are depicted in Fig. 1-c. Slightly later, another study of $Eu_{1-x}Y_xMnO_3$ using

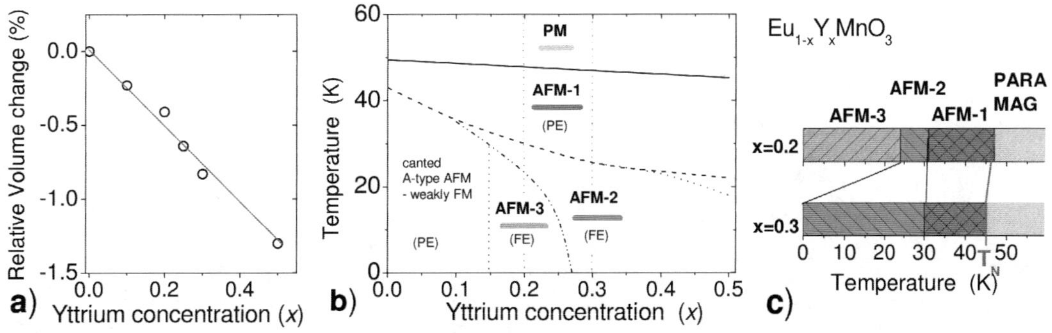


**Figure 1.** a) Change of unit cell volume with Y concentration (after [2]). b) Phase diagram of $Eu_{1-x}Y_xMnO_3$ according to [2, 3]. c) The antiferromagnetic states for $x = 0.2$ and 0.3 in detail.

magnetization, dielectric and structural measurements [4] led to an alternate phase diagram where the main difference concerns the $x = 0.2$ compound. The weakly ferromagnetic (canted A-type antiferromagnet) state that is found in [2] as the second magnetic state only for $x \leq 0.1$ is extended to $x \approx 0.27$. In consequence, this state which must be paraelectric on structural arguments, is then the ground state for $Eu_{0.8}Y_{0.2}MnO_3$. The aim of this study is to gain additional information on the magnetic states of $Eu_{0.8}Y_{0.2}MnO_3$ and $Eu_{0.7}Y_{0.3}MnO_3$ from the study of internal fields. We shall discuss in section 6 our findings from $\mu$SR and Mössbauer spectroscopy with respect to the results of [2, 3] and [4] as well as additional studies by [5, 6].

## 2. Experimental

The powder material used was prepared using standard solid state reaction procedures [7]. The samples consisted of non-oriented small single crystalline grains prepared from the the same batch of materials used in [2]. Powder X-ray diffraction revealed single phase material with proper crystallographic structure. For the $\mu$SR measurements, the powder was pressed into tablets which were mounted between thin aluminized Mylar foils in the helium gas flow of a variable temperature cryostat. Data were recorded at the 'DOLLY' set-up of the Swiss Muon Source in zero field (ZF) and weak (10 mT) transverse field (TF). The VETO mode was enabled for suppressing the "background signal" from muons stopped outside the sample. Its absence was verified by TF measurements.

## 3. $\mu$SR data for $Eu_{0.8}Y_{0.2}MnO_3$

As depicted in Fig.2-a, the spectra change around 47 K from an oscillatory pattern (in TF) or a simple exponential decay (in ZF), both typical for the paramagnetic state, to the characteristic pattern for polycrystalline long-range ordered (LRO) magnets given by:

$$A_{LRO}(t) = A_0[(2/3)\exp(-\lambda_{tr}\, t)\cos(2\pi\nu_\mu t + \varphi) + (1/3)\exp(-\lambda_{lg}\, t)] \qquad (1)$$

with $\nu_\mu = \gamma_\mu B_\mu / 2\pi$ where $B_\mu$is the mean field at the muon site, and $\gamma_\mu$ the muon gyromagnetic ratio. The transverse relaxation rate is a measure of the static field distribution width ($\Delta B_\mu = \lambda_{tr}/\gamma_\mu$), while the longitudinal rate $\lambda_{lg}$ gives the fluctuation rate of the ordered atomic moments (here the $Mn^{3+}$moments) with $\lambda_{lg} \propto 1/\tau_{Mn}$. Eq. (1) is strictly valid only in ZF, but also holds for weak TF ($B_{TF} \leq B_\mu$). Between 47 K and 50 K the $\mu$SR spectra consist of the sum of the paramagnetic and the LRO signals (see spectrum at 47.5 K in Fig.2-a). This means that the sample is a heterogeneous mixture of paramagnetic and LRO fractions. The temperature dependence of the paramagnetic fraction is shown in Fig.2-b. The transition into the LRO state

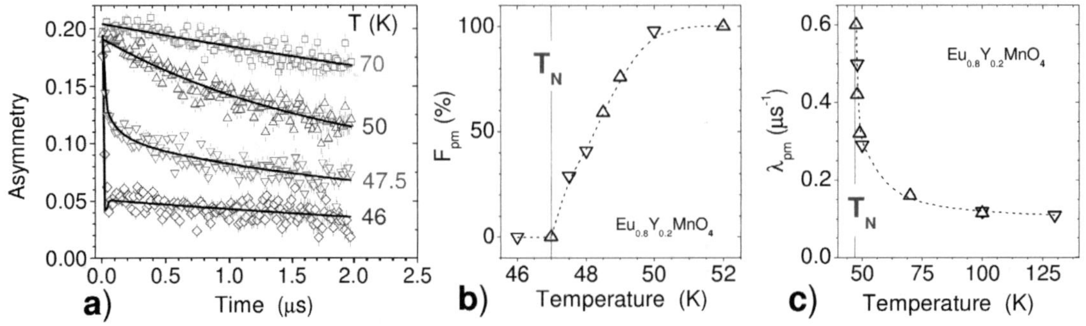


**Figure 2.** a) ZF spectra around $T_N$. The lines are fits discussed in text. Temperature dependences of b) the paramagnetic fraction $F_{pm}$ and c) the paramagnetic relaxation rate $\lambda_{pm}$. $\bigtriangledown$ =ZF data, $\bigtriangleup$ = TF data. The broken lines are guides to the eye.

is spread out over ~3 K. Taking as Néel temperature the first appearance of the LRO pattern gives $T_N = 47\pm0.5$ K in agreement with bulk magnetic data [2]. The paramagnetic relaxation rate exhibits the typical sharp rise on approach to a second order phase transition (Fig.2-c).

Fig. 3-a displays the early time parts of typical ZF spectra below $T_N$, and Fig. 3-b the temperature dependences of the magnetic parameters. The spectra at 47, 46 and 43 K, characteristic for the AFM-1 state, can be fitted well using Eq. (1). They are fully consistent with a commensurate modulated collinear spin structure. The spin modulation causes a large field distribution width $\Delta B_\mu / B_\mu(0) \sim 1$, provided the modulation wavelength is not too short. A short wavelength would result in a structured field distribution that is not observed. It cannot be excluded that small distortions of the collinear spin structure on a local scale also contribute to $\Delta B_\mu$. At 27 K we obtain a spectrum similar to those recorded at 20 K (not shown) and 10 K (i.e., in AFM-3). The $\mu$SR data indicate a direct transition from the AFM-1 state into the AFM-3 ground state.

Fitting the spectra of the AFM-3 state in the same manner as the AFM-1 spectra lead to unreasonably large values of the phase $\varphi$ given in Eq. (1). This feature is indicative for the need

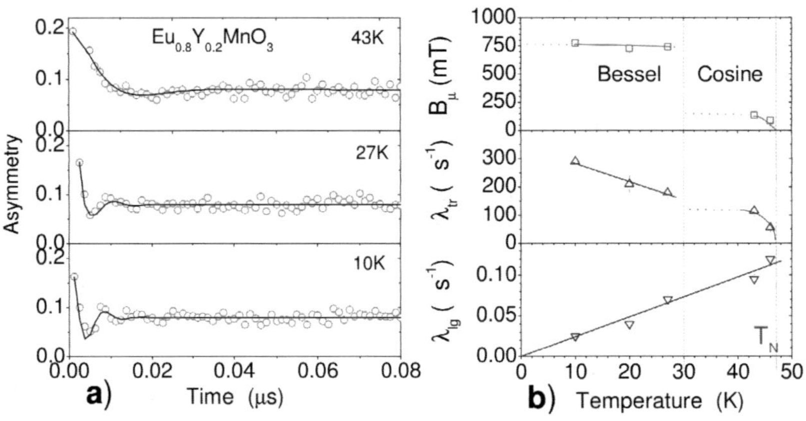


**Figure 3.** a): Selected ZF spectra of $Eu_{0.8}Y_{0.2}MnO_3$ at different temperatures. The lines are fits discussed in text. b): Temperature dependences of the local field $B_\mu$, the transverse ($\lambda_{tr}$) and longitudinal ($\lambda_{lg}$) relaxation rates. The lines trough the data points are guides to the eye.

to replace the cosine function by the first order Bessel function $J_0$ [8]. At least for simple cases like a sinusoidally modulated spin structure, it has been shown [8, 9] that the Bessel function is the result of incommensurability of the modulation. $\mu$SR is not a reciprocal space probe like neutron diffraction and hence gives no direct information of the spatial arrangements in LRO spin structures. But quite generally, the appearance of of Bessel oscillations in the $\mu$SR spectra can be taken as an indication for incommensurate spin modulation (see examples in [9]). Neutron data on the spin structure are not reported. The Bessel function takes care of the field distribution caused by the incommensurate spin modulation [8]. The frequency of Bessel oscillations is given by the maximal field of this field distribution, in contrast to the cosine oscillations where $\nu_\mu$ is given by the mean field. For clarity we use in the figures $B_\mu$ for both situations. $B_\mu$ increases sharply at the 30 K transition. The transverse relaxation rate is still large in AFM-3 ground state. Since the Bessel function takes care of the field distribution caused by the incommensurate spin modulation, the substantial distribution width given by $\lambda_{tr}$ points to some additional random local disturbances in the AFM-3 spin structure. The spin fluctuations reflected by $\lambda_{lg}$ show the expected move toward the static limit for $T \to 0$. They are not affected by the changes in spin structure, an observation also made in other magnetic compounds (see, for example [9]). It indicates that the slow electronic moment fluctuations seen uniquely by $\mu$SR do not exhibit critical behavior at the transit to a different spin arrangement.

## 4. $\mu$SR data for $Eu_{0.7}Y_{0.3}MnO_3$

Using the same procedures as those employed for $x = 0.2$ leads to $T_N = 45 \pm 0.5$ K. Again one observes a small spread over $\sim$3 K of the transition regime and the strong rise in paramagnetic relaxation rate (from 0.15 $\mu s^{-1}$ at 60K to 1.5 $\mu s^{-1}$ at 46K) on approach to $T_N$, confirming here as well the second order nature of the transition at $T_N$.

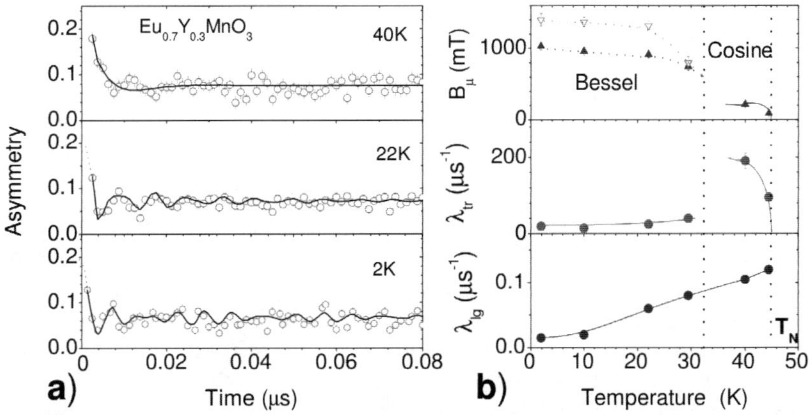


**Figure 4.** a): Selected ZF spectra of $Eu_{0.7}Y_{0.3}MnO_3$ at different temperatures. The lines are fits discussed in text. b): Temperature dependences of the local field $B_\mu$, the transverse ($\lambda_{tr}$) and longitudinal ($\lambda_{lg}$) relaxation rates. The lines through the data points are guides to the eye.

The early time parts of selected ZF spectra taken below $T_N$ is presented in Fig. 4-a, and the temperature dependences of the magnetic parameters in Fig. 4-b. Spectra within the AFM-1 regime (e.g., at 40 K) were successfully treated in the same manner as previously for the $x = 0.2$ material, showing that the characteristics of the AFM-1 state are the same in both compounds. Fits of the spectra in the AFM-2 state (taken at 30, 22, and 2 K ) reveal a two frequency Bessel oscillatory pattern. At 30 K the two frequency requirement is marginal, but

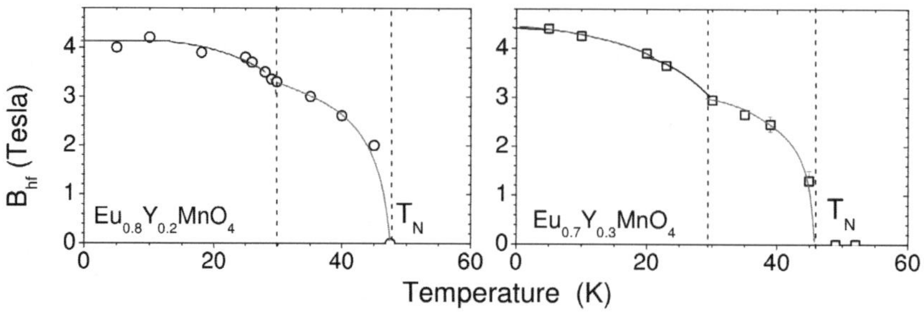


**Figure 5.** The magnetic field at the $^{151}$Eu nucleus ($B_{hf}$) in $Eu_{0.8}Y_{0.2}MnO_3$ and $Eu_{0.7}Y_{0.3}MnO_3$ versus temperature. The solid lines indicate the extrapolation of $B_{hf}$ to $T \rightarrow 0$.

it becomes distinctive at lower temperatures. Since the paramagnetic spectra confirm a single interstitial muon stopping site, the two frequencies must arise from the field distribution of the LRO spin arrangement. It reflects, according to [10], an incommensurate spiral antiferromagnetic structure. We have previously observed in $FeCr_2S_4$ [11] a similar type of spectra and interpreted them the same way. Present in $Eu_{0.7}Y_{0.3}MnO_3$ is also the large jump in $B_\mu$ at the transition point and the continuous reduction of $\lambda_{lg}$ with cooling. In contrast, $\lambda_{tr}$ is very small meaning the spiral structure is not modulated and well developed on short-range and long-range scales.

## 5. Mössbauer data

Only a brief summary is presented, more details will be found in a forthcoming publication [13]. The 21 keV resonance of $^{151}$Eu was used. Technical features and typical applications of this resonance can be found in [14]. The parameter of concern is the hyperfine field $B_{hf}$ at the Eu nucleus. In standard Mössbauer studies (e.g., $^{57}$Fe) it is generated by an unfilled electron shell in the resonant ion. Here, in the resonant ion ($Eu^{3+}$) all electron orbits are closed and $B_{hf}$ originates from the interaction with the moment on neighboring $Mn^{3+}$ ions, somewhat in analogy to $B_\mu$. Yet, the information contained in $B_{hf}$ is different: (a) The Eu nucleus as the probe is fixed onto a regular lattice site, while the muon sits on an (unknown) interstitial site. (b) $B_{hf}$ is not the raw field of the $Mn^{3+}$ moments at the Eu lattice site. Rather, the field from Mn moments polarizes the Eu electron shell and the field at the nucleus is amplified by this shell distortion and thus enhanced. (c) The coupling between the muon and the Mn field is predominantly dipolar and hence highly sensitive to the geometrical arrangement of the Mn moments. Due to the dominant enhancement by the polarized electron shell, this is not the case for $B_{hf}$. (d) The $^{151}$Eu nucleus has a sizable quadrupole moment and responds also to electrical interactions. Since the magnetic coupling is dominant, and due to the limited resolution resulting from the low size of $B_{hf}$, the quadrupolar interaction was neglected in first approximation.

$B_{hf}$ versus temperature is plotted in Fig. 5. The slightly higher field in the AFM-1 state of $x = 0.3$ is confirmed. Most noticeable is the much smaller increase of $B_{hf}(0)$ between the AFM-1 state and the magnetic ground states (AFM-3 or AFM-2) when compared to $B_\mu$. This confirms that the large jump in $B_\mu$ arises from a change in spin arrangement, rather than from a change in Mn moment. The size of $B_{hf}(T \rightarrow 0)$ is only $\sim$4 T, a rather small value compared to $B_{hf}(0)$ produced by an open electron shell in Eu (typically around 30 T [14]). Even for an induced field the size is low which indicates that the distortion of the Eu shell, i.e. the mixing of the non-magnetic ground state with magnetic higher crystalline electric field electronic states, is small and points furthermore to a weak Mn-Eu exchange coupling, estimated to be $\sim$0.5 K.

Journal of Physics: Conference Series **551** (2014) 012014 doi:10.1088/1742-6596/551/1/012014

## 6. Discussion

We now discuss the results of our $\mu$SR and Mössbauer studies on the different magnetic states in $Eu_{0.8}Y_{0.2}MnO_3$ and $Eu_{0.7}Y_{0.3}MnO_3$ in the light of existing bulk magnetic data.

**a)** The Néel temperatures derived agree well with [2] and confirm their slight decrease with increasing $x$. The $\mu$SR spectra show the coexistence of paramagnetic and antiferromagnetic fractions over a range of 3 K around $T_N$ and confirm further the second order nature of the transition at $T_N$.

**b)** The spin structure in the uppermost magnetic state (AFM-1) is similar for $x = 0.2$ and 0.3. The $\mu$SR spectra are consistent with a commensurate modulated collinear spin arrangement, as proposed in [2, 4]. The observed wide distribution of the local field requires a not too short modulation wave length (see also [4]). The saturation internal field $B_\mu(T \to 0)$ is somewhat larger in $x = 0.3$ than in $x = 0.2$ as the result of the reduced unit cell volume.

**c)** The $\mu$SR results for the magnetic states following the AFM-1 state on reduced temperature (AFM-2, AFM-3) are characteristic for incommensurate spin structures. The question of incommensurate structures in the ferroelectric ground states had remained open on the basis of bulk measurements [2]. It is, however, not decisive with respect to the multiferroic properties. A sudden rise of $B_\mu$ occurs when crossing the lower AFM-1 phase boundary. The corresponding Mössbauer data show that the magnitude of the ordered $Mn^{3+}$ moment does not change much, if at all. The sudden increase of $B_\mu$ then reflects a substantial change in spin arrangement.

**d)** The $\mu$SR spectral features of the ground state in $x = 0.3$ (AFM-2) point toward a well developed incommensurate antiferromagnetic spiral structure, in accordance with earlier suggestions on the basis of the ferroelectric properties (see, for example [12]).

**e)** The $\mu$SR data for $x = 0.2$ show that the magnetic ground state (AFM-3) follows directly the AFM-1 state without going through an intermediate phase (AFM-2).

**f)** As mentioned already in section 1, the nature of the magnetic ground state for $x = 0.2$ is controversial. In [2] the $Eu_{0.8}Y_{0.2}MnO_3$ material shows definitely ferroelectric polarization. The same result is also reported in [5, 6]. It completely contradicts the canted A-type AFM structure claimed by [4] as the $x = 0.2$ ground state, since that state cannot support ferroelectricity on structural grounds. Indeed it is listed as a paraelectric state in the phase diagram of [4]. Two scenarios have been suggested to remedy the situation: (1) The $x = 0.2$ material is heterogeneous having undergone phase separation. It consists of a weakly ferromagnetic fraction having the canted A-type AFM spin structure established for $x \leq 0.1$ and a non-ferromagnetic fraction featuring the antiferromagnetic spin structure of the $x = 0.3$ magnetic ground state. This approach is discussed, in particular, in [6] and [4]. (2) The formation of a new weakly ferromagnetic AFM phase having a spin structure which breaks inversion symmetry. This approach is presented in [2] with the proposal of a non collinear probably cone-like AFM spin structure. A basically similar approach is given in [5].

The $\mu$SR data definitely exclude a two phase heterogeneous material. If two fractions with substantially different spin structures would exist, the spectrum should consist of a superposition of two different signals, similar to the coexistence of a paramagnetic and a antiferromagnetic phase around $T_N$ (see Fig. 2-a). The rather involved spectrum of AFM-2 would lead to a very complex sum spectrum. This is not the case. The spectrum can well be fitted by a single $\mu$SR signal. Furthermore, a ground state having the canted A-type AFM structure not only contradicts the electric polarization data, but the $\mu$SR spectrum as well, which should consist of a simple cosine oscillation and not the Bessel oscillation observed. On the other hand, a non-collinear cone-like incommensurately modulated AFM structure is fully supported by $\mu$SR.

## Acknowledgments

The $\mu$SR studies were performed at the Swiss Muon Source of the Paul Scherrer Institute (PSI). We thank R. Scheuermann and H. Luetkens for their assistance. This work was supported in part by the Deutsche Forschungsgememeinschaft (DFG) via the Transregional Collaborative Research Center TRR80 (Augsburg, Munich, Stuttgart)

## References

[1] Khomski D I 2006 *J. Magn. Magn. Mat.* **306** 1
[2] Hemberger J *et al.* 2007 *Phys. Rev.* B**75** 035118
[3] Noda K *et al.* 2006 *J. Appl. Phys.* **99** 08S905
[4] Yamasaki Y *et al.* 2007 *Phys. Rev.* B **76** 184418
[5] Ivanov V Yu *et al.* 2006 *J. Magn. Magn. Mat.* **300** e130
[6] Danjoh S *et al.* 2009 *Phys. Rev.* B **80** 180408(R)
[7] Balbashov M *et al.* 1996 *J. Cryst. Growth* **67** 365
[8] Yaouanc A and Dalmas de Réotier P 2011 *Muon Spin Rotation, Relaxation, and Resonance* (Oxford: Oxford University Press)
[9] Kalvius G M, Noakes D R and Hartmann O 2001 in *Handbook on the Physics and Chemistry of Rare Earth* vol 32 chapter 206 (Amsterdam: Elsevier Science)
[10] Noakes D R 2002 *An Ornamental Garden of Field Distributions* http//musr.org/intro/ppt/GardenExport
[11] Kalvius G M *et al.* 2010 *J. Physics: Condens. Matter* **22** 052205
[12] Mostovoy M 2006 *Phys. Rev. Letters* **96** 067601
[13] Litterst F J *et al.* to be published
[14] Grandjean F and Long G J 1989 in *Mössbauer Spectroscopy Applied to Inorganic Chemistry* vol 3 (New York: Plenum press) p 513

Journal of Physics: Conference Series **551** (2014) 012015 doi:10.1088/1742-6596/551/1/012015

# A $\mu$SR study of the ruthenium perovskites $ACu_3Ru_4O_{12}$ with A = Ca, Pr, Nd

**G M Kalvius[1], O Hartmann[2], A Günther[3], A Krimmel[3], A Loidl[3] R Wäppling[2], K Sedlak[4] and R Scheuermann[4]**

[1] Physics Department, Technical University Munich, 85747 Garching, Germany
[2] Department of Physics and Astronomy, Uppsala University, 75120 Uppsala, Sweden
[3] Experimental Physics V, University of Augsburg, 86159 Augsburg, Germany
[4] Laboratory for Muon Spin Spectroscopy, Paul Scherrer Institute, CH-5232 Villigen-PSI, Switzerland

E-mail: kalvius@ph.tum.de

**Abstract.** The metallic ruthenium perovskites $ACu_3Ru_4O_{12}$ with A = Ca, Pr, Nd were investigated by zero field (ZF) and weak transverse field (TF) muon spin rotation/relaxation ($\mu$SR) spectroscopy. The ZF spectra for $CaCu_3Ru_4O_{12}$ show pure static Gaussian Kubo-Toyabe relaxation arising from the interaction between the muon spin and the nuclear moments of the Cu ions. This confirms that no atomic magnetic moment exists. A sudden increase of the lattice parameter $a$ when heating above ~150 K had previously been detected by neutron diffraction. The root mean square field at the muon site $B_{rms}$ was found to be 0.15 mT independent of temperature, in particular around 150 K. Also, no change of spectral parameters is seen in the weak TF data in that temperature range. Those findings imply that the structural change around 150 K takes place without noticeably shifting atomic positions. The spectra for $PrCu_3Ru_4O_{12}$ and $NdCu_3Ru_4O_{12}$ are dominated by the interaction with the dynamic rare earth moments. The analysis requires the use of the electron-nuclear double relaxation formalism. The electronic part shows simple exponential relaxation typical for a paramagnet. It features, with reducing the temperature a steep increase of paramagnetic relaxation rate as is characteristic for an approach to a magnetic spin freezing transition from above. That result suggests that the magnetic ground states of $PrCu_3Ru_4O_{12}$ and $NdCu_3Ru_4O_{12}$ are spin frozen states, although bulk magnetic data give no direct evidence in that direction.

## 1. Introduction

The ruthenates $ACu_3Ru_4O_{12}$ with A = Ca, Pr, Nd are metallic compounds having a modified perovskite structure [1]. This modification is described by a tilting of the $RuO_6$ octahedra along the < 111 > axis mainly arising from the bonding between the new C-site and oxygen [2] accompanied by the lowering of the Ru–O–Ru bond angle and the appearance of an almost square-planar coordinated C-site, while leaving the basic cubic structure intact. From this structural variation several unusual physical properties emerge and show a high sensitivity in case of chemical substitution at any ionic position.

The compound $CaCu_3Ru_4O_{12}$ exhibits heavy fermion [3] and non-Fermi liquid properties [4] and shows characteristics of a classical intermediate valence (IV) system having low-energy charge fluctuations [5]. Neutron powder diffraction revealed a sudden increase of lattice constant $a$ (i.e., an increase of unit cell volume) around 150 K with the basic crystal structure remaining

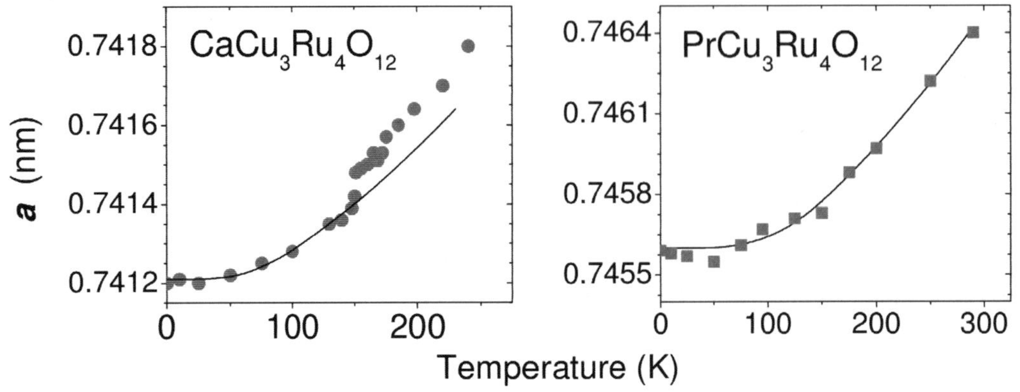


**Figure 1.** Temperature dependence of the lattice constant $a$ from powder neutron diffraction measurements for $CaCu_3Ru_4O_{12}$ (left) and for $PrCu_3Ru_4O_{12}$ (right). The solid lines are anharmonic fits to the thermal expansion. After [5, 6].

conserved [5]. That feature is not seen in the rare earth ruthenates [6]. Fig. 1 depicts the situation in $CaCu_3Ru_4O_{12}$ and $PrCu_3Ru_4O_{12}$. Since the volume increase is only seen in $CaCu_3Ru_4O_{12}$, that feature must be connected to its IV properties and be of electronic nature. For example, a transition from localized $4d$ electrons above ~150 K to itinerant $4d$ band states below ~150 K is proposed [5].

The rare earth homologues $PrCu_3Ru_4O_{12}$ and $NdCu_3Ru_4O_{12}$ also show heavy-fermion-like properties [7]. However, their magnetic behavior is largely determined by the localized $4f$ moments of the rare earth constituents [6]. In order to suppress the influence of the $d$-electrons of Cu and Ru, the susceptibility of the analogous ($4f$ states free) $LaCu_3Ru_4O_{12}$ was subtracted. The pure $4f$ inverse susceptibility $\chi^{-1}_{\mathrm{Pr,Nd}}$ shows proper Curie–Weiss (CW) behavior with moments close to the free ion values of $Pr^{3+}$ and $Nd^{3+}$ down to approximately 200 K. At lower temperatures the data on $1/\chi$ deviate from the linear temperature dependence. This is attributed to crystalline electric field (CEF) interactions. At very low temperatures ($T$ <10 K) the slope of the inverse susceptibility increases strongly. This is ascribed to a decreasing Curie constant implying, especially for $Pr^{3+}$, a ground state with a small magnetic moment. No indication for spin freezing is apparent. The extensive study of the rare earth ruthenates [6] includes electrical resistivity, heat capacity, NMR and NQR data, as well as inelastic neutron scattering work on the Pr compound. In all cases the results can consistently be explained within a CEF states model.

## 2. Experimental

The polycrystalline samples were synthesized by solid-state reactions as described in [8]. The crystal structures of the samples were verified by x-ray and neutron powder diffraction. For the $\mu$SR measurements the powders were put into an envelope made of thin aluminized Mylar tape and then mounted with similar tape on aluminum forks of the cryostat stick. The Dolly surface beam facility was used with the vertical Oxford variable temperature cryostat (VARIOX). Zero field (ZF) and transverse field (TF) measurements were carried out between 2 K and 200 K. The sample was large compared to the beam spot and also the VETO was enabled, so no background signal was to be expected. This was verified by TF measurements. Transverse fields were restricted to 5mT to avoid beam bending and with it a change of spectral baseline (alpha-parameter). Alpha was found to be quite constant and only weakly temperature dependent.

The asymmetry spectra were least squares fitted using the WKM program (Technical University Braunschweig).

## 3. Results

### *3.1. $CaCu_3Ru_4O_{12}$*

In weak TF (5 mT) a single Gaussian damped oscillatory pattern:

$$A_{TF}(t) = A_0 \exp(-\sigma^2 t^2)\cos(2\pi\nu_\mu t + \phi) \tag{1}$$

with full and temperature independent initial asymmetry ($A_0 = 0.237\pm0.001$) is seen. The pure Gaussian relaxation means that the local field at the muon is static on the $\mu$SR timescale. The precession frequency $\nu_\mu$ corresponds within the limits of error to the applied transverse field. This means that the muon stops at only one interstitial lattice site. Also independent of temperature is the relaxation rate $\sigma = 0.085\pm0.003\ \mu s^{-1}$, revealing that the muon rests stationary at its single stopping site.

The ZF spectra are fully consistent with the TF data. Two typical examples of ZF spectra are depicted in Fig. 2-left. The time dependence of the asymmetry ($A$(t)) clearly shows Gaussian decay. An initial Gaussian relaxation in a ZF pattern is the signature of static Gaussian Kubo-Toyabe (sGKT) relaxation [9]:

$$A_{sgkt}(t) = A_0[(2/3)(1-\Delta^2 t^2)\exp(-\Delta^2 t^2/2) + (1/3)] \tag{2}$$

The sole parameter of physical relevance is the static relaxation rate $\Delta$ which is related to the the width of the distribution of the field at the muon site. The mean field of the gaussian distribution in Eq.(2) is $\langle B\rangle = 0$. The distribution of the of magnitude of the field $|B|$

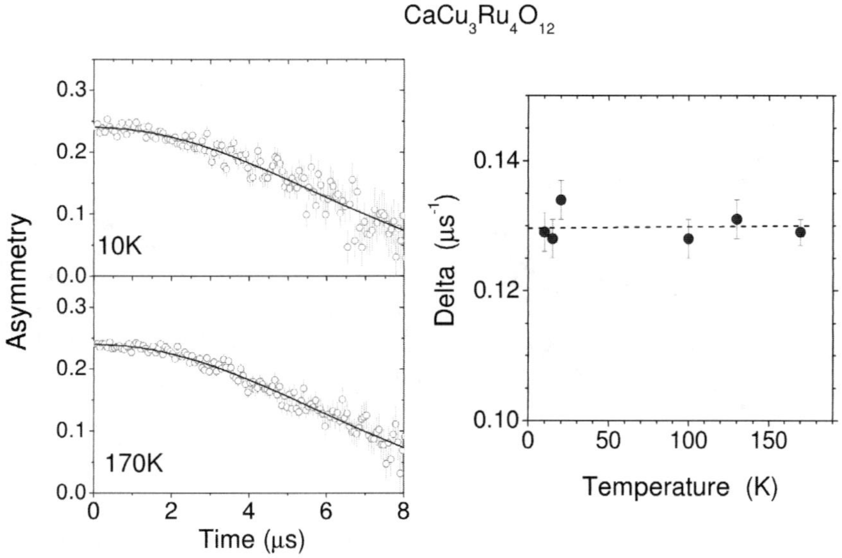


**Figure 2.** Left: ZF spectra of $CaCu_3Ru_4O_{12}$ at 10 K and 170 K. The solid lines are fits to a static Gaussian Kubo-Toyabe (sGKT) function. Right: Temperature dependences of the static width $\Delta$. The dashed line is a linear fit to the data points.

is Maxwellian with $|B|^2 = \Delta^2/\gamma_\mu{}^2$. One commonly uses the root mean square field $B_{rms} = \Delta/\gamma_\mu$ to descibe the distibution of the local field.

The sGKT function in total is not a simple Gaussian decay of asymmetry. After its initial Gaussian decay, A(t) reaches a minimum around t$\approx$1.7/$\Delta$. It then recovers to the time independent value $A_0/3$ for t$\geq$ 3.5/$\Delta$. In the present case these features are outside the time window of the $\mu$SR spectrometer due to the small value of $\Delta = 0.13\ \mu s^{-1}$(see below). Hence Fig. 2 is able to display only the very initial part of the sGKT function. This leads to a problem in fitting procedure. Using as fit function Eq.(2) with the parameters $A_0$ and $\Delta$ unrestricted, the system is unable to find the baseline. One needs to fix the initial asymmetry to the value obtained from the TF spectra ($A_0 = 0.237$, see above). This produces consistent results, in particular with respect to $\Delta$. The results are shown in Fig. 2-right.

One finds $\Delta$ to be independent of temperature with a mean value of $\Delta = 0.130 \pm 0.002$, corresponding to $B_{rms} \approx 0.15$ mT. This extremely low value points to nuclear dipole moments as the origin of the local field. Of the constituents of $CaCu_3Ru_4O_{12}$, oxygen is nearly 100% $^{16}O$ which has no magnetic dipole moment. In Ca the only isotope having a magnetic moment (-1.3$\mu_N$) is $^{43}Ca$, which, however, has a very low abundance (0.13%). Ru has two isotopes of interest ($^{99}Ru$ and $^{101}Ru$) with moderate abundances (13% and 17%) but with very low moment magnitudes (-0.6 and -0.7$\mu_N$). Cu consists to 100% of the two isotopes $^{63}Cu$ and $^{65}Cu$, both having sizable nuclear moments (2.2 and 2.4$\mu_N$). Thus, the dominant source of the sGKT relaxation is the coupling between the muon spin and the Cu nuclear moments. Nuclear dipole moments are of course not really static, but their fluctuation rate is rather slow (often in the ms regime) and thus they appear static on the $\mu$SR time scale.

The pure sGKT spectra confirm that no ion with a magnetic moment is present in $CaCu_3Ru_4O_{12}$. From the constancy of $B_{rms}(T)$, in particular in the range of the transition region around 150 K, two additional conclusions can be drawn. Firstly, it confirms the conclusions of [5] that lattice symmetry is maintained and that the source of the transition is most likely a change in electron structure. Secondly, such a change in electron structure does not lead to the creation of even a very small atomic magnetic moment.

### 3.2. $PrCu_3Ru_4O_{12}$ and $NdCu_3Ru_4O_{12}$

The central feature determining the $\mu$SR spectra of the rare earth ruthenates is the coupling of the the localized $4f$ moments to the muon spin. As long as no strong spin-spin correlations between the $4f$ moments exist (e.g., in the absence of short-range or long-range magnetic order) the moments will fluctuate rapidly (free paramagnetic moments) leading to a simple exponential decay of asymmetry with rate $\lambda$. This rate is inversely proportional to the fluctuation rate $1/\tau_{4f}$ of the rare earth moments. It may become quite small in the regime of fast fluctuation, meaning that the sGKT relaxation of the Cu nuclear is no longer negligible with respect to the exponential relaxation by the electronic $4f$ moments. In a proper fit of the spectra of $PrCu_3Ru_4O_{12}$ and $NdCu_3Ru_4O_{12}$, both relaxation channels have to be taken into account. In case the two interactions are independent of each other and occur on significantly different time scales, the total relaxation function can be well approximated by the product of the electronic and nuclear relaxation functions (electron-nuclear double relaxation):

$$A_{dr}(t) = A_0\{\exp(-\lambda t) \times [(2/3)(1 - \Delta^2 t^2)\exp(-\Delta^2 t^2/2) + (1/3)]\} \quad (3)$$

The conditions for applying Eq.(3) are well fulfilled here. The electronic moments fluctuate rapidly and rest on different ions ($Pr^{3+}$ or $Nd^{3+}$) than the static nuclear moments of Cu.

Typical ZF spectra of the two rare earth ruthenates least squares fitted with Eq.(3) are shown in Fig. 3. For reasons discussed above, we had to fix either the initial asymmetry ($A_0 = 0.24$ for $PrCu_3Ru_4O_{12}$) or the baseline parameter alpha ($\alpha = 0.4$ for $NdCu_3Ru_4O_{12}$), taking these values from the TF spectra. The nuclear static relaxation rate $\Delta$ is 0.13 $\mu s^{-1}$ as in the case of

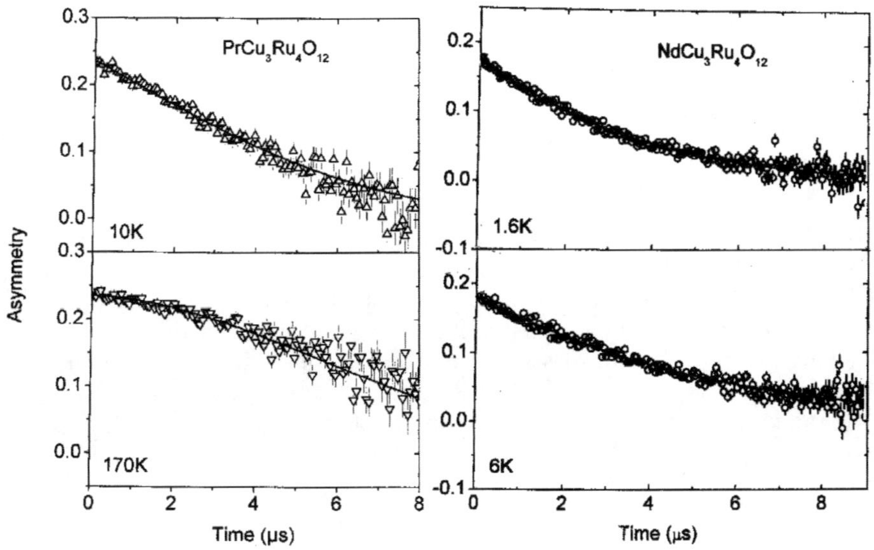


**Figure 3.** Typical ZF spectra of $PrCu_3Ru_4O_{12}$ (left) and $NdCu_3Ru_4O_{12}$ (right). The solid lines are fits to the electron-nuclear double relaxation discussed in text. The $NdCu_3Ru_4O_{12}$ sample was slightly smaller, hence the initial asymmetry is somewhat lower.

$CaCu_3Ru_4O_{12}$. The important result is the temperature dependence of the electronic relaxation rate $\lambda$ as displayed in Fig. 4

Looking first at the situation in $PrCu_3Ru_4O_{12}$, one observes the typical behavior of $\lambda(T)$ for a free paramagnetic ion. With rising temperature the fluctuation rate of the $Pr^{3+}$ moments increases leading to a continuous reduction in magnitude of $\lambda$. The smoothness of the curve $\lambda(T)$ indicates that no change of magnitude in the $Pr^{3+}$ moments (e.g., by CEF effects) occurs in the temperature range covered. The steep rise of $\lambda(T)$ seen at low temperatures is characteristic for critical slowing down of paramagnetic moments on approach to a spin freezing transition. The data of Fig. 4-left then point to a transition into a quasi static spin state (either long-range or

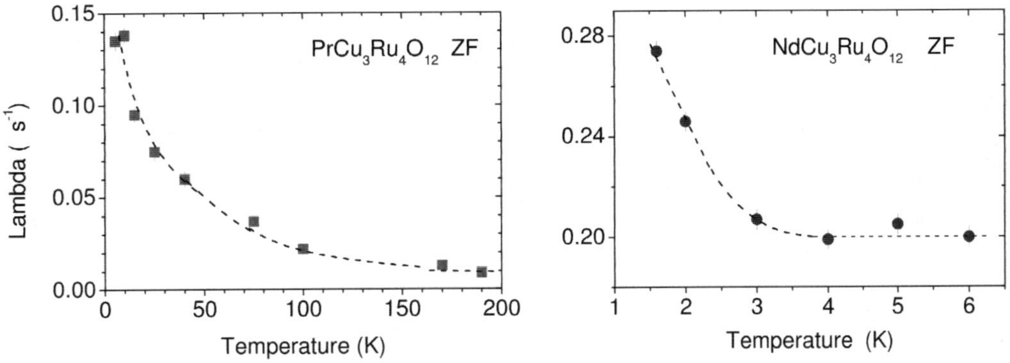


**Figure 4.** Temperature dependences of the electronic relaxation rates for $PrCu_3Ru_4O_{12}$ (left) and $NdCu_3Ru_4O_{12}$ (right). The dotted lines are guides to the eye only. Note the difference in temperature range.

short-range correlated) at low temperatures.

The situation in $NdCu_3Ru_4O_{12}$ is alike as far as the steep increase of $\lambda\,(T)$ at low temperatures is concerned. Again the immediate approach to a spin freezing transition is indicated. Unusual, and at present not understood, is the leveling-off of $\lambda\,(T)$ at a fairly high value of $\lambda$ only $\sim$2 K above the onset of critical slowing down.

In general, the relaxation rate $\lambda$ in $NdCu_3Ru_4O_{12}$ is about a factor of two larger in comparison to $PrCu_3Ru_4O_{12}$, although the free ion magnetic moments are nearly the same. The likely reason is the different CEF behavior of the two rare earth ions. $Nd^{3+}$ is a Kramers ion, while $Pr^{3+}$ is not. In the former case one expects the CEF ground state to be the maximum J doublet ($\pm 9/2$), well isolated from excited states formed by lower J doublets. Spin fluctuation between well separated CEF states will be comparatively slow. In the non Kramers ion $Pr^{3+}$ is the situation more complex. Its exact CEF ground state is not known, but expected to be an admixture of different J states, with the various possible CEF states not largely separated. While this may explain the difference in the magnitude of $\lambda$, the more normal behavior of $\lambda\,(T)$ in CEF sensitive $PrCu_3Ru_4O_{12}$ remains curious.

In summary, the nature of the magnetic ground states in the two rare earth ruthenates remains enigmatic. $\mu$SR points toward a spin frozen state, while magnetization data give no direct evidence for spin freezing. On the other hand, a dynamic state of correlated spins for $T \to 0$ is not easily compatible with the present $\mu$SR data. Since the discussion of the susceptibility curve [6] is based on the influence of CEF interactions, one has to keep in mind, when comparing it to the $\mu$SR data, that the presence of the charged muon may lead to disturbances of the local symmetry around the magnetic ion. This in turn can induce a change in the CEF level scheme by the stopped muon. Such an effect is most likely to occur in $Pr^{3+}$ compounds with a non-magnetic or weakly magnetic ground state. The situation of a "muon induced effect" has been observed previously in some cases, for example in $PrNi_5$ [10]. However, in $Nd^{3+}$ materials the CEF levels (usually Kramers doublets) are more sturdy and unlikely to respond significantly to the symmetry disturbance by the presence of the muon. Clearly, more extensive measurements are needed for both methods in order to clarify the complex magnetic situation in the low temperature range.

## Acknowledgments

The $\mu$SR studies were performed at the Swiss Muon Source of the Paul Scherrer Institute (PSI). We thank A. Amato and H. Luetkens for their assistance. This work was supported in part by the Deutsche Forschungsgemeinschaft (DFG) via Research unit 960 "Quantum Phase Transitions", DFG Contract No. SCHE487/7 and the project COST P16 ECOM of the European Union.

## References

[1] Labeau M et al. 1980 *J. Solid State Chem.* **33** 257
[2] Schwingenschlögel U et al. 2003 *Chem. Phys. Letters* **370** 719
[3] Kobayashi W et al. 2004 *J. Phys. Soc. Japan* **73** 2373
[4] Krimmel A et al. 2008 *Phys. Rev. B* **78** 165126
[5] Krimmel A et al. 2009 *Phys. Rev. B* **80** 121101
[6] Günther A et al. 2013 preprint (december 2013)
[7] Buettgen N et al. 2010 *J. Low Temp. Phys.* **161** 148
[8] Ebbinghaus S G Weidenkaff A and Cava R J 2002 *J. Solid State Chem.* **167** 126
[9] Kubo R 1981 *Hyperfine Interactions* **8** 731
[10] Feyerherm R et al. 1995 *Z. Phys. B* **99** 3

Journal of Physics: Conference Series **551** (2014) 012016 doi:10.1088/1742-6596/551/1/012016

# Spin-polaron band in the ferromagnetic heavy-fermion superconductor $UGe_2$

**V G Storchak[1], J H Brewer[2], D G Eshchenko[3], P W Mengyan[4], O E Parfenov[1] and D Sokolov[5]**

[1] Kurchatov Institute, Kurchatov Sq. 1, Moscow 123182, Russia
[2] Department of Physics and Astronomy, University of British Columbia, Vancouver, BC, Canada V6T 1Z1
[3] Bruker BioSpin AG, Industriestrasse 26, 8117 Fällanden, Switzerland
[4] Department of Physics, Texas Tech University, Lubbock, Texas 79409-1051, USA
[5] School of Physics and CSEC, The University of Edinburgh, Edinburgh EH9 3JZ, UK

E-mail: `mussr@triumf.ca`

**Abstract.** It has long been believed that coexistence among ferromagnetic ordering, superconductivity or heavy-fermion behaviour is impossible, as the former supports parallel spin alignment while the latter two phenomena assume a spin-singlet configuration. This understanding has recently been challenged by a number of observations in uranium intermetallic systems where superconductivity (SC) is found within a ferromagnetic state and both ordering phenomena are facilitated by the same set of comparatively heavy quasiparticles which bind into spin-triplet pairs in the SC state. Within the heavy-fermion scenario, this mechanism necessarily assumes that the magnetism has a band character. This band is expected to be responsible for all three phenomena — heavy-fermion behaviour, ferromagnetism and superconductivity — although its nature and the nature of the heavy quasiparticles have so far remained unclear. Our high-field muon spin rotation measurements are indicative of spin polarons of subnanometer size in $UGe_2$. These spin polarons behave as heavy quasiparticles made of $5f$ electrons. Once coherence is established, they may form a narrow spin-polaron band which thus may provide a natural reconciliation of itinerant ferromagnetism with spin-triplet superconductivity and heavy-fermion behaviour.

## 1. Introduction

Within the BCS theory of superconductivity (SC), it became clear long ago [1] that pairing of electrons in the spin-singlet state is effectively destroyed by an exchange mechanism arising from strong Coulomb interactions between the valence electrons. In a ferromagnetically (FM) ordered state, this exchange interaction tends to align the spins of electrons within a Cooper pair in parallel, thereby effectively preventing the pairing. Likewise, within the standard heavy-fermion (HF) approach, the Kondo effect quenches the on-site magnetic moment by spin fluctuations, thereby destroying pairs.

Although these three phenomena — magnetism, superconductivity and HF behaviour — have been considered in the past to be mutually antagonistic, the following findings clearly establish their possible coexistence (for a recent review, see [2]). A distinctive class of $f$-electron Ce- or U-based systems [3] convincingly shows that HF behaviour may coexist with SC. Most importantly, the SC pairing occurs among the heavy quasiparticles rather than within a band of light electrons: the SC energy gap opens up within the band of heavy quasiparticles. The attractive interaction between the quasiparticles is probably not provided by the electron-phonon interaction as in ordinary BCS superconductors, but rather calls for an alternative mechanism which is offered by various spin-fluctuation models [4] of magnetically mediated SC.

In these HF materials SC may coexist and couple with magnetism: in fact, many f-electron HF systems exhibit SC deep within magnetically ordered states, suggesting that magnetism may promote rather than destroy the superconductivity. It is remarkable that the same set of heavy quasiparticles apparently supports both the magnetism and superconductivity [3].

In the theory of magnetically mediated superconductivity, it is important that the ferromagnetism itself is of itinerant character, similar to that in the canonical d-electron ferromagnets Fe, Co or Ni. Conventionally, electrons in solids are classified as either itinerant or localized. In strongly correlated electron systems, specifically HF systems, such a clear distinction is often obscured, since signatures of both pictures appear. The possibility that the same electrons might simultaneously exhibit both localized and itinerant characteristics due to strong Coulomb interactions has developed into a duality problem — are they localized, itinerant or of a dual nature (partially localized and partially itinerant)? The electronic duality proposed for many different HF systems exhibiting simultaneous SC and magnetism would require the same f-electron to display both localized and itinerant nature *simultaneously*. The 5f electron duality problem is still debated in $UGe_2$, which is proposed to be viewed as a two-subset electronic system, where some of the 5f electrons are localized and responsible for the ferromagnetic moment and huge magnetocrystalline anisotropy, while the remaining 5f electrons are itinerant and responsible for unconventional SC [5].

Here we propose a specific concept that may supply the necessary requirement of simultaneously itinerant and localized electrons: formation of a spin-polaron band in which quasiparticle excitations of a low energy scale (several meV) around the Fermi energy are responsible for HF behaviour, SC and magnetism.

The standard theory for the formation of the extremely narrow high mass bands characteristic of heavy fermion metals starts from a set of strongly localized f-electrons. The appearance of a new (low) energy scale in this approach results from hybridization with the delocalized conduction states and strong correlations within the f-shells. A different approach, that we wish to develop here, starts with a delocalized band carrier whose transport depends upon the strength of its coupling with excitations of the medium. This is similar to the case of a lattice polaron (LP), where an electron accompanied by lattice modes forms a quasiparticle in which a local distortion of the crystal structure follows the charge carrier adiabatically and whose bandwidth is reduced by up to 4 orders of magnitude relative to that of ordinary electrons in conventional metals [6]. At low temperature, as long as the polaron is still much lighter than the atoms composing the medium, the charge is then delocalized within the LP band. A remarkable collapse of LP band at higher temperature marks a crossover from coherent band dynamics to incoherent hopping of localized states, analogous to the so-called dynamic destruction of the band for the tunnelling dynamics of heavy particles, such as protons, isotopic defects, or muons and muonium [7].

In close analogy, the exchange interaction (J) between a free carrier and local spins can cause electron localization into a FM droplet on the scale of the lattice spacing in a paramagnetic (PM) or AFM sea [8]. This charge carrier, accompanied by reorientations of local spins, forms a spin polaron [6, 8] (SP) with a large composite spin. As in the case of the LP, formation of a SP profoundly renormalizes the bare electron band into an extremely narrow ($\sim 0.001 - 0.1$ eV) spin-polaron band, which will favour coherent SP band dynamics at low temperature as long as spin fluctuations are suppressed. A remarkable result is that heavy f-like quasiparticles (SP) become part of the Fermi surface. We specifically note that within this framework the heavy quasiparticle is not an f-electron but the composite quasiparticle, a SP, formed as a result of s(p)-f interaction. Here we present spectroscopic indication, obtained by muon spin rotation, for spin polarons in $UGe_2$, confined within $R = 0.25(1)$ nm, with a high spin of $S = 4.3 \pm 0.3$. At low temperature, SP tend to form a narrow spin-polaron band in the vicinity of $E_F$, profoundly modifying the magnetic, transport, optical and thermodynamic properties of the host.

## 2. Results and discussion

Single crystals of $UGe_2$ for the current studies were grown by the Czochralski technique under purified Ar atmosphere with a radio-frequency heating. Single crystals of a disk shape were oriented using a white beam X-ray backscattering Laue method, sparkcut and etched to remove the oxidised surface. The electrical resistivity shows a residual resistivity ratio RRR=62 and FM transition at $T_c$= 52.6 K.

Time-differential $\mu^+$SR experiments, using 100% spin-polarized positive muons implanted into these samples, were carried out on the M15 surface muon channel at TRIUMF using the HiTime spectrometer. At high temperature, Fourier transforms of the time spectra in a magnetic field (B) transverse to the initial muon spin polarization direction and parallel to the easy magnetization direction (a axis) exhibit a single peak at the muon frequency $\nu_\mu = \gamma_\mu B/2\pi$ (where $\gamma_\mu = 2\pi \times 135.53879$ MHz/T is the muon magnetogyric ratio). However, below T = 100 K the $\mu^+$SR spectra change abruptly to reveal two peaks (Fig. 1). The evolution of these signals with temperature is presented in Fig. 2. These two peaks are also shifted to lower frequencies relative to the single peak (not shown in Fig. 1) detected in a reference sample ($CaCO_3$), which occurs at the bare muon frequency.

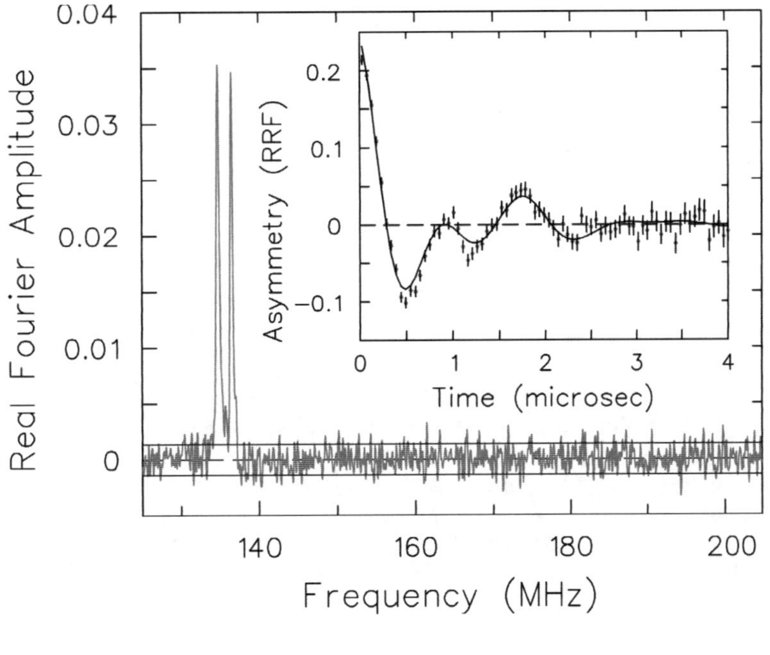


**Figure 1.** Frequency spectrum of muon spin precession in $UGe_2$ in a transverse magnetic field of H=1 T at T=75 K. Inset: same spectrum in the time domain in a rotating reference frame at 135.53333 MHz. The two-frequency precession pattern characteristic of a localized electron hyperfine-coupled to a muon is clearly apparent in both domains.

Observation of two peaks in the Fourier spectra prompted the authors of [9] to suggest two magnetically inequivalent sites occupied by the positive muon in $UGe_2$. Although such

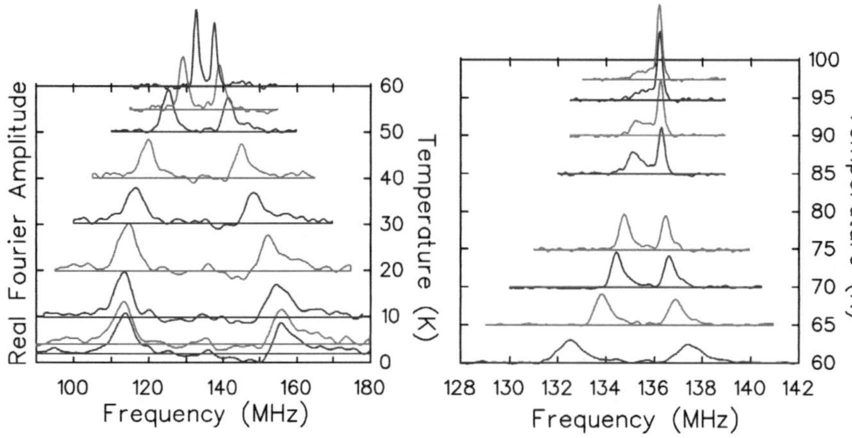


**Figure 2.** Fourier transforms of the muon spin precession signal in UGe2 in a transverse external magnetic field of 1T at different temperatures. The characteristic SP lines appear below about 100 K and persist through the FM transition (Tc = 52.6 K) down to the lowest measured temperature.

an approach constitutes the conventional assignment, the two-peak signal may alternatively be interpreted as indication of a muon-electron bound state.

In particular, the two lines shown in Fig. 1 and Fig. 2 may indicate the characteristic signature of a coupled muon-electron spin system in high magnetic field [10]. We suggest that the observed bound state is a spin polaron. In a PM or metallic environment, the strong pair exchange interaction of the bound electron with itinerant spins would result in rapid spin fluctuations of this electron, averaging the hyperfine interaction to zero which, in turn, would result in a collapse of the doublet into a single line at $\nu_\mu$ [11], if the local FM ordering mediated by this electron did not hold the electron's spin orientation locked. In metals, however, even the protective local FM environment of a SP does not ensure observation of the doublet unless the SP spin (S) is decoupled from its magnetic environment [12, 13]. Such decoupling is possible in high B when the Zeeman energy of $S$ exceeds an exchange interaction ($I$) between local spins [10]. This is the case in magnetic insulators where the SP doublet is detected up to very high temperature. In metals, RKKY interactions make $I$ much stronger, so that decoupling would require a very high magnetic field that is inaccessible in the current experiment. In $UGe_2$, above about 100 K, spin exchange with the magnetic environment effectively averages the muon-electron hyperfine interaction, causing the collapse of the doublet into a single peak as presented in Fig. 2. Such a collapse does not necessarily mean that the SP does not form in $UGe_2$ above 100 K; we just may not see its fingerprint, which is a characteristic doublet [11]. The abrupt appearance of a SP doublet below about 100 K is possible due to another effective decoupling mechanism the opening of a spin gap due to crystal field splitting of the U ions spin excitation, characteristic of U compounds [5]. This is consistent with optical conductivity measurements which exhibit a dramatic reduction of the spin scattering rate below 120 K [14].

Temperature and magnetic field dependencies of the signal frequency splitting, $\Delta\nu$, provide information on the characteristic size through the hyperfine coupling ($A$) and determine the composite spin of the SP [10, 11]. Within a mean field approximation, $\Delta\nu$ is proportional to a Brillouin function. For $g\,\mu_B B \ll k_B T$, $\Delta\nu$ is a linear function of both $B$ and $1/(T - T_c)$ [10]:

$$\Delta\nu = A\left[\frac{g\,\mu_B B}{3k_B(T - T_c)}\right](\mathcal{S} + 1)\ . \qquad (1)$$

At low $T$ and high $B$, however, this equation is no longer valid, as the composite spin $S$ is fully polarized. Therefore, in a magnetic field high enough that the muon Zeeman energy exceeds the hyperfine coupling, $\Delta\nu$ saturates at the value of A [10, 11]. In $UGe_2$, $\Delta\nu$ saturates as a function of both inverse temperature (in a magnetic field $H = 1$ T) and $H$ (at $T = 5$ K) at the same value, $A = 41 \pm 2$ MHz.

This hyperfine coupling in the SP is about 100 times less than that for Mu in vacuum, implying that the SP radius is 0.25(1) nm. This is consistent with the muon being centred between two U atoms (x=0.5, y=0.5, z=0.5) giving a muon-U distance of 0.214 nm. Fitting Eq. (1) to the data yields $S = 4.3(3)$ and $T_c = 52(1)$ K, consistent with our estimate of $S$ from the SP radius and $T_c$ of our sample, respectively.

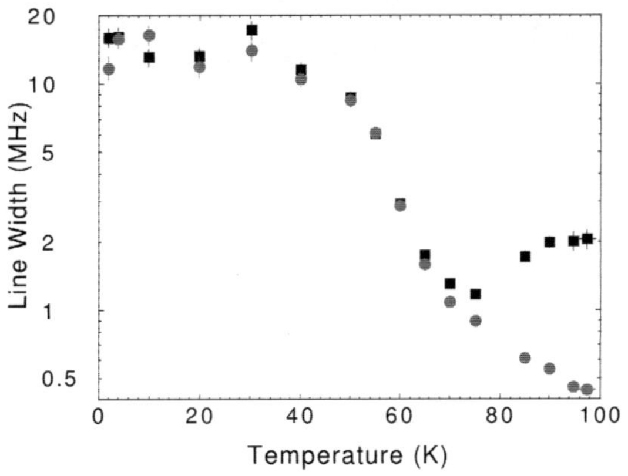


**Figure 3.** Temperature dependences of the SP line widths of the signals with higher (red circles) and lower (black squares) frequencies in $UGe_2$ in a magnetic field of $H = 1$ T.

The exchange interaction governs SP formation and dynamics in $UGe_2$, since the Coulomb interaction is effectively screened. Therefore the role of the muon, which may be important for SP formation in magnetic semiconductors and insulators [10], is reduced to that of an innocent bystander in metals. The appearance of the mid-infrared feature in optical conductivity experiment in $UGe_2$ is attributed to a polaron response [14]. We argue that once the host lattice is populated by free spin polarons, one of them may be captured by the muon to reveal the fingerprint of a bound muon-electron state the characteristic doublet.

In $UGe_2$, the dynamics of such SP clearly shows a qualitative change around 80 K. At high temperature, the asymmetric doublet shape (Fig. 2, right panel) indicates the nearly static character of the SP [12]. A remarkable crossover occurs at about 80 K, where not only do the line widths become equal (Fig. 3), but also the spectral weights (corresponding state populations) of the two lines within the doublet effectively equilibrate, clearly indicating the onset of the effective spin-exchange mechanism within the SP system, as other spin-exchange channels are rather ineffective [15]. This marks a crossover from the static SP to itinerant polaron behaviour. At high temperature, the characteristic length (in Angstroms) of the ferromagnetic fluctuations determined from neutron measurements, $\xi_{FM} = 3.45/(T/T_c - 1)^{1/2}$, is much less than the SP size, 2R, which causes strong SP localization due to the significant energy shift of the nearest equivalent SP positions within the lattice (so-called static destruction of the band [7]). At lower temperature, once $\xi_{FM}$ exceeds the SP size, energy levels equilibrate within the length scale of $\xi_{FM}$, which causes SP delocalization. A simple estimate shows that $\xi_{FM}$ becomes equal to 2R at

Journal of Physics: Conference Series **551** (2014) 012016 doi:10.1088/1742-6596/551/1/012016

77 K, which agrees well with the experiment. The amplitudes or spectral weights of both lines are temperature independent below about 80 K (Fig. 2), which clearly indicates temperature independent populations of the spin-up and spin-down states which in turn is inconsistent with a localized SP but rather signifies its itinerant nature. Thus, our model suggests that the SP, captured by the muon, stays localized and is protected from spin exchange above 80 K; however, below 80 K it remains localized but undergoes active spin exchange with free itinerant SP.

However, the remarkable itinerancy acquired by SP below 80 K does not mean they form a band at 80 K. Instead, SP transport in $UGe_2$ just below 80 K is determined by hopping (diffusive) dynamics similar to the hopping dynamics of muon or Mu states in the regime dominated by static destruction of the band when dynamic fluctuations of the environment (phonons in the case of Mu) makes such hopping dynamics possible [7]. In fact, coherent SP transport is only possible in a spin-polarized band where spin-flip scattering is suppressed due to the absence of SP with the opposite spin state. In a FM state, such coherent transport of SP becomes possible due to splitting of the majority and minority spin subbands. This splitting may be viewed as Zeeman splitting due to spontaneous magnetization. In $UGe_2$, such splitting of spin subbands occurs at $T^*$, which separates the high-temperature weakly polarized phase (FM1) from the low-temperature strongly polarized phase (FM2) and shows up as anomalies in resistivity, heat capacity, magnetization, magnetoresistivity and de Haas-van Alphen measurements [5].

Formation of the spin polaron band may provide an explanation for a list of mysteries of $UGe_2$ as well as other HF and strongly correlated materials, such as itinerant versus localized carriers in a duality problem when local moments acquire itinerancy, small Fermi surface versus large Fermi surface including both conduction electrons and local spins, huge anisotropic magnetoresistance, Fermi surface reconstruction at T* and the nature of the mysterious FM transition within the FM phase.

## References

[1] Berk N F & Schrieffer J R 1966 *Phys. Rev. Lett.* **17** 433
[2] Pfleiderer C H 2009 *Rev. Mod. Phys.* **81** 1551
[3] Fisk Z *et al.* 1988 *Science* **239** 33
[4] Monthoux P, Pines D & Lonzarich G G 2007 *Nature* **450** 1177
[5] Troc R, Gajek Z. & Pikul A 2012 *Phys. Rev.* B **86** 224403
[6] Nagaev E L 2002 *Magnetic Semiconductors* (Imperial College, London).
[7] Storchak V G and Prokof'ev N V 1998 *Rev. Mod. Phys.* **70** 929
[8] Mott N F 1990 *Metal-insulator transitions* (Taylor & Francis, Bristol).
[9] Sakarya S *et al.* 2010 *Phys. Rev.* B **81** 024429
[10] Storchak V G *et al.* 2009 *Phys. Rev.* B **80** 235203
[11] Storchak V G *et al.* 2011 *Phys. Rev.* B **83** 077202
[12] Storchak V G *et al.* 2010 *Phys. Rev. Lett.* **105** 076402
[13] Storchak V G *et al.* 2011 *Phys. Rev.* B **83** 140404(R)
[14] Guritanu V *et al.* 2008 *Phys. Rev.* B **78** 172406
[15] Storchak V G *et al.* 2013 *ArXiv:1309.6407*

# Magnetic order and muon motion in $VO_2$

**P W Mengyan, R L Lichti, B B Baker and G Jayarathna**

Department of Physics, Texas Tech University, Lubbock, TX 79409-1051 USA

E-mail: rick.mengyan@ttu.edu

**Abstract.** We report the first transverse, longitudinal and zero field measurements performed on Vanadium Dioxide ($VO_2$) where we find a significant local magnetic field below $T = 34.7 \pm 0.16$ K at the muon site and $Mu_d$ motion above 340 K. Distinct shifts in the muon relaxation rates and Kubo-Toyabe delta parameters, near the metal to semiconductor and structural transition ($T_{MST} \approx 340$ K), show that $\mu^+$SR measurements may provide additional means to further characterize the mechanisms contributing to these transitions that are yet to be understood.

## 1. Introduction

Vanadium dioxide ($VO_2$) has an ultra-fast, reversible metal-semiconductor phase transition (MST) at $T_{MST} \approx 340$ K that coincides with a structural phase transition (rutile $T > T_{MST}$; monoclinic $T < T_{MST}$) below which the bandgap changes from nominally 0 eV to about 1 eV and the conductivity decreases by several orders of magnitude [1, 2, 3]. This transition can be triggered thermally, optically, electrically or with application of pressure. Incorporating just a few percent of H reduces $T_{MST}$ to near 200 K while tremendously reducing the change in properties, such as conductivity and bandgap [4]. Dopants such as titanium, tungsten, niobium or gold only reduce $T_{MST}$ to between 280 K and 320 K whilst having minimal effect on the electronic or optical properties of the host material [5]. $T_{MST}$ increases upwards of 400 K when elements such as fluorine, aluminum or chromium are introduced to the system. Many studies have looked specifically at the effects that some dopants and impurities have on material characteristics when crossing the MST [5] as there are a plethora of potential technological applications, but there is still no consensus as to whether the fundamental mechanism responsible for the transition is related to electron-phonon interactions or strong electron-electron interactions (i.e. Peierls mechanism or Mott-Hubbard transition) [6]. Furthermore, the role impurities play has yet to be determined. An understanding of the mechanism governing the transitions is required for device application as control and optimization of the material properties is critical.

In the metallic phase ($T > T_{MST}$), $VO_2$ is described as rutile with a tetragonal body-centered unit cell where each V is surrounded by an octahedral arrangement of oxygen atoms. Below $T_{MST}$, in the semiconducting phase, the system has a doubled unit cell in which V atoms pair along the $c$-axis where one V atom (of each pair) shifts in the $a$-$b$ plane as well as closer to its partner (parallel to the $c$-axis). The net result is a zig-zag chain of vanadium atoms. Whilst in the rutile structure, each $V^{4+}$ cation has an electron near the Fermi level occupying the lowest $3d$ level. Asymmetry in the crystal field causes the $3d$ states to be split so that the lowest orbital is predominantly aligned along the $c$-axis ($3d_{||}$). In the semiconducting phase, dimerization of the $V^{4+}$ ions result in pairs of these $3d$ electrons being forced into a singlet state ($S_{pair} = 0$) and hence contributes to the high resistivity and non-zero bandgap. Impurities (dopants, structural defects, vacancies etc) in the $VO_2$ system do affect resulting properties of the transitions such as the temperature at which it occurs, change in conductivity, transition rate, variation in optical absorption, transmission and reflectivity. If impurities modify $VO_2$ properties by disrupting the dimerization process, an outcome of this partial dimerization may be the development of a non-zero net spin.

Here we present some of our initial findings from the first investigation of $VO_2$ compounds via the muon spin research technique where the muon ($\mu^+$) is used both as an experimentally accessible analog to the hydrogen impurity as well as a sensitive local magnetic probe to reveal properties not previously reported. For oxides in general, an implanted $\mu^+$ typically bonds with oxygen, forming the $[OMu]^-$ complex (c.f. $[OH]^-$), which we specifically refer to as $Mu_d$ throughout this paper since the bound complex is diamagnetic.

## 2. Experimental Details

A bulk sintered 7x7x1 $mm^3$ sample of $VO_2$, acquired from a commercial source, was used in these experiments. XRD measurements were completed at the materials characterization lab at the STFC ISIS facility (Didcot, UK) and show no obvious chemical phases beyond what is expected for $VO_2$. Transverse field muon spin rotation (TF-$\mu^+$SR) measurements were conducted using the HiTime spectrometer on the M15 surface muon channel at TRIUMF (Vancouver, Canada) in temperatures from 2.5 K to 300 K and magnetic fields up to 6.5 T applied perpendicular to the initial muon spin polarization direction. Low TF measurements ($B_{TF}$ = 100G) were completed on the EMU instrument at the STFC ISIS facility (Didcot, UK). Both of these measurements are aimed at characterizing the $\mu^+$ stopping site, investigating the possibility of $Mu^0$ ($\mu^+e^-$) formation and explore the possibility of local magnetism through the MST and well into each phase. Zero field muon spin relaxation measurements (ZF-$\mu^+$SR) were primarily conducted on EMU (ISIS) in a temperature range of 8 K to 560 K. The main goal of these measurements were to characterize motional dynamics of $Mu_d$ and any variations in the local field throughout each phase. Supplemental ZF measurements were completed with the HiTime and Helios spectrometers on the M15 and M20C surface muon channels (respectively) at TRIUMF, below 300 K, where the higher frequency resolution capabilities (c.f. properties of the continuous wave source and instrument design) allow for additional characterization of the local magnetic environment. Longitudinal field muon spin relaxation measurements (LF-$\mu^+$SR) were conducted with EMU (ISIS) and Helios (M20C, TRIUMF) with magnetic fields up to 4.5 T applied parallel to the initial muon spin polarization direction and the goal of characterizing any dynamics assocaited with the observed magnetic features.

## 3. Results and Discussion

$\mu^+$SR data reveal three distinct regimes of $Mu_d$ behavior in $VO_2$. For ZF data analysis, the baseline is determined at each temperature by applying a low transverse field ($B_{TF}$ = 100G) and adjusting the alpha parameter so that the precession signal is centered about zero asymmetry. This corrects for detector efficiency and any anisotropy in the experimental geometry. The ZF data between 100 K and 300 K fit well to a static Kubo-Toyabe function [7] (sKT) with $\Delta = 0.171 \pm 0.004$ $\mu s^{-1}$ and a non-relaxing component (Figure 1a).

Above 300 K there is some slight variation in $\Delta$ between 320 K and 350 K where there is a small but definite decrease which leads to $\Delta = 0.165 \pm 0.005$ $\mu s^{-1}$ above 350 K. Above 450 K the signal character in the time domain changes dramatically (Figure 1b), as a drop in $\Delta$ when initially characterized with a static Kubo-Toyabe function. The signal above 340 K is appropriately fit using a dynamic Kubo-Toyabe function [7] (dKT) where the $\Delta$ is fixed at the stable value determined between 350 K and 450 K and the hop rate ($\nu$) is then allowed to

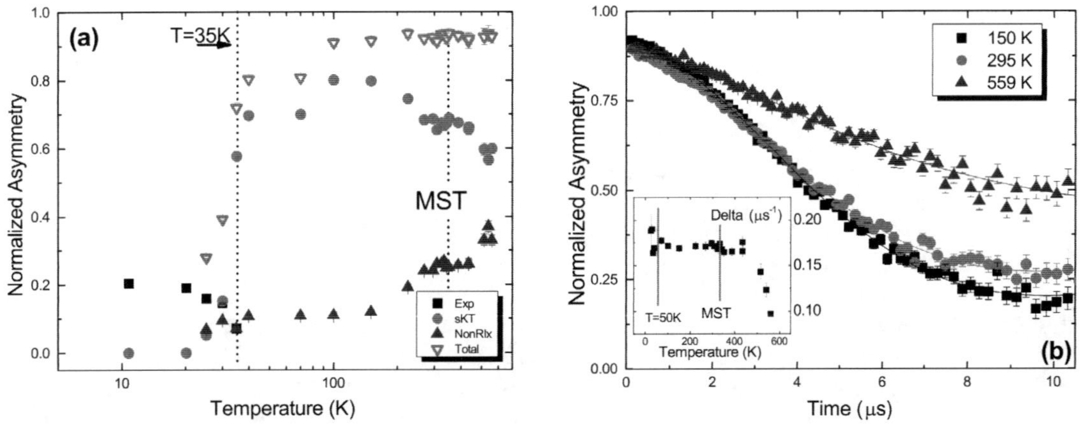


**Figure 1. (a):** ZF Asymmetry from 8 K to 600 K where the sharp decrease in asymmetry near 35 K is due to the abrupt ($\delta T < 5$ K) development of a magnetic phase. The slight decrease in static Kubo-Toyabe (sKT) asymmetry near 100 K is attributed to the development of highly localized magnetic environment or magnetic fluctuations as a precursor to the significant local field development below 35 K. Features around 340 K relate to the MST. **(b):** ZF time-domain signal data from EMU at ISIS. *Inset:* Temperature dependence of sKT $\Delta$. The sharp drop above 450 K suggests the onset of motion.

vary. These dynamics can be traced down to, at least, 340 K ($T_{MST}$) and possibly to 300 K. Considerable uncertainty in $\nu$ arises below 340 K since the change in local field distribution at the $Mu_d$ site is a result of the structural transition which contributes to a real variation in $\Delta$. Additionally, the high correlation between $\Delta$ and $\nu$, in the function itself, mandates that one determine $\Delta$ in a region where dynamics are sufficiently slow so that a static function is appropriate before applying a dynamic function to extract $\nu$. Therefore, any variation in $\Delta$ between 300 K and 340 K directly contributes to the uncertainty of $\nu$, in this region. Assuming the structural transition provides a real change in $\Delta$ and that the stability of $\Delta$ between 350 K and 450 K accurately reflects the second moment of the field distribution at the $Mu_d$ site in the rutile state ($T > T_{MST}$), we can assign $\Delta$ ($0.165 \pm 0.005$ $\mu s^{-1}$) to the $Mu_d$ site in this phase. With $\Delta$ fixed, analysis reveal dynamics begin slightly above 340 K and continue to above 560 K (Figure 2) which we assign to the onset of $Mu_d$ motion. A simple Arrhenius fit $[\nu(T) \sim \nu_0 exp(-E_a/KT)]$ to the hop rates above 400 K (where fits are more reliable) result in an activation energy of $E_A \approx 60 \pm 10$ meV and prefactor of $\nu_0 = 1.3 \pm 0.15 \mu s^{-1}$. The hop rate leveling off at higher temperatures and hence the extremely low prefactor is consistent with local motion triggered by a wag mode or a phonon assisted tunneling process. Below 100 K there is a slight drop in the ZF sKT component's asymmetry accompanied by a comparable rise in the asymmetry of the non-relaxing component. Near 35 K the overall character of the time domain signal changes significantly as the sKT quickly vanishes as an exponentially relaxing signal (as observed at ISIS) then dominates the spectra implying either a fast depolarization or frequencies beyond the detection limit.

Weak ($B_{TF}$ = 100G) transverse field measurements from 8 K to 560 K, conducted at ISIS, were fit to a single relaxing oscillation and a non-oscillating component ($P(t) = A_1 exp(-\lambda_1 t) cos(\omega_L t + \phi) + A_2$) and reveal significant features at 35 K, 100 K and 340 K. Crossing over 340 K (MST) from below, we observe a decrease in $A_2$ and a sharp minimum in $\lambda_1$ followed by a peak in $\lambda_1$ near 450 K (Figure 3). When traversing 100 K from above, we observe an increase in $A_2$ and correlated decrease in $A_1$. At the lowest temperature, $A_2$ is nearly

Journal of Physics: Conference Series **551** (2014) 012017 doi:10.1088/1742-6596/551/1/012017

1/3 of the total maximum asymmetry (measured at $T = 560$ K).

Higher TF measurements ($B_{TF} = \{0.03, 0.1, 3.0\}T$) between 2.6 K and 300 K, at TRIUMF, show qualitatively comparable behavior to what is observed at ISIS in low field (Figure 3). Accounting for the difference in measurement limits of the instruments at ISIS and TRIUMF, one can understand why below 100 K at TRIUMF we resolve two different regimes of $Mu_d$ behavior that were only hinted at in the data from ISIS. Specifically, a high frequency oscillation with a small amplitude develops between 35 K and 100 K (4), and below 35 K a significant local field develops (5). Above 120 K, these high TF data show a single, well defined oscillation with a frequency that matched the signal found in the non-magnetic $CaCO_3$ reference material mounted behind the $VO_2$ sample suggesting that between 120 K and 300 K, the effective field at the muon site ($B_{eff}$) is simply equal the externally applied field ($B_{ext}$). Peaks in the relaxation rate of this diamagnetic component are centered around 100 K and 35 K, with a minimum near 70 K (not shown) and corresponds with the onset of the change in signal character. Below 35 K a minimum of two oscillating components are required, as a second frequency with growing asymmetry and

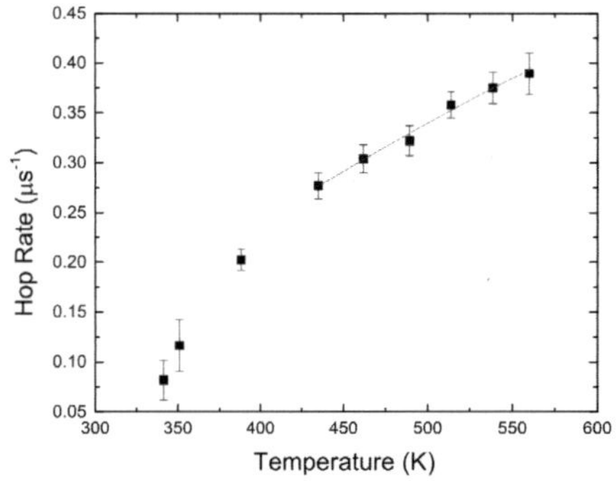


**Figure 2.** Hop rates extracted from fits to time domain data with a dynamic Kubo-Toyabe function (dKT). An Arrhenius fit to the higher temperature hop rates yields $E_A = 60 \pm 10$meV and $\nu_0 = 1.3 \pm 0.15\mu s^{-1}$. The very low hop rates, $\nu_0$ and $E_A$ are characteristic of local motion.

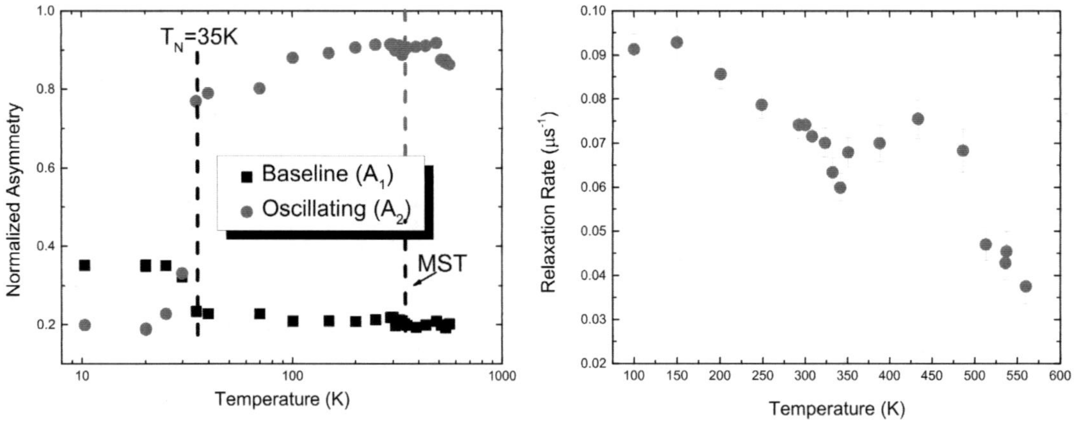


**Figure 3.** $B_{TF} = 100$G fit asymmetry for oscillating ($A_1$) and baseline ($A_2$) components (*left*) and the relaxation rate ($\lambda_1$) associated with the oscillating component (*right*).

relaxation rate develops rapidly below 35 K (Figure 5). A third higher frequency component to account for the frequency present between 35 K and 100 K (Figure 4) should be included but has not yet been folded into this preliminary analysis of the spectra, below 35 K. The growth in asymmetry of this shifted signal is complemented by the reduction in diamagnetic asymmetry. The fit frequency and relaxation rate of the shifted signal (below 35 K) follow a critical power law ($f(T) \sim f_0(1 - [T/T_c]^\alpha)^\beta$) with $\alpha = 1.09 \pm 0.2$, $\beta = 0.35 \pm 0.07$ and $f_0 \approx 22 \pm 1$ MHz (Figure 4). Since $B_{ext} = 102.05 \pm 0.04$ mT (determined via fits to diamagnetic component) and $f_0 \to B_{0,eff} = 162 \pm 7$ mT, there is clearly a non-zero local field ($B_{loc} = 60 \pm 7$ mT) contribution. The source of $B_{loc}$ has not yet been determined. Similar results are found at the other measured fields. Note that this low temperature behavior is not consistent with $Mu^0$ formation but instead developing magnetic order and alignment with the externally applied field.

These measurements on $VO_2$ are the first part of a large scale project geared towards understanding the $Mu_d$ behavior, local electronic and magnetic field structure and an attempt to probe the fundamental mechanisms responsible for the various transitions occurring in $VO_2$ compounds. Part of this project includes investigating $VO_2$ compounds doped with a variety of elements that alter the MST by either raising or lowering $T_{MST}$ as well as stabilizing the low temperature magnetic phase by significantly disrupting the dimerization. Additional measurements and some modeling are required to better determine the source of the magnetism and driving mechanisms for each transition.

As of now, we suggest that a disruption in the V–V dimers introduce a non-zero net spin and hence lead to the observed magnetic features. Additional $\mu^+$SR measurements coupled with other techniques, such as neutron scattering and bulk magnetization measurements, are required to determine properties such as moment direction, size and further characterize the low temperature magnetic phase. We are furthering this effort by continuing to investigate $VO_2$ compounds with different impurities and varying concentrations with the goal of also mapping

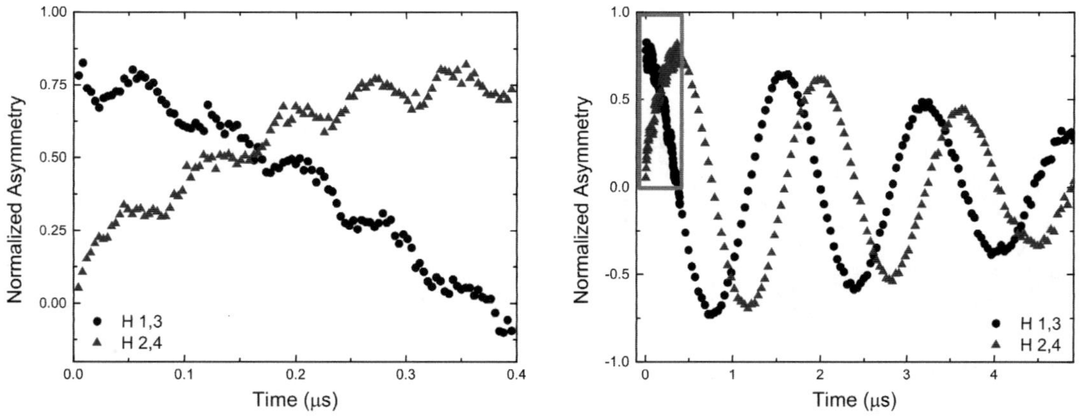


**Figure 4.** TF-$\mu^+$SR spectra with $B_{TF} = 100$mT where both pair of opposing counters (labeled *H 1,3* and *H 2,4*) are shown in the time domain with a rotating reference frame frequency set at $f_{RRF} = 13.25$ MHz. These plots show a region where the non-magnetic phase and a small fraction of the magnetism is observed. *Left*: Highlighting the small amplitude oscillation that fits to a field that roughly matches the 162 mT field observed at 5 K. At 5 K, this component dominates the spectra. *Right*: Precession at a frequency that matches the $\mu^+$ signal in non-magnetic reference material and what one expects for the 100 mT externally applied field. The red rectangle highlights the region enlarged in the left frame.

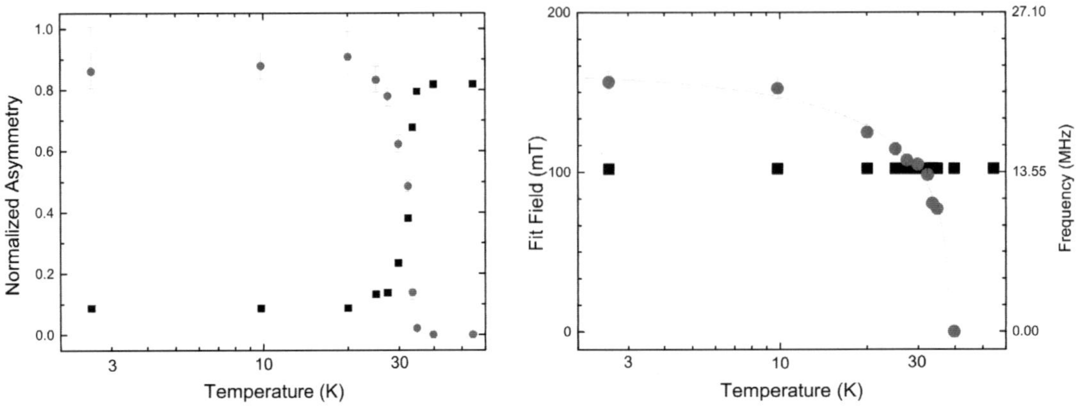


**Figure 5.** Normalized asymmetry (*left*) associated fit frequencies [fields] (*right*) from two component analysis on TF-$\mu^+$SR data in an applied field of 100 mT utilizing a rotating reference frame. The developing internal field ($T_c = 34.7 \pm 0.12$ K) follows a critical power law (fit shown, discussed in text) and at its maximum, $B_{loc} = 60 \pm 7$ mT.

out the role that the impurities play in modifying the local environment.

## 4. Summary

From this first analysis, we report the detection of a significant local magnetic field that develops in stoichiometric $VO_2$ below $T_c \approx 35$ K with a maximum value of $B_{loc} = 60 \pm 7$ mT and that the muon detects features that are consistent with short range magnetic correlations that begin below 100 K and persist down to $T_c$. Between 100 K and 300 K, we find no significant dynamics in either $Mu_d$ motion or variations in local fields. We find that the muon is, in fact, sensitive to the MST and structural transition around 340 K. In the metallic phase ($T > T_{MST}$), we find that $\Delta = 0.165 \pm 0.005\ \mu s^{-1}$ is slightly different from the $\Delta = 0.171 \pm 0.004\ \mu s^{-1}$ measured in the semiconducting phase ($T < T_{MST}$). We find that $Mu_d$ dynamics, either local motion triggered by a wag mode or phonon assisted tunneling, are significant above 450 K ($E_A \approx 60 \pm 10$meV, $\nu_0 = 1.3 \pm 0.15\mu s^{-1}$) but can be traced down to 340 K.

## Acknowledgments

Funding for this project is provided by the Welch Foundation (grant D-1321). We would also like to recognize and thank the MuSR groups at both ISIS and TRIUMF for their time, effort and support.

## References

[1] Imada M *et al.* 1998 *Rev. Mod. Phys.* **70** 1039
[2] Goodenough J B 1971 *J. Solid State Chem.* **3** 490
[3] Morin F J 1959 *Phys. Rev. Lett.* **3** 34
[4] Wu C *et al.* 2011 *J. Am. Chem. Soc.* **133** 13798
[5] See e.g.: Kiri P *et al.* 2010 *Adv. Mat. Lett.* **1** 86; Burkhardt *et al.* 1999 *Thin Solid Films* **345** 229; Tang C *et al.* 1985 *Phys. Rev.* B **31** 1000
[6] See e.g.: Qazilbash M M *et al.* 2007 *Science* **318** 1750; Laad M S *et al.* 2006 *Phys. Rev.* B **73** 195120; Pergament A 2003 *J. Phys: Condens. Matt.* **15** 3217; Cavalleri A *et al.* 2001 *Phys. Rev. Lett.* **87** 237401
[7] Hayano R S *et al* 1979 *Phys. Rev.* B **20** 850

# Muon spin precession in ferromagnetic $DyAl_2$

**O Hartmann**[1], **G M Kalvius**[2] **and R Wäppling**[1]

[1] Department of Physics and Astronomy, Uppsala University, 75120 Uppsala, Sweden
[2] Physics Department, Technical University Munich, 85747 Garching, Germany

E-mail: ola.hartmann@fysik.uu.se

**Abstract.** The Laves phase intermetallics $REAl_2$ were studied by $\mu$SR over several decades. Results had mainly been obtained for the temperature and field dependencies of relaxation rates in the paramagnetic state. However, it turned out that spontaneous precession signals in the ordered magnetic state were very difficult to observe. We report here the observation a weak precession signal in a single crystalline sample of $DyAl_2$. The precession signal seen has low asymmetry and is heavily damped. This is understood with the notion that the magnetic structure of $DyAl_2$ splits the internal field at the interstitial muon stopping site into several components, and that of those only one of the field components leads to the visible precession signal. The observed temperature dependence of the precession frequency is compatible with the change of easy axis of magnetization at $T = 40$ K.

## 1. Introduction

Like all rare earth (RE) di-aluminates, $DyAl_2$ crystalizes in the C15 Laves phase structure. Other members of this crystallographic group are the analogeous compounds with transition metals and actinides. A fair number of these Laves phase intermetallics have been studied by $\mu$SR [1] on account of their fundamental magnetic behavior. This holds in particular for the $REAl_2$ intermetallics, mainly because (with the exception of RE = Ce) they show only one (second order) magnetic phase transition into a simple ferromagnetic state. Nearly all of the earlier $REAl_2$ studies concentrated on the paramagnetic range, since the precession signals in the ferromagnetic state could not be seen (again with the exception of $CeAl_2$ and one recently found case). Those extensive studies of $REAl_2$ compounds in the paramagnetic range [2, 3, 4], including prominently $DyAl_2$, were undertaken with the aim to elucidate the fluctuations of paramagnetic 4f-moments, especially on approach to the Curie point. The $\mu$SR spectra show single pure exponential decay of asymmetry with rate $\lambda_{pm}$. Foremost established was the divergent rise in $\lambda_{pm} \propto \tau_{4f}$ (with $1/\tau_{4f}$ being the fluctuation rate of the 4f moments) over a fairly wide temperature range above $T_C$, i.e., much wider than the common 'critical range' in mean field theory. This established regime of strong rise in $\lambda_{pm}$ is commonly termed 'critical slowing down' and reflects the growing size of spin clusters due to increasing spin-spin correlations. The larger the spin cluster, the longer the time needed to turn the full cluster by thermal agitation (see also the discussion in [5] relating to iron and some of its intermetallics). Also explored in the previous study of $REAl_2$ materials was the temperature dependence of the paramagnetic frequency shift which can have magnitudes of several percent and shows as well a strong rise on approaching $T_C$, quite compatible with the picture for $\lambda_{pm}$. The above mentioned recent exception where a signal in the ferromagnetic state has been seen is $SmAl_2$ [6], which has a comparatively small magnetic moment. Its $\mu$SR spectra display a clear precession pattern with normal asymmetry.

In this report we present new results on $DyAl_2$ using a spherical single crystalline sample. We were able to observe a weak precession signal in the ferromagnetic regime ($T_C \approx 65$ K), a first in a high moment $REAl_2$ compound.

## 2. Experimental

We have carried out our studies on single crystalline $DyAl_2$ at different facilities. With surface muons at M13 of TRIUMF and DOLLY of PSI as well as with backward muons at GPD of PSI. Zero field (ZF) and (in the paramagnetic range) weak tranverse field (TF) data were taken from 190 K down to 10 K in different orientations between [100] and [111]. A previous run sequence with the single crystalline sample at ISIS extended the higher temperature region to 1000 K. The low temperature range $T<T_C$ includes in particular the change in easy axis of magnetization being [111] close to $T_C$ and [100] at low temperatures which, according to [7], occurs around 40 K.

## 3. Results

### *3.1. Paramagnetic range*

The $\mu$SR spectra for $T > T_C$ show in ZF as well as in TF single simple exponential decay of muon spin polarization with full initial asymmetry confirming a unique single interstitial muon stopping site. This conclusion is further corroborated by high field (up to 300 mT) TF measurements which still show a single signal shifted by a well defined paramagnetic Knigt shift [2]. The single site approach holds for all members of the $REAl_2$ series and also in $\mu$SR studies of other C15 Laves phase intermetallics [1], especially for the results in magnetically ordered $CeAl_2$ and $SmAl_2$. Examples of paramagnetic ZF spectra taken within this work are depicted in Fig. 2-left (top curves).

The temperature dependence of the paramagnetic relaxation rate $\lambda_{pm}$ is displayed in Fig. 1

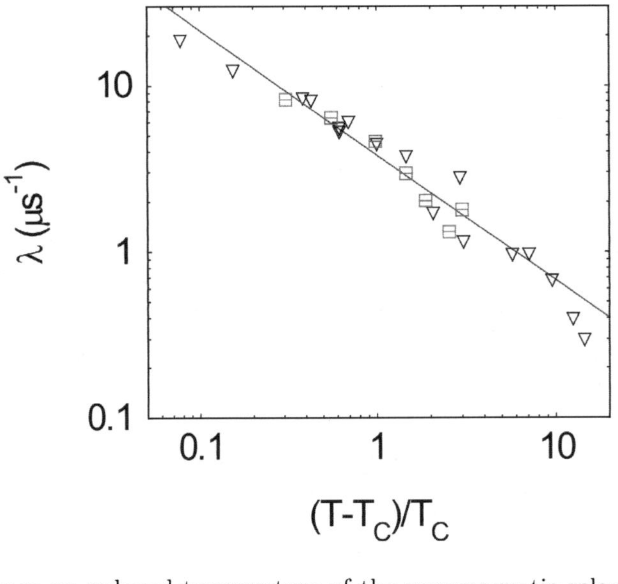


**Figure 1.** Dependence on reduced temperature of the paramagnetic relaxation rate in $DyAl_2$ from ZF and low TF measurements. Triangles: ZF data from the single crystalline sample; Squares: Older weak TF data from a polycristalline sample. The line is a least squares fit explained in text.

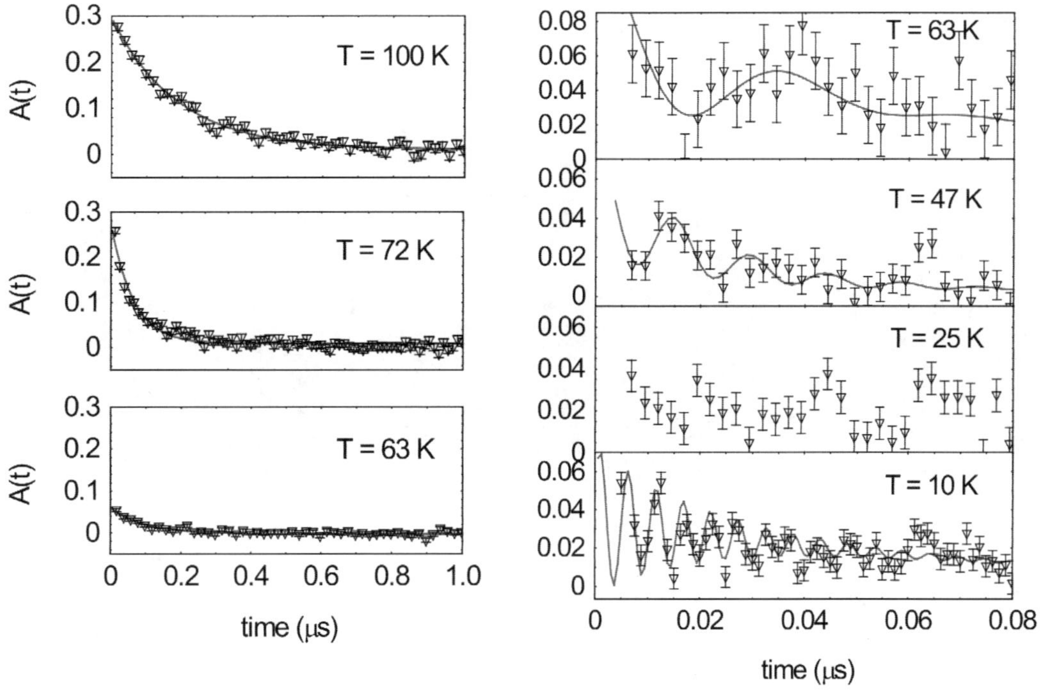


**Figure 2.** Left: ZF spectra at 100 K, 72 K (paramagnetic) and 63 K (ferromagnetic). Right: Muon precession patterns in the ferromagnetic regime at T = 63 K, 47 K, 25 K, and 10 K. Note the different scales of the $x$ and $y$ axes between the left and right panels. The lines are least squares fits discussed in text. The 25 K spectrum shows no distict precession pattern and cannot be fitted in the usual manner as discussed in text

including earlier weak TF data [2] on a polycrystalline sample. The new and the old data agree well with each other. The solid line is a least squares fit giving $\lambda_{\rm pm} \propto [(T\text{-}T_C)/T_C]^{-w}$ with $w = 0.76 \pm 0.06$, in agreement with values found in earlier studies [3, 4].

### *3.2. Ferromagnetic range*

A muon signal is visible in all the three sets of data introduced in section 2. Spectra at select temperatures are shown in Fig. 2-b. They all have a low initial asymmetry of 0.03-0.05, in contrast to the paramagnetic spectra with $A_0 \sim 0.2$. Clearly, only a part of the full muon signal is observable below $T_C$. The remaining signal relaxes extremely fast and is in most cases hidden within the initial dead time (approximately between 5 and 10 ns) of the $\mu$SR spectrometer. In the majority of cases, a spontaneous muon spin precession signal could be seen, yet the oscillations were heavily damped (up to $\sim$50 $\mu s^{-1}$). The spectra were fitted with a single cosine oscillation having exponential damping. It cannot be excluded that some spectra contain more than one oscillatory frequency. Some of the data points can be succesfully fitted with a set of free parameters, but in others, expecially for the GPD data, the phase and the initial asymmetry (set to 0.04) had to be kept fixed. In the temperature region between 40-15 K the data cannot be fitted reliably in that way, neither with a single oscillation nor with two oscillation frequencies. The signal is just too heavily damped for the fit procedure to lock on to oscillations. Nearly all of the signal is within the dead time range and not observable. A typical example is the spectrum at 25 K in Fig. 2-right. The high damping rate also prevents a location of precession

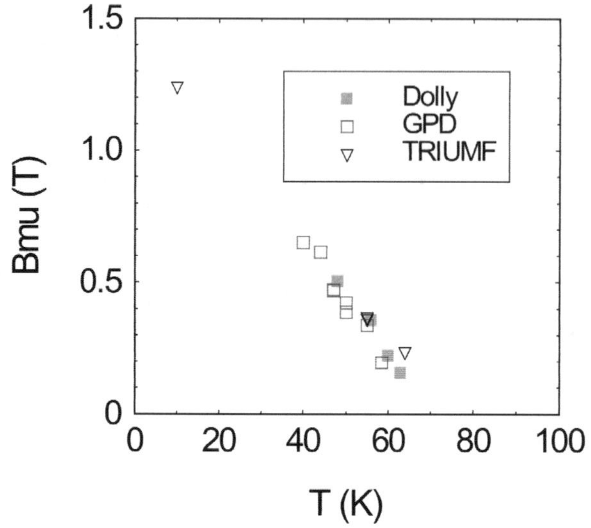


**Figure 3.** Observed interstitial fields in $DyAl_2$. The region between 40 K and 15 K is excluded since reliable fits could not de obtained (see text)

frequencies in a FFT analysis. The interstitial fields derived from the observed frequencies in the cosine oscillation fits are summarized in Fig 3.

## 4. Discussion

As said in the introductory chapter, the exponent $w$ of the temperature dependence of the paramagnetic relaxation rate cannot be related directly to theoretical treatments of phase transitions since it is not the critical parameter in their sense. This issue is discussed extensively in [8], and also in [5]. The value of $w$ varies markedlly throughout the literature and often turns out to be different for different samples of the same compound. Clearly more macroscopic

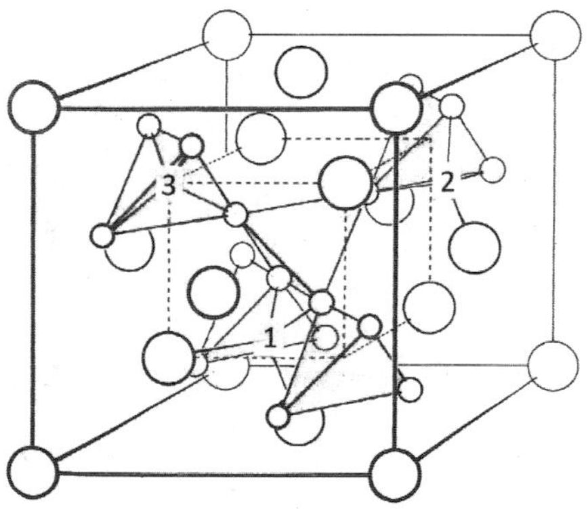


**Figure 4.** Structure of the C15 $REAl_2$ lattice (RE = large circles, Al = small circles). The 3 types of interstitial sites are marked 1 ,2 ,3 in the figure with 1 being the likely muon position (2-2 site).

parameters enter like crystal faults, grain sizes and domain volume enter. It is usualy also method sensitive, even within the various nuclear methods.

In all $REAl_2$ compounds we expect the muons to reside at the 2-2 site (g-site) of the C15 Laves phase structure that is shown in Fig. 4. It is the preferred site for low concentration of hydrogen (see discussion in [2]), and was also found for muons in $CeAl_2$ [9]. Furthermore the muons are likely to be stationary below $T = 80$ K as seen in non-magnetic $LaAl_2$ [10].

That only a part of the muon signal is observed in the ferromagnetic range is presumably due to the fact that the magnetic structure of the compound will split the local fields at the interstitial sites into several components. As shown in Fig. 4, the 2–2-site forms a tetrahedron with 2 RE and 2 Al atoms, but the distances from the muon to the RE and the Al atoms are not identical. There are 96 different 2–2 sites in the unit cell and the distribution of dipolar fields on these sites depends on the magnetization direction. Calculations of the magnitude of the internal fields are shown in Fig. 5 for muons on the 2–2 sites, and also for a slightly shifted 2–2-site where the distances to the 4 neighboring atoms are all equal. In the standard 2–2 site there are 4 or 6 different fields, but they are pairwise only of opposite signs, reducing the numbers to 2 and 3 respectively. If the muon position is offset even further splittings occur as displayed in Fig. 5-c and -d.

The saturation value of the interstitial field $B_\mu$ is on the order of 1 T a value considerably larger than that found [6] in $SmAl_2$ (~0.3 T), reflecting mainly the much larger free ion moment of $Dy^{3+}$. It is, in fact, in roughly the same range as $B_\mu$ observed in ferromagnetic dysprosium metal [11, 1]

The temperature dependence of $B_\mu$ shown in Fig. 3 is unusual since it shows essentially a linear behavior. Clearly, this relies on the single low temperature data point (10 K), yet this data point is characterized by a fairly clear oscillatory pattern (see Fig. 2-right) defining its frequency rather well. In contrast, the temperature dependence of $B_\mu$ in $SmAl_2$ can well be reproduced [6] by

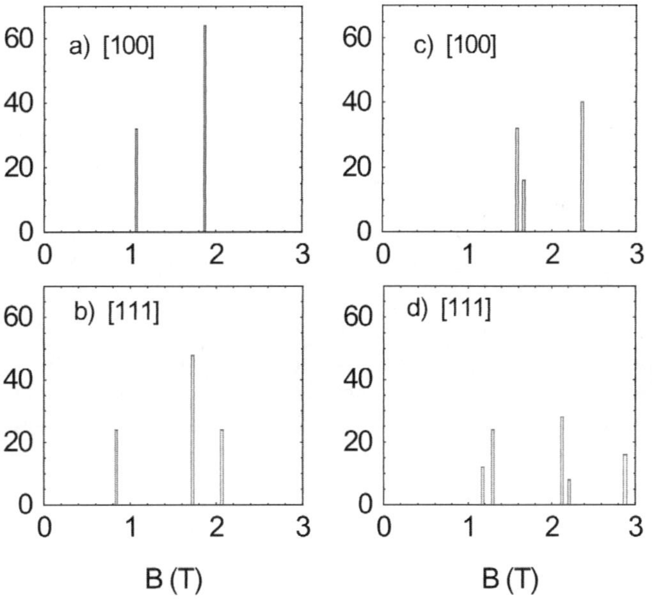


**Figure 5.** Relative magnitude of calculated dipolar fields at T=0 in $DyAl_2$ for [100] and [111] magnetization directions in regular 2-2 sites (a and b), and muon shifted 2-2-sites (c and d). See text for details.

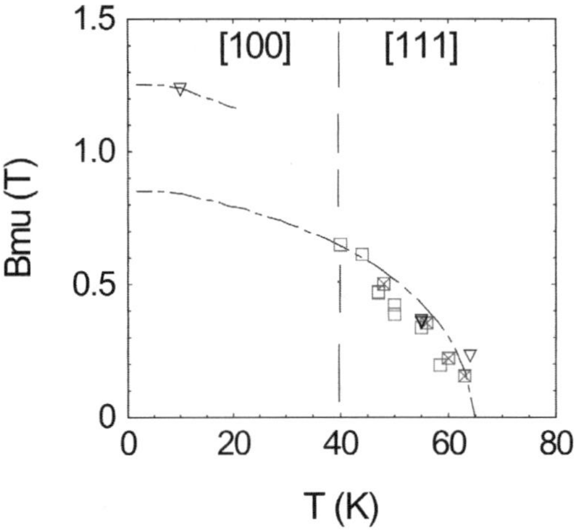


**Figure 6.** Possible explanation of the dependence of the local field on temperature (see text for details). The broken lines are theoretical magnetization functions for $J = 15/2$.

the common Bloch spin wave approach (Brillouin-like behavior). A nearly linear dependence of $B_{\rm hf}$ could possibly result from strong crystalline electric field (CEF) interaction. In a high moment Kramers ion as $Dy^{3+}$ CEF interactions are small. On the other hand, a reorientation transition of the easy axis of magnetization from [111] to [100] at 40 K has been established for $DyAl_2$ from magnetization and heat capacity data on a single crystal [7]. The calculations of the dipolar field component show a shift to higher values when changing the axis of magnetization from [111] to [100], as seen when comparing panels b) and a) in Fig. 5. At 40 K, corresponding to $T/T_C \approx 0.6$, one expects the value of $B_\mu$ to be close to its saturation value $B_\mu(T \to 0)$ and no large rise in field should occur when lowering the temperature further. The observation of the opposite behavior then supports the interpretation that the rise in field for T<40 K is the outcome of the change of magnetization direction. Assumed in this conclusion is, however, that we only observe the lowest precession frequencies and that the contribution from the contact field at the muon site is small compared to the dipole field. The latter assumption in particular is quite reasonable (see [1]).

One may finally interpret the dependence of $B_\mu$(T) shown in Fig. 3 as consisting of two 'normal' (i.e., Brillouin-like) temperature dependences of the interstitial field as displayed in Fig. 6. It has been said before, that unfortunately reliable fits of the spectra for temperatures between 40 and 15 K in terms of one or more cosine frequencies were not possible because of excessive damping rates. The most probable explanation is that the reorientation of magnetization directions is distributed in temperature and consequently, the spectra consist of an overlay of various precession signals. Such a spread of spin turning transition has been oserved previously, for example in the spin reorientation transition in Gd metal [12], albeit over a smaller temperature range.

**Acknowledgments**
The $\mu$SR studies were performed in part at TRIUMF (Vancouver BC, Canada) and in part at the Swiss Muon Source of the Paul Scherrer Institute (PSI). We thank B. Hitti, R. Scheuermann and H. Luetkens for their assistance during the measurements.

## References

[1] Kalvius G M, Noakes D R and Hartmann O 2001 *Handbook on the Physics and Chemistry of Rare Earth* vol 32 (Amsterdam: Elsevier Science) p 55
[2] Hartmann O *et al.* 1986 *J. Phys. (Metal Phys.)* F **16** 1573
[3] Hartmann O *et al.* 1984 *Hyperfine Interactions* **17-19** 491
[4] Kalvius G M *et al.* 1984 *Hyperfine Interactions* **17-19** 497
[5] Herlach D *et al.* 1986 *Hyperfine Interactions* **31** 287
[6] Pratt F L *et al.* 2006 *Physica* B **374 - 375** 34
[7] Lima A L *et al.* 2005 *Phys. Rev.* B **72** 024403
[8] Hohenemser C *et al.* 1982 *Hyperfine Interactions* **49** 267
[9] Hartmann O *et al.* 1989 *Hyperfine Interactions* **51** 995
[10] Hartmann O *et al.* 1990 *Hyperfine Interactions* **64** 369
[11] Ekström M *et al.* 1997 *Hyperfine Interactions* **104** 281
[12] Hartmann O *et al.* 1994 *Hyperfine Interactions* **85** 251

# Spin dynamics of Mn pyrochlore lattice in $YMn_2Zn_{20-x}In_x$

**M Miyazaki[1], R Kadono[1,2], M Hiraishi[1], I Yamauchi[1], A Koda[1,2], K M Kojima[1,2], I Kawasaki[3], I Watanabe[3], Y Okamoto[4†] and Z Hiroi[4]**

[1] Muon Science Laboratory and Condensed Matter Research Center, Institute for Materials Structure Science, High Energy Accelerator Research Organization (KEK), Tsukuba, Ibaraki 305-0801, Japan

[2] Department of Materials Structure Science, The Graduate University for Advanced Studies (SOKENDAI), Tsukuba, Ibaraki 305-0801, Japan

[3]Advanced Meson Science Laboratory, RIKEN Nishina Center, Wako, Saitama 351-0198, Japan

[4] Institute for Solid State Physics, University of Tokyo, Kashiwa, Chiba 277-8581, Japan

E-mail: `mmiya@post.kek.jp`

**Abstract.** It is shown in a new class of pyrochlore compound $YMn_2Zn_{20-x}In_x$ (with $x = 2.36$) that the spin fluctuation rate ($\nu$) is proportional to temperature ($\nu \propto T$) below $T_0 \sim$10 K where heavy-fermion(HF)-like behavior is observed. Such a linear $T$ dependence, commonly found in Y(Sc)$Mn_2$ and $LiV_2O_4$, is expected for intersecting quasi-1D Hubbard chains, suggesting that the quasi-1D character of the $t_{2g}$ band associated with the pyrochlore lattice plays an important role in the emergence of the HF-like state, where $T_0$ may correspond to intra-chain antiferromagnetic interaction energy between Mn spins.

## 1. Introduction

Geometrical frustration in the electronic degrees of freedom such as spin, charge, and orbit, which is often realized on the stages of highly symmetric crystals, has been one of the major topics in the field of condensed matter physics. In particular, the heavy fermion (HF) behavior in $d$-electron transition metal compounds such as $Y_{1-x}Sc_xMn_2$ [1, 2, 3] and $LiV_2O_4$ [4, 5, 6] has been under the spotlight of broad interest, where such a local electronic correlation specific to highly symmetric pyrochlore structure may be of direct relevance to the formation of heavy quasiparticle (QP) state. However, despite decades of studies, the microscopic mechanism how such a local correlation is transmuted into the heavy QP mass of itinerant $d$-electrons in those compounds still remains controversial.

One of the bottlenecks for the investigation of the $d$-electron HF-like state is the limited variety of actual candidate compounds. Recent development of a ternary intermetallic compound, $YMn_2Zn_{20-x}In_x$ [7], might be a relief to this situation. It exhibits a Sommerfeld coefficient $\gamma \sim 200$ mJ/K$^2$mol [8] which is much larger than that calculated by the band theory for uncorrelated electrons, $\gamma_{\rm band} = 34$ mJ/K$^2$mol in $YMn_2Zn_{20}$ [7].

[4] † Present address: Department of Applied Physics, Nagoya University, Chikusa-ku, Nagoya 464-8603, Japan.

The compound crystallizes in the cubic $CeCr_2Al_{20}$ structure with the space group $Fd\bar{3}m$ (space group no. 227) shown in figure 1(a), where the Y and Mn atoms respectively occupy the $8a$ and $16d$ sites forming diamond and pyrochlore lattices common to the cubic Laves-phase $YMn_2$. Meanwhile, In atoms occupy the $16c$ site forming pyrochlore lattice, and Zn

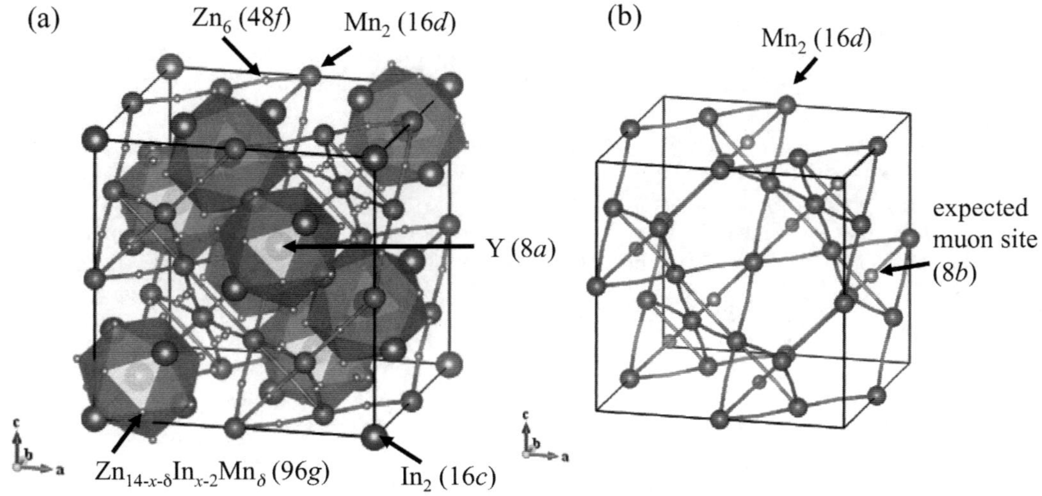


**Figure 1.** (a) Crystal structure of $YMn_{2+\delta}Zn_{20-x-\delta}In_x$, where Y, Mn, and Zn atoms are indicated by blue, red, and white spheres, respectively. A part of Zn at $96g$ site is substituted by In to stabilize the $CeCr_2Al_{20}$ structure. The actual compound exhibits a small amount ($\delta$) of excess Mn that occupies the $96g$ site to substitute for Zn. The Mn-Zn tetrahedra form a network similar to the pyrochlore lattice, where the local interaction between Mn atoms is relatively weaker than that in $Y_{1-x}Sc_xMn_2$ due to Zn atoms situated between Mn atoms. (b) Expected muon site at the center of Mn tetrahedra at $8b$ site is shown.

atoms at the $48f$ site are located between Mn atoms, and those atoms at the $96g$ site are located around Y atoms, so that the pyrochlore lattice composed of the Mn atoms is almost doubly expanded in comparison with that in $Y_{1-x}Sc_xMn_2$ without modifying the tetrahedral symmetry. Although the compound with the ideal $YMn_2Zn_{20}$ composition has not yet been obtained, partial substitution of Zn with In is known to be effective for stabilizing the structure.

## 2. Experiment

We have investigated the spin dynamics of Mn $d$-electrons by means of muon spin relaxation ($\mu$SR) under a zero/longitudinal field (ZF/LF) in a sample whose chemical composition is more precisely expressed as $YMn_{2+\delta}Zn_{20-x-\delta}In_x$ with $x = 2.36$, where the influence of excess Mn seems to be minimal [$\delta = 0.11(1)$]. The sample was a mosaic of single crystals glued with varnish on a sample holder made of high purity silver (12 mm$\phi$ disc). The experiment was carried out at RIKEN-RAL port 2 using the ARGUS spectrometer with "flypast" setup that allows low background measurements for small samples. $\mu$SR measurements were performed using two cryostats with different gas refrigerants, one with $^3$He covering a temperature range from 50 K to 300 mK and another with $^4$He from ambient temperature to 4.2 K, respectively.

$\mu$SR measurements for 300 K–4.2 K were made under LF = 10 mT to quench the depolarization due to random local fields from nuclear magnetic moments. Meanwhile, those for 50 K–300 mK were made under LF = 395 mT to discern depolarization due to excess Mn at very low temperature, where the respective yield of signals from pyrochore Mn and that from excess Mn was estimated from field dependence of LF-$\mu$SR spectra.

## 3. Results and analysis

Typical examples of ZF/LF-$\mu$SR spectra obtained at 4.2 K and 300 mK are shown in figure 2. The depolarization rate ($\lambda$) at 4.2 K is mostly independent of the magnitude of external field (mostly unchanged between 10 mT and 395 mT), indicating that the spectra are in the limit of motional narrowing. These spectra were analyzed by curve fits using a form,

$$A(t) = A_0 G_z^{\mathrm{KT}}(t) \exp(-\lambda t) + A_b \tag{1}$$

where $A(t)$ is the positron asymmetry, $A_0$ is the initial positron decay asymmetry, $G_z^{\mathrm{KT}}(t)$ is the Kubo-Toyabe relaxation function, and $A_b$ is the time-independent background mainly coming from muons stopped in the sample holder made by silver. The signal to background ratio ($A_0/A_b$) was ~3.8 for the $^4$He cryostat and ~3 for the $^3$He cryostat over a time range of 0-20 µs, allowing reliable deduction of $\lambda$ at small values ($\sim 10^{-2}$ µs$^{-1}$)[figure 3].

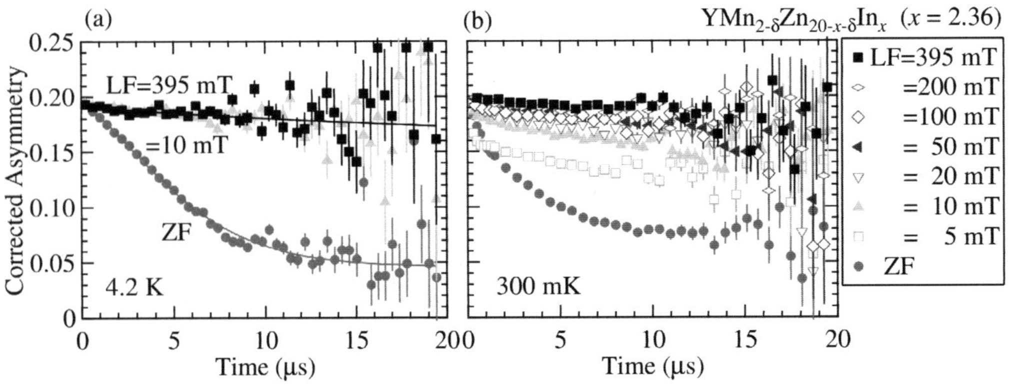


**Figure 2.** ZF/LF-$\mu$SR time spectra in $YMn_{2+\delta}Zn_{20-x-\delta}In_x$ with $x$=2.36 ($\delta$=0.11) measured at (a) 4.2 K and (b) 300 mK. Solid curves are best fits using eq.(1). The asymmetry of the spectrum under LF = 395 mT at 4.2 K is adjusted for comparison with the spectrum under LF = 10 mT.

The spin fluctuation rate $\nu$ at a given temperature $T$ is deduced using the following relationship between $\lambda$ and $\nu$ in metals,

$$\lambda \approx \frac{1}{T_1} = \frac{2\delta_\mu^2 \nu}{\nu^2 + \gamma_\mu^2 H_0^2} \cdot \frac{\chi k_B T}{N_A \mu_B^2} \tag{2}$$

where $T_1$ is spin-lattice relaxation time, $\chi(T)$ is the magnetic susceptibility, $N_A$ is the Avogadro number, $\delta_\mu$ is the muon hyperfine parameter, $\gamma_\mu = 2\pi \times 135.54$ MHz/T is the muon gyromagnetic ratio, and $H_0$ is the external field. Assuming that muons located at the center of Mn tetrahedra (8*b* site, shown in figure 1(b)), $\delta_\mu$ is calculated by the muon-Mn (electronic spin) dipolar tensor $A_i^{\alpha,\beta}$ as

$$\delta_\mu^2 = \gamma_\mu^2 \sum_{i,\alpha,\beta} \left[ (A_i^{\alpha,\beta}) \mu_i \right]^2 = \gamma_\mu^2 \sum_{i,\alpha,\beta} \left[ \frac{\mu_i}{r_i^3} (\delta_{\alpha,\beta} - \frac{3 r_i^\alpha r_i^\beta}{r_i^2}) \right]^2 \quad (\alpha = x, y,\ \beta = x, y, z), \tag{3}$$

where $r_i$ is the distance between muon and the $i$-th Mn atom, yielding $\delta_\mu = \gamma_\mu \times 19.59$ mT/$\mu_B = 16.68$ μs$^{-1}/\mu_B$. The nuclear dipolar width can be estimated using the above equation by substituting $\mu_i$ with those of the nuclear magnetic moments to yield $\Delta = \gamma_\mu \times 0.1257$ mT $=$ 0.107 μs$^{-1}$. The calculation of $\delta_\mu$ includes contribution from Mn atoms situated at the nearest and next nearest neighboring tetrahedra. These values are in good agreements with those deduced from the curve fits, i.e., $\Delta = 0.128(3)$ μs$^{-1} = \gamma_\mu \times 0.150(4)$ mT at 4.2 K as a parameter of $G_z^{KT}(t)$ in eq.(1) and 15.9(28) μs$^{-1} = \gamma_\mu \times 18.7(33)$ mT/$\mu_B$ determined by longitudinal field dependence of $\lambda$ at 0.303 K using eq.(2).

The fluctuation rate of the hyperfine field is deduced by modifying eq.(2),

$$\nu = \frac{2\delta_\mu^2}{\lambda} \cdot \frac{\chi k_B T}{N_A \mu_B^2} + \sqrt{\left(\frac{2\delta_\mu^2}{\lambda} \cdot \frac{\chi k_B T}{N_A \mu_B^2}\right) - \gamma_\mu^2 H_0^2}, \tag{4}$$

where $\lambda$ and $\chi$ are the quantities determined experimentally. Here, in order to avoid influence of excess Mn, we rely on $\lambda$ obtained from data under LF with magnitude greater than 100 mT and $\chi$ evaluated for the Mn atoms forming pyochlore lattice, i.e., $C = 0.2$ cm$^3$K/mol and $\theta_W =$ −11 K for $\delta \to 0$ [8]. Taking $\delta_\mu = 16.68$ μs$^{-1}$, we obtain $\nu$ vs $T$ as shown in figure 4.

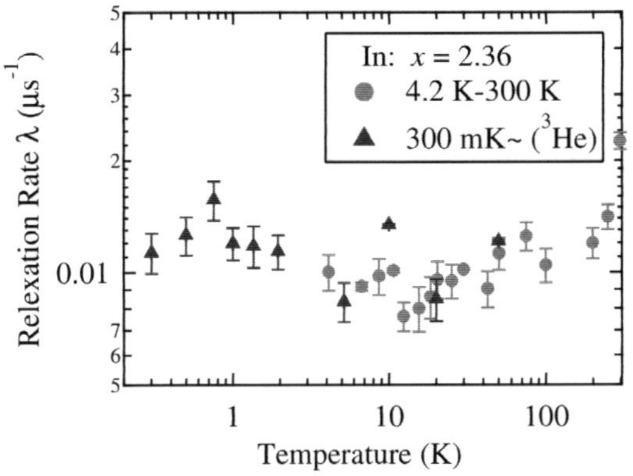


**Figure 3.** Temperature dependences of muon spin depolarization rate ($\lambda$) in $YMn_{2+\delta}Zn_{20-x-\delta}In_x$ with $x = 2.36$ ($\delta = 0.11$), where two sets of symbols correspond to the data obtained using different cryostats (see text).

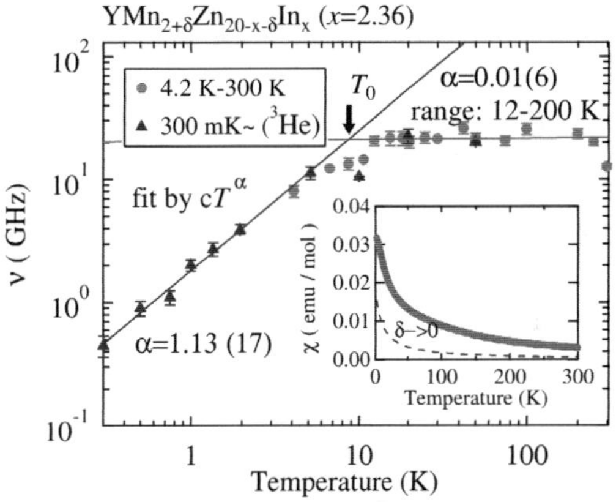


**Figure 4.** The fluctuation rate ($\nu$) of the hyperfine field vs $T$ in $YMn_{2+\delta}Zn_{20-x-\delta}In_x$ with $x = 2.36$ ($\delta = 0.11$). Solid lines are results of curve fits using a power-law ($\nu = cT^\alpha$), with $c$ and $\alpha$ being free parameters, for the respective range of temperature. $T_0$ is the suggested onset temperature for the spin correlation of quasi-1D Hubbard chains. Inset: Susceptibility of $YMn_{2+\delta}Zn_{20-x-\delta}In_x$ with $x = 2.36$ ($\delta = 0.11$, red circles) which includes contribution from excess Mn other than the intrinsic $\chi$ for $\delta = 0$ (blue dashed line)[8].

## 4. Discussion

It is noticeable that $\nu$ is linearly dependent on temperature ($\nu \propto T$) below $T_0 \sim 10$ K. Such a behavior is in marked resemblance with those observed in $Y_{1-x}Sc_xMn_2$ [9] and $LiV_2O_4$ [10, 11], which is understood as a characteristic property of spin-spin correlation predicted for intersecting Hubbard chains as a theoretical model for pyrochlore lattice[12]. This implies the crucial role of $t_{2g}$ orbitals as one-dimensional (1D) chains that are under a strong geometrical constraint of pyrochlore lattice structure, and further suggests the dimensional crossover due to coupling between these chains as a possible origin of the heavy-fermion state [13]. Then, the onset temperature $T_0$ might be interpreted as the energy scale for an effective intra-chain exchange interaction which would be thermally disrupted above $T_0$.

It is interesting to note that $T_0$ seems to correspond to the temperature below which the HF-like behavior sets in, as suggested from the temperature dependence of the Sommerfeld coefficient ($\gamma$) [8]. The relatively limited range of temperature for the HF-like behavior in $YMn_{2+\delta}Zn_{20-x-\delta}In_x$ may be understood as a consequence of weaker Mn-Mn interaction in comparison with that in $Y_{1-x}Sc_xMn_2$ due to nearly doubled Mn-Mn distance.

The behavior of $\nu$ above $T_0$ might evoke the "Korringa relation", i.e., $\nu \propto 1/T_1T \rightarrow$ *const.* in normal metals, where one would expect $\nu_F \approx 1/\hbar N(E_F) = 2\pi^2 k_B^2 N_A/3\hbar\gamma$. However, $\nu_F$ is estimated to be $\approx 10^{13-14}$ s$^{-1}$ from $\gamma =$ 80-200 mJ/K$^2$mol, which is three orders of magnitude smaller than that in normal metals. Thus, the observed spin fluctuation above $T_0$ does not seem to be directly related with the Pauli paramagnetism.

## 5. Summary

Spin fluctuation rate of Mn spins in a *d*-electron heavy fermion like compound $YMn_{2+\delta}Zn_{20-x-\delta}In_x$ has been deduced from ZF/LF-$\mu$SR measurements. The temperature dependence of $\nu$ below $T_0 \sim 10$ K shows $T$-linear dependence that is expected for the spin fluctuation in quasi-1D Hubbard chains as a model of the pyrochlore lattice. It comprises the third example of such a behavior next to $Y_{1-x}Sc_xMn_2$ and $LiV_2O_4$. The HF-like behavior seems to set in below $T_0$, in good correspondence with the emergence of quasi-1D-like spin fluctuation. The relatively small $\gamma$ compared with $Y_{1-x}Sc_xMn_2$ may be understood as due to a small effective intra-chain exchange energy suggested by low $T_0$.

## Acknowledgments

This work was partially supported by the KEK-MSL Inter-University Research Program for Oversea Muon Facilities (2011A0011) and by a Grant-in-Aid for Creative Scientific Research on Priority Areas from the Ministry of Education, Culture, Sports, Science and Technology, Japan. A part of experiment was carried out as a recovery program from the East Japan Earthquake in 2011 (R-1011), and we appreciate RIKEN-RAL and ISIS for sparing beam time to this program. We would like to thank the RIKEN-RAL and ISIS staff for their technical support during $\mu$SR experiment.

## References

[1] Wada H, Shiga M and Nakamura Y 1987 *J. Magn. Magn. Mat.* **70** 17
[2] Shiga M 1988 *Physica* B **149** 293-305
[3] Wada H, Shiga M and Nakamura Y 1989 *Physica B* **161** 197-202
[4] Kondo S, Johnston D C, Swenson C A, Borsa F, Mahajan A V, Miller L L, Gu T, Goldman A I, Maple M B, Gajewski D A, Freeman E J, Dilley N R, Dickey R P, Merrin J, Kojima K, Luke G M, Uemura Y J, Chmaissem O and Jorgensen J D 1997 *Phys. Rev. Lett.* **78** 3729
[5] Takagi H, Urano C, Kondo S, Nohara M, Ueda Y, Shiraki T and Okubo T 1999 *Mater. Sci. Eng.* **63** 147-150
[6] Urano C, Nohara M, Kondo S, Sakai F, Takagi H, Shiraki T and Okubo T 2000 *Phys. Rev. Lett.* **85** 1052
[7] Okamoto Y, Shimizu T, Yamaura J, Kiuchi Y and Hiroi H 2010 *J. Phys. Soc. Jpn.* **79** 093712
[8] Okamoto Y, Shimizu T, Yamaura J and Hiroi H 2012 *J. Solid State Chem.* **191** 246-56

[9] Miyazaki M, Kadono R, Hiraishi M, Masuda T, Koda A, Kojima K M, Yamazaki T, Tabata Y and Nakamura H 2011 *J. Phys. Soc. Jpn.* **80** 063707
[10] Kadono R, Koda A, Higemoto W, Ohishi K, Ueda H, Urano C, Kondo S, Nohara M and Takagi H 2012 *J. Phys. Soc. Jpn.* **81** 014709
[11] Lee S H, Qiu Y, Broholm C, Ueda Y and Rush J J 2001 *Phys. Rev. Lett.* **86** 5554
[12] Lee J D 2003 *Phys. Rev.* B **67** 153108
[13] Fujimoto S 2002 *Phys. Rev.* B **65** 155108

# Magnetism and magnetic order in the pyrochlore iridates in the insulator-to-metal crossover region

**M J Graf[1], S M Disseler[2], C Dhital[1], T Hogan[1], M Bojko[1], A Amato[3], H Luetkens[3], C Baines[3], D Margineda[4], S R Giblin[4], M Jura[5] and S D Wilson[1,6]**

[1] Department of Physics, Boston College, Chestnut Hill, MA 02467, USA
[2] Center for Neutron Research, National Institute for Standards and Technology, Gaithersburg, MD 20899 USA
[3] Laboratory for Muon Spin Spectroscopy, Paul Scherrer Institut, CH-5232 Villigen PSI, Switzerland
[4] School of Physics and Astronomy, Cardiff University, Cardiff CF24 3AA, UK
[5] Rutherford Appleton Laboratory, Didcot, Oxfordshire OX11 0QX, UK

E-mail: grafm@bc.edu

**Abstract.** We present resistivity, magnetization, and zero field muon spin relaxation ($\mu$SR) data for the pyrochlore iridate materials $Nd_{2-x}Ca_xIr_2O_7$ ($x = 0, 0.06$, and $0.10$) and $Sm_2Ir_2O_7$. While $Nd_2Ir_2O_7$ (Nd227) is weakly conducting, $Sm_2Ir_2O_7$ (Sm227) has a slowly diverging resistivity at low temperature. Nd227 and Sm227 exhibit magnetic anomalies at $T_M = 105K$ and $137K$, respectively. However, zero-field $\mu$SR measurements show that long-range magnetic order of the $Ir^{4+}$ sublattice sets in at much lower temperatures ($T_{LRO} \sim 8K$ for Nd227 and $70K$ for Sm227); both materials show heavily damped muon precession with a characteristic frequency near 9 MHz. The magnetic anomaly at $T_M$ in Nd227 is not significantly affected by the introduction of hole carriers by Ca-substitution in the conducting $Nd_{2-x}Ca_xIr_2O_7$ samples, but the muon precession is fully suppressed for both.

## 1. Introduction

Strongly correlated transition metal oxides exhibit a wide-range of important and interesting phenomena, including high temperature superconductivity and colossal magnetoresistance. While iridium oxides have a much reduced Ir-Ir correlation energy due to their extended 5d orbitals, the strong spin orbit interaction (SOI) energy makes possible a SOI-assisted Mott-like insulating state with $J_{eff} = 1/2$, as observed in $Sr_2IrO_4$ [1]. The pyrochlore iridates $A_2Ir_2O_7$ ($A$ = lanthanide or Y), or $A227$, consist of interpenetrating corner-sharing tetragonal structures for the $A^{3+}$ and $Ir^{4+}$ ions, suggesting that frustration may also play a role in determining the magnetic ground state. For all members of the family (apart from $A$ = Pr), a transition in the DC magnetic susceptibility associated with the Ir sublattice is observed at a temperature $T_M$, which ranges from near 200 K ($A$ = Y) to near 30 K ($A$ = Nd). The deviation of field-cooled (FC) from zero-field-cooled (ZFC) data below $T_M$ bears a resemblance to a spin-glass transition, or possibly very weak ferromagnetism. For small $A$-ion radii $R_A$ an insulating low temperature ground state is observed, and for several of species of $A$ a concomitant metal-insulator transition has been

[6] Present address: Materials Dept., Univ. of California Santa Barbara, Santa Barbara CA 93106 USA

reported at temperature $T_{MI} \sim T_M$ [2]. Zero field muon spin rotation ($\mu$SR) measurements confirm long-range magnetic ordering of the Ir sublattice at temperatures $T_{LRO} = 130K, 150K$ and $120K$ for the insulators Yb227, Y227 [3], and Eu227 [4], respectively. The muon precession frequencies are near 14 MHz, corresponding to an $Ir^{4+}$ moment $\mu_{Ir} \sim 0.3\mu_B$ [5], and are lightly damped. The moments order in an all-in/all-out (AIAO) magnetic structure, where the $Ir^{4+}$ moments are pointing either all in towards or away from the center of a given tetrahedron, as indicated by $\mu$SR measurements [3,5] and resonant inelastic x-ray scattering measurements [6].

In contrast, Pr227, with a large $R_A$, has a conducting ground state with no magnetic transition down to $T = 25mK$ [7]. This system is proposed to be a chiral spin liquid at low temperatures [8]. Factors that may play a role in the crossover from magnetically ordered insulator to a conductor lacking long-range order include the $A$-ion radius and magnetic moment, the carrier density, geometric frustration effects, and structural and stoichiometric disorder. This crossover region, i.e., near $A =$ Nd and Sm, is of particular interest. Theoretical work predicts topologically non-trivial ground states in this region that may require the presence of magnetic order to break time-reversal symmetry [9]. The Nd227 ground state is very sensitive to the details in preparation, as expected in a crossover region: samples of comparable purity and quality can be either insulating or weakly conducting. The weakly conducting samples have a relatively high $T_M = 105K$ [10] compared to the insulating samples with $T_{MI} \sim T_M = 33K$ [2]. However, regardless of whether the low temperature state is insulating or conducting, long range magnetic order in the $Ir^{4+}$ sublattice is observed in Nd227 with a muon precession frequency of roughly 9 MHz and with very strong damping [10,11]. For the insulating samples $T_{LRO} \sim T_M = 33K$, while for the weakly conducting samples $T_{LRO} \sim 8K \ll T_M$, suggesting an extended region of a magnetic phase with no long-range order that appears to be correlated with the presence of additional charge carriers. The magnetic state of the $Nd^{3+}$ ions remains an open question: neutron scattering measurements indicate the Nd-sublattice also orders in an AIAO structure [12], but $\mu$SR [10], Hall, and magnetization measurements [13] suggest a spin-ice like state.

In this work we present preliminary results on the effects of carrier density on the the magnetic transitions at temperatures $T_M$ and $T_{LRO}$ in Nd227 by introducing hole carriers through Ca-substitution. Additionally, we present the first $\mu$SR measurements on Sm227.

## 2. Samples

Polycrystalline samples of Nd227 (Sm227) were synthesized by reacting stoichiometric amounts of 99.99% purity $Nd_2O_3$ ($Sm_2O_3$) and $IrO_2$ (99.9%) in air to minimize variations in the oxygen stoichiometry. Powders were pelletized using an isostatic cold press and reacted between 900 C to 1125 C over six days with several intermediate grindings. The samples were determined to be phase-pure with the exception of two minor impurity phases of $IrO_2$ and $Nd_2O_3$ ($Sm_2O_3$) each comprising less than 1% of the total volume fraction. Powder x-ray diffraction measurements yielded narrow lineshapes comparable to the best previously reported data on Nd227 and Sm227, demonstrating the high homogeneity of the sample stoichiometry. $Nd_{2-x}Ca_xIr_2O_7$ polycrystalline samples with $x = 0.06$ and $0.10$ were prepared by reacting stoichiometric amounts of $Nd_2O_3$, $CaCo_3$ and $IrO_2$ in air as described above. Several EDS measurements taken on the nominal $x = 0.10$ sample over areas of 100 $\mu m^2$ show that the average Ca concentration was $x = 0.08$. X-ray refinement yields a slight decrease in the lattice parameter from $a_{Nd} = 10.375$ to $10.351\mathring{A}$ for the nominal $x = 0.10$ sample.

## 3. Resistivity

The temperature dependent resistivity was measured in the range $0.3 < T < 300K$ via a four-terminal AC resistance bridge. The samples were pressed sintered pellets, with an error in the resistivity of approximately 20% based on uncertainties in the sample dimensions. Excitation level dependence studies at the lowest temperatures showed there was no self heating of the samples; we note that Sm227 has a very long thermal equibration time. In Figure 1, we show the temperature dependent resistivity for Sm227 and Nd227, along with previously published data for Yb227 [3]. Yb227 is insulating and exhibits a temperature dependence consistent with variable range hopping at low temperatures. Nd227 exhibits a Kondo-like behavior, with a logarithmic upturn below 10 K and saturation at very low temperatures. Sm227 exhibits a behavior intermediate between Yb227 and Nd227, and is qualitatively similar to earlier results [2], with a crossover from high temperature conducting behavior to an insulating behavior at low temperatures. The crossover from conductor to insulator occurs at $T \sim 70K < T_M$ (see next section), with a resistivity that is slowly divergent down to 0.3 K. This does not follow the expected variable range hopping law, but rather an inverse power law $T^{-\beta}$ with $\beta = 0.52(2)$.

Also shown in Figure 1 are data for $Nd_{1.94}Ca_{0.06}Ir_2O_7$ and $Nd_{1.9}Ca_{0.1}Ir_2O_7$. Assuming one hole carrier per Ca ion, the expected hole carrier density in the $x = 0.06$ sample should be approximately twice the estimated upper limit for the undoped Nd227 carrier density from Hall effect measurements [13]; Hall measurements are now underway to confirm the actual carrier densities. We observe a reduction in both the overall value of the resistivity and in the low temperature upturn relative to the room temperature resistivity, with the unexpected result that the room temperature resistivity of $x = 0.06$ being lower than that of the $x = 0.10$ sample. The unusual behavior is illustrated in the inset to Figure 1, where the fractional change of resistance (relative to room temperature) versus temperature is shown. At present it is not clear if this is due to intrinsic effects, such as enhanced scattering by Ca impurities, or extrinsic effects (e.g. subtle differences in pressure or temperature while pressing the powders).

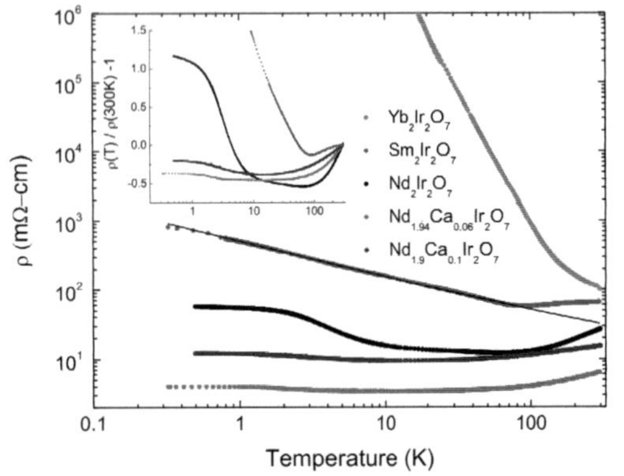


**Figure 1.** The temperature dependent resistivity for several pyrochlore iridates; data for Yb227 is taken from Ref. 3. The solid line is a fit to the Sm227 data at low temperatures (see text).

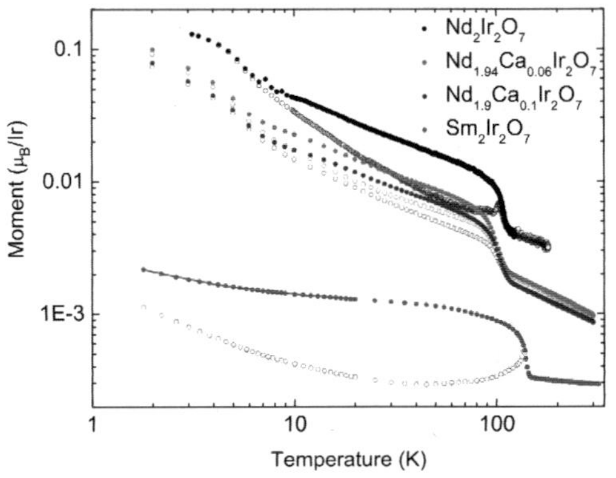


**Figure 2.** The temperature dependent magnetization in fields of 1000 Oe for several pyrochlore iridates (open circles: zero field cooled; filled circles: field cooled); solid line is a fit to the Sm227 data (see text).

## 4. Magnetization

Data were taken in Quantum Design MPMS SQUID magnetometers located in Cardiff, Didcot, and Chestnut Hill. In Figure 2 we show the temperature-dependent DC magnetization for the Nd-based samples and Sm227 (taken in fields of 1000 Oe). It is immediately evident that $T_M = 105K$ is not significantly modifed by the Ca-substitution. The hysteresis between the FC and ZFC curves for the Nd-based samples decreases at low temperatures; this may reflect an interplay between the $Nd^{3+}$ and $Ir^{4+}$ sublattices, and is the subject of an ongoing study. For

Sm227, $T_M = 137K$, higher than the value 118 K observed in Ref. 2. The $Sm^{3+}$ contribution to the observed magnetization is significantly smaller than that of $Nd^{3+}$. Assuming that the ordered $Ir^{4+}$ moments give a nearly temperature independent contribution to the DC susceptibility at low temperature in addition to the paramagnetic $Sm^{3+}$ contribution, we fit the zero-field cooled data below 20 K to a Curie-Weiss form including an additive constant. This yields an (upper bound) effective moment of $\mu_{Sm} = 0.23(1)\mu_B$ and a Curie-Weiss temperature of $T_{CW} = -0.6(1)K$, indicative of very weak $Sm^{3+}$ antiferromagnetic correlations. This value of $\mu_{Sm}$ is well below the free ion value $0.8\mu_B$, but not unusual [14, 15].

## 5. Zero-field $\mu$SR

Data were taken at PSI using the GPS and LTF spectrometers and at ISIS on the EMU spectrometer. In Figure 3 we show zero-field muon depolarization curves at $T = 25mK$ for the Nd-based and Sm227 samples. Also shown for comparison is the depolarization of insulating Yb227 [3]; the large moment ($\mu_{Yb} = 3.3\mu_B$) moderately damps the muon precession at low temperatures. Three points are immediately obvious: (1) the depolarization in the Nd227 and Sm227 are nearly identical; (2) these are both at significantly lower frequency and are more strongly damped than oscillations observed in insulating $A$227 materials (e.g., Yb227), and (3) there are no signs of muon precession in the Ca-substituted samples.

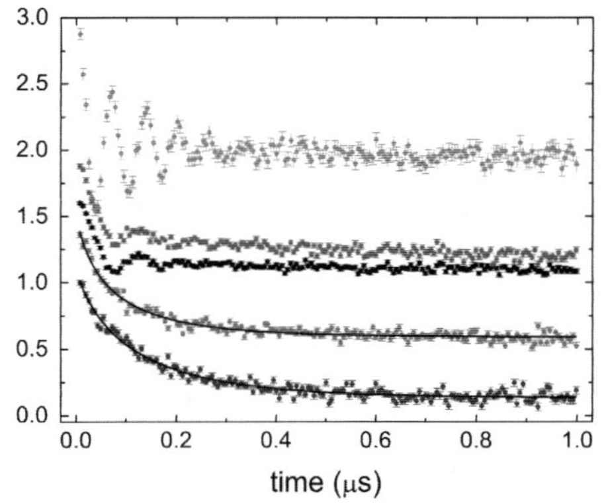


**Figure 3.** Zero field depolarization with time at $T = 25mK$. From top to bottom: Yb227, Sm227, Nd227, $Nd_{1.94}Ca_{0.06}Ir_2O_7$, and $Nd_{1.9}Ca_{0.1}Ir_2O_7$. Curves are offset by constant amounts except for the Yb227 curve. Solid lines are fits as described in the text.

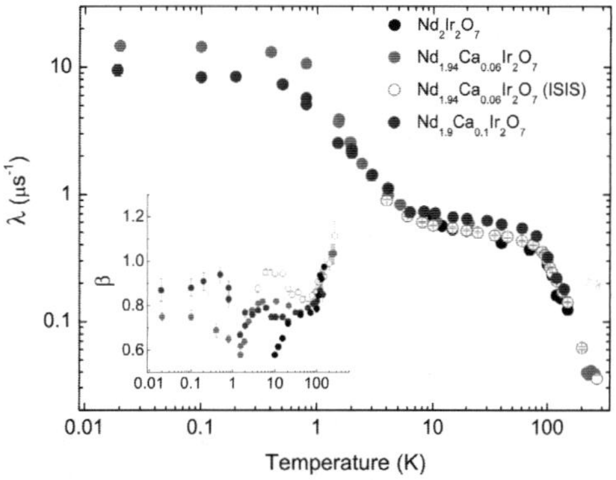


**Figure 4.** Temperature dependence of the stretched exponential depolarization rate for Nd227 [10], $Nd_{1.94}Ca_{0.06}Ir_2O_7$, and $Nd_{1.9}Ca_{0.1}Ir_2O_7$. Inset: temperature dependence of the exponent.

The strikingly different depolarization of Sm227 and Nd227 compared to that of the insulating $A$227 materials is not yet understood, and may be due to reduced $Ir^{4+}$ moment [15], a different magnetic ordering structure, a different muon stopping site, or the presence of magnetic $A$-ions. The last possibility is unlikely, given that the $Yb^{3+}$ has a larger moment than both the $Nd^{3+}$ ($\mu_{Nd} = 2.8\mu_B$) or the $Sm^{3+}$. The muon stopping site is also unlikely to be different: preliminary computer calculations [5, 10] have been made for the electrostatic potential in Sm227, and yield nearly identical results as for Y227. A reduction in the Ir moment is an appealing scenario, but is hard to reconcile with the very small change in lattice parameters concomitant with the dramatic change in magnetic behavior between Eu227 and Sm227. Recent reanalysis of the Nd227 $\mu$SR results shows that a Bessel depolarization function yields a superior fit than the phenomenological function used in Ref. 10, consistent with an incommensurate, rather than AIAO, ordering of the $Ir^{4+}$ sublattice for both Nd227 and Sm227. This leads us to believe that the delicately balanced magnetic interactions shift between Eu227 and Sm227, producing a different (and possibly incommensurate) ordering structure. A detailed analysis will be presented in a future publication. The change in magnetic structure is not solely determined by the presence of charge carriers, since both weakly conducting [10] and insulating [11] samples of Nd227 have similar muon precession frequencies and very strong damping.

From the Ca-substituted data we infer that adding hole carriers strongly suppresses the onset of magnetic ordering simultaneous to destabilizing the spin-orbit Mott phase. This is consistent with the earlier observation that while insulating Nd227 shows spontaneous muon precession near 33 K, for weakly conducting Nd227 the onset temperature is roughly 8 K. In contrast, the addition of hole carriers has little effect on the magnetic transition at $T_M$, as shown in Figure 2. The depolarization of the Ca-substituted samples has been fit to a stretched exponential function $A(t) = A_0 exp(-(\lambda t)^{\beta})$ (solid lines in Figure 3), and the temperature dependences of $\lambda$ and $\beta$ are shown in Figure 4. Also shown is the stretched exponential fit for Nd227 for $T \geq 10K > T_{LRO}$, taken from Ref. 10. In the Ca-substituted samples the $Ir^{4+}$ moments freeze near 100 K, but do not order to low temperatures, so we associate the increase in $\lambda$ below 10 K with $Nd^{3+}$ fluctuation rate slowing down into the muon time window. Preliminary measurements on the $x = 0.10$ sample at $2K$ show that fast relaxation persists in longitudinal fields up to 0.5 T, strongly suggesting that the low temperature depolarization is dynamical.

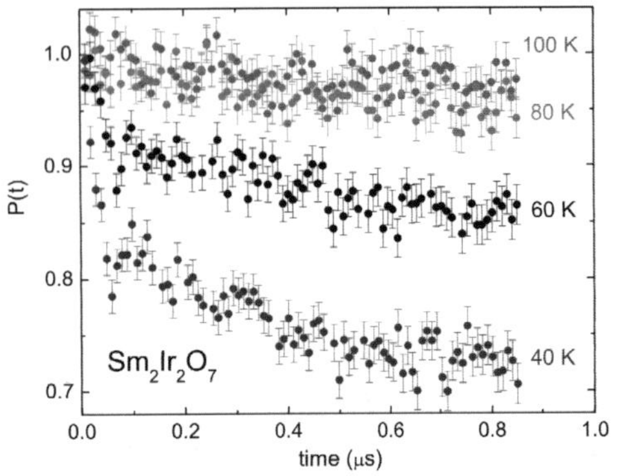


**Figure 5.** Evolution of the zero field depolarization of Sm227 with temperature in the vicinity of $T_{LRO}$.

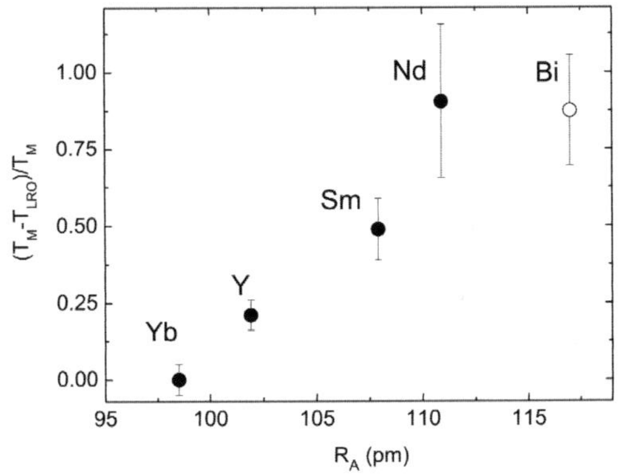


**Figure 6.** The variation of the fractional suppression of $T_{LRO}$ with respect to $T_M$ with $A$-ion radius. Bi227 data taken from Ref. 16.

Tracking the temperature dependence of the Sm227 ZF-$\mu$SR depolarization shows that an oscillatory-like time dependence sets in below $T = 80K$, as shown in Figure 5. Based solely on this data we cannot determine if this is connected to true long-range magnetic ordering in the Ir sublattice, or simply due to Ir-moment spin freezing. However, taken in context of the Nd227 measurements for which the oscillatory behavior is more clearly distinguished from Kubo-Gauss depolarization [10], we assume that it is indeed due to the onset of long-range magnetic order, i.e. $T_{LRO} \sim 70K$ for Sm227. This is well below $T_M = 137K$, and is consistent with our trend reported earlier: as the $A$-ion radius is increased a short range ordered magnetic phase is stabilized such that $T_{LRO} < T_M$. In Figure 6 we have plotted the fractional suppression $1 - T_{LRO}/T_M$ versus $A$-ion radius, as extracted from our data for Yb227, Y227, Sm227, and Nd227; also included is the fractional difference in the two magnetic transition temperatures reported for Bi227 [16], although the exact nature of those two transitions is not clear. The correlation is evident, though the underlying mechanism remains unclear.

## 6. Summary and Conclusions

Addition of carriers to Nd227 via Ca substitution has little effect on the magnetization anomaly observed at $T_M$, but completely suppresses the onset of long-range magnetic order; it would be of interest to study the effects of Ca-substitution on $T_M$ for insulating samples of Nd227 and other insulating pyrochlore iridates [17]. Additionally, Sm227 has magnetic properties that are very similar to Nd227: $T_{LRO} < T_M$ and heavily damped muon precession near 9 MHz. The underlying mechanisms causing the stabilization of a magnetic phase lacking long-range order and the abrupt change in magnetic ordering between Eu227 and Sm227 remain unknown. Detailed studies of the changes in stoichiometry [18,19] and possible distortions of the local $A$-ion environment with $A$-ion radius are necessary to fully understand these behaviors.

## Acknowledgments

$\mu$SR measurements were performed at the Swiss Muon Source and the ISIS Muon Facility. This work was supported in part by the National Science Foundation grants DMR-1056625 (SDW) and DMR-1337576 (SDW and MJG), the National Research Council (SMD), and EPSCR (SRG).

## References

[1] Kim B J, Jin H, Moon S J, Kim J-Y, Park B-G, Leem C S, Yu J, Noh T W, Kim C, Oh S-J, Park J-H, Durairaj V, Cao G and Rotenberg E 2008 *Phys. Rev. Lett.* **101** 076402
[2] Matsuhira K, Wakeshima M, Nakanishi R, Yamada T, Nakamura A, Kawano W, Takagi S and Hinatsu Y 2007 *J. Phys. Soc. Jpn.* **76** 04370 ; Matsuhira K, Wakeshima M, Hinatsu Y and Takagi S 2011 *J. Phys. Soc. Jpn.* **80** 094701
[3] Disseler S M, Dhital C, Amato A, Giblin S R, de la Cruz C, Wilson S D and Graf M J 2012 *Phys. Rev.* B **86** 014428
[4] Zhao S, Mackie J M, MacLaughlin D E, Bernal O O, Ishikawa J J, Ohta Y and Nakatsuji S 2011 *Phys. Rev.* B **83** 180402R
[5] Disseler S M 2014 *Phys. Rev.* B **89** 140413R
[6] Sagayama H, Uematsu D, Arima T, Sugimoto K, Ishikawa J J, E. OFarrell and S. Nakatsuji 2013 *Phys. Rev.* B **87** 100403R
[7] MacLaughlin D E, Ohta Y, Machida Y, Nakatsuji S, Luke G M, Ishida K, Heffner R H, Shu L and Bernal O O 2008 *Physica* B **404** 667
[8] Machida Y, Nakatsuji S, Onoda S, Tayama T and Sakakibara T 2010 *Nature* **463** 210
[9] Witczak-Krempa W, Chen G, Kim Y B and Balents L 2014 *Annu. Rev. Condens. Matter Phys.* **5** 57
[10] Disseler S M, Dhital C, Hogan T C , Amato A, Giblin S R, de la Cruz C, Daoud-Aladine A, Wilson S D and Graf M J 2012 *Phys. Rev.* B **85** 174441
[11] Guo H, Matsuhira K, Kawasaki I, Wakeshima M, Hinatsu Y, Watanabe I and Xu Z 2013 *Phys. Rev.* B **88** 060411
[12] Tomiyasu K, Matsuhira K, Iwasa K, Watahiki M, Takagi S, Wakeshima M, Hinatsu Y, Yokoyama M, Ohoyama K and Yamada K 2012 *J. Phys. Soc. Jpn.* **81** 034709
[13] Disseler S M, Giblin S R, Dhital C, Lukas K C, Wilson S D and Graf M J 2013 *Phys. Rev.* B **87** 060403R
[14] Singh S , Saha S, Dhar S K , Suryanarayanan R, Sood A K and Revcolevschi A 2008 *Phys. Rev.* B **77** 054408
[15] Malkin B Z, Lummen T T A , van Loosdrecht P H M, Dhalenne G and Zakirov A R 2010 *J. Phys. Condens. Matter* **22** 276003
[16] Baker P J, Möller J S, Pratt F L, Hayes W, Blundell S J, Lancaster T, Qi T F and Cao G 2013 *Phys. Rev.* B **87** 180409R
[17] Fukazawa H and Maeno Y 2002 *J. Phys. Soc. Jpn.* **71** 2578
[18] Ishikawa J J, T. OFarrell E C and Nakatsuji S 2012 *Phys. Rev.* B **85** 245109
[19] See for example Section II.C in Gardner J S, Gingras M J P and Greedan J E 2010 *Rev. Mod. Phys.* **82** 53

# Low temperature crystal structure and local magnetometry for the geometrically frustrated pyrochlore $Tb_2Ti_2O_7$

**P Dalmas de Réotier[1,2], A Yaouanc[1,2], A Bertin[1,2], C Marin[1,2], S Vanishri[1,2], D Sheptyakov[3], A Cervellino[4], B Roessli[3] and C Baines[5]**

[1] Univ. Grenoble Alpes, INAC-SPSMS, F-38000 Grenoble, France
[2] CEA, INAC-SPSMS, F-38000 Grenoble, France
[3] Laboratory for Neutron Scattering and Imaging, Paul Scherrer Institute, CH-5232 Villigen-PSI, Switzerland
[4] Swiss Light Source, Paul Scherrer Institute, CH-5232 Villigen-PSI, Switzerland
[5] Laboratory for Muon-Spin Spectroscopy, Paul Scherrer Institute, CH-5232 Villigen-PSI, Switzerland

E-mail: pierre.dalmas-de-reotier@cea.fr

**Abstract.** We report synchrotron radiation diffraction and muon spin rotation ($\mu$SR) measurements on the frustrated pyrochlore magnet $Tb_2Ti_2O_7$. The powder diffraction study of a crushed crystal fragment does not reveal any structural change down to 4 K. The $\mu$SR measurements performed at 20 mK on a mosaic of single crystals with an external magnetic field applied along a three-fold axis are consistent with published a.c. magnetic-susceptibility measurements at 16 mK. While an inflection point could be present around an internal field intensity slightly above 0.3 T, the data barely support the presence of a magnetization plateau.

The study of geometrically frustrated magnetic systems reveals a large variety of new magnetic phases. Among the frustrated materials, the rare-earth pyrochlore oxides which crystallize in a cubic structure ($Fd\bar{3}m$ space group) form a family for which different exotic ground states have been found [1]. Focusing on the insulators, we mention (i) the spin-ice ground state of $Ho_2Ti_2O_7$ and $Dy_2Ti_2O_7$ [2, 3], (ii) the spin-liquid ground state of $Yb_2Ti_2O_7$ [4] characterized by a pronounced peak in the specific heat for powder samples [5, 6], although an exotic ordered magnetic state has been reported for a single crystal [7, 8], (iii) the unconventional dynamical ground state of $Tb_2Sn_2O_7$ for which magnetic Bragg reflections are observed by neutron diffraction [9] while no spontaneous magnetic field is found by the zero-field positive muon spin relaxation ($\mu$SR) technique [10, 11], (iv) the persistent spin dynamics detected in the ordered states of $Gd_2Sn_2O_7$, $Gd_2Ti_2O_7$ and $Er_2Ti_2O_7$ [12, 13, 14, 15, 16], and (v) the splayed ferromagnet $Yb_2Sn_2O_7$, i.e. essentially a ferromagnetic compound [17, 18, 19], with an emergent gauge field [20].

The most mysterious compound of the rare-earth pyrochlore oxide family might be $Tb_2Ti_2O_7$ for which no long-range magnetic order is detected down to 20 mK, far below the absolute value of its Curie-Weiss temperature $\Theta_{CW} = -19$ K [21, 22]. Two theoretical ground states have been suggested: a quantum spin ice ground state proposed in Ref. [23] and a Jahn-Teller like distorted ground state based on specific heat measurements [24]. In analogy with the spin-ice systems, a magnetization plateau is expected at low temperature for an external magnetic field $\mathbf{B}_{\rm ext}$ applied along a [111] crystal direction if the former ground state is reached [25]. While static magnetization measurements have not found any signature of the predicted plateau down to 43 mK [26, 27, 28], the presence of a weak magnetization plateau below about 0.05 K has been proposed from a.c. susceptibility experiments for $0.06 < B_{\rm ext} < 0.6$ T [29]. Pointing out to the complexity of the physics involved, an anomaly in the static magnetic response has been observed around 0.15 K [30, 22, 26, 27] which strangely enough corresponds to a minimum rather than a maximum in the specific heat [22]. Pinch points in the neutron scattering intensity have been observed [31] with dispersive excitations emerging from them [32, 33], suggesting a strong magnetoelastic coupling in the Coulomb phase of $Tb_2Ti_2O_7$. In addition, a neutron scattering intensity measured at the $\left(\frac{1}{2}, \frac{1}{2}, \frac{1}{2}\right)$ position in reciprocal space has attracted some attention [31, 34, 35, 36]. However, its characterization and origin are a subject of discussions.

Given the dependence of some physical properties of $Tb_2Ti_2O_7$ on the sample preparation [24, 37, 22, 6, 26, 36], here we shall first present a high-resolution synchrotron radiation diffraction study on a crushed fragment of a single crystal. Thereafter we shall report on frequency shifts as measured on a crystal by the transverse-field (TF)-$\mu$SR technique with an external field $\mathbf{B}_{\rm ext}$ applied along a three-fold axis. They give access to the thermal and field dependences of a physical quantity proportional to the static magnetization. These measurements, for which the influence of a small amount of an impurity phase, if any, should be minimum,[1] barely support the existence of magnetization plateau at 20 mK.

For the two experiments we used parts of the crystal denoted C in Refs. [24, 37, 26]. This crystal is characterized by a very small residual entropy [22], as expected for a sample in thermodynamical equilibrium [38]. We briefly recall the procedure for the crystal synthesis. A polycrystalline $Tb_2Ti_2O_7$ powder has been first prepared from commercial $Tb_2O_3$ and $TiO_2$ powders of respective purity 4N and 4N5. An initial heat treatment of these powders to 1200°C has been followed by a second treatment up to 1350°C with an intermediate grinding and compaction so as to obtain a dense rod. The crystal has been subsequently prepared from the $Tb_2Ti_2O_7$ rod using the traveling floating zone technique under oxygen gas and with a translational velocity of 7 mm/h. For the $\mu$SR measurements, plates whose normal axis is a [111] axis have been cut from the single crystal: their thickness is about 1/3 mm and their lateral size is up to 6 mm.

The synchrotron radiation measurements have been performed at the high resolution powder diffractometer of the Material Science (MS) beamline at the Swiss Light Source (Paul Scherrer Institut, Switzerland). An x-ray beam of wavelength 0.49646 Å, corresponding to an energy $\simeq 25.0$ keV, has been used. The x-ray flux is maximum at this energy at the MS beamline [39]. A crushed fragment of the $Tb_2Ti_2O_7$ crystal and $\simeq 15$ wt.% of silicon powder have been mixed and ground to obtain a homogeneous mixture. The specimen has been loaded into a 0.3 mm diameter glass capillary. The presence of silicon helps in reducing the $Tb_2Ti_2O_7$ sample x-ray absorption and provides a convenient calibration. The data have been taken from room temperature down to 4 K.

Figure 1 displays an example of a diffraction diagram. The solid line results from a Rietveld fit. The quality of the fit is excellent. In Figs. 2 and 3 we compare the profiles of the $(12, 0, 0)$ and $(8, 8, 0)$ Bragg profiles recorded at 20 and 4 K. In contrast to the result of Ruff *et al.* [40] who found a profile broadening below 20 K, no shape change of the profiles is observed at low temperature. To further check for the behavior of our $Tb_2Ti_2O_7$ specimen, we determined the temperature dependence of the lattice parameter $a$. The compound exhibits the expected smooth thermal contraction as it is cooled down with a plateau below $\approx 25$ K to $a \simeq 10.1368$ Å. In Ref. [40] an expansion was reported below 20 K. Therefore, our measurements do not support

[1] As an example, see A. Yaouanc *et al.*, to be published.

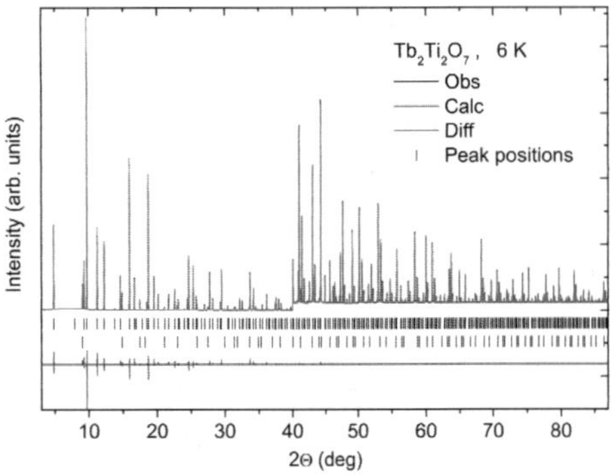


**Figure 1.** Rietveld refinement plot of synchrotron x-ray powder diffraction data of $Tb_2Ti_2O_7$ collected at 6 K with a photon energy of 25 keV. Observed points, calculated profile and difference curve are shown. Ticks below the graph show the calculated peak positions for $Tb_2Ti_2O_7$ and Si (upper and lower rows respectively). The intensities beyond $2\Theta = 40°$ have been enlarged by a factor of 10 in order to illustrate the agreement quality at higher angles.

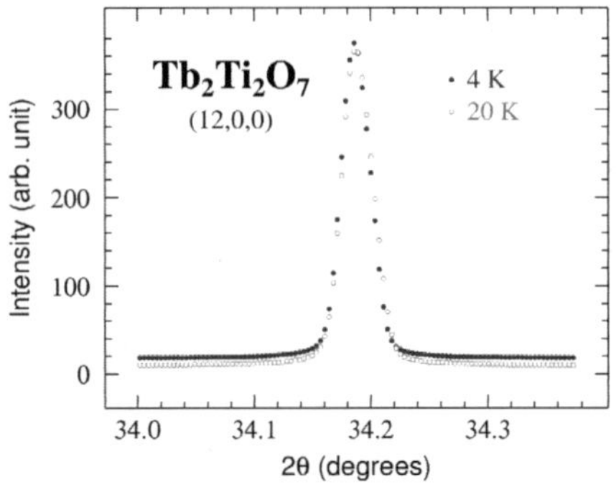


**Figure 2.** Comparison of the $(12, 0, 0)$ x-ray Bragg peak profiles measured at 20 and 4 K for our $Tb_2Ti_2O_7$ powder. The full width at half maximum of the Bragg peak is $9 \times 10^{-3}$ in reciprocal lattice units.

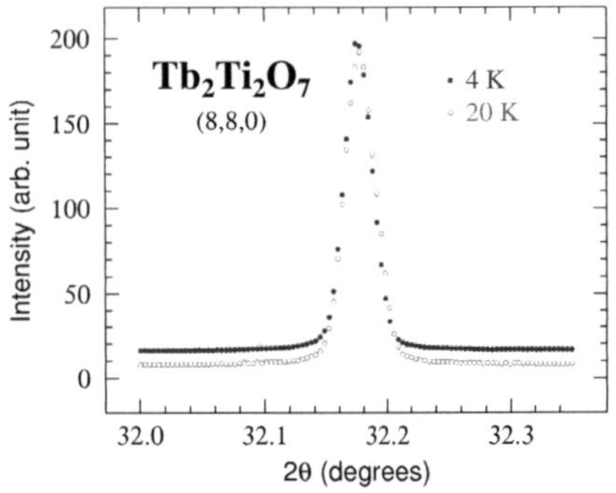


**Figure 3.** Comparison of the $(8, 8, 0)$ x-ray Bragg peak profiles measured at 20 and 4 K for our $Tb_2Ti_2O_7$ powder.

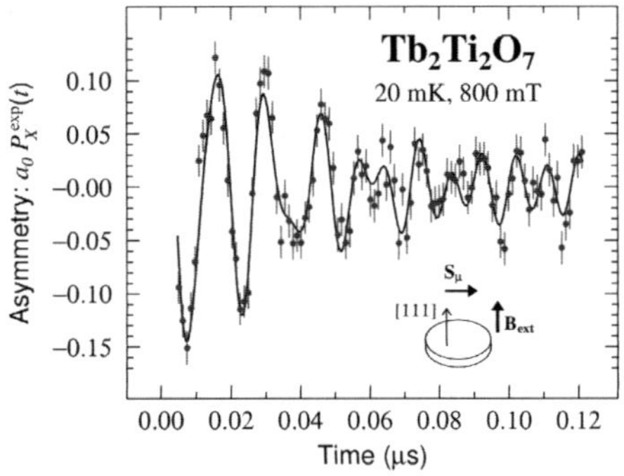


**Figure 4.** A typical TF-$\mu$SR asymmetry time spectrum recorded at 20 mK for a mosaic of $Tb_2Ti_2O_7$ crystals with $\mathbf{B}_{\rm ext}$ applied along a three-fold axis and $B_{\rm ext} = 800$ mT. The solid line results from a fit as explained in the main text.

the previous claim of the presence of structural fluctuations below $\approx 20$ K in $Tb_2Ti_2O_7$. They are in fact fully consistent with the results of Goto *et al.* [41]. We measure for the lattice parameter $a = 10.15735\,(10)$ Å at 295 K. This is slightly larger than $a = 10.15529\,(1)$ Å recently reported [33]. In fact, as pointed out in this reference, literature values of the lattice parameter of $Tb_2Ti_2O_7$ are clustered around 10.154 Å, but outlying values do exist. This casts doubt on the assignment of composition based on the lattice parameter [36].

Finally the value for the position parameter of the oxygen ion at position $48f$ is $x = 0.32783\,(3)$

for temperatures below 50 K.[2] Above 200 K we find that it slightly decreases: $x = 0.3270\,(1)$. These values are in agreement with the literature [42].

The $\mu$SR study has been performed at the Low Temperature Facility of the Swiss Muon Source (Paul Scherrer Institute, Switzerland) with a disk made of a mosaic of crystal plates previously described. It has covered the temperature range from 20 to 500 mK. We have set $\mathbf{B}_{\rm ext}$ perpendicular to the sample disk and therefore parallel to a $Tb_2Ti_2O_7$ [111] axis, and the initial muon polarization $\mathbf{S}_\mu$ perpendicular to it. By definition $\mathbf{B}_{\rm ext}$ and $\mathbf{S}_\mu$ are parallel to the $Z$ and $X$ axes, respectively. The geometry of the experiment is illustrated in the inset of Fig. 4.

The quantity of interest is the TF-$\mu$SR asymmetry time spectrum $a_0 P_X^{\rm exp}(t)$, where $P_X^{\rm exp}(t)$ describes the evolution of the muon polarization under $\mathbf{B}_{\rm ext}$. The positron detectors are located perpendicular to the $Z$ axis [43]. In Fig. 4 a typical asymmetry spectrum is displayed. It can be described as the weighted sum of two beating damped oscillating components: the first accounting for the muons implanted in the sample and the second modeling the response for the muons stopped in the sample surroundings, essentially the silver sample holder:

$$a_0 P_X^{\rm exp}(t) \quad = \quad a_1 \exp(-\lambda_{X,1} t)\cos(2\pi\nu_1 t + \varphi) + a_2 \exp(-\lambda_{X,2} t)\cos(2\pi\nu_2 t + \varphi). \qquad (1)$$

The analysis of the presented spectrum yields $a_1 = 0.192\,(13)$ and $a_2 = 0.028\,(2)$, i.e. only $\approx 13\%$ of the muons are stopped in the surroundings with $\nu_2 = 108.46\,(1)$ MHz. This latter value is very close to the precession frequency $\nu_{\rm ext} = \gamma_\mu B_{\rm ext}/(2\pi) = 108.43$ MHz expected for muons submitted to a field $B_{\rm ext} = 800$ mT. Here $\gamma_\mu = 851.6$ Mrad s$^{-1}$ T$^{-1}$ is the muon gyromagnetic ratio.

The interest is not in $\nu_1$, but rather in the frequency shift $\Delta\nu = \nu_1 - \nu_2$, or more conveniently the normalized frequency shift $K_{\rm exp} = \Delta\nu/\nu_2$ which, to a good approximation, can be written as $K_{\rm exp} = \Delta\nu/\nu_{\rm ext}$. This quantity, already known to be negative for $Tb_2Ti_2O_7$ [44, 22], is displayed in Fig. 5 versus $B_{\rm ext}$ for $T = 500$ and 20 mK. Relative to the data at 500 mK, at 20 mK an extra contribution to $K_{\rm exp}$ appears below $\approx 0.6$ T. This is consistent with the first magnetization curves recorded at 500 and 57 mK which also differentiate themselves only below 0.5 T [26]. Following the difficulties recently pointed out to correct for demagnetization [45], we shall refrain to do it for our data. However, the qualitative consistency of our data and the magnetizations displayed in Ref. [26], for which demagnetization corrections were not required, suggests our field scale to be correct.

Since $Tb_2Ti_2O_7$ is an insulator, only the dipole field from the terbium magnetic moments contributes to the frequency shift. For the sake of derivation, assuming the sample to be a sphere, i.e. the relevant demagnetization coefficient $N^Z = 1/3$, $K_{\rm exp} = K'_{\rm dip} = \mathcal{D}^{ZZ}(\mu_0 M/B_{\rm ext})$ where $K'_{\rm dip}$ accounts for the shift from the dipole field arising from the magnetic moments inside the Lorentz sphere, $\mathcal{D}^{ZZ}$ is a dipole tensor element, and $M$ is the magnetization along a three-fold axis. Note that if we were in the linear regime of the magnetization, the term in parentheses would be the magnetic susceptibility $\chi$ and we would recover the usual formula [43] $K_{\rm exp} = K'_{\rm dip} = \mathcal{D}^{ZZ}\chi$.[3] Hence, we expect the product $-B_{\rm ext}K_{\rm exp}$ to be proportional to $M$. In Fig. 6 we plot this product versus $B_{\rm ext}$. If there was a definitive plateau in the magnetization, the product would be field independent in a finite field range. This is not observed. However, as indicated by the up-arrow, a weak inflection point is present for the 20 mK data at $B_{\rm ext} \simeq 0.66$ T. It has disappeared at 500 mK.

Yin *et al.* have performed a.c. magnetic-susceptibility measurements for a $Tb_2Ti_2O_7$ crystal [29] with $\mathbf{B}_{\rm ext}$ parallel to a three-fold axis. The data presented in their Fig. 1d for $T = 16$ mT have been recorded with an a.c. field amplitude of 0.94 mT and $B_{\rm ext}$ extending up to 1.5 T.

[2] The $x$ values are given for the description of the space group with the origin at site of symmetry $.\bar{3}m$. The origin of the lattice is occupied by a titanium atom.

[3] Here we have neglected the anisotropy of the susceptibility.

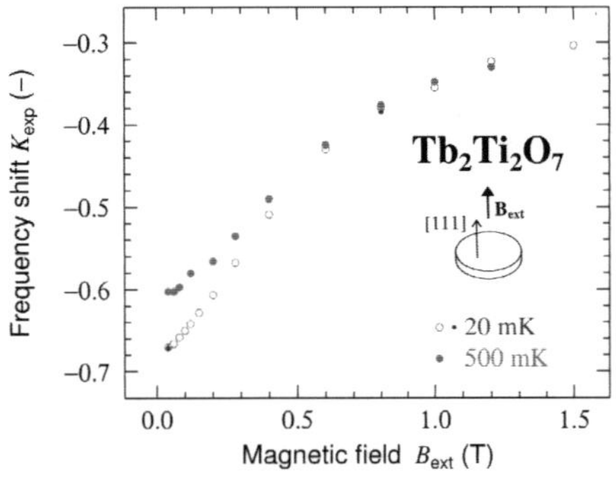


**Figure 5.** Field intensity dependence of the normalized frequency shift $K_{\rm exp}$ for our mosaic of $Tb_2Ti_2O_7$ crystals. $\mathbf{B}_{\rm ext}$ has been applied along a three-fold axis. The shifts have been measured at 20 and 500 mK. The error bars are smaller than the symbols. The data at 20 mK shown by open circles have been measured after zero-field cooling by increasing $B_{\rm ext}$ up to 1.5 T. Further data (small bullets) recorded after decreasing $B_{\rm ext}$ to 800 and 40 mT show no hysteretic effect. In contrast, an hysteretic response at 60 mT and low temperature has been observed in a temperature scan [22].

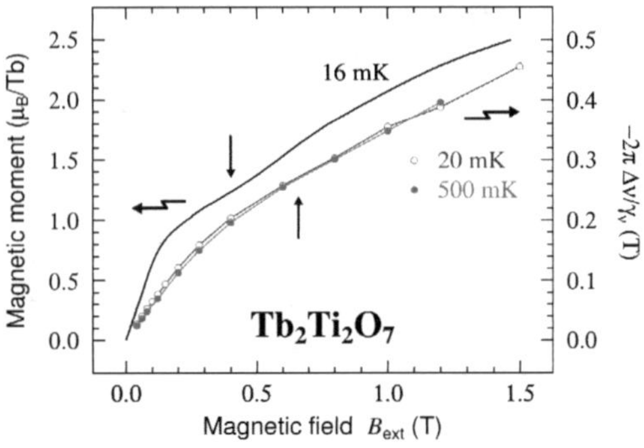


**Figure 6.** Open circles and bullets: product $-B_{\rm ext}K_{\rm exp} = -2\pi\Delta\nu/\gamma_\mu$ deduced from the 20 and 500 mK data displayed in Fig. 5, versus $B_{\rm ext}$. The experimental points are linked by segments. Solid line: field dependence of the terbium magnetic moment measured at 16 mK for $\mathbf{B}_{\rm ext}$ applied along a [111] crystal direction. As explained in the main text, the latter curve is computed from the data published by Yin *et al* [29].

With these experimental conditions the measurements give access to $\mu_0 {\rm d}\chi'_Z/{\rm d}B_{\rm ext}$ rather than $\chi'_Z$, where $\chi'_Z$ is the real part of the susceptibility. This means that $M$ can be deduced from the data of Yin *et al.* by field integration. Normalizing the result by the volume per terbium ion and dividing by the electronic Bohr magneton, we obtain the magnetic moment per terbium ion versus $B_{\rm ext}$ displayed in Fig. 6. An inflection point occurs at $B_{\rm ext} \simeq 0.40$ T as indicated by the down-arrow. In Fig. 1d of Ref. [29] the data were not corrected for the demagnetization field. However, it is relatively small: about 60 mT around 0.6 T; see Note 35 in Ref. [29]. Because of our experimental geometry, see Fig. 5, it is expected to be larger in our case. Assuming a reasonable demagnetization field of $\approx 0.30$ T at 0.66 T, the position of the inflection point inferred from our data and from Ref. [29] would coincide. Expressed in terms of the internal magnetic field, the inflection point would be located slightly above 0.3 T.

Therefore our $\mu$SR measurements and a.c. susceptibility data suggest an inflection point in the magnetization could be present slightly above 0.3 T at low temperature. As deduced from Fig. 6, there is an apparent discrepancy between the two data sets at lower field. However, since the initial slope of the magnetization curve at 57 mT is about $10\,\mu_{\rm B}/{\rm T}$ [26], we compute a susceptibility of $\approx 1.8$, a quite large value, in agreement with the literature [29]. Our frequency shift data at low field are therefore mainly controlled by the shape of our mosaic sample. From Ref. [26] we estimate $\mu_0 M/B_{\rm ext}$ to be about three times smaller at 0.4 T. This much smaller value explains that the field dependence of the two data sets are similar at high field.

The authors of the a.c. susceptibility measurements have stressed that $\chi'_Z$ displays a bump at $B_{\rm ext} \simeq 0.5$ T for $T = 40$ mK only if the field is swept sufficiently slowly, i.e for 0.625 mT/min but not for 3.75 mT/min; see Fig. S2 of the supplemental material to Ref. [29]. Our measurement sequences have consisted of changing $B_{\rm ext}$ step by step, taking between 10 to 20 min for a field change. This is therefore quicker than 0.625 mT/min. However, when at a given $B_{\rm ext}$ value, we stood there without any field change about 1 hour for data taking. This is in contrast to the a.c. susceptibility measurements done with an a.c. field of frequency 9.6 Hz for the data of Fig. S2 and 0.21 Hz for the measurements presented in Fig. 1d.

In summary, we have reported a high resolution synchrotron radiation diffraction investigation of a piece of a crushed $Tb_2Ti_2O_7$ single crystal, and frequency shift $\mu$SR measurements at 20 mK on a mosaic of single crystals of the same compound for an external field applied along a three-fold axis. The x-ray study has not revealed any structure change down to 4 K, confirming the high quality of our crystals. The $\mu$SR measurements have given us access to a quantity proportional to the terbium magnetic moment. While an inflection point seems to be present slightly above 0.3 T for the field dependence of the moment at the lowest temperature, in agreement with previously published a.c. susceptibility data, there is barely any signature of a magnetization plateau.

**Acknowledgments**
PDR gratefully acknowledges partial support of Prof. H. Keller from the University of Zurich for the $\mu$SR measurements. Part of this work has been performed at the Swiss Light Source and the Swiss Muon Source, Paul Scherrer Institute, Villigen, Switzerland.

**References**
[1] Gardner J S, Gingras M J P and Greedan J E 2010 *Rev. Mod. Phys.* **82** 53
[2] Harris M J, Bramwell S T, McMorrow D F, Zeiske T and Godfrey K W 1997 *Phys. Rev. Lett.* **79** 2554–2557
[3] Ramirez A P, Hayashi A, Cava R J, Siddharthan R and Shastry B S 1999 *Nature* **399** 333
[4] Hodges J A, Bonville P, Forget A, Yaouanc A, Dalmas de Réotier P, André G, Rams M, Królas K, Ritter C, Gubbens P C M, Kaiser C T, King P J C and Baines C 2002 *Phys. Rev. Lett.* **88** 077204
[5] Yaouanc A, Dalmas de Réotier P, Marin C and Glazkov V 2011 *Phys. Rev.* B **84** 172408
[6] Ross K A, Yaraskavitch L R, Laver M, Gardner J S, Quilliam J A, Meng S, Kycia J B, Singh D K, Proffen T, Dabkowska H A and Gaulin B D 2011 *Phys. Rev.* B **84** 174442
[7] Yasui Y, Kanada M, Ito M, Harashina H, Sato M, Okumura H, Kakurai K and Kadowaki H 2002 *J. Phys. Soc. Jpn.* **71** 599
[8] Chang L J, Onoda S, Su Y, Kao Y J, Tsuei K D, Yasui Y, Kakurai K and Lees M R 2012 *Nat. Commun.* **3** 992
[9] Mirebeau I, Apetrei A, Rodríguez-Carvajal J, Bonville P, Forget A, Colson D, Glazkov V, Sanchez J P, Isnard O and Suard E 2005 *Phys. Rev. Lett.* **94** 246402
[10] Dalmas de Réotier P, Yaouanc A, Keller L, Cervellino A, Roessli B, Baines C, Forget A, Vaju C, Gubbens P C M, Amato A and King P J C 2006 *Phys. Rev. Lett.* **96** 127202
[11] Bert F, Mendels P, Olariu A, Blanchard N, Collin G, Amato A, Baines C and Hillier A D 2006 *Phys. Rev. Lett.* **97** 117203
[12] Bertin E, Bonville P, Bouchaud J P, Hodges J A, Sanchez J P and Vulliet P 2002 *Eur. Phys. J.* B **27** 347–354
[13] Yaouanc A, P Dalmas de Réotier, Glazkov V, Marin C, Bonville P, Hodges J A, Gubbens P C M, Sakarya S and Baines C 2005 *Phys. Rev. Lett.* **95** 047203
[14] Chapuis Y, Dalmas de Réotier P, Marin C, Yaouanc A, Forget A, Amato A and Baines C 2009 *Physica* B **404** 686
[15] Lago J, Lancaster T, Blundell S J, Bramwell S T, Pratt F L, Shirai M and Baines C 2005 *J. Phys.: Condens. Matter* **17** 979
[16] Dalmas de Réotier P, Yaouanc A, Chapuis Y, Curnoe S H, Grenier B, Ressouche E, Marin C, Lago J, Baines C and Giblin S R 2012 *Phys. Rev.* B **86** 104424
[17] Yaouanc A, Dalmas de Réotier P, Bonville P, Hodges J A, Glazkov V, Keller L, Sikolenko V, Bartkowiak M, Amato A, Baines C, King P J C, Gubbens P C M and Forget A 2013 *Phys. Rev. Lett.* **110** 127207
[18] Dun Z L, Choi E S, Zhou H D, Hallas A M, Silverstein H J, Qiu Y, Copley J R D, Gardner J S and Wiebe C R 2013 *Phys. Rev.* B **87** 134408
[19] Lago J, Živković I, Piatek J O, Álvarez P, Hüvonen D, Pratt F L, Díaz M and Rojo T 2014 *Phys. Rev.* B **89** 024421
[20] Savary L and Balents L 2012 *Phys. Rev. Lett.* **108** 037202

[21] Gardner J S, Dunsiger S R, Gaulin B D, Gingras M J P, Greedan J E, Kiefl R F, Lumsden M D, MacFarlane W A, Raju N P, Sonier J E, Swainson I and Tun Z 1999 *Phys. Rev. Lett.* **82** 1012–1015
[22] Yaouanc A, Dalmas de Réotier P, Chapuis Y, Marin C, Vanishri S, Aoki D, Fåk B, Regnault L P, Buisson C, Amato A, Baines C and Hillier A D 2011 *Phys. Rev.* B **84** 184403
[23] Molavian H R, Gingras M J P and Canals B 2007 *Phys. Rev. Lett.* **98** 157204
[24] Chapuis Y, Yaouanc A, Dalmas de Réotier P, Marin C, Vanishri S, Curnoe S H, Vâju C and Forget A 2010 *Phys. Rev.* B **82** 100402(R)
[25] Molavian H R and Gingras M J P 2009 *J. Phys.: Condens. Matter* **21** 172201
[26] Lhotel E, Paulsen C, Dalmas de Réotier P, Yaouanc A, Marin C and Vanishri S 2012 *Phys. Rev.* B **86** 020410(R)
[27] Legl S, Krey C, Dunsiger S R, Dabkowska H A, Rodriguez J A, Luke G M and Pfleiderer C 2012 *Phys. Rev. Lett.* **109** 047201
[28] Sazonov A P, Gukasov A, Cao H B, Bonville P, Ressouche E, Decorse C and Mirebeau I 2013 *Phys. Rev.* B **88** 184428
[29] Yin L, Xia J S, Takano Y, Sullivan N S, Li Q J and Sun X F 2013 *Phys. Rev. Lett.* **110** 137201
[30] Luo G, Hess S T and Corruccini L R 2001 *Phys. Lett.* A **291** 306–310
[31] Fennell T, Kenzelmann M, Roessli B, Haas M K and Cava R J 2012 *Phys. Rev. Lett.* **109** 017201
[32] Guitteny S, Robert J, Bonville P, Ollivier J, Decorse C, Steffens P, Boehm M, Mutka H, Mirebeau I and Petit S 2013 *Phys. Rev. Lett.* **111** 087201
[33] Fennell T, Kenzelmann M, Roessli B, Mutka H, Ollivier J, Ruminy M, Stuhr U, Zaharko O, Bovo L, Cervellino A, Haas M K and Cava R J 2014 *Phys. Rev. Lett.* **112** 017203
[34] Petit S, Bonville P, Robert J, Decorse C and Mirebeau I 2012 *Phys. Rev.* B **86** 174403
[35] Fritsch K, Ross K A, Qiu Y, Copley J R D, Guidi T, Bewley R I, Dabkowska H A and Gaulin B D 2013 *Phys. Rev.* B **87** 094410
[36] Taniguchi T, Kadowaki H, Takatsu H, Fåk B, Ollivier J, Yamazaki T, Sato T J, Yoshizawa H, Shimura Y, Sakakibara T, Hong T, Goto K, Yaraskavitch L R and Kycia J B 2013 *Phys. Rev.* B **87** 060408
[37] Chapuis Y 2009 *Frustration géométrique, transitions de phase et ordre dynamique* Ph.D. thesis Université Joseph Fourier Grenoble
[38] Revell H M, Yaraskavitch L R, Mason J D, Ross K A, Noad H M L, Dabkowska H A, Gaulin B D, Henelius P and Kycia J B 2013 *Nature Phys.* **9** 34
[39] Willmott P R, Meister D, Leake S J, Lange M, Bergamaschi A, Böge M, Calvi M, Cancellieri C, Casati N, Cervellino A, Chen Q, David C, Flechsig U, Gozzo F, Henrich B, Jäggi-Spielmann S, Jakob B, Kalichava I, Karvinen P, Krempasky J, Lüdeke A, Lüscher R, Maag S, Quitmann C, Reinle-Schmitt M L, Schmidt T, Schmitt B, Streun A, Vartiainen I, Vitins M, Wang X and Wullschlegera R 2013 *J. Synchrotron Rad.* **20** 667–682
[40] Ruff J P C, Gaulin B D, Castellan J P, Rule K C, Clancy J P, Rodriguez J and Dabkowska H A 2007 *Phys. Rev. Lett.* **99** 237202
[41] Goto K, Takatsu H, Taniguchi T and Kadowaki H 2012 *J. Phys. Soc. Jpn* **81** 015001
[42] Helean K B, Ushakov S V, Brown C E, Navrotsky A, Lian J, Ewing R C, Farmer J M and Boatner L A 2004 *J. Solid State Chem.* **177** 1858
[43] Yaouanc A and Dalmas de Réotier P 2011 *Muon Spin Rotation, Relaxation, and Resonance: Applications to Condensed Matter* International Series of Monographs on Physiscs 147 (Oxford: Oxford University Press)
[44] Ofer O, Keren A and Baines C 2007 *J. Phys.: Condens. Matter* **19** 145270
[45] Bovo L, Jaubert L D C, Holdsworth P C W and Bramwell S T 2013 *J. Phys.: Condens. Matter* **25** 386002

Journal of Physics: Conference Series **551** (2014) 012022 doi:10.1088/1742-6596/551/1/012022

# Search for ALCR-$\mu^+$SR spin polaron resonances in $Cd_2Re_2O_7$ and $FeGa_3$

**J H Brewer**[1], **V G Storchak**[2], **D G Eshchenko**[3], **R L Lichti**[4], **P W Mengyan**[4] **and D J Arseneau**[5]

[1] Department of Physics and Astronomy, University of British Columbia, Vancouver, BC, Canada V6T 1Z1

[2] Kurchatov Institute, Kurchatov Sq. 1, Moscow 123182, Russia

[3] Bruker BioSpin AG, Industriestrasse 26, 8117 Fällanden, Switzerland

[4] Department of Physics, Texas Tech University, Lubbock, Texas 79409-1051, USA

[5] TRIUMF, 4004 Wesbrook Mall, Vancouver, BC, Canada V6T 2A3

E-mail: jess@triumf.ca

**Abstract.** Certain magnetic materials may contain spin polarons (SP) — tiny, stable ferromagnetic "droplets" in a paramagnetic or weakly ordered "sea" of more random spins. The positive muon may become associated with SP, giving a unique window on their properties. In this experiment the $\mu^+$SR spectroscopy of "avoided level crossing resonance" ($\mu$ALCR) was used on several candidate materials to further explore this model. The results leave many questions unanswered.

## 1. Introduction

Numerous semiconducting[1, 2, 3] and metallic[4] magnetic materials have been found to exhibit characteristic two-frequency $\mu^+$SR precession signals in high transverse magnetic field (HTF-$\mu^+$SR), but skepticism remains over the assignment of these spectra to muons associated with de Gennes' legendary magnetic polarons (MP)[5] or spin polarons (SP). This is understandable, since the SP picture is a radical departure from "conventional wisdom" about both muonium and magnetism. It is therefore incumbent upon both advocates and adversaries of this interpretation to present as much spectroscopic evidence as possible in support or contradiction of the SP picture.

In studies of muonium (Mu = $\mu^+e^-$) in solids and muonated radicals in liquids, the hyperfine frequency of the species containing the muon is routinely measured in high transverse magnetic field (HTF-$\mu^+$SR) as the splitting $A$ between two precession frequencies on opposite sides of the muon Larmor frequency $\nu_\mu$, as shown in Fig. 1. In a SP, the electron's spin is locked to that of the local ferromagnetic (FM) "droplet" by a very strong exchange interaction, but the same sort of splitting still appears. Of course, muons in an antiferromagnetic (AFM) crystal with local fields parallel or antiparallel to the applied field will produce a similar pattern, so other considerations must help determine which scenario applies.

The basic mechanism of muon avoided level crossing resonance ($\mu$ALCR) is depicted in Fig. 2. Usually it is assumed that $J \ll A$. The opposite is the case in a SP, if $J$ is the exchange interaction between the SP electron and the paramagnetic ions comprising the SP. However, if

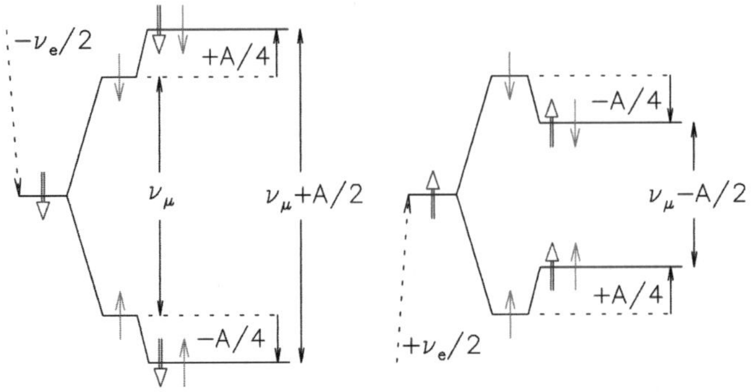


**Figure 1.** Muonium energy level diagram in the limit of high magnetic field (muon Larmor frequency $\nu_\mu \gg A$, the hyperfine frequency).

$J$ is the *nuclear* hyperfine (NHF) coupling between the SP electron and a nearby *nuclear* spin, the same situation may apply as in ordinary semiconductors[6] or radicals [7], for which $\mu$ALCR spectroscopy [6, 7, 8] has yielded a plethora of information about the location, structure and effective spin hamiltonian of the paramagnetic center.

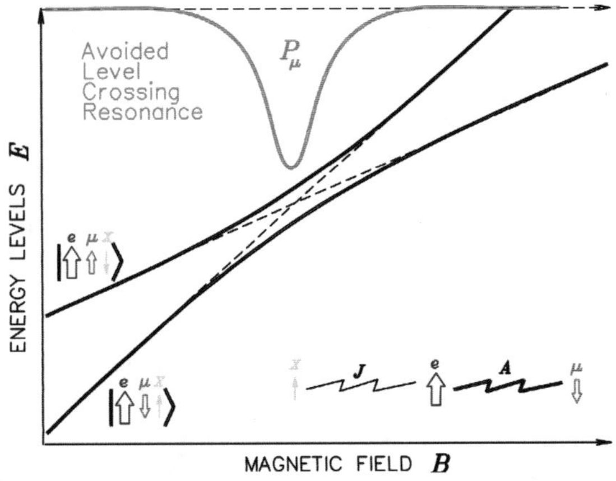


**Figure 2.** Avoided level crossings occur when "flipping" a third spin $X$ that couples to the muon and/or electron causes the same change in energy as "flopping" the muon spin at a certain applied field $B_0$. A resonant loss of muon polarization results at $B_0$. This method is widely used to identify the partner $X$, its interaction strength $J$ and, from that, the site and environment of the muon.

It is therefore natural (even mandatory) to explore the $\mu$ALCR spectroscopy of candidate SP systems in the same way. We performed an initial study of several nominally dissimilar materials with nearly identical SP-like HTF-$\mu^+$SR spectra.

## 2. SP$\mu$ Spectroscopy

The MP or SP is formed when a free electron in a paramagnetic or weakly magnetic host has a huge exchange interaction $J$ with neighboring magnetic ions, causing their spins to be locked together (along with that of said electron) into a small ($\sim$ 1 unit cell) ferromagnetic (FM) "droplet" — the SP — widely believed to be the root cause of various important phenomena such as metal-insulator transitions in magnetic semiconductors (MS)[9]. The energy decrease due to this local FM ordering compensates for the increased kinetic energy of the SP electron caused by its localization. In some cases the additional electrostatic binding energy of said electron to the $\mu^+$ assists in the localization; in other cases a SP population exists independent of the muon but the negatively charged SP attracts (and is bound to) the $\mu^+$ in a neutral "SP$\mu$" complex, analogous to muonium or muonated radicals.

The ability of HTF-$\mu^+$SR to probe the SP depends upon the muon's incorporation into the SP in a consistent way; thus the system in question is often described as a *bound* magnetic polaron (BMP). However, this designation is usually taken to imply a SP that is not free to move through the lattice, which may be a poor picture of the SP$\mu$ system in cases where the muon is "bound" to the SP only by the latter's negative charge, providing only a small extra impedance to the mobility of this exotic charge carrier.

The two SP candidate materials chosen for this $\mu$ALCR study are the metallic, magnetically frustrated pyrochlore $Cd_2Re_2O_7$ and the magnetic semiconductor $FeGa_3$, both of whose HTF-$\mu^+$SR frequency spectra are shown in Fig. 3.

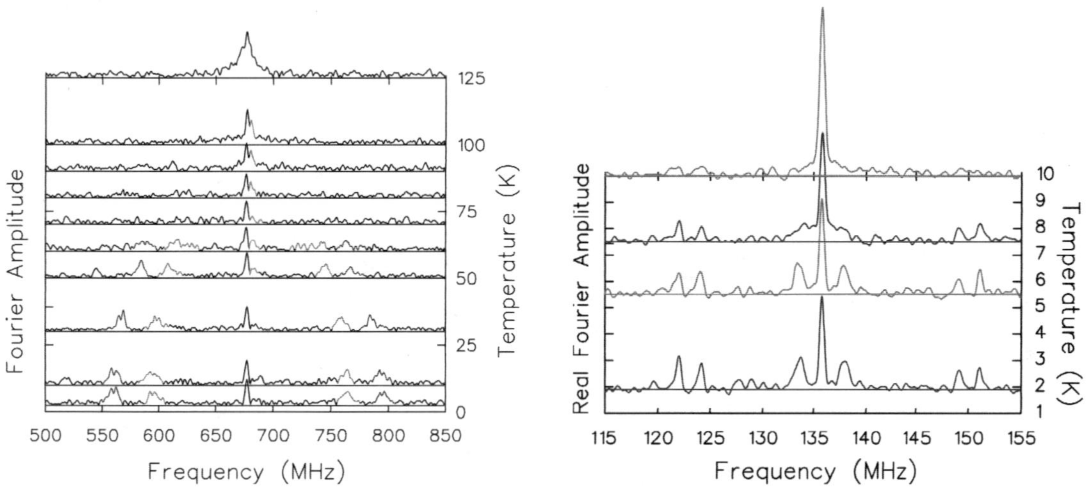


**Figure 3.** Left: Frequency spectra of HTF-$\mu^+$SR signals in $Cd_2Re_2O_7$ at $H = 1$ T[4]. The splittings do not change appreciably with $H$ (not shown). Right: HTF-$\mu^+$SR frequency spectra at $H = 1$ T in the magnetic semiconductor $FeGa_3$. Again the splittings do not change significantly with $H$.

$Cd_2Re_2O_7$ is a weakly metallic pyrochlore with geometrically frustrated magnetism, a heavy electron effective mass and a superconducting transition at $\sim 1$ K. In the region below $\sim 60$ K, where its resistivity follows a $T^2$ dependence (characteristic of a Fermi liquid regime), the HTF-$\mu^+$SR frequency spectrum at $H = 5$ T shows the splittings characteristic of a SP with a large hyperfine coupling[4].

$FeGa_3$ is a narrow-gap diamagnetic semiconductor for which the gap formation is attributed to strong electron correlations within a narrow $3d$ band. Below about 10 K an extremely narrow SP band is thought to form; in the same range, characteristic SP splittings are observed in the HTF-$\mu^+$SR frequency spectra, with splittings independent of field up to 5 T[3].

Both materials have plentiful nuclear spins which might couple to the SP electron, possibly producing a rich $\mu$ALCR spectrum.

The HTF-$\mu^+$SR "signature" of SP$\mu$, like that of Mu or a muonated radical, is a pair of frequencies split about the free muon Larmor frequency ($\nu_\mu$[MHz] $= 135.54 B_\mu$[T], where $B_\mu$

is the local magnetic field at the muon site) by $+\Delta\nu_+$ and $-\Delta\nu_-$ with $\Delta\nu_+ - \Delta\nu_- = A$, the effective frequency of the $\mu$-$e$ HF interaction. This similarity, plus the fact that SP$\mu$ spectra were first discovered in MS, may tempt one to attribute such spectra to normal Mu atoms; but this cannot be the case, first because (as in $Cd_2Re_2O_7$, shown in Fig. 3[left]) SP$\mu$ is seen in *metallic* hosts where the Coulomb attraction of the $\mu^+$ is screened by conduction electrons, and second because the fields of large nearby moments in the paramagnetic phase would instantly relax the Mu electron if it were not tightly locked to the spins in the FM "droplet". Very similar spectra are seen in the MS $FeGa_3$ (Fig. 2) at low temperature. It is remarkable that the signature of SP$\mu$ should be observed in such different magnetic systems; usually one does not expect a hitherto undiscovered phenomenon to be quite so ubiquitous. In each instance one can often find alternative explanations for the observed pairs of lines — two sites with different Knight shifts, or one site in an antiferromagnetic environment, or other more exotic interpretations invoking modification by the $\mu^+$ of the local magnetic properties of the host. However, such alternative-hunting fails to explain the remarkable consistency of the HTF-$\mu^+$SR spectra observed in such diverse magnetic metals, semiconductors and insulators.

## 3. SP-$\mu$ALCR Experiments

We therefore measured $\mu$ALCR spectra over a very large field range in the correlated metallic pyrochlore $Cd_2Re_2O_7$ at 30 K and in the semiconductor $FeGa_3$ at 10 K. The results are shown in Figs. 4 and 5.

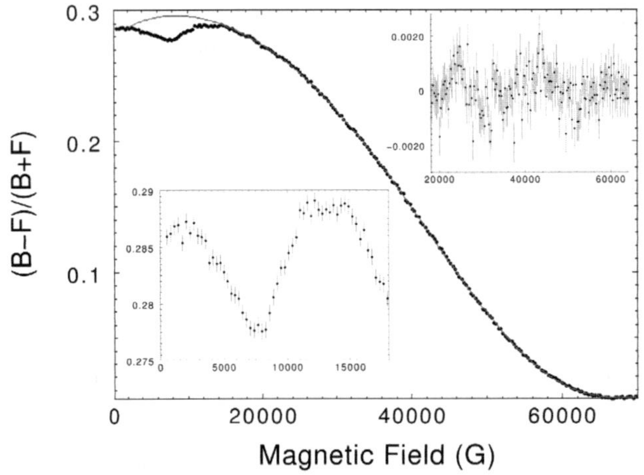


**Figure 4.** $\mu$ALCR spectrum from 0 to 7 T in metallic $Cd_2Re_2O_7$ at 30 K, taken on the M15 muon channel at TRIUMF with the *HiTime* $\mu$SR spectrometer. Upper right inset: residuals from a polynomial fit (red line) to the systematic decrease at high field due to curling up of electron orbits. Lower left inset: detail of a broad resonance at $\sim$ 0.75 T ($\nu_\mu \approx$ 100 MHz $\approx A_{SP}/2$) and a hint of a smaller resonance at $\sim$ 1.3 T.

In each case the measurements were made (as usual) by scaling the Forward (F) and backward (B) raw positron count rates and calculating the asymmetry as $(B-F)/(B+F)$. As is often the case, this "raw" asymmetry is susceptible to huge systematic longitudinal field (LF) dependences as positron orbits "curl up" in high fields. Another common problem is stochastic variations due to instabilities in beamline power supplies or the proton beam drifting on the production target;

that effect requires use of differential method where the applied LF is rapidly "toggled" between $H - \delta H$ and $H + \delta H$. This latter method was not required for these experiments on M15 and the newly-rebuilt M20, whose power supplies are very stable. Therefore the huge systematic changes in the "baseline" asymmetry are evident in Figs. 4 and 5.

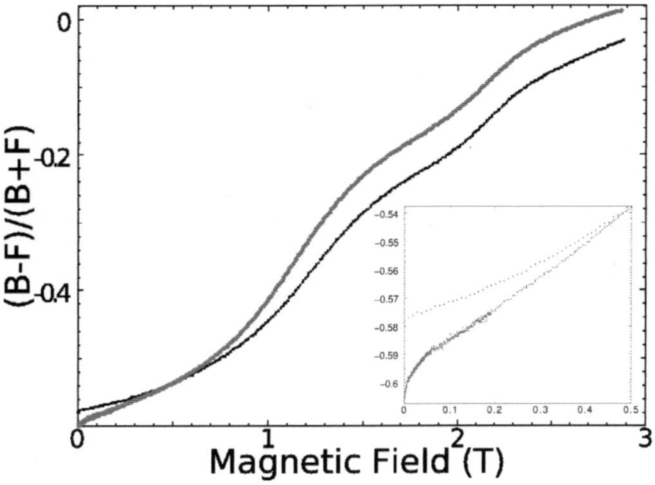


**Figure 5.** $\mu$ALCR spectra from 0 to 3 T at 5 K in $FeGa_3$ (upper curve, red circles) and Ag (lower curve, black squares), taken on the new M20 muon channel at TRIUMF with the *Helios* $\mu$SR spectrometer. If there is any resonant "dip" in the $FeGa_3$ polarization curve, it is either extremely broad or extremely narrow. The decrease in the $FeGa_3$ curve at very low field (see inset, lower right) is presumably just decoupling from local dipolar fields.

The large, broad resonance in $Cd_2Re_2O_7$ at $\sim$ 0.75 T (corresponding to a frequency $\nu_\mu \approx 100$ MHz) is consistent with a zero-crossing resonance (ZCR) for muons at a site with a local field of that magnitude antiparallel to the applied field, as in $MnF_2$ [10]. This would imply that the SP hyperfine splitting $A_{SP}$ in $Cd_2Re_2O_7$ at low $T$ is not the large ($\Delta\nu_{\rm big} \approx 200$ MHz) splitting shown in Fig. 3 but the smaller ($A_{SP} = \Delta\nu_{\rm small} \sim 30$ MHz) splitting between blue and red peaks in each of the groups on either side of $\nu_\mu$, while $\Delta\nu_{\rm big}$ comes from local magnetic ordering at the muon site. If there is actually a smaller resonance at $\sim 1.3$ T, it might be a true $\mu$ALCR, presumably with $X =$ a Re nuclear moment.

Presumably the $H$ dependence of the asymmetry in Ag is due entirely to systematics and represents a calibration for full asymmetry, except for a possible offset due to slight geometrical differences. The Ag curve also serves to isolate the rather dramatic effects due to refocussing of the muon beam, "curl-up" of positron orbits and other "systematics". It is therefore unsurprising that the $FeGa_3$ asymmetry has the same shape at high LF (above about 1.6 T) except for a geometrical offset. At lower fields, however, considerable asymmetry is lost in $FeGa_3$ (relative to Ag), in a monotonic drop of about 0.066, nearly half of a typical full asymmetry in such LF measurements. This is consistent with Mu decoupling in a semiconductor with nearly 100% Mu formation and a nearly vacuum-like Mu HF coupling. The additional drop-off as $H \to 0$ is presumably due to decoupling from local nuclear dipolar fields. The HTF-$\mu^+$SR splittings observed for $FeGa_3$ in Fig. 3 are certainly not consistent with a HF coupling of close to $A_0 = 4463$ MHz for a majority of muons forming vacuum-like Mu atoms.

## 4. Conclusions

Our search for rich $\mu$ALCR structure in these SP candidate materials has been generally unsuccessful. The NHF couplings between the SP electron and neighboring nuclei may be either too weak or too anisotropic to engender resonances. There might be extremely *narrow* resonances that would require scanning the appropriate field range with much smaller step sizes. (We used steps of 4, 10, 50 and 100 G in low, medium and high field regions.) Such a survey would benefit enormously from a good guess of where such resonances might be expected to appear. Such guesses might benefit in turn from DFT analysis.

## References

[1] Storchak V G *et al.* 2009 *Phys Rev.* B **80** 235203
[2] Storchak V G *et al.* 2009 *Phys. Rev.* B **79** 193205
[3] Storchak V G *et al.* 2012 *J. Phys. Condens. Matt.* **24** 185601
[4] Storchak V G *et al.* 2010 *Phys. Rev. Lett.* **105** 76402
[5] de Gennes P G 1960 *Phys. Rev.* B **118** 141
[6] Kiefl R F *et al.* 1987 *Phys. Rev. Lett.* **58** 1780
[7] Kiefl R F *et al.* 1986 *Phys. Rev.* A **34** 681
[8] Kreitzman S R *et al.* 1986 *Phys. Rev. Lett.* **56** 181
[9] Nagaev E L 2002 Colossal Magnetoresistance and Phase Separation in *Magnetic Semiconductors* (London: Imperial College Press)
[10] Kiefl R F *et al.* 1987 *Phys. Rev.* B **35** 2079

Journal of Physics: Conference Series **551** (2014) 012023 doi:10.1088/1742-6596/551/1/012023

# $\beta$-NMR Study of a buried Mn $\delta$-doped layer in a silicon host

**D Cortie[1,2], T Buck[2], R M L McFadden[3], J Xiao[3], C D P Levy[4], M Dehn[5], M Pearson[4], G D Morris[4], S R Dunsiger[5], R F Kiefl[2], F J Rueß[6], A Fuhrer[6] and W A MacFarlane[1,3]**

[1] Quantum Matter Institute, University of British Columbia, Vancouver, British Columbia V6T 1Z4, Canada
[2] Department of Physics and Astronomy, University of British Columbia, 6224 Agricultural Road, Vancouver, BC V6T 1Z1, Canada
[3] Chemistry Department, University of British Columbia, Vancouver, BC, V6T 1Z1, Canada
[4] TRIUMF, 4004 Wesbrook Mall, Vancouver, BC V6T 2A3, Canada
[5] Physik-Department, Technische Universität München, James-Franck-Straße 1, 85748 Garching bei München, Germany
[6] IBM Research-Zurich, Säumerstrasse 4, 8803 Rüschlikon, Switzerland

E-mail: cortie@phas.ubc.ca

**Abstract.** Low temperature growth methods were used to encapsulate a buried Mn $\delta$-doping layer into a silicon host. A $\beta$-NMR investigation was performed of the magnetic properties in the temperature range 10 – 300 K using spin-polarized $^8Li^+$. A depth-dependent broadening and shift of the NMR resonance was detected that is consistent with internal fields distributed at depths of 10 – 30 nm beneath the surface. At low temperatures, a negative relative shift occurred and the resonance was significantly broadened. At 300 K the line-shape could be described by a single Gaussian line, however, at 10 K the line is best approximated by a two component Lorentzian shape consisting of a broad and narrow component as anticipated for a diluted magnetic alloy. The overall magnitude of the resonance shift at both temperatures is small suggesting a weak interaction between the $^8Li^+$ and the magnetic Mn environment.

## 1. Introduction

Precise methods to confine magnetic impurities in silicon are a key step to establishing spin-based functionality in this highly compatible semiconductor. Past work reported high-temperature ferromagnetic behavior in Mn-implanted Si [1], however direct verification of an intrinsic mechanism is complicated by the phase-segregation that occurs during high temperature annealing [2]. By employing low temperature growth methods based on molecular beam epitaxy, it is possible to form buried Mn $\delta$-doped layers embedded in silicon with a tightly confined spatial distribution [3]. We report here on the study of a single buried Mn $\delta$-doped layer encapsulated in silicon. The local magnetic effect of the narrow Mn-distribution was investigated with the sensitive depth-resolved probe of $\beta$-detected nuclear magnetic resonance ($\beta$-NMR) using a beam of radioactive $^8Li^+$ ions [4, 5]. Based on past $\beta$-NMR experiments on other magnetic systems, including Mn-doped GaAs [6] and $Mn_{12}$-molecular monolayers [8], it was predicted that a shifted resonance frequency and a broadened lineshape would provide a sensitive measure of paramagnetic or ferromagnetic behavior in the impurity layer. Here we report on the $^8Li^+$ NMR resonance at various implant depths and proximities to the $\delta$-doped layer at 10 and 300 K. The results are consistent with a depth dependent magnetic distribution as expected from confined Mn moments in a $\delta$-layer. However, the observed signatures are also substantially different from the ones observed in ferromagnetic Mn-doped GaAs films [6].

## 2. Experiment

A dual-chamber Omicron VT AFM/STM UHV system was used to deposit surface Mn monolayers and encapsulate these beneath a Si capping layer [3]. The deposition was performed on chemically cleaned p-type B-doped (10$\Omega$ cm) Si(100) substrates with carrier densities of approximately 2x$10^{15}$ $cm^{-3}$. A low concentration of Mn equivalent to 0.2 – 0.3 monolayers was encapsulated at a depth of 10 – 20 nm in a silicon host by using the two step process consisting of low-temperature surface deposition followed by the addition of a Si capping layer. $\beta$-NMR was used to study the buried magnetic $\delta$-doped region at 10 K and 300 K using ion beams of the nuclear-spin polarized isotope $^8Li^+$ implanted at variable depths in the range 1 – 100 nm. Monte Carlo simulations using the SRIM package were used to predict the penetration depth of the ions into the sample. The experimental study was performed under a high field of 6.55 T applied normal to the substrate surface.

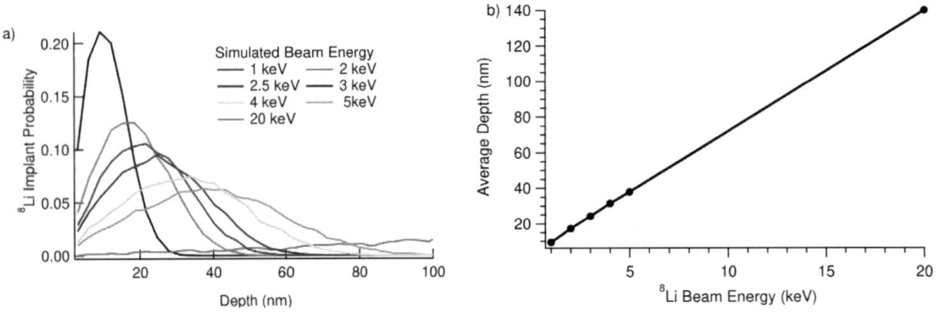


**Figure 1.** a) Predicted $^8Li^+$stopping range distribution from Monte Carlo simulations at different implant energies and b) Average stopping depth of $^8Li^+$ into Si.

## 3. Results and Discussion

Figure 1 shows the simulated implantation profile for the $^8Li^+$ calculated using Monte Carlo simulations in the SRIM 2012 software [10]. The addition of a small amount of Mn in the simulated layer profile was found to have negligible effects on the overall distribution which is dominated by the stopping power of silicon, although it should be noted that secondary ion-channeling effects are not modelled by SRIM. It is clear that, for lower ion-beam energies (1 – 5 keV), a substantial portion of Li is expected to stop near the nominal depth of the Mn monolayer (10 – 20 nm). On the other hand, at the highest possible energy (20 keV) there is negligible overlap between the impurity layer and the $^8Li^+$ profile. In general, one would expect the maximum magnetic signal to arise when the probing profile overlaps the magnetic layer. For a magnetic layer buried at 10 – 20 nm, the signal is therefore expected to be maximized at 1 – 3 keV. A common limitation of NMR and $\beta$-NMR in strongly magnetic materials, however, is the wipeout effect where fast relaxation destroys the observable asymmetry signal. This is a feature of strongly ordered ferromagnetic or antiferromagnetic materials. For this reason it is conceivable that a missing fraction of asymmetry may correspond to $^8Li^+$ located within a few lattice spacings of Mn moments, however as the $\delta$-doped layer is so thin, this fraction will not be large at any energy.

Figure 2 a) displays the observed nuclear magnetic resonance at 300 K at various beam energies. The observed resonances are remarkably *narrow*, partly as a result of the absence of nuclear dipolar broadening due to the scarcity of spin-bearing $^{29}$Si. In contrast the linewidth is greater than 2 kHz in GaAs due in part to the Ga and As nuclear moments [6]. The narrow line in Si is ideal for measuring small resonance shifts. The resonance shows no quadrupolar splittings indicating that the $^{8}Li^{+}$ occupies a site of cubic symmetry, likely the tetrahedral interstitial [7]. It is found that the peak shape can be well described using a single Gaussian. The amplitude of the peak asymmetry for the Mn-doped Si resonance is also quite large even at

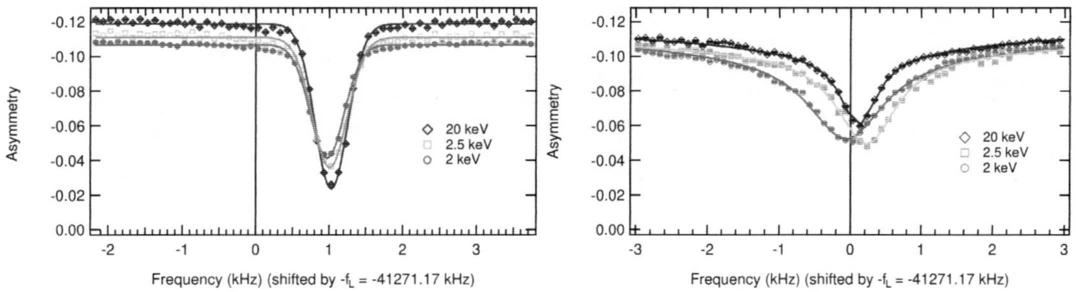


**Figure 2.** NMR resonance of $^{8}Li^{+}$ in Mn-doped Si at a) 300 K and b) 10 K at various probing beam energies. All frequencies have been shifted by the Larmor frequency ($f_L$) taken from a reference sample. Markers are the data. Solid lines are fits to the data.

very low energies of 2 keV. A reduction in amplitude is tied to line broadening from a magnetic origin, but could also result from an increased missing fraction of asymmetry or back-scattering of low energy ions. All of the resonances are shifted to higher frequencies by approximately 20 parts-per-million (ppm) with respect to the reference frequency, however, it is clear that there are also smaller relative shifts in the resonance frequency corresponding to a depth dependent feature for Li stopping in the range of 10 – 30 nm beneath the surface, at some proximity to the Mn layer. The relative shifts between different implantation energies are very small ($< 0.1$ kHz or $\approx 2$ ppm). Past $\beta$-NMR work on $Ga_{1-x}Mn_xAs$ observed two well-separated resonance peaks corresponding to the magnetic and non-magnetic local environments surrounding the $^{8}Li^{+}$ probe [6]. In contrast to that sample, where the magnetic layer was very thick (180 nm), we do not expect a significant fraction of the $^{8}Li^{+}$ in the thin $\delta$-doped layer at any energy. Even at the nominally optimal stopping ranges given by 1 – 2 keV beam energy (see Fig. 1), less than 20 % of the $^{8}Li^{+}$ could be considered to be in the $\delta$-doped layer. Instead, in the neighborhood of the narrow Mn-doped layer, the $^{8}Li^{+}$ is expected to sense a distribution of medium-range dipolar fields due to the variable distance between the implant site and the magnetic layer. This would result in an inhomogeneously broadened resonance line. It is clear that the peak width, described by the full-width-at-half-maximum (FWHM) of the Gaussian peaks at 300 K is extremely similar for all beam energies 2 – 20 keV (see also Figure 3). We discount the 1 keV data because there is reason to suspect that at these low energies the signal is artificially broadened from a known background signal related to $^{8}Li^{+}$ backscattering. The absence of a substantial magnetic effect on the linewidth in this range suggests that any Mn local moment magnetism is highly dynamic and lies in the paramagnetic regime as might be expected at 300 K (e.g. from the behavior of the helimagnet MnSi), or that antiferromagnetic Mn-Mn interactions reduce the net local moment. Past work showed the tendency for Mn to cluster and this cannot be excluded here [3], so that it is possible that nearest-neighbor Mn-Mn interactions play an important role in determining the overall magnetic behavior.

Given the relatively weak magnetic features at room temperature, it is useful to compare the

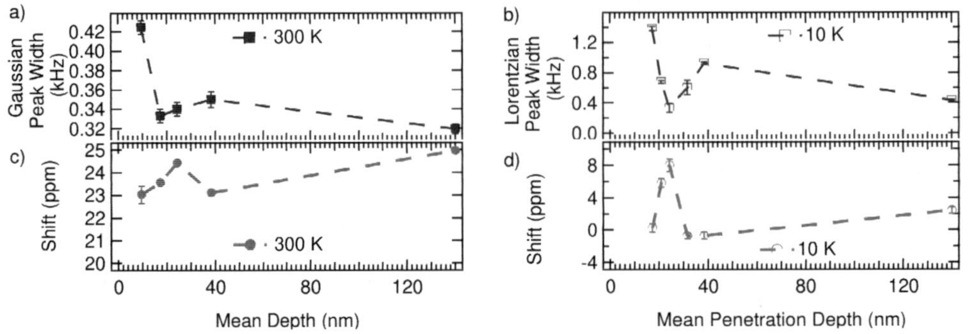


**Figure 3.** (Top) Summary of the peak-fits of the peak-width at a) 300 K and b) 10 K for various implant depths corresponding to different implant energies. (Bottom) Resonance frequency at 300 K (c) and 10 K (d) for various implant depths.

behavior of the same sample at 10 K. If the sample is paramagnetic or ferromagnetic with a Curie temperature $\ll$ 300 K, one would anticipate a very strong change in magnetic moment with temperature. The magnetic effects should scale as $H/T$ if the susceptibility is Curie-like. Figure 2 b) shows the observed resonance at 10 K under various different energies with identical RF conditions to those in Figure 2 a). Significant modifications to the line shape and frequency shift are observed at lower temperatures. The resonance overall at low temperature is broadened compared to the room temperature data, and takes on widths approaching 1 kHz. It is found that, unlike the 300 K data, a Gaussian line shape does not accurately describe the 10 K resonance shape, and the best description is a dual-component Lorentzian consisting of a broad and narrow component. A quasi-Lorentzian lineshape is commonly found in NMR studies of dilute magnetic alloys, where a wide range of local magnetic environments reflects the range of distances between the probe nuclei and the dilute magnetic ions [9]. The two Lorenzian components used to fit the 10 K data were found to share a common average frequency and shift within experimental uncertainty. The FWHM of the narrower component was found to be nearly constant ($\approx$ 0.4 kHz) at all beam energies, whereas the wider component varied strongly (0.4 – 1.2 kHz). The narrow line is ascribed to the intrinsic response of the silicon, whereas the broader component is a feature of the distribution of proximal magnetic fields detected at the $^{8}Li^{+}$ site.

Figures 3 a) and b) summarize the width at various implant depths at room temperature and low temperature respectively. It is clear that the maximum resonance broadening occurs at energies corresponding to depths of 20 – 30 nm suggesting a buried magnetic layer. At 300 K, the FWHM of the peak is nearly independent of the probing depth for energies greater than 1 keV (i.e. for depths greater than 10 nm where backscattering plays little role). However, at lower temperature, the peak-width is strongly depth dependent. Collectively, the room temperature and low temperature data implies that $^{8}Li^{+}$ nuclei located at a depth of $\approx$ 20 nm are sensitive to a magnetic perturbation provided by the Mn impurity which is largely paramagnetic at room temperature and becomes more polarized at lower temperature. The line broadening with different beam energy is quite dramatic at low temperature, and it is likely that the major source is the presence of an inhomogeneous magnetic dipolar field associated with the magnetic layer. It is known, however, that even undoped Si has a slightly broadened resonance at low temperature with a shift that depends on carrier doping [8] and bare GaAs substrates occasionally show a weak depth dependency. Therefore, in order to quantify the magnitude of the Mn-related magnetic field, careful control measurements on our specific doping regime over a range of beam energies are required. Figure 3 c) and d) shows a summary of the fitted frequency at 300 K and

10 K for various penetration depths corresponding to the beam energies 2 – 20 keV. A similar depth-dependent trend is noted in the room temperature and low temperature data. At 300 K and 10 K, the maximum shift and minimum broadening coincide at a penetration depth near 20 nm, corresponding to a beam energy of 2 keV. The relative shifts near the buried feature are far larger at 10 K (+8 ppm) near the magnetic depth profile, than in the room temperature state (+2 ppm) which implies an increased internal magnetic field. It is not yet completely clear whether this distribution is paramagnetic or ferromagnetic in nature, as this will be established using a study of the full temperature dependence. The slight modification of the detection depth between 300 K and 10 K implies the $^8Li^+$ may experience a subtly different magnetic stray field distribution in the room temperature and low temperature state associated with non-random order. Overall, the shift of the resonance deep within the Si substrate is decreased at low temperature, showing that a negative shift relative to the room temperature data is intrinsic to the silicon substrate. This may reflect electron localization resulting in a lower contact hyperfine field at the $^8Li^+$ implant site. The interplay between this effect and magnetic shifts needs to be taken into account and complicates interpretation. The length-scale where the Mn impurity is detected agrees with structural studies on past samples [3], although it is perhaps slightly deeper than the value expected for the deposition times used in the present study (10 – 20 nm). It is possible that the slight difference originates if the SRIM profiles overestimate the implantation depth. Given the relatively high local concentration of the Mn at peak, it is also conceivable that the signal is completely lost for $^8Li^+$ arriving too nearby to the $\delta$ layer.

## 4. Summary

Our initial results show that it is feasible to detect a minute amount of buried magnetic impurity in a silicon host. This is made clear through the systematic beam-energy and temperature dependency of the nuclear magnetic resonance. At room temperature, we observed a single resolved peak that depends on implant energy with maximum shifts appearing at beam energies of 2.5 keV, corresponding to a mean ion range of $\approx$ 20 nm beneath the surface. At lower temperature, the peak was broadened and consisted of a narrow and broad Lorentzian component. This is consistent with a strong temperature dependence in the sample associated with a buried magnetic feature. Future work will aim to investigate Mn-layers buried more deeply into the sample (50 nm) over a wide range of temperatures to track the magnetic transition and ascertain whether it is paramagnetic in nature. In particular, comparison with the blank Si control samples will facilitate quantitative analysis of the internal field distribution.

## References

[1] Bolduc M *et al.* 2005 *Phys. Rev.* B **71** 033302
[2] Zhou S *et al.* 2007 *Phys. Rev.* B **75** 085203
[3] Rueß F J *et al.* 2013 *Appl. Phys. Lett.* **102** 082101
[4] MacFarlane W A *et al.* 2013 *Phys. Rev.* B **88** 144424
[5] Salman Z *et al.* 2012 *Phys. Rev. Lett.* **109** 257207
[6] Song Q *et al.* 2011 *Phys. Rev.* B **84** 054414
[7] Wahl U 1997 *Physics Reports* **280** 145
[8] Salman Z *et al.* 2007 *Nano. Lett.* **7** 1551
[9] Walstedt R E and Walker L R 1974 *Phys. Rev.* B **9** 4857
[10] Biersack J P and Haggmark L G 1980 *Nucl. Inst. Meth.* **174** 257

Journal of Physics: Conference Series **551** (2014) 012024 doi:10.1088/1742-6596/551/1/012024

# Local magnetic order in $La_2CuO_4$ seen via $\mu^+$SR spectroscopy

**V G Storchak[1], J H Brewer[2], D G Eshchenko[3], P W Mengyan[4], O E Parfenov[1], A M Tokmachev[1], P Dosanjh[2] and S N Barilo[5]**

[1] Kurchatov Institute, Kurchatov Sq. 1, Moscow 123182, Russia
[2] Department of Physics and Astronomy, University of British Columbia, Vancouver, BC, Canada V6T 1Z1
[3] Bruker BioSpin AG, Industriestrasse 26, 8117 Fällanden, Switzerland
[4] Department of Physics, Texas Tech University, Lubbock, Texas 79409-1051, USA
[5] Institute of Solid State and Semiconductor Physics, National Academy of Sciences of Belarus, Minsk 220072, Belarus

E-mail: mussr@triumf.ca

**Abstract.** The origin of a pseudogap in the underdoped phase of cuprates has become one of the leading issues in understanding the mechanism(s) of high-$T_c$ superconductivity. Several experiments (i.e. polarized neutron diffraction studies) support theoretical models based on the circulating current (CC) picture. These CC models suggest a novel ordered phase formed by orbital magnetic moments caused by spontaneous orbital currents. However to date, $\mu$SR experiments in cuprate superconductors have not revealed any magnetic field higher than 0.2 G associated with orbital moments. Here we present high magnetic field $\mu$SR spectroscopy of a stoichiometric $La_2CuO_4$ crystal which shows antiferromagnetic splittings and additional 45 G line splittings, consistent with orbital currents.

## 1. Introduction

The origin of the pseudogap (PG) and its relation to the nature of high-$T_c$ superconductivity has been highly debated for over 20 years. In the underdoped regime of the phase diagram, cuprate superconductors show features of a correlated metal exhibiting a remarkable departure from the standard Fermi liquid behavior. Their transport, magnetic and thermodynamic properties suggest strong reduction of the electronic density of states (DOS) below a temperature $T^*$, though the DOS does not reach zero at the lowest temperature and the system remains metallic [1]. However, ARPES experiments [2] suggest opening of a $q$-dependent real gap in a one-particle excitation spectrum, supported by other spectroscopies [3].

Certain theoretical models consider the PG state as an ordered phase with a well defined order parameter and a related broken symmetry [4, 5, 6]. In this scenario, the PG temperature $T^*$ is a phase transition temperature to an ordered state formed by orbital magnetic moments caused by spontaneous orbital currents flowing in specific patterns within or around $CuO_2$ plaquettes. The fluctuations associated with the broken symmetry are considered to be responsible for both the superconductivity, playing role of a pairing glue, and the non-Fermi-liquid behavior below the $T^*$ phase transition line on the phase diagram.

The key experimental evidence of a novel ordered state comes from spin-polarized neutron scattering experiments which reported commensurate magnetic peaks below $T^*$ in YBCO and H1201 [7, 8], consistent with ARPES data, which point to time-reversal symmetry breaking (TRSB) that nevertheless preserves lattice translation invariance. Sizeable magnetic moments on the order of 0.1-0.2 $\mu_B$ are reported in the PG state [7, 8].

However, $\mu$SR experiments which have convincingly demonstrated their sensitivity to TRSB fields in a number of weak magnetic systems, produced no evidence of a magnetic order in the PG state. At first, spontaneous static magnetic fields were reported in YBCO, which were later reinterpreted as being due to spatial charge inhomogeneities [9]. No TRSB is reported in LSCO where $\mu$SR experiments set an upper limit of $\sim$ 0.2 G for magnetic field at the muon site while the expected local field is estimated to amount about 40 G [10].

This discrepancy between neutron and muon experiments has been attributed to screening of the charge density in the metallic-plane unit cells in the vicinity of the muon [11]. On the other hand, orbital currents and associated magnetic moments may be present in the limiting case of the underdoped regime — in the insulating stoichiometric $La_2CuO_4$, should CC be an intimate feature of chemical bonding in the $CuO_2$ plane.

Here we present $\mu$SR spectroscopy of stoichiometric $La_2CuO_4$ crystal (accompanied by SQUID measurements), in which the magnetic field at the muon should not be affected by charge density screening.

## 2. Results and discussion

High-quality $La_2CuO_{4+x}$ single crystal grown from CuO flux have been used for current studies. Oxygen over its stoichiometric content has been removed by annealing in vacuum. The lattice parameters of the crystal correspond to the low temperature orthorhombic stoichiometric $La_2CuO_4$ [12]. Néel temperature measured using SQUID amounts to 320 K.

Time-differential $\mu$SR experiments, using 100% spin-polarized positive muons implanted into these samples, were carried out on the M15 surface muon channel using the *HiTime* spectrometer. At low temperature, zero field $\mu$SR spectra consist of two (small and large amplitude) components, well known from numerous previous studies and indicative of two inequivalent muon sites in antiferromagnetic $La_2CuO_4$. Figure 1 shows temperature dependences of the corresponding frequencies. Above 150 K, the low-frequency component (signal with a small amplitude) disappears below the background. The Néel temperature $T_N \approx 320$ K and magnetic fields at the muon sites $B = 428.7$ G and $B = 111.8$ G determined from the temperature dependences shown in Figure 1 are consistent with numerous earlier measurements of stoichiometric $La_2CuO_4$.

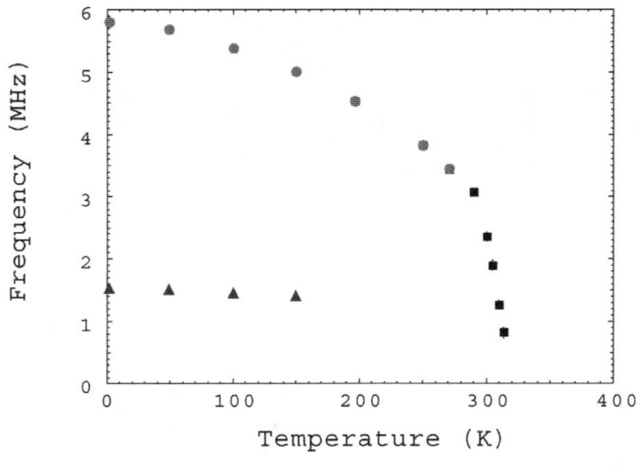


**Figure 1.** Temperature dependences of the muon spin precession frequencies in stoichiometric $La_2CuO_4$ in zero applied magnetic field. Red circles and blue triangles: frequencies obtained via a 2-component fit of $\mu$SR spectra at low temperature. Black squares: frequencies obtained via a 1-component fit of $\mu$SR spectra at higher temperature.

In high magnetic field $B$ transverse to the initial muon spin polarization and parallel to the $c$-axis of $La_2CuO_4$ crystal, muon spin precession spectra exhibit at least 7-component pattern. The characteristic Fourier transform is shown in Figure 2. Central line at about 135.6 MHz coincides with the single line detected in $CaCO_3$, a non-magnetic reference sample. We attribute this line to those muons (small portion of the muon flux) which missed the sample and stopped in a non-magnetic environment. Two small peaks positioned symmetrically around the central line correspond to the AFM splitting of the low-frequency (low-amplitude) signal observed in zero magnetic field. Our main interest is focused on large-amplitude signals again positioned symmetrically around the central line. We attribute these signals to high-frequency (large-amplitude) signal in Figure 1.

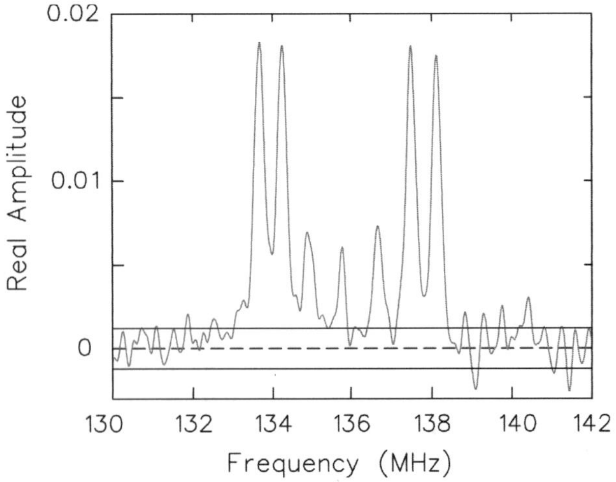


**Figure 2.** Frequency spectrum of muon spin precession in $La_2CuO_4$ in a transverse magnetic field of $H = 1$ T at $T = 100$ K. Central line frequency coincides with that of a single peak detected in the reference non-magnetic sample ($CaCO_3$).

In order to uncover the origin(s) of the magnetic field at the muon site, which may come from different sources, we must exactly determine the muon positions in the $La_2CuO_4$ lattice.

Stoichiometric $La_2CuO_4$ is a collinear antiferromagnetic material with four sublattices. Magnetic moment on Cu atoms is about 0.5 $\mu_B$ directed along the diagonal of the $CuO_2$ plaquette [13]. A peculiar feature of magnetic ordering in $La_2CuO_4$ is the presence of a weak ($\sim$0.002 $\mu_B$) ferromagnetic coupling of spins within $CuO_2$ layers [13, 14] which is believed to originate from Dzyaloshinskii-Moriya exchange interaction. These moments are orthogonal to $CuO_2$ layers and have opposite directions in the neighbouring planes. They are coupled to in-plane magnetic moments on Cu via tilting of $CuO_6$ octahedra. Therefore, $CuO_2$ is an antiferromagnet with hidden, weak ferromagnetism. A spin-flop transition to this hidden phase (weak ferromagnetism) takes place in high magnetic field applied along the $c$-axis [13, 14]. Within this phase, the direction of ferromagnetic moments in every $CuO_2$ plane becomes the same. Our SQUID measurements (Figure 3) at 250 K show that the spin-flop transition is observed at magnetic field of about 3.5-4 T. The magnetization jump $\sim$0.002 $\mu_B$ is consistent with literature data [13, 14].

The spin-flop transition discussed above is clearly seen in $\mu$SR measurements at 250 K (Figure 4), where half of the lines in the spectra vanish at a magnetic field of about 4 T. A remarkable consequence of the spin-flop transition is that the magnetic unit cell becomes equivalent to the crystallographic unit cell; the number of magnetic sublattices decrease from 4 to 2. From symmetry considerations this means that tilting of the $CuO_6$ octahedra forces the muon to occupy only half of the otherwise crystallographically equivalent positions. However, in order to go beyond general considerations, possible muon positions should be clearly established.

The muon stopping site(s) in $La_2CuO_4$ have been discussed in several papers [15]. Information

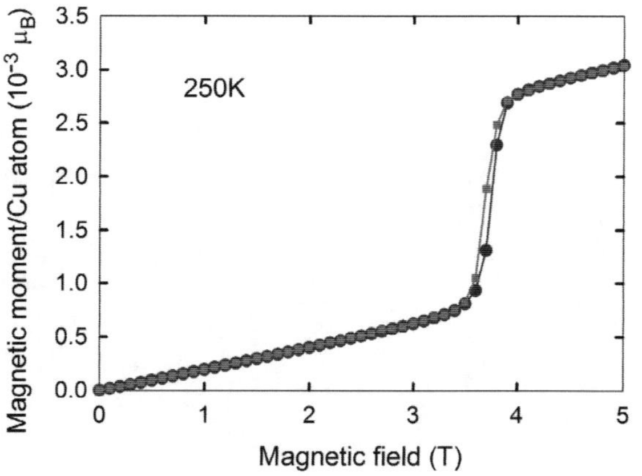


**Figure 3.** Magnetic field dependence of magnetic moment per Cu atom along the $c$-axis. Blue color: measurements in increasing magnetic field; red color: measurements in decreasing magnetic field.

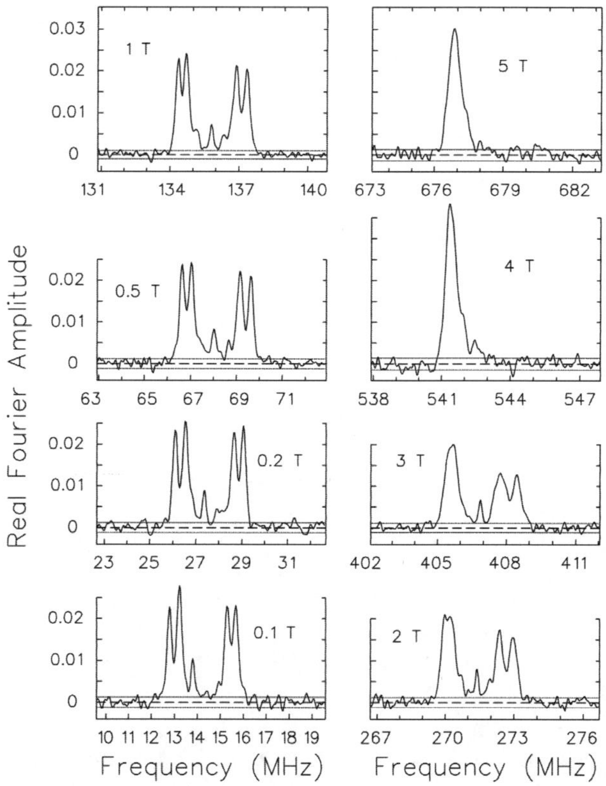


**Figure 4.** Fourier transforms of the muon spin precession signal in $La_2CuO_4$ at $T$ = 250 K in different transverse external magnetic fields.

obtained from zero-field $\mu$SR spectra is insufficient and additional data are typically sought from electronic structure calculations. It is common to predict the muon position to be in the vicinity of the apical oxygen atom but the exact positions are quite different in a number of publications leading to much confusion. In contrast, high-field $\mu$SR measurements with different setups provide data on projections of the magnetic field on the muon which correspond perfectly to the frequencies determined from the zero-field experiments.

To find muon positions based on this extended set of experimental data, we performed calculations based on the dipole-field approximation [16] assuming the periodic antiferromagnetic structure with the known magnetic moments on Cu atoms. There are only two types of inequivalent muon positions for the high-frequency high-amplitude signal in Figure 1: one is close to the apical oxygen and the other is very close to La atom. From chemical bonding considerations the first type is far more stable. Similar analysis is possible for the low-frequency signal and only one of the two positions is chemically substantiated. Figure 5 shows possible muon positions for both low- and high-frequency signals. The tilting of the $CuO_6$ octahedra produces two types of chemically inequivalent positions differing by the distance between the muon and the apical oxygen atom. Only one type of these positions is occupied by muons. Our calculations confirm that it is consistent with vanishing half of the spectral lines at the spin-flop transition (accompanied by 10% decrease of the line shift from the free muon signal).

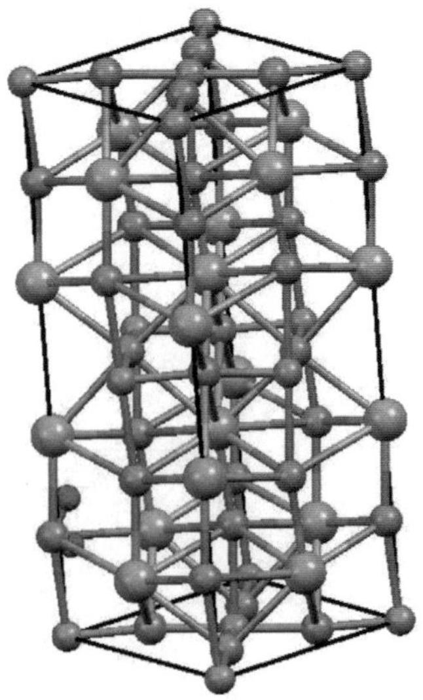

**Figure 5.** Unit cell for the low temperature orthorhombic phase of $La_2CuO_4$: green - Cu atoms, grey - La atoms, blue - O atoms, red - muon positions (high- and low-frequency signals come from the muon residing near Cu and La, respectively); atomic positions are shown for pristine $La_2CuO_4$, the perturbation caused by the muon is not taken into account.

One should expect 4 distinct signals in the high-field $\mu$SR spectra from two independent muon positions known from zero-field $\mu$SR experiments and the antiferromagnetic nature of the sample. Instead, at least 6 peaks have been observed in high magnetic fields of different directions. That means that apart from the AFM field, there is *an additional source of magnetic field on the muon* where some muons experience magnetic field larger than that coming from AFM while the other feel smaller magnetic field which causes the characteristic splitting of the large-amplitude signals (see Figure 4). The apparent absence of splittings in low-frequency signals can be explained by the increased distance from the $CuO_2$ planes leading to much smaller splittings which are not well resolved in the experiments.

There are two types of models explaining intra-unit-cell magnetic order in cuprates which may reveal the nature of the additional source of magnetic field on the muon. One relies upon spins being ascribed to oxygen atoms while the other is based on circulating orbital currents. Polarized neutron experiments have not been able to differentiate the existing models beyond establishing TRSB [7, 8]. It seems to be possible using high-field $\mu$SR experiments. While the strength of additional magnetic fields on the muon (which causes the observed splitting in the first place) is not known *a priori*, the ratio of signal splittings for different directions of the external magnetic field can be used to rule out certain models; although, a lot can be inferred from simple symmetry considerations.

The first class of models is based on the antiferromagnetic ordering of spins on in-plane oxygen atoms. When the spins are directed along the *b*- or *c*-axis of the crystal [17] the spectra are not described even qualitatively due to absence of splitting along the *c*-axis. The case of oxygen spins directed along the $a$-axis [7] is less obvious as the symmetry is correct but the ratio of the splittings along the $b$- and $c$-axes is about 1.6, while our experiments determine it to be close to 1.

Magnetic orders based on CC provide a happy alternative. The model dubbed $CC\text{-}\Theta_I$ and formed by 4 orbital current loops for each $CuO_2$ plaquette [4] is not consistent with polarized neutron experiments as well as with our $\mu$SR data complemented by symmetry considerations. The most widely used model is $CC\text{-}\Theta_{II}$ with 2 opposite current loops O-Cu-O per $CuO_2$ plaquette [5]. This model has correct symmetry properties and predicts the ratio of the splittings along the $b$- and $c$-axes to be about 1.1, which is close to the observed value. The orbital magnetic moment extracted from our data amounts about 0.04 $\mu_B$, which is comparable with the theoretical value on the order 0.1 $\mu_B$ [4]. An alternative staggered orbital current phase with the same symmetry as $CC\text{-}\Theta_{II}$ but Cu-Cu-Cu currents [18] is also unlikely because the ratio of the splittings along the $b$- and $c$-axes is too large.

In summary, high magnetic field $\mu$SR studies in stoichiometric $La_2CuO_4$ reveal additional source of magnetic field on the muon consistent with Varma model of circulating currents [5].

## Acknowledgments

This work was supported by the Kurchatov Institute, NSERC of Canada and the U.S. DOE, Basic Energy Sciences (Grant DE-SC0001769).

## References


[1] Norman M R and Pépin C 2003 *Rep. Prog. Phys.* **66** 1547
[2] Kaminski A *et al.* 2002 *Nature* **416** 610
[3] Timusk T and Statt B 1999 *Rep. Prog. Phys.* **62** 61
[4] Varma C M 1997 *Phys. Rev.* B **55** 14554
[5] Varma C M 2006 *Phys. Rev.* B **73** 155113
[6] Chakravarty S, Laughlin R B, Morr D K and Nayak C 2001 *Phys. Rev.* B **63** 094503
[7] Fauqué B, Sidis Y, Hinkov V, Palihès S, Lin C T, Chaud X and Bourges P 2006 *Phys. Rev. Lett.* **96** 197001
[8] Mook H A, Sidis Y, Fauqué B, Balédent V and Bourges P 2008 *Phys. Rev.* B **78** 020506
Li Y, Balédent V, Barišić N, Cho Y, Fauqué B, Sidis Y, Yu G, Zhao X, Bourges P and Greven M 2008 *Nature* **455** 372
[9] Sonier J E *et al.* 2001 *Science* **292** 1692
Sonier J E *et al.* 2002 *Phys. Rev.* B **66** 134501
[10] MacDougall G J, Aczel A A, Carlo J P, Ito T, Rodriguez J, Russo P L, Uemura Y J, Wakimoto S and Luke G M 2008 *Phys. Rev. Lett.* **101** 017001
[11] Shekhter A, Shu L, Aji V, MacLaughlin D E and Varma C M 2008 *Phys. Rev. Lett.* **101** 227004
[12] Parfenov O E, Nikonov A A and Barilo S N 2002 *JETP Letters* **76** 616
[13] Kastner M A, Birgeneau R J, Shirane G and Endoh Y 1998 *Rev. Mod. Phys.* **70** 897
[14] Thio T, Thurston T R, Preyer N W, Picone P J, Kastner M A, Jenssen H P, Gabbe D R, Chen C Y, Birgeneau R J and Aharony A 1988 *Phys. Rev.* B **38** 905
[15] Hitti B, Birrer P, Fisher K, Gygax F N, Lippelt E, Maletta H, Schenk A and Weber M 1990 *Hyperf. Int.* **63** 287
McMullen T, Jena P and Khanna S N 1991 *Int. J. Mod. Phys.* B **5** 1579
Saito R, Kamimura H and Nagamine K 1991 *Physica* C **185-189** 1217
Sulaiman S B, Srinivas S, Sahoo N, Hagelberg F, Das T P, Torikai E and Nagamine K 1994 *Phys. Rev.* B **49** 9879
Suter H U, Stoll E P and Meier P F 2003 *Physica* B **326** 329
Huang W, Pacradouni V, Kennett M P, Komiya S and Sonier J E 2012 *Phys. Rev.* B **85** 104527
Adiperdana B, Dharmawan I A, Siregar R E, Sulaiman S, Mohamed-Ibrahim M I and Watanabe I 2013 *AIP Conf. Proc.* **1554** 214
[16] Yaouanc A and Dalmas de Réotier P 2011 *Muon Spin Rotation, Relaxation and Resonance* (Oxford: Oxford University Press)
[17] Moskvin A S 2012 *JETP Letters* **96** 385
[18] Stanescu T D and Phillips P 2004 *Phys. Rev.* B **69** 245104

# Interplay of magnetism and superconductivity in $EuFe_2(As_{1-x}P_x)_2$ single crystals probed by muon spin rotation and $^{57}$Fe Mössbauer spectroscopy

**T Goltz[1], S Kamusella[1], H S Jeevan[2], P Gegenwart[2,4], H Luetkens[3], P Materne[1], J Spehling[1], R Sarkar[1] and H-H Klauss[1]**

[1] Technical University of Dresden, Institute of Solid State Physics, Germany
[2] I. Physik. Institut, Georg-August Universität Göttingen, Germany
[3] Laboratory for Muon Spin Spectroscopy, Paul Scherrer Institut, Villigen, Switzerland

E-mail: goltz@physik.tu-dresden.de

**Abstract.** We present our results of a local probe study on $EuFe_2(As_{1-x}P_x)_2$ single crystals with $x$=0.13, 0.19 and 0.28 by means of muon spin rotation and $^{57}$Fe Mössbauer spectroscopy. We focus our discussion on the sample with $x$=0.19 *viz.* at the optimal substitution level, where bulk superconductivity ($T_{SC}$ = 28 K) sets in above static europium order ($T^{Eu}$ = 20 K) but well below the onset of the iron antiferromagnetic (AFM) transition (~100 K). We find enhanced spin dynamics in the Fe sublattice closely above $T_{SC}$ and propose that these are related to enhanced Eu fluctuations due to the evident coupling of both sublattices observed in our experiments.

## 1. Introduction

The interplay of magnetism and superconductivity is one of the central topics in the contemporary studies on ferropnictides. Notably, the $AFe_2As_2$-based compounds ($A$=Ba [1], Sr [2] and Eu [3]) have been widely studied since reasonably good single crystals can be obtained for chemical substitution on all three sites. Of particular interest within this so-called '122-family' is the superconducting $EuFe_2(As_{1-x}P_x)_2$ system for two reasons. Firstly, the substitution of As by P is (nominally) isovalent thus superconductivity is not introduced by extra charge carriers and secondly, it contains a magnetic rare earth element on the $A$-site giving rise to magnetic order of the local $Eu^{2+}$ $4f$ electrons in addition to the *itinerant* antiferromagnetic iron order of the $3d$ conduction band electrons.

For $EuFe_2(As_{1-x}P_x)_2$, previous studies reported that the Fe AFM ordering and the accompanying structural transition from tetragonal to orthorhombic is suppressed upon P substitution and eventually vanishes *prior* to the appearance of a superconducting dome [4, 5, 6, 7]. In contrast, by measuring resistivity on a single crystal with $x$=0.13 under hydrostatical pressure, Tokiwa et al. [8] demonstrated the presence of a *precursory* structural and Fe AFM transition above $T_{SC}$ between $p = 0.4 - 0.8$ GPa (refering to $x_P = 0.15 - 0.20$). Likewise, a $\mu$SR pressure study on powdered samples by Guguchia et al. [9] evidenced static magnetic order above the onset of superconductivity for similar pressures but they also conclude that the SDW

[4] Present address: Exp. Physics VI, Center for Electronic Correlations and Magnetism, University of Augsburg, Germany

ground state is differently affected by $x$ and $p$. Only recently, Nandi et al. [10] showed the existence of a finite orthorhombic splitting reminiscent of weak Fe order [11] below 50 K in a superconducting ($T_{SC} = 25$ K) single crystal with $x_P = 0.15$ at ambient conditions.

Up to now, no comprehensive microscopic study of the ($T$-$x_P$) electronic phase diagram on $EuFe_2(As_{1-x}P_x)_2$ single crystals without any explicit symmetry-breaking forces (*viz.* at zero external field and ambient pressure) is available to the best of our knowledge. Some microscopic studies on polycrystalline material were done by Nowik et al. [7] and Guguchia et al. [9] but the reported transition temperatures from macroscopic measurements differ systematically comparing single crystalline [4, 5, 12] and polycrystalline [6, 7] samples. We only found one local probe study on a P-substituted $EuFe_2(As_{1-x}P_x)_2$ single crystal with $x$=0.3 ($T_{SC}^{mid}$=10.5 K) by Munevar et al. [13] but unfortunately they focussed on low temperatures and did not investigated temperatures above 50 K.

In view of this gap this work emphazises, that further microscopic studies of single crystalline $EuFe_2(As_{1-x}P_x)_2$ in the full temperature range are needed to get a better understanding of the precursory ($T > T^{Eu}$) Fe order and its possible importance for the appearance of superconductivity. Due to the lenght restriction of this article, we can only present the main results from the sample with $x$=0.19. A more detailed discussion of the obtained ($T$-$x_P$) phase diagram shown in Fig. 3 is in preparation [14].

## 2. Experimental Methods

Single crystals of $EuFe_2(As_{1-x}P_x)_2$ were grown by the FeAs self-flux method. The homogeneity and actual composition of the three samples with $x$=0.13, 0.19 and 0.28 was confirmed within $\Delta x = 0.01$ error by EDX microprobe analysis on several points of the sample. Thermodynamic properties were determined by resistivity, magnetization and specific heat measurements according to [4] indicating bulk superconductivity for the sample with $x$=0.19 whereas the samples with $x$=0.13 and 0.28 are non-superconducting. $^{57}$Fe Mössbauer spectroscopy (MS) was performed in a standard transmission geometry setup using a $^{57}$Co/Rh source with an experimental line width (HWHM) of $\omega = 0.135(5)$ mm/s. An Oxford LLD1 cryostat with a standard VTI was used to stabilize temperatures between 2 K and 300 K. Mössbauer spectra were evaluated by diagonalizing the full static hyperfine Hamiltonian including electric quadrupole and magnetic hyperfine interaction using the maximum entropy method (MEM) option to extract the iron hyperfine field distribution $\rho$(B) provided by the MössFit package [15]. Muon spin rotation measurements were performed using the GPS spectrometer at the $\pi$M3 beamline of the Swiss Muon Source at the Paul Scherrer Institut, Switzerland. The data was analyzed with the MUSRFIT package [16]. In all $\mu$SR experiments, the plate-like single crystals were mounted with the crystal $c$-axis parallel to the muon beam. The initial muon spin polarization was rotated by $-42°$ with respect to the beam and the $z$-direction of the laboratory framework as determined by transerse field (TF) experiments. In this arrangement, we can simultaneously measure the zero field (ZF) $\mu$SR time spectra for $\mu$ spin $\perp c$ and $\mu$ spin $\parallel c$ by evaluating the asymmetry signal from the up-down (UD) and forward-backward (FB) detector-pairs seperately.

## 3. Results

### *3.1. $x$=0.13 single crystal (batch no. HS03T2)*

Iron AFM ordering sets in below 115(5) K and the magnetically ordered volume fraction gradually increases at lower temperatures and saturates at ~65% for T=40 K as determined by weak TF-$\mu$SR. We define the AFM transition temperature for this sample at the midpoint of the very broad transition leading to $T_N^{Fe} = 71(3)$ K. Consistently, $^{57}$Fe MS spectra show (i) a significant line broadening between 140 and 80 K and (ii) a clear change in the hyperfine field distribution $\rho$(B) shifting about 75% of the spectral weight to discernible hyperfine field values ($\rho$(B) >1.25 T) below 40 K. This change in $\rho$(B) is reminiscent to the modification of

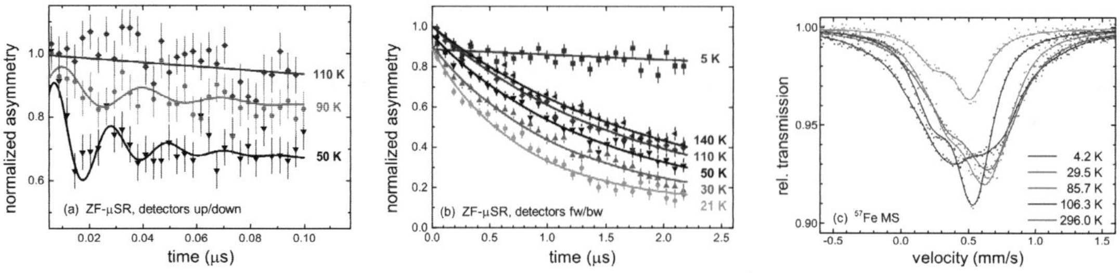


**Figure 1.** Selective ZF-$\mu$SR and $^{57}$Fe Mössbauer spectra of $EuFe_2(As_{1-x}P_x)_2$ with $x$=0.19.

the hyperfine field distribution W(B)[1] due to the alteration of the SDW shape from sinoidal to rectangular in $EuFe_2As_2$ between 191 and 145 K reported by Błachowski et al. [17]. Our MS data yield a mean hyperfine field value of $\langle B_{\rm hf}^{\rm Fe}\rangle = 4.1(3)$ T and a total spectral shift of IS=0.50(1) mm/s relative to metallic iron at $T$=4.2 K. Independent of temperature, the local field is canted by 15(5)° out of the ab-plane ($\theta^{\text{c-axis}} = 75(5)^\circ$).

### *3.2. x=0.19 single crystal (batch no. HS03E)*

ZF-$\mu$SR and $^{57}$Fe Mössbauer spectra are shown in Fig. 1. The results of our analysis are compiled in Fig. 2. The ZF-$\mu$SR spectra at high temperatures (T>150 K) show a very weakly damped exponential muon spin depolarization in both detector pairs. Below 100 K, a strongly damped, cosine shaped precession signal becomes visible in the UD detector pair ($\mu$ spin $\perp$ $c$) at short times marking the onset of static *iron* magnetic ordering. The amplitude becomes more pronounced upon further cooling down to 30 K showing a gradual increase of the magnetic volume fraction $V^{\rm osc}$, see Fig. 2(c). Analogue to [9] we used the muon spin depolarization function

$$P_{\rm UD}(t) = V_{\rm UD}^{\rm osc}\,\cos\left(2\pi f_\mu\cdot t+\phi\right)e^{-\lambda_{\rm UD}^{T}\cdot t} + V_{\rm UD}^{\rm relax}\,e^{-\lambda_{\rm UD}^{L}\cdot t} \quad (1)$$

to fit our data. The total initial asymmetry $P_{\rm UD}(t=0)$ is normalized to 1 at 150 K. In the FB detector pairs ($\mu$ spin $\|$ $c$), no oscillations can be observed for all temperatures. Instead, only a slow exponentially relaxing signal according to the second term in Eq. (1) with a small decrease in the initial value of the normalized asymmetry $P_{\rm FB}(t=0)$ is found. $P_{\rm FB}(t=0)$ was likewise normalized to 1 at 150 K and decreases continiously below 100 K to 0.9 at 21 K ($T > T^{\rm Eu}$), see Fig. 1(b). From this follows, that the local field at the muon site due to static Fe ordering points essentially parallel $c$ beeing suggestive of an ordered iron moment within the a/b plane. This finding is corroborated by our results from $^{57}$Fe Mössbauer spectroscopy which yield a tilting angle of $\theta^{\text{c-axis}} = 78(5)^\circ$ for the static iron hyperfine field with respect to the c-axis in the same temperature range as shown in Fig. 2(d).

Below 50 K, we observe additional damping of the oscillatory signal in the ZF-$\mu$SR spectra. The transverse relaxation rate $\lambda_{\rm UD}^{T}$ increases from 50 to 75 $\mu s^{-1}$ closely above $T_{\rm SC}$ and gets back to 50 $\mu s^{-1}$ below $T^{\rm Eu}$, see Fig. 2(b).

Below 20 K, additive Eu FM ordering is evidenced by a much faster (>100 MHz) precessing signal than for the iron order in the UD detector pair ($\mu$ spin $\perp$ $c$) and a nearly constant muon spin depolarization in the FB detector pair ($\mu$ spin $\|$ $c$). From the absolute asymmetry values, $V_{\rm UD}^{\rm osc}$ and $P_{\rm FB}$(t=0) in the ordered ($T$=1.6 K) and normal state ($T$>150 K), we calculated [14] the c-axis tilting angle of the local field at the muon site to be 9(2)° at $T$=1.6 K. The temperature dependence of $B_{\rm hf}^{\mu}$ is shown in Fig. 2(a) together with $\langle B_{\rm hf}^{\rm MS}\rangle$ from $^{57}$Fe Mössbauer spectroscopy.

[1] Note that when using the maximum entropy method, $\rho$(B) is deduced directly from the raw data whereas W(B) is constructed from the SDW shape, thus a priori includes an implicit premise about the underlying physics.

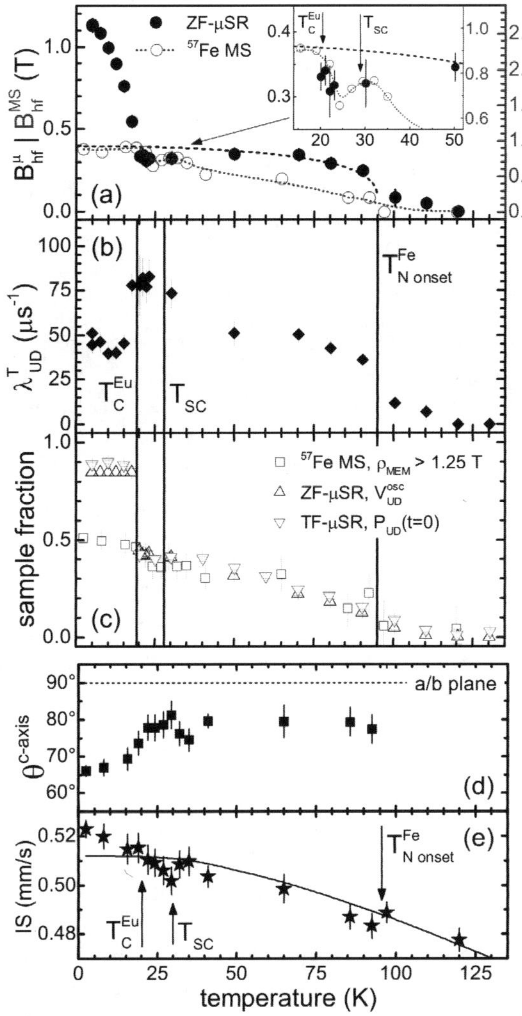


**Figure 2.**
Compilation of the main results for $EuFe_2(As_{1-x}P_x)_2$ with $x$=0.19:

(a) Local field at the muon site $B^{\mu}_{\rm hf}$ (black dots, left scale) and on the $^{57}$Fe nucleus $B^{\rm MS}_{\rm hf}$ (blue dots, right scale). The black dashed line is a fit to a phenomenological order parameter function [9] given by $B(T) = B_0 \cdot \left[1 - (T/T^{\rm Fe}_{\rm N\,onset})^{\alpha}\right]^{\beta}$ yielding $\alpha$=2, $\beta$=0.27(5) and $B_0$=0.4 T. The dotted blue line is a guide to the eye, emphazising the small but reproducible decrease of $B^{\rm MS}_{\rm hf}$ below the superconducting transition temperature.

(b) $\mu$SR transverse relaxation rate $\lambda^{T}_{\rm UD}$

(c) Sample fractions associated to the static Fe magnetic ordered volume fraction: ZF-$\mu$SR oscillatory signal $V^{\rm osc}_{\rm UD}$ (black triangles), total initial asymmetry $P_{\rm UD}$(t=0) from weak TF-$\mu$SR (red triangles) and integrated spectral weight with discernable hyperfine field values $\rho_{\rm MEM} > 1.25$ T from $^{57}$Fe MS (blue squares).

(d) Tilting angle $\theta^{\rm c\text{-}axis}$ of the $^{57}$Fe MS hyperfine field with respect to the c-axis.

(e) $^{57}$Fe MS total spectral shift (IS) relative to metallic iron. The line is a standart Debye-fit yielding $\theta_D$=260(110) K and $M_{\rm eff}$=68(11) u. Note, that a slight decrease of IS below the onset of Fe order is also reported for single crystalline $EuFe_2As_2$ [17].

### *3.3. x=0.28 single crystal (batch no. HS33/34/35)*

For $EuFe_2(As_{1-x}P_x)_2$ with $x$=0.28, we observe very weak iron magnetism below 120(5) K as deduced from significant line broadening of the $^{57}$Fe MS spectra using a pseudo-paramagnetic fit. More clearly, the ZF-$\mu$SR time spectra show weak Fe order seen by additional strong electronic relaxation ($\approx$50 $\mu s^{-1}$) at short times for $T$=100 K and below, which is not present at 150 K. The relative intensity for this fast relaxation signal can be well separated and accounts for the magnetic volume fraction. It gradually increases from $\sim$4% at 100 K close to 20% above the onset of static Eu ordering.

Below $T^{\rm Eu}$=20 K, ZF-$\mu$SR time spectra of $EuFe_2(As_{1-x}P_x)_2$ with $x$=0.28 are similar to those of the $x$=0.19 sample. The local field at the muon site lies essentially parallel to the c-axis and the Eu magnetism covers the complete sample volume. $^{57}$Fe MS spectra show an increasing transferred hyperfine field up to 1.2(1) T at 4.2 K which is accompagnied by a continious spatial reorientation of the (total) hyperfine field. The tilting of $\langle B^{\rm MS}_{\rm hf}\rangle$ with respect to the c-axis changes from $\theta^{\rm c\text{-}axis}$=60° at 19 K to 40° at $T$=4.2 K.

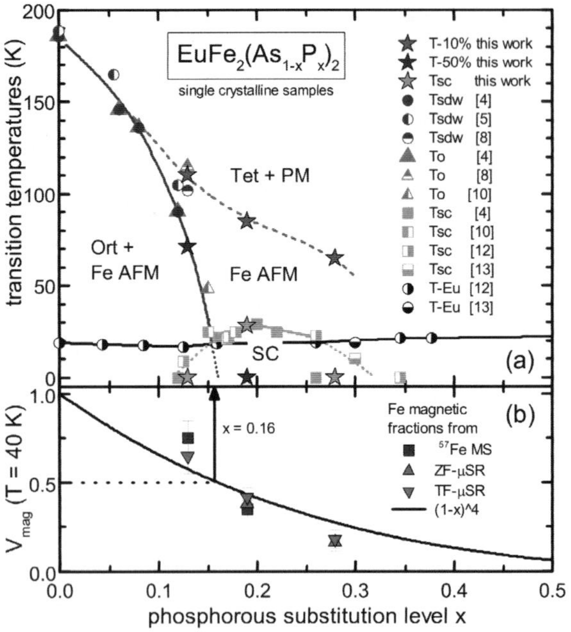


**Figure 3.**
(a) ($T$-$x_P$) phase diagram for single crystalline $EuFe_2(As_{1-x}P_x)_2$ under ambient conditions. Our finding of precursory Fe magnetism for $x$=0.19 and 0.28 is illustrated by $T_{10\%}$ (turquois). Structural (red triangles), Fe SDW (blue dots), static Eu (black dots) and superconducting (orange squares) phase transition temperatures from macroscopic methods are taken from the corresponding references. $T_{10\%}$ and $T_{50\%}$ are defined in the main article; Tet=tetragonal, Ort=orthorhombic.
(b) Fe AFM magnetic volume fraction at $T$=40 K ($T > T^{\mathrm{Eu}}$) in comparision to $P(x)=(1-x)^4$ (see main article).
Bulk superconductivity emerges close to $x$=0.16 where $V_{\mathrm{mag}}$ falls below the value of 50% according to $(1-x)^4$ and consequently, $T_{50\%}(=T_{\mathrm{SDW}})$ vanishes.

## 4. Discussion

### *4.1. Precursory Fe order ($T > T_{SC}$, $T > T^{Eu}$) in $EuFe_2(As_{1-x}P_x)_2$ with $x$=0.13, 0.19 and 0.28*

For $x$=0.13 and $x$=0.19, we find static iron magnetism below 115(5) and 95(5) K. The Fe hyperfine field $\langle B_{\mathrm{hf}}^{\mathrm{MS}} \rangle$ as well as the magnetic volume fraction ($V_{\mathrm{mag}}$) increases gradually upon cooling. At 40 K (fairly above $T^{\mathrm{Eu}}$), we find $V_{\mathrm{mag}}$=70% and 40% along with $B_{\mathrm{hf}}^{\mathrm{MS}}$=4.0 and 0.5 T, respectively. For $x$=0.28, $V_{\mathrm{mag}}$ does not exceed 20% above $T^{\mathrm{Eu}}$ and it remains unclear whether or not it is static on the $\mu$SR and Mössbauer timescale. In order to be sure not to overestimate the magnetic anomalies for the compilation of the ($T$-$x_P$) phase diagram (Fig. 3), we define firstly $T_{10\%}$ as the temperature, where at least 10% of our raw data signal displays magnetic behaviour. Secondly, since a thermodynamic phase transition temperature cannot be well-defined from the very gradual magnetic transition we find for our $EuFe_2(As_{1-x}P_x)_2$ samples, we relate $T_{\mathrm{SDW}}$ from macroscopic measurements to $T_{50\%}$, defined by the temperature for which $V_{\mathrm{mag}}$=50% (percolation threshold). In this context, we emphasize that for $x$=0.19 and $x$=0.28, $V_{\mathrm{mag}}$ is always smaller than 50%, so the observed weak iron magnetism might not be seen by resistivity or other macroscopic methods.

The obtained values for $V_{\mathrm{mag}}$ above $T^{\mathrm{Eu}}$ lie close to the probalistic expression $P(x)=(1-x)^4$ which accounts for the number of Fe atoms surrounded by four As atoms, see Fig. 3(b). This connection was pointed out by Nowik et al. [7] who interpreted the two parts in terms of commensurate ($P(x)$) and incommensurate ($1-P(x)$) magnetic components assuming $V_{\mathrm{mag}}$=100% below a given temperature. Nevertheless, considering that $\mu$SR is a much more sensitive probe for small ($< 0.1\,\mu_B$) magnetic moments, we can self-consistently substanciate our interpretation of the $^{57}$Fe MS spectra in terms of magnetic and non-magnetic sample fractions to our weak TF-$\mu$SR data. Furthermore, the cosine shaped ZF-$\mu$SR precession signal we observe for $x$=0.19 typically accounts for commensurate magnetic order. However, without contradiction to [7], we conclude that on an atomic lenght scale, *coherent* and presumably static magnetism is related to iron atoms which are surrounded by As atoms only.

*4.2. Superconductivity and magnetism in $EuFe_2(As_{1-x}P_x)_2$ with $x$=0.19 ($T<T_{SC}$, $T>T^{Eu}$)*
For $x$=0.19, we find a small but clearly discernible decrease of $\langle B_{\rm hf}^{\rm MS}\rangle$ and constant $V_{\rm mag}$ below the superconducting transition temperature $T_{\rm SC}$ as shown in Fig. 2(a) and (c). This strongly suggests that magnetism and superconductivity compete for the same electrons at least in parts of the sample volume thus pointing to coexistence of superconductivity and Fe magnetism. The observed decrease in the Mössbauer hyperfine field is accompagnied by an increase of the field distribution width in $\rho_{\rm MEM}$. Consistently, the ZF-$\mu$SR transverse relaxation rate $\lambda^T$ increases from 50 to 75 $\mu s^{-1}$ above $T_{\rm SC}$ (see Fig. 2(b)). Since $\lambda^T$ contains static and dynamic components, we conclude that close to $T_{\rm SC}$ enhanced spin dynamics weaken the static character of the Fe order. A change in the dynamics of the Fe magnetic moment closely above $T_{\rm SC}$ was reported by Munevar et al. for a superconducting $x$=0.3 single crystal [13]. However, for our $x$=0.19 sample we find that below $T^{\rm Eu}$, the value of $\lambda^T$ is steplike-wise restored to 50 $\mu s^{-1}$, indicating that the Fe sublattice magnetization is (re-)stabilized by the static Eu order. This leads us to the conclusion that the enhanced spin dynamics of the Fe sublattice below 40 K should be rather referred to enhanced Eu fluctuations. An interplay of both sublattices is clearly evidenced by the increase of the $^{57}$Fe MS total spectral shift below $T^{\rm Eu}$ in Fig. 2(e).

To summarize, we investigated the ($T$-$x_{\rm P}$) electronic phase diagram for single crystalline $EuFe_2(As_{1-x}P_x)_2$ by a local probe study of three samples with $x$=0.13, 0.19 and 0.28 under ambient conditions. Fe magnetism was found for all samples and the magnetic volume fractions are found to be related to iron atoms which are surrounded by As atoms only. For $x$=0.19, our data suggests that magnetism coexists with superconductivity.

## Acknowledgments

This work was financially supported by the German Research Foundation (DFG) within SPP 1458 (projects KL 1086/10-1 and GE 1640/4-2), GRK 1621 and grant no. SA 2426/1-1. Part of this work was performed at the Swiss Muon Source (Villigen, Switzerland). T.G. thanks Shibabrata Nandi, Shuai Jiang and Sina Zapf for valuable discussions.

## References

[1] Rotter M, Tegel M and Johrendt D 2008 *Phys. Rev. Lett.* **101** 107006
[2] Leithe-Jasper A, Schnelle W, Geibel C and Rosner H 2008 *Phys. Rev. Lett.* **101** 207004
[3] Tegel M, Rotter M, Weiss V, Schappacher F M, Pöttgen R and Johrendt D *Journal of Physics: Condensed Matter* **20** 452201
[4] Jeevan H S, Kasinathan D, Rosner H and Gegenwart P 2011 *Phys. Rev.* B **83** 054511
[5] Zapf S, Wu D, Bogani L, Jeevan H S, Gegenwart P and Dressel M 2011 *Phys. Rev.* B **84** 140503
[6] Cao G, Xu S, Ren Z, Jiang S, Feng C and Xu Z A *Journal of Physics: Condensed Matter* **23** 464204
[7] Nowik I, Felner I, Ren Z, Cao G H and Xu Z A *Journal of Physics: Condensed Matter* **23** 065701
[8] Tokiwa Y, Hübner S H, Beck O, Jeevan H S and Gegenwart P 2012 *Phys. Rev.* B **86** 220505
[9] Guguchia Z, Shengelaya A, Maisuradze A, Howald L, Bukowski Z, Chikovani M, Luetkens H, Katrych S, Karpinski J and Keller H 2013 *Journal of Superconductivity and Novel Magnetism* **26** 285–295
[10] Nandi S, Jin W T, Xiao Y, Su Y, Price S, Shukla D K, Strempfer J, Jeevan H S, Gegenwart P and Brückel T 2014 *Phys. Rev.* B **89** 014512
[11] Goltz T, Zinth V, Johrendt D, Rosner H, Pascua G, Luetkens H, Materne P and Klauss H H 2014 *Phys. Rev.* B **89** 144511
[12] Zapf S, Jeevan H S, Ivek T, Pfister F, Klingert F, Jiang S, Wu D, Gegenwart P, Kremer R K and Dressel M 2013 *Phys. Rev. Lett.* **110** 237002
[13] Munevar J, Micklitz H, Alzamora M, Argello C, Goko T, Ning F, Munsie T, Williams T, Aczel A, Luke G, Chen G, Yu W, Uemura Y and Baggio-Saitovitch E 2014 *Solid State Communications* **187** 18–22
[14] Goltz T Ph.D. Thesis, TU Dresden (in preparation)
[15] Kamusella S MössFit: A Free Framework for $^{57}$Fe Mössbauer Data Analysis (private communication)
[16] Suter A and Wojek B 2012 *Physics Procedia* **30** 69–73
[17] Błachowski A, Ruebenbauer K, Żukrowski J, Rogacki K, Bukowski Z and Karpinski J 2011 *Phys. Rev.* B **83** 134410

# CDW order and unconventional s-wave superconductivity in $Ba_{1-x}Na_xTi_2Sb_2O$

**Sirko Kamusella[1], Phuong Doan[2], Til Goltz[1], Hubertus Luetkens[3], Rajib Sarkar[1], Arnold Guloy[2] and Hans-Henning Klauss[1]**

[1] Institut für Festkörperphysik, Technische Universität Dresden, 01062 Dresden, Germany
[2] TCSUH and Department of Chemistry, University of Houston, Texas, 772034-5003, USA
[3] Laboratory for Muon Spin Spectroscopy, Paul Scherrer Institut, 5232 Villigen, Switzerland

E-mail: `sirko.kamusella@tu-dresden.de`

**Abstract.** Due to its anticuprate $Ti_2O$ layer and its fascinating phase diagram with a large coexistence area of superconductivity and a density wave phase, the new class of titanium based superconductors attracts great scientific interest. In this paper we report μSR investigation on powder samples of $Ba_{1-x}Na_xTi_2Sb_2O$ ($x = 0, 0.15, 0.25$). Our results exhibit both the presence of a charge density wave and superconductivity in $Ba_{1-x}Na_xTi_2Sb_2O$. The superconducting order parameter, extracted from a vortex state analysis using the numeric Ginzburg-Landau model, is compatible with a s-wave symmetry. In the universal Uemura classification of superconductors this compound is at the verge of unconventional superconductivity.

## 1. Introduction

The observation of superconductivity with critical temperatures ($T_c$) up to 5.5 K in the $Ba_{1-x}Na_xTi_2Sb_2O$ [1] raised much interest in the compounds of the $CeCr_2Si_2C$-type, which both from the structural and magnetic point of view, may be considered as an unusual superconductor.

The nominal valence state $3d^1$ of Ti is comparable to $3d^9$ state of Cu in the cuprates [2, 3]. Compared to the cuprates the oxygen spacing is strongly increased and the parent compound is a metal. In contrast to the planar coordination of Cu in the cuprates, the Ti ions exhibit an extra 90° Ti-Sb-Ti coupling. This effects an octahedral coordination of the Ti atoms between four Sb and two oxygen atoms. In the oxochalcogenides the impact of this structural feature on the order of localized transition metal moments is controversially discussed [4]. For the $Ba_{1-x}Na_xTi_2Sb_2O$ instead the Fermi surface consists of cylinders [5] enabling density wave (DW) instabilities which were uncovered by ARPES [6] in the related $BaTi_2As_2O$. In $BaTi_2Sb_2O$ such a phase transition was seen by $\chi(T)$ and $\rho(T)$ measurements [1, 2]. NMR investigations [3] assigned it to a charge density wave (CDW) order, although a commensurate spin density wave (SDW) scenario could not be ruled out. The interstitial muon can conclusively discriminated both phenomena.

Numerous approaches have been taken, such as charge doping, structural modification and chemical pressure (table 1), to suppress the DW phase and thus increase the $T_C = 1.2$ K [2] of the parent compound. However, an isovalent Sb/Bi substitution, as reported by Yajima *et al.* [7], deserves particular attention. It results in a two-dome structure in $T_c$ with two distinct superconducting phases revealing a second mechanism of suppression of superconductivity.

In $Ba_{1-x}Na_xTi_2Sb_2O$ the sodium substitution leads to effective hole doping within the $Ti_2Sb_2O$-layers. The parent compounds from both ends of the phase diagram, $BaTi_2Sb_2O$ and

**Table 1.** Substitution schemes and structural modification of the $BaTi_2Sb_2O$. Chemical pressure ("p") as well as hole doping ("h") and interlayers ("IL") can increase $T_C$. If no $T_{DW}$ is stated, the density wave state is no longer traceable. If additionally marked with a star *, $T_{DW}$ went to 0 K (convex) with increasing substitution level, otherwise the signal disappeared at finite temperatures before reaching optimal substitution (concave).

| original | none | Sb | Sb | Sb | Ba | Ba | Ba | Ba | Ba | Ba |
|---|---|---|---|---|---|---|---|---|---|---|
| substituent | none | As | Bi | Sn | Na | Rb | K | $(FeAs)_2$ | $(SrF)_2$ | $(SmO)_2$ |
| | [1,7] | [11] | [2] | [12] | [1] | [13] | [14] | [15] | [16] | [16] |
| mechanism | - | p | p | h | h | h | h | IL | IL | IL |
| $T_{DW}$ [K] | 54 | 200 | -* | 30 | 30 | -* | - | 125 | 198 | 230 |
| $T_{C,max}$ [K] | 1.2 | - | 4.6 | 2.5 | 5.5 | 5.4 | 6.1 | 21.5 | - | - |

$Na_2Ti_2Sb_2O$, show DW transition temperatures of 50 K [7] and 125 K [8,9], respectively. The sodium compound differers firstly in the nominal interlayer atom count and secondly displays staggered $Ti_2Sb_2O$-layers and thus reduced c-axis spacing. However X-ray diffraction pattern [1] place our samples ($x \leq 0.25$) close to the Ba parent compound.

By means of TF-μSR we studied the field and temperature dependence of the flux line lattice to extract superconducting correlation lengths, upper critical fields and the symmetry of the superconducting wave function. For all three Na levels we rule out static magnetic order from our ZF-μSR measurements, and thus a SDW state. This confirms the CDW scenario [7,10].

## 2. Experimental

Phase pure $Ba_{1-x}Na_xTi_2Sb_2O$ polycrystalline specimens were synthesised as described previously [1]. Air sensitive powder samples were filled into kapton sealed bags of sticky tape and measured by means of μSR at the GPS (x = 0, 0.15, beamline $\pi$M3.2) and Dolly (x = 0.25, beamline $\pi$E1) instruments at the Paul Scherrer Institut (PSI) Villigen, Switzerland. Due to small sample mass of $\approx$ 20 mg the spectrometers were run in veto mode. The measurements comprised zero field (ZF) temperature scans in the DW temperature range and transverse field (TF) measurements (except for x = 0) up to 0.52 T. The data were analysed using the free software *Musrfit* [17].

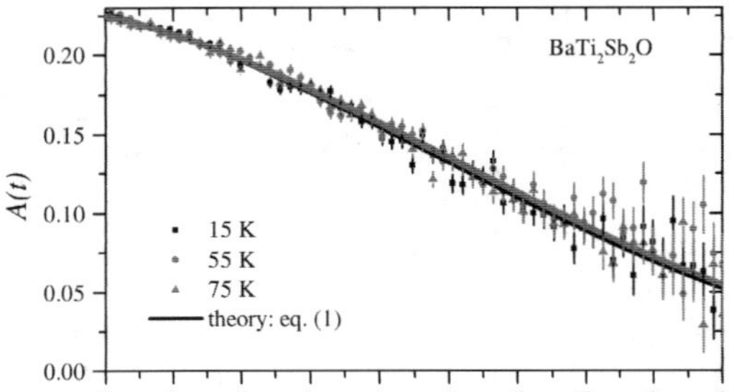


**Figure 1.** Zero field (ZF) spectra of all substitution levels show no indications of static magnetic order caused by a SDW.

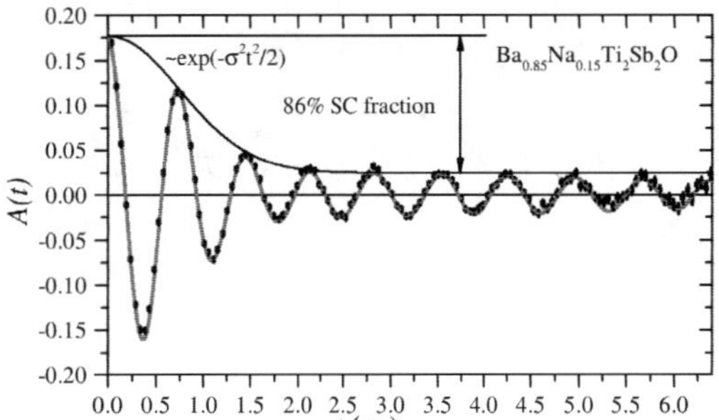


**Figure 2.** Transverse field (TF) (0.01 T) spectrum at 1.6 K to determine superconducting volume fraction and damping rate $\sigma_{SC}$.

## 3. Results and Discussion

Susceptibility measurements indicated superconducting $T_c$s of 5.5 K and 4.8 K for the x = 0.15 and x = 0.25 samples. For x = 0 a $T_c < 1.5$ K falls below experimental limits [18].

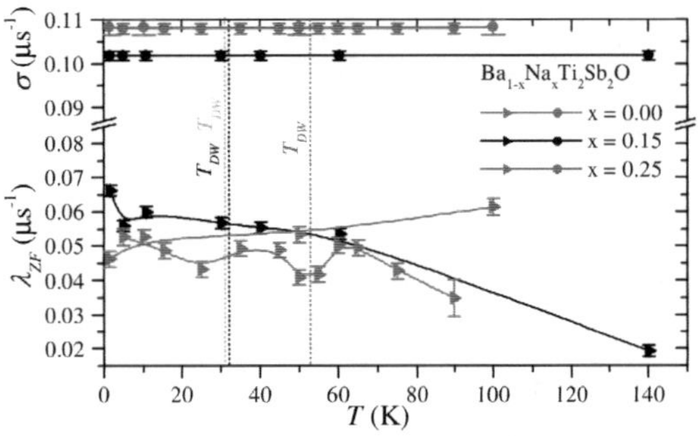


**Figure 3.** The zero field (ZF) μSR time spectra were described by equation (1). As $\sigma$ and $\lambda_{ZF}$ are highly correlated in this model, $\sigma$ was fitted globally over temperature. This nearly substitution independent damping rate is due to randomly distributed nuclear moments. In the exponential relaxation there is neither a significant feature indicating the $T_{DW}$, nor does $\lambda$ above and below $T_{DW}$ differ. The latter contradicts the SDW scenario.

ZF μSR measurements (Fig. 1) were performed to study the density wave transition. The asymmetry $A(t)$ shows the relaxation of the muon spin polarization due to randomly distributed nuclear moments ($\sigma$) and a exponential relaxation ($\lambda_{ZF}$). It was modelled by the following function:

$$\frac{A(t)}{A_0} = \left(\frac{1}{3} + \frac{2}{3}(1 - \sigma^2 t^2) e^{-\frac{\sigma^2 t^2}{2}}\right) e^{-\lambda_{ZF} t} \tag{1}$$

Temperature dependencies of $\sigma$ and $\lambda_{ZF}$ are presented and discussed in Fig. 3. The data is consistent with the CDW scenario, but not with the SDW scenario.

To estimate the superconducting volume fraction $f_{sc}$ and to determine the superconducting order parameters, TF μSR experiments were carried out on the samples x = 0.15 and x = 0.25 (see Fig. 2). Type II superconductors exhibit a characteristic flux line lattice when exposed to an external magnetic field $B_{ex}$. At temperatures above $T_c$, the muon spins polarized perpendicular to the external magnetic field precesses with the frequency $\frac{\gamma}{2\pi} B_{ex} = 135.5 \frac{\text{MHz}}{\text{T}} B_{ex}$. As the field is compressed into flux lines, there will be a unique field distribution $\rho(B_{ex})$ depending on the London penetration depth $\lambda$, the superconducting coherence length $\xi$ and the applied magnetic field $B_{ex}$. Often it is sufficient to treat this distribution in a powder sample as a Gaussian distribution [19] with the second moment $\sigma_B$, which is associated with a damping of the precession signal with the rate $\sigma_{SC} = \gamma\sigma_B$. This most probably is the case for $Ba_{1-x}Na_xTi_2Sb_2O$: a Fourier transform of the TF μSR time spectra reveals a paramagnetic line with precession frequency $\omega_0 = \gamma B_{ex}$ and a broadened line corresponding to $f_{sc}$ of the sample.

The μSR spectra were fitted using the following theory for the asymmetry:

$$\frac{A(t)}{A_0} = f_{sc} \cos(\gamma(B_{ex} + \Delta B)t + \varphi) e^{-(\sigma_{sc}^2 + \sigma_{nc}^2) t^2/2} + (1 - f_{sc}) \cos(\gamma B_{ex} t + \varphi) e^{-\sigma_{nc}^2 t^2/2} \tag{2}$$

The damping rate $\sigma_{nc}$ of the normal state was determined for all $B_{ex}$ at 10 K. From this theory the superconducting volume fraction $f_{sc}$ can be deduced (Fig. 4).

Figure 4 indicates that $f_{sc}$ of the x = 0.15 sample increases more sharply compared to the one of the x = 0.25 sample, where a more gradual enhancement is observed. The field and temperature dependence of $\sigma_{sc}(B,T)$ were further analysed using the numerical Ginzburg-Landau (NGL) model introduced by Brandt [20] (see Fig. 5). Although the often applied analytical approximation for the triangular flux line lattice

$$\sigma_{sc}^{\approx} = \gamma\sigma_B^{\approx} = \gamma \frac{0.172\Phi_0}{2\pi} \left(1 - \frac{B}{B_{c2}}\right) \left(1 + 1.21 \left(1 - \sqrt{\frac{B}{B_{c2}}}\right)^3\right) \frac{1}{\lambda^2} \tag{3}$$

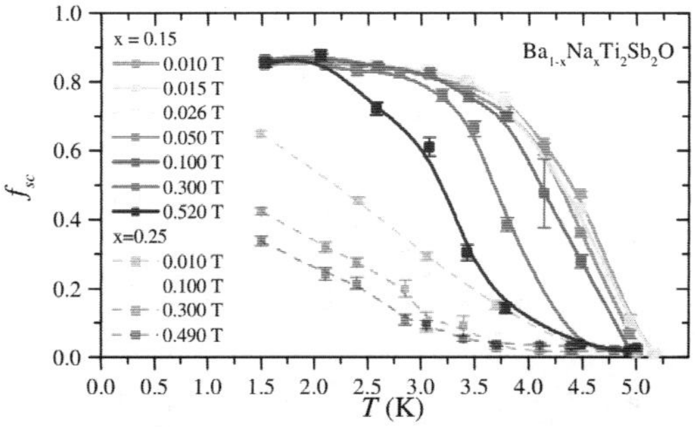


**Figure 4.** Superconducting volume fraction $f_{sc}$ extracted from TF data show a less pronounced transition of x = 0.25 compared to the x = 0.15 sample.

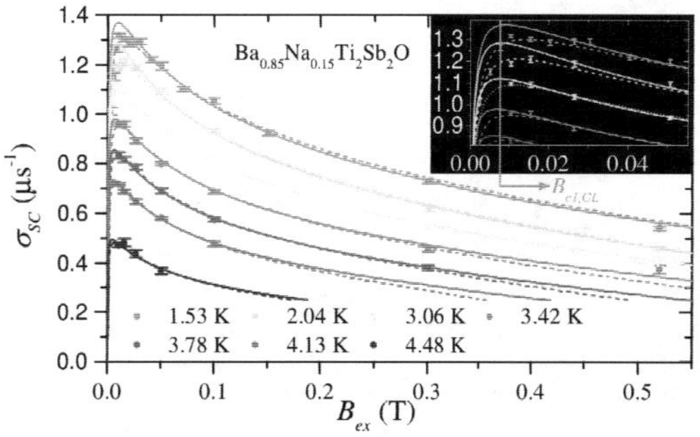


**Figure 5.** The complete $\sigma_{SC}(B_{ex}, T)$ is well described within the NGL theory [20]. The quadratic flux line lattice (dashed) is to be favoured over the triangular (solid).

gives a reasonable fit to the data for $B \geq 0.015\,\mathrm{T}$, we rather followed Brandt's more exact three step scheme (for details see [20]). In this numerical model the flux line lattice $B(\vec{R})$ is described in a cosine series and the Fourier coefficients $b_{\vec{K}}$ are modified to fulfil Ginzburg-Landau theory.

$$B(\vec{R}) = B_{ex} + \sum_{\vec{K} \neq \vec{0}} b_{\vec{K}} \cos(\vec{K}\vec{R}) \tag{4}$$

From the coefficients $b_{\vec{K}}$ one calculates the second moment $\sigma_B$ of the distribution of magnetic flux $\rho(B)$, which is proportional to the superconducting damping rate $\sigma_{SC}$ measured in the µSR TF experiment.

$$\sigma_{sc} = \gamma\sigma_B = \gamma\sqrt{\left\langle (B(\vec{R}) - B_{ex})^2 \right\rangle_{\vec{R}}} = \gamma\sqrt{\frac{1}{2}\sum_{\vec{K}} b_{\vec{K}}^2} \tag{5}$$

A fitting routine based on Gnu Scientific Libary (GSL) [21] was written to extract the effective penetration depth $\lambda_{eff}(T)$ and the upper critical field $B_{c2}(T)$ from the field dependence of $\sigma_{SC}$.

In Ginzburg Landau (GL) theory the upper critical field $B_{c2}$ and the superconducting coherence length $\xi$ are related by $\xi^2 = \Phi_0/2\pi B_{c2}$ leading to a coherence length of less than 20 nm. With the lower critical field $B_{c1,GL} = \mu_0\Phi_0/4\pi\lambda_{eff}^2 \cdot \ln(\kappa + const.)$ calculated from the already deduced data, it can be shown that only two data points lie outside the model range $B_{c1} < B_{ex} < B_{c2}$ (Fig. 5, inset). All superconducting properties are summarized in table 2.

To describe the superfluid density, which is proportional to $\lambda_{eff}^{-2}$, we adopted the isotropic s-wave model [22] from von Rohr [10] and Nozaki [23], who measured at $He^3$-temperatures. The quality of the fit does not depend on the flux line lattice symmetry which was used to extract $\sigma_{SC}$. Superconducting gaps $\Delta_0$ of 0.77(1) meV and 0.8(1) meV for x = 0.15 and x = 0.25 are obtained from the model. These values are in accordance with heat capacity measurements [18].

Although µSR is usually not the method of choice to determine upper critical fields, the NGL automatically produces these values. The temperature dependence of $B_{C2}$ is sensitive to the choice of flux line lattice symmetry (see Fig. 7). Only the quadratic flux line lattice data points are well described by a WHH model and in line with susceptibility data $B_{C2,\chi}(T)$ [18]. The WHH model yields Maki parameters of 0.6(8) and 1(1) for x = 0.15 and x = 0.25 respectively.

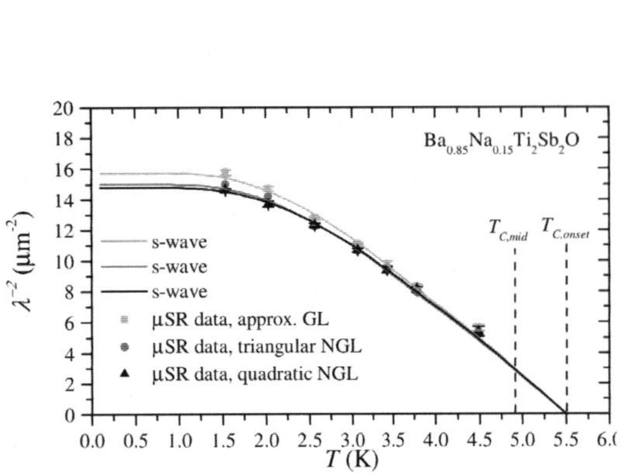


**Figure 6.** A s-wave model is sufficient to describe the temperature dependence of the superfluid density.

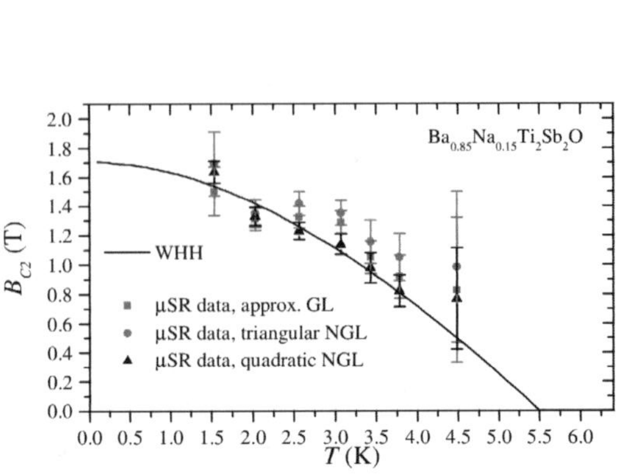


**Figure 7.** The quadratic NGL model yields most reasonable results for the temperature dependence of $B_{C2}$ compared to $B_{C2,\chi}$ [18].

**Table 2.** Summary of superconducting properties (index "0" $\equiv T$=0-value). $\Delta_0$ and $\lambda_0$ are extracted from the s-wave-fits of $\lambda_{eff}^{-2}$, $\kappa$ is averaged over all temperature values in the NGL model, $B_{c20}$ is extrapolated from the WHH model [calculated from $\lambda_0$ and $\kappa$ according to GL theory], $\xi_0$ and $B_{c10}$ are calculated from $B_{c20}$ and $\lambda_0$ and $\kappa$ respectively, following the GL theory.

| Na content | $B_{c20}$ [T] | $\Delta_0$ [meV] | $\lambda_0$ [nm] | $\xi_0$ [nm] | $\kappa$ | $B_{c10}$ [T] |
|---|---|---|---|---|---|---|
| 15% | 1.7(1) [1.60(0.16)] | 0.77(1) | 258(6) | 14.3(7) | 18.2(8) | 0.007(3) |
| 25% | 2.6(3) [2.40(0.87)] | 0.78(15) | 323(17) | 12(2) | 28(3) | 0.005(3) |

This means a Pauli limitation of this superconductor. Although the $\sigma_{SC}(B_{ex}, T)$ data (Fig. 5) is better matched by the quadratic flux line lattice, we stress that our μSR data alone are not able to discriminate between the different flux line lattice symmetries. In order to validate such a statement, single crystal measurements of the probably anisotropic $\sigma_{SC}$ are essential.

The universal weak coupling BCS behaviour of the gap over critical temperature ratio $\Delta/k_BT_c = 1.764$ is matched by the 25% Na sample (1.9(4)). The more precise value of 1.66(5) for 15% Na lies rather below the BCS prediction, but is still closer to the BCS value than to the value of 1.45 which was extracted by von Rohr for the x=0.15 substitution [10].

With the facts accumulated so far the $Ba_{1-x}Na_xTi_2Sb_2O$ indeed seems to be an ordinary weak coupled BCS superconductor.

Nevertheless we like to draw attention to Uemura's universal classification of unconventional superconductors sharing a common condensation mechanism and/or thermodynamic description different from the conventional superconductors [24]. This difference manifests itself in a relatively low Fermi energy. Following Uemura's formula for the Fermi temperature for two dimensional systems $T_F = \frac{\hbar^2\pi}{k_B}\frac{c}{\mu_0 e^2}\frac{1+\xi/l}{\lambda^2}$ with the mean free path $l \gg \xi$ and the lattice parameter c, $Ba_{1-x}Na_xTi_2Sb_2O$ emerges at the verge of unconventional superconductivity (Fig. 8).

## 4. Conclusion

ZF-μSR studies of $Ba_{1-x}Na_xTi_2Sb_2O$ support the commensurate CDW scenario [3] without static magnetic order in that compound. Our TF-μSR data were analysed using the NGL model [20]. All results are consistent with the ones obtained by an independent μSR study [10]. Following the Uemura scheme $Ba_{1-x}Na_xTi_2Sb_2O$ was found in the vicinity of unconventional superconductors.

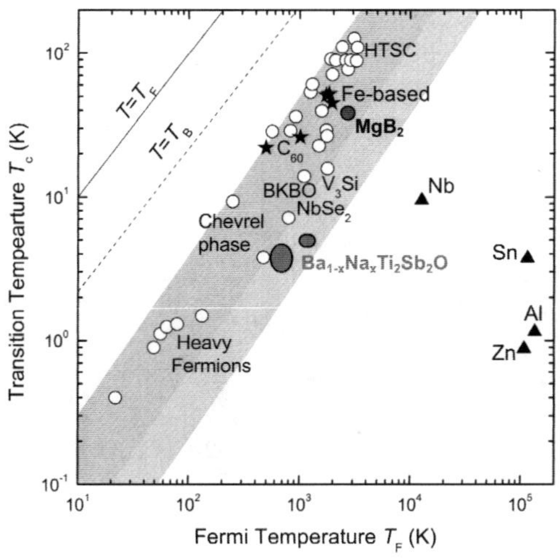


**Figure 8.** Plot adapted from [25]. Although the universality of the Uemura classification was criticized recently [26] we emphasize its significance to group compounds owing to a common, unusual condensation mechanism and an unconventional thermodynamic description of the superconducting state. The emergence of the strong electron-phonon coupled $MgB_2$ in this plot illustrates, that phonon mediated pairing is no criteria of exclusion within this classification. The vicinity of $Ba_{1-x}Na_xTi_2Sb_2O$ to unconventional superconductors is rather attributed to the multiband electron [5] and layered crystal structure (as in $MgB_2$), to the presumably important 3d-electron correlations and to the competing CDW order (the latter two similar in $NbSb_2$).


## Acknowledgments

We appreciate the financial support by the Deutsche Forschungsgemeinschaft (DFG) through the priority program SPP 1458 (KL 1086/10-1), the Research Training Group GRK 1621 and the grant number SA 2426/1-1. Support from TCSUH and the R. A. Welch Foundation (E-1297) is also gratefully acknowledged. The experiment was performed at the Swiss Muon Source (Villigen, Switzerland).


## References


[1] Doan P, Gooch M *et al.* 2012 *J. Am. Chem. Soc.* **134** 16520–16523
[2] Yajima T, Nakano K, Takeiri F *et al.* 2013 *J. Phys. Soc. Jpn.* **82** 013703
[3] Kitagawa S, Ishida K, Nakano K, Yajima T and Kageyama H 2013 *Phys. Rev.* B **87** 060510
[4] Fuwa Y, Wakeshima M and Hinatsu Y 2010 *J. Phys.: Condens. Matter* **22** 346003
[5] Singh D J 2012 *New J. Phys.* **14** 123003
[6] Xu H C, Xu M, Peng R, Zhang Y, Ge Q Q, Qin F, Xia M, Ying J J *et al.* 2014 *Phys. Rev.* B **89** 155108
[7] Yajima T, Nakano K, Takeiri F, Nozaki Y, Kobayashi Y and Kageyama H 2013 *J. Phys. Soc. Jpn.* **82** 033705
[8] Adam A and Schuster H U 1990 *Z. Anorg. Allg. Chem.* **584** 150–158
[9] Liu R H, Tan D, Song Y A, Li Q J, Yan Y J, Ying J J *et al.* 2009 *Phys. Rev.* B **80** 144516
[10] von Rohr F, Schilling A, Nesper R, Baines C and Bendele M 2013 *Phys. Rev.* B **88** 140501
[11] Wang X, Yan Y, Ying J, Li Q, Zhang M, Xu N and Chen X 2010 *J. Phys.: Condens. Matter* **22** 075702
[12] Nakano K, Yajima T, Takeiri F, A Green M, Hester J, Kobayashi Y and Kageyama H 2013 *J. Phys. Soc. Jpn.* **82** 074707
[13] von Rohr F, Nesper R and Schilling A 2014 *Phys. Rev.* B **89** 094505
[14] Pachmayr U and Johrendt D 2013 *Solid State Sci.* **28** 21–34
[15] Sun Y L, Ablimit A, Bao J K, Jiang H, Zhou J and Cao G H 2013 *Sci. Tech. Adv. Mater.* **14** 055008
[16] Liu R, Song Y, Li Q, Ying J, Yan Y, He Y and Chen X 2010 *Chem. Mater.* **22** 1503–1508
[17] Suter A and Wojek B 2012 *Physics Procedia* **30** 69–73
[18] Gooch M, Doan P, Tang Z, Lorenz B, Guloy A M and Chu P C 2013 *Phys. Rev.* B **88** 064510
[19] Maeter H 2012 *Muon spin relaxation and rotation of iron-based superconductors* Ph.D. thesis TU Dresden
[20] Brandt E H 2003 *Phys. Rev.* B **68** 054506
[21] Galassi M *et al.* 2011 *Gnu Scientific Library: Reference Manual, Edition 1.15, for GSL Version 1.15*
[22] Prozorov R 2008 *Supercond. Sci. Technol.* **21** 082003
[23] Nozaki Y, Nakano K, Yajima T, Kageyama H, Frandsen B, Liu L *et al.* 2013 *Phys. Rev.* B **88** 214506
[24] Uemura Y, Le L, Luke G, Sternlieb B, Wu W, Brewer J *et al.* 1991 *Phys. Rev. Lett.* **66** 2665–2668
[25] Khasanov R, Luetkens H, Amato A, Klauss H H *et al.* 2008 *Phys. Rev.* B **78** 092506
[26] Dordevic S, Basov D and Homes C 2013 *Scientific Reports* **3** 1713

# Field dependence of the superconducting gap in $YPd_2Sn$: A $\mu$SR and NMR study

**E Morenzoni[1], H Saadaoui[1], A Amato[1], C Baines[1], H Luetkens[1], E Pomjakushina[2], M Pikulski[1,3] and T Shiroka[1,3]**

[1] Laboratory for Muon Spin Spectroscopy, Paul Scherrer Institute, CH-5232 Villigen PSI, Switzerland

[2] Laboratory for Developments and Methods, Paul Scherrer Institut, CH-5232 Villigen PSI, Switzerland

[3] Laboratorium für Festkörperphysik, ETH-Hönggerberg, CH-8093 Zürich, Switzerland

E-mail: `elvezio.morenzoni@psi.ch`

**Abstract.** We have performed muon spin rotation/relaxation and $^{119}$Sn nuclear magnetic resonance (NMR) measurements to study the vortex state of polycrystalline samples of $YPd_2Sn$ ($T_c = 5.4$ K), over a wide range of applied magnetic fields up to $B_{c2}(T)$. Measurements in the vortex state provide the temperature dependence of the effective magnetic penetration depth $\lambda(T)$ and the field dependence of the superconducting gap $\Delta(0)$. The results are consistent with a very dirty $s$-wave BCS superconductor with $\lambda(0) = 212(1)$ nm, a gap $\Delta(0) = 0.85(3)$ meV, and a Ginzburg-Landau coherence length $\xi_{GL}(0) \cong 23$ nm. The $\mu$SR data in a broad range of applied fields are well reproduced by taking into account a field-related reduction of the effective superconducting gap. Interestingly, the ratio $2\Delta(0)/(k_B T_c)$ appears to a good approximation to be field-independent, with a value at low field of 3.85(9), implying a field dependence of the gap $\Delta(T = 0, B) = \Delta(T = 0, B = 0)\sqrt{1 - \frac{B}{B_{c2}(0)}}$. We discuss the significance of this result.

The compound $YPd_2Sn$ is a so-called full Heusler compound with general formula $AT_2M$ ($M$ main group element, $T$ transition element, and $A$ either a rare earth or another transition element). Out of many hundreds of Heusler compounds only less than 30 (with Pd, Ni, or Au at the T site) are superconducting at ambient pressure. $YPd_2Sn$ is the one with the highest superconducting transition temperature, $T_c = 5.4$ K.

Besides superconductivity Heusler materials [1] display a large variety of interesting electronic properties such as different types of magnetic order (ferro-, antiferro-, and ferrimagnetism), heavy fermion behavior or half metallic ferromagnetism [2]. Some materials of this class may also exhibit properties typical of a topological insulator [3]. Their electronic tunability and multifunctionality make them attractive candidates for spintronics applications.

Recent transport and thermodynamic studies have identified strong electron-phonon coupling as the most important factor leading to superconductivity in these families. However, in contrast to the simple BCS theory, no conventional dependence of $T_c$ as a function of the BCS parameters, such as $N(0)$ or the Debye temperature, was found [4].

Unlike conventional BCS superconductors, several Pd-based compounds, such as $YbPd_2Sn$ [5] and $ErPd_2Sn$ [6], display co-existing superconductivity and long-range antiferromagnetic order. A clear and complete understanding of the origin of superconductivity, magnetism, and especially of their coexistence or interplay in the full Heusler compounds, is still missing.

In $YPd_2Sn$, earlier $\mu$SR data also showed an anomaly of possible magnetic origin setting in at $T_c$. For a non-magnetic superconductor such as $YPd_2Sn$, the large variation of $T_c$'s reported in the literature is also striking. We have therefore characterized the magnetic and superconducting properties of $YPd_2Sn$ at microscopic level also to clarify the role of Pd, which due to its large Stoner factor is near to ferromagnetism. The detailed results of these investigations are reported in [7].

Here we focus on investigations of the vortex state and we discuss the possible field dependence of the energy gap in a conventional BCS superconductor. The low $B_{c2}(0) = 0.57$ T of this material allowed to study the $B - T$ phase diagram in the vortex state over a large range of fields up to the phase boundary and to investigate how the BCS gap is affected by a large magnetic field. We analyzed the $\mu$SR data by using a sum of Gaussian functions. This has the advantage to be model independent and provides also the best fit to the data over the entire range of applied fields [8]. After subtraction of the background contribution, the time spectra were fitted with the following expression

$$A(t) = e^{-\sigma_n^2 t^2/2} \sum_{i=1}^{N} A_i e^{-\sigma_i^2 t^2/2} \cos(\gamma_\mu B_i t + \varphi). \tag{1}$$

Here $\varphi$ is the initial muon spin phase, while $A_i$, $\sigma_i$, and $B_i$ are the amplitude, relaxation rate and first moment of the internal field of the $i$-th Gaussian component, respectively. $\sigma_n$ is the small contribution to the field distribution arising from the nuclear moments, independent of temperature and determined well above $T_c$.

The standard deviation of the multi-Gaussian internal field distribution is then given by:

$$\langle \Delta B^2 \rangle = \frac{\sigma_s^2}{\gamma_\mu^2} = \sum_{i=1}^{N} \frac{A_i}{A_1 + \ldots + A_N} \left[ (\frac{\sigma_i}{\gamma_\mu})^2 + (B_i - \langle B \rangle)^2 \right], \tag{2}$$

with

$$\langle B \rangle = \sum_{i=1}^{N} \frac{A_i B_i}{A_1 + A_2 + \ldots + A_N}. \tag{3}$$

In the fit the number of components $N$ was increased until the $\chi^2$ of the fit did not change significantly. Typically, $N$ is between 1 (high-field data) and 3 (low-field data).

The ability of $\mu$SR to measure the field distribution and its moments in the vortex state is a powerful tool to determine characteristic length scales of superconductors such as the magnetic penetration depth and the coherence length, and electronic properties such as the superconducting gap and its symmetry. The ideally periodic flux line lattice (FLL) of known symmetry has been computed from various theories, see [9, 10, 11, 12] and references therein. The field distribution of the vortex lattice contains detailed information about the superconducting state. Recently, for instance, the diffraction of Cooper pairs on the vortex lattice of a very clean superconductor predicted by the Delrieu treatment of Gorkov theory has been detected in Nb crystals by a careful analysis of the TF spectra [12, 13].

The London and Ginzburg-Landau theories have the practical advantage to provide relatively simple expressions for quantities such as the field variance $\langle \Delta B^2 \rangle = \frac{\sigma_s^2}{\gamma_\mu^2}$ and are therefore widely used to analyse the $\mu$SR data. However, the original London model neglects the vortex core, corresponding to the case of an extreme type-II superconductor $\kappa = \frac{\lambda}{\xi} \gg 1$ and small applied fields $B$ or, equivalently, small $b \equiv \langle B \rangle / B_{c2}$, $\langle B \rangle$ average local field, which is generally close to the applied field; in our experiment equal to it within 1%. On the other hand the Ginzburg-Landau theory, which takes into account the vortex core, is actually valid only close to $T_c$.

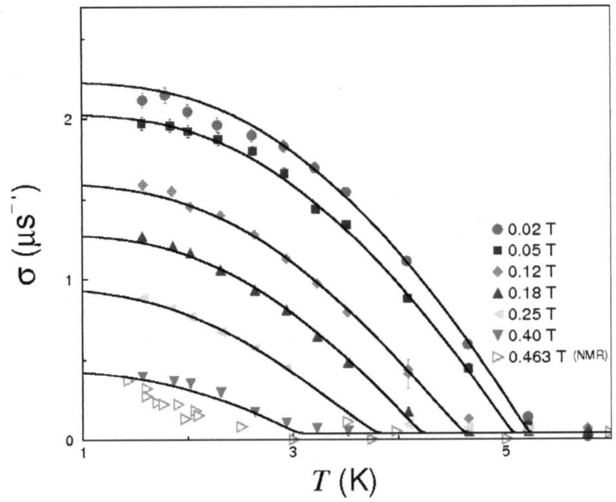


**Figure 1.** Temperature dependence of the muon spin depolarization rate in the vortex state of $YPd_2Sn$, measured after cooling below $T_c$ in different applied fields. Also plotted are the NMR linewidths measured at 0.463 T and converted into equivalent $\mu$SR relaxation rates. The fit lines were obtained using the procedure explained in the text.

In the range $0.13/\kappa^2 \ll b \ll 1$ the London model gives the well known field-independent $\langle \Delta B^2 \rangle = 0.00371 \frac{\Phi_0^2}{\lambda^4}$ [14], or equivalently $\sigma_{sc,L}(T)[\mu s^{-1}] = \frac{1.0728 \times 10^5}{\lambda^2(T)[\text{nm}^2]}$, where the subscript $L$ indicates the London limit.

If the applied field is not small with respect to $B_{c2}$, the standard deviation becomes field dependent, because of the effect of the finite size of the vortex cores and of the reduced intervortex distance. The effect of the finite size of the vortex cores and the expected field dependence of the field variance have been taken into account in the London model by introducing a Gaussian cutoff in the Fourier coefficients of the local field [9, 11]. Ginzburg-Landau theory provides a natural description of the vortex cores of the FLL [10].

One finds that the standard deviation in an applied field is a function of $b$ and can be written as

$$\sigma_s(T, b) = \sigma_{s,L} f(b). \tag{4}$$

Analytical expressions for $f(b)$ have been computed from the London model with cutoff (LC) [9] and from a numerical solution of the Ginzburg-Landau equation (GL) [10]. Ref. [11] also presented an approximate analytical solution and numerical calculations of this function. In the first two cases $f(b)$ simplifies to:

$$f_{LC}(b) = 0.45175(1 - b)\sqrt{1 + 3.9(1 - b^2)} \tag{5}$$

$$f_{GL}(b) = 0.45249(1 - b)\left[1 + 1.21(1 - \sqrt{b})^3\right] \tag{6}$$

where the coefficient have been chosen so that for $b = 0$, which formally corresponds to the London limit, we obtain $\sigma_{s,L}$. Both expressions have been computed for a wide range of fields in the mixed state, the first for $0 < b < 1$ and the second for $0.25/\kappa^{1.3} \lessapprox b \leqslant 1$ for $\kappa \geqslant 5$.

Eq. 4 has been used to analyze data in several cases over the full temperature range, where a decrease in the width of the internal field distribution when increasing the field toward $B_{c2}$ has

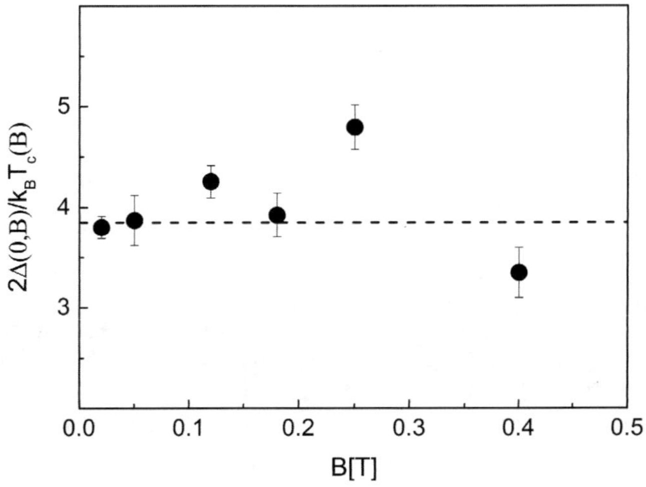


**Figure 2.** Field dependence of the gap-to-$T_c$ ratio $2\Delta(0,B)/(k_B T_c(B))$.

been observed (see for instance [15]). Generally, for each temperature the $\sigma_s$ values obtained at various $b$ are fitted with Eq. 5 or 6 by leaving the parameters $\lambda$ and $B_{c2}$ free and finally obtaining from the global temperature data $B_{c2}(T)$ and $\lambda(T)$. This procedure assumes no field dependence of $\lambda$. This is true and justified at $T = 0$ for a conventional superconductor with a single isotropic gap, to which the calculations of Ref. [9, 10] are related, but not at $T > 0$; i.e., the temperature dependence of $\lambda$ cannot be considered as field independent. This follows from the fact that the magnetic penetration depth diverges at the critical temperature, which in turn depends on the applied field $T_c(B)$. Since the phase transition from the normal state to the vortex state is a second-order transition, this implicitly introduces for all $T > 0$ a field dependence of $\lambda$.

The temperature dependence of the magnetic penetration depth is a measure of the superfluid density $n_s \propto 1/\lambda^2$. Within the BCS approach it is determined by the temperature dependence of the BCS gap. To take the above described field effects on the magnetic penetration depth into account we introduce a field-dependent effective superconducting gap $\Delta(0,B)$. Physically, this can be understood by considering that fields of not negligible magnitude with respect to $B_{c2}$ act as pair breakers, smear out the sharp edges of the spectroscopic gap, and lead to an effective reduction of $\Delta(0)$.

Following Ref. [16] we parametrize the BCS gap as

$$\Delta(T,B) = \Delta(0,B)\tanh\left[\frac{\pi k_{\mathrm{B}} T_c(B)}{\Delta(0,B)}\sqrt{\frac{T_c(B)}{T} - 1}\right]. \tag{7}$$

Our set of data of $\sigma_s(T,b)$ can be well fitted using the expression for a dirty superconductor with a single $s$-wave gap [17] (see Fig. 1)

$$\frac{n_s(T)}{n_s(0)} = \frac{\lambda^2(0)}{\lambda^2(T)} = \frac{\Delta(T,B)}{\Delta(0,B)}\tanh\frac{\Delta(T,B)}{2k_{\mathrm{B}}T}. \tag{8}$$

The dirty character is confirmed by our independent determination of the normal and the superconducting state parameters via transport and magnetization measurements [7], where we find that, while remaining a good metal with $k_F\ell \simeq 60$, $YPd_2Sn$ has a very low value of $\ell/\xi_0 = 0.008$ ($\xi_0$, BCS coherence length). Fig. 1 also shows the NMR line width measured at 0.463 T scaled to the muon data by taking into account the different gyromagnetic ratios ($\gamma_{^{119}Sn}/\gamma_\mu = 0.1178$). It is interesting to note that the magnitude of the field broadening obtained by the two techniques agrees fairly well. However, the NMR data seem to be affected by additional broadening effects at low temperatures, probably due to the proximity to the phase boundary. This reflects the limitations of NMR with respect to $\mu$SR when studying the vortex state of superconductors with low $B_{c2}$ values.

From the fit we obtain $T_c(B)$ and $\Delta(0, B)$. The field dependence of the critical temperature obtained in such a way is in good agreement with the $B_{c2}(T)$ curve obtained from magneto-resistivity and ac magnetization measurements and is well reproduced by the prediction of the Werthamer-Helfand-Hohenberg (WHH) expression [7]. This agreement supports our analysis. Fig. 2 shows the field dependence of the ratio $2\Delta(0)/(k_BT_c)$. Interestingly, it appears to a good approximation to be field-independent, with an average value of 3.85(9). From the usual approximation $B_{c2}(T) = B_{c2}(0)[1 - (\frac{T}{T_c})^2]$, one obtains $T_c(B) = T_c(0)\sqrt{(1 - \frac{B}{B_{c2}(0)}}$. Therefore, a constant gap-to-$T_c$ ratio implies a field dependence of the gap $\Delta(0, B) = \Delta(0, 0)\sqrt{1 - \frac{B}{B_{c2}(0)}}$. Assuming a field independent gap-to-$T_c$ ratio, the field and temperature dependence of the gap can be described by the compact formula

$$\Delta(T, B) = \Delta(0, 0)\sqrt{1 - \frac{B}{B_{c2}(0)}} \tanh\left[R\sqrt{\frac{T_c(B)}{T} - 1}\right]. \tag{9}$$

In our specific case $R = 1.6152$, and for an ideal (weak limit) BCS superconductor, $R = 2\pi/3.5278 = 1.781$. In the microscopic BCS theory for superconductivity the external magnetic field is treated as a weak perturbation. Therefore, e.g., the solution of the Meissner effect is a low-field limit solution and the gap has no field dependence. In Ginzburg-Landau theory the spatially averaged Ginzburg-Landau order parameter has a field dependence $\langle|\Psi_{GL}(\vec{r}, B)|^2\rangle = \Psi_0^2 1[-\frac{B}{B_{c2}(0)}]$. Since the Gorkov gap (or pair) potential $\Delta_G(\vec{r}) \propto \Psi_{GL}(\vec{r})$, it is interesting to note that, although the BCS (energy) gap is not necessarily the same as the Gorkov gap potential (see ,e.g., gapless superconductivity), a field reduction of the BCS gap following $\sqrt{1 - \frac{B}{B_{c2}(0)}}$ is obtained if we identify the BCS gap with the spatially averaged Gorkov gap potential $\Delta(T = 0, B)^2 = \langle|\Delta_G(\vec{r})|^2\rangle \propto \langle|\Psi_{GL}(\vec{r}, B)|^2\rangle = \Psi_0^2[1 - \frac{B}{B_{c2}(0)}]$.

In conclusion, we find that the $\mu$SR data of $YPd_2Sn$ are consistently reproduced in a broad range of applied fields with a *field-dependent* isotropic superconducting gap, which decreases with field with a similar functional dependence as the critical temperature. The ratio $2\Delta(0)/(k_BT_c) = 3.85(9)$ is consistent with the presence of an important electron-phonon coupling in this compound. The effective magnetic penetration depth is $\lambda(0) = 212(1)$ nm. Assuming a field independent gap-to-$T_c$ ratio a simple parametrization of the gap can be used to take into account the reduction of the superfluid density (for $T \neq 0$) of a conventional $s$-wave BCS superconductor in a magnetic field.

## References


[1] Heusler F 1903 *Verh. Dtsch. Phys. Ges.* **5**, 219
[2] Graf T *et al.* 2011 *Prog. Solid State Chem.* **39** 1
[3] Butch N P *et al.* 2011 *Phys. Rev.* B **84** 220504(R)
[4] Klimczuk T *et al.* 2012 *Phys. Rev.* B **85** 174505
[5] Kierstead H A *et al.* 1985 *Phys. Rev.* B **32** 135
[6] Shelton R N *et al.* 1986 *Phys. Rev.* B **34** 199
[7] Saadaoui H *et al.* 2013 *Phys. Rev.* B **88** 094518

[8] Maisuradze A *et al.* 2009 *J. Phys.: Condens. Matter* **21** 075701
[9] Brandt E H 1988 *Phys. Rev.* B **37** 2349
[10] Brandt E H 2003 *Phys. Rev.* B **68** 054506
[11] Yaouanc A, Dalmas de Réotier P and Brandt E H 1997 *Phys. Rev.* B **55** 11107
[12] Maisuradze A and Yaouanc A 2013 *Phys. Rev.* B **87** 134508
[13] Maisuradze A *et al.* 2013 *Phys. Rev.* B **88** 140509(R)
[14] Barford W and Gunn J M F 1988 *Physica* C **156** 515
[15] Shermadini Z *et al.* 2010 *Phys. Rev.* B **82** 144527
[16] Gross F *et al.* 1986 *Z. Phys.* B **64** 175
[17] Tinkham M 2004 *Introduction to Superconductivity* 2nd ed. (New York: Dover Publications)

# Magnetic order in the 2D Heavy-Fermion system $CePt_2In_7$ studied by $\mu^+SR$

**M Månsson[1,2], K Prša[1,3], Y Sassa[3], P H Tobash[4], E D Bauer[4], C Rusu[5], D Andreica[5], O Tjernberg[6], K Sedlak[7], M Grioni[8], T Durakiewicz[4] and J Sugiyama[9]**

[1] Laboratory for Quantum Magnetism, École Polytechnique Fédérale de Lausanne (EPFL), CH-1015 Lausanne, Switzerland
[2] Laboratory for Neutron Scattering & Imaging, Paul Scherrer Institute, CH-5232 Villigen PSI, Switzerland
[3] Laboratory for Solid state physics, ETH Zürich, CH-8093 Zürich, Switzerland
[4] Los Alamos National Laboratory, Los Alamos, New Mexico 87545, USA
[5] Faculty of Physics, Babes-Bolyai University, 400084 Cluj-Napoca, Romania
[6] Materials Physics, Royal Institute of Technology KTH, S-164 40 Kista, Sweden
[7] Laboratory for Muon Spin Spectroscopy, Paul Scherrer Institute, CH-5232 Villigen PSI, Switzerland
[8] Institute of Condensed Matter Physics, École Polytechnique Fédérale de Lausanne (EPFL), CH-1015 Lausanne, Switzerland
[9] Toyota Central Research & Development Laboratories, Inc., 41-1 Yokomichi, Nagakute, Aichi 480-1192, Japan

E-mail: `martin.mansson@epfl.ch`

**Abstract.** The low-temperature microscopic magnetic properties of the quasi-2D heavy-fermion compound, $CePt_2In_7$ are investigated by using a positive muon-spin rotation and relaxation ($\mu^+SR$) technique. Clear evidence for the formation of a commensurate antiferromagnetic order below $T_N \approx 5.40$ K is presented. The magnetic order parameter is shown to fit well to a modified BCS gap-energy function in a strong-coupling scenario.

## 1. Introduction

$CePt_2In_7$ is a recently discovered heavy fermion material belonging to the same family as the rather famous $CeIn_3$ [1] and Ce$M$In$_5$ ($M$ = Co, Rh, Ir) [2, 3] compounds. However, the spacing between Ce-In planes in $CePt_2In_7$ is drastically increased [4] relative to its Ce$M$In$_5$ cousins. As might be expected, the Fermi surface is more two-dimensional [5]. This compound crystallizes in the tetragonal space group $I4/mmm$ and represents a novel structure type [Pearson symbol $tI20$, $a = 4.6093$ Å, $c = 21.627$ Å, $Z = 2$] as shown in Fig. 1(a). The magnetic susceptibility, $\chi(T)$ [6, 7] exhibits a Curie-Weiss behavior for $T \geq 150$ K, with an effective moment $\mu_{\rm eff} = 2.93\ \mu_{\rm B}$, which is a bit higher than what can be expected for the $Ce^{3+}$ free ion moment ($p = 2.54$). $\chi(T)$ displays a maximum at 8 K, and an inflection at $T_N \approx 5$ K, signaling the onset of antiferromagnetism (AF). Further, resistivity data [7], $\rho(T)$, show metallic behavior with inflections at $T_{\rm Infl} = 25$ K and $T_{\rm Inf2} = 100$ K, and finally a sharp suppression below 8 K. The origin of these two high-temperature inflection points is presently unknown, but may be related to either a crystalline electric field excitation and/or the onset of coherence of the Kondo lattice [7, 8]. The specific

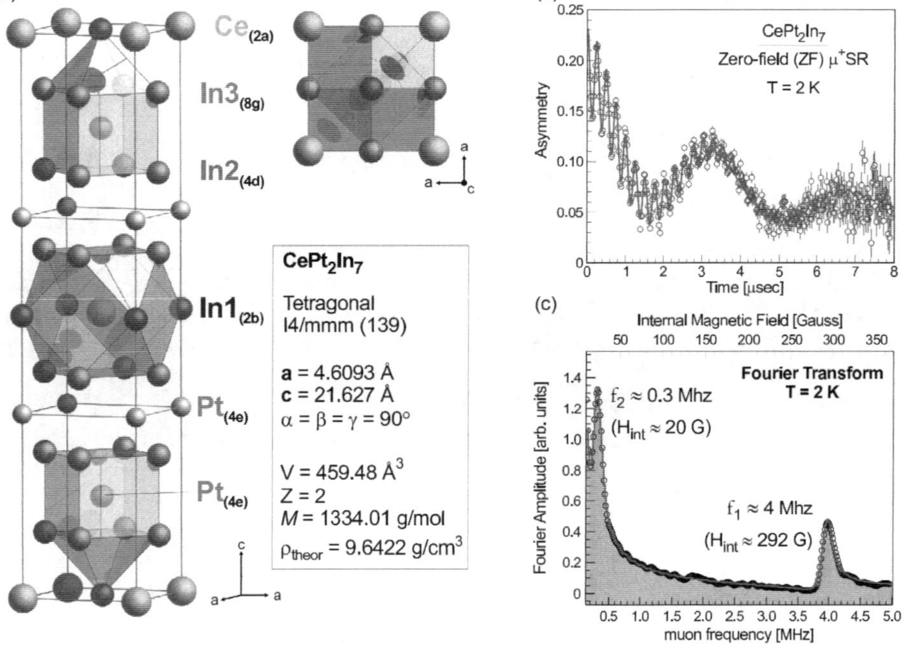


**Figure 1.** (Color online) (a) Side-view and top-view of the crystal structure of $CePt_2In_7$. (b) Zero-field (ZF) $\mu^+$SR time-spectrum acquired at $T = 2$ K. (c) Fourier transform of the time-spectrum at $T = 2$ K, clearly revealing the two frequencies ($f_1$ & $f_2$). Solid red lines in (b) and (c) represent the fit according to Eq. 1.

heat divided by temperature ($C/T$) exhibits a main peak at $T_N \approx 5.3$ K associated with the AF transition [6, 7]. Noteworthy is also the large value of $C/T \approx 450$ mJ/mol·K$^2$ [7], which indicates a substantial enhancement of the effective mass, $m^*$. Further, nuclear quadrupolar resonance (NQR) studies have indicated that the AF structure could be both commensurate [7] as well as incommensurate [9], the details of which are strongly dependent on how the polycrystalline or single crystalline samples were prepared. This suggests that strain may affect the magnetic properties of $CePt_2In_7$. NQR measurements indicate that strong AF fluctuations could be present in the paramagnetic state all the way up to $4T_N$ [7].

Electrical resistivity and ac-calorimetry measurements under pressure were recently carried out on single crystals of $CePt_2In_7$ [10]. These experiments reveal a quantum critical point at a critical pressure $P_c \approx 3.2$ GPa where the AF order is completely suppressed, in which the electrical resistivity exhibits a powerlaw $T$-dependence near $P_c$. In addition, a narrow dome of superconductivity is observed around $P_c$, suggesting that critical AF fluctuations may mediate the Cooper pairing. In contrast, a broad superconducting dome is found in poylcrystalline samples [11], indicating that strain at the grain boundaries also affects the superconductivity. NMR/NQR measurements under pressure on single crystals [12] reveal that a localized-itinerant crossover of the Ce-4$f$ electron occurs within the AF state at $P^* \approx 2.5$ GPa, approximately the pressure where superconductivity first emerges.

The positive muon-spin rotation and relaxation ($\mu^+$SR) technique has during the last 30 years been established as a unique and powerful method to study local magnetic/dipole fields as well as spin order and dynamics in a wide range of materials [13, 14, 15, 16, 17]. By implanting highly spin polarized muons ($S = \frac{1}{2}$) into a compound and monitoring the angular and temporal decay into positrons, the muon acts as an *ultra-sensitive* and *highly localized* magnetic probe. In this paper, we present a first principal study of microscopic magnetic order and transitions in the quasi-2D heavy fermion system $CePt_2In_7$ using mainly zero-field (ZF) $\mu^+$SR.

## 2. Experimental Details

Approximately 1.5 grams of polycrystalline $CePt_2In_7$ sample in the ratio 1:2:7.5 were prepared by arc-melting on a Cu hearth under UHP Ar atmosphere with a Zr getter. The resulting compound was subsequently wrapped in Ta foil and annealed under vacuum at 500°C for 2 weeks to stabilize the 1:2:7 phase. The sample preparation procedure is described in greater detail elsewhere [6, 18]. For the $\mu^+$SR experiment the sample was placed in a small envelope made of 50 $\mu$m thin Al-coated Mylar tape and then attached to a low-background fork-type sample holder. In order to make certain that the muons stopped primarily inside the sample, we ensured that the side facing the muon beamline was covered only by a single layer of Mylar tape. Subsequently, $\mu^+$SR spectra were recorded at the Swiss Muon Source (S$\mu$S), Paul Scherrer Institut, Villigen, Switzerland. By using the DOLLY spectrometer at the surface muon beamline $\mu$E1, zero-field (ZF) and weak transverse-field (wTF) data were collected for 0.25 K $\leq T \leq$ 20 K. The experimental setup and techniques were described in detail elsewhere [19].

## 3. Results and Discussion

$\mu^+$SR spectra recorded in ZF at $T = 2$ K [see Fig. 1(b-c)] show clear oscillations indicating that the muon-spins are precessing around spontaneous internal magnetic fields created, at the muon site, by the AF ordering. The fast Fourier transform (FFT) of the raw time spectrum clearly shows that the oscillating signal is composed of two different frequencies $f_1 \approx 4$ MHz and $f_2 \approx 0.3$ MHz. Accordingly, the ZF muSR spectra recorded at temperatures well below $T_N$ were well fitted by the combination of two exponentially damped cosine oscillations corresponding to an average of 2/3 of the muons that "see" a magnetic field perpendicular to the muon's initial polarization, a slowly relaxing non-oscillatory signal corresponding to the remaining 1/3 of the muons for which $H_{int}$ is parallel to their initial polarization, and a small background (BG) signal from a minor fraction of the muons stopping in the sample holder:

$$\begin{aligned} A_0\, P_{ZF}(t) &= A_1 \cos(2\pi \cdot f_1 t + \phi_1) \exp(-\lambda_1 t) \\ &+ A_2 \cos(2\pi \cdot f_2 t + \phi_2) \exp(-\lambda_2 t) \\ &+ A_{tail} \exp(-\lambda_{tail} t) \\ &+ A_{BG} \exp(-\lambda_{BG} t) \\ &+ A_{KT}\, G^{SGKT}(\Delta_{KT}, t). \end{aligned} \quad (1)$$

Here, $A_0$ is the initial ($t = 0$) asymmetry, $P_{ZF}(t)$ is the muon-spin polarization function in ZF, $f_1$ and $f_2$ are the two muon Larmor frequencies related to the internal fields as $f_i = (\gamma_\mu/2\pi) \times B_{int}$, $\phi_1$ & $\phi_2$ are the initial phases and $\lambda_1$ & $\lambda_2$ are the exponential relaxation rates of the precessing signals. Further, $\lambda_{tail}$ is the exponential relaxation rate of the tail signal, while $\Delta_{KT}$ is the static field distribution width of the isotropic dipolar field from the frozen (nuclear) magnetic moments. Finally, $A_1$, $A_2$, $A_{tail}$ and $A_{BG}$ are the asymmetries (i.e. volume fraction) of the four individual signals. As seen from Eq. 1, also a fifth component is added with a static Gaussian Kubo-Toyabe (KT) function [20] [$G^{SGKT}(\Delta_{KT}, t)$] that describes the presence of randomly oriented (frozen) spin system with a static moment. This last component only becomes important in the cross-over region close to $T_N$ where the fit function evolves from Eq. 1 to a simple Kubo-Toyabe function as the transition from AF to PM order occurs.

As seen from Fig. 2(a) and from the temperature dependencies of $f_1$ and $f_2$ shown in Fig. 2(b), as $T$ increases the oscillation frequencies are gradually suppressed and finally disappear around

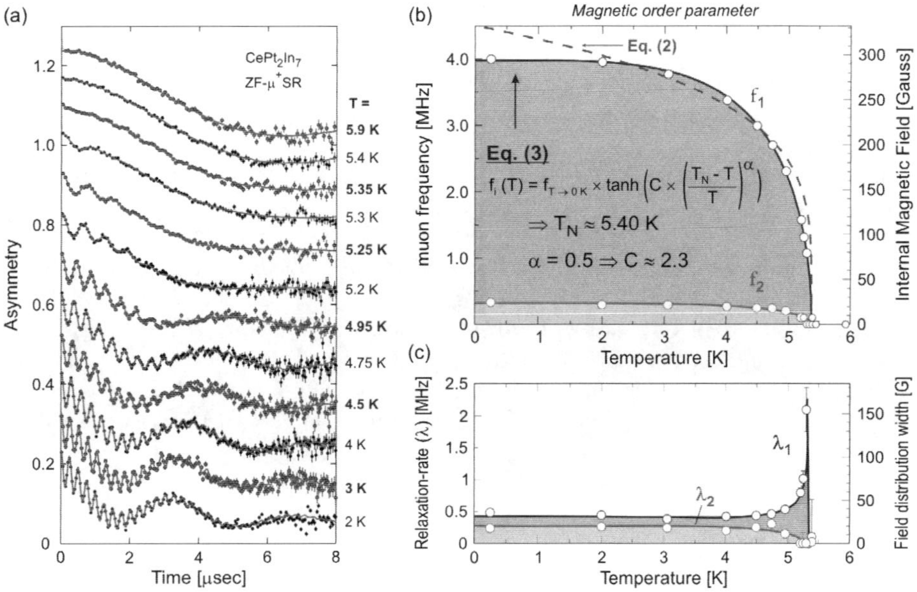


**Figure 2.** (Color online) (a) Temperature dependent ZF $\mu^+$SR spectra where the solid (red) lines represent the fit result using Eq. (1). Each spectrum is offset by 0.1 in asymmetry for clarity of display. (b) Magnetic order parameter fitted to Eq. 2 (dashed line) and Eq. 3 (solid lines), respectively. (c) Temperature-dependence of the relaxation rates, $\lambda_i$. Solid lines in (c) are only guides to the eye.

$T_N^{\mu} \approx 5.40$ K. The magnetic order parameter shown in Fig. 2(b) could only be rather poorly fitted in the complete temperature-range by a basic mean-field theory approach:

$$f_{\rm i} = f_{\rm T\to 0K} \cdot \left(\frac{T_{\rm N} - T}{T_{\rm N}}\right)^{\beta}. \quad (2)$$

As seen from the dashed line in Fig. 2(b) such function generates a clearly inadequate fit for both the high and low-temperature regime. Even when fitted only to the critical temperature region in the vicinity of $T_{\rm N} = 5.40$ K the critical exponent $\beta \approx 0.25$, which is in the intermediate range between the predictions for a 2D and 3D Ising model ($\beta = 0.125$ and 0.3125, respectively) [21, 22], which is unreasonable since $CePt_2In_7$ is expected to have a strong 2D character. The metallic conductivity of this material instead imply that the magnetic order parameter [Fig. 2(b)] could be fitted by the BCS gap-energy function that is commonly used for a spin-density wave (SDW) order parameter [23]:

$$f_{\rm i} = f_{\rm T\to 0K} \cdot \tanh\left[C \cdot \left(\frac{T_{\rm N} - T}{T}\right)^{\alpha}\right]. \quad (3)$$

Here $C = 1.74$ and $\alpha = 1/2$ are for the conventional weak-coupling BSC scenario while a higher value of $C$ is a sign for the presence of strong-coupling effects [24, 25]. Fitting the magnetic order parameter to Eq. 3 in the weak-coupling scenario, yields a very poor result. However, when allowing also $C$ to be fitted a very good result if obtained for $T_N^{\mu} \approx 5.40$ K, $C \approx 2.3$ and $\alpha = 1/2$ (*fixed*), as shown in Fig. 2(b) (solid lines). This is a clear indication that a strong-coupling scenario is more appropriate for this compound, as could be expected from the reported high effective mass, $m^*$ resulting from the interactions of the conduction electrons with the magnetic Ce ions [26].

The initial phases of the oscillations, $\phi_1$ and $\phi_2$, are equal and close to zero in the whole temperature range. Such fact together with a rather symmetric field-distribution shown in the FFT spectrum [Fig. 1(c)], suggest that the AF order is commensurate to the crystallographic lattice, as has been partly proposed by NQR [7]. However, it should be emphasized that neutron diffraction would be the appropriate tool, more suitable than both NQR and $\mu^+$SR, for robustly determining the detailed spin-structure of $CePt_2In_7$. Further, that $f_1$ and $f_2$ show the same temperature dependence and are equally well fitted to Eq. 3, clearly suggests that the multiple frequencies are not caused by the coexistence of two different phases in the sample but a direct result of two magnetically inequivalent interatomic muon stopping sites within the crystal lattice.

Figure 2(c) shows the temperature dependence of the relaxation rate (field distribution width) for the fast oscillation ($\lambda_1$), which display a critical behavior close to the transition. This is caused by that the AF fluctuations become stronger and stronger with increasing temperature, until the static AF order is completely destroyed at $T_N$. For the slow oscillation, $\lambda_2$ appears to display a different temperature dependence close to the transition. However, this is most likely merely a fitting-effect caused by a problem to disentangle the slow oscillation from the evolving KT function close to $T_N$.

## 4. Conclusions

By the use of a positive muon-spin rotation and relaxation ($\mu^+$SR) technique we find clear evidence for the formation of a commensurate antiferromagnetic order in $CePt_2In_7$ below $T_N \approx 5.40$ K. The magnetic order parameter can be well fitted to a modified BSC gap-energy function in a strong-coupling scenario.


## Acknowledgments

This work was performed at the Swiss Muon Source (S$\mu$S), Paul Scherrer Institut (PSI), Villigen, Switzerland and we are thankful to Robert Scheuermann, Bastian Wojek and Matthias Elender for assistance with the $\mu^+$SR experiments. This work is partially supported by the Swiss National Science Foundation (SNSF), by Grant-in-Aid for Scientific Research on Innovative Areas "Ultra Slow Muon" (No. 23108003) of the Ministry of Education, Culture, Sports, Science, and Technology, Japan, MEXT KAKENHI Grant No. 23108003, JSPS KAKENHI Grant No. 26286084, and the Swedish Research Council (VR). Work at LANL was performed under the auspices of the U.S. DOE, Office of Basic Energy Sciences, Division of Materials Sciences and Engineering. D.A. acknowledges financial support from the Romanian UEFISCDI project No. PN-II-ID-PCE-2011-3-0583 (85/2011). All images involving crystal structure were made with DIAMOND and the $\mu^+$SR data was fitted using `musrfit` [27].


## References


[1] Mathur N D, Grosche F M, Julian S R, Walker I R, Freye D M, Haselwimmer R K W and Lonzarich G G 1998 *Nature* **394** 39

[2] Park T, Ronning F, Yuan H Q, Salamon M B, Movshovich R, Sarrao J L and Thompson J D 2006 *Nature* **440** 65

[3] Gerber S *et al.* 2014 *Nature Physics* **10** 126

[4] Kurenbaeva Zh M , Murashova E V, Seropegin Y D, Noël H and Tursina A I 2008 *Intermetallics* **16** 979

[5] Altarawneh M M, Harrison N, McDonald R D, Balakirev F F, Mielke C H, Tobash P H, Zhu J -X, Thompson J D, Ronning F and Bauer E D 2011 *Phys. Rev.* B **83** 081103(R)

[6] Bauer E D, Sidorov V A, Lee H, Kurita N, Ronning F, Movshovich R and Thompson J D 2010 *J. Phys.: Conf. Ser.* **200** 012011

[7] apRoberts-Warren N, Dioguardi A P, Shockley A C, Lin C H, Crocker J, Klavins and Curro N J 2010 *Phys. Rev.* B **81** 180403(R)

[8] Matsumoto M, Han M J, Otsuki J, and Savrasov S Yu 2012 *Phys. Rev.* B **82** 180515(R)

[9] Sakai H, Tokunaga Y, Kambe S, Lee H-O, Sidorov V A, Tobash P H, Ronning F, Bauer E D and Thompson J D 2011 *Phys. Rev.* B **83** 140408(R)
[10] Sidorov V A, Lu X, Park T, Lee H, Tobash P H, Baumbach R E, Ronning F, Bauer E D and Thompson J D 2013 *Phys. Rev.* B **88** 020503(R)
[11] Bauer E D, Lee H O, Sidorov V A, Kurita N, Gofryk K, Zhu J- X, Ronning F, Movshovich R, Thompson J D and Park T 2012 *Phys. Rev.* B **81** 180507(R)
[12] Sakai H, Tokunaga Y, Kambe S, Ronning F, Bauer E D and Thompson J D 2014 *Phys. Rev. Lett.* **112** 206401
[13] Denison A B, Graf H, Kündig W and Meier P F 1979 *Helvetica Phys. Acta* **52** 460
[14] Cox S F J 1987 *J. Phys. C: Solid State Phys.* **20** 3187
[15] Kalvius G M, Noakes D R and Hartmann O 2001 *Handbook on the Physics and Chemistry of Rare Earths* (Amsterdam: Elsevier Science) vol 32 chapter 206
[16] Sugiyama J, Mukai K, Ikedo Y, Nozaki H, Månsson M and Watanabe I 2009 *Phys. Rev. Lett.* **103** 147601
[17] Månsson M and Sugiyama J 2013 *Physica Scripta* **88** 068509
[18] Tobash P H, Ronning F, Thompson J D, Scott B L, Moll P J W, Batlogg B and Bauer E D 2012 *J. Phys.: Condens. Matter* **24** 015601
[19] Yaouanc A and Dalmas de Réotier P 2011 *Muon Spin Rotation, Relaxation, and Resonance* (Oxford: Oxford University Press)
[20] Kubo R and Toyabe T 1967 *Magnetic Resonance and Relaxation* (Amsterdam: North-Holland)
[21] Collins M F 1989 *Magnetic critical scattering* (Oxford: Oxford University Press)
[22] Stanley H E 1971 *Introduction to Phase Transitions and Critical Phenomena* (Oxford: Clarendon)
[23] Grüner G 1994 *Density Waves in Solids* (Reading: Addison-Wesley-Longmans)
[24] Tinkham M 1996 *Introduction to Superconductivity* (New York: McGraw-Hill Inc.)
[25] Betouras J and Joynt R 1995 *Europhys. Lett.* **31** 119
[26] Fisk Z, Hess D W, Pethick C J, Pines D, Smith J L, Thompson J D and Willis J O 1988 *Science* **239** 33
[27] Suter A and Wojek B M 2012 *Physics Procedia* **30** 69

# Superconducting properties of $Ca_3Ir_4Sn_{13}$: a $\mu$SR study

**P K Biswas[1], A Amato[1], Kefeng Wang[2], C Petrovic[2], R Khasanov[1], H Luetkens[1] and E Morenzoni[1]**

[1] Laboratory for Muon Spin Spectroscopy, Paul Scherrer Institute, CH-5232 Villigen PSI, Switzerland

[2] Condensed Matter Physics and Materials Science Department, Brookhaven National Laboratory, Upton, New York 11973, USA

E-mail: pabitra.biswas@psi.ch, elvezio.morenzoni@psi.ch

**Abstract.** Muon spin relaxation and rotation ($\mu$SR) measurements have been performed to study the superconducting properties of $Ca_3Ir_4Sn_{13}$. Zero-field $\mu$SR data shows no sign of any magnetic anomaly in $Ca_3Ir_4Sn_{13}$ at the superlattice transition temperature, $T^*$ or in the superconducting ground state. Transverse-field $\mu$SR measurements in the vortex state provided the temperature dependence of the magnetic penetration depth $\lambda$. The dependence of $\lambda^{-2}$ with temperature is consistent with the existence of a single $s$-wave energy gap in the superconducting state of $Ca_3Ir_4Sn_{13}$ with a gap value of 1.51(5) meV at absolute zero temperature. The magnetic penetration depth at zero temperature $\lambda(0)$ is 351(4) nm. The ratio $\Delta(0)/k_BT_c = 2.41(8)$ indicates that $Ca_3Ir_4Sn_{13}$ is a strong-coupling superconductor.

## 1. Introduction

The ternary intermetallic stannide compounds, $M_3Ir_4Sn_{13}$, where $M$ = Ca, Sr, etc. are of particular interest because they exhibit many exotic physical properties such as superconductivity, magnetic or charge order, and structural instabilities [3, 4, 5]. $Ca_3Ir_4Sn_{13}$ one of the most important material among these compounds shows a superconducting ground state below 7 K [1, 2]. Although this material was synthesized some 30 years ago, only very recently it has regained particular attention due to the possible coexistence or competition between superconducting and ferromagnetic spin fluctuation or charge density wave (CDW) states.

Resistivity and susceptibility measurements on $Ca_3Ir_4Sn_{13}$ show a broad peak like anomaly at $T^* \approx 33$ K. A similar anomaly has also been observed in the isoelectric sister compound $Sr_3Ir_4Sn_{13}$ at $T^* \approx 147$ K. Initially these $T^*$ anomalies were attributed to ferromagnetic (FM) fluctuation. The claim was made with the observation that the resistivity follows a non-Fermi liquid temperature dependence, but that the Fermi liquid behaviour can be restored by applying a high magnetic field [3]. Under an applied hydrostatic pressure, the $T_c$ of $Ca_3Ir_4Sn_{13}$ increases up to 4 GPa then falls for higher pressures [4].

Recently, single crystal x-ray diffraction studies [4] showed that the $T^*$ anomaly in $M_3Ir_4Sn_{13}$ ($M$ = Ca, Sr) is related to a second order superlattice transition from simple cubic parent phase, the $I$-phase, to a superlattice variant, the $I'$-phase, with a lattice parameter twice that of the $I$-phase. It has been further argued that this superlattice transition is associated with a charge density wave transition of the conduction electron system. At low temperatures, thermal conductivity and specific heat measurements indicate weakly correlated nodeless superconductivity in $Ca_3Ir_4Sn_{13}$ [5, 6]. On the other hand, recent $\mu$SR study on $Ca_3Ir_4Sn_{13}$ [7] determined a very high gap-to-$T_c$ ratio value $\Delta(0)/(k_BT_c) = 5$, which is unusually large even for a very strongly coupled BCS superconductor. In this contest, it is therefore important to reveal the true nature of the superconducting gap structure in $M_3Ir_4Sn_{13}$ ($M$ = Ca, Sr).

Here, we use the $\mu$SR technique as a microscopic local magnetic probe to investigate the magnetic properties of the superconducting ground state and determine the superconducting gap structures in $Ca_3Ir_4Sn_{13}$.

## 2. Experimental details

Single crystal samples of $Ca_3Ir_4Sn_{13}$ were prepared using a high temperature self-flux method [1, 2]. Sample characterization measurements were performed at the Brookhaven National Laboratory. The transverse-field (TF) and zero-field (ZF) $\mu$SR experiments were carried out at the $\pi$E1 and $\pi$E3 beam lines at the Paul Scherrer Institute (Villigen, Switzerland). The sample was cooled to the base temperature (1.6 K) at $H = 0$ during ZF-$\mu$SR experiments and in series of fields ranging from 2 mT to 0.5 T in TF-$\mu$SR experiments. The typical counting statistics were $\sim$ 8 million events per data point. The ZF and TF $\mu$SR data were analyzed by using the free software package MUSRFIT [8].

## 3. Results and discussion

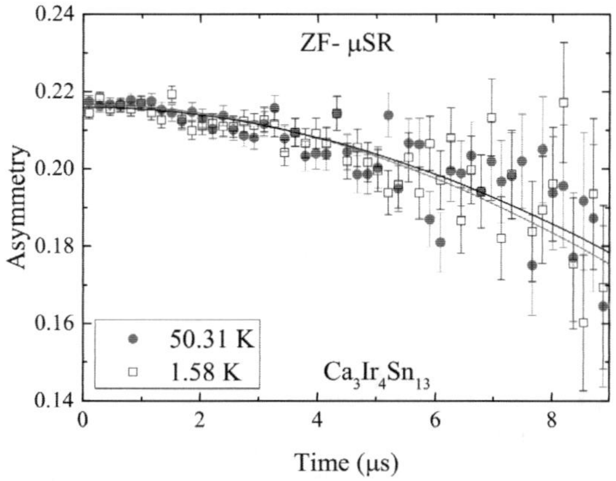


**Figure 1.** (Color online) ZF-$\mu$SR spectra of $Ca_3Ir_4Sn_{13}$ taken at 1.58 and 50.3 K. The solid lines are fits to the data using Eq. 1.

Figure 1 compares the ZF-$\mu$SR signals collected above $T^*$ and below $T_c$. Both signals are practically identical, indicating that no additional magnetic moments (either static or dynamic) appear below $T^*$ and also below $T_c$. ZF-$\mu$SR data can be well described using a Gaussian Kubo-Toyabe relaxation function, [9]

$$A(t) = A(0)\left\{\frac{1}{3} + \frac{2}{3}\left(1 - \sigma^2t^2\right)\exp\left(-\frac{\sigma^2t^2}{2}\right)\right\}, \tag{1}$$

where $A(0)$ is the initial asymmetry and $\sigma$ describes the muon spin relaxation rate due to the presence of static nuclear moments in $Ca_3Ir_4Sn_{13}$. The values of $\sigma$ obtained from the fits are 0.048(2) and 0.051(2) $\mu s^{-1}$ for 1.58 and 50.3 K, respectively.

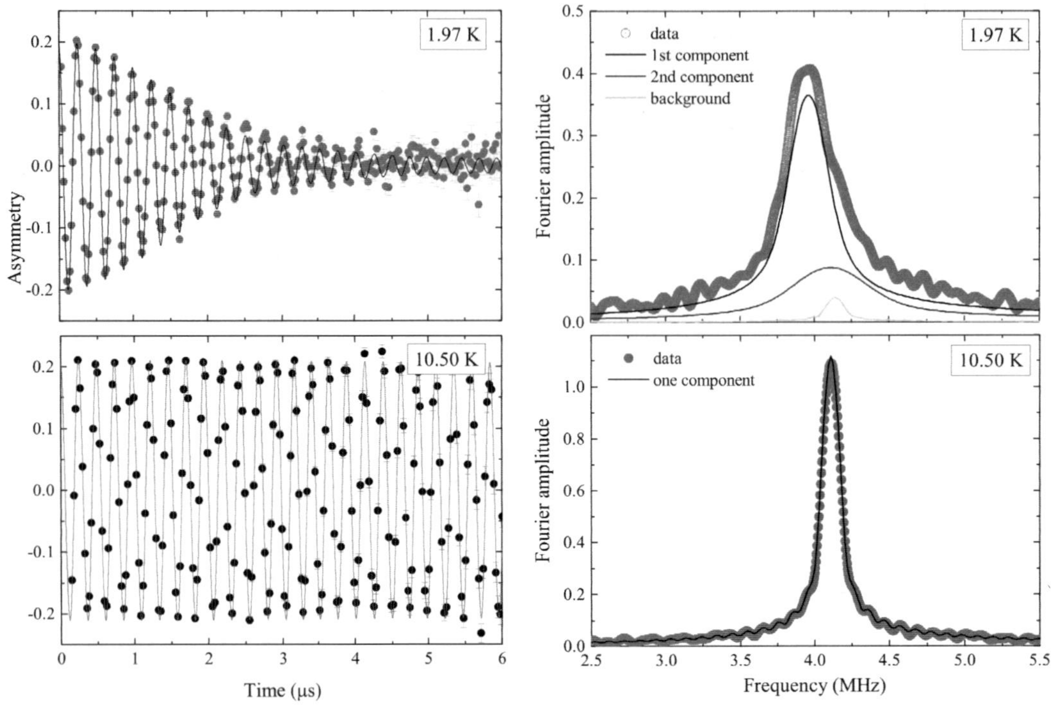


**Figure 2.** (Color online) (Left panel) TF-$\mu$SR signals for $Ca_3Ir_4Sn_{13}$ measured at 1.97 and 10.5 K in an applied field of 300 Oe. The solid lines are the fits to the data using Eq. 2. (Right panel) Corresponding TF spectra plotted in the frequency domain.

The left panel of Fig. 2 shows the TF-$\mu$SR precession signals of $Ca_3Ir_4Sn_{13}$ taken at 1.97 and 10.5 K in an applied field of 300 Oe. The right panel of Fig. 2 shows corresponding TF spectra plotted in the frequency domain. The signal in the normal state exhibit almost no damping due to homogeneous magnetic field distribution inside the sample. However, the signal taken at 1.97 K (below $T_c$) decays very quickly due to the inhomogeneous field distribution generated by the superconducting vortex lattice [10].

We can determine the second moment of the magnetic field distribution associated with the vortex state from the TF-$\mu$SR spectra. To do that the TF-$\mu$SR spectra were fitted using a multi-component Gaussian curve [11, 12]:

$$A(t) = \sum_{i=1}^{N} A_i \exp(-\sigma_i^2 t^2/2)\cos(\gamma_\mu B_i t + \phi) + A_{bg}\cos(\gamma_\mu B_{bg} t + \phi), \tag{2}$$

where $\phi$, $A_i$, $\sigma_i$, and $B_i$ are the initial phase, asymmetry, relaxation rate, and mean field (first moment) of the $i$th Gaussian component, respectively. $A_{bg}$ and $B_{bg}$ are the asymmetry and field, respectively due to the background contribution. We found that two Gaussian components ($N = 2$) are sufficient to fit the muon time spectra data. For $N = 2$, the first and second moments of $P(B)$ are given by

$$\langle B \rangle = \sum_{i=1}^{2} \frac{A_i B_i}{A_1 + A_2}, \tag{3}$$

and

$$\left\langle \Delta B^2 \right\rangle = \frac{\sigma^2}{\gamma_\mu} = \sum_{i=1}^{2} \frac{A_i}{A_1 + A_2} \left\{ (\sigma_i/\gamma_\mu)^2 + [B_i - \langle B \rangle]^2 \right\}, \tag{4}$$

where $\gamma_\mu = 2\pi \times 135.5388$ MHz/T is the muon gyromagnetic ratio and $\sigma$ the muon depolarization rate. Fig. 3 shows the temperature dependence of $\sigma$ of $Ca_3Ir_4Sn_{13}$ for an applied field of 300 Oe.

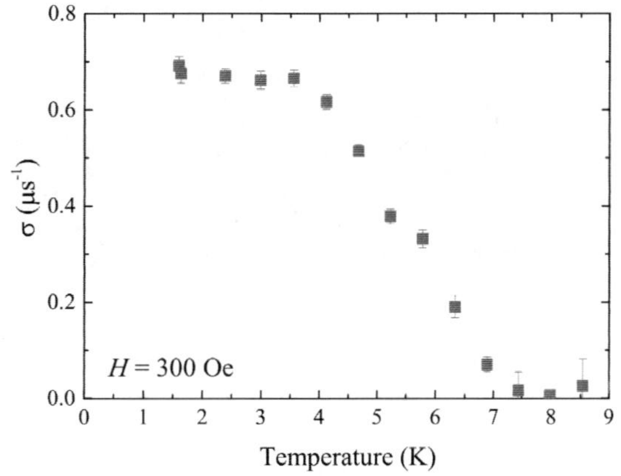


**Figure 3.** (Color online) Temperature dependence of the muon depolarization rate $\sigma$ in $Ca_3Ir_4Sn_{13}$ collected in an applied magnetic field of 300 Oe.

The superconducting contribution to $\sigma$ is obtained by subtracting the nuclear moment contribution (measured above $T_c$) as ${\sigma_{sc}}^2 = \sigma^2 - {\sigma_{\rm nm}}^2$. For an isotropic type-II superconductor

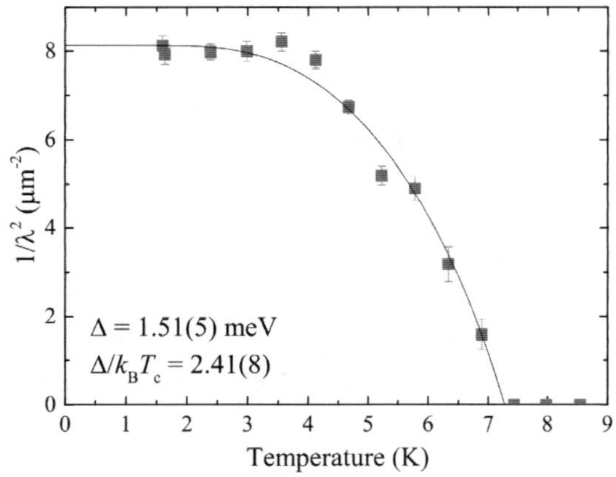


**Figure 4.** (Color online) The temperature dependence of $\lambda^{-2}(T)$. The solid line is a fit to the data using a single $s$-wave BCS model.

with a hexagonal Abrikosov vortex lattice described by Ginzburg-Landau theory, the magnetic penetration depth $\lambda$ is related to $\sigma_{sc}$ by the equation [10]:

$$\sigma_{sc}(b)[\mu s^{-1}] = 4.854 \times 10^4 (1-b)[1 + 1.21(1-\sqrt{b})^3]\lambda^{-2}[\text{nm}^{-2}], \tag{5}$$

Here $b = \langle B \rangle / B_{c2}$ is a reduced magnetic field. Fig. 4 shows the temperature dependence of $\lambda^{-2}$, i.e. the effective superfluid density, which is nearly flat below 4 K. This suggests that $Ca_3Ir_4Sn_{13}$ is a nodeless fully gapped superconductor. The solid line is a fit to the data using a single gap BCS $s$-wave model [13, 14]:

$$\frac{\lambda^{-2}(T)}{\lambda^{-2}(0)} = 1 + 2\int_{\Delta(T)}^{\infty} \left(\frac{\partial f}{\partial E}\right) \frac{EdE}{\sqrt{E^2 - \Delta(T)^2}}. \tag{6}$$

Here $\lambda(0)$ is the zero-temperature value of the magnetic penetration depth, and $f = [1 + \exp(E/k_BT)]^{-1}$ is the Fermi function. We approximated the temperature dependence of the gap with $\Delta(T) = \Delta(0)\tanh\{1.82[1.018(T_c/T - 1)]^{0.51}\}$ [15]. The fit yields $\lambda(0) = 351(4)$ nm, and $\Delta(0) = 1.51(5)$ meV. The gap to $T_c$ ratio $\Delta(0)/k_BT_c = 2.41(8)$ is higher than the BCS value of 1.76, suggesting that $Ca_3Ir_4Sn_{13}$ is a strong-coupling superconductor. Our values of $\Delta(0)$ and gap-to-$T_c$ ratio obtained by a microscopic measurements are in good agreement with those obtained from specific heat measurements [16], whereas they are nearly 50 % lower than the values obtained in the previous $\mu$SR measurements by S. Gerber et al. [7].

## 4. Conclusions

In conclusion, ZF-$\mu$SR results do not find evidence of any magnetism in $Ca_3Ir_4Sn_{13}$ below $T^*$ as well as in the superconducting state. TF-$\mu$SR results indicate for a fully gapped superconducting state in $Ca_3Ir_4Sn_{13}$. $\lambda(T)$ can be well fitted using a single BCS $s$-wave model with gap value, $\Delta(0) = 1.51(5)$ meV and penetration depth, $\lambda(0) = 351(4)$ nm. The value of the gap to $T_c$ ratio, 2.41(8) is higher than the BCS value of 1.76 and suggests that $Ca_3Ir_4Sn_{13}$ is a strong-coupling superconductor.

## Acknowledgments

The $\mu$SR experiments were performed at the Swiss Muon Source (S$\mu$S), Paul Scherrer Institute (PSI, Switzerland). Work at Brookhaven is supported by the US DOE under Contract No. DE-AC02-98CH10886.

## References

[1] Remeika J, Esponosa G, Cooper A, Barz H, Rowell J, McWhan D, Vandenberg J, Moncton D, Fisk Z, Woolf L, Hamaker H, Maple M, Shirane G and Thomlinson M 1980 *Solid State Commun.* **34** 923
[2] Espinosa G P 1980 *Mater. Res. Bull.* **15** 791
[3] Yang J, Chen B, Michioka C and Yoshimura K 2010 *J. Phys. Soc. Jpn.* **19** 113705
[4] Klintberg L E, Goh S K, Alireza P L, Saines P J, Tompsett D A, Logg P W, Yang J, Chen B, Yoshimura K and Grosche F M 2012 *Phys. Rev. Lett.* **109** 237008
[5] Zhou S Y, Zhang H, Hong X C, Pan B Y, Qiu X, Dong W N, Li X L and Li S Y 2012 *Phys. Rev.* B **86** 064504
[6] Wang K and Petrovic C 2012 *Phys. Rev.* B **86** 024522
[7] Gerber S, Gavilano J L, Medarde M, Pomjakushin V, Baines C, Pomjakushina E, Conder K and Kenzelmann M 2013 *Phys. Rev.* B **88** 104505
[8] Suter A and Wojek B M 2012 *Physics Procedia* **30** 69
[9] Kubo R 1981 *Hyperfine Interact.* **8** 731
[10] Brandt E H 2003 *Phys. Rev.* B **68** 054506
[11] Weber M, Amato A, Gygax N, Schenck A, Maletta H, Duginov V N, Grebinnik V G, Lazarev A B, Olshevsky V G, Pomjakushin V Yu, Shilov S N, Zhukov V A, Kirillov B F, Pirogov A V, Ponomarev A N, Storchak V G, Kapusta S and Bock J 1993 *Phys. Rev.* B **48** 13022
[12] Maisuradze A, Khasanov R, Shengelaya A and Keller H 2009 *J. Phys.: Condens. Matter* **21** 075701
[13] Tinkham M 1975 *Introduction to Superconductivity* (New York: McGraw-Hill)
[14] Prozorov R and Giannetta R W 2006 *Supercond. Sci. Technol.* **19** R41
[15] Carrington A and Manzano F 2003 *Physica* C **385** 205
[16] Kase N, Hayamizu H and Akimitsu J 2011 *Phys. Rev.* B **83** 184509

Journal of Physics: Conference Series **551** (2014) 012030 doi:10.1088/1742-6596/551/1/012030

# TF-$\mu^+$SR in overdoped $Tl_2Ba_2CuO_{6+x}$: simplicity still elusive

**J H Brewer[1], S L Stubbs[1], R Liang[1], D A Bonn[1], W N Hardy[1], W A MacFarlane[2] and D C Peets[3]**

[1] Department of Physics and Astronomy, The University of British Columbia, Vancouver, BC, Canada V6T 1Z1

[2] Department of Chemistry, The University of British Columbia, Vancouver, BC, Canada V6T 1Z1

[3] Center for Correlated Electron Systems, Institute for Basic Science, Seoul National University, Seoul 151-747, Korea

E-mail: jess@triumf.ca

**Abstract.** Encouraged by predictions of simple Fermi liquid behavior in the overdoped versions of high temperature superconductors (H$T_c$SC) based on $CuO_2$ planes, we have performed TF-$\mu^+$SR experiments on several high quality oriented crystal mosaics of $Tl_2Ba_2CuO_{6+x}$ ("Tl-2201") in the vortex-lattice state where the magnetic penetration depth $\lambda$ can be extracted from the characteristic superconducting (SC) lineshape by $\chi^2$-minimization fitting in the time domain. The results show evidence for a surprising degree of disorder, causing some "smearing" of the characteristic lineshape features, but we were able to extract $\lambda$ as a function of temperature, revealing a remarkably consistent $T$-dependence of the SC carrier density $n_s \propto \lambda^{-2}$ that suggests this "role-model cuprate" is anything but "simple". The extracted superfluid density for all dopings exhibits a clear inflection point with temperature, suggesting a vortex lattice transition. Such a transition would have important implications for other measurements.

In the cuprate H$T_c$SCs, charge doping of the $CuO_2$ planes engenders superconductivity (SC) between the limits of an undoped insulating antiferromagnet (AFM) and what is believed to be a conventional Fermi liquid (FL) state at high dopings. One would like to understand how the unconventional SC state evolves out of these more conventional electronic ground states. However, overdoped cuprates are rare and few have highly ordered $CuO_2$ planes. The $Tl_2Ba_2CuO_{6+x}$ (Tl-2201) system offers tunability throughout the overdoped regime with highly-ordered, isolated, flat $CuO_2$ planes, doped *via* dilute interstitial oxygen atoms in the distant TlO layers [1]; the overdoping arises from substitution of copper on the thallium site at the $\sim 8\%$ level [2, 3, 4] or self-doping to eliminate a predicted electron pocket at the $\Gamma$ point [5].

Previous data on powder samples [6] was used to extend the "Uemura plot" [7] to overdoped cuprates, but such samples do not produce characteristic lineshapes allowing detailed analysis.

We recently reported the first TF-$\mu^+$SR results on high-quality single crystal mosaics of Tl-2201 for a range of dopings[8]; additional details are provided here. From the magnetic field distribution in the vortex state, we extract the temperature dependence of the SC order parameter and the magnetic penetration depth $\lambda$. Surprisingly, these do not appear to follow the simple $d$-wave temperature dependence found in under- and optimally-doped cuprates[9], exhibiting instead an unexpected inflection point in a qualitatively different screening response.

Single crystals of Tl-2201 were grown in gold-sealed alumina crucibles by an encapsulated copper-rich self-flux method as described elsewhere [10]. It was necessary to vary the degree of sealing during the growth process, to allow evolved gases to escape, then to contain volatile and toxic $Tl_2O$ during growth. The oxygen content was set by annealing under controlled oxygen partial pressures and temperatures [11]; two different annealing schemes were employed depending on the desired oxygen content [10]. Crystals were assembled in mosaics on substrates of aluminized mylar or GaAs, with the $c$-axis perpendicular to the substrate.

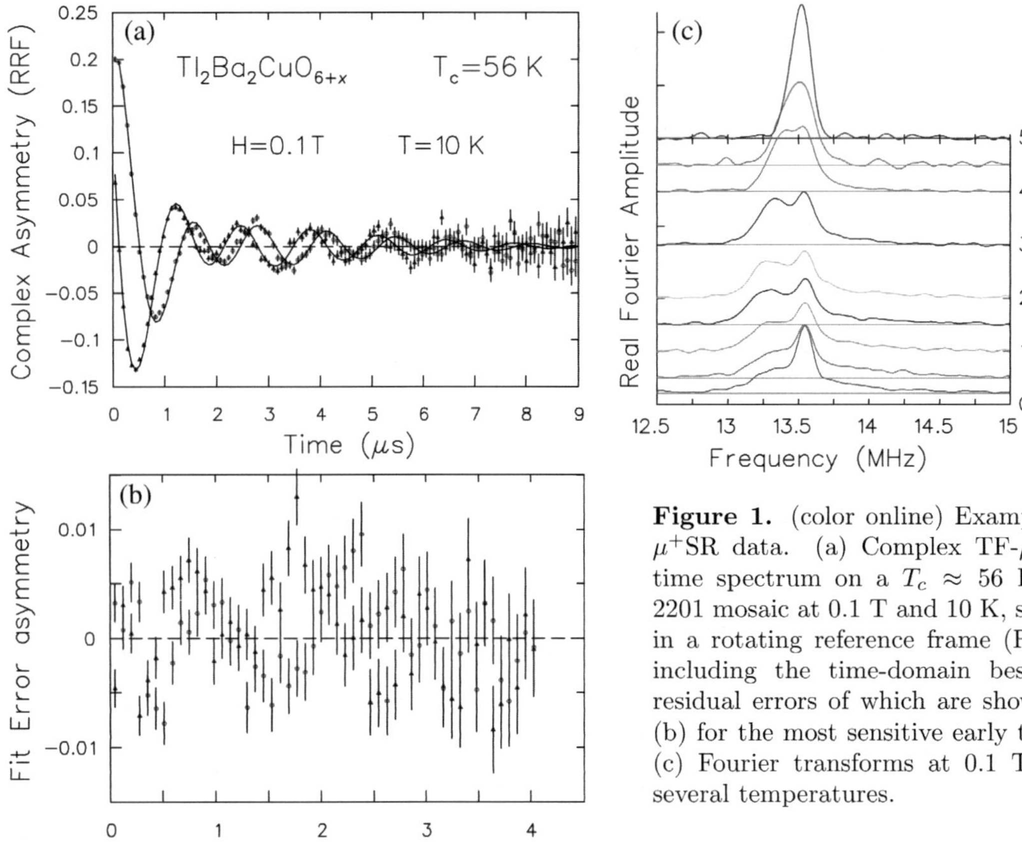


**Figure 1.** (color online) Example of $\mu^+$SR data. (a) Complex TF-$\mu^+$SR time spectrum on a $T_c \approx 56$ K Tl-2201 mosaic at 0.1 T and 10 K, shown in a rotating reference frame (RRF), including the time-domain best fit, residual errors of which are shown in (b) for the most sensitive early times. (c) Fourier transforms at 0.1 T and several temperatures.

The main advantage of TF-$\mu^+SR$ in studies of Type-II SC is its ability, under optimal circumstances, to determine the *absolute value* of the magnetic penetration depth $\lambda$ and its inverse square, which is proportional to the density $n_s$ of SC carriers[9]. Unfortunately, "optimal circumstances" do not yet apply in the case of $Tl_2Ba_2CuO_{6+x}$, where only small improvements of global $\chi^2$-minimization fits distinguish between the broadening $\sigma_d$ due to vortex lattice disorder [which should scale with $\lambda^{-2}(T)$] and the $T$-independent broadening $\sigma_0$ due to nuclear dipoles and crystal defects. The amplitude $A_B$ of the background signal due to muons stopping outside the sample is also known only by finding the best global fit; like $\sigma_0$, the value of this parameter is subtly coupled to that of $\lambda^{-2}$.

An example of the data collected, in this case at 10 K on a $T_c = 56$ K mosaic, is shown in Fig. 1, including the time domain fit in panel (a) and its residuals in panel (b). Time-domain fits to a numerically generated vortex lattice field distribution [9] converged very well and fully reproduce the data on all mosaics at all fields and temperatures. Fourier transforms corresponding to the field distribution are also shown for a variety of temperatures — an additional peak just above the "cusp" is attributable to muons stopping outside the sample; the time-domain fits account for this contribution. The high-frequency "cutoff" is indistinct in these mosaics, precluding a quantitative analysis of the coherence length $\xi$, but the absolute magnetic penetration depth $\lambda$, manifested here in the linewidth, may be reliably extracted to within $\sim 10\%$.

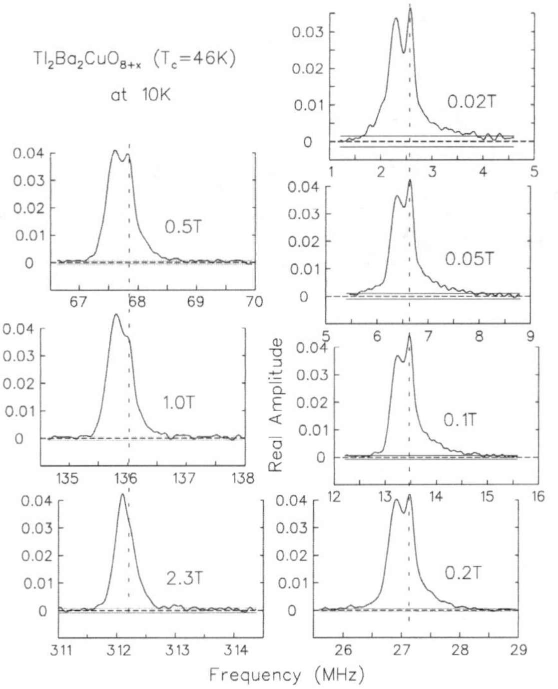


**Figure 2.** Field dependence of TF-$\mu^+$SR lineshape in a $T_c \approx 46$ K Tl-2201 mosaic at 10 K. At lower fields the lineshape develops a low-frequency tail, possibly due to stronger pinning disorder. At high fields the high-frequency "tail" is suppressed. At all fields the sharp features such as the "cusp" (van Hove singularity from saddle points between vortices) and high-frequency "cutoff" (due to the finite size of vortex cores) are "smeared out", presumably due to disorder in the vortex lattice.

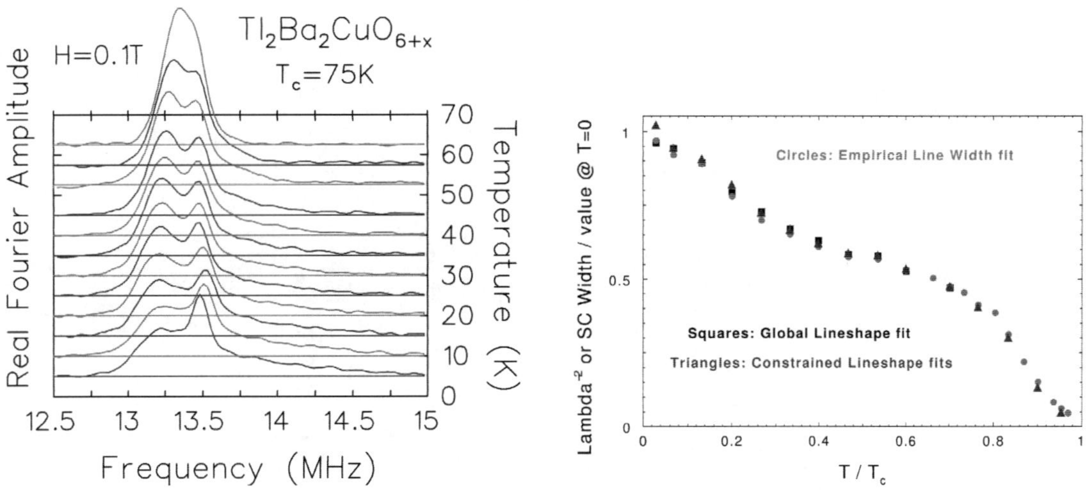


**Figure 3.** (color online) LEFT: temperature dependence of frequency spectra at $H = 0.1$ T for a $T_c = 75$ K Tl-2201 mosaic. RIGHT: normalized $\lambda^{-2}(T)/\lambda^{-2}(0)$ (squares and triangles) and similarly normalized empirical linewidth (circles) *vs.* reduced temperature $T/T_c$ for the same sample.

For the $T_c = 75$ K mosaic at $H = 0.1$ T, fits were also made to an empirical lineshape model composed of a Gaussian "cusp" signal, an exponential "tail" signal and an exponential "breadth" signal whose positions and widths are scaled in fixed ratios. This model gives excellent fits in a tiny fraction of the time required for the numerical simulations, but (a) the overall width (scaling factor) cannot be directly interpreted in terms of $\lambda^{-2}$; and (b) the normalized $T$-dependence of the "linewidth" usually has a different shape than that of $\lambda^{-2}$. Here (see Fig. 3) the two match perfectly, suggesting that in some cases this economical method may have validity; but it must always be "spot-checked"! All results shown in later Figures (including those for the $T_c = 75$ K mosaic) are from rigorous numerical lineshape simulation fits, *not* from the empirical model.

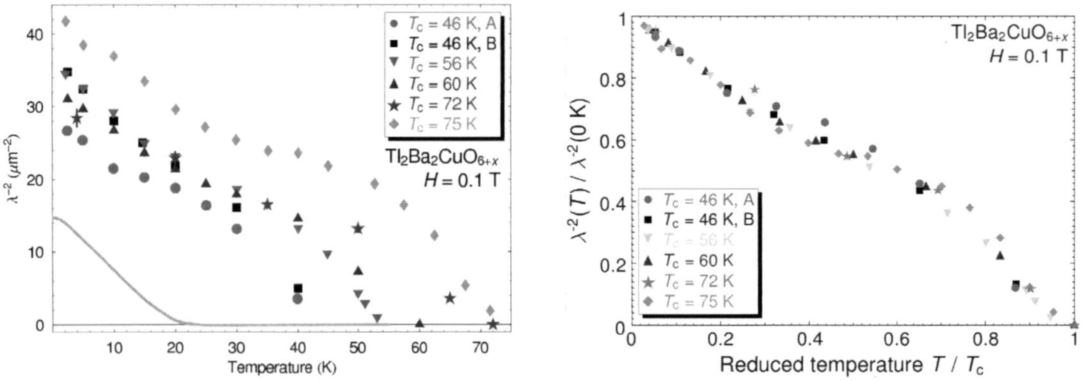


**Figure 4.** (color online) LEFT: temperature dependence of fitted $\lambda^{-2}$ values at $H = 0.1$ T for all Tl-2201 mosaics studied. The smooth curve is the Meissner phase microwave data[12] on a $T_c = 25$ K Tl-2201 crystal at 2.497 GHz. RIGHT: same data plotted as normalized $\lambda^{-2}(T)/\lambda^{-2}(0)$ *vs.* reduced temperature $T/T_c$. All dopings exhibit essentially the same $T/T_c$-dependence of the normalized superfluid density, with an inflection point at $T \sim 0.5T_c$. The fact that the susceptibility shows no such inflection point indicates that it is intrinsic to the vortex phase.

**Table 1.** Zero-temperature magnetic penetration depths in 0.1 T for overdoped Tl-2201 mosaics having various $T_c$s, from a linear extrapolation of $\lambda^{-2}(T)$ at low temperatures, with estimated uncertainties in parentheses.

| $T_c$ (K) | 46(1) (A) | 46(1) (B) | 56(1) | 60(1) | 72(1) | 75(1) |
|---|---|---|---|---|---|---|
| $\lambda_{ab}(0)$ (nm) | 187(2) | 165(2) | 166(1) | 175(1) | 182(2) | 153(2) |

Figure 4 (left) shows the extracted $\lambda^{-2}(T)$ for the six mosaics measured. A rather similar, and highly unusual, $T$-dependence is immediately apparent. The extent of this similarity is more striking in Fig. 4 (right), where $\lambda^{-2}$ is normalized to its linearly extrapolated $T = 0$ value (see Table 1) and plotted against the reduced temperature $T/T_c$ — the *relative* temperature dependence is almost identical. Its most intriguing feature, exhibited clearly in all six mosaics, is upward curvature around $\frac{1}{2}T_c$. This unusual temperature dependence is essentially model independent and is evident in any measure of the linewidth. The $T$-dependence of the SC carrier density in a single-gap $s$- or $d$-wave superconductor exhibits downward curvature over the entire temperature range $0 - T_c$.

The $d$-wave symmetry of the SC order parameter in Tl-2201 has been conclusively established by observation of half-integer flux quanta at crystal boundaries in films[13], and the presence of line nodes is evident in microwave measurements[14, 15, 12] and thermal transport[16]. Indeed, our data also show linear $T$-dependence below $\sim 0.2T_c$, characteristic of line nodes. Near $T_c$, our $T$-dependence is also not unusual.

The scaling of the $T$-dependence over a wide doping range on different mosaics strongly suggests that it is intrinsic to the vortices, rather than a phase transition within the SC dome, such as an extension of the $T^*$ pseudogap transition; moreover, the similarity with much more disordered overdoped LSCO[17] and cleaner, slightly underdoped $YBa_2Cu_4O_8$[18] implies that it is not peculiar to $Tl_2Ba_2CuO_{6+x}$.

A number of possible explanations for the self-similar $T$-dependence of $\lambda^{-2}$ are explored in a forthcoming publication[8]: multiband SC with independent gaps[19], dilute paramagnetic impurities[20] and phase transitions in the vortex lattice [21] are all ruled out by various arguments; a transition from a triangular to a square vortex lattice [22] remains a possibility, but preliminary fits to a square lattice show no improvement of $\chi^2$ relative to the usual triangular lattice model.

A 3D-2D "vortex lattice melting" transition would produce much more dramatic lineshape changes then are seen in these data; thermal "depinning", on the other hand, would produce a small *drop* in the apparent linewidth, if anything.

We conclude that the upward curvature arises from some fundamental property intrinsic to the $d$-wave vortices themselves. Franz *et al.*[23] predicted a small but ubiquitous admixture of $s$-wave order in the vicinity of a $d$-wave vortex core where $d$-wave order varies spatially due to the vortex lattice. Because the shielding current is only gapless in certain directions, the supercurrent path cannot be cylindrical about the vortex core, altering the field distribution. Intervortex interactions and the fraction of muons near vortex cores both increase with temperature as $\xi$ and $\lambda$ expand, and with field as the vortex density increases, but the reduced gradients of the order parameter mean that less $s$-wave component should be induced. An anomalous excess broadening of the $\mu^+$SR lineshape would indeed be introduced, but detailed modeling is required to determine whether this can successfully explain the observed temperature dependence.

The nature of SC in the cuprates remains an open question; the overdoped regime offers the promising prospect of understanding the normal state from which SC emerges. However, our $\mu^+$SR data indicate that the story here is not simple. An unusual $T$-dependence of the penetration depth, seen now in at least three distinct material families and inconsistent with microwave measurements of nominally the same quantity on equivalent crystals of Tl-2201[12], is most readily explained by a generic phase transition of the vortex lattice. Aside from the vortex physics and what we may learn about the gap symmetry of the material, this has important ramifications for other techniques. $\mu$SR is uniquely suited to extracting the absolute penetration depth, and the values obtained from this technique underpin results in other techniques which can't measure it in absolute terms (or at all). These absolute values, however, are model-dependent. A square vortex lattice at low temperature, for instance, would require recalculating a number of previous results throughout the field of $HT_c$SC, and would serve as a caution to researchers working on other systems that may have line nodes.

## Acknowledgments

This work was supported by the Canadian Institute for Advanced Research and the Natural Sciences and Engineering Research Council of Canada. We thank Ch. Bernhard, J.E. Sonier, D.M. Broun, and M. Franz for stimulating discussions, D. Deepwell and D.M. Broun for microwave data, and the TRIUMF CMMS staff for their assistance.

## References

[1] Wagner J L, Chmaissem O, Jorgensen J D, Hinks D G, Radaelli P G, Hunter B A and Jensen W R 1997 *Physica* C **277** 170–182
[2] Shimakawa Y 1993 *Physica* C **204** 247–261
[3] Liu R S, Hughes S D, Angel R J, Hackwell T P, Mackenzie A P and Edwards P P 1992 *Physica* C **198** 203–208
[4] Peets D C, Mottershead J D F, Wu B, Elfimov I S, Liang R, Hardy W N, Bonn D A, Raudsepp M, Ingle N J C and Damascelli A 2007 *New J. Phys.* **9** 28
[5] Platé M, Mottershead J D F, Elfimov I S, Peets D C, Liang R, Bonn D A, Hardy W N, Chiuzbaian S, Falub M, Shi M, Patthey L and Damascelli A 2005 *Phys. Rev. Lett.* **95** 077001
[6] Niedermayer C, Bernhard C, Binninger U, Glückler H, Tallon J, Ansaldo E and Budnick J 1993 *Phys. Rev. Lett.* **71** 1764
[7] Uemura Y J *et al* 1989 *Phys. Rev. Lett.* **62** 2317–2320
[8] Brewer J H, Stubbs S L, Liang R, Bonn D, Hardy W, MacFarlane W A and Peets D C 2014 Signatures of new *d*-wave vortex physics in overdoped $Tl_2Ba_2CuO_{6+x}$ revealed by TF-$\mu^+$SR (*Preprint arXiv:1404.4727*)
[9] Sonier J E, Brewer J H and Kiefl R F 2000 *Rev. Mod. Phys.* **72**(3) 769–811
[10] Peets D C, Liang R, Raudsepp M, Hardy W N and Bonn D A 2010 *J. Cryst. Growth* **312** 344–350
[11] Opagiste C, Triscone G, Couach M, Jondo T K, Junod A, Khoder A F and Muller J 1993 *Physica* C **213** 17–25
[12] Deepwell D, Peets D C, Truncik C J S, Murphy N C, Kennett M P, Huttema W A, Liang R, Bonn D A, Hardy W N and Broun D M 2013 *Phys. Rev.* B **88** 214509
[13] Tsuei C C, Kirtley J R, Ren Z F, Wang J H, Raffy H and Li Z Z 1997 *Nature* **387** 481–483
[14] Broun D M, Morgan D C, Ormeno R J, Lee S F, Tyler A W, Mackenzie A P and Waldram J R 1997 *Phys. Rev.* B **56** R11443–R11446
[15] Özcan S, Turner P J, Waldram J R, Drost R J, Kes P H and Broun D M 2006 *Phys. Rev.* B **73** 064506
[16] Hawthorn D *et al* 2007 *Phys. Rev.* B **75** 104518
[17] Khasanov R, Shengelaya A, Maisuradze A, Mattina F L, Bussmann-Holder A, Keller H and Muller K A 2007 *Phys. Rev. Lett.* **98** 057007
[18] Khasanov R, Shengelaya A, Karpinski J, Bussmann-Holder A, Keller H and Mller K 2008 *J. Supercond. Nov. Magn.* **21** 81–85 ISSN 1557-1939
[19] Zehetmayer M 2013 *Supercond. Sci. Technol.* **26** 043001
[20] Sonier J *Rep. Prog. Phys.* **70** 1717
[21] Zheng D, Li J, Lin C, Yan S, Fang L, Ni Y and Zhao Z 2000 *Physica* C **341–348** 1373–1374
[22] White J *et al* 2011 *Phys. Rev.* B **84** 104519
[23] Franz M, Kallin C, Soininen P I, Berlinsky A J and Fetter A L 1996 *Phys. Rev.* B **53** 5795

# $\mu$SR study of Al-0.67%Mg-0.77%Si alloys

**K Nishimura[1], K Matsuda[1], R Komaki[1], N Nunomura[2], S Wenner[3], R Holmestad[3], T Matsuzaki[4], I Watanabe[4], F L Pratt[5] and C D Marioara[6]**

[1] Faculty of Engineering, University of Toyama, Toyama, 930-8555, Japan
[2] Information Technology Center, University of Toyama, Toyama, 930-8555, Japan
[3] Department of Physics, NTNU, Hogskoleringen 5, Trondheim NO-7491, Norway
[4] Advanced Meson Science Laboratory, RIKEN Nishina Center for Accelerator Based Science, RIKEN, Wako, 351-0198, Japan
[5] ISIS Facility, Rutherford Appleton Laboratory, Chilton, OX11 0QX, UK
[6] Materials and Chemistry, SINTEF, Trondheim, NO-7491, Norway

E-mail: nishi@eng.u-toyama.ac.jp

**Abstract.** Zero-field muon spin relaxation measurements were carried out with Al-0.67%Mg-0.77%Si alloys in the temperature range from 20 K to 300 K. Observed relaxation spectra were compared with the relaxation functions calculated by a Monte Carlo simulation with four fitting parameters: the dipolar width, trapping rate, detrapping rate and fraction of initially trapped muons. From the fitting, the temperature variations of the trapping rates reveal that there are three temperature regions concerning muon kinetics. In the low temperature region below 120 K, muons appeared to be trapped in a shallow potential yielded by dissolved Mg atoms, and thus little effect of heat treatment of the samples was observed, while in the mid and high temperature regions, the trapping rates clearly depended on the heat treatment of the samples suggesting muon-cluster and/or muon-vacancy interactions.

## 1. Introduction

High energy-efficiency is a common requirement in industries, along which Al-Mg-Si alloys, commercially known as 6xxx series aluminum alloys, are in high demand as a material for vehicles because of their low weight, excellent formability and age-hardenability. It is a peculiarly attractive feature of 6xxx aluminum alloys that only 1 % Mg + Si solute atoms increase the mechanical hardness by a factor of approximately 5 from the pure aluminum after a proper heat treatment[1, 2]. This motivates intensive studies of physical properties and microstructures to facilitate a practical use of the alloys. Previous investigations with electrical resistivity[3, 4], transmission electron microscope (TEM)[5, 6], atom probe tomography (APT)[7, 8, 9], differential scanning calorimetry (DSC) [10] etc., illustrate a broadly acceptable relationship that a large number of small size precipitations of solute elements result in a high strength, but a small number of large size precipitations lessen the strength. A key factor has been considered to be a vacancy behavior which stimulates diffusion of Mg and Si and nucleation of clusters. The Mg-Si-vacancy clusters could lead to initial precipitations of Mg and Si. The vacancies and clusters are important, but they are too small to be observed directly. The positron annihilation spectroscopy (PAS) has been sometime used to monitor a number of vacancies in materials[11, 12].

We have commenced a muon spin relaxation spectroscopy to study a behavior of vacancies in Al-Mg-Si alloys. The observed muon spin relaxation functions give us rich information on muon hopping between vacancies and defects in Al alloys. Our recent work with Al-1.6%$Mg_2$Si, Al-0.5%Mg, and Al-0.5%Si alloys has demonstrated that the muon trapping rates ($\nu_t$) depended on the heat treatment and solute concentrations: 1) the concentration of dissolved Mg dominates the $\nu_t$ values at lower temperatures, 2) the $\nu_t$ values around 200 K reflect a number of clusters, 3) a natural aging (alloys are stored at room temperature) effect is clearly observed with Al-0.5%Si[12, 13, 14]. This contribution will report experimental results of muon spin relaxation versus temperature with Al-0.67%Mg-0.77%Si (balanced concentrations of Mg and Si, noted as 1.4B). Observed zero-field relaxation spectra were compared with the simulated functions by the Monte Carlo technique.

## 2. Experimental

Samples were prepared using pure elements of Al, Mg and Si, of which the purities are more than 99.99%. The materials were cast, rolled to 1 mm thickness sheets, and cut to dimensions of 25 × 25 × 1 mm plates. All the samples underwent a heat treatment at 575 °C for 1 hour and subsequently quenching in ice water (usually called as solution heat treatment, SHT). After SHT, three kinds of processes were employed: 1) $\mu$SR measurement started immediately after the quenching (within approximately 15 min). The data set with this process is noted as 1.4B-AQ. 2) A sample was stored at room temperature for 163 days before measurement (noted as 1.4B-163d). 3) Soon after SHT, a sample was annealed again at 70 °C for 1000 minutes (noted as 1.4B-70C).

Muons have been known to diffuse very fast in Al; hence several tens of ppm of impurities, such Mg, Mn, Cu, Ag, etc., produce significant influence on the depolarization rate[15, 16]. Since the base aluminum used in this study carries a few ppm of Si, Fe, Cu, Mn and Mg, we measured a temperature dependence of $\mu$SR spectra with the base Al to estimate a background from impurities. The $\mu$SR experiment was carried out at the RIKEN-RAL Muon Facility in Rutherford-Appleton laboratory[17]. We used the ARGUS muon spectrometer with a helium flow cryostat to control the temperature of the samples. The 20 to 60 million positron counts (events) were recorded at a constant temperature in the range from 20 to 300 K, increasing the temperature.

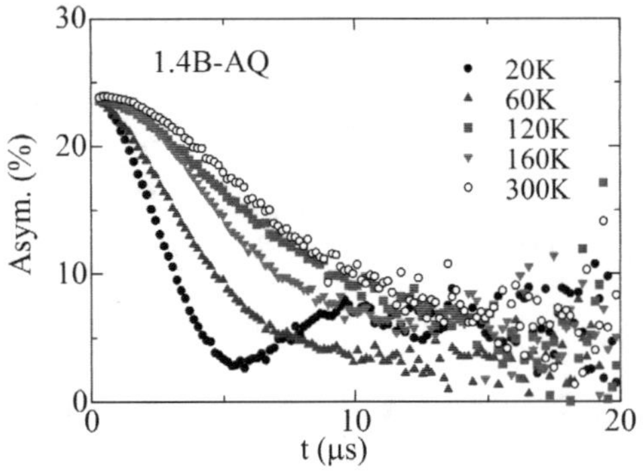


**Figure 1.** Zero-field muon spin relaxation spectra with the quenched Al-0.67%Mg-0.77%Si sample observed at selected temperatures from 20 to 300 K.

## 3. Results and discussions

Figure 1 shows the zero-field relaxation spectra at selected temperatures of the sample, 1.4B-AQ. It is obvious that the muon spin depolarization rate depended on temperature. The spectrum at 20 K appears a typical shape described by the Kubo-Toyabe function [18]:

$$g(t) = \frac{1}{3} + \frac{2}{3}\left(1 - \Delta^2 t^2\right)\exp\left(-\frac{1}{2}\Delta^2 t^2\right).$$

In the present samples, Al nuclei dominantly generate dipole fields causing muon spin relaxation; contributions from Mg and Si nuclei are negligibly small. At the present experimental temperatures, the nuclear magnetic moments of Al are randomly oriented, and thus we assume that the resultant averaged field acting on trapped muons in a certain site has a Gaussian distribution, of which the field vector oriented randomly. The parameter $\Delta$ is the dipolar width of the Gaussian distributed field. As the temperature was raised, the depolarization rate was gradually decreased due to detrapping of muons by thermal agitation. The diffusion of muons effects not only on a correlation time between the trapped muons and the magnetic field, but also the dipolar width, which would decrease as motional narrowing; consequently data analysis by only the Kubo-Toyabe function is difficult. It is worth mentioning that the depolarization at 120 K is slower than that at 160 K as seen in the relaxation spectra in Fig. 1.

We have successfully interpreted the observed relaxation spectra using a Monte Carlo simulation, in which four variable parameters were employed: the dipolar width ($\Delta$), trapping rate ($\nu_t$), detrapping rate ($\nu_d$), and fraction of initially trapped muons ($p_0$)[13, 14]. This method has been introduced in the literature[19, 20]. An ensemble of 60 million muons are simulated to produce a five dimensional relaxation function $f$ ($\Delta$, $\nu_t$, $\nu_d$, $p_0$, *time*), in which a muon spin is assumed to depolarize only when it is trapped, but no relaxation occurs during diffusion (two state model). Simulated relaxation functions are compared with the experimental ones to extract the best fit parameters.

The results of fitting are shown in Figs. 2 $\sim$ 5. The deduced $\Delta$ values in Fig. 2 of the alloy samples are larger than that of base Al, which is most likely ascribed to the solute Mg and Si elements since randomly substituting Mg and Si for Al sites lead to inhomogeneous magnetic fields at muon sites. The fraction of initially trapped muons in Fig. 3 is almost independent of temperature, and there is little variation among the present samples. The temperature variations of the trapping rates of the three 1.4B samples in Fig. 4 closely resemble each other in the low temperature region; the $\nu_t$ decreases as temperature increases. It means that the heat treatments of the samples made little influence on the trapping rates below 120 K. This property of $\nu_t$ at low temperatures has been also observed with the alloys of Al+1.6%$Mg_2Si$ and Al+0.5%Mg, but

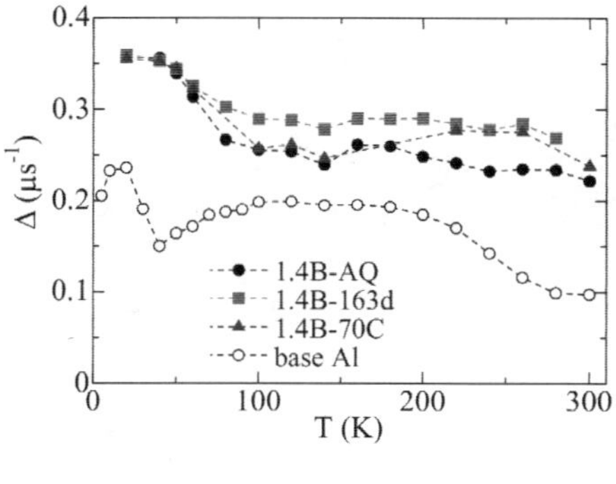


**Figure 2.** Temperature variations of dipolar width for Al-0.67%Mg-0.77%Si and base Al samples.

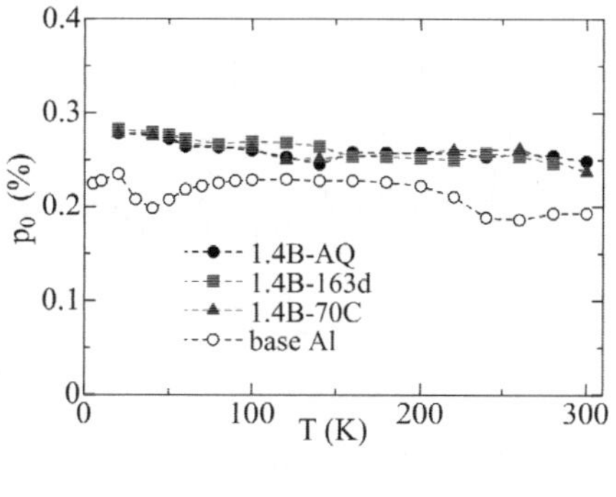


**Figure 3.** Temperature variations of fraction of initially trapped muons for Al-0.67%Mg-0.77%Si and base Al samples.

Journal of Physics: Conference Series **551** (2014) 012031 doi:10.1088/1742-6596/551/1/012031

not with Al+0.5%Si alloys[14]. The $\nu_t$ variations below 120 K are, therefore, related with the dissolved Mg atoms via Mg-muon interactions. There are possibly a large number of dissolved Mg atoms which provide shallow potentials attracting positive muons. Oppositely the detrapping rates at low temperature in Fig. 5 gradually increase with increasing temperature, implying a thermal energy makes trapped muons to escape from the shallow potential.

According to our previous studies[13, 14], the trapping rates around 200 K were found to be dominated by the Mg-Si-vacancy clusters since the heat treatment of samples significantly affected the extracted $\nu_t$ values. The sample heat-treated at a proper temperature and a proper period of time would have a high density of small Mg-Si-vacancy clusters, resulting in a high $\nu_t$ value as seen with the 1.4B-70C sample in Fig. 4; while the trapping rates near 300 K were relatively large like as the quenched 1.4B-AQ sample. Similar phenomena have been found with the other quenched samples of Al-Mg-Si, Al-Mg, and Al-Si[13, 14]. It is therefore reasonable to consider that the high $\nu_t$ values around 300 K with the 1.4B-AQ are associated with vacancies via muon-vacancy interactions.

In order to make the above interpretations clearer, the trapping rates associated with solute elements of Mg and Si were estimated by subtracting the $\nu_t$ values of base Al from those of 1.4B

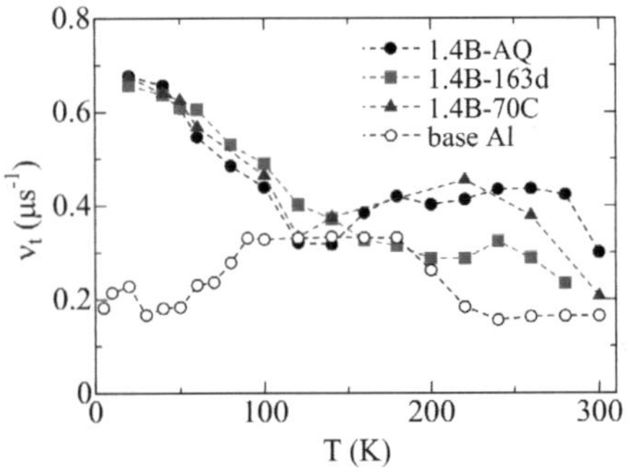


**Figure 4.** Temperature variations of trapping rate for Al-0.67%Mg-0.77%Si and base Al samples.

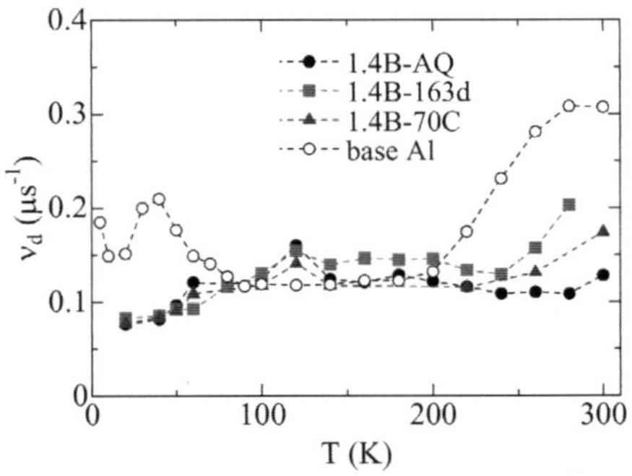


**Figure 5.** Temperature variations of detrapping rate for Al-0.67%Mg-0.77%Si and base Al samples.

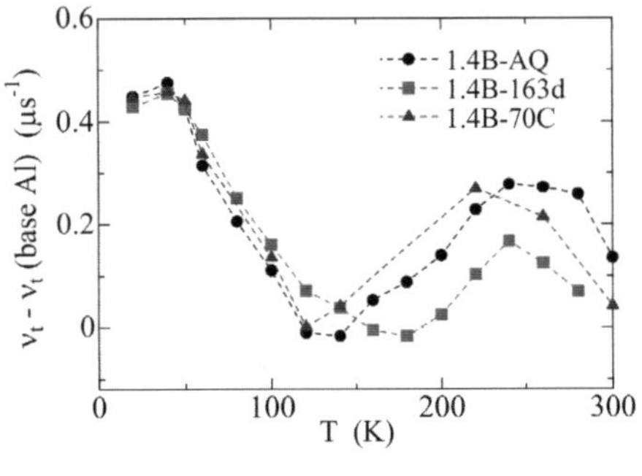


**Figure 6.** Temperature variations of trapping rate caused by solute Mg and Si atoms for Al-0.67%Mg-0.77%Si.

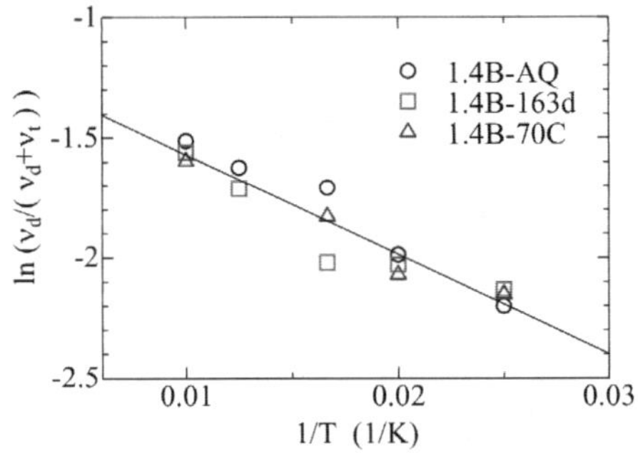


**Figure 7.** Arrhenius plot of the equilibrium fraction of free muons in the temperature range 40 - 100 K.

samples at each temperature point in Fig. 6. The trapping rates at low temperature appear overlapping considerably. A density of Mg-Si-vacancy clusters would reflect on the trapping rates at 220 K; that is, the largest with the heat treated sample, 1.4B-70C. The larger trapping rates with the 1.4B-AQ above 260 K most likely result from a larger number of vacancies. It is interesting that the 1.4B-163d alloy stored at room temperature forms a less number of the Mg-Si-vacancy clusters, as it is clear from the trapping rates around 200 K. This explanation would be supported by the study of atom probe tomography; an appropriately heat treated Al-Mg-Si alloy produced about five times more clusters than that stored at room temperature[9]. It should be noted that absolute values of density and size of clusters largely depends on the solute Mg and Si concentrations, as well as a heat treatment condition[10].

It is meaningful to investigate binding energies of muons in Al-Mg-Si alloys to clarify structure of Mg-Si-vacancy clusters. Although a typical Arrhenius plot is unlikely accessible in the mid and high temperature regions, the data points in Fig. 6 at low temperatures yield a reasonably linear Arrhenius plot, as shown in Fig. 7. Since little effect of the heat treatment on the trapping rates was noticeable in the low temperature region, we fit all the data from 40 K to 100 K to a straight line, which provides a activation energy of 3.6 ($\pm$0.4) meV. This value is comparable to 6.9 meV found with Al-1.6%$Mg_2Si$ in the similar temperature region[14]. The activation energy with a dilute Mg and well annealed alloy (Al-250ppmMg) has been reported to be 11 meV[19]. The order-of-magnitude agreement between these energy values supports our interpretation that muons are trapped by a shallow potential by dissolved Mg atoms in the low temperature region.

The present results of trapping rates suggested that a storage at room temperature of Si rich 6xxx Al alloys would bring about a significant negative effect on the mechanical strength because the trapping rate with 1.4B-163d related with a number of clusters is rather smaller than those of Al-1.6%$Mg_2Si$ and Al-1.0%$Mg_2Si$ stored in the same condition[14]. If a long period of room temperature storage is inevitable after SHT, it is desirable to control the concentration ratio of Mg/Si to be around two.

## Acknowledgments
One of the authors (K. Nishimura) acknowledges JSPS KAKENHI No.25289260.

## References
[1] Matsuda K, Taniguchi S, Kido K, Uetani Y and Ikeno S 2002 *Mater. Trans.* **43** 2789
[2] Moons T, Ratchev P, Smet P D, Verlinden B and Houtte P V 1996 *Scripta Mater.* **35** 939
[3] Seyedrezai H, Grebennikov D, Mascher P and Zurob H S 2009 *Mater. Sci. Eng.* A **525** 186
[4] Gaffar M A, Gaber A, Mostafa M S and Zeid E F A 2007 *Mater. Sci. Eng.* A **465** 274
[5] Matsuda K, Gamada H, Fujii K, Uetani Y, Sato T, Kamio A and Ikeno S 1998 *Metall. Mater. Trans.* A **29** 1161
[6] Marioara C D, Nordmark H, Andersen S J and Holmestad R 2006 *J. Mater. Sci.* **41** 471
[7] Murayama M and Hono K 1999 *Acta Mater.* **47** 1537
[8] Serizawa A, Hirosawa S and Sato T 2008 *Matall. Mater. Trans* A **39** 243
[9] Torsæter M, Hasting H S , Lefebvre W, Marioara C D, Walmsley J C and Andersen S J and Holmestad R 2010 *J Appl. Phys.* **108** 073527
[10] Kim S, Kim J Tezuka H, Kobayashi E and Sato T 2013 *Mater. Trans.* **54** 297
[11] Banhart J, Lay M D H, Chang C S T and Hill A J 2011 *Phys. Rev.* B **83** 014101
[12] Wenner S, Matsuda K, Nishimura K, Banhart J, Matsuzaki T, Tomono D, Pratt F L, Liu M, Yan Y, Marioara C D and Holmestad R 2012 *13th International Conference on Aluminum Alloys (Pittsburgh)* ed Heiland W, Rollett A D and Cassada W (Wiley) pp 37-42
[13] Wenner S, Holmestad R, Matsuda K, Nishimura K, Matsuzaki T, Tomono D, Pratt F L and Marioara C D 2012 *Phys. Rev.* B **86** 104201
[14] Wenner S, Nishimura K, Matsuda K, Matsuzaki T, Tomono D, Pratt F L, Marioara C D and Holmestad R 2013 *Acta Mater.* **61** 6082
[15] Kehr K W, Richter D, Welter J -M, Hartmann O, Karlsson E, Norlin L O, Niinikoski T O, Yaouanc A 1982 *Phys. Rev.* B **26** 567

[16] Hartmann O, Karlsson E, Wäckelgård E, Wäppling R, Richter D, Hempelmann R, Niinikoski T O 1988 *Phys. Rev.* B **37** 4425
[17] Matsuzaki , Ishida K, Nagamine K, Watanabe I, Eaton G H and Williams W G 2001 *Nucl. Instr. Methods* A **465** 365
[18] Hayano R S, Uemura Y J, Imazato J, Nishida N, Yamazaki T and Kubo R 1979 *Phys. Rev.* B **20** 85
[19] Sato E, Hatano T, Suzuki Y, Imafuku M, Sunaga M, Doyama M, Morozumi Y, Suzuki T and Nagamine K 1984 *Hyper. Inter.* **17-19** 203
[20] Hatano T, Suzuki Y, Doyama M, Uemura Y J, Yamazaki T and Brewer J H 1984 *Hyper. Inter.* **17-19** 211

# $\beta$-NMR of $^8Li^+$ in rutile $TiO_2$

**R M L McFadden[1], D L Cortie[1,2,3], D J Arseneau[4], T J Buck[2], C-C Chen[5], M H Dehn[6], S R Dunsiger[6], R F Kiefl[2], C D P Levy[4], C Li[7], G D Morris[4], M R Pearson[4], D Samuelis[5], J Xiao[1], J Maier[5] and W A MacFarlane[1]**

[1] Department of Chemistry, University of British Columbia, 2036 Main Mall, Vancouver, BC V6T 1Z1, Canada
[2] Department of Physics and Astronomy, University of British Columbia, 6224 Agricultural Road, Vancouver, BC V6T 1Z1, Canada
[3] Quantum Matter Institute, University of British Columbia, Vancouver, BC V6T 1Z4
[4] TRIUMF, 4004 Wesbrook Mall, Vancouver, BC V6T 2A3, Canada
[5] Max-Planck-Institut für Festkörperforschung, D-70569 Stuttgart, Germany
[6] Physik-Department, Technische Universität München, James-Franck-Straße 1, 85748 Garching bei München, Germany
[7] Shanghai Institute of Ceramics, Chinese Academy of Sciences, 1295 Dingxi Road, Shanghai, P.R. China 200050

E-mail: wam@chem.ubc.ca

**Abstract.** We report preliminary low-energy $\beta$-NMR measurements of $^8Li^+$ implanted in single crystal rutile $TiO_2$ at an applied field of 6.55 T and 300 K. We observe a broad 12 kHz wide quadrupole split resonance with unresolved features and a sharp component at the Larmor frequency. The line broadening may be caused by overlapping multi-quantum transitions or motion of $^8Li^+$ on the scale of its lifetime (1.21 s). We also find spin-lattice relaxation that is relatively fast compared to other wide band gap insulators. The origin of this fast relaxation is also likely quadrupolar and may be due to anisotropic $^8Li^+$ diffusion.

## 1. Introduction

Like the many $TiO_2$ polymorphs, rutile has applications as an anode material in lithium-ion batteries [1, 2]. Its tetragonal structure is comprised of edge-sharing $TiO_6$ octahedra that extend along the $c$-axis, forming natural channels which Li can occupy during electrochemical intercalation. For incorporation into functional devices, a detailed understanding of the diffusivity of $Li^+$ is necessary. Macroscopic methods (e.g., impedance spectroscopy) are often used to probe ionic diffusion, but are unable to distinguish between different conductive pathways. The applicability of NMR to probe dynamics in solids is well known [3], and can be especially useful in resolving multiple diffusive pathways activated in different thermal regions [4]. Li dynamics in $TiO_2$ polymorphs (and related compounds) have been studied by conventional NMR [5, 6, 7, 8, 9, 10], but no such investigation has been undertaken for rutile. We report here a preliminary look at $Li^+$ implanted (i.e., not doped) in rutile titania using beta-detected NMR ($\beta$-NMR) at room temperature and high magnetic field.

Journal of Physics: Conference Series **551** (2014) 012032 doi:10.1088/1742-6596/551/1/012032

**Figure 1.** Cleaved rutile $TiO_2$ (100) substrate from Crystal GmbH mounted in the $\beta$-NMR sample holder. The pale yellow colour is indicative of a minor presence of defects.

## 2. Experimental

A 10 mm × 10 mm single crystal rutile $TiO_2$ (100) substrate was purchased from Crystal GmbH. The rutile polymorph is tetragonal ($a = b = 0.4584$ nm, $c = 0.2953$ nm) with space group $P4_2/mnm$. The substrate appears transparent pale yellow in colour, qualitatively indicating a minor presence of defects [11]. The sample was cleaved for mounting on a cold finger cryostat (see Figure 1).

$\beta$-NMR experiments were performed at TRIUMF's Isotope Separator and Accelerator (ISAC) facility using a low-energy ($\sim$ 20 keV) radioactive beam of spin-polarized $^8Li^+$ with a typical flux of $\sim 10^6$ ions $s^{-1}$ and spot size of $\sim$ 2 mm in diameter. The probe nucleus, $^8Li$, has spin $I = 2$, a quadrupole moment $Q = +31.4$ mb, a gyromagnetic ratio $\gamma = 6.3015$ MHz $T^{-1}$, and a mean lifetime $\tau = 1.21$ s. A large nuclear spin-polarization ($\sim$ 70%) is achieved by collinear optical pumping with resonant circularly polarized laser light [12] such that the direction of spin-polarization is aligned parallel or antiparallel to the static applied magnetic field, $B_0$. A high-voltage bias is used to decelerate the beam prior to implantation. Depth-profiles were estimated from Monte Carlo simulations of $10^6$ ions using SRIM [13] (see Figure 2). Details of the spectrometer can be found elsewhere [14, 15].

Resonance measurements were performed using a continuous $^8Li^+$ beam with a transverse RF-field $H_1$ operated in continuous wave (CW) mode. Application of an RF-field at the appropriate

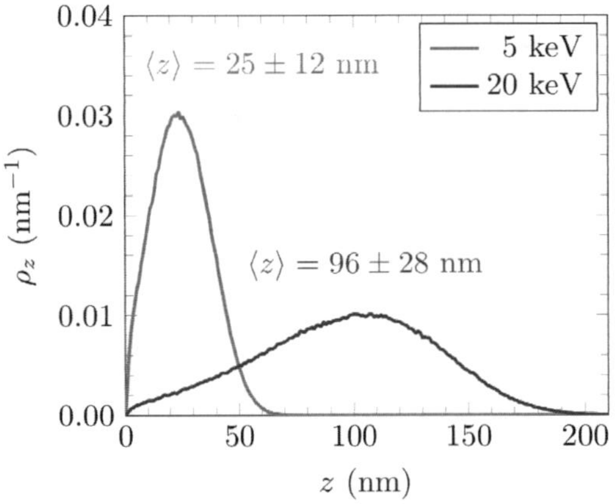


**Figure 2.** SRIM stopping profiles for $10^6$ $^8Li^+$ ions implanted in $TiO_2$, histogrammed into 1 nm wide bins. Here, $z$ is the distance from the substrate surface and $\rho_z$ is the stopping probability density. Mean and root mean square implantation depths are inset over the profiles.

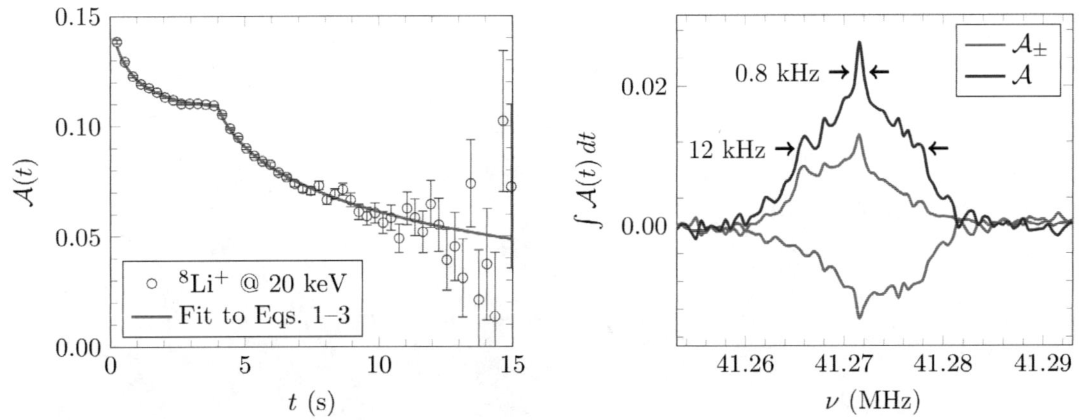


**Figure 3.** $\beta$-NMR spectra of $^8Li^+$ implanted in rutile $TiO_2$ (100) at 20 keV and 300 K. *Left*: Spin-lattice relaxation — time-differential asymmetry showing the temporal evolution of $^8Li^+$ spin-polarization. *Right*: CW resonance spectra — time-integrated asymmetry of the positive and negative helicities, $\mathcal{A}_\pm$, and the combined asymmetry, $\mathcal{A}$, from both helicities.

resonant frequency causes rapid $^8$Li spin-precession perpendicular to the $H_1$ field-axis, destroying $^8$Li polarization, observed as a change in time-integrated asymmetry. Spin-lattice relaxation (SLR) measurements were performed using a pulsed beam and no RF-field. The time-differential asymmetry, which is proportional to $^8$Li spin-polarization, is monitored both during and following a 4 second $^8Li^+$ beam pulse. All measurements were performed in a static magnetic field $B_0 = 6.55$ T $\parallel$ the rutile $TiO_2$ (100) (i.e., the $a$-axis).

## 3. Results & Discussion

Figure 3 (*left*) shows the spin-lattice relaxation (SLR) spectrum at 300 K of $^8Li^+$ implanted in rutile $TiO_2$ at 20 keV. The time-evolution of $^8Li^+$ asymmetry follows [16]:

$$\mathcal{A}(t) = \frac{\int_0^t \exp\left[-\left(t-t'\right)/\tau\right] f\left(t,t';T_1\right)\,dt'}{\int_0^t \exp\left[-t'/\tau\right]\,dt'} \qquad t \leq \Delta \tag{1}$$

$$= \frac{\int_0^\Delta \exp\left[-\left(\Delta-t'\right)/\tau\right] f\left(t,t';T_1\right)\,dt'}{\int_0^\Delta \exp\left[-t'/\tau\right]\,dt'} \qquad t > \Delta \tag{2}$$

where $\tau$ is the mean lifetime of $^8$Li, and $\Delta$ is the duration of the beam pulse, corresponding to the pronounced kink at $t = 4$ s. $f\left(t,t';T_1\right)$ is the longitudinal relaxation function phenomenologically defined as a stretched exponential:

$$f\left(t,t';T_1\right) = \mathcal{A}_0 \exp\left[-\left\{\left(t-t'\right)/T_1\right\}^\beta\right], \tag{3}$$

where $\mathcal{A}_0$, $T_1$, and $\beta$ are the initial asymmetry, SLR relaxation time, and stretching exponent ($0 \leq \beta \leq 1$). A best fit of Eqs. 1–3 to the SLR spectra in Figure 3 (*left*) yields: $\mathcal{A}_0 = 0.166 \pm 0.004$, $\beta = 0.386 \pm 0.020$, and $T_1^{-1} = 0.141 \pm 0.005$ s$^{-1}$. Such a relaxation rate is relatively high compared to other insulating metal oxides [17], where $^8Li^+$ can be nearly non-relaxing.[1] As rutile is non-cubic and virtually without nuclear moments ($^{47}$Ti: $I = 3/2$, 7.28%; $^{49}$Ti: $I = 5/2$,

[1] See e.g., W A MacFarlane et al. and Z Salman et al. in these proceedings.

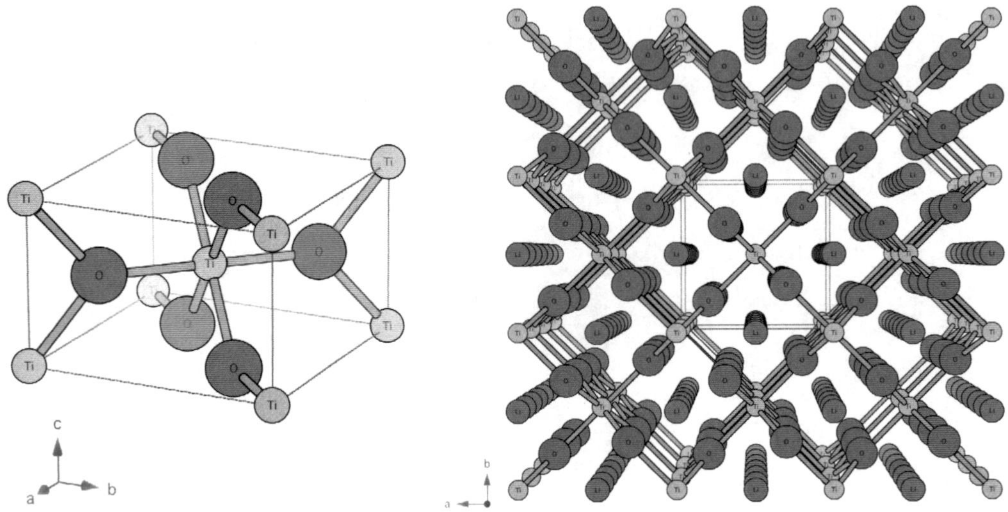


**Figure 4.** Crystal structure of rutile $TiO_2$ — oxygen and titanium atoms appear as red and light blue spheres, with bonded atoms connected by grey cylinders. *Left*: Unit cell. *Right*: View along the $c$-axis of a $3 \times 3$ rutile supercell. Octahedral Li sites are shown as dark blue spheres and the dashed blue line encloses the unit cell.

5.51%, $^{17}O$: $I = 5/2$, 0.038%), it is likely that the relaxation is predominantly quadrupolar in nature. It is worth noting that a previous $\beta$-NMR investigation of $^8Li^+$ in rutile $TiO_2$ reported a 30% preservation of nuclear spin-polarization at low magnetic fields in stark contrast to 100% retention for other implanted nuclear probes (e.g., $^{12}B$) [18]. This is consistent with our observation of appreciable $^8Li^+$ relaxation. Additionally, we find a depth dependence to the spin-lattice relaxation rate at 300 K, with $T_1^{-1}$ decreasing by a factor of $\sim 5$ when the implantation energy is decreased from 20 keV to 5 keV. The origin of this depth-dependence is not well-understood at this time.

A CW resonance spectrum of 20 keV $^8Li^+$ implanted in rutile $TiO_2$ at 300 K is shown in Figure 3 (*right*). The asymmetric time-integrated asymmetries from spin-polarization with different helicities ($\mathcal{A}_\pm$) are characteristic of a quadrupole split resonance [17, 19]. From the combined asymmetry ($\mathcal{A}$), we find a broad, $12.48 \pm 0.12$ kHz wide quadrupole split resonance with unresolved features. This is in stark contrast to the $^8Li^+$ resonance in Bi, where the four quadrupole satellite and three double-quantum transitions are well-resolved,[2] or in the perovskite $SrTiO_3$, where the satellite splitting is greater by about an order of magnitude [17]. X-ray diffraction and reflectometry measurements do not reveal the presence of impurity phases, though static disorder (e.g., from oxygen vacancies, as indicated by the discolouration of the crystal, or transition metal impurities) is a possible cause for the broadening. A sharp, large amplitude, $0.76 \pm 0.05$ kHz wide resonance is also observed at the Larmor frequency, $\nu_0 = 41.2715$ MHz. This may be a multi-quantum transition (e.g., the double-quantum $|\Delta m| = 2$ transition between $m = \pm 1$ states). Such transitions appear as large sharp (narrow) enhancements in the resonance spectrum and can be especially prevalent in CW $\beta$-NMR, where strong RF fields are often used [20], as is the case here. This can easily be tested by comparing spectra at different

[2] See W A MacFarlane et al. in these proceedings.

values of the RF magnetic field amplitude $H_1$. A multiquantum resonance will be attenuated very strongly by a reduced $H_1$.

Another possibility is that the observed pattern is due to Li residing in multiple sites with overlapping resonances. Molecular dynamics (MD) and density functional theory (DFT) calculations of lithiated rutile ($Li_xTiO_2$: $0 < x \leq 1$) [21, 22, 23, 24, 25] predict preferential sites for lithium at octahedral and tetrahedral oxygen-coordinated interstitial sites. Figure 4 shows the octahedral site for Li along the rutile $c$-axis, which is predicted to be energetically favourable over the tetrahedral site by $\sim 0.8$ eV [21, 22].

It is also possible that the sharp component of the resonance in Figure 3 (*right*) is from a motionally narrowed fraction of $^8Li^+$, as it experiences an average electric field gradient (EFG) from site hopping during its lifetime. MD and DFT calculations also predict fast, anisotropic 1D diffusion of $Li^+$ along the rutile $c$-axis with a low hopping-barrier of $\sim 0.05$ eV [21, 22, 24, 25].[3] These barriers are significantly lower than what is observed experimentally from macroscopic methods [26, 27], though such measurements are done far from the limit of infinitely dilute lithium, where the potential energy landscape is likely altered by lattice distortions and $Li^+$–$Li^+$ interactions [22, 28]. The broad component of the resonance may be due to fast-diffusion of $^8Li^+$ to Frenkel pairs, formed from irradiation, before they heal. Motion of $^8Li^+$ during its lifetime may also explain the relatively large SLR rate observed and measuring its temperature dependence would be a way of testing this. It will be interesting to see if the predicted low-barrier hopping can be probed with $\beta$-NMR, which is uniquely suited for diffusion measurements at such dilute lithium concentrations.

## 4. Summary

In a preliminary study of $^8Li^+$ in rutile $TiO_2$ at 300 K and $B_0 = 6.55$ T $\parallel$ the rutile (100) using $\beta$-NMR, we find a relatively fast relaxation compared to other wide band gap insulators and an unresolved quadrupole split resonance in contrast to well-resolved quadrupolar patterns in other materials. We propose that these may be due to motion of $^8Li^+$ during its lifetime. A full temperature scan of SLR rates and resonance lineshapes (at different RF powers) in future experiments will help determine the relaxation mechanism and elucidate the behaviour of $Li^+$ in rutile titania.

## References

[1] Hu Y S, Kienle L, Guo Y G and Maier J 2006 *Adv. Mater.* **18** 1421–1426
[2] Reddy M V, Subba Rao G V and Chowdari B V R 2013 *Chem. Rev.* **113** 5364–5457
[3] Kuhn A, Kunze M, Sreeraj P, Wiemhöfer H D, Thangadurai V, Wilkening M and Heitjans P 2012 *Solid State Nucl. Magn. Reson.* **42** 2–8
[4] Kuhn A, Sreeraj P, Pöttgen R, Wiemhöfer H D, Wilkening M and Heitjans P 2011 *J. Am. Chem. Soc.* **133** 11018–11021
[5] Wagemaker M, van de Krol R, Kentgens A P M, van Well A A and Mulder F M 2001 *J. Am. Chem. Soc.* **123** 11454–11461
[6] Wagemaker M, Kentgens A P M and Mulder F M 2002 *Nature* **418** 397–399
[7] Wilkening M, Amade R, Iwaniak W and Heitjans P 2007 *Phys. Chem. Chem. Phys.* **9** 1239–1246
[8] Wilkening M, Iwaniak W, Heine J, Epp V, Kleinert A, Behrens M, Nuspl G, Bensch W and Heitjans P 2007 *Phys. Chem. Chem. Phys.* **9** 6199–6202
[9] Wilkening M, Lyness C, Armstrong A R and Bruce P G 2009 *J. Phys. Chem.* C **113** 4741–4744
[10] Bottke P, Ren Y, Hanzu I, Bruce P G and Wilkening M 2014 *Phys. Chem. Chem. Phys.* **16** 1894–1901
[11] Diebold U 2003 *Surf. Sci. Rep.* **48** 53–229
[12] Levy C D P, Pearson M R, Kiefl R F, Mané E, Morris G D and Voss A 2014 *Hyperfine Interact.* **225** 165–172
[13] Ziegler J F, Ziegler M and Biersack J 2010 *Nucl. Instrum. Methods Phys. Res.* B **268** 1818–1823
[14] Morris G D, MacFarlane W A, Chow K H, Salman Z, Arseneau D J, Daviel S, Hatakeyama A, Kreitzman

[3] Barriers for diffusion in the $ab$-plane are predicted to be larger by an order of magnitude.

S R, Levy C D P, Poutissou R, Heffner R H, Elenewski J E, Greene L H and Kiefl R F 2004 *Phys. Rev. Lett.* **93** 157601
[15] Morris G D 2014 *Hyperfine Interact.* **225** 173–182
[16] Salman Z, Kiefl R F, Chow K H, Hossain M D, Keeler T A, Kreitzman S R, Levy C D P, Miller R I, Parolin T J, Pearson M R, Saadaoui H, Schultz J D, Smadella M, Wang D and MacFarlane W A 2006 *Phys. Rev. Lett.* **96** 147601
[17] MacFarlane W A, Morris G D, Chow K H, Baartman R A, Daviel S, Dunsiger S, Hatakeyama A, Kreitzman S, Levy C, Miller R, Nichol K, Poutissou R, Dumont E, Greene L H and Kiefl R F 2003 *Physica* B **326** 209–212
[18] Ogura M, Minamisono K, Sumikama T, Nagatomo T, Iwakoshi T, Miyake T, Hashimoto K, Kudo S, Arimura K, Ota M, Akutsu K, Sato K, Mihara M, Fukuda M, Matsuta K, Akai H and Minamisono T 2001 *Hyperfine Interact.* **136–137** 195–199
[19] Parolin T J, Shi J, Salman Z, Chow K H, Dosanjh P, Saadaoui H, Song Q, Hossain M D, Kiefl R F, Levy C D P, Pearson M R and MacFarlane W A 2009 *Phys. Rev.* B **80** 174109
[20] Dubbers D, Dörr K, Ackermann H, Fujara F, Grupp H, Grupp M, Heitjans P, Körblein A and Stöckmann H J 1977 *Z. Phys.* A **282** 243–248
[21] Koudriachova M V, Harrison N M and de Leeuw S W 2001 *Phys. Rev. Lett.* **86** 1275–1278
[22] Koudriachova M V, Harrison N M and de Leeuw S W 2002 *Phys. Rev.* B **65** 235423
[23] Koudriachova M V, Harrison N M and de Leeuw S W 2003 *Solid State Ionics* **157** 35–38
[24] Kerisit S, Rosso K M, Yang Z and Liu J 2009 *J. Phys. Chem.* C **113** 20998–21007
[25] Yildirim H, Greeley J P and Sankaranarayanan S K R S 2012 *Phys. Chem. Chem. Phys.* **14** 4565–4576
[26] Johnson O W 1964 *Phys. Rev.* **136** A284–A290
[27] Bach S, Pereira-Ramos J and Willman P 2010 *Electrochim. Acta* **55** 4952–4959
[28] Koudriachova M V, Harrison N M and de Leeuw S W 2002 *Comput. Mater. Sci.* **24** 235–240

Journal of Physics: Conference Series **551** (2014) 012033 doi:10.1088/1742-6596/551/1/012033

# $^8Li^+$ $\beta$-NMR in the Cubic Insulator MgO

**W A MacFarlane[1], T J Parolin[1], D L Cortie[1,2,3], K H Chow[4], M D Hossain[2], R F Kiefl[2], C D P Levy[5], R M L McFadden[1], G D Morris[5], M R Pearson[5], H Saadaoui[2,6], Z Salman[5,6], Q Song[2] and D Wang[2]**

[1] Chemistry Department, University of British Columbia, Vancouver, BC, V6T 1Z1, Canada
[2] Department of Physics and Astronomy, University of British Columbia, Vancouver, BC, V6T 1Z1, Canada
[3] Quantum Matter Institute, University of British Columbia, Vancouver, BC, V6T 1Z1, Canada
[4] Department of Physics, University of Alberta, Edmonton, AB, T6G 2E1, Canada
[5] TRIUMF, Vancouver, B.C. Canada, V6T 2A3

E-mail: wam@chem.ubc.ca

**Abstract.** We present extensive high magnetic field $\beta$-NMR measurements of $^8Li^+$ implanted in single crystals of MgO. The narrow resonance, consistent with a cubic $^8Li^+$ site, likely the tetrahedral interstititial, is used routinely as a reference for shift measurements. We show the intrinsic linewidth is on the order of 200 Hz, allowing a frequency determination to an accuracy of a few Hz. We find no implantation energy dependence of the resonance within a few ppm, but there is evidence of slow spin dynamics in hole-burning measurements. The spin lattice relaxation is slow. The temperature dependence reveals interesting changes at low temperature whose origin remains uncertain.

## 1. Introduction

The high field $^8Li^+$ $\beta$-NMR resonance in rocksalt MgO is very narrow, with minimal nuclear dipolar broadening and no quadrupolar splitting, the latter implying a site with cubic symmetry for the implanted ion. The resonance in MgO is commonly used as a $\beta$-NMR frequency reference, from which the resonance shifts in other materials are measured. It will also serve as a useful reference for $^{31}Mg$ $\beta$-NMR[1] currently being developed at both ISAC and ISOLDE. Here the implanted ion will be isotopic with the stable Mg in the host. MgO is often used as a film growth substrate, and may have interesting magnetic properties when doped[2] or at surfaces[3]. It is also an important comparison with other rocksalt compounds such as EuO and LiF. Li doped MgO (MgO:Li) is also useful as a heterogeneous catalyst[4], and aside from chemical doping routes, $Li^+$ ion implantation has also been studied recently as a means to modify MgO[5]. In this context, local information on the Li site and defect electronic structure for truly isolated $Li^+$ would be a valuable addition to understanding MgO:Li. Here we report extensive characterization of the $^8Li^+$ $\beta$-NMR in MgO in high magnetic field.

[6] Present address: Paul Scherrer Institute, Laboratory for Muon Spin Spectroscopy, 5232 Villigen PSI, Switzerland

Journal of Physics: Conference Series **551** (2014) 012033 doi:10.1088/1742-6596/551/1/012033

## 2. Experiment

$\beta$-NMR experiments were carried out at the Isotope Separator and Accelerator (ISAC) facility at TRIUMF between 2002 and 2013. The $^8Li^+$ has nuclear spin $I = 2$, radioactive lifetime $\tau = 1.21$ s, gyromagnetic ratio $\gamma = 6.3015$ MHz/T, and a small electric quadrupole moment $Q = +31.4$ mb. A highly polarized beam of $^8Li^+$ was implanted in the sample with a typical flux of $10^6$/s into a beam spot $\sim$ 2 mm in diameter. The implantation energy was varied by biasing the spectrometer to high positive voltage, thus electrostatically decelerating the incident $^8Li^+$ beam from its transport energy (typically 20 or 30 keV). The nuclear polarization is monitored through the anisotropic $\beta$-decay of the highly polarized $^8Li$ in two fast plastic scintillation detectors up- (B) and down- (F) stream of the sample. Resonances were measured with a continuous beam by introducing a transverse RF magnetic field, $B_1$, either as CW or modulated pulses, and scanning the frequency through the Larmor frequency at a rate slow compared to $\tau$. On resonance, the $^8Li^+$ precess about $B_1$ and the beta decay asymmetry is consequently reduced. To measure the spin lattice relaxation rate, the incoming $^8Li^+$ beam was pulsed with a fast electrostatic kicker in the absence of any RF, and the relaxation of the beta decay asymmetry was monitored during and after the beam pulse. In each type of measurement both spin directions of the $^8Li^+$ (parallel and antiparallel) relative to the large static field $B_0$, produced by a high homogeneity superconducting solenoid, were collected by alternating the helicity of the polarizing laser light. This has a number of advantages, including determination of the baseline of the helicity combined asymmetry. The field $B_0$ ranged from 1.3 to 6.55 T.

We present data on a number of epitaxially polished MgO single crystals labelled F and R (supplier: Crystal GmBH, Berlin), M (supplier: MTI Corp., Richmond, CA), C1 and C2 (capped in situ with epitaxial Ag layers 19 and 29 nm thick).

## 3. Results and Discussion

The NMR of $^8Li^+$ implanted in MgO consists of a single narrow resonance near the Larmor frequency in the applied field, with no evidence of quadrupolar splitting, implying a site of cubic symmetry. Power broadening (familiar from RF$\mu$SR[6]) is clearly evident in the resonances with CW RF in Fig. 1. The Lorentzian linewidth (HWHM throughout), shown in Fig. 2, can be

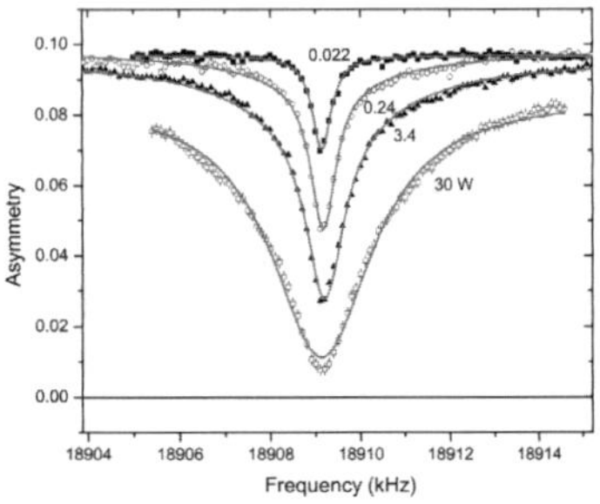


**Figure 1.** The CW resonance of $^8Li^+$ implanted into MgO (3 T, 300 K, 30 keV) at different RF powers.

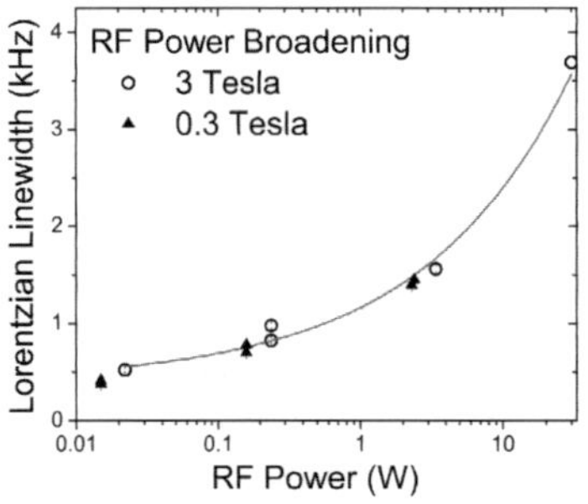


**Figure 2.** The Lorentzian linewidths of the data of Fig. 1 as a function of the RF power.

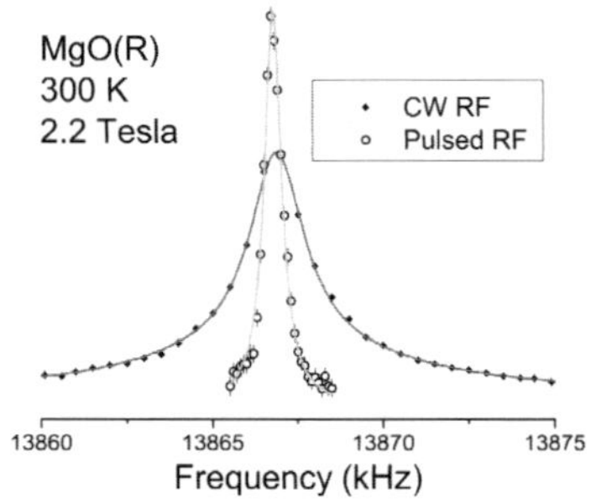


**Figure 3.** CW vs pulsed resonances at 19 keV into MgO sample R.

extrapolated to the zero power limit of $w_0 = 407(160)$ Hz, independent of $B_0$[7]. This intrinsic linewidth is small on the scale of typical nuclear dipolar widths since MgO has small nuclear

[7] The resonance is better fit by the sum of two Lorentzian lines sharing the same frequency, but with different widths

moments in low abundance. However, use of a pulsed RF mode leads to a substantially narrower resonance (Fig. 3) of width 190(16) Hz, independent of $B_0$ up to 6.55 Tesla. The difference is likely the result of spin dynamical effects in the CW resonance that are suppressed by the short RF pulses (see below).

We use the narrow resonance in MgO as a frequency reference for shifts of in other materials, so it is interesting to consider how close the resonance frequency is to the free Larmor frequency $\nu_L$. We expect the $^8Li^+$ chemical shift to be small (on the order of a few ppm) as it is generally for Li NMR, e.g. see Ref. [7]. We should, however, consider demagnetization. Approximating the sample as a thin plate in a perpendicular field, the shift *correction* due to demagnetization is[8]

$$K_d = \frac{8\pi}{3}\chi, \tag{1}$$

where $\chi$ is the dimensionless CGS volume susceptibility. MgO is weakly diamagnetic[9], with $\chi = -1.38 \times 10^{-6}$, yielding $K_d = -11.5$ ppm, i.e. the resonance in MgO is shifted slightly up in frequency from $\nu_L$. This correction is important for determining small *absolute* shifts, e.g. the Knight shift in a metal where an accurate zero is required, especially if that shift is used in forming the Korringa ratio.

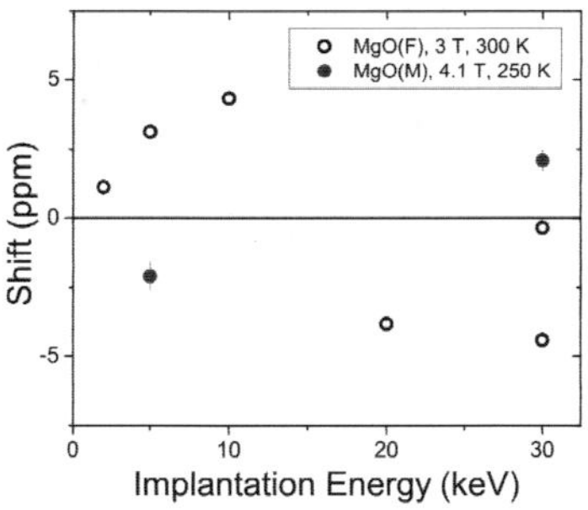


**Figure 4.** The shift of the resonance as a function of $^8Li^+$ implantation energy, referred to zero on average.

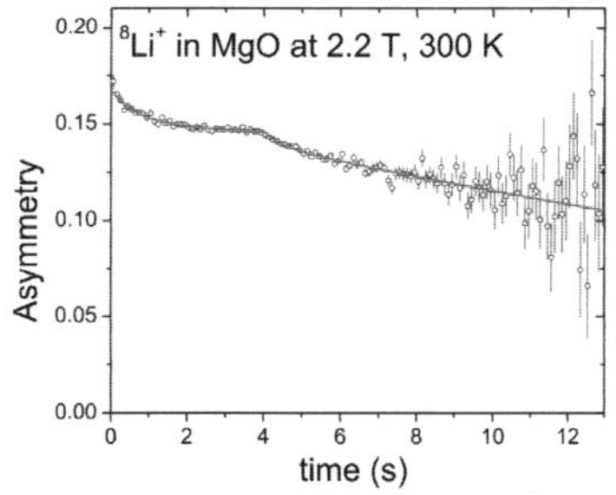


**Figure 5.** Spin lattice relaxation of $^8Li^+$ in MgO(R) at 2.2 Tesla, 300 K, 20 keV.

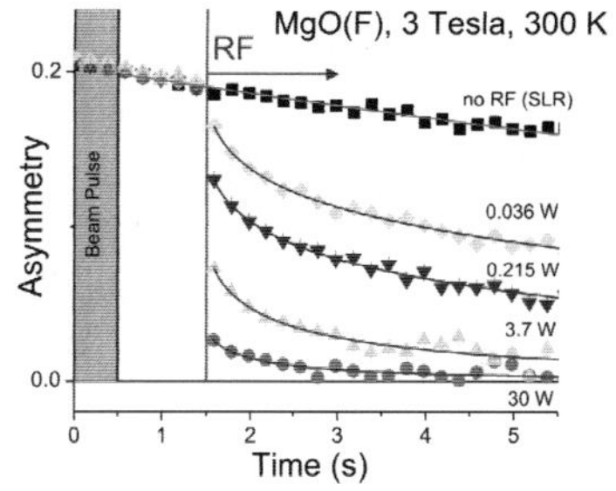


**Figure 6.** Hole burning measurements in MgO(F), see text.

A key strength of implanted ion $\beta$-NMR is the ability to vary the implantation energy $E$ and hence implantation depth of the probe ions, thus obtaining *depth resolution*. A scan of $E$ in MgO, shown in Fig. 4, reveals the resonance shift is depth independent (within 5 ppm), establishing an important comparison case for nontrivial *topological* insulators, such as some Bi compunds[10, 11], where depth dependence due to topological surface states is expected. Note that the statistical errors from the fits are much smaller than the scatter in the data, indicating that systematics are dominating the uncertainty in the resonance position. This is typically the case in these high rate experiments. Systematic drifts in the baseline asymmetry can result from fluctuations of the $^8Li^+$ beam.

Spin lattice relaxation (SLR) of $^8Li^+$ in MgO at high magnetic fields is slow. Fig. 5 shows data for a 4 s pulse of beam fit to a biexponential relaxation. We attribute the large slow relaxing signal to MgO, here with rate $\lambda = 1/T_1 = 0.0312(29)$ s$^{-1}$. The small fast component is likely a background signal.

While no RF is used in the SLR measurement, we can apply RF at a specific time during such a measurement. If the RF is on resonance, it causes a sudden downward step in the asymmetry, of height proportional to the fraction of the line affected by the RF. A higher RF power thus leads to a larger step. However, if there are spin dynamics in the resonance, i.e. $^8$Li correspond to more than one unique frequency (within the resonance) during their lifetimes, then the RF has further effects. Spectral dynamics (diffusion) can be caused by $^8Li^+$ motion and/or host nuclear spin fluctuations. With the RF applied continuously at a fixed frequency, one continues to destroy the polarization of any $^8Li^+$ that stray into the "hole" that the RF "burns" into the resonance, leading to another mechanism of relaxation in the presence of the RF. Hole-burning measurements, where the RF is applied continuously 1 second after a 0.5 s pulse of beam are shown in Fig. 6. At the highest power (30 W), almost all the $^8Li^+$ asymmetry is destroyed instantly. The step is reduced with reduced power, but there also appears a new time dependent relaxation that is much faster than the intrinsic SLR (also shown), caused by spectral diffusion. This effect has been used in conventional NMR, for example, to study slow diffusion[12]. A simple small frequency step random walk model predicts a relaxation proportional to $\exp(-\sqrt{\Gamma t})$. The relaxation after the RF is turned on is fit to a fraction relaxing at the intrinsic $1/T_1$ plus such a term (blue curves). The hole-burning relaxation rate $\Gamma$ is found to be 1.2(7) $s^{-1}$. This dynamic relaxation acts to increase the effective width of a resonance under CW RF, while pulsed RF, to the extent that the pulses are fast on the time scale of $\Gamma$ is unaffected. This likely accounts for the substantially smaller pulsed RF linewidth mentioned above.

We now consider possible mechanisms for spectral diffusion. $^{25}$Mg is a low moment nucleus unfavourable for NMR, but some limited results are available[13, 14]. Its $T_1$ is about 50 s at 300 K[14], much slower than $\Gamma$, but it is possible that the transverse relaxation time $T_2$ is significantly faster and might account for the timescale $\Gamma$. Alternatively, $^8Li^+$ may hop between equivalent lattice sites on the second timescale at 300 K. The distinction between these two mechanisms may be made from the temperature dependence, with the hop rate exhibiting an activated behaviour while $T_2$ should be roughly $T$ independent. As has been found in conventional NMR, hole-burning may provide a useful probe of very slow motion of the probe $^8Li^+$.

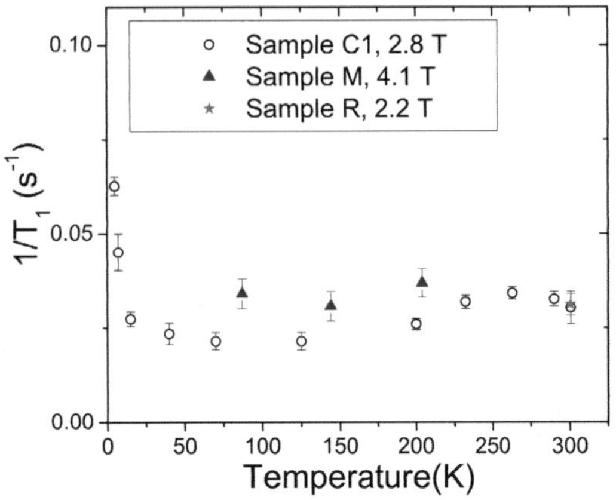


**Figure 7.** Spin lattice relaxation of $^8Li^+$ at high fields as a function of temperature in several samples.

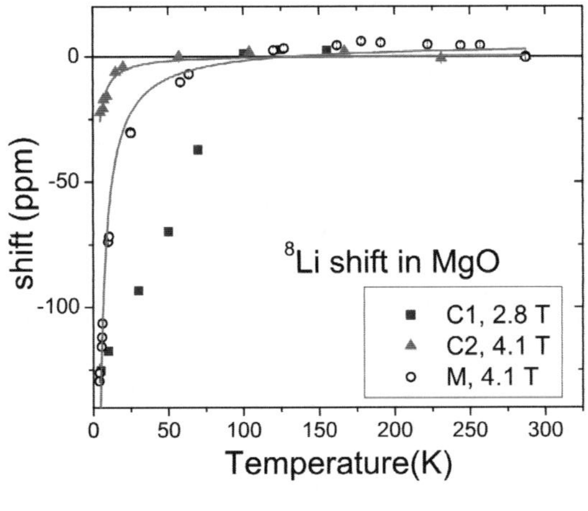


**Figure 8.** The temperature dependence of the shift of the $^8Li^+$ resonance with respect to 300 K. The fits shown are to a (negative) Curie law plus a constant.

We have also measured the temperature dependence of both SLR and the shift (Figs. 7 and 8) that reveal an increase in $1/T_1$ at low temperature and a negative shift of the resonance below 100 K that increases as temperature is lowered. The origin of these changes at low temperature is not clear, but here we consider some possibilities.

Magnetic impurities in the MgO would cause a negative demagnetization shift via Eq. (1) with a uniform paramagnetic $\chi$; however, the concentration of such defects required to account for the observed shift seems unreasonably high (at the % level). Moreover, such a concentration should cause significant broadening of the resonance, while only a modest broadening is found.

Studies of Li doped MgO find that substitutional $Li'_{Mg}$ localizes a hole, i.e. the $[Li]^0$ centre[15]. As $Li^+$ is negatively charged relative to the $Mg^{++}$ that it replaces, there is a Coulomb attraction for positive mobile charges, and at low temperatures an oxygen $2p$ hole is localized on the near neighbours of the Li defect. The holes are provided by high energy irradiation or by quenching[16]. EPR[15, 17, 18] and ENDOR[19] find a negative hyperfine field on the Li nucleus[20, 21], and a significant anisotropy due to relaxation of the lattice with respect to the $Li^+$–$O^-$ axis, breaking local cubic symmetry[15, 19]. The isotropic part of the hyperfine interaction with the hole spin is found to be $-4.539$ MHz[19]. The disappearance of anisotropy with increasing temperature was interpreted as a result of slow hopping of the hole among the neighbouring oxygen ions[18, 19]. This picture was supported by several theoretical calculations[21, 22, 23] and is quite different from a symmetric localized state predicted more recently for Ca vacancies in CaO[24].

There are a number of problems attributing the negative $^8Li$ shift at low temperature to such a localized hole. In particular, we see no evidence of a noncubic environment at the $^8Li^+$, and, rather than a large hyperfine field, we have only a small shift on the scale of 100 ppm compared to the hyperfine field from ENDOR which corresponds to more than $10^5$ ppm at 4.1 T. This is somewhat reminiscent of Mu in heavily $n$ type Si[25], where the shift is the remnant of the Muonium hyperfine field reduced by averaging due to rapid exchange of the Mu electron with the surrounding conduction electrons at densities on the order of $10^{18}/cm^3$. In MgO, there is no such bath of mobile electrons, unless there is a significant transient population of carriers related to the implantation itself. The density of implantation-related carriers would have to be quite high in the vicinity of the stopped $^8Li^+$ to make this mechanism feasible. The sample dependence in Fig. 8 is also not easy to explain in this picture.

Finally, it is not even clear that the implanted $^8Li^+$ is at the substitutional site. In many materials, we find substitutional $^8Li^+$ only at high temperature. At low temperatures, $^8Li^+$ is generally in a metastable interstitial site. Such a site is available here, the tetrahedral interstitial ($T$) site. Based on effective ionic radii in oxides, the $T$ site is too small to accommodate $Li^+$, and significant outward relaxation of the 4 neighbouring $O^{2-}$ would be necessary. Our initial density functional calculations suggest that this is a higher energy site for $^8Li^+$ than the Mg site, but in the absence of a Mg vacancy, this may well be the only available cubic site. Unlike $Li'_{Mg}$, $Li^{\cdot}_T$, is a positive defect that would tend to localize an electron (rather than a hole), and this is not likely on the neighbouring $O^{2-}$.

Further experiments and calculations are necessary to characterize the electronic structure of the implanted $^8Li^+$ in MgO.

## 4. Summary

We have presented extensive results on the $\beta$-NMR of $^8Li^+$ implanted in cubic MgO, where it occupies a cubic site, likely the tetrahedral interstitial site. We routinely use the narrow resonance as a reference for resonance shifts, and we show here that the intrinsic linewidth is on the order of 200 Hz, which allows a reference frequency to be measured to an accuracy of a few Hz. We find no implantation energy dependence of the resonance within a few ppm, but we find evidence of slow spin dynamics at 300 K from hole-burning measurements. The temperature dependence reveals interesting changes at low temperature whose origin remains uncertain.

## Acknowledgments
We thank S. Hak and T. Hibma of the University of Groningen for samples C1 and C2; the TRIUMF CMMS and ISAC facility for technical support; and S.X.Y. Wang for the spectral diffusion modeling.

## References
[1] Kowalska M, Yordanov D T, Blaum K, Himpe P, Lievens P, Mallion S, Neugart R, Neyens G and Vermeulen N 2008 *Phys. Rev.* C **77** 034307
[2] Elfimov I S, Rusydi A, Csiszar S I, Hu Z, Hsieh H H, Lin H J, Chen C T, Liang R and Sawatzky G A 2007 *Phys. Rev. Lett.* **98** 137202
[3] Gallego S, Beltrán J I, Cerdá J and Muñoz M C 2005 *J. Phys. Cond. Matt.* **17** L451
[4] Arndt S, Laugel G, Levchenko S, Horn R, Baerns M, Scheffler M, Schlögl R and Schomäcker R 2011 *Catalysis Reviews* **53** 424–514
[5] Tardío M M, Ramírez R, González R and Chen Y 2002 *Phys. Rev.* B **66** 134202
[6] Kreitzman S 1991 *Hyp. Int.* **65** 1055–1069
[7] Böhmer R, Jeffrey K and Vogel M 2007 *Prog. NMR Spectrosc.* **50** 87–174
[8] Xu M, Hossain M, Saadaoui H, Parolin T, Chow K, Keeler T, Kiefl R, Morris G, Salman Z, Song Q, Wang D and MacFarlane W 2008 *J. Magn. Reson.* **191** 47–55
[9] Candea R M, Hudgens S J and Kastner M 1978 *Phys. Rev.* B **18** 2733–2738
[10] MacFarlane W *et al.* In preparation
[11] Koumoulis D *et al.* In preparation
[12] Kuhns P L and Conradi M S 1982 *J. Chem. Phys.* **77** 1771–1778
[13] Dupree R and Smith M E 1988 *J. Chem. Soc., Chem. Commun.* 1483–1485
[14] Fiske P, Stebbins J and Farnan I 1994 *Physics and Chemistry of Minerals* **20** 587–593
[15] Schirmer O 1971 *J. Phys. Chem. Sol.* **32** 499–509
[16] Abraham M M, Chen Y, Boatner L A and Reynolds R W 1976 *Phys. Rev. Lett.* **37** 849–852
[17] Halliburton L E, Kappers L A, Cowan D L, Dravnieks F and Wertz J E 1973 *Phys. Rev. Lett.* **30** 607–610
[18] Sanjuan M L and Orera V M 1986 *J. Phys.* C **19** 67
[19] Abraham M M, Unruh W P and Chen Y 1974 *Phys. Rev.* B **10** 3540–3545
[20] Bass R 1984 *Phys. Rev.* B **30** 5334–5335
[21] Lichanot A, Larrieu C, Zicovich-Wilson C, Roetti C, Orlando R and Dovesi R 1998 *J. Phys. Chem. Sol.* **59** 1119–1124
[22] Zuo J, Pandey R and Kunz A B 1991 *Phys. Rev.* B **44** 7187–7191
[23] Shluger A L, Kotomin E A and Kantorovich L N 1986 *J. Phys.* C **19** 4183
[24] Elfimov I S, Yunoki S and Sawatzky G A 2002 *Phys. Rev. Lett.* **89** 216403
[25] Chow K H, Kiefl R F, Hitti B, Estle T L and Lichti R L 2000 *Phys. Rev. Lett.* **84** 2251–2254

# $\beta$-detected nuclear quadrupole resonance and relaxation of $^8Li^+$ in sapphire

**Z Salman[1], K H Chow[2], M D Hossain[3], R F Kiefl[3,4], C D P Levy[4], T J Parolin[5], M R Pearson[4], H Saadaoui[1], D Wang[3] and W A MacFarlane[5]**

[1] Laboratory for Muon Spin Spectroscopy, Paul Scherrer Institute, CH-5232 Villigen PSI, Switzerland
[2] Department of Physics, University of Alberta, Edmonton T6G 2G7, AB, Canada
[3] Department of Physics and Astronomy, University of British Columbia, Vancouver V6T 1Z1, BC, Canada
[4] TRIUMF, 4004 Wesbrook Mall, Vancouver V6T 2A3, BC, Canada
[5] Department of Chemistry, University of British Columbia, Vancouver V6T 1Z1, BC, Canada

E-mail: zaher.salman@psi.ch

**Abstract.** We report detailed behaviour of low energy $^8$Li implanted near the surface of $\alpha$-$Al_2O_3$ single crystal, as revealed by beta-detected NQR of $^8$Li. We find that the implanted $^8$Li occupies at least two sites with non-cubic symmetry in the $Al_2O_3$ lattice. In both sites the $^8$Li experiences axially symmetric electric field gradient, with the main principal axis along the $c$-crystallographic direction. The temperature and field dependence of the spin lattice relaxation of $^8$Li in $\alpha$-$Al_2O_3$, indicate that the $^8$Li diffusion is negligible on the scale of its lifetime, 1.21 s.

## 1. Introduction

Sapphire ($\alpha$-$Al_2O_3$) is an important material with many technological applications, for example in optoelectronics, catalysis and environmental chemistry [1]. It is an exceptionally stable oxide with a very large formation enthalpy, high mechanical strength and thermal conductivity, which is often used as a substrate for thin film deposition. Point defects can strongly modify the properties of sapphire, which may affect its use, for example in proposed gravity wave detectors [2]. Therefore, they are of considerable interest, and have been the subject of numerous theoretical investigations [3, 4, 5]. Ion implantation is an important method to create point defects in materials and thereby modify their properties as has been studied extensively in sapphire [6], including as a synthetic route to included metallic nanoparticles [7]. There are, however, few detailed investigations of the crystallographic sites of low dose and low energy implanted ions in the crystalline matrix [8, 6].

In addition to the importance of point defects, it has been shown that mixtures of sapphire with Li ionic conductor nanoparticles can enhance $Li^+$ conductivity in these materials [9, 10]. This effect was attributed to the heterointerface in this nanocomposite material. However, the contribution of sapphire to this enhancement and its microscopic details are still not well understood.

In this paper we present zero magnetic field $\beta$-detected nuclear quadrupole resonance ($\beta$-NQR) of $^8Li^+$ implanted near the surface of sapphire. We find that $^8$Li occupies at least two inequivalent crystallographic sites, distinguished by their local quadrupolar interactions. Both sites exhibit an axially symmetric electric field gradient (EFG), with the main principal axis (MPA) along the *c*-crystallographic direction. We also measure the temperature dependence of the resonance spectra and the spin lattice relaxation down to liquid helium temperatures. The temperature and field dependence of the spin lattice relaxation rate indicate that the main source of spin relaxation of $^8$Li in $Al_2O_3$ is fluctuations of the EFG. These results rule out $^8$Li diffusion in sapphire, even near its surface at and below 300 K.

## 2. Experimental

$\beta$-NQR is a magnetic resonance technique similar to nuclear magnetic resonance and muon spin relaxation ($\mu$SR). The local spin probe used in these experiments is $^8$Li. A low energy (28 keV) beam of radioactive $^8Li^+$ is produced at the isotope separator and accelerator (ISAC) at TRIUMF in Vancouver, Canada. It is then spin-polarized using a collinear optical pumping method, yielding nuclear polarization as high as 70%, and subsequently implanted into the studied sample. Since the implanted beam energy is relatively low, the $^8$Li stops near the surface of the sample, e.g. at an average depth of $\sim$ 1500 Å in $Al_2O_3$ at the implantation energy used here. The nuclear polarization, and its time evolution, is the quantity of interest in our experiments. It can be measured through the $\beta$-decay asymmetry, where an electron is emitted preferentially opposite to the direction of the nuclear polarization at the time of decay [11] and detected by appropriately positioned scintillation counters. $^8$Li is a spin $I = 2$ nucleus with a small electric quadrupole moment $Q = +31.4$ mB and gyromagnetic ratio $\gamma = 6.3015$ MHz/T. The magnetic resonance is detected by monitoring the time-averaged nuclear polarization as a function of a small radio frequency (RF) magnetic field applied perpendicular to the initial polarization direction. Alternatively, the spin lattice relaxation of the $^8$Li nuclear spin can be measured by implanting a short pulse of beam (e.g. 4 seconds) and measuring the polarization as a function of time during and after the beam pulse with no RF field. More details about the technique can be found in Refs. [12, 13].

In the experiments reported here, the sample is an epitaxially polished single crystal substrate of $\alpha$-$Al_2O_3$. The *c*-axis oriented crystal was 12.5mm square and 0.5 mm thick. It was mounted on the cold finger of a He flow cryostat. In all the measurements, the *c*-axis was oriented 45° relative to the Li nuclear polarization. The experiments were performed in zero and low magnetic field, applied along the initial nuclear polarization of the implanted $^8$Li.

## 3. Theory

Typically, the implanted $^8$Li resides at well-defined crystalline lattice sites which can be interstitial or substitutional. Sites with non-cubic symmetry are characterized by an EFG which couples to the electric quadrupole moment of $^8$Li giving rise to a zero field (ZF) splitting in the nuclear spin levels [14]. This splitting and the resulting nuclear quadrupole resonance spectrum are thus a fingerprint of the $^8$Li site. Assuming the EFG tensor is axially symmetric about $z$, the quadrupolar spin Hamiltonian is [15]:

$$H_q = h\nu_q[I_z^2 - 2] \tag{1}$$

where $\nu_q = e^2qQ/8$, $eq = V_{zz}$ is the EFG along $z$, and $Q$ is the electric quadrupole moment of the nucleus. The energy eigenvalues, $E_m = h\nu_q(m^2 - 2)$, are a function of the azimuthal quantum number $m$ where $I_z|m\rangle = m|m\rangle$. Small perturbations, such as small magnetic fields or slight deviations in the quadrupolar interaction, can have a surprisingly large effect on the nuclear polarization and $\beta$-NQR measurements. This may be traced to the fact that in zero magnetic field the $|\pm m\rangle$ states are degenerate and therefore easily mixed by small perturbations. Crystal imperfections, resulting in, for example, non-axial terms in the EFG, also lead to $|\pm m\rangle$ state mixing. Consider for example a small term which breaks the axial symmetry of the EFG [14, 16]:

$$H_\eta = \eta h\nu_q(I_x^2 - I_y^2). \tag{2}$$

Here $\eta$ is the conventional dimensionless EFG asymmetry parameter [14, 16]. $H_\eta$ mixes the $m = \pm 1$ levels in first order with a characteristic frequency splitting between the mixed levels. We have shown [13, 17] that even small values of $\eta$ (e.g. 0.01) will produce strong mixing of the $|\pm m\rangle$ states and a resulting fast drop of the initial nuclear polarization. Therefore, our $\beta$-NQR measurements are effectively insensitive to $^8$Li sites with non-axial EFG, such as sites near a defect or vacancy. In what follows we will assume $\eta = 0$.

Another perturbation that can lead to mixing between different $|m\rangle$ states is a small magnetic field perpendicular to $z$ (the MPA). Consider first a small magnetic field which is oriented along $z$, $\mathbf{B} = B\hat{z}$. The resulting Zeeman interaction lifts the degeneracy between $|\pm m\rangle$ states in first order but does not mix the states since it commutes with $H_q$. Although the measured resonance positions are shifted by a small amount there is no change in the amplitudes. For example, the resonance corresponding to the $|2\rangle \to |1\rangle$ transition shifts by $\gamma B$. On the other hand if $\mathbf{B} = B\hat{x}$ is applied in the $x$ direction the Zeeman interaction does not commute with $H_q$. Such a perturbation mixes the different $m$ levels and consequently has a dramatic effect on the nuclear polarization. In general, the Hamiltonian of the $^8$Li spin in an applied field $\mathbf{B}$ is

$$H = H_q - h\nu\left(I_z\cos\theta + I_x\sin\theta\right), \tag{3}$$

where $\nu = \gamma B$ and $\theta$ is the angle between the MPA and the applied field direction. Using perturbation theory the dimensionless mixing $\Delta_{m,n}$ between states $|m\rangle$ and $|n\rangle$ can be written as (see e.g. [13])

$$\begin{aligned}
\Delta_{2,1} &= \left(\frac{\nu\sin\theta}{3\nu_q - \nu\cos\theta}\right)^2 &&\simeq \frac{\sin^2\theta}{9}\left(\frac{\nu}{\nu_q}\right)^2 \\
\Delta_{-2,-1} &= \left(\frac{\nu\sin\theta}{3\nu_q + \nu\cos\theta}\right)^2 &&\simeq \frac{\sin^2\theta}{9}\left(\frac{\nu}{\nu_q}\right)^2 \\
\Delta_{1,0} &= \frac{3}{2}\left(\frac{\nu\sin\theta}{\nu_q - \nu\cos\theta}\right)^2 &&\simeq \frac{3\sin^2\theta}{2}\left(\frac{\nu}{\nu_q}\right)^2 \\
\Delta_{-1,0} &= \frac{3}{2}\left(\frac{\nu\sin\theta}{\nu_q + \nu\cos\theta}\right)^2 &&\simeq \frac{3\sin^2\theta}{2}\left(\frac{\nu}{\nu_q}\right)^2 .
\end{aligned} \tag{4}$$

Note that $\Delta_{m,n}$ for all other $m$ and $n$ values vanish in first order. Provided this mixing is larger or comparable to the $^8$Li decay rate, it will affect both the measured initial nuclear polarization and its relaxation rate.

## 4. Results

In the process of carrying out the resonance measurements, we found that the $^8$Li$^+$ beam caused the sapphire crystal to scintillate (see Fig. 1), likely due to the $\beta$-emission produced in its decay

**Figure 1.** Demonstration of sapphire's use as a scintillator. The image of the $^8$Li$^+$ beam impinging on sapphire at a rate of $10^7$ per second, taken with a CCD camera over a 10 second exposure. The dashed line highlights a 8x8 mm$^2$ area.

as well as the subsequent $\alpha$ decay of its daughter $^8$Be. This property is not surprising in view of sapphire's well-known luminescent properties [18, 19], but it does provide an interesting application as an ultra-high vacuum compatible, thermally conductive backing material for $\beta$-NMR samples, which enables tracking and positioning of the beam accurately on small samples.

Initially we tried to measure $\beta$-NQR spectra, while implanting $^8$Li with its polarization perpendicular to the $c$-axis of the $Al_2O_3$ crystal. Under these conditions there was no measurable asymmetry in ZF, possibly due to (i) all the implanted $^8$Li occupy a site with a very small nuclear quadrupole interaction, (ii) the MPA at $^8$Li site is along the $c$-axis, or (iii) a large $\eta$ which produces a large mixing between the different $|m\rangle$ states and a subsequent loss of polarization. However, when we rotated the $Al_2O_3$ crystal by 45°, two resonances were observed at room temperature with a non-vanishing asymmetry off resonance (Fig. 2). This clearly indicates that

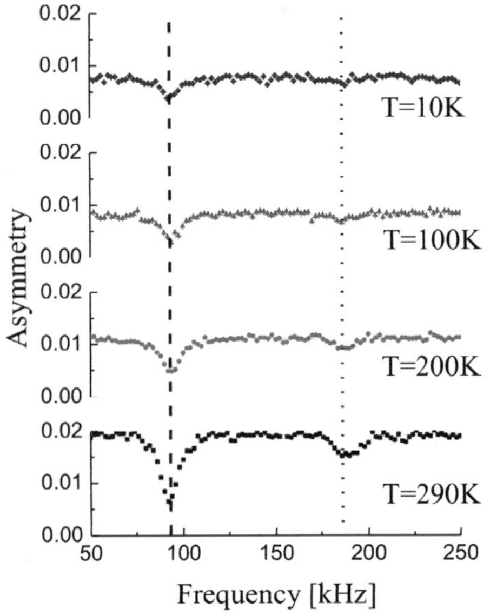


**Figure 2.** $\beta$-NQR spectra at various temperatures and in ZF. There is no significant temperature dependence of the resonance frequency, amplitude or broadening. Note however, a reduction of the baseline asymmetry is observed as the temperature is lowered. The solid lines are fits to Lorentzian lineshapes.

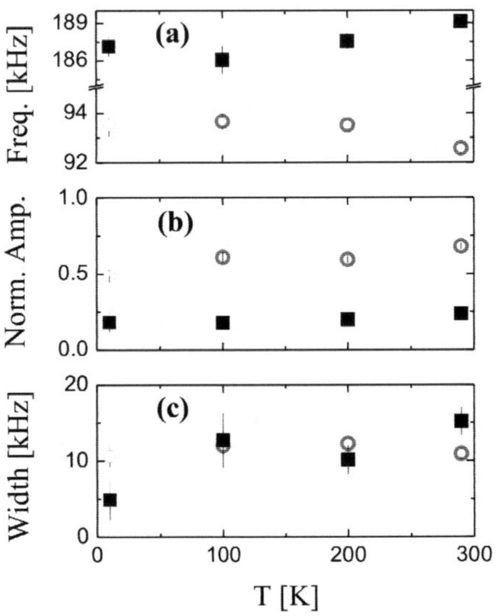


**Figure 3.** The resonance frequencies (a), normalized amplitudes (b) and widths (c) as a function of temperature, obtained from fitting the $\beta$-NQR spectra to Lorentzian lineshapes. The circles and squares indicate low and high frequency resonance parameters, respectively.

at least a large fraction of the $^8$Li occupy sites with a nearly axial EFG which is oriented along the $c$-axis. The two resonances are attributed to two distinct sites, each with a different EFG tensor. Since the ratio between the resonance frequencies of the two lines is $\sim 2$, they cannot be attributed to the $|\pm 2\rangle \rightarrow |\pm 1\rangle$ and $|\pm 1\rangle \rightarrow |\pm 0\rangle$ transitions which should have a ratio of 3 in ZF. As can be seen in Figs. 2 and 3, there is no significant temperature dependent shift or broadening of the resonances, such as that seen in the $\beta$-NQR signal in $SrTiO_3$[17]. However, the measured asymmetry off resonance (i.e. the baseline of the spectra) decreases by 50% from 290 K down to 10K. This decrease can be attributed to an increase in the spin lattice relaxation of $^8$Li as the temperature is decreased (see below).

The observed resonances fit very well to Lorentzian lineshapes. From these fits we extract the resonance frequencies, their amplitudes (normalized by the baseline), and widths as shown in Fig. 3. The first resonance frequency is between 92-94 kHz while the second falls between 186-189 kHz, with corresponding normalized amplitudes of $\sim 0.6$ and $\sim 0.2$ and an almost equal width of $\sim 10$ kHz. We do not observe a strong temperature dependence, indicating that there are no sudden changes in the structural/electronic properties of sapphire or of the $^8Li^+$ site. The small variations as a function of temperature could be due to thermal expansion/contraction. Note that the sum of normalized amplitudes is $\sim 0.8$ leaving a possibility that a small fraction of $^8$Li is implanted in sites with nearly cubic symmetry or non-axially symmetric EFG.

Now we turn to the discussion of the spin lattice relaxation of the $^8$Li nuclear polarization. As mentioned earlier, this is measured by implanting the $^8$Li beam for a period of 4 seconds, and measuring the polarization as a function of time, $p_z(t)$, during and after the beam pulse. Typical polarization curves measured at various temperatures in ZF are shown in Fig. 4(a). Note the gradual increase of the relaxation rate as the temperature is decreased. In contrast,

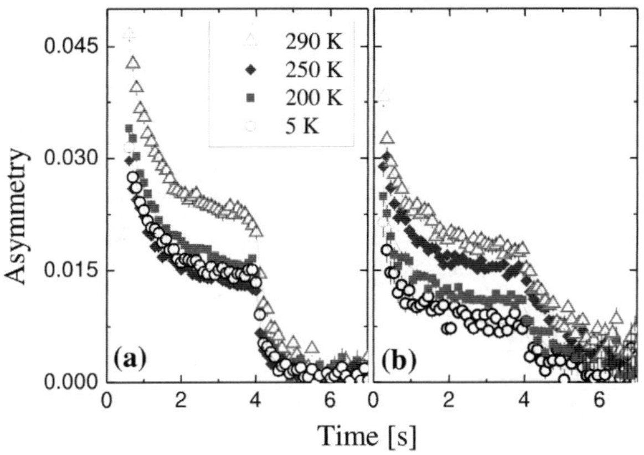


**Figure 4.** $p_z(t)$ measured in (a) ZF and and (b) 15 mT at various temperatures. The angle between the $c$-axis of the $Al_2O_3$ crystal and the initial polarization is 45°. The solid lines are fits to stretch exponential relaxation function (see text).

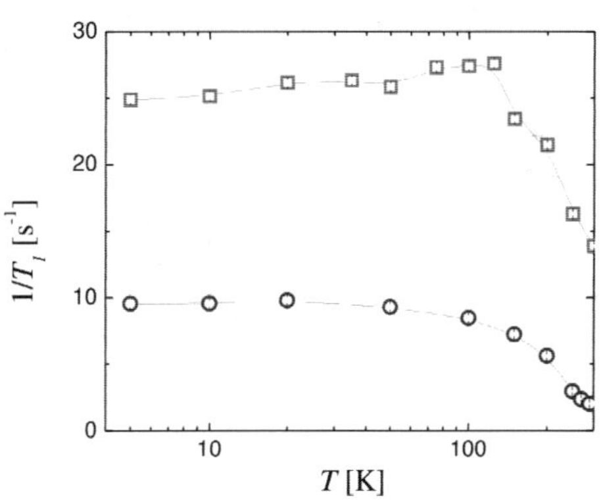


**Figure 5.** The temperature dependence of the spin lattice relaxation rate as a function of temperature at zero field (circles) and at 15 mT (squares). The solid lines are a guide for the eye.

measurements of $p_z(t)$ in a magnetic field of 15 mT, applied along the direction of the $^8$Li initial nuclear spin, show that in this case the relaxation rate does not increase monotonically with decreasing temperature [Fig. 4(b)]. Rather, it peaks at $\sim 125$ K and then decreases gradually as the temperature is decreased further. $p_z(t)$ in these measurements is determined by both the $^8$Li spin-lattice relaxation rate $1/T_1$ and its radioactive lifetime $\tau = 1.21$s. Assuming a general spin relaxation function $f(t, t_p : 1/T_1)$ for the fraction of $^8$Li implanted in the sample at $t_p$, the polarization follows [17]

$$p_z(t) = \begin{cases} \dfrac{\int_0^t e^{-(t-t_p)/\tau} f(t,t_p:1/T_1)dt_p}{\int_0^t e^{-t/\tau}dt} & t \leq T \\ \dfrac{\int_0^T e^{-(T-t_p)/\tau} f(t,t_p:1/T_1)dt_p}{\int_0^T e^{-t/\tau}dt} & t > T. \end{cases} \tag{5}$$

The data in Figs. 4(a) and (b) were fit to Eq. (5) with a phenomenological stretch exponential form,

$$f(t, t_p : 1/T_1) = A_0 e^{-[(t-t_p)/T_1]^\beta}. \tag{6}$$

In the fits we assume that the initial asymmetry, $A_0$ (which is proportional to the initial polarization), and the exponent $\beta$ are common for all temperatures, which yields $A_0 = 0.53(6)$ in ZF and 0.200(6) in 15 mT with $\beta = 0.43(2)$ for both fields. The relaxation rates obtained from the fits are shown in Fig. 5. As expected, $1/T_1$ in ZF increases gradually below 290 K and saturates below 100 K, while in 15 mT it increases sharply below 290 K, peaks at $\sim$ 125 K and then decreases gradually as temperature is decreased further. Note, in the 15 mT measurements the applied field enhances the relaxation rate. This is due to mixing, $\Delta_{m,n}$, between the $|m\rangle$ and $|n\rangle$ states produced by the magnetic field component, $B_x$ perpendicular to the MPA of the EFG. The effect of the applied field on the relaxation may also be responsible for the difference in the temperature dependence of the relaxation rate between the ZF and 15 mT measurements. We should also point out here that the measured $1/T_1$ in this case reflects the weighted average of the $1/T_1$ of both sites.

## 5. Conclusions

The stopping site of $^8$Li is consistent with the results from channelling experiments on Er implanted sapphire [8]. There it was found that at room temperature $\sim$ 70% of the Er occupy an interstitial site near the free octahedral site while $\sim$ 20% occupy the tetrahedral site. Both sites exhibit an axially symmetric EFG, with the MPA along the *c*-crystallographic direction. The temperature dependence of $1/T_1$ resembles that of the lattice constants in sapphire [20]. This is strong indication that the main source of spin relaxation of $^8$Li in $Al_2O_3$ is small fluctuations of the EFG, which also scale with the lattice constants. These results rule out $^8$Li diffusion in sapphire, even near its surface, as a contributor to the enhanced Li conductivity in mixtures of sapphire with Li ionic conductor nanoparticles [9, 10] .

## Acknowledgments

We acknowledge the technical assistance of R. Abasalti, B. Hitti, D. Arseneau, S. Daviel, and funding from NSERC Canada. TRIUMF is funded in part by NRC Canada.

## References

[1] Al-Abadleh H and Grassian V 2003 *Surf. Sci. Rep.* **52** 63
[2] Yan Z *et al.* 2004 *J. Opt. A* **6** 635
[3] Jacobs P W M and Kotomin E A 1994 *J. Am. Ceram. Soc.* **77** 2505–2508
[4] Jacobs P W M and Kotomin E A 1992 *Phys. Rev. Lett.* **69** 1411
[5] Carrasco J, Gomes J R B and Illas F 2004 *Phys. Rev.* B **69** 064116
[6] McHargue C J 1998 *Mat. Sci. Eng. A* **253** 94–105
[7] Stepanov A L and Khaibullin I B 2005 *Rev. Adv. Mater. Sci.* **9** 109129
[8] Alves E, daSilva M F, vandenHoven G N, Polman A, Melo A A and Soares J C 1995 *Nucl. Instrum. Meth.* B **106** 429–432
[9] Ardel G, Golodnitsky D, Peled E, Wang Y F, Wang G, Bajue S and Greenbaum S 1998 *Solid State Ionics* **115** 477–485
[10] Wilkening M, Indris S and Heitjans P 2003 *Phys. Chem. Chem. Phys.* **5** 2225–2231
[11] Crane S G, Brice S J, Goldschmidt A, Guckert R, Hime A, Kitten J J, Vieira D J and Zhao X 2001 *Phys. Rev. Lett.* **86** 2967
[12] Morris G D, MacFarlane W A, Chow K H, Salman Z, Arseneau D J, Daviel S, Hatakeyama A, Kreitzman S R, Levy C D P, Poutissou R, Heffner R H, Elenewski J E, Greene L H and Kiefl R F 2004 *Phys. Rev. Lett.* **93** 157601
[13] Salman Z, Reynard E P, MacFarlane W A, Chow K H, Chakhalian J, Kreitzman S R, Daviel S, Levy C D P, Poutissou R and Kiefl R F 2004 *Phys. Rev.* B **70** 104404
[14] Cohen M and Reif F 1957 *Solid State Physics* **5** 321–438
[15] Slichter C P 1990 *Principles of Magnetic Resonance* 3rd ed (New York: Springer-Verlag)
[16] Das T and Hahn E L 1958 *Nuclear Quadrupole Resonance Spectroscopy* (Academic Press Inc.)
[17] Salman Z, Kiefl R F, Chow K H, Hossain M D, Keeler T A, Kreitzman S R, Levy C D P, Miller R I, Parolin T J, Pearson M R, Saadaoui H, Schultz J D, Smadella M, Wang D and MacFarlane W A 2006 *Phys. Rev. Lett.* **96** 147601

[18] Caulfield K J, Cooper R and Boas J F 1993 *Phys. Rev.* B **47** 55
[19] Kotomin E, Tale I, Butlers P and Kulis P 1989 *J. Phys. Cond. Mat.* **1** 6777–6785
[20] Lucht M, Lerche M, Wille H C, Shvyd'ko Y V, Rüter H D, Gerdau E and Becker P 2003 *J. Appl. Cryst.* **36** 1075–1081

Journal of Physics: Conference Series **551** (2014) 012035 doi:10.1088/1742-6596/551/1/012035

# Na-ion dynamics in Quasi-1D compound $NaV_2O_4$

**M Månsson[1,2], I Umegaki[3], H Nozaki[3], Y Higuchi[3], I Kawasaki[4], I Watanabe[4], H Sakurai[5] and J Sugiyama[3]**

[1] Laboratory for Quantum Magnetism, École Polytechnique Fédérale de Lausanne (EPFL), CH-1015 Lausanne, Switzerland
[2] Laboratory for Neutron Scattering & Imaging, Paul Scherrer Institute, CH-5232 Villigen PSI, Switzerland
[3] Toyota Central Research & Development Laboratories, Inc., 41-1 Yokomichi, Nagakute, Aichi 480-1192, Japan
[4] Advanced Meson Science Laboratory, RIKEN, 2-1 Hirosawa, Wako, Saitama 351-0198, Japan
[5] National Institute for Materials Science, Namiki, Tsukuba, Ibaraki 305-0044, Japan

E-mail: `martin.mansson@epfl.ch`

**Abstract.** We have used the pulsed muon source at ISIS to study high-temperature Na-ion dynamics in the quasi-one-dimensional (Q1D) metallic antiferromagnet $NaV_2O_4$. By performing systematic zero-field and longitudinal-field measurements as a function of temperature we clearly distinguish that the hopping rate increases exponentially above $T_{\rm diff} \approx 250$ K. The data is well fitted to an Arrhenius type equation typical for a diffusion process, showing that the Na-ions starts to be mobile above $T_{\rm diff}$. Such results make this compound very interesting for the tuning of Q1D magnetism using atomic-scale ion-texturing through the periodic potential from ordered Na-vacancies. Further, it also opens the door to possible use of $NaV_2O_4$ and related compounds in energy related applications.

## 1. Introduction

The interest for Quasi-one-dimensional (Q1D) magnets has been extensive in both the experimental, as well as theoretical research communities [1, 2, 3]. The reason is that these materials display a variety of fascinating phenomena, such as unconventional magnetic ground states [4] and quantum effects [5]. The physics behind these phenomena is governed by the strong spin-spin interaction along the 1D direction, together with a much weaker coupling along the other directions. Further, it is well known that an ideal 1D AF spin system does not show long-range ordering (LRO) above $T = 0$ K due to strong quantum spin fluctuation, causing a series of unconventional phenomena. Also Q1D metals have been in the center of attention for scientists worldwide since decades [6], due to their tendency toward Fermi surface instabilities, *e.g.* density waves, and for their ability to form non-Fermi-liquid ground states. At present day, big efforts are still being conducted to understand spurious Q1D phenomena and to theoretically predict the ground states. *e.g.* the superconducting (non-BCS) state that occur when applying pressure to the $\beta$-$Na_{0.33}V_2O_5$ compound [7, 8]. Another example is $BaVS_3$ which showed Luttinger liquid behavior in its photoemission spectra [9].

Very recently, a new member of the Q1D vanadates, $NaV_2O_4$, was discovered and synthesized by our collaborators at the National Institute for Materials Science (NIMS) in Japan [10] as well as by another group [11]. This compound is very interesting since it is something as unusual as

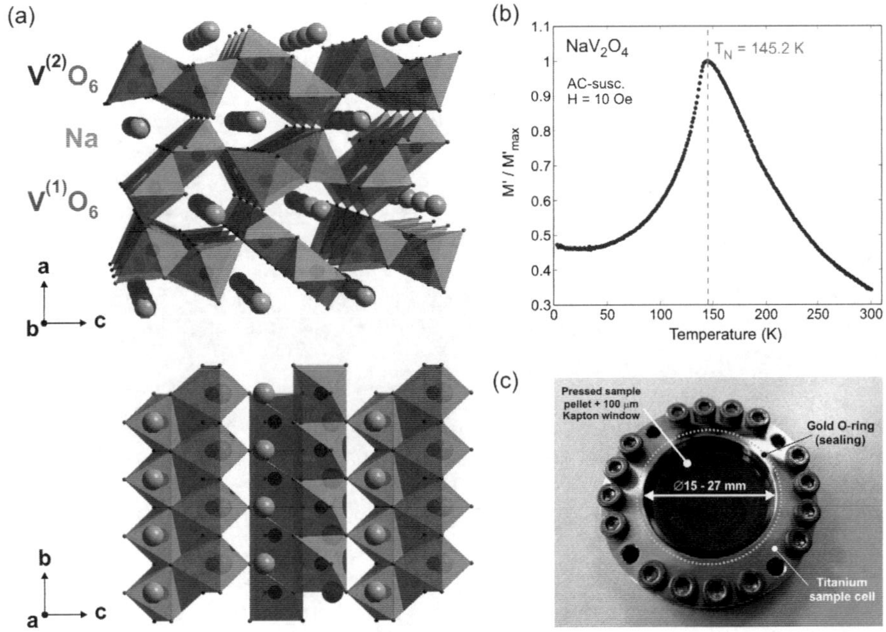


**Figure 1.** (Color online) (a) Crystallographic structure of $NaV_2O_4$ for two different viewing directions that clearly shows the quasi-1D Na-ion diffusion channels along the **b**-axis. (b) Normalized AC-susceptibility showing how $NaV_2O_4$ enters into an antiferromagnetic long-range order below $T_N \approx 145.2$ K. (c) Sample powder cell made of titanium showing the thin (100 $\mu$m) Kapton entrance window covering the pressed sample pellet. The yellow dashed line schematically shows the underlying gold O-ring that is used for sealing the cell inside a helium glove-box to avoid sample degradation.

a Q1D metallic antiferromagnet. $NaV_2O_4$ belongs to a $CaFe_2O_4$-type orthorhombic structure [see Fig. 1(a)] having a *Pnma* space group. In this structure $V_2O_4$ double (*zigzag*) chains are formed by a network of edge-sharing $VO_6$ octahedra aligned along the **b**-axis so as to make irregular hexagonal 1D channels. The sodium ions are located in the centre of such channels and are thought to be highly mobile in the direction along the **b**-axis. Our group has already studied $NaV_2O_4$ and related compounds extensively using both neutron scattering [12] and muon spin rotation/relaxation ($\mu^+$SR) [13, 14]. Such studies, however, focused on clarifying their unconventional magnetic ground states that arise from the anisotropic Q1D spin interactions. However, in this brief report we instead wish to focus on the high-temperature ion-dynamics connected to the Q1D diffusion channels.

## 2. Experimental Details

A polycrystalline sample of $NaV_2O_4$ was prepared by a solid-state reaction technique under a pressure of 6 GPa using $Na_4V_2O_7$ and $V_2O_3$ powders as starting materials. A mixture of the two powders was packed in an Au capsule, then heated at 1300°C for 1 h, and finally quenched to ambient $T$. A powder XRD analysis showed that the samples were single phase with an orthorhombic system of space group *Pnma* at ambient $T$. DC-$\chi(T)$ data from the present sample reproduce the prior measurements showing the entrance into an AF state at $T_N \approx 145.2$ K [see Fig. 1(b)]. The preparation and characterization of our samples were described in greater detail elsewhere [10].

For the $\mu^+$SR experiment, approximately 2 grams of $NaV_2O_4$ sample was pressed into a disc with a 24 mm diameter and 1.5 mm thickness. Inside a helium glove-box the disc was packed into a Au-sealed (gold O-ring) powder cell made of pure titanium using a thin (100 $\mu$m) Kapton film as 'entrance window' for the muons [see Fig 1(c)]. In addition, a silver mask with a hole matching the sample diameter was mounted onto the cell to ensure that the any minor background signal is non-relaxing in a wide temperature range. The cell was mounted onto the Cu end-plate of a liquid-He flow-type cryostat in the temperature range between 10 and 500 K. Subsequently, ZF-, weak transverse-field (wTF) and LF-$\mu^+$SR spectra were collected using the RIKEN-RAL / ARGUS spectrometer at the pulsed muon source ISIS/RAL in UK. The experimental techniques are described in more detail elsewhere [15].

## 3. Results and Discussion

In order to investigate the diffusive properties of Na-ions in the $NaV_2O_4$ compound we have used our previously presented $\mu^+$SR technique for studying solid state ion diffusion in battery related materials [16, 17]. In similarity to our previous extensive measurements of both Li-ion [16, 18, 19, 20, 21, 22, 23] and Na-ion [17] diffusion, a series of ZF, wTF = 30 G as well as LF = 5 G and 10 G $\mu^+$SR spectra were acquired in the temperature range between 140 and 500 K [see Fig. 2(a-b)]. At each temperature, the ZF and two LF spectra were fitted by an exponentially relaxing dynamic Kubo-Toyabe (KT) function plus a small background signal from the fraction of muons stopped mainly in the silver mask:

$$A_0\, P(t) \;=\; A_{\rm KT}G^{\rm DGKT}(\Delta,\nu,t)\exp(-\lambda t) + A_{\rm BG}. \qquad (1)$$

Furthermore, a global fitting procedure was employed over the entire temperature range using a common alpha ($\alpha$) and background asymmetry ($A_{\rm BG}$), but with temperature dependent field fluctuation rate ($\nu$) and relaxation rate ($\lambda$). The static width of the local field distribution ($\Delta$) was from individual fitting found to be more or less independent of temperature [$\Delta \approx 0.2{\cdot}10^6$ s$^{-1}$] and was finally also treated as a common parameter for the global fit over the entire temperature range.

Below ambient temperature, the spectra display a clear static behavior as shown for instance by the $T = 150$ K spectra in Fig. 2(a) and with moderately increasing temperature the spectra remains more or less the same. However, just below ambient temperature a clear dynamic contribution sets in, as shown in the $T = 475$ K spectra [Fig. 2(b)]. The temperature dependence of the Na-ion hopping rate [$\nu(T)$] obtained from fitting the data to Eq. 1 is shown in Fig. 2(c). As can be seen $\nu(T)$ displays a clear diffusive behavior above $T_{\rm diff} \approx 250$ K. By fitting $\nu(T)$ to an Arrhenius type equation [dashed line in Fig. 2(c)], it is possible to extract the activation energy $E_{\rm a} \approx 225$ meV.

Our results, hence, indicate that either Na-ions or muons starts to be mobile above $T_{\rm diff}$ in the $NaV_2O_4$ compound. Unfortunately there are no detailed high-temperature $^{23}$Na-NMR studies of this compound but there is a short manuscript [24] stating that some effects are seen around $T = 200$ K that could be interpreted as a sign for changes at the Na-sites. Further, the fact that $\Delta$ does not change much with temperature could also be taken as an indication that the muons form rather strong bonds to the oxygen and remains static in this compound up to $T = 500$ K. To further clarify this issue more detailed studies and modelling need to be performed. However, if we for now assume that Na-ions are indeed diffusing, it is intriguing to point out that they only become static just above the magnetic transition temperature $T_{\rm N} \approx 145.2$ K. Such a situation is rather similar to the layered frustrated magnet $Na_xCoO_2$ (NCO). NCO is a rather famous compound with an intriguing phase-diagram displaying superconducting [25, 26], magnetic [27, 28, 29, 30], as well as thermoelectric [31, 32] phases that strongly depend on the Na-content ($x$). It is well known that in NCO Na-vacancies can be ordered in many different

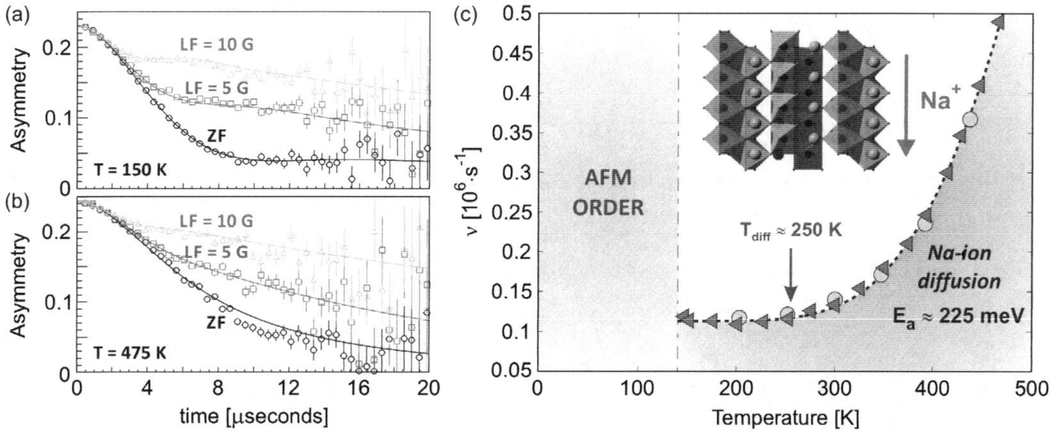


**Figure 2.** (Color online) Zero-field (ZF) and Longitudinal-field (LF = 5 and 10 G) $\mu^+$SR time spectra collected at (a) $T = 150$ K and (b) $T = 475$ K. (c) Temperature dependence of the hopping-rate, $\nu(T)$ showing the exponential increase above $T_{\rm diff} \approx 250$ K. Dashed line is fit to an Arrhenius type equation that yields the activation energy $E_{\rm a} \approx 225$ meV for the Na-ion diffusion process. Filled circles and triangles, respectively, represent data from two different experiments and samples that were performed in order to verify the reproducibility of our results.

2D and 3D configurations [33, 34, 35] and also undergo order-disorder transitions [36, 37, 38] at several different temperatures depending on the Na content ($x$). It was lately demonstrated that the electronic properties of the $CoO_2$ layers are not only governed by the number of conduction electrons on the triangular Co layers (i.e. $x$) but also the (periodic) Coulomb potential caused by (ordered) Na vacancies [34]. In fact, Na ordering and its effect on the Fermi surface was found as the key factor that separates the Curie-Weiss metal (x > 0.5) and Pauli metal (x < 0.5) [39]. The reconstruction of the Fermi surface with small Fermi-surface pockets caused by Na superstructures has later been demonstrated by the Shubnikov de Haas (SdH) oscillation effect [40].

Further, Schultze *et al.* proved by muon spin relaxation ($\mu^+$SR) and susceptibility measurements that not only the Na order but also the Na dynamics is of great importance [41]. They observed a new magnetic phase transition at $T = 8$ K for $x = 0.8$ samples, that could be connected to thermal history, i.e. Na rearrangement/diffusion, in a narrow temperature range around 200 K. This was the first clear proof that the sodium movement at high temperature could strongly influence the low-temperature magnetic properties. Later, other members of the same laboratory followed up this study with an $^{23}$Na NMR investigation of the same samples [42]. Their results show a rapidly increasing mobility and diffusion of Na ions above 200 K, and finally at $T_m = 291$ K, the Na layers melts and enters into a 2D liquid state. Above $T_m$, the NMR response is similar to what has been observed only in superionic conductors containing Na layers [43]. Finally, we have within our collaboration very recently revealed new detailed information regarding subtle structural transitions that unlocks the diffusion pathways [44] as well as presented novel data on the Na-diffusion in NCO [17]. This has opened up intriguing possibilities for tuning fundamental physical properties in correlated electron systems by controlling the dynamic properties and through that the self-assembling structure on a sub-nanoscale [45, 46, 28, 47].

From our current $\mu^+$SR data it seems very promising that a similar situation could be present also in $NaV_2O_4$. First of all, the low-temperature magnetic structure of $NaV_2O_4$ is Q1D, rather complex and there have been some questions concerning details of the incommensurate (IC) magnetic nature [12, 14]. Further, the ion-dynamic region is in very close vicinity to the magnetic phase, in fact, much closer than for the above mentioned NCO system. This makes it very plausible that a tuning of the low-temperature Q1D incommensurate magnetic order as well as electronic properties can be achieved through a careful control of the Na-vacancy landscape. Currently there are no investigations available concerning Na-deficient $Na_xV_2O_4$ samples. However, it would be of high interest for us to investigate how the Na-content ($x$) affects both the low-temperature magnetic properties as well as how this can be connected to the ion-dynamic phase at temperatures just above the magnetic transition.

Finally, the fact that Na-ions are mobile around room-temperature also opens the door to possible use of $NaV_2O_4$ and related compounds in next generation energy devices/applications. Such materials could not only function *e.g.* as an ionic conductor but the direction of conduction would be strongly anisotropic as determined by the Q1D crystal structure. This is very similar to the $LiFePO_4$ battery compound that presently is one of the most promising and investigated cathode materials [19, 48]. However, further investigations of Na-deficient samples [$Na_xV_2O_4$] are needed using both electrochemical methods as well as miscroscopic studies of ion-diffusion and changes in the atomic structure with $x$. For the current sample our previous neutron and x-ray diffraction studies [12] indicate that the sample is stoichiometric i.e. if the sodium ions are diffusing it has to be through some interstitial sites.

**Acknowledgments**
We are grateful to the staff of ISIS and RIKEN-RAL for their assistance with our $\mu^+$SR experiments. This research was financially supported by the Swiss National Science Foundation (SNSF), MEXT KAKENHI Grant No. 23108003 and JSPS KAKENHI Grant No. 26286084.

## References

[1] Masuda T, Zheludev A, Manaka H, Regnault L -P, Chung J -H and Qiu Y 2006 *Phys. Rev. Lett.* **96** 047210
[2] Mihály, Balázs Dóra, András Ványolos, Helmuth Berger and László Forró 2006 *Phys. Rev. Lett.* **97** 067206
[3] Zvyagin A A and Drechsler S -L 2008 *Phys. Rev.* B **78** 014429
[4] Sugiyama J *et al.* 2006 *Phys. Rev. Lett.* **96** 197206
[5] Simutis G *et al.* 2013 *Phys. Rev. Lett.* **111** 067204
[6] Grioni M, Pons S and Frantzeskakis E 2009 *J. Phys.: Cond. Mat.* **21** 023201
[7] Yamauchi T, Ueda Y and Môri N 2002 *Phys. Rev. Lett.* **89** 057002
[8] Okazaki K, Fujimori A, Yamauchi T and Ueda Y 2004 *Phys. Rev.* B **69** 140506
[9] Nakamura M, Sekiyama A, Namatame H, Fujimori A, Yoshihara H, Ohtani T, Misu A and Takano M 1994 *Phys. Rev. B* **49**, 16191
[10] Sakurai H 2008 *Phys. Rev.* B **78** 094410
[11] Yamaura K, Arai M, Sato A, Karki A B, Young D P, Movshovich R, Okamoto S, Mandrus D and Takayama-Muromachi E 2007 *Phys. Rev. Lett.* **99** 196601
[12] Nozaki H, Sugiyama J, Månsson M, Harada M, Pomjakushin V, Sikolenko V, Cervellino A, Roessli B and Sakurai H 2010 *Phys. Rev.* B **81** 100410(R)
[13] Sugiyama J, Ikedo Y, Goko T, Ansaldo E J, Brewer J H, Russo P L, Chow K H and Sakurai H 2008 *Phys. Rev.* B **78** 224406
[14] Ofer O, Ikedo Y, Goko T, Månsson M, Sugiyama J, Ansaldo E J, Brewer J H, Chow K H and Sakurai H 2010*Phys. Rev.* B **82** 094410
[15] Yaouanc A and Dalmas de Réotier P 2011 *Muon Spin Rotation, Relaxation, and Resonance* (Oxford: Oxford University Press)
[16] Sugiyama J, Mukai K, Ikedo Y, Nozaki H, Månsson M and Watanabe I 2009 *Phys. Rev. Lett.* **103** 147601
[17] Månsson M and Sugiyama J 2013 *Physica Scripta* **88** 068509
[18] Sugiyama J *et al.* 2010 *Phys. Rev.* B **82** 224412
[19] Sugiyama J *et al.* 2011 *Phys. Rev.* B **84** 054430
[20] Sugiyama J *et al.* 2012 *Phys. Rev.* B **85** 054111
[21] Sugiyama J, Mukai K, Nozaki H, Harada M, Månsson M, Kamazawa K, Andreica D, Amato A and Hillier A D 2013 *Phys. Rev.* B **87** 024409

[22] Sugiyama J, Nozaki H, Mukai K, Harada M, Månsson M, and Hillier A 2014 *Solid State Ionics* **262** 901
[23] Sugiyama J *et al.* 2012 *Phys. Proc.* **30** 105
[24] Takeda H, Itoh M and Sakurai H 2011 *Journal of Physics: Conference Series* **273** 012142
[25] Takada K, Sakurai H, Takayama-Muromachi E, Izumi F, Dilanian R A and Sasaki T 2003 *Nature* **422** 53
[26] Sugiyama J *et al.* 2010 *Phys. Rev.* B **82** 214505
[27] Sugiyama J, Brewer J H, Ansaldo E J, Itahara H, Tani T, Mikami M, Mori Y, Sasaki T, Hébert S and Maignan A 2004 *Phys. Rev. Lett.* **92** 017602
[28] Foo M L, Wang Y, Watauchi S, Zandbergen H W, He T, Cava R J and Ong N P 2004 *Phys. Rev. Lett.* **92** 247001
[29] Bayrakci S P, Bernhard C, Chen D P, Keimer B, Kremer R K, Lemmens P, Lin C T, Niedermayer C and Strempfer J 2004 *Phys. Rev.* B **69**, 100410(R)
[30] Mendels P *et al.* 2005 *Phys. Rev. Lett.* **94** 136403
[31] Terasaki I, Sasago Y and Uchinokura K 1997 *Phys. Rev. B* **56** R12685(R)
[32] Lee M, Vici L, Li L, Wang Y, Foo M L, Watauchi S, Pascal Jr R A, Cava R J and Ong N P 2006 *Nature Mater.* **5** 537
[33] Wang Y, Ding Y and Ni J 2009 *J. Phys.: Cond. Mat.* **21** 035401
[34] Roger M *et al.* 2007 *Nature* **445** 631
[35] Zhang P, Capaz R B, Cohen M L, and Louie S G 2005 *Phys. Rev.* B **71** 153102
[36] Morris D J P *et al.* 2009 *Phys. Rev.* B **79** 100103(R)
[37] Hinuma Y, Meng Y S, and Ceder G 2008 *Phys. Rev.* B **77** 224111
[38] Igarashi D, Miyazaki Y, Kajitani T, and Yubuta K 2008 *Phys. Rev.* B **78** 184112
[39] Bobroff J, Lang G, Alloul H, Blanchard N, and Collin G 2006 *Phys. Rev. Lett.* **96** 107201
[40] Balicas L, Jo Y J, Shu G J, Chou F C and Lee P A 2008 *Phys. Rev. Lett.* **100** 126405
[41] Schulze T F, Häfliger P S, Niedermayer Ch, Mattenberger K, Bubenhofer S and Batlogg B 2008 *Phys. Rev. Lett.* **100** 026407
[42] Weller M, Sacchetti A, Ott H R, Mattenberger K and Batlogg B 2009 *Phys. Rev. Lett.* **102** 056401
[43] Villa M and Bjorkstam J L 1980 *Phys. Rev.* B **22** 5033
[44] Medarde M *et al.* 2013 *Phys. Rev. Lett.* **110** 266401
[45] McElroy K, Lee J, Slezak J A, Lee D -H, Eisaki H, Uchida S and Davis J C 2005 *Science* **309** 1048
[46] Zhang J, Ismail, Moore R G, Wang S -C, Ding H, Jin R, Mandrus D and Plummer E W 2006 *Phys. Rev. Lett.* **96** 066401
[47] Julien M -H *et al.* 2008 *Phys. Rev. Lett.* **100** 096405
[48] Wang Y, He P and Zhou H 2011 *Energy Environ. Sci.* **4** 805

# In situ $\mu^+$SR measurements on the hydrogen desorption reaction of magnesium hydride

**I Umegaki[1*], H Nozaki[1], M Harada[1], Y Higuchi[1], T Noritake[1], M Matsumoto[1], S-i Towata[1], E J Ansaldo[2], J H Brewer[2,3], A Koda[4], Y Miyake[4] and J Sugiyama[1**]**

[1] Toyota Central Research & Development Laboratories, Inc., 41-1 Yokomichi, Nagakute, Aichi 480-1192, Japan
[2] TRIUMF, 4004 Wesbrook Mall, Vancouver, BC, V6T 2A3 Canada
[3] Department of Physics & Astronomy, University of British Columbia, Vancouver, BC, V6T 1Z1 Canada
[4] Muon Science Laboratory, Institute of Materials Structure Science, KEK, Tsukuba, Ibaraki 305-0801, Japan

E-mail: *e4774@mosk.tytlabs.co.jp, **e0589@mosk.tytlabs.co.jp

**Abstract.** In order to clarify the reason why the hydrogen desorption temperature ($T_d$) of $MgH_2$ is lowered by milling, we have studied the change in a local nuclear magnetic field with temperature by means of $\mu^+$SR. We have found a very clear oscillation in the ZF-spectrum at 2 K for the "milled" and "milled with $Nb_2O_5$" samples, while such oscillation is weaker for the "as prepared" $MgH_2$. It was also found that the oscillation signal is stable up to 250 K and is assigned mainly due to the formation of a H-$\mu$-H system. At temperatures above ambient $T$, we also found that the ZF-$\mu^+$SR spectrum exhibits a static Kubo-Toyabe behavior due to the nuclear magnetic field of $^1$H. Furthermore, it was clarified that rapid H diffusion starts well below $T_d$ only in the milled samples, leading to the conclusion that the consequent enhanced diffusion rate in $MgH_2$ is essential to accelerate the desorption reaction and to decrease $T_d$.

## 1. Introduction

Despite huge efforts for searching and developing hydrogen storage materials [1], it is still a great challenge for materials science and engineering to make a safe, light-weight, high-capacity hydrogen storage system. Unfortunately, all the alloys are too heavy to carry on automobiles due to their small gravimetric $H_2$ density. On the other hand, complex hydrides are possible candidates for future onboard hydrogen storage materials [2, 3], if their hydrogen desorption temperature ($T_d$) is reasonably lowered, ideally down to ambient temperature [4, 5]. Among many complex hydrides, magnesium hydride, $MgH_2$ (Fig. 1), has attracted special attention due to its relatively large gravimetric $H_2$ density. Furthermore, since $T_d$ is decreased down to 500 K by milling $MgH_2$ with 5 wt% $Nb_2O_5$ [6], for reasons currently unknown [7, 8], $MgH_2$ is considered to be one of the most promising hydrogen storage materials.

Recent $\mu^+$SR experiments on the other complex-type hydrides, $NaAlH_4$ [9] and $M(BH_4)_n$ [10], proposed a significant role of H-diffusion in solids on determining $T_d$ through the observation of H-$\mu$-H and/or H-$\mu$ systems. This is a promising future of $\mu^+$SR, because such three-spin and two-spin systems are formed only when muons are implanted into samples. However, the oscillation signals due to the formation of such systems become weaker and weaker with increasing temperature. As a result, there was no $\mu^+$SR study of such compounds above ambient temperature. Even if the implanted muons are mobile by thermal activation, muons are still expected to *see* local nuclear magnetic fields via a dynamic Kubo-Toyabe behavior. We have, therefore, measured $\mu^+$SR spectra on $MgH_2$ in a wide temperature range in order to investigate the mechanism on the decrease in $T_d$ by milling with $Nb_2O_5$.

## 2. Experiment

A coarse $MgH_2$ powder, which is called as an "as prepared" sample, was purchased from Avocado. The average grain diameter ($d_g$) of the "as prepared" sample is 100 nm. Then, the "as prepared" sample was packed in a ball mill container in an Ar-filled glove-box, and then, milled for 24 hours by a planetary ball mill in an Ar atmosphere in order to reduce $d_g$ down to 8-9 nm (a "milled" sample). Using the same milling procedure, but together with 5 wt% $Nb_2O_5$ powder, a "milled with $Nb_2O_5$" sample was also prepared. $T_d$ of the three samples was determined by temperature-programmed desorption (TPD) measurements, in which the desorption gas was detected by a mass spectrometer (TP-5000, Ohkura-Riken) during heating with a rate 5 K/min in a He flow with a flow rate 60 ml/min. Such measurements revealed that the amount of desorbed hydrogen does not change with milling.

For the $\mu^+$SR measurements, each powder sample was pressed into a pellet, and then, the pellet was packed into a titanium cell sealed with a gold O-ring in a glovebox, because $MgH_2$ is sensitive to moisture. The $\mu^+$SR spectra were measured at TRIUMF in the temperature range between 2 and 300 K, and at J-PARC in the temperature range between 300 and 750 K. Particularly in J-PARC, the pressure of the sample space, which is evacuated by a turbo-molecular pump, was monitored during the $\mu^+$SR measurements. In order to release $H_2$ in the Ti cell to the sample space, small holes were made on the Ti foil window just before the measurements.

## 3. Results

Figure 2 shows the TPD profile of the present three samples. It is confirmed that $T_d = 710$ K for the "as prepared" sample, but $T_d = 620$ K for the "milled" sample and 530 K for the "milled

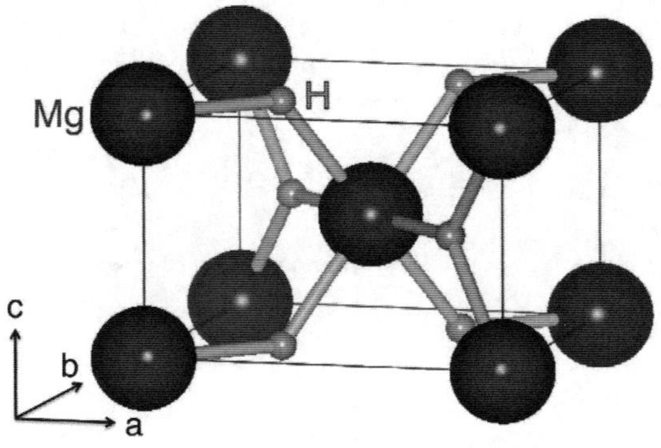


**Figure 1.** Crystal structure of tetragonal $MgH_2$ with space group $P4_2/mnm$, $a$ = 4.5180 Å and $c$ = 3.0211 Å. Mg ion occupies the center and corner sites in a unit cell, and H ions are connected to Mg ions with the bond length 1.95 Å[11].

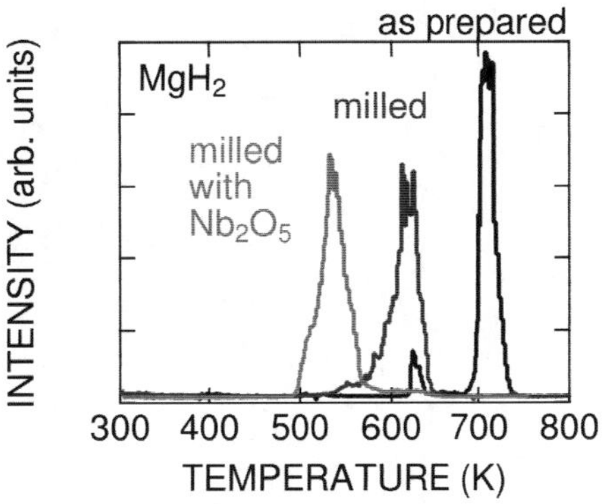


**Figure 2.** Temperature-programmed desorption (TPD) profile for the "as prepared", "milled", and "milled with $Nb_2O_5$" $MgH_2$ samples.

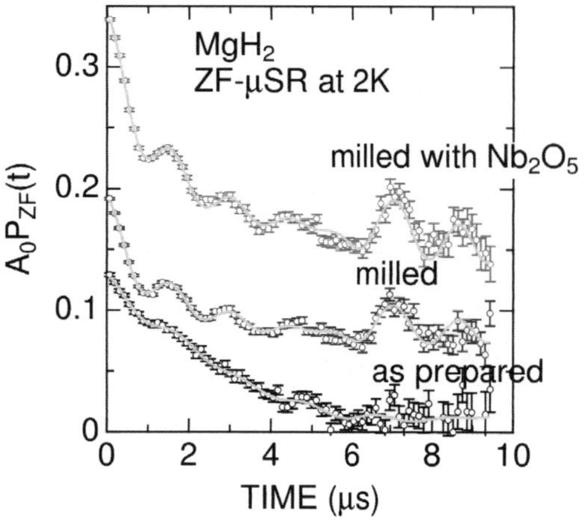


**Figure 3.** ZF-$\mu^+$SR spectra for the three kinds of $MgH_2$ measured at 2 K in TRIUMF. The spectra for the "milled" and "milled with $Nb_2O_5$" samples are shifted upward by 0.05 and 0.1 for clarity of display. The oscillations indicate the formation of the H-$\mu$-H system.

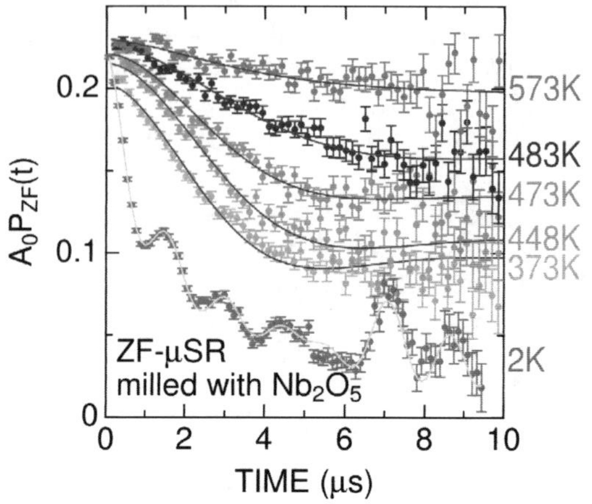


**Figure 4.** Temperature variation of ZF-$\mu^+$SR spectrum for the "milled with $Nb_2O_5$" sample obtained in J-PARC. The data at 2 K, which is the same to that in Fig. 3, is also plotted for comparison, but shifted downward by 0.05 for clarity of display.

with $Nb_2O_5$" sample.

### *3.1. Below ambient temperature*

Figure 3 shows the ZF-$\mu^+$SR spectra obtained at 2 K for the three samples. The ZF-spectra for the two milled samples exhibit a very clear oscillation, indicating the formation of the H-$\mu$-H and/or H-$\mu$ systems. In fact, the ZF-spectra for both samples were reasonably fitted by a combination of an exponentially relaxing H-$\mu$-H signal, two exponentially relaxing cosine signals probably due to a localized state with three (or more) spins, and two non-oscillatory relaxing signals. The other four signals besides the H-$\mu$-H signal suggest the presence of multiple muon sites in the $MgH_2$ lattice. We will call this complex situation as the formation of a "H-$\mu$-H like" system.

Back to the H-$\mu$-H signal, the muon-spin polarization function ($P_{\mathrm{H}\mu\mathrm{H}}$) is given by [12];

$$\begin{aligned} P_{\mathrm{H}\mu\mathrm{H}}(t) &= \frac{1}{2} + \frac{1}{6}\cos(\sqrt{3}\omega_{\mathrm{d}}t) \\ &+ \frac{1+1/\sqrt{3}}{6}\cos(\frac{3+\sqrt{3}}{2}\omega_{\mathrm{d}}t) \\ &+ \frac{1-1/\sqrt{3}}{6}\cos(\frac{3-\sqrt{3}}{2}\omega_{\mathrm{d}}t)\ , \end{aligned} \quad (1)$$

and [12, 13]

$$\omega_d \equiv 2\pi f_d = \frac{\mu_0 \hbar \gamma_\mu \gamma_{\mathrm{H}}}{4\pi r^3}. \quad (2)$$

Here, $\gamma_\mu$ is the nuclear gyromagnetic ratio of $\mu$, $\gamma_{\mathrm{H}}$ is the nuclear gyromagnetic ratio of $^1$H, and $r$ is the distance between $\mu$ and H in the H-$\mu$-H system. Since the fit provided that $f_{\mathrm{d}} = 0.25$ MHz for both "milled sample" and "milled with $Nb_2O_5$" samples, $r$ is estimated as 1.16 Å, which is close to a half of the distance between the two nearest neighboring hydrogens (2.52 Å). This means that some of the implanted muons (about 25%) sit at the middle of the two neighboring hydrogens, resulting in the formation of a H-$\mu$-H system.

Note that even for the "as prepared" sample, the ZF-spectrum shows a weaker, but clearly evident oscillation. Therefore, the formation of the H-$\mu$-H like system is found to be an intrinsic phenomenon for $MgH_2$. The enhancement of the oscillatory signal by milling suggests the role of fresh surface on the formation of the H-$\mu$-H like system. That is, for the "as prepared" sample, hydrogens in the surface layer are expected to bind to oxygens, while it is stored in Ar atmosphere. As a result, the H-$\mu$-H like signal is a minor component in the ZF-spectrum for the "as prepared" sample, together due to the large particle size of the sample (100 nm). This also proposes a significant role of surface to determine $T_d$ for $MgH_2$.

In brief summary, although the $\mu^+$SR experiment at low temperatures clarified the variation of static nature of $MgH_2$ by milling, there is still no information on dynamics of hydrogens. We therefore measured $\mu^+$SR spectra for the three samples at high temperatures, i.e. from ambient temperature to $T_d$ and above.

### *3.2. Above ambient temperature*

Figure 4 shows the ZF-$\mu^+$SR spectra at high temperatures for the "milled with $Nb_2O_5$" sample. At 373 K, the ZF-spectrum exhibits a static Gaussian Kubo-Toyabe (KT) behavior due to a nuclear magnetic field of $^1$H. As temperature increases from 373 K, the KT relaxation becomes weaker and weaker, and finally, the spectrum looks almost time-independent at 573 K.

In order to know the change in the $\mu^+$SR parameters with temperature, Figs. 5(b), 5(c) and 5(d) show the temperature dependences of the relaxation rate ($\lambda_{TF}$) of the wTF-spectrum, which corresponds to a spin-spin relaxation rate ($1/T_2$), i.e., the distribution width of the nuclear magnetic field ($\Delta$), and the field fluctuation rate ($\nu$) for the three samples. For the "as prepared sample", $\lambda_{TF}$, $\Delta$ and $\nu$ are almost temperature independent up to 630 K, and then become very small, eventually zero at temperatures above $T_d$. This means that the hydrogen desorption reaction abruptly occurs and completes at $T_d$. In fact, at temperatures above $T_d$, the wTF-spectrum exhibited a non-relaxing oscillatory signal from the Ti-cell and Mg. Furthermore, the wTF-spectrum obtained in cooling mode from 750 K did not change even below $T_d$, suggesting that all the hydrogen atoms are removed from the sample. For such situation, since the ZF-spectrum exhibits an exponentially relaxing behavior with a very small relaxation rate, it is impossible to evaluate $\Delta$ and $\nu$. This is the reason why we show the $\lambda_{TF}(T)$ curve in Fig. 5.

For the "milled" sample, $\lambda_{TF}$ gradually decreases with temperature above 470 K, while $\nu$ apparently increases with temperature also above 470 K. This suggests that the H-desorption reaction starts to occur far below $T_d$ (=620 K) for the "milled" sample, because $1/T_2$ decreases with decreasing the number density of H around the muon. Moreover, the increase in $\nu$ implies rapid diffusion of H and/or muon in the sample, which is proposed to be a significant factor to decrease $T_d$ for the other hydrogen storage materials, such as $LiAlH_4$ [9] and $M(BH_4)_n$ [10].

For the "milled with $Nb_2O_5$" sample, as temperature increases from ambient temperature, $\nu$ starts to increase above 450 K as for the "milled" sample, but the slope ($d\mu/dT$) is rather steep compared with that for the "milled" sample and $\nu$ reaches a maximum ($\sim 0.5 \times 10^6$ s$^{-1}$) at 550 K. The maximum value of $\nu$ is larger by 2.5 times than the highest $\nu$ for the "milled" sample. Simultaneously, the $\lambda_{TF}(T)$ curve shows a rapid decrease down to zero at around 500 K. These changes are good agreement with the TPD curve [Fig. 5(a)]. It should be emphasized that, even for the "milled" sample, $\nu$ also starts to increase above 470 K. In addition, we measured the wTF-spectra for the "milled with $Nb_2O_5$" sample with decreasing $T$ above $T_d$. Since the wTF-spectrum does not exhibits a relaxation even below $T_d$ due to the absence of H, $\lambda_{TF}$ measured on cooling is almost zero down to the lowest $T$ measured (=433 K) [Fig. 5(b)].

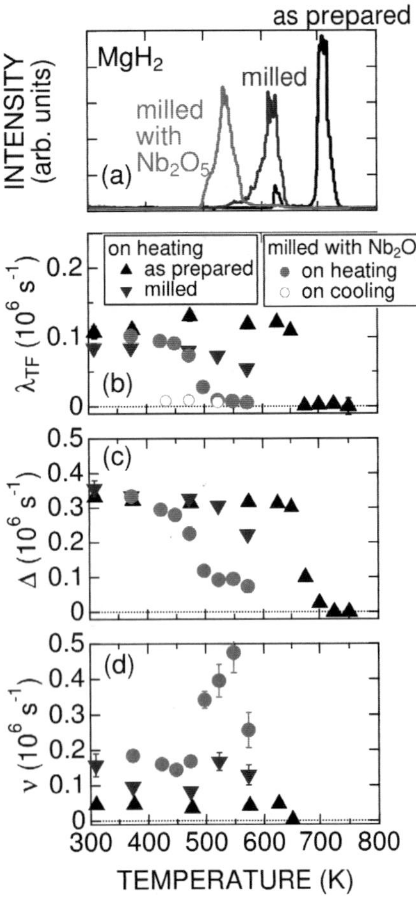


**Figure 5.** (a) TPD result and temperature dependences of (b) the exponential relaxation rate of the weak transverse field spectrum ($\lambda_{\rm TF}$) and (c) the filed distribution width ($\Delta$) (d) the filed fluctuation rate ($\nu$) for the "as prepared", "milled", and "milled with $Nb_2O_5$" samples. (a) is the same to Fig. 2.

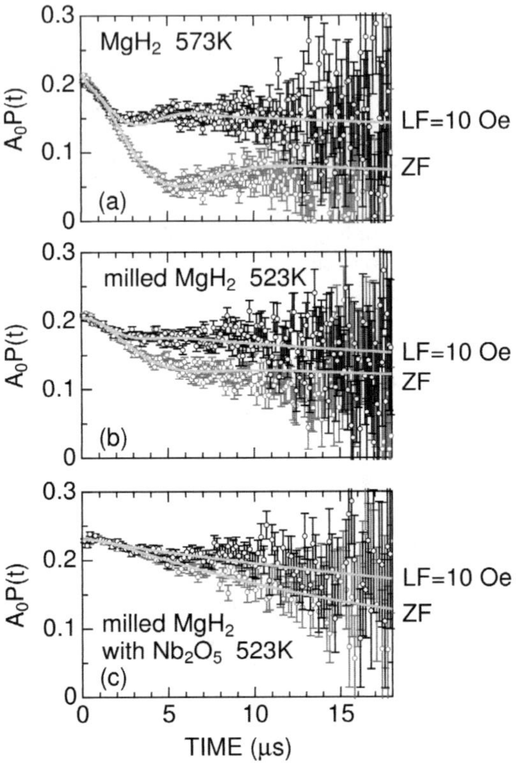


**Figure 6.** ZF- and LF-spectra for (a) the "as prepared" sample at 573 K, (b) the "milled" sample at 523 K, and (c) the "milled with $Nb_2O_5$" sample. The applied LF was 10 Oe. These spectra were obtained in J-PARC.

After the measurements above $T_{\rm d}$, the residue in the Ti cell was checked by XRD analyses in air. It was found that the residue consists of mainly Mg and small amount of MgO, $Mg(OH)_2$, and $MgH_2$ (below 3%). This means that almost all the hydrogen were released from the sample by the hydrogen desorption reaction.

## 4. Discussion

Here, we discuss the relationship between the magnitude of $T_{\rm d}$ and the obtained $\mu^+$SR parameters. The most remarkable difference of the $\mu^+$SR parameters among the three samples is that $\nu$ starts to increase above 470 K both for the "milled" and "milled with $Nb_2O_5$" samples, but DOES NOT even at $T_{\rm d}$ for the "as prepared" sample. The increase in $\nu$ is naturally caused by H- and/or $\mu$-diffusion in the $MgH_2$ lattice. Furthermore, since $\mu$ is static even at 573 K in the "as prepared" sample (Fig. 6), the increase in $\nu$ (decrease in $\lambda_{TF}$) is likely due to H-diffusion, which also induces $\mu$-diffusion. Then, why H- and $\mu$-diffusion are observed only for the two milled samples? As mentioned in Sec. 3.1, each particle in the two milled samples is surrounded by fresh surface with greater surface area to particle volume ratio, across which the liberated hydrogen is released to atmosphere. The fresh surface, which is evidenced by the presence of the H-$\mu$-H like system, is naturally thought as a starting point of the hydrogen desorption reaction, meaning the formation of defects at the surface. When such defects are formed at the surface, the liberated hydrogen is easily moved to the surface through H-diffusion in solids. This would be a reason for the decrease in $T_d$ by milling. Since the milling with $Nb_2O_5$ does not change the onset temperature of the hydrogen desorption reaction, $Nb_2O_5$ is most likely to function as catalyst, leading to the enhancement of the reaction rate.

Finally, we wish to mention the tentative result of an in situ $\mu^+$SR measurement. During the $\mu^+$SR measurement on the "milled with $Nb_2O_5$" sample, we also monitored the pressure ($P$) in the sample space. When $P$ starts to increase due to the hydrogen desorption reaction, $\lambda_{TF}$ starts to decrease, whereas $\nu$ starts to increase. This is, to our knowledge, a first **"in situ $\mu^+$SR measurement"** to *see* the change in real time local magnetic environments. Such in situ measurements will provide significant information on the desorption reaction in hydrogen storage materials.

## Acknowledgments

We appreciate the staff of TRIUMF and J-PARC for their help to our experiments. This work was supported by Grant-in-Aid for Scientific Research on Innovative Areas "Ultra Slow Muon Microscope" (No. 26108722, 23108003) of the Ministry of Education, Culture, Sports, Science, and Technology, Japan.

## References

[1] Schlapbach L and Züttel A 2001 *Nature* **414** 353
[2] Orimo S, Nakamori Y, Eliseo J R, Züttel A, and Jensen C M 2007 *Chem. Rev.* **107** 4111
[3] Züttel A, Borgschulte A and Orimo S 2007 *Scrip. Mater.* **56** 823
[4] Orimo S, Nakamori Y, Ohba N, Miwa K, Aoki M, Towata S-i and Züttel A 2006 *Appl. Phys. Lett.* **89** 021920
[5] Hwang S J, Bowman R C, Reiter Jr J W, Rijssenbeek J, Soloveichik G L, Zhao J-C, Kabbour H and Ahn C C 2008 *J. Phys. Chem. Lett.* C **112** 3164
[6] Hanada N, Ichikawa T and Fujii H 2007 *J. Alloys Compounds* **446-447** 67
[7] Liang G, Huot J, Boily S, Van Neste A and Schulz R 1999 *J. Alloys Compounds* **292** 247
[8] Bobet J-L, Chevalier B, Song M Y, Darriet B, and Etourneau J 2002 *J. Alloys Compounds* **336** 292
[9] Kadono R, Shimomura K, Satoh K H, Takeshita S, Koda A, Nishiyama K, Akiba E, Ayabe R M, Kuba M and Jensen C M 2008 *Phys. Rev. Lett.* **100** 026401
[10] Sugiyama J, Ikedo Y, Noritake T, Ofer O, Goko T, Månsson M, Miwa K, Ansaldo E J, Brewer J H, Chow K H and Towata S I 2010 *Phys. Rev.* B **81** 092103
[11] Noritake T, Aoki M, Towata S-i, Seno Y, Hirose Y, Nishibori E, Takata M and Sakata M 2009 *Appl. Phys. Lett.* **81** 2008
[12] Brewer J H, Kreitzman S R, Noakes D R, Ansaldo E J, Harshman D R and Keitel R 1986 *Phys. Rev.* B **33** 7813(R)
[13] Månsson M, Ikedo Y, Nozaki H, Sugiyama J, Russo P L, Andreica D, Shizuya M, Isobe M and Takayama-Muromachi E 2010 *Solid State Commun.* **150** 307

Journal of Physics: Conference Series **551** (2014) 012037 doi:10.1088/1742-6596/551/1/012037

# Lithium Diffusion & Magnetism in Battery Cathode Material $Li_xNi_{1/3}Co_{1/3}Mn_{1/3}O_2$

**M Månsson[1,2], H Nozaki[3], J M Wikberg[4], K Prša[1,5], Y Sassa[5], M Dahbi[6], K Kamazawa[7], K Sedlak[8], I Watanabe[9] and J Sugiyama[3]**

[1] Laboratory for Quantum Magnetism, École Polytechnique Fédérale de Lausanne (EPFL), CH-1015 Lausanne, Switzerland
[2] Lab. for Neutron Scattering & Imaging, Paul Scherrer Institute, CH-5232 Villigen PSI, Switzerland
[3] Toyota Central Research & Development Laboratories, Inc., 41-1 Yokomichi, Nagakute, Aichi 480-1192, Japan
[4] Department of Physics and Astronomy, Molecular and Condensed Matter Physics, Uppsala University, Box 516, S-75120 Uppsala, Sweden
[5] Laboratory for Solid state physics, ETH Zürich, CH-8093 Zürich, Switzerland
[6] Department of Chemistry, Ångström Laboratory, Uppsala University, Box 538, S-75121 Uppsala, Sweden
[7] Comprehensive Research Organization for Science and Society (CROSS), Tokai, Ibaragi 319-1106, Japan
[8] Laboratory for Muon Spin Spectroscopy, Paul Scherrer Institute, CH-5232 Villigen PSI, Switzerland
[9] Advanced Meson Science Laboratory, RIKEN, 2-1 Hirosawa, Wako, Saitama 351-0198, Japan

E-mail: `martin.mansson@epfl.ch`

**Abstract.** We have studied low-temperature magnetic properties as well as high-temperature lithium ion diffusion in the battery cathode materials $Li_xNi_{1/3}Co_{1/3}Mn_{1/3}O_2$ by the use of muon spin rotation/relaxation. Our data reveal that the samples enter into a 2D spin-glass state below $T_{SG} \approx 12$ K. We further show that lithium diffusion channels become active for $T \geq T_{diff} \approx 125$ K where the Li-ion hopping-rate $[\nu(T)]$ starts to increase exponentially. Further, $\nu(T)$ is found to fit very well to an Arrhenius type equation and the activation energy for the diffusion process is extracted as $E_a \approx 100$ meV.

## 1. Introduction

One of the main obstacles for a general breakthrough of electric automobiles is the development of a high-capacity, cheap and safe rechargeable battery. The most widely used cathode material is by far $LiCoO_2$ [1, 2], however, cobalt is very expensive and there is a strong driving force to find new cheaper and more environmental friendly cathode materials. One of the big problems for battery application is the structural distortions that occur [3, 4] when large amounts of Li is deintercalated during charge / discharge cycles of the battery [see Fig. 1]. Some efforts have been made to minimize these effects *e.g.* by combining Co and Ni into a $LiNi_{1-x}Co_xO_2$ solid solution. However, this compound suffers from strong intermixing of Ni into the Li layers for the highly delithiated state [5]. Recently, the mixed transition metal oxides (MTMO's) of the form $Li_xNi_{1-y-z}Co_yMn_zO_2$ have become the center of attention [6, 7]. In particular has the

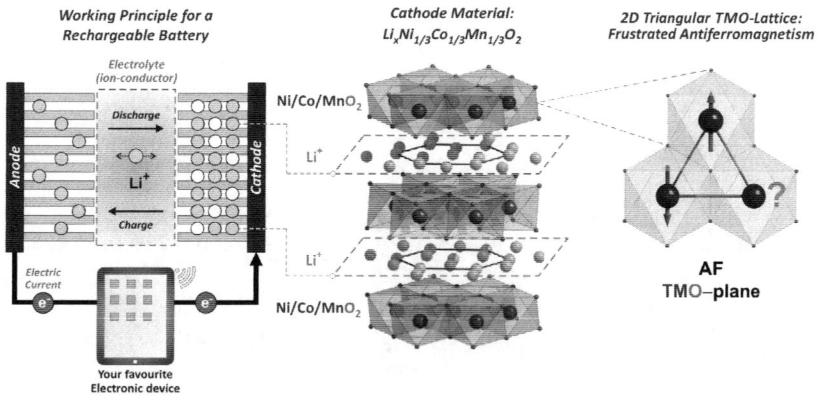


**Figure 1.** Schematic figure showing the operational principle for a standard rechargeable battery (left). Middle panel show the archetypical atomic structure for a layered battery cathode materials. Finally, right panel show the triangular transition metal oxide (TMO) planes that creates the possibility for 2D frustrated antiferromagnetism (AF) within the planes. This figure highlights that the very same layered TMO compounds can be of interest for energy applications as well as low-temperature fundamental condensed matter physics.

$Li_xNi_{1/3}Co_{1/3}Mn_{1/3}O_2$ compound [8] been put forward as one of the most promising battery electrode materials for the future. The structure of these compounds is the same as for the fundamental $LiCoO_2$ i.e. a rhomohedral lattice (space group $R\bar{3}m$) where Ni/Co/$MnO_2$ planes are stacked between nonmagnetic Li layers along the $c$-axis [see Fig. 1]. In similarity with $LiNiO_2$ some intermixing of Ni in the Li layers can be expected. However, in this particular case, this actually helps to stabilize the structure and improve capacity retention [9].

In addition to their extensive use in energy related applications, these materials also display very interesting and complex magnetic properties at lower temperatures. This is due to the 2D triangular antiferromagnetic (AF) lattice [10] (see also Fig. 1) created by the transition metal ions. For the $Li_xNi_{1/3}Co_{1/3}Mn_{1/3}O_2$ compounds the situation is even more complex since a combination of three different transition metals is present within the geometrically frustrated magnetic planes [11, 12, 13]. In this paper we have investigated both the low-temperature magnetic properties as well as the high-temperature ion-diffusion for two $Li_xNi_{1/3}Co_{1/3}Mn_{1/3}O_2$ samples ($x = 1$ and 0.3) using muon spin rotation/relaxation ($\mu^+$SR).

## 2. Experimental Details

Powder samples of $Li_xNi_{1/3}Co_{1/3}Mn_{1/3}O_2$ compound were prepared at Uppsala University, Sweden. For the low-temperature magnetic measurements, two samples ($x = 1$ and 0.3) of approximately 1 g each were sealed inside small envelopes made of very thin Al-coated Mylar tape under helium atmosphere and then attached to a low-background, fork-type, sample holder. Subsequently, high-resolution $\mu^+$SR spectra were measured at the (continuous, DC) Swiss Muon Source (S$\mu$S), Paul Scherrer Institute, Villigen, Switzerland. By using the muon beamline $\pi$E1 with the Dolly spectrometer, zero-field (ZF) and weak transverse-field (wTF) spectra were collected for $2\text{ K} \leq T \leq 150$ K. For the ion-diffusion experiments, approximately 2 grams of the same samples were pressed into two discs with a 24 mm diameter and 1.5 mm thickness. Inside a helium glove-box the discs were packed into a Au-sealed (gold O-ring) powder cells made of pure titanium using a thin (100 $\mu$m) Kapton film as 'entrance window' for the muons. In addition, a silver mask with a hole matching the samples diameter was mounted onto the cell to ensure that the any minor background signal is non-relaxing over a wide temperature range. The cell was mounted onto the Cu end-plate of a liquid-He flow-type cryostat in the temperature range between 10 and 500 K. Subsequently, ZF-, wTF- and LF-$\mu^+$SR spectra were collected using the RIKEN-RAL / ARGUS spectrometer at the pulsed muon source ISIS/RAL in UK. The experimental techniques are described in more detail elsewhere [14].

## 3. Results and Discussion: Low-temperature Magnetic Properties

To obtain a first insight into the nature of the magnetic order in these compounds, we acquired zero-field (ZF) data at the lowest temperature ($T = 2$ K). The $\mu^+$SR time spectrum show no indication of spontaneous muon precession, but rather only a fast exponentially decaying signal for both samples [see Fig. 2(a-b)]. Such signal indicate either a wide field-distribution, which is typical for spin-glasses (SG) or the presence of dynamical correlations. However, recent magnetometry measurements of our samples ($x = 1$ and $x = 0.3$) using zero-field-cooled (ZFC) and field-cooled (FC) protocols [15] clearly show a divergence around $T = 10$ K, which is a clear indication for a magnetically frustrated SG state. Further, information obtained also from related compounds using magnetometry as well as $\mu^+$SR [11, 12, 13] clearly favours the static SG scenario . The ZF data was found to be well fitted to the sum of a very fast exponentially relaxing signal and a slower decaying stretched-exponential function:

$$A_0\, P_{\rm ZF}(t) = A_{\rm fast} \cdot e^{-\lambda_{\rm fast} t} + A_{\rm slow} \cdot e^{[(-\lambda_{\rm slow} t)^n]}. \tag{1}$$

This kind of stretched exponential relaxation is found to appear in a range of different systems, however, they are commonly connected to SG systems [16] containing geometrical frustration *e.g.* the title compound. In similarity to our previous $\mu^+$SR studies of the closely related $Li(Ni_{0.8}Co_{0.1}Mn_{0.1})O_2$ compound [13], we find that also the present compounds enter into a disordered spin-glass (SG) state at lowest temperature ($T = 2$ K). Further, from the fits to Eq. 1 the critical exponent $n$ reaches a value $n \leq 1/2$ [$n_{x=1} \approx 0.41$ and $n_{x=0.3} \approx 0.44$], possibly indicating a decreased dimensionality, however, this is only a phenomenological parametrisation of the data and other experimental methods needs to be employed. From ZF measurements at $T = 30$ K as well as $T = 100$ K it is evident that both samples are in a paramagnetic state above at least $T = 30$ K [see insets Fig. 2(a-b)].

To further investigate the PM to SG phase-transition we also performed detailed weak-transverse field (wTF =50 G) measurements as a function of temperature. The wTF $\mu^+$SR time spectra for selected temperatures are shown in Fig. 2(c-d). Around $T = 10$ K a total suppression of the signal from the externally applied field is seen as annihilation of the wTF asymmetry ($A_{\rm TF}$). The data is found to be well fitted by a combination of an exponentially relaxing cosine oscillation and a fast relaxing component (the latter becomes dominant at and below the transition):

$$A_0\, P_{\rm TF}(t) = A_{\rm TF} \cos(2\pi f_{\rm TF} \cdot t + \phi_{\rm TF}) \cdot e^{-\lambda_{\rm TF} t} + A_{\rm fast} \cdot e^{-\lambda_{\rm fast} t}. \tag{2}$$

From such fits, the temperature dependence of $A_{\rm TF}$ can be extracted and a better picture of the transition is obtained [Fig. 2(e-f)]. For both samples it is clear that a magnetic phase transition into the previously mentioned SG state occurs at $T_{\rm SG} \approx 12$ K where $A_{\rm TF}$ drops to zero very rapidly. This is in fact rather different from the $Li(Ni_{0.8}Co_{0.1}Mn_{0.1})O_2$ compound [13] where we found a gradual transition over a 50 K wide temperature range. Hence, contrary to the cluster glass like state found for the previous compound, $Li_xNi_{1/3}Co_{1/3}Mn_{1/3}O_2$ is believed to enter into a simple SG phase below $T_{\rm SG}$. This could possibly be explained by the equal distribution of transition metals creating a higher degree of disorder on the TMO planes. Worth noticing is also that the $x = 1$ sample seems to have a slightly wider transition than $x = 0.3$. This could be an artefact related to too low statistics for the wTF data. Another explanation could be that stoichiometric $x = 1$ samples are well known to be notoriously difficult to synthesize.

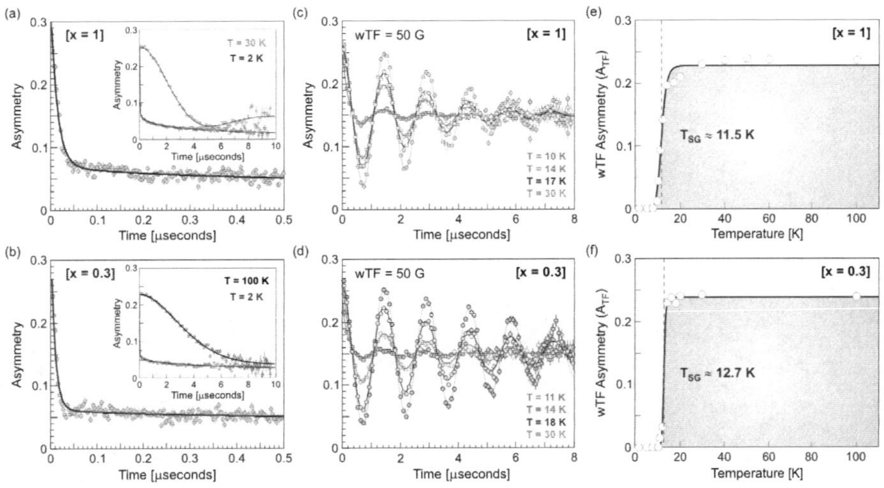

**Figure 2. Low-temperature magnetic measurements at S$\mu$S continuous muon source.** (a-b) Zero-field (ZF) $\mu^+$SR spectra at $T = 2$ K from the $x = 1$ and $x = 0.3$ samples, respectively. Data are shown in the very short time-scale in order to emphasize the fast relaxing signal. Solid line is a fit to Eq. 1. Insets show data in the longer time-scale together with a spectra at higher temperature, which in both cases display clear Kubo-Toyabe behaviour. (c-d) weak-transverse field (wTF $= 50$ G) data from the $x = 1$ and $x = 0.3$ samples, respectively, for a series of temperatures across the magnetic transition. Solid lines are fits to Eq. 2. (e-f) Temperature dependence of the wTF asymmetry ($A_{\rm TF}$) from the two samples. A clear entrance into a magnetic state is observed in both cases for $T_{\rm SG} \approx 12$ K.

## 4. Results and Discussion: High-temperature Li-ion Diffusion

In order to investigate Li-ion diffusion in these materials we have used our previously presented $\mu^+$SR technique for studying solid state ion diffusion in battery related materials [17, 18]. In similarity to our previous investigations of both Li-ion [17, 19, 20, 21, 22, 23, 24] and Na-ion [18] diffusion, a series of ZF, wTF $= 30$ G as well as LF $= 5$ G and 10 G $\mu^+$SR spectra were acquired in the temperature range between 50 and 500 K. At each temperature, the ZF and two LF spectra were fitted by an exponentially relaxing dynamic Kubo-Toyabe (KT) function plus a small background signal from the fraction of muons stopped mainly in the silver mask:

$$A_0\, P(t) \quad = \quad A_{\rm KT} G^{\rm DGKT}(\Delta, \nu, t) \exp(-\lambda t) + A_{\rm BG}. \tag{3}$$

Here the exponential relaxation of the KT is associated with the rapid electronic/paramagnetic fluctuations mainly from the TMO planes. Furthermore, a global fitting procedure was employed over the entire temperature range using a common alpha ($\alpha$) and background asymmetry ($A_{\rm BG}$), but with temperature dependent field fluctuation rate ($\nu$). The static width of the local field distribution ($\Delta$) and relaxation rate ($\lambda$) were from individual fitting found to be more or less independent of temperature and were finally also treated as common parameters for the global fit over the entire temperature range for each sample [$\Delta_{x=1} \approx 0.35 \cdot 10^6$ s$^{-1}$ and $\Delta_{x=0.3} \approx 0.24 \cdot 10^6$ s$^{-1}$ along with $\lambda \approx 0.025 \cdot 10^6$ s$^{-1}$ for both samples].

Below $T \approx 120$ K, the spectra display a clear static behavior as shown for instance by the $T = 50$ K spectra for the $x = 1$ samples in Fig. 3(a) [except for the minor relaxation due to paramagnetic fluctuations as mentioned above]. However, above $T_{\rm diff} \approx 125$ K clear dynamic contribution sets in, which increases strongly with temperature, as shown in the $T = 450$ K

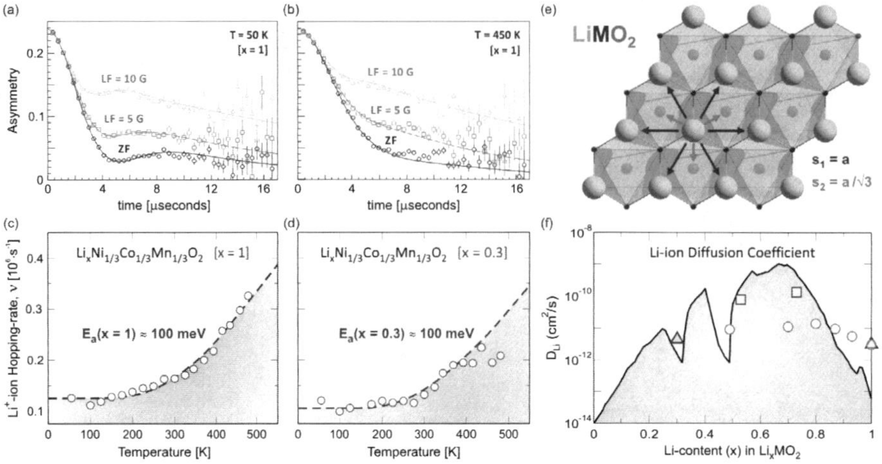


**Figure 3. High-temperature diffusion measurements at ISIS pulsed muon source.** Zero-field (ZF) and Longitudinal-field (LF = 5 and 10 G) $\mu^+$SR time spectra collected at (a) $T = 50$ K and (b) $T = 450$ K for the $x = 1$ sample. Solid lines are fits to Eq. 3. (c-d) Temperature dependence of the Li-ion hopping-rate [$\nu(T)$] obtained from our $\mu^+$SR experiments for $x = 1$ and $x = 0.3$ samples, respectively. (e) $Li^+$-ion jump-paths and distances within the lithium planes. (f) Resulting Li-ion diffusion coefficient ($D_{\rm Li}$) for the current two $Li_xNi_{1/3}Co_{1/3}Mn_{1/3}O_2$ samples shown as filled blue triangles. Also shown are results from previous $\mu^+$SR measurements on similar samples with different $x$ [25] (open circles) as well as from the $Li_xCoO_2$ compound [17] (filled yellow squares) together with a calculated curve for $D_{\rm Li}$ in $Li_xCoO_2$ as solid black line [3].

spectra for the $x = 1$ sample [Fig. 3(b)]. The temperature dependence of the Li-ion hopping rate [$\nu(T)$] obtained from fitting both data sets to Eq. 3 are shown in Fig. 3(c-d) for the $x = 1$ and $x = 0.3$ samples, respectively. As seen, $\nu(T)$ displays a clear diffusive behavior above $T_{\rm diff} \approx 150$ K for both samples. By fitting $\nu(T)$ to an Arrhenius type equation [dashed line in Fig. 3(c-d)], it is possible to extract the activation energy $E_{\rm a} \approx 100$ meV for both samples. As seen in Fig. 3(d), for the $x = 0.3$ sample there is a small deviation at higher temperatures. One could argue that this is due to the onset of muon diffusion in parallel to the Li-ion diffusion. However, this is highly unlikely since then a similar effect could be expected also for the $x = 1$ sample. Instead we attribute this feature to the fact that this was the last sample we measured and had to cut down on the statistics for the last couple of temperature points. If we assume that $\nu(T)$ is a direct measure of the Li-ion hopping rate we can then express the diffusion coefficient $D_{\rm Li}$ as [26]:

$$D_{\rm Li} = \sum_{i=1}^{n} \frac{1}{N_i} Z_i s_i^2 \nu. \tag{4}$$

Here, $N_i$ is the number of possible Li jump sites for the $i$:th path, $Z_i$ is the vacancy fraction, $s_i$ is the jump distance and $\nu$ is the hopping rate obtained from our $\mu^+$SR experiments. For the current samples there are in similarity to $Li_xCoO_2$ samples 2 different jumping paths to either a partly occupied Li-ion site ($N_1 = 6$, $Z_1 = 1 - x$ and $s_1 = $ **a**-axis $= 2.86347(2)$ Å) or to an empty interstitial site ($N_2 = 3$, $Z_2 = 1$ and $s_2 = $ **a**-axis$/\sqrt{3}$) [see also Fig. 3(e)]. As a result we can calculate $D_{\rm Li}$ at 300 K for our two samples as $D_{\rm Li}^{300{\rm K}}[x = 1] = 3.5 \cdot 10^{-12}$ cm$^2$s$^{-1}$ and

$D_{\text{Li}}^{300\text{K}}[x = 0.3] = 5 \cdot 10^{-12}$ cm$^2$s$^{-1}$, respectively. If we insert such values in a graph showing $D_{\text{Li}}$ from previous measurements of $\text{Li}_x\text{Ni}_{1/3}\text{Co}_{1/3}\text{Mn}_{1/3}\text{O}_2$ samples [25] we clearly see that the value for $x = 1$ is fully reproducible. Our new data for $x = 0.3$ falls where one can expect considering the previous measurements of $x = 0.49$ samples. Please note that a different sample provider was used for the current study as compared to previously published data [25].

An interesting detail appears when looking at the first principle calculations for $D_{\text{Li}}$ in the closely related $\text{Li}_x\text{CoO}_2$ compound [3] [see solid black line in Fig. 3(f)], several sharp minima at *e.g.* $x = 0.33$ and $x = 0.5$ are visible. Such minima are related to different type of Li-vacancy ordering. However, from the present and previous measurements on $\text{Li}_x\text{Ni}_{1/3}\text{Co}_{1/3}\text{Mn}_{1/3}\text{O}_2$ for a series of different Li-content, $x$, it seems clear that no such minima are present in this compound. This is probably reasonable since the disorder in the MTMO planes would create an unfavorable environment for the spontaneous formation of any Li-vacancy ordering. Such disorder is most likely also the cause for the overall lower diffusion rate in the MTMO compared to the $\text{LiCoO}_2$ compound causing for instance the interstitial sites to be less favourable. Another possibility is that we have an intermixing between the Li and Ni sites. As we have already found from our previous measurements on the $\text{LiNiO}_2$ compound [19] the diffusion rate is drastically lowered if such intermixing is present. To reveal the details of such issues we will need to perform further measurements using *e.g.* neutron diffraction and/or quasi-elastic neutron scattering (QENS).

## 5. Summary

By the use of a positive muon-spin rotation and relaxation ($\mu^+$SR) technique we have investigated the mixed transition metal oxide battery cathode material $\text{Li}_x\text{Ni}_{1/3}\text{Co}_{1/3}\text{Mn}_{1/3}\text{O}_2$. Low-temperature ($T = 2 - 100$ K) measurements show that this compounds enters into a spin-glass like magnetic state below approximately $T = 12$ K for both lithium contents $x = 1$ and $x = 0.3$, respectively. High-temperature ($T = 50 - 500$ K) investigations show details regarding the Li-ion diffusion process in these compounds. Above $T_{\text{diff}} \approx 125$ K the lithium ions start to diffuse as seen from the exponential increase of the hopping rate $\nu(T)$. We are able to extract the activation energy for the diffusion mechanism as $E_{\text{a}} \approx 100$ meV, which seems to be approximately independent of the lithium content $x$.

## Acknowledgments

This work was performed using the DOLLY spectrometer at the Swiss Muon Source (S$\mu$S) of the Paul Scherrer Institut (PSI), Villigen, Switzerland as well as the RIKEN-RAL / ARGUS spectrometer at the pulsed muon source ISIS/RAL in UK. We are very grateful for the support obtained during our experiments from the beamline and technical staff of the two facilities. This research was financially supported by the Swedish Research Council (VR), Swedish Energy Agency, Knut and Alice Wallenberg Foundation (KAW), the Swiss National Science Foundation (SNSF), MEXT KAKENHI Grant No. 23108003 and JSPS KAKENHI Grant No. 26286084.

## References

[1] Mizushima K, Jones P C, Wiseman P J and Goodenough J B 1980 *Mater. Res. Bull.* **15** 783
[2] Li H, Wang Z, Chen L and Huang X 2009 *Adv. Mater.* **21** 4593
[3] Van der Ven A, Aydinol M K, Ceder G, Kresse G and Hafner J 1998 *Phys. Rev.* B **58** 2975
[4] Seguin L, Amatucci G, Anne M, Chabre Y, Strobel P, Tarascon J M and Vaughan G 1999 *J. Power Sources* **81/82** 604
[5] Chebiam R V, Prado F and Manthiram A 2001 *J. El. Chem. Soc.* **148** A49
[6] Yoshio M, Noguchi H, Itoh J -I, Okada M and Mouri T 2000 *J. Pow. Sour.* **90** 176
[7] Choi J and Manthiram A 2006 *J. Pow. Sour.* **162** 667
[8] Ohzuku T and Makimura Y 2001 *Chem. Lett.* **7** 642
[9] Dahbi M, Wikberg J M, Saadoune I, Gustafsson T, Svedlindh P and Edström K 2009 *Electrochimica Acta* **54** 3211

[10] Anderson P W 1973 *Mater. Res. Bull.* **8**, 153
[11] Wikberg J M, Dahbi M, Saadoune I, Gustafsson T, Edström K and Svedlindh P 2010 *Phys. Rev.* B **81** 224411
[12] Wikberg J M, Dahbi M, Saadoune I, Gustafsson T, Edström K and Svedlindh 2010 *J. Appl. Phys.* **180** 083909
[13] Wikberg J M, Månsson M, Dahbi M, Kamazawa K and Sugiyama J 2012 *Phys. Proc.* **30** 202
[14] Yaouanc A and Dalmas de Réotier P 2011 *Muon Spin Rotation, Relaxation, and Resonance* (Oxford: Oxford University Press)
[15] Wikberg J M 2014 *Private Communication*
[16] Phillips C 1996 *Rep. Prog. Phys.* **59** 1133
[17] Sugiyama J, Mukai K, Ikedo Y, Nozaki H, Månsson M and Watanabe I 2009 *Phys. Rev. Lett.* **103** 147601
[18] Månsson M and Sugiyama J 2013 *Physica Scripta* **88** 068509
[19] Sugiyama J *et al.* 2010 *Phys. Rev.* B **82** 224412
[20] Sugiyama J *et al.* 2011 *Phys. Rev.* B **84** 054430
[21] Sugiyama J *et al.* 2012 *Phys. Rev.* B **85** 054111
[22] Sugiyama J, Mukai K, Nozaki H, Harada M, Månsson M, Kamazawa K, Andreica D, Amato A, and Hillier A D 2013 *Phys. Rev.* B **87**, 024409
[23] Sugiyama J, Nozaki H, Mukai K, Harada M, Månsson , and Hillier A 2014 *Solid State Ionics* **262** 901
[24] Sugiyama J *et al.* 2012 *Phys. Proc.* **30** 105
[25] Sugiyama J, Mukai K, Harada M, Nozaki H, Miwa K, Shiotsuki T, Shindo Y, Giblin S R and Lord J S 2013 *Phys. Chem. Chem. Phys.* **15** 10402
[26] Borg R J and Dienes G J 1988 *An Introduction to Solid State Diffusion* (San Diego: Academic)

Journal of Physics: Conference Series **551** (2014) 012038 doi:10.1088/1742-6596/551/1/012038

# Investigating the compatibility of PEEK polymer for the fabrication of sample cells for use in muon spin spectroscopy

**L Chandrasena[1], I McKenzie[1,2], J-C Brodovitch[1], M Mozafari[1], S P Cottrell[3] and P W Percival[1,2]**

[1] Department of Chemistry, Simon Fraser University, Burnaby, B.C. Canada V5A 1S6
[2] Centre for Molecular and Materials Science, TRIUMF, Vancouver, B.C. Canada V6T 2A3
[3] ISIS Facility, STFC Rutherford Appleton Laboratory, Chilton, Didcot, Oxon, U.K. OX11 0QX

E-mail: iain.mckenzie@triumf.ca

**Abstract.** Polyether ether ketone (PEEK) is a thermoplastic polymer with a wide range of applications due to its chemical inertness and thermal stability, and for these reasons sample cells for gas and liquid phase $\mu$SR have been constructed from PEEK. Muon level-crossing resonance ($\mu$LCR) studies of PEEK revealed a broad, strong $\mu$LCR signal that, we hypothesize, is due to multiple overlapping resonances from the various muonium (Mu) adducts of PEEK. To investigate this, two monomer units from PEEK (4,4′-dihydroxybenzophenone and *para*-dimethoxybenzene) were studied in solution using transverse-field muon spin rotation (TF-$\mu$SR) and $\mu$LCR. Two different muoniated radicals were formed by Mu addition to 4,4′-dihydroxybenzophenone and one radical was formed in *para*-dimethoxybenzene. The $\mu$SR spectra were assigned by comparing the experimentally measured muon and proton hyperfine coupling constants with values calculated for the possible structures using Gaussian-09 software with the B3LYP functional and 6-31G basis set. Good agreement was found for cyclohexadienyl-type radicals formed by Mu addition to the benzene rings of the monomer units. We can also infer that these radicals are being formed in PEEK, and based on this we conclude that sample cells made of PEEK are unsuitable for many types of $\mu$SR experiment.

## 1. Introduction

Materials used for $\mu$SR sample cell construction should have a number of important properties. They should be chemically inert and non-magnetic, withstand high pressure and temperature conditions, yet allow the fabrication of thin enough entrance windows to facilitate muon penetration with minimum stops and scattering in the window itself to optimize muon stops inside the cell cavity containing the target material. In addition, for radio-frequency experiments the sample cell should be electrically non-conducting to prevent the RF cavity from interacting with the cell material [1].

In this context polyether ether ketone (PEEK) was considered an ideal material for sample cell construction. It is a thermoplastic polymer with a wide range of application due to its chemical inertness and thermal stability up to 533 K [2]. Muons stop in this material with only a small amount of diamagnetic polarization ($P_{\rm D}$ ~0.1 % with about 90% missing fraction) and cells have been fabricated for gas samples up to 50 bar of pressure [1].

In a study of muon implantation in inert gases, a large (∼30%) muonium (Mu) signal was detected in a sample of high purity neon in a PEEK sample cell. This was a magnitude larger than previous measurements on neon carried out over a large pressure range. It was suggested that a charge neutralization reaction occurs with an unknown impurity outgassing from the PEEK polymer cell: $NeMu^+ + X \rightarrow NeX^+ + Mu$. In addition, longitudinal-field repolarization studies suggested the formation of a radical state in the cell body.

The objective of the study reported here was to determine whether muoniated radicals are formed in PEEK, which would make it an unsuitable material for sample cells of experiments studying Mu and muoniated radicals, and to identify the structures of the resulting radicals. PEEK is a polymer consisting of alternating aromatic ring structures, a 4,4′-substituted benzophenone and a *para*-substituted benzene, with ether linkages. It is well established that Mu adds to substituted benzene rings to produce cyclohexadienyl radicals [3], and that Mu also adds to ketone groups [4]. We have performed muon level-crossing resonance ($\mu$LCR) measurements on PEEK and transverse-field muon spin rotation (TF-$\mu$SR) and $\mu$LCR measurements on two monomer units from PEEK (4,4′-dihydroxybenzophenone and *para*-dimethoxybenzene) in solution. The structures of these molecules are shown in Figure 1.

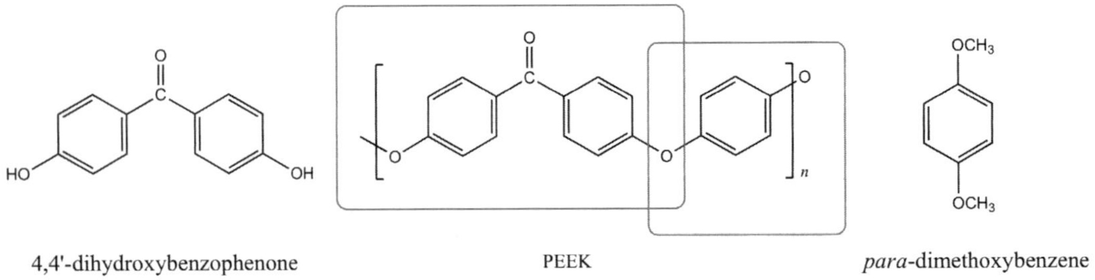


**Figure 1.** The structure of the polyether ether ketone (PEEK) repeating unit and the sub-units of *para*-dimethoxybenzene and 4,4′-dihydroxybenzophenone.

## 2. Experimental

### *2.1. Sample Preparation*

A 6.5 mm thick sheet of PEEK was purchased from Goodfellow Inc and a 4 cm × 4 cm sample was cut from this sheet. *Para*-dimethoxybenzene and 4,4′-dihydroxybenzophenone were purchased from Sigma-Aldrich and dissolved in anhydrous tetrahydrofuran and ethanol, respectively. The sample solutions were deoxygenated by the freeze-pump-thaw technique and sealed in stainless-steel cells fitted with windows made of 25 $\mu$m thick foils of the same material.

### *2.2. TF-$\mu$SR and $\mu$LCR Experiments*

The $\mu$LCR measurement on PEEK was performed with the ALC spectrometer on the $\pi$E3 beam line of the Paul Scherrer Institute, Villigen AG, Switzerland. This measurement was made at 360 K and no field modulation was used. $\mu$SR experiments on *para*-dimethoxybenzene and 4,4′-dihydroxybenzophenone were performed using the HELIOS spectrometer on the M20 surface muon channel at the TRIUMF cyclotron facility in Vancouver, Canada. The magnetic field was aligned along the beam direction, and the beam was tuned to give transverse spin-polarization for TF-$\mu$SR experiments, and longitudinal spin-polarization for $\mu$LCR experiments. Square-wave field modulation (approx. ±80 G amplitude) was used to suppress baseline fluctuations in $\mu$LCR. These measurements were made at 298 K.

## 3. Results and Discussion

The raw $\mu$LCR spectrum of PEEK at 360 K is shown in Figure 2(left). There is a broad feature between 14 and 23 kG that is superimposed on a non-linear background. This background can be attributed to the magnetic field effect on the muon beam spot and positron trajectories in the ALC spectrometer [5]. In keeping with previous practice [6, 7] the background was modelled with a polynomial, which in this case was $4^{\text{th}}$ order. The background-subtracted $\mu$LCR spectrum of PEEK is shown in Figure 2(right). Our hypothesis is that this spectral feature is due to multiple overlapping resonances of cyclohexadienyl radicals with restricted rotation. To confirm this we turn to the study of the PEEK sub-units.

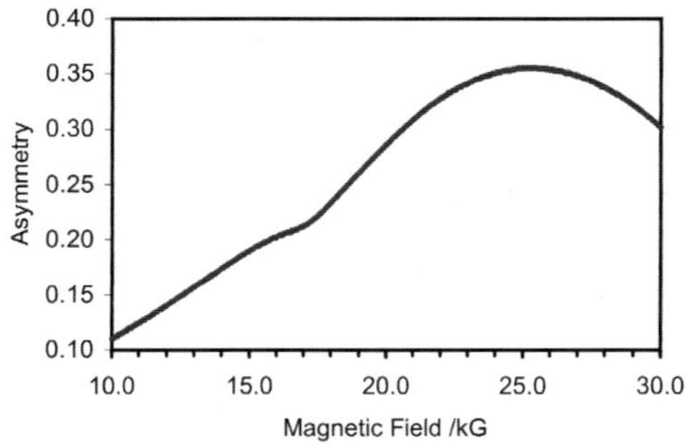


**Figure 2.** (left) Raw $\mu$LCR spectrum of PEEK at 360 K and (right) background-subtracted $\mu$LCR spectrum of PEEK at 360 K.

The TF-$\mu$SR results are displayed as Fourier power spectra in Figure 3. The signal at 131 MHz arises from muons in diamagnetic environments whereas muoniated radicals give rise to a pair of precession frequencies almost symmetrically placed about the diamagnetic peak. Data was collected from positron detectors in phase quadrature, so that negative precessions can be distinguished via complex Fourier transformation. The signal at -131 MHz in the right-hand spectrum is an artifact (ghost frequency of the strong diamagnetic signal caused by imperfect quadrature). A single radical was detected from the *para*-dimethoxybenzene sample and two from 4,4′-dihydroxybenzophenone. The precession frequencies were determined by fitting the original $\mu$SR time spectra, and the muon hyperfine constants were calculated either from the separation of radical precession frequencies or from the stronger radical precession and the diamagnetic frequency, using standard formulae:

$$A_\mu = \nu_{\text{R2}} - \nu_{\text{R1}} \tag{1}$$

$$A_\mu = 2\frac{(\nu_{\text{e}} + \nu_{\text{R1}})(\nu_\mu - \nu_{\text{R1}})}{(\nu_{\text{e}} - \nu_\mu + 2\nu_{\text{R1}})} \tag{2}$$

where $\nu_{\text{R1}}$ and $\nu_{\text{R2}}$ are the radical precession frequencies, and $\nu_{\text{e}}$ and $\nu_\mu$ are the electron and muon Larmor frequencies. All three radicals exhibit large hyperfine constants (Table 1) characteristic of substituted cyclohexadienyl radicals. There was no evidence to support the formation of a ketyl radical by Mu addition to the ketone group of the 4,4′-dihydroxybenzophenone.

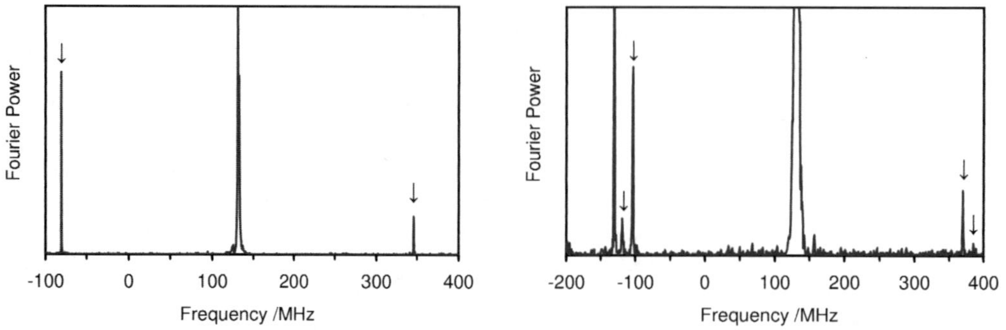


**Figure 3.** Transverse-field $\mu$SR spectra at 9.65 kG obtained from (left) 1.5 M *para*-dimethoxybenzene in tetrahydrofuran; and (right) 1.05 M 4,4′-dihydroxybenzophenone in ethanol. Radical precession signals are marked with arrows.

**Table 1.** Muon hyperfine constants ($A_\mu$) determined from TF-$\mu$SR and calculated $\Delta_1$ avoided level-crossing fields ($B_{\rm LCR}^{\Delta_1}$).

| Parent molecule | $A_\mu$ / MHz | $B_{\rm LCR}^{\Delta_1}$ / kG |
|---|---|---|
| para-dimethoxybenzene | 426.8(1) | 15.668(4) |
| 4,4′-dihydroxybenzophenone | 473.8(1) | 17.394(4) |
| 4,4′-dihydroxybenzophenone | 504.2(3) | 18.510(11) |

The $\mu$LCR results also support the conclusion that one cyclohexadienyl-type radical is formed from the *para*-dimethoxybenzene and two such radicals from 4,4′-dihydroxybenzophenone. The resonances displayed in Figure 4 are typical for the methylene (-CHMu-) fragment of a cyclohexadienyl. Proton hyperfine constants were calculated from the resonance field positions using the standard formula [8],

$$B_{\rm LCR} = \frac{1}{2}\left[\frac{(A_\mu - A_{\rm p})}{(\gamma_\mu - \gamma_{\rm p})} - \frac{(A_\mu + A_{\rm p})}{\gamma_{\rm e}}\right] \tag{3}$$

and are included in Table 2. In principle there is ambiguity in assigning the two resonances

**Table 2.** Proton hyperfine constants ($A_{\rm p}$) determined from $\Delta_0$ avoided level-crossing fields ($B_{\rm LCR}^{\Delta_0}$), and the hyperfine isotope effect.

| Parent molecule | $B_{\rm LCR}^{\Delta_0}$ / kG | $A_{\rm p}$ / MHz | 0.31413 $A_\mu/A_{\rm p}$ |
|---|---|---|---|
| *para*-dimethoxybenzene | 16.999(2) | 109.0(1) | 1.23 |
| 4,4′-dihydroxybenzophenone | 19.064(4) | 117.4(1) | 1.27 |
| 4,4′-dihydroxybenzophenone | 20.402(8) | 122.8(4) | 1.29 |

from dimethoxybenzene to the two radicals detected by TF-$\mu$SR. In practice the relative signal

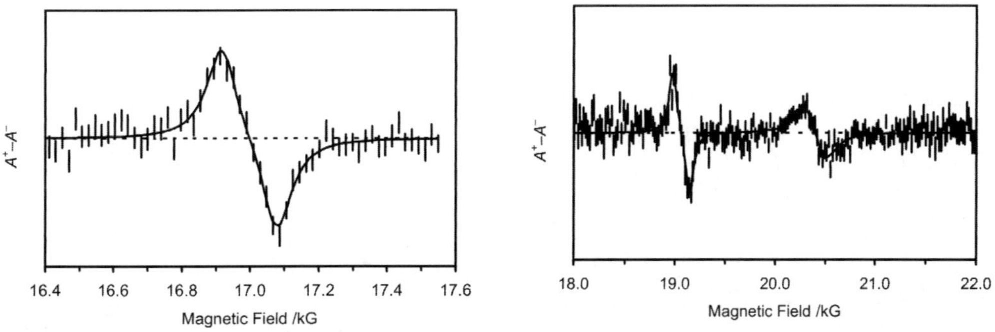


**Figure 4.** Avoided level-crossing spectra obtained from (left) 1.5 M *para*-dimethoxybenzene in tetrahydrofuran; and (right) 1.05 M 4,4′-dihydroxybenzophenone in ethanol.

amplitudes and the consistent ratio of muon to proton hfcs leaves little doubt that the assignment of Table 2 is correct.

DFT calculations were carried out to help assign structures to the three radicals detected experimentally. The Gaussian 09 suite of programs was used to optimize molecular geometries of potential muoniated radicals and compute their isotropic hyperfine constants, using the unrestricted B3LYP hybrid density functional and the 6-31G basis set. The muonium atom was explicitly included in the calculations as an isotope of H with mass 0.113429 u and magnetic moment 8.6578226 $\mu_N$. It is thus possible to take account of isotope effects by calculating vibrationally averaged hyperfine constants. Table 3 lists hyperfine constants calculated for the reference geometry (identical for all H isotopes) and vibrationally averaged at 298 K. In the final column the experimentally determined values are compared with calculated values for the reference geometry, because that is the isotope effect most often quoted in the literature. The structures of the radicals are displayed in Figure 5.

**Table 3.** Muon and proton hyperfine constants calculated for the reference geometry ($A_0$) and as vibrationally averaged at 298 K ($A_{298}$).

| Parent molecule | Radical | Nucleus | $A_0$ / MHz | $A_{298}$ / MHz | $A_{exp}/A_0$ |
|---|---|---|---|---|---|
| *para*-dimethoxybenzene | 1 | $\mu$ | 425.6 | 520.9 | 1.00 |
| *para*-dimethoxybenzene | 1 | p | 138.0 | 132.3 | 0.79 |
| 4,4′-dihydroxybenzophenone | 2 | $\mu$ | 425.5 | 538.7 | 1.11 |
| 4,4′-dihydroxybenzophenone | 2 | p | 138.1 | 133.1 | 0.85 |
| 4,4′-dihydroxybenzophenone | 3 | $\mu$ | 448.5 | 510.7 | 1.12 |
| 4,4′-dihydroxybenzophenone | 3 | p | 154.9 | 139.3 | 0.79 |

The similarity of isotope effects for Mu adducts of 4,4′-dihydroxybenzophenone supports the assignment of structures implicit in the ordering of data in Table 3. Calculations were also carried out for the radical formed by Mu addition to the benzophenone ketone. As expected, the muon hyperfine constant is much smaller than those determined by TF-$\mu$SR, nor could such a radical account for the LCR resonances. Finally, the Mu addition to ketones is known to be much slower than addition to aromatic rings [9]. This possibility is therefore rejected.

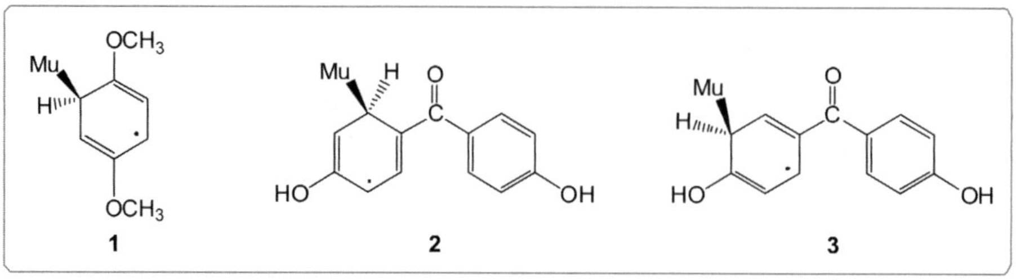


**Figure 5.** Muoniated radicals identified in this work.

Inspection of Tables 1-3 shows that there is reasonable agreement between the calculated hyperfine constants and the experimental results. However, although the inclusion of vibrational averaging addresses the isotope effect, there is still not a perfect match. Possible factors for the discrepancy include the use of a relatively modest basis set, the neglect of solvent effects, and inaccurate calculation of vibrational frequencies, which leads to overestimation of the temperature effect (cf. $A_0$ and $A_{298}$ in Table 3). Preliminary investigations suggests that this last factor is the major source of the discrepancy.

## 4. Conclusions

Extrapolating the results of this investigation of the monomer subunits of PEEK, it is clear that overlapping resonances from radicals formed by Mu addition to benzene rings in the polymer can account for the broad $\mu$LCR signal observed in PEEK. It is therefore concluded that PEEK sample cells are unsuitable for many types of $\mu$SR experiment, in particular those involving muonium or muoniated free radicals, because the PEEK polymer can produce signals which could interfere with those of the sample under study.

## Acknowledgments

Part of this research was carried out with the aid of the staff and facilities of the TRIUMF Centre for Molecular and Materials Science (CMMS). TRIUMF is funded through the National Research Council of Canada; the M20 beam line was built with funds from Canada Foundation for Innovation and British Columbia Knowledge Development Fund; and CMMS is supported by a Major Resources Support grant from the Natural Sciences and Engineering Research Council of Canada. The computational work made use of resources provided by WestGrid and Compute/Calcul Canada. Support from the staff at the Laboratory for Muon Spin Spectroscopy at the Paul Scherrer Institute is gratefully acknowledged. This research project has been supported by the European Commission under the seventh Framework Programme through the "Research Infrastructures" action of the "Capacities" Programme, Contract No. CP-CSA-INFRA-2008-1.1.1 Number 226507-NMI3.

## References

[1] Johnson C, Cottrell S P, Ghandi K and Fleming D G 2005 *J. Phys. B: At. Mol. Opt. Phys.* **38** 119
[2] Patel P, Hull T R, McCabe R W, Flath D, Grasmeder J and Percy M 2010 *Polym. Degrad. Stab.* **95** 709
[3] Roduner E, Brinkman G A and Louwrier P W F 1982 *Chem. Phys.* **73** 117
[4] Cox S F J, Geeson D A, Rhodes C J, Roduner E, Scott C A and Symons M C R 1986 *Hyp. Int.* **32** 763
[5] Stoykov A, Scheuermann R, Sedlak K, Shiroka T and Zhuk V 2009 *Physica* B **404** 986
[6] McKenzie I, Scheuermann R and Sedlak K 2012 *J. Phys. Chem.* A **116** 7765

[7] McKenzie I, Cammidge A N, Gopee H, Dilger H, Scheuermann R, Stoykov A and Jayasooriya U A 2013 *Phys. Rev.* E **87** 012504
[8] McKenzie I 2013 *Annu. Rep. Prog. Chem., Sect. C: Phys. Chem.* **109** 65
[9] Barnabas M V and Walker D C 1991 *Can. J. Chem.* **69** 1252

Journal of Physics: Conference Series **551** (2014) 012039 doi:10.1088/1742-6596/551/1/012039

# A Brief Survey of $\beta$-Detected NMR of Implanted $^8Li^+$ in Organic Polymers

**F H McGee,[1] I McKenzie,[2,3] T Buck,[1] C R Daley,[4] J A Forrest,[4] M Harada,[5] R F Kiefl,[1,6] C D P Levy,[2] G D Morris,[2] M R Pearson,[2] J Sugiyama,[5] D Wang[1] and W A MacFarlane[7]**

[1] Department of Physics, University of British Columbia, Vancouver, BC, V6T 1Z1, Canada
[2] TRIUMF, Vancouver, B.C. Canada, V6T 2A3
[3] Department of Chemistry, Simon Fraser University, Burnaby, B.C. Canada, V5A 1S6
[4] Department of Physics and Astronomy, University of Waterloo, Waterloo, Ont. Canada N2L 3G1
[5] Toyota Central Research and Development Laboratories Inc., Nagakute, Aichi 480-1192, Japan
[6] Canadian Institute for Advanced Research, Canada
[7] Chemistry Department, University of British Columbia, Vancouver, BC, V6T 1Z1, Canada

E-mail: wam@chem.ubc.ca, iain.mckenzie@triumf.ca

**Abstract.** Unlike the positive muon, we expect the chemistry of the implanted $^8Li^+$ $\beta$-NMR probe in organic polymers to be simply that of the monovalent ion, but almost nothing is known about the NMR of isolated $Li^+$ in this context. Here, we present a brief survey of $^8Li^+$ $\beta$-NMR in a variety of insulating polymers at high magnetic field, including polyimide, PET, polycarbonate, polystyrene and polyethylene oxide. In all cases, we find a large-amplitude, broad Lorentzian resonance near the Larmor frequency, consistent with the expected diamagnetic charge state. We also find remarkably fast spin-lattice relaxation rates $1/T_1$. There is very little dependence of either linewidth or $1/T_1$ on the proton density, the main source of nuclear dipolar magnetic fields, leading us to conclude the main contribution to both broadening and spin relaxation at room temperature is *quadrupolar* in origin. This behaviour is very different from crystalline insulators such as MgO and $Al_2O_3$, and suggests that $^8Li^+$ $\beta$-NMR will be an important probe of polymer dynamics. Additionally, we note dramatically different behaviour of one sample above its glass transition, motivating the construction of a high temperature spectrometer to enable further exploration at elevated temperature.

## 1. Introduction

Our understanding of polymers has made great strides in recent years thanks to the emergence of new analysis techniques, but many of their properties remain a mystery. One area where this is particularly true is the study of depth-dependent properties of polymers. $\beta$-detected NMR ($\beta$-NMR) has a depth resolution on the order of a few nanometers in the surface layer, allowing detailed studies of depth dependence. This technique, analogous to low energy $\mu$SR, uses the anisotropic beta decay to measure magnetic resonance and relaxation of extremely dilute implanted radioactive probes. At TRIUMF's Isotope Separator and Accelerator (ISAC) facility [1] a dedicated system for producing high intensity spin polarized beams of $\beta$-NMR probes is routinely used to study depth dependent phenomena in crystalline solids [2], but only recently has this capability been applied to more disordered materials such as glassy organic polymers. Here we present the results of preliminary experiments conducted on several polymers to assess the viability the $\beta$-NMR technique (using the $^8Li^+$ probe) as a tool to study their properties across a wide range of temperatures and conditions.

## 2. Experimental

$\beta$-NMR measurements for this experiment were performed at the ISAC facility at TRIUMF, using an optically pumped, highly polarized beam of radioactive $^8Li^+$ to implant ions into the sample at varying energies in the range 0.1-30 keV. The $^8$Li nucleus has spin $I = 2$, gyromagnetic ratio $\gamma = 6.3015$ MHz/T, and electric quadrupole moment $Q = +31.4$ mb. The spin is polarized by in-flight optical pumping to polarization $\sim 70$ %. Resonances were acquired with continuous wave (CW) transverse radiofrequency (RF) magnetic field with frequency stepped slowly relative to the $^8$Li lifetime of 1.21 s through a frequency range around the $^8Li^+$ Larmor frequency. A static magnetic field of 6.55 T was applied perpendicular to the sample surface (parallel to both the beam and the spin polarization).

Three of the samples (Kapton (polyimide), PC (polycarbonate) and PET (polyethylene terphthalate)) were freestanding polymer foils used as templates for nanowire growth in swift heavy ion tracks produced at the GSI [3]. The polystyrene (PS) films were spin-coated on $Al_2O_3$ substrates [4]. Polystyrene-d3 and d5 are deuterated along the molecular backbone and phenol side-groups respectively, while polystyrene-d8 is fully deuterated. Polyethylene oxide (PEO) was also spin-coated on $Al_2O_3$[5]. All of the thin films were approximately 200 nm thick. At the implantation energies used, we expect none of the $^8Li^+$ to penetrate through to the substrate. The polymer properties are summarized in Table 1 and the structures are shown in Figure 1. All data was taken at room temperature, except the PEO which was somewhat colder at 260 K.

**Table 1.** Summary of samples, including the calculated protons density $\rho_p$ and glass transition temperature $T_g$ from the literature.

| Sample | $\rho_p$ (mol/cm$^3$) | $T_g$ (K) |
|---|---|---|
| Polystyrene-d8 (0.9 kg/mol) film | 0.000 | 270 |
| Polystyrene-d5 (25.3 kg/mol) film | 0.030 | 373 |
| Polyimide (Kapton 50HN) foil | 0.037 | 658 |
| Polystyrene-d3 (22.3 kg/mol) film | 0.050 | 373 |
| PET (Hostaphan RN) foil | 0.058 | 342 |
| Polycarbonate foil | 0.066 | 420 |
| Polystyrene (1.140 kg/mol) film | 0.081 | 293 |
| Polyethylene oxide film | 0.103 | 223 |

## 3. Results and Discussion

$\beta$-NMR spectroscopy provides information about the local environment of lithium ions, which will be interacting with the different functional groups of the polymers. The lithium ion will preferentially stop in regions of high electron density. We propose stopping sites based on the structures of the polymers. The main areas where $^8Li^+$ is likely found would be adjacent to a phenyl or aromatic ring system, where it is bound via the cation-$\pi$ interaction, bound to the oxygen of a carbonyl group and bound to an ether or ester oxygen. The final location of $^8Li^+$ will be a delicate balance of the interactions with all the neighbouring functional groups.

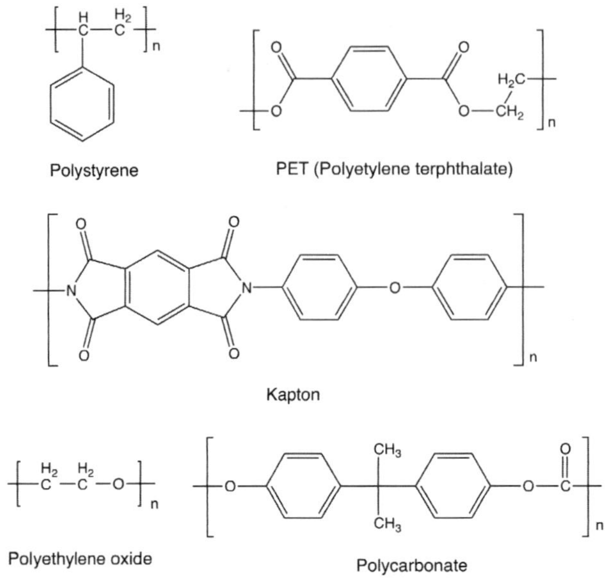


**Figure 1.** Structures of the polymers studied with $\beta$-NMR

We have used DFT calculations (RB3LYP/6-311+G(d,p)) of model compounds (benzene, dimethyl ether, acetone and ethylene carbonate) to estimate the binding energy ($BE$) for $Li^+$ bound to the different functional groups present in the polymers (Figures 1,2). The calculations were performed with the Gaussian 09 package. $BE = [E(\mathrm{Li}^+) + E(\mathrm{X})] - E(\mathrm{Li}^+ - \mathrm{X})$, where $E$ is the internal energy and X is the molecule to which the lithium ion is bound. The calculated $BE$ values are shown in Table 2 and are in good agreement with experimental values and previous calculations by Blint [6]. All of the sites have quite large $BE$, although the interaction with carbonyl groups is preferred. In PS it is likely that $Li^+$ is sandwiched between two adjacent phenyl rings with a total $BE$ of 265.3±15.1 kJ/mol [7]. In the other polymers $Li^+$ is likely to interact with multiple functional groups, so that the effective $BE$ will be larger than that of a single molecule. The preferred interaction with carbonyl groups means that in most of the polymers studied here the $^8Li^+$ is not likely to be close to protons.

**Table 2.** Summary of experimental and calculated $Li^+$ binding energies for model compounds

| Molecule | Experimental kJ/mol | HF/6-31G$^a$ kJ/mol | B3LYP/6-311+G(d,p)$^b$ kJ/mol |
|---|---|---|---|
| Benzene | 161.1±13.5$^c$ | – | 160.6 |
| Dimethyl ether | 165.0±10.6$^d$ | 165 | 170.1 |
| Acetone | 186$^e$ | 200 | 202.6 |
| Ethylene carbonate | – | 218 | 215.8 |

$^a$Ref [6] $^b$Present work $^c$Ref. [7] $^d$Ref. [8] $^e$Ref. [9]

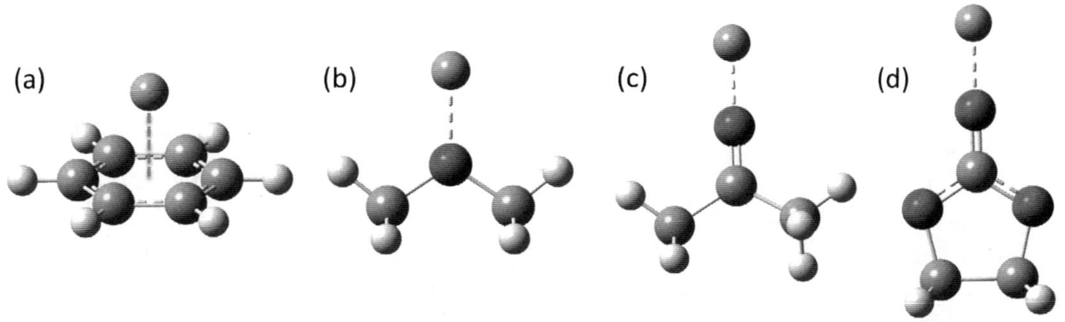


**Figure 2.** Structures of lithium ions (purple) bound to (a) benzene, (b) dimethyl ether, (c) acetone and (d) ethylene carbonate. Carbon = grey; oxygen = red; hydrogen = white.

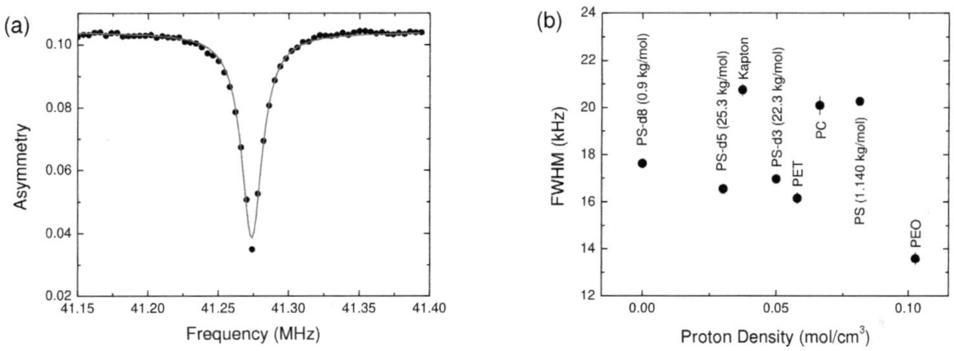


**Figure 3.** (a) $\beta$-NMR resonance spectrum of $^8Li^+$ in the fully deuterated PS film at 300 K and 6.55 T and (b) the resonance linewidth in the full set of polymer samples as a function of the proton density. The measurements were made at 300 K except for PEO, which was studied at 260 K.

An example of a time-integral resonance spectrum is shown in Figure 3a. All of the resonance spectra were fit with a single Lorentzian function and the linewidth does not appear to depend on the proton density (Figure 3b). An example of a time-differential relaxation measurement is shown in Figure 4a. The relaxation was fit to the stretched-exponential function $e^{-(t/T_1)^\beta}$ with $\beta$ fixed to 0.5. The root exponential relaxation is probably the result of a distribution of relaxation times due to the microscopic inhomogeneity of the $^8Li^+$ stopping sites. Stretched exponential relaxation has been observed in dielectric relaxation [10] and NMR [11, 12] studies of glassy materials. The relaxation is remarkably fast in all the polymer samples compared to wide band gap crystalline insulators we have studied at comparably high magnetic fields (e.g. $Al_2O_3$ and MgO). $1/T_1$ does not depend significantly on the proton density except for PEO, where the spin-lattice relaxation is much faster than in the other polymers.

The spin relaxation rate will depend on the EFG as well as on the distance between $^8Li$ and surrounding protons. The linewidth of the resonance does not depend on the proton density and this is most dramatically apparent among the polystyrene samples with varying levels of

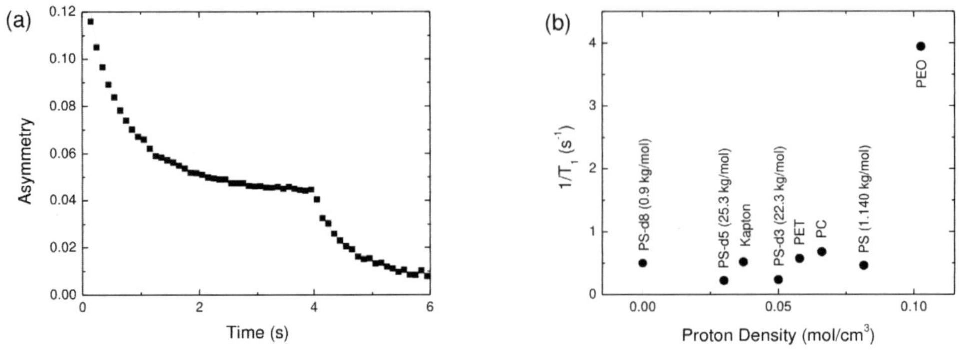


**Figure 4.** (a) Spin-lattice relaxation measurement of PEO film at 300 K and 6.55 T and (b) the spin-lattice relaxation rate versus proton density for the full set of polymer samples.

deuteration, which have almost identical FWHM values. This suggests that the nuclear magnetic dipolar field from the protons (i.e. dipolar broadening) plays a negligible role in broadening the resonance. Contributions from other nuclear moments ($^{13}$C etc.) are even smaller, so we conclude the linewidth is unrelated to the magnetic dipolar interaction.

One possible source of magnetic fields that could contribute to both the resonance and relaxation is long-lived paramagnetic radical species produced via radiolysis during the $^{8}Li^{+}$ implantation. Electronic magnetic moments would produce a large hyperfine field on the $^{8}$Li. In the simplest case of isolated nearby radicals, however, such fields would behave as an isolated local moment with net fields scaling as $B/T$. Temperature dependent studies in PEO [5] and PS [4] do not reveal such temperature dependence.

Instead, we suspect that the quadrupolar interaction between the $^{8}$Li 's small nuclear electric quadrupole moment and the local electric field gradient (EFG) at the $^{8}Li^{+}$ site must be the dominant source of *both* resonance broadening and spin lattice relaxation. For a single well-defined site, the quadrupolar interaction splits the resonance into a pattern of four quadrupole satellites centred on the undisturbed resonance frequency, e.g. in bismuth [13] or $Al_2O_3$ [14]. In a noncrystalline glassy material such as a polymer, the EFG varies based on the local configuration of polymer surrounding $^{8}Li^{+}$, resulting instead in a single broadened quadrupolar "powder pattern". The EFG is expected to be somewhat material dependent, but would be independent of proton density as observed. This hypothesis will be tested by measuring the magnetic field dependence of the linewidth. In the case of quadrupolar broadening, the width should be field independent in contrast to the electronic magnetic fields of radicals.

Fluctuations of the EFG due to motion of the polymer chains or $^{8}Li^{+}$ itself will give rise to spin relaxation. Quadrupolar relaxation is important in conventional NMR of quadrupolar nuclei [15]. The similar behaviour in Kapton, PET, PC and high-molecular-weight ($M_w$) PS is perhaps not surprising as the measurements were made significantly below $T_g$ where the polymer chains are not highly mobile. The spin relaxation is dominated by the static characteristics of the local environments of $^{8}Li^{+}$ which are likely similar. The strength of the interaction between $^{8}Li^{+}$ and the functional groups of the polymers suggests that $^{8}Li^{+}$ is not diffusing at the temperatures where the measurements were made.

The measurements on low-$M_w$ PS were made above $T_g$ and suggest that the polymer dynamics are faster than the high-$M_w$ polymer [4]. We suggest that $^{8}Li^{+}$ is not diffusing above $T_g$ because of the strength of the interaction with the neighbouring phenyl rings and that these phenyl rings cause it to be partially encapsulated. We have recently measured the depth dependence of the average spin-lattice relaxation rate and shown that there is a layer near the surface $\sim$ 10 nm thick with faster dynamics than the bulk [4].

PEO is well known to be a $Li^+$ ionic conductor [16] and the measurements were made above $T_g$. $^8Li^+$ diffusion is likely important in PEO because the lithium ion is not encapsulated as in PS and motion of the polymer chains will bring suitable sites close to $^8Li^+$ and make it favourable to hop. The hopping of $^8Li^+$ results in a fluctuating EFG and leads to the much faster $1/T_1$. The temperature dependence of the average spin-lattice relaxation time up to 317 K was used to determine the activation energy for hopping in pure PEO and PEO with 30 wt.% of $LiCF_3SO_3$ [5].

## 4. Conclusions

We believe that $^8Li^+$ will be an important probe of the dynamics of polymeric materials, particularly in thin films and near the free surface where other techniques (notably conventional NMR) have very limited applicability. It will be particularly informative to make measurements closer to $T_g$ but the maximum temperature of the current spectrometer is $\sim$320 K. We are currently designing a high temperature spectrometer to enable measurements through the glass transition in all the polymers in the current study.

## Acknowledgments

We thank C. Trautmann and M. E. Toimil-Molares of the Materials Research Department of the Gesellschaft für Schwerionenforschung (GSI), Darmstadt for providing the polymer foils.

## References

[1] Dilling J, Krücken R and Merminga L (eds) 2014 *ISAC and ARIEL: The TRIUMF Radioactive Beam Facilities and the Scientific Program* (Heidelberg: Springer)
[2] Morris G 2014 *Hyperfine Interact.* **225** 173
[3] Toimil-Molares M E 2012 *Beilstein J. Nanotechnol.* **3** 860
[4] McKenzie I, Daley C R, Kiefl R F, Levy C D P, MacFarlane W A, Morris G D, Pearson M R, Wang D and Forrest J A 2014 *Submitted to ACS Macro Letter*
[5] McKenzie I, Harada M, Kiefl R F, Levy C D P, MacFarlane W A, Morris G D, Ogata S I, Pearson M R and Sugiyama J 2014 *J. Am. Chem. Soc.* **136** 7833
[6] Blint R J 1995 *J. Electrochem. Soc.* **142** 696
[7] Amicangelo J C and Armentrout P B 2000 *J. Phys. Chem.* A **104** 11420
[8] Rodgers M T and Armentrout P B 2000 *Mass Spectrom. Rev.* **19** 215
[9] Staley R H and Beauchamp J L 1975 *J. Am. Chem. Soc.* **97** 5920
[10] Lindsey C P and Patterson G D 1980 *J. Chem. Phys.* **73** 3348
[11] Spiess H W 2007 *Multidimensional Solid State NMR of Polymers* (New York: John Wiley & Sons, Ltd) pp 1–15
[12] Rössler E, Stapf S and Fatkullin N 2013 *Current Opinion in Colloid & Interface Science* **18** 173
[13] MacFarlane W A, Levy C D P, Pearson M R, Buck T, Chow K H, Hariwal A N, Kiefl R F, McGee F H, Morris G D and Wang D 2014 *J. Phys.: Conf. Ser.* **551** 012059
[14] Salman Z, Chow K H, Hossain M D, Kiefl R F, Levy C D P, Parolin T J, Pearson M R, Saadaoui H, Wang D and MacFarlane W A 2014 *J. Phys.: Conf. Ser.* **551** 012034
[15] Suter A, Mali M, Roos J and Brinkmann D 1998 *J. Phys.: Condens. Matter* **10** 5977
[16] Brinkmann D 1992 *Progress in Nuclear Magnetic Resonance Spectroscopy* **24** 527

# Paramagnetic muon states in mesoporous carbon materials

**R M Macrae[1], C Upchurch[1], D K Rose[1], Y Miyake[2,3], A Koda[2,3], J S Lord[4], P S Shuttleworth[5,6]**

[1] School of Mathematics and Sciences, Marian University, 3200 Cold Spring Road, Indianapolis, IN 46222, USA
[2] Muon Science Section, Materials and Life Science Division, J-PARC Center, 2-4 Shirane, Shirakata, Tokai-mura, Naka-gun, Ibaraki 319-1195, Japan
[3] Institute of Materials Structure Science, KEK, Tsukuba, Ibaraki 305-0801, Japan
[4] ISIS Facility, Rutherford Appleton Laboratory, Chilton, Oxfordshire, OX11 0QX, UK
[5] Green Chemistry Centre of Excellence, Department of Chemistry, University of York, York YO10 5DD, UK

E-mail: rmacrae@marian.edu

**Abstract.** We present results of longitudinal field repolarisation measurements carried out at J-PARC and ISIS on the "green" functional carbon materials Starbon 300 and Starbon 800, synthesized using starch as a template and subjected to pyrolysis treatments at different temperatures (300°C and 800°C respectively); pyrolysis at low temperature may be expected to yield a material retaining more of the "hydrophilic" properties of the original starch material in its chemically active voids, while high temperature pyrolysis may be expected to lead to "hydrophobic" voids and a more graphitic material. The hydrophilic material shows a larger repolarising fraction than the hydrophobic material, with a hyperfine constant on the order of 200-300 MHz. This is likely to be a superposition of the repolarisation of multiple radicals. Several candidate and model species are investigated through accompanying density functional theory calculations.

## 1. Introduction

Given the approaching world bottlenecks in energy and material resources, interest has recently taken hold in the idea of using carbon biomass as feedstock for the production of low-cost, "green" functional carbon materials. Approaches such as silica-templating [1] and hydrothermal carbonization (HTC) [2] are effective in the morphology-controlled synthesis of carbonaceous nanostructures from carbohydrate parent materials such as starch, cellulose, cyclodextrins, and simple sugars. Such materials are expected to find a variety of technological uses ranging from nanoengineering [3] to spintronics [4] to ultracapacitors [5]. In the last of these, where the storage capacity of electrical double-layer capacitors can be optimized through the use of a nanostructured material, there has recently been interest in replacing activated charcoal (a granular form of carbon in which the interparticle voids constitute the storage space) with better-characterized high-tech forms of carbon such as aligned single-walled nanotubes [6]. Fundamental studies of such materials using a probe which can give some local insight into

[6] Present Address: Polymer Physics Group, CSIC, Madrid, 28006, Spain

the physical and chemical properties of the voids and of the bulk material itself are thus of some importance.

Muon spin rotation studies have been carried out on diamond [7, 8], graphite [9, 10, 11], amorphous carbon [12], fullerenes [13], and nanotubes [14], and are summarized in the recent review by Cox [15]. While in diamond both bond-centered muonium and interstitial muonium are observed [7], in graphite the paramagnetic state formed on muon implantation undergoes rapid spin or charge exchange with the conduction electrons leading to a collapse of the radical line and a temperature-dependent "paramagnetic shift" in the muon Larmor frequency [9, 10, 11]. Insulating amorphous carbon presumably contains a large variety of paramagnetic sites (not experimentally resolvable), but was observed by Cox and Davis to have a diamagnetic fraction of around 40% [12]. In single-walled nanotubes MacFarlane and co-workers observed a high fraction (about 40%) of vacuum-like muonium [14]. Fullerenes show distinct states corresponding to muons stopping respectively inside and outside the cages [13].

Thus, it is evident that $\mu$SR methods distinguish clearly between a variety of different bonding types exhibited by forms of pure carbon; however, to date there have been no studies in the published literature of muon implantation in nanostructured carbonaceous materials of catalytic importance.

## 2. Starbons

Activated carbons are carbonaceous materials with broad industrial applications ranging from biomedical filtration devices to supercapacitors. Typical activated carbons are made using high-temperature inert atmosphere pyrolysis of a hydrocarbon starting material to form a *char*, followed by *activation* via high temperature reaction with $O_2$, $CO_2$, or steam. The meso/microporosity of the material (up to about 500 $m^2/g$ internal pore surface area) is generated by the activation process. The degree of control over the pore structure of the material is typically low.

Recently an alternative strategy for preparation of activated carbon materials has been developed, based on the use of expanded starch as a templating material and relatively mild pyrolysis conditions, with no separate activation step [16]. Properties of these Starbon® materials are "tunable" in a manner dependent on the temperature and duration of pyrolysis; pyrolysis at low temperature may be expected to yield a material retaining more of the "hydrophilic" properties of the original starch material in its chemically active voids, while high temperature pyrolysis may be expected to lead to "hydrophobic" voids and a more graphitic material. Two samples were obtained from the Green Chemistry Centre of Excellence at the University of York, representing the hydrophilic (Starbon 300) and hydrophobic (Starbon 800) extremes of the preparation method. (The numerical suffixes represent the pyrolysis temperature in degrees Celsius.)

## 3. Experimental

Complex amorphous materials are not amenable to study using muon spectroscopy methods that rely upon precise measurement of spin precession frequencies or detection of sharp RF or level crossing resonances, as the spin polarisation is likely to be distributed over a broad variety of muon addition sites; we therefore employed the bluntest instrument in the $\mu$SR arsenal, longitudinal field muon spin repolarisation spectroscopy (LF$\mu$SR). In this method, paramagnetic species are detected via the recovery of muon spin polarisation against an applied longitudinal field, yielding semiquantitative information on the magnitude of the isotropic hyperfine coupling $A_\mu$ (which can distinguish between muonium, high-coupling radicals, or low-coupling radicals), the presence and type of hyperfine anisotropy (axial or complete), and the presence or otherwise of sources of spin relaxation (electron mobility, chemical reactions).

### *3.1. J-PARC measurements*

Room temperature longitudinal field repolarisation measurements were carried out on foil-encapsulated powder samples using the D1 instrument at J-PARC over the field range 0-1600 G; the results appear in Figures 1 (Starbon 300) and 2 (Starbon 800). In Starbon 800 (Figure 2) only a small fraction exhibits repolarisation, while there is a much larger repolarising fraction in Starbon 300. This suggests that the "hydrophilic" material retains substantial isolated radical-forming double bond character whereas the "hydrophobic" material is substantially graphitized. Relaxation rates are modest ($\lambda < 0.5\mu s^{-1}$) over the entire field range.

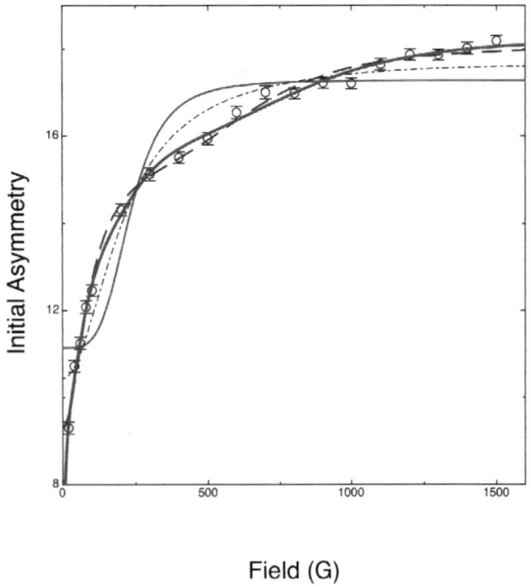


**Figure 1.** Room temperature LF repolarisation curve for Starbon 300. The curves represent fits to the data described in the text.

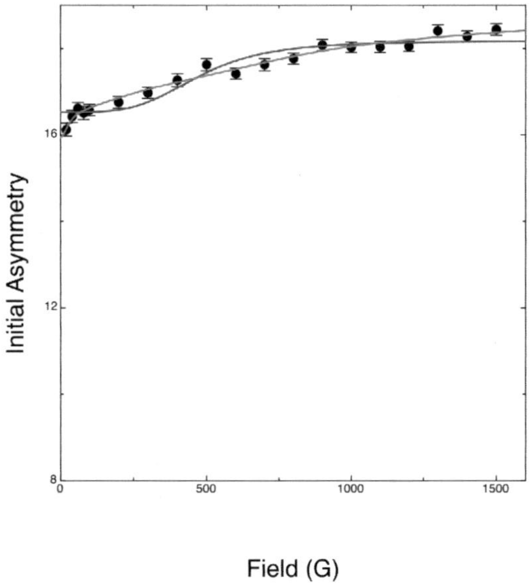


**Figure 2.** Room temperature LF repolarisation curve for Starbon 800. The curves represent fits to the data described in the text.

The solid and dashed curves in Figures 1 and 2 represent fits to the data. One or more repolarising paramagnetic fractions are considered, together with a non-repolarising background. In the hydrophilic sample (Figure 1), a single isotropic radical (solid red curve, $\chi^2 = 12.00$) yields a poor fit to the experimental points, with an optimum $A_\mu$ of 375 MHz. Better fits are obtained by allowing for an anisotropic hyperfine tensor or for more than one radical. A fit to two isotropic radicals yielded a much improved fit (dashed blue curve, $\chi^2 = 0.69$), corresponding to $A_\mu$ values of 258 MHz and 154 MHz. Together, these account for almost all of the observed asymmetry, which is inconsistent with the approximately 1/3 of the total asymmetry observed to precess under transverse field. (This is most likely due to muons stopping in the sample holder; studies with a holmium sample offering the same cross-sectional area as the Starbon sample showed that about 2/3 of the polarisation should be attributed to muons stopping in the sample.) A more appropriate fitting function for a polycrystalline or amorphous solid would make use of one or more radicals with a powder-averaged fully anisotropic hyperfine interaction [18]. With a single repolarising radical fraction in which both the isotropic component and the dipolar part (considered axial) are left as free parameters, the best fit yielded $A_\mu = 574$ MHz and $D_{\|} = 112$ MHz (black dot-dashed curve, $\chi^2 = 4.47$), with the paramagnetic fraction accounting for about 1/3 of the total asymmetry; the fractional hyperfine anisotropy in this fit is somewhat too large to be plausible for a typical muonium adducts to a C=C double bond. Choosing instead to restrict the anisotropic term to 10% of the magnitude of the isotropic term, but allowing two radicals, yielded $A_\mu$ = 228 MHz (54%) and $A_\mu$ = 298 MHz (24%) (purple curve, $\chi^2$ = 0.43); this approach gave the best quantitative fit, but a slightly larger estimated total asymmetry. (Note that the polarisation recovery is not quite completed at 1600 G in this case.) The red and violet curves in Figure 2 represent isotropic single radical and anisotropic two-radical fits to the Starbon 800 data; the small fraction of recovering polarisation makes reliable fits problematic in this case, but the patterns are consistent with a coupling of some few hundred MHz.

### *3.2. ISIS measurements*

As a check on the high-field component of the polarisation recovery, and to search for any level crossings that may be detectable, additional repolarisation scans were carried out on the Starbon 300 sample using the Hi-Fi instrument at ISIS, over a field range up to about 10 kG at 300 K, and up to 5 kG at 20 K. The powder samples were mounted in an Al foil packet, masked with a silver plate. Thus, some part of the non-repolarising fraction probably corresponds to muons stopping in silver. A field scan up to 12 kG was also carried out using an Al foil dummy sample of equivalent stopping density in order to correct for field-dependent background due to changing decay positron trajectories. No sign of any level crossings was seen at fields up to 10 kG, excluding the detection of well-defined zero-crossings from radicals with $A_{\rm iso}$ up to 300 MHz. Fits to the data between 50 G and 5 kG are shown in Figure 3. The repolarisation behavior changes very little between 300 K (solid dots) and 20 K (dots inside circles), indicating that the formation of the paramagnetic species either has a low barrier or is non-thermal, and also that the radicals are likely to be in a rather rigid environment, not strongly subject to dynamical averaging of $A_{\rm iso}$ itself. (That there may be some dynamical modulation of the dipolar part of the coupling is evident from the differences observable at low field.) The fit parameters in each case represent two radicals, each with $D_\parallel$ fixed to $0.1A_\mu$. A relaxing signal component with $\lambda$ on the order of 3 $\mu s^{-1}$ is decoupled in a few tens of gauss. Observed relaxation effects are most likely related to slow desorption of oxygen and other gases from these microporous materials.

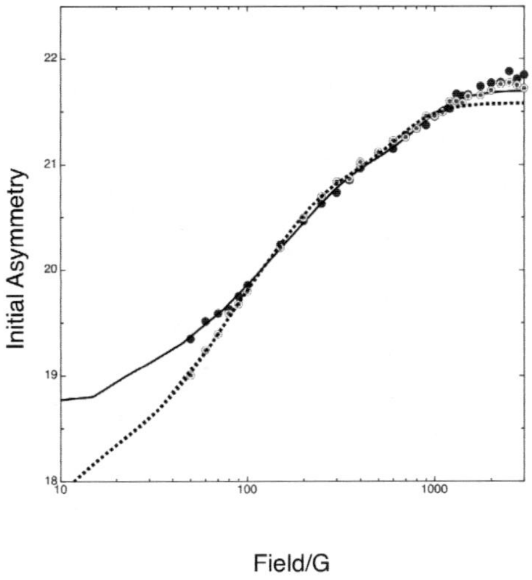


**Figure 3.** LF repolarisation data obtained at 300 K (solid circles, bold solid curve) and 20 K (dots inside circles, bold dotted curve) on the Starbon 300 sample using the Hi-Fi instrument at ISIS. Curves are fits to two anisotropic radials as described above.

Fit parameters:
$A_\mu$ = 297 MHz (14 %) and 178 MHz (6 %), $\chi^2$ = 0.569 (300 K data); $A_\mu$ = 212 MHz (21 %) and 203 MHz (6 %), $\chi^2$ = 0.655 (20 K data).

Simulations using *Quantum* [17] suggest that an alternative model might be a single radical with $A_\mu \sim$ 300 MHz with the low-field part of the polarization recovery then due to decoupling of a *geminal* proton (for which $A_H$ can be expected to be close to $A_\mu/3.1833$); in this case, the difference between the 300 K and 20 K datasets would be due to a change in the thermally-

Journal of Physics: Conference Series **551** (2014) 012040 doi:10.1088/1742-6596/551/1/012040

averaged dipolar part of the hyperfine interaction. Such a picture, however, requires the room temperature hyperfine tensor to be close to isotropic.

## 4. Computational results

Starbons and similar materials are likely to have a complex structure intermediate between that of their parent material (usually starch or cellulose) and graphitic carbon. Thus, there is probably no single dominant type of radical-forming site present in these materials. Nonetheless, it is possible to obtain some information about likely site chemistry from a study of site energetics and the magnitudes of hyperfine coupling constants.

### *4.1. Muonium adducts to hydroxymethylfurfural*

Hydroxymethylfurfural (HMF) is a compound produced in the dehydration of hexose sugars, recently reviewed for its potential as a potential renewable "platform chemical", a starting material in industrial synthesis of polymers, pharmaceuticals, and other products [19]. It has been suggested by Titirici *et al.* as a small-molecule intermediate in the HTC process [20]. Under conditions of reasonably low temperature (below about 300°C), hydrolysis of the starch or cellulose template is followed by dehydration of hexoses into HMF (and of pentoses into furfural), whereupon a complex series of ring-openings, condensations, and aromatisation reactions leads to the final product, a polymeric furanic structure with aliphatic bridges and terminal carbonyl and hydroxyl groups.

All calculations used the computational package *Gaussian 09* [21], at the B3LYP/6-311G(d,p) level of density functional theory. Optimized geometries of the six radical adducts considered are shown in Figure 4, while their relative energies and the corresponding Fermi contact hyperfine couplings (proton values scaled by $\frac{\mu_\mu}{\mu_p}$) appear in Table 1. As Mu addition to C=C is typically more rapid than to C=O, radical 2 may be a reasonable model species for the repolarising fraction in Starbon 300. (Computational studies of more realistic oligomeric models are in progress.)

**Table 1.** Hyperfine coupling constants for muonium adducts to hydroxymethylfurfural calculated as described in the text. No corrections are made for zero-point stretching or vibrational averaging. Pairs of $A_\mu$ values correspond to situations where Mu can add in either of two, slightly inequivalent, ways.

| Radical | Energy/kJ mol$^{-1}$ | $A_\mu$/MHz |
|---|---|---|
| 1 | 0 | -25 |
| 2 | 1.29 | 369/442 |
| 3 | 24.85 | 312 |
| 4 | 43.36 | 318/324 |
| 5 | 79.14 | 251 |
| 6 | 99.00 | 356/662 |

### *4.2. Muonium addition to graphitic analogues*

The use of polycyclic aromatic hydrocarbons as molecular analogs of graphite (edge-terminated with hydrogen) is widespread in the theoretical literature. We utilise the molecule tribenzo[adg]coronene ($C_{36}H_{18}$) as a model of an "all-armchair" nanographene. There are six distinct muonium addition sites to this species: three outer edge sites, two "bay" sites, and an interior site. Addition at edge sites is 50-70 kJ mol$^{-1}$ more favorable than addition at interior

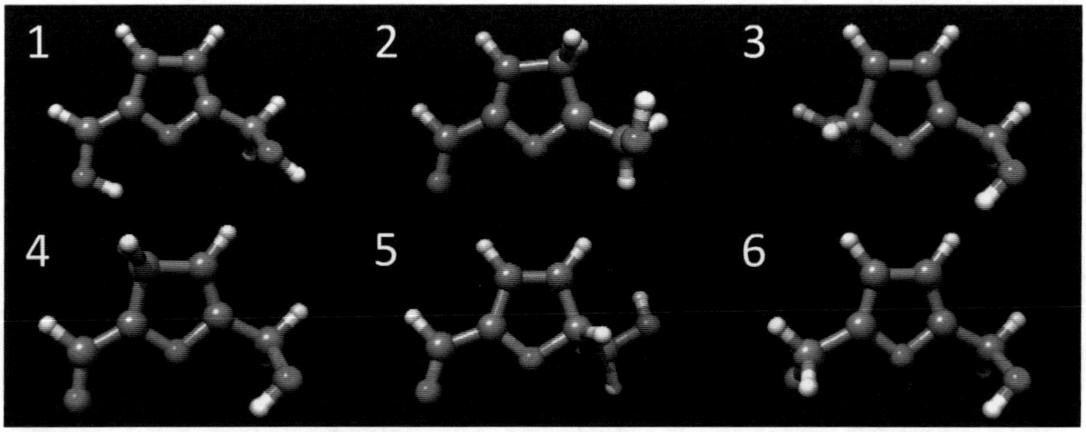


**Figure 4.** Optimized structures of the six muonium adducts to hydroxymethylfurfural. Oxygen atoms are red, carbon atoms grey, hydrogens white. The muon site is indicated in yellow.

or bay sites. Fermi contact couplings for the radicals (scaled to muon values, but uncorrected for the C-Mu zero-point stretch) range from 337 MHz for the lowest-energy edge radical to 583 MHz for the interior radical, which is the least energetically-favorable structure, involving substantial geometrical distortion from the planarity of the parent structure. As an "all-zigzag" model nanographene we consider dibenzo[bc,kl]coronene ($C_{30}H_{14}$), which possesses one terminal site, three edge sites, two "fissure" sites, and three interior sites. In this case the zig-zag edge sites exhibit both the lowest energies overall (lower by about 150 kJ mol$^{-1}$ than the interior or fissure sites), as well as unusually small Fermi contact couplings (down to 80 MHz) occasioned by long-range migration of spin density. A more complete report describing calculations on these species is in preparation and will appear elsewhere.

## Acknowledgments

The authors thank the Green Chemistry Centre of Excellence at the University of York for providing the Starbon samples. The computational part of this work used the Extreme Science and Engineering Discovery Environment (XSEDE), which is supported by National Science Foundation grant number ACI-1053575. The primary computational resource used was the Trestles system at the San Diego Supercomputer Center. PS gratefully acknowledges the Ministerio de Ciencia e Innovación for the concession of a Juan de la Cierva fellowship (JCI-2011-10836). Additionally, RMM and CU thank Marian University for travel support.

## References

[1] Ryoo R, Joo S H and Jun S 1999 *J. Phys. Chem.* B **103** 7743-6
[2] Titirici M-M 2013 *Sustainable Carbon Materials from Hydrothermal Processes* (Chichester: Wiley)
[3] Nakayama N and Akita S 2003 *New J. Phys.* **5** 128
[4] Zhao B, Mönch I, Vinzelberg H, Mühl T and Schneider C M 2002 *Appl. Phys. Lett.* **80** 3144
[5] Stoller M D, Park S, Zhu Y, An J and Ruoff R S 2008 *Nano Lett.* **8** 3498
[6] MIT LEES laboratory project; see e.g `http://lees.mit.edu/lees/posters/RU13_signorelli.pdf`
[7] Holzschuh E, Kündig W, Meier P F, Patterson B D, Sellschop J P F, Stemmet M C and Appel H 1982 *Phys. Rev.* A **25** 1272
[8] Patterson B D 1988 *Rev. Mod. Phys.* **60** 69
[9] Hutchinson D P, Menes J, Shapiro G, Patlach A M and Penman S 1961 *Phys. Rev. Lett.* **7** 129
[10] Chakhalian J, Kiefl R F, Dunsiger S R, MacFarlane W A, Miller R, Sonier J E and Fischer J E 2002 *Phys. Rev.* B **66** 155107

[11] Cox S F J, Cottrell S P, Charlton M, Donnelly P A, Ewels C, Heggie M and Hourahine B 2001 *J. Phys.: Condens. Matter* **13** 2169
[12] Davis E A and Cox S F J 2000 ISIS Annual Report, cited in [15]
[13] Ansaldo E J, Niedermayer C and Stronach C E 1991 *Nature* **353** 121
[14] MacFarlane W A, Chakhalian J, Kiefl R F, Dunsiger S R, Miller R I, Sonier J E, Thess A, Rinzler A G, Smalley R E and Fischer J E 2000 *Physica* B **289-290** 589
[15] Cox S F J 2009 *Rep. Prog. Phys.* **72** 116501
[16] Budarin V, Clark J H, Hardy J J E, Luque R, Milkowski K, Tavener S J and Wilson A J 2006 *Angew. Chem. Int. Ed.* **45** 3782
[17] Lord J S 2006 *Physica* B **374-375** 472-474
[18] Pratt F L 1997 *Phil. Mag. Lett.* **75** 371
[19] van Putten R J, van der Waal J C, de Jong E, Rasrendra C B, Heeres H J and de Vries J G 2013 *Chem. Rev.* **113** 1499-597
[20] Titirici M-M, White R J, Falco C and Sevilla M 2012 *Energy. Environ. Sci.* **5** 6796-822
[21] Gaussian 09, Revision D.01, Frisch M J *et al.*, Gaussian, Inc., Wallingford CT, 2013

Journal of Physics: Conference Series **551** (2014) 012041 doi:10.1088/1742-6596/551/1/012041

# Molecular Dynamics of Transient Radicals in Solution Probed by Longitudinal-Field Muon Spin Relaxation

**I McKenzie**[1,2] **and S P Cottrell**[3]

[1] Centre for Molecular and Materials Science, TRIUMF, Vancouver, B.C. Canada V6T 2A3
[2] Department of Chemistry, Simon Fraser University, Burnaby, B.C. Canada V5A 1S6
[3] ISIS Facility, STFC Rutherford Appleton Laboratory, Chilton, Didcot, Oxon, U.K. OX11 0QX

E-mail: iain.mckenzie@triumf.ca

**Abstract.** Longitudinal field muon spin relaxation (LF-$\mu$SR) has been used to measure the magnetic field dependence of the longitudinal field muon spin relaxation rates ($\lambda_\mu = 1/T_1^\mu$) of two muoniated radicals, 2-muoxyprop-2-yl radical (formed by Mu addition to acetone) and the muoniated 1,2-dicarboxyvinyl radical dianion (formed by Mu addition to the 1,2-dicarboxyacetylene dianion), in aqueous solution. The rotational correlations time of the radicals in solution at 298 K were determined from the magnetic field dependence of $\lambda_\mu$ and were found to be $32 \pm 2$ ps for the 2-muoxyprop-2-yl radical and $55 \pm 3$ ps for the muoniated 1,2-dicarboxyvinyl radical dianion.

## 1. Introduction

The measurement of nuclear and electron spin relaxation rates as a function of magnetic field and temperature can give useful information about molecular dynamics of free radicals in the liquid-phase and the interaction between the radical and the solvent [1, 2, 3]. Electron paramagnetic resonance is the main technique used to study free radicals; however, measurements are often difficult in solution due to the high reactivity of the radical species. In such cases, there can be important advantages in using muon spectroscopic techniques to study muoniated isotopomers of the radicals. These methods and their role in free radical chemistry have been extensively reviewed [4, 5], so only a basic description is given here. A beam of spin-polarized positive muons is injected into a sample and the positron produced by the decay of each muon detected. Muons can pick up a radiolytic electron and form Mu, which is considered to be a light isotope of hydrogen. This species can then react with unsaturated molecules to produce a muoniated radical, where the muon is a ~100% polarized spin label. The parity-violating decay of the muon provides a convenient way to monitor the evolution of the muon spin polarization, and its lifetime (2.2 $\mu$s) is comparable to many molecular processes.

In this paper we report measurements of the magnetic field dependence of the longitudinal field muon spin relaxation rates ($\lambda_\mu = 1/T_1^\mu$) of two muoniated radicals, 2-muoxyprop-2-yl radical (formed by Mu addition to acetone [6]) and the muoniated 1,2-dicarboxyvinyl radical dianion (formed by Mu addition to the 1,2-dicarboxyacetylene dianion [7]), in aqueous solution. The structures of the muoniated radicals are shown in Fig. 1. The electron spin-lattice relaxation

Journal of Physics: Conference Series **551** (2014) 012041 doi:10.1088/1742-6596/551/1/012041

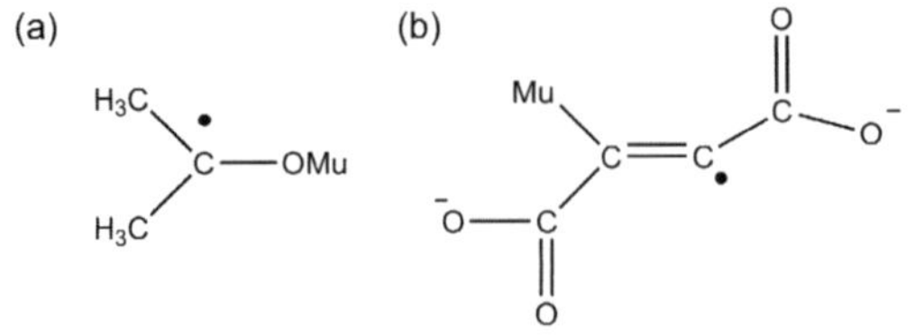


**Figure 1.** Structures of (a) the 2-muoxyprop-2-yl radical and (b) the muoniated 1,2-dicarboxyvinyl radical dianion.

times ($T_1^e$) of the non-muoniated isotopomers of the radicals have been previously measured using EPR spectroscopy [2, 3] but only at one field, which provides insufficient data to determine the rotational correlation time. The rotational correlation time of a muoniated radical can be determined by measuring the magnetic field dependence of $\lambda_\mu$ [8]. The goal of these experiments was to determine whether the reorientational dynamics of an open shell species in solution are significantly different from the closed shell analogues.

## 2. Experimental

Two samples were used in these experiments; a 1.0 M aqueous solution of acetone and a 0.5 M aqueous solution of sodium acetylenedicarboxylate ($Na_2C_4O_4$). Acetone was purchased from Sigma-Aldrich and dissolved in Millipore water. Acetylenedicarboxylic acid was purchased from Fluka. The 0.5 M aqueous solution of $Na_2C_4O_4$ was prepared by dissolving acetylenedicarboxylic acid in Millipore water and titrating with NaOH. Molecular oxygen was removed from the solutions by repeated cycles of freezing, pumping and thawing; a necessary step to avoid Heisenberg spin-exchange between the $O_2$ and the Mu or radical species contributing to the spin relaxation rate.

The samples were sealed in a titanium cell with a 100 $\mu$m titanium window, through which surface muons (28 MeV $c^{-1}$) can penetrate. The sample cells were mounted in a Closed Cycle Cryocooler to control the temperature at 298 K. Measurements were made using the EMU spectrometer at the ISIS Pulsed Muon Facility, STFC Rutherford Appleton Laboratory [9, 10].

The LF-$\mu$SR technique involves measuring the asymmetry of the muon decay, $a(t)$, as a function of time. The asymmetry is proportional to the muon polarization along the $z$-axis (i.e. the beam axis). The asymmetry is obtained by counting the number of positrons detected in the forward ($N_F$) and backward ($N_B$) counters (with respect to the muon beam momentum) as a function of time after the implantation of the muons.

$$a(t) = \frac{N_F(t) - \alpha N_B(t)}{N_F(t) - \alpha N_B(t)} \tag{1}$$

where $\alpha$ is a measure of the relative efficiencies of the forward and backward detectors, and was determined by measuring the spin rotation spectrum in a small transverse field (20 G). Decay events from each detector are accumulated over many muon pulses and formed into separate histograms each of total length 32 $\mu$s with a timing resolution of 16 ns. The maximum asymmetry of the EMU spectrometer, $a_{max}$, was determined by carrying out a similar spin rotation measurement for muons stopped in a silver plate - where all the muons are in a diamagnetic environment.

Spin relaxation in a multilevel system, as for the radicals studied here, is properly described by the sum of several exponential terms [11]. Properties of the system, such as a distribution of

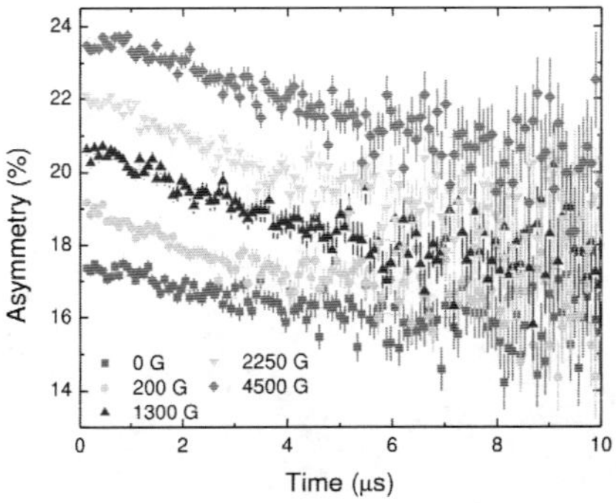


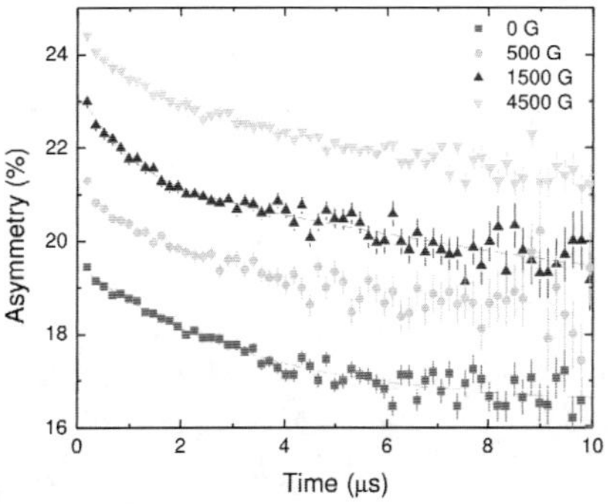


**Figure 2.** (Left) LF-$\mu$SR spectra of 1.0 M aqueous acetone solution and (Right) LF-$\mu$SR spectra of 0.5 M aqueous $Na_2C_4O_4$ solution as a function of longitudinal magnetic field.

correlation times, local environments or, perhaps, hindered molecular dynamics may, however, complicate this analysis, and a form of the stretch exponential (Eq. 2) can frequently be expected. The time dependence of the asymmetry was modelled with the following equation,

$$a(t) = a_D + a_R e^{-(\lambda_\mu t)^\beta} \tag{2}$$

where $a_D$ is the asymmetry due to muons in diamagnetic environments and is equal to the precession amplitude in a 20 G transverse field, $a_R$ is the asymmetry due to muons in a paramagnetic species, $\lambda_\mu$ is the muon spin lattice relaxation rate ($= 1/T_1^\mu$) and $\beta$ is the stretch exponent. In our case, data for the $(CH_3)_2COMu$ radical could be adequately modelled with $\beta$ fixed at 1; in contrast, however, the vinyl radical dianion displays a root exponential time dependence at fields above 10 G. The initial polarization of the paramagnetic species is given by:

$$P_R = \frac{a(0) - a_D}{a_{max} - a_D} \tag{3}$$

### 3. Results and Discussion

We obtained LF-$\mu$SR spectra of all samples as a function of longitudinal magnetic field and examples for both $(CH_3)_2COMu$ and $C_4O_4Mu^{2-}$are shown in Figure 2.

The identity of the paramagnetic species was confirmed by measuring the initial polarization as a function of the applied magnetic field. This is typically called a repolarization curve and the inflection point is related to $A_\mu$ as described by the following equation:

$$P_z = \frac{\frac{1}{2} + x^2}{1 + x^2} \tag{4}$$

where $x = \gamma_e B/A_\mu$ and $\gamma_e$ is the electron gyromagnetic ratio.

The outcome is unambiguous for muons stopped in water (shown as the blue line in Fig. 3), where the repolarization of the prompt Mu state yields a curve with a single well defined inflection point that gives a fitted value for $A_\mu = 4836 \pm 156$ MHz. This value can be contrasted with an accurate measurement made using spin rotation methods that gives $A_\mu = 4432.4 \pm 1$ MHz [12], suggesting that large errors might be expected for coupling constants determined by repolarization.

In comparison to the repolarization curve measured for Mu in aqueous solution, the curves measured for the transient radicals (also shown in Fig. 3) clearly exhibit a more complex polarization recovery with field, with both curves apparently containing two components. Multiple components to a repolarization curve can indicate either the formation and coexistence of multiple paramagnetic species or that the lifetime of the Mu precursor is long with respect to $1/A_\mu \sim 0.2$ ns. The latter case is expected for the present system since the molecular species being studied are in dilute solution. The rate of Mu addition to acetone in a concentrated aqueous solution was determined to be $(8.5 \pm 2.1) \times 10^7$ M$^{-1}$s$^{-1}$ [13] so we expect the lifetime of Mu in the 1.0 M acetone solution to be ~12 ns.

A phenomenological model that describes the latter situation was first proposed by Cox, for deep and shallow-donor states in semiconductors [14] and, more recently, developed by Fleming et al. during a study of Mu adducts of bromine [15]. The model suggests a modified function for the repolarization curve, of the form:

$$P_{\mathrm{R}} = f_O + f_S \times \frac{\frac{1}{2} + x_{\mathrm{Mu}}^2}{1 + x_{\mathrm{Mu}}^2} \times \frac{\frac{1}{2} + x_{\mathrm{R}}^2}{1 + x_{\mathrm{R}}^2} \tag{5}$$

where $x_{\mathrm{Mu}} = \gamma_e B / A_\mu^{\mathrm{Mu}}$ models the precursor Mu state and $x_{\mathrm{R}} = \gamma_e B / A_\mu^{\mathrm{R}}$ the final radical state. Nuclear couplings are accommodated to a limited extent by an 'offset' parameter, $f_O$, with errors due to changes in $\alpha$ with magnetic field accounted for by a 'scaling factor', $f_S$.

Using this equation to fit the curves obtained for the transient radicals, with the muon hfcc of Mu fixed at 4432.4 MHz [12], provides an estimate for the muon hfcc of the final state for each radical species. Values obtained are shown in Table 1 together with muon hfccs determined using spin rotation methods for identical systems in high concentration. Considering that repolarization curves are a notoriously poor method of determining muon hfccs and the given limitations of the fitting model, large errors in determining the hfccs are likely - certainly much larger than the statistical errors determined from the fit and quoted in the table. The results obtained from the repolarization measurements are therefore of limited quantitative value, but importantly provide a good indication that final state radical species are formed as anticipated in this system.

**Table 1.** Comparison of muon hyperfine coupling constants determined by TF-$\mu$SR and repolarization curves.

| Sample | Radical | $A_\mu^{\mathrm{TF-\mu SR}}$ / MHz | $B_0$ / G | $A_\mu^{\mathrm{repol}}$ / MHz |
|---|---|---|---|---|
| 1.0 M acetone | $(H_3C)_2COMu$ | ~22[a] | 19±2 | 53±6 |
| 0.5 $Na_2C_4O_4$ | $C_4O_4Mu^{2-}$ | 493.8±0.2[b] | 132±23 | 370±64 |

[a] Ref. [6], [b] Ref. [7]

Fitted values for the relaxation rate, $\lambda_\mu$, in both $(H_3C)_2COMu$ and $C_4O_4Mu^{2-}$ are shown in Fig. 4, and were found to depend on the applied magnetic field in a similar manner. Fleming et al. thoroughly investigated the magnetic field dependence of $\lambda_\mu$ of the muoniated ethyl radical in the gas phase [8]. There are several relaxation processes, which dominate at different values of the applied magnetic field. A muon spin flip can be induced directly or indirectly by flipping the electron spin, which is coupled to the muon via the hyperfine interaction, or by coupled muon-electron "flip-flop" or "flip-flip" processes. The field dependence of $\lambda_\mu$ for both $(H_3C)_2COMu$ and $C_4O_4Mu^{2-}$ could best be fit using the expression given by Fleming et al. for

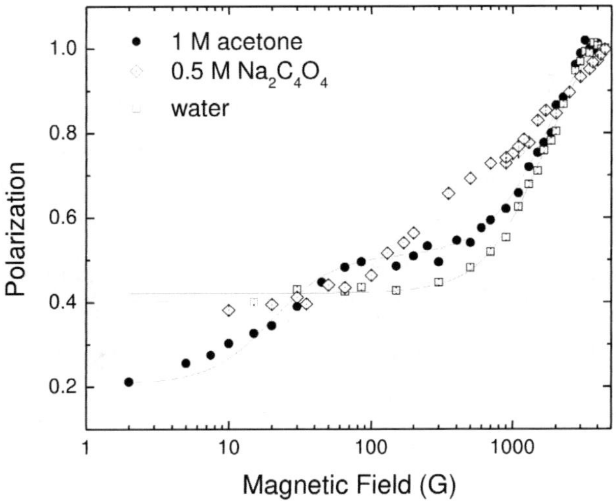


**Figure 3.** Repolarization curves for Mu formed in water, 0.5 M $Na_2C_4O_4$ in water and 1.0 M acetone in water. The solid lines for $(H_3C)_2COMu$ and $C_4O_4Mu^{2-}$ are fits using equation 5.

coupled processes with the addition of a field-independent relaxation term, $\lambda_0$:

$$\lambda_\mu = \lambda_0 + \frac{2\Delta^2_{\rm ME}\tau_c}{1+(\gamma_e\tau_c)^2\cdot B^2} \tag{6}$$

where $\Delta_{\rm ME}$ is the matrix element and $\tau_c$ is the motional correlation time. From the magnetic field dependence of the relaxation data we have determined that $\tau_c$ of $(H_3C)_2COMu$ is $32\pm2$ ps and $\tau_c$ of $C_4O_4Mu^{2-}$ is $55\pm3$ ps.

For molecules in $H_2O$ at room temperature ($\eta = 1$ cP) $\tau_c$ can be approximated as

$$\tau_c(\text{ps}) \simeq 0.4 \times M_w \tag{7}$$

where $M_w$ is the molecular weight of the radical [3]. Thus the expected $\tau_c$ of $(H_3C)_2COMu$ is ~23 ps and the expected $\tau_c$ of $C_4O_4Mu^{2-}$ is ~45 ps. The experimentally determined $\tau_c$ are therefore in good agreement with the predictions. The value of $\tau_c$ of $C_4O_4Mu^{2-}$ at 302 K reported in a previous publication from measurements of $\lambda_\mu$ in zero magnetic field (1.1±0.1 ns [7]) appears to be off by a factor of ~25 from the current measurement. In part this discrepancy may be due to the need to develop Eq. 6 to properly account for the root exponential behaviour of the data; however, a review of recent NMR literature [16] suggests this is likely to have minimal impact on the results obtained. Another source of error may be the value of the hyperfine anisotropy, derived from DFT calculations, that was used to determine $\tau_c$ in the zero field measurements. Further work is required to properly understand the origin of this discrepancy.

## 4. Conclusions

We have demonstrated that the rotational correlation times of transient radicals in solution can be determined using longitudinal field muon spin relaxation measurements. The measured

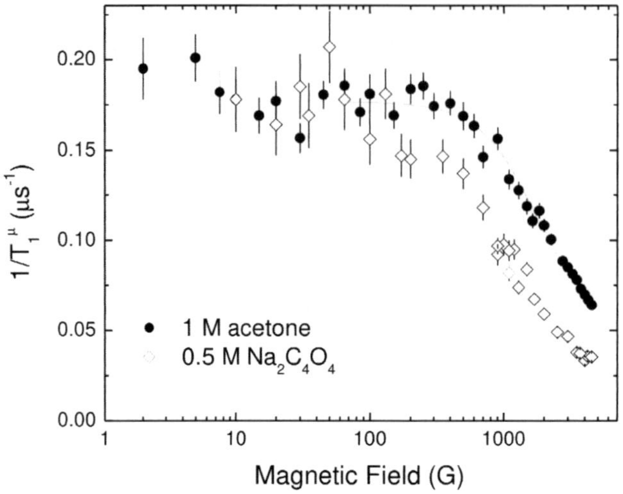


**Figure 4.** Magnetic field dependence of the longitudinal field muon spin relaxation rate for $(H_3C)_2COMu$ and $C_4O_4Mu^{2-}$. The solid lines for $(H_3C)_2COMu$ and $C_4O_4Mu^{2-}$ are fits using Eq. 6.

rotational correlation times are in good agreement with predictions based on approximations made for closed shell molecules. The rotational correlation time of $C_4O_4Mu^{2-}$ determined from the magnetic field dependence of $\lambda_\mu$ differs by ~25 from the value determined from the temperature dependence of $\lambda_\mu$ in zero field.

## References

[1] Paul H 1975 *Chem. Phys. Lett.* **32** 472
[2] Fessenden R W, Hornak J P and Venkataraman B 1981 *J. Chem. Phys.* **74** 3694
[3] Bartels D M, Lawler R G and Trifunac A D 1985 *J. Chem. Phys.* **83** 2686
[4] McKenzie I and Roduner E 2009 *Naturwissenschaften* **96** 873
[5] McKenzie I 2013 *Annu. Rep. Prog. Chem.* C **109** 65
[6] Hill A, Symons M C R, Cox S F J, De Renzi R, Scott C A, Bucci C and Vecli A 1985 *J. Chem. Soc. Faraday Trans. 1* **81** 433
[7] McKenzie I 2011 *Phys. Chem. Chem. Phys.* **13** 1168
[8] Fleming D G, Pan J J, Senba M, Arseneau D J, Kiefl R F, Shelley M Y, Cox S F J, Percival P W and Brodovitch J-C 1996 *J. Chem. Phys.* **105** 7517
[9] Eaton G H, Clarke-Gayther M A, Scott C A, Uden C N and Williams W G 1994 *Nucl. Instr. Meth. Phys. Res.* A **342** 319
[10] Giblin S R, Cottrell S P, King P J C, Tomlinson S, Jago S J S, Randall L J, Roberts M J, Norris J, Howarth S, Mutamba Q B, Rhodes N J and Akeroyd F A 2014 *Nucl. Instr. Meth. Phys. Res.* A **751** 70
[11] Cox S F J 1998 *Solid State Nucl. Magn. Reson.* **11** 103
[12] McKenzie I, Dilger H, Roduner E, Scheuermann R and Zimmermann U 2008 *J. Phys. Chem.* B **112** 3070
[13] Venkateswaran K, Kiefl R F, Barnabas M V, Stadlbauer J M, Ng B W and Walker D C 1988 *Chem. Phys. Lett.* **145** 289
[14] Cox S F J 2009 *Rep. Prog. Phys.* **72** 116501
[15] Fleming D G, Cottrell S P, McKenzie I and Macrae R M 2012 *Phys. Chem. Chem. Phys.* **14** 10953
[16] Kuhn A, Kunze M, Sreeraj P, Wiemhöfer H-D, Thangadurai V, Wilkening M and Heitjans P 2012 *Solid State Nucl. Magn. Reson.* **42** 2

Journal of Physics: Conference Series **551** (2014) 012042 doi:10.1088/1742-6596/551/1/012042

# Measurement of hyperfine coupling constants of muoniated radicals in small molecule semiconductors

**L Schulz[1,†], K Wang[1,2], M Willis[1,2], L Nuccio[2,3], P Murahari[2], S Zhang[1], F L Pratt[4], J S Lord[4], N A Morley[5], C Bernhard[3] and A J Drew[1,2,4,‡]**

[1] Sichuan University, School of Physical Science and Technology, Chengdu, Sichuan 610064, China

[2] Queen Mary University of London, Mile End Road, London E1 4NS, UK

[3] Department of Physics and Fribourg Center for Nanomaterials-Frimat, University of Fribourg, Chemin du Musée 3, 1700 Fribourg, Switzerland

[4] ISIS Muon Facility, Rutherford Appleton Laboratory, STFC, Harwell Science Campus, Chilton, Didcot OX11, UK

[5] Department of Materals Engineering and Science, University of Sheffield, Mappin Street, Sheffield, S1 3JD, UK

E-mail: † lschulz@scu.edu.cn, ‡ a.j.drew@qmul.ac.uk

**Abstract.** We report the hyperfine coupling constants of muoniated radicals formed in a number of organic semiconductors, via transverse field measurements taken in the Paschen Back limit, and compare the results to avoided level crossing resonances. Five muoniated radicals are found in tetracene, despite there only being three potential non-equivalent bonding sites, and we suggest that this might be down to crystal packing effects. For 6,13-bis(triisopropylsilylethynyl) pentacene and 6,13-bis(trimethlsilylethynyl)-pentacene, we demonstrate that the transverse field data supports the previously published avoided level crossing resonances.

## 1. Introduction

The investigation of organic materials (OSC) is vital for the advancement of organic electronics. Owing to the ability to probe at a microscopic (local) level, the muon spin rotation/relaxation/resonance (MuSR) techniques have been extensively employed for studying this material class. In particular, it has been used to quantitatively determine the electron spin relaxation rate and the electron spin relaxation mechanisms in organic semiconductors by using the sensitivity to spin dynamics in the system at so-called avoided-level crossings (ALC) [1, 2, 3]. A more specialized technique, the low-energy muon spin rotation technique, allowed for implanting muons at buried interfaces of fully functional organic-based spintronic devices, thereby enabling the measurement of spin injection into the organic spacer layer [4]. This resulted in a better understanding of their functionality [4] and how the spintronic property of these devices can be engineered [5]. In these latter studies of spin dynamics in organic semiconductors, it is of critical importance to characterise the muoniated radicals, which are very much material specific. Here we report avoided level crossing (ALC) and transverse field (TF) measurements taken in the Paschen Back limit to determine the hyperfine coupling (HFC) constant of muoniated radicals in a number of organic semiconductors [6, 7, 8].

## 2. Theoretical background

A considerable fraction of the muons that are implanted into an organic material can usually capture an electron to form muonium, which to all intense purposes, is analogous to hydrogen in an organic semiconductor. The muon and electron spins interact via the hyperfine interaction, resulting in a four level system (one spin singlet and three spin triplet states). After a chemical reaction of the muonium with a molecule, a molecular radical (molecule with an unpaired electron) is formed, typically with a significantly lower HFC constant, $A$, than the vacuum value. At the time of formation, the coupled spin system can be with equal probability either in the $|\uparrow_\mu\uparrow_e\rangle$ or $|\uparrow_\mu\downarrow_e\rangle$ state, the latter being a combination of the triplet $|\uparrow_\mu\downarrow_e\rangle + |\downarrow_\mu\uparrow_e\rangle$ and singlet $|\uparrow_\mu\downarrow_e\rangle - |\downarrow_\mu\uparrow_e\rangle$ states. The four energy eigenvalues of the system, corresponding to the three spin triplets and one singlet, are [11]

$$E_{|\uparrow_\mu\uparrow_e\rangle} = E_1 = \frac{\hbar}{4}\left[A + 2B\left(\gamma_e - \gamma_\mu\right)\right] \quad (1)$$

$$E_{|\uparrow_\mu\downarrow_e\rangle+|\downarrow_\mu\uparrow_e\rangle} = E_2 = -\frac{\hbar}{4}\left[A - 2\sqrt{A^2 + B^2\left(\gamma_e + \gamma_\mu\right)^2}\right] \quad (2)$$

$$E_{|\downarrow_\mu\downarrow_e\rangle} = E_3 = \frac{\hbar}{4}\left[A - 2B\left(\gamma_e - \gamma_\mu\right)\right] \quad (3)$$

$$E_{|\uparrow_\mu\downarrow_e\rangle-|\downarrow_\mu\uparrow_e\rangle} = E_4 = -\frac{\hbar}{4}\left[A + 2\sqrt{A^2 + B^2\left(\gamma_e + \gamma_\mu\right)^2}\right]. \quad (4)$$

Upon application of an external transverse magnetic field, $B$, the muon spin in isolated muonium precesses at a frequency that corresponds to the energy difference $E_i$ - $E_j$. In high transverse magnetic fields ($B \gg A/\left(\gamma_e + \gamma_\mu\right)$), only the muon spin precession frequencies governed by the energy differences $E_1$ - $E_2$ and $E_3$ - $E_4$ are observable. The other precessions are quenched out, i.e. their amplitude is zero [9, 11]. This is the so-called Paschen-Back regime. The two observable muon spin precession frequencies are [10]

$$f_{12} = \frac{1}{2}A + \frac{1}{2}\left(\gamma_e - \gamma_\mu\right)B - \frac{1}{2}A\sqrt{1 + \left(\frac{2B\left(\gamma_e + \gamma_\mu\right)}{A}\right)^2} \quad (5)$$

and

$$f_{34} = \frac{1}{2}A - \frac{1}{2}\left(\gamma_e - \gamma_\mu\right)B + \frac{1}{2}A\sqrt{1 + \left(\frac{2B\left(\gamma_e + \gamma_\mu\right)}{A}\right)^2}. \quad (6)$$

The sum of the two frequencies amounts to the isotropic hyperfine coupling constant [9, 10]

$$A = f_{12} + f_{34} \quad (7)$$

and practically in high TF experiment, these frequencies can be accessed either by conventional Fourier transforms or by fitting the time dependent asymmetry $A_0$ directly to a sum of relaxed oscillations (two frequencies $f_{12}$ and $f_{34}$ per muonium site plus the diamagnetic fraction).

## 3. Results and Discussion

The MuSR experiment was performed on polycrystalline samples of the organic semiconductors tetracene, 6,13-bis(triisopropylsilylethynyl)pentacene (TIPS-pentacene) and 6,13-bis(trimethlsilylethynyl)-pentacene (TMS-pentacene). The transverse field (TF) measurements were taken in the Paschen-Back regime at a temperature of 10 K using the GPS spectometer of the Paul-Scherrer-Institute, Switzerland, and the avoided level crossing (ALC) spectra were taken on the HiFi instrument at the ISIS Pulsed Muon Source, UK. For each TF data set, a total of 250 million events were measured, and we took 10 million events for each magnetic field in the ALCs .

### *3.1. Tetracene*

Figures 1a-f shows the measured forward-backward asymmetry and Fourier transforms at 100 mT, 150 mT, and 200 mT. The Fast Fourier Transforms (FFT) were obtained by apodising the time spectra with an exponential function with a characteristic time of 0.3 $\mu$s prior to using the FFTW routine [13]. The truncated peak at about 13.5 MHz in the 100 mT data corresponds to the interstitial free muons that precess in the externally applied magnetic field. Many satellite peaks are observable in the spectra, in pairs where the sum of the two frequencies corresponds to the HF coupling constant. The two most clear satellite peaks correspond to a HFC of $A_{\mathrm{Tc}} = 136.3 \pm 0.3$ MHz. The quite complex Fourier spectra indicate there are additional components that are difficult to resolve, which is also clear in the raw time spectra. To aid in interpretation of this data, we performed an Avoided Level Crossing (ALC) measurement on the same sample using the HiFi instrument at ISIS, which is shown in Figure 1g. Multiple ALC resonances are clear, particularly at 10K, indicated by the black arrows. We modelled the spectrum using the freely available Quantum software [14] and were able to resolve five independent ALCs. Details of the hyperfine coupling constants extracted from the 10 K data are given in Table 1; further details of the ALC spectra on tetracene and other heteroacenes will be given elsewhere [15]. It is worth noting here, however, that from the molecular structure shown in Figure 1h, only three non-equivalent sites should be present in tetracene. Given that the $\Delta_0$ ALC lines are unlikely to be large or wide enough to be observable in these spectra [2], we believe that the crystal packing breaking the equivalency of the different sites is a more likely explanation, resulting in more $\Delta_1$ ALC lines than expected from the three sites. It is also worth noting that the various lines in the frequency spectra will be broadened by the anisotropic HFC, even in the Paschen-Back limit, making them weak and indistinct in the Fourier spectra.

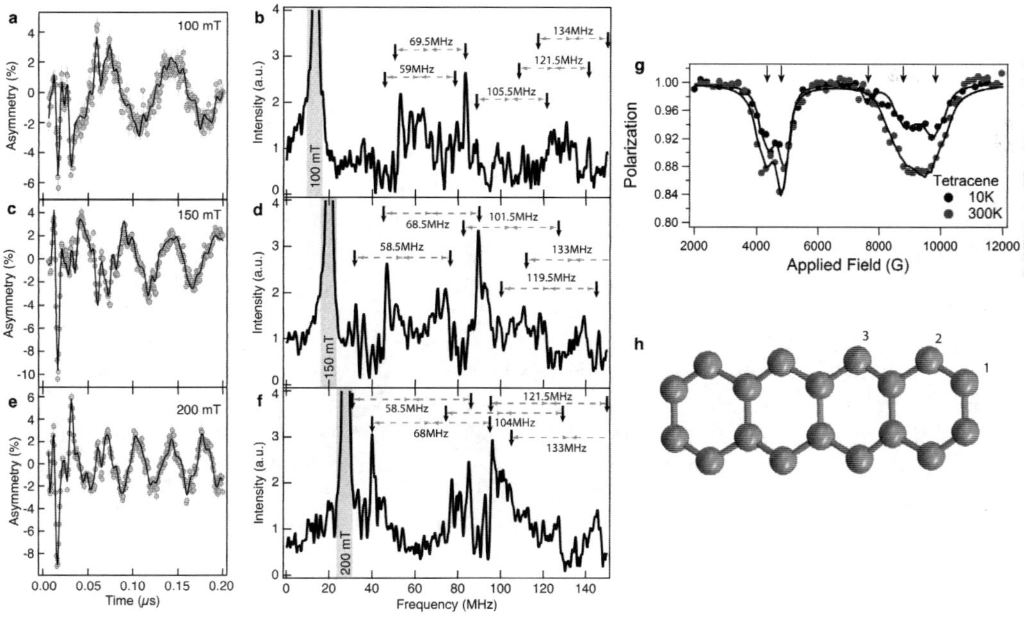


**Figure 1.** Transverse magnetic field data for tetracene using a,b) 100 mT, c,d) 150 mT and e,f) 200 mT. Data is shown in the a,c,e) time domain data (points) and fits (lines). b,d,f) show frequency domain, with the black arrows indicating the peaks predicted by the time domain fits. g) An ALC spectra for tetracene at 10 and 300 K. The lines show the results of the modelling, from which the HFCs can be extracted. h) The molecular structure of tetracene with H omitted.

**Table 1.** Extracted hyperfine parameters for the five different muoniated radicals in tetracene, from the ALC modelling and time domain fits for the three different fields. The satellite peaks expected from the time domain fits are indicated on the frequency spectrum in Figure 1b), d) and f). Frequencies are in MHz and have been rounded to the nearest integer. Errors have been rounded up.

| | **ALC** | **TD** | **0.2 T** | | **TD** | **0.15 T** | | **TD** | **0.1 T** | |
|---|---|---|---|---|---|---|---|---|---|---|
| | HFC | HFC | $f_{12}$ | $f_{34}$ | HFC | $f_{12}$ | $f_{34}$ | HFC | $f_{12}$ | $f_{34}$ |
| 1 | 120 | 117±2 | 31±2 | 86±1 | 117±9 | 39±1 | 78±9 | 118±2 | 45±2 | 73±1 |
| 2 | 135 | 136±1 | 40±1 | 96±1 | 137±1 | 47±1 | 90±1 | 139±1 | 55±1 | 84±1 |
| 3 | 208 | 208±3 | 75±1 | 133±3 | 203±8 | 82±8 | 121±1 | 211±2 | 92±1 | 119±2 |
| 4 | 243 | 243±1 | 96±1 | 147±1 | 239±1 | 100±1 | 139±1 | 243±10 | 108±10 | 135±1 |
| 5 | 267 | 266±3 | 104±2 | 162±2 | 266±1 | 111±1 | 155±1 | 268 ±4 | 119±2 | 149±3 |

Using the information on the HFC extracted from the ALCs. we fit the time series spectra to an eleven component fit; ten frequencies from the muonium (five lots of $f_{12}$ and $f_{34}$ per muonium site) plus the diamagnetic fraction. The resultant fits are shown in Figure 1a,c, and e, and it is clear that the complex time domain fits are in extremely good agreement with the experimental data. The results of the fitted frequencies are shown in Table 1, and they compare very favourably to the values obtained from the ALC modelling. This indicates that the five ALC lines are all from $\Delta_1$ transitions.

### *3.2. TIPS-Pentacene*

Figures 2a-f shows similar data sets for applied fields of 200, 250 and 500 mT for the case of TIPS-pentacene. The characteristic frequencies are once more indicated by the black arrows in Figures 2b,d and f. Interestingly, the spectra are significantly less noisy, or less complex, than tetracene. However, the satellite peaks appear relatively broad, which could be either from the anisotropic terms of the hyperfine tensor or from more than one site with a slightly different isotropic hyperfine coupling constant, such that the multiple satellite peaks are partially merged and difficult to unambiguously distinguish from each other. Previously, maximum entropy analysis of the ALC measurements and TF data (shown in Figure 2g and h respectively) indicated there were two muoniated radicals with similar HFCs, [5]. These two sites may originate from different muonium bonding positions (or binding angles to the same site) with a slightly different HFC constants. DFT calculations in the original work suggested that sites 3 and 5, shown in Figure 2i, have very close HFCs resulting in overlapping ALCs. The TF measurements presented here are consistent with the previously reported data. The separation of the satellite peaks, $2S$, increases as a function of the external field. Notice that one frequency is negative signifying a reversal of the precession direction. As negative and positive precession frequencies cannot be distinguished in a power spectrum, both appear as positive values. However, the distance, $S$, is reflected from the ordinate axis as can be seen in Figure 2e, thereby revealing that the value is indeed negative. The average value for the isotropic HF interaction constant extracted from a three-component fit to the time domain data is $A_{\mathrm{TIPS}} = 81.4 \pm 0.6$ MHz. Unfortunately, since the two HFCs are so close, it is not possible to fit a five component function to the data, although the maximum entropy, FFTs and the ALC data suggest it is present.

Journal of Physics: Conference Series **551** (2014) 012042 doi:10.1088/1742-6596/551/1/012042

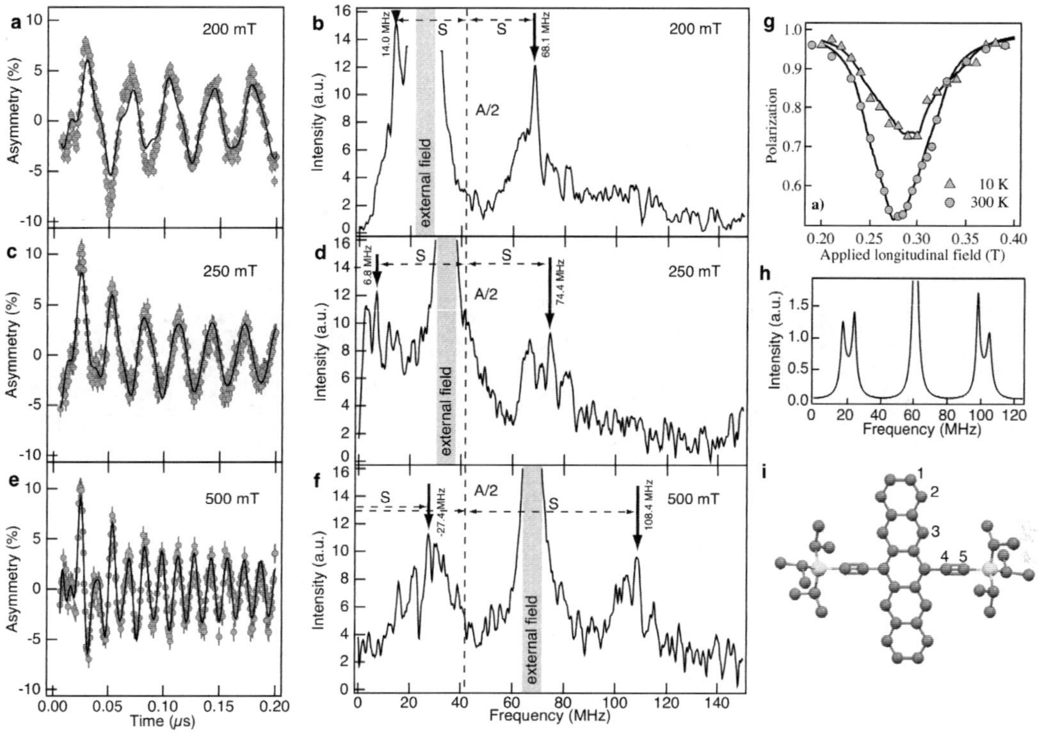


**Figure 2.** a,c,e) Time-domain asymmetry plots of transverse field measurement of TIPS-pentacene at 200 mT, 250 mT and 500 mT. The solid lines are fits to the data. b,d,f) The corresponding power spectra in the frequency domain. The satellite peaks are pointed out by the arrows. g) Muon's polarisation as a function of magnetic field, where two distinct ALC resonances are observable. h) Maximum entropy analysis of the TF data shown in f, where two very distinct peaks are observable. i) Molecular structure of TIPS-Pentacene, with the muonium bonding sites indicated. All data were taken at a temperature of 10 K.

### *3.3. TMS-Pentacene*

An equivalent data set for TMS-pentacene for the applied fields of 100 and 250 mT are shown in Figure 3. Similar to TIPS-Pentacene, the spectra are significantly less noisy or less complex than tetracene. However, it can be seen that there is also a skewness to the satellite peaks. Previously reported ALC measurements indicated that three sites with similar isotropic HFCs exist in TMS-pentacene [3], as shown by the black arrows in Figure 3f, suggesting that the skewness observed in Figure 3 is indeed related to the multiple positions of the muonium. However, we are unable to completely rule out that it is down to anisotropy. Due to the difficulties resolving these additional peaks, the pair of satellite peaks with the highest intensity is followed in the FFTs, where it can be seen that $S$ increases with increasing external field, while $A$ and hence $A/2$ remain almost constant. The isotropic HF coupling constant $A_{\mathrm{TMS}} = 84.0 \pm 0.6$ MHz extracted from fits to the time domain is very close to the value (82 MHz) extracted from the most intense ALC shown in Figure 3f [3]. As discussed above, only two ALC resonances were observed in TIPS-pentacene at this field, despite the similarity in molecular structure with TMS-Pentacene. The additional site in TMS-pentacene could be related to solid-state packing, as TMS-pentacene crystalises with a herringbone packing [3] whereas TIPS-pentacene crystallises with a two-dimensional brickwork arrangement, with close $\pi - \pi$ overlap [2].

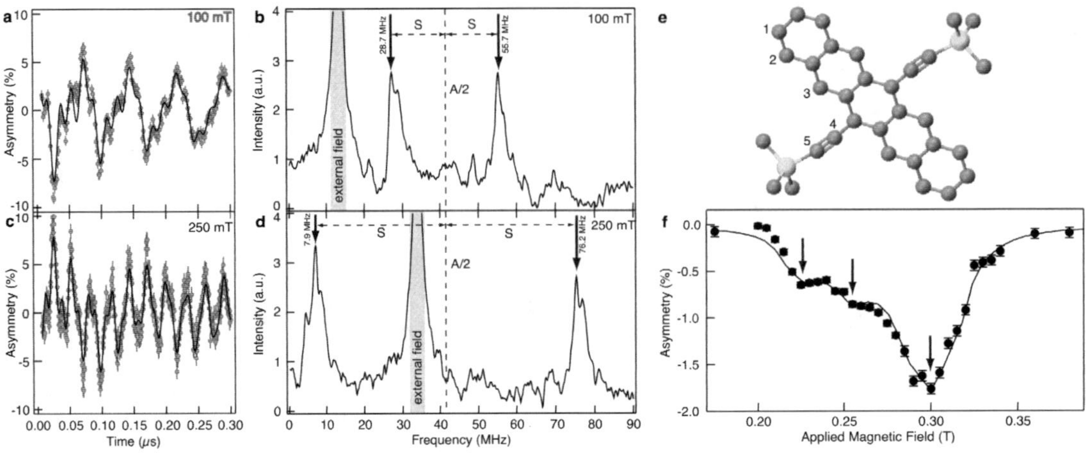

**Figure 3.** a-d) The TF data for TMS-pentacene. e) The molecular structure of TMS-Pentacene, with five potential muonium adducts shown. d) The asymmetry as a function of field, where three clear ALC resonances are observable. All data were taken at a temperature of 10 K.

## 4. Conclusions

We have reported HFC constants for muoniated radicals in three organic semiconductors, using a combination of TF and ALC techniques. In all cases, the extracted constants from the TF and ALC measurements are within the margin of error of each other, and their values are within the expected range in molecules such as these [2]. Interestingly, in tetracene, we observe two ALC resonances more than there are unique sites on the molecule. TF data suggests these ALC resonances are not $\Delta_0$ lines, since the data appears consistent with multiple components. The eleven component fit (five HFCs plus diamagnetic) give HFCs that are very similar to those extracted from the ALCs. We conclude these additional HFCs are related to the crystal packing, either by lifting the inequivalency of bonding sites or by enabling different bonding angles on the same site. We also observe more ALC resonances in TMS-pentacene than in TIPS-pentacene, despite a similar molecular structure. We conclude that this is also due to crystal packing.

## References

[1] Nuccio L *et al.* 2013 *Phys. Rev. Lett.* **110** 216602
[2] Schulz L *et al.* 2011 *Phys. Rev.* B **84** 085209
[3] Nuccio L *et al.* 2011 *J. Phys.: Conf. Ser.* **292** 012004
[4] Drew A J *et al.* 2009 *Nat. Mat.* **8** 109
[5] Schulz L *et al.* 2011 *Nat. Mat.* **10** 39
[6] Duarte J P *et al.* 2009 *Physica* B **404** 933
[7] Duarte J P *et al.* 2003 *Physica* B **326** 94
[8] Ohira S *et al.* 2000 *Physica* B **289** 123
[9] Patterson B D 1988 *Rev. Mod. Phys.* **60** 69
[10] Kiefl R F *et al.* 1984 *Phys. Rev. Lett.* **53** 90
[11] Blundell S J 2004 *Chem. Rev.* **104** 5717
[12] Roduner E 1993 *Chem. Soc. Rev.* **22** 337
[13] Frigo M and Johnson S G 2005 *Proc. IEEE* **93** 216
[14] Lord J S 2006 *Physica* B **374** 472
[15] Wang K *et al.* to be published

Journal of Physics: Conference Series **551** (2014) 012043 doi:10.1088/1742-6596/551/1/012043

# Muonium response to oxygen content in biological aqueous solutions for cancer research

**A D Pant[1], K Nagamine[2,3,4], I Shiraki[1], E Torikai[1], K Shimomura[4], F L Pratt[5], H Ariga[6], K Ishida[7] and J S Schultz[8]**

[1] Interdisciplinary Graduate School of Medicine and Engineering, University of Yamanashi, 4-3-11 Takeda, Kofu, 400-8511, Japan
[2] Atomic Physics Laboratory, RIKEN, 2-1 Hirosawa, Wako, Saitama 351-0198, Japan
[3] Physics and Astronomy, University of California, Riverside, USA
[4] Muon Science Laboratory, IMSS. KEK, Oho, Tsukuba, Ibaraki, Japan
[5] ISIS, Rutherford Appleton Laboratory, Oxford, UK
[6] Catalysis Research Centre, Hokkaido University, Sapporo, Japan
[7] Advanced Meson Science Laboratory, RIKEN, 2-1Hirosawa, Wako, Japan
[8] Department of Bio-engineering, University of California, Riverside, USA

E-mail: kanetada.nagamine@kek.jp

**Abstract.** Muonium (Mu), which is known to exhibit a characteristic concentration dependent spin relaxation change when impacting molecular oxygen dissolved in water, was found to show a similar behaviour in aqueous solutions of Tris Buffered Saline (TBS), albumin, serum, and hemoglobin (Hb). These effects, along with the interaction of Mu with deoxy-Hb (which is modulated by oxygen by the formation of oxy-Hb) suggest that the muon method can be applied to systematic studies of oxygen dependent effects in biological systems. This muon method may particularly be applicable at low (hypoxic) oxygen levels, an important consideration in the radiation treatment of cancer.

## 1. Introduction

Hypoxia, or low oxygenation, is known as an important factor in tumor biology and the response of tumors to radiation treatment. In cancer patients, an accurate measurement of hypoxia in specific regions may have an important predictive value in the management of treatment and outcome of the disease [1, 2, 3, 4]. The National Cancer Workshop (Tatum, et al 2006 [4]) on hypoxia imaging techniques pointed out the need for improved $O_2$ detection methods for cancer treatment. Here, we propose the use of the polarized positive muon ($\mu^+$) produced at the accelerator facility as a new sensitive method to probe existence of the paramagnetic $O_2$ in the cancer of eventually human body.

There have been trials employing PET, MRI and EPR for this purpose but with major limitations as summarized in the followings [4].

**PET** $^{18}$F-labelled fluoro-misonidazole ($^{18}$F-FMISO) tracer is widely used in PET for monitoring hypoxia. It does not directly measure molecular $O_2$ in the tumor, but the tracer retention affected primarily by $O_2$. It images hypoxic cells and re-oxygenation following radiation therapy, which is usually measured in off-line manner after radiation therapy.

Journal of Physics: Conference Series **551** (2014) 012043 doi:10.1088/1742-6596/551/1/012043

**MRI** Blood oxygen level-dependent (BOLD) MRI does not directly measure $O_2$ molecule in blood but rather detects deoxy-Hb. Also, special attention is required for use of the high magnetic fields for in-situ MRI.

**EPR** Two unpaired electrons in molecular $O_2$ cause spin relaxation of the surrounding paramagnetic constituents. Usually, infusible tracers are used to detect line width broadening due to interactions with molecular $O_2$. Thus, the EPR method is not truly non-invasive.

On the other hand, since late 70s, there have been experimental studies on the effects of dissolved oxygen on the spin relaxation of Mu in pure water [5, 6, 7]. The relaxation rate constant change $\lambda_{Mu}/[O_2]$ of Mu is known to be $(1.8 \pm 0.1) \times 10^{10}$ L mol$^{-1}$ s$^{-1}$ [7]. The origin of this relaxation is ascribed to electron spin exchange interaction with dissolved paramagnetic molecular $O_2$. The sensitivity range to the spin relaxation of a pulsed muon is $10^4$ to $10^7$ s$^{-1}$ for an intense muon beam now available at accelerator laboratories like ISIS (UK) or J-PARC (Japan). Therefore, the sensitivity of the muon spin probes for $pO_2$(Mu) becomes $0.5 \times 10^{-6}$ to $0.5 \times 10^{-3}$ mol L$^{-1}$ based on the experimental data mentioned above.

The solubility of $O_2$ in water at 23 °C is known as $1.29 \times 10^{-5}$ mol L$^{-1}$ kPa$^{-1}$ [8]. Thus, under 1 atmosphere, the oxygen solubility $pO_2$(s.l.) at 23 °C is $1.3 \times 10^{-3}$ mol L$^{-1}$. (Here s.l. is used for solubility limit.) This means that the measurable partial pressure, which is $pO_2$(Mu) divided by $pO_2$(s.l.) becomes $0.4 \times 10^{-3}$ to 0.4. These values can be expressed in units of mmHg so that the oxygen tension in water sensed by the muon spin probe becomes $3.0 \times 10^{-1}$ to $3.0 \times 10^2$ mmHg perfectly matching to the hypoxia range of values.

The proposed Mu relaxation method is able to detect and measure molecular $O_2$ concentration in tissues directly. It is able to monitor non-invasively in a small area of human tissue at any temperature and without any strong magnetic fields. A particularly desirable feature of the present muon method is a capability of non-invasive and sensitive imaging; spatial resolution of the mm level at the depth of up to 20 cm and $O_2$ concentration sensitivity down to a $pO_2$ of 0.1 mmHg.

The problem to be solved before serious application to hypoxia studies is a possible existence of the background signals from other magnetic molecules in human tissues. Magnetic property of Hb in blood [9] is a concern. At the same time chemical reaction between Mu and biological molecules may disturb the expected observations. In the present study, we conducted several test experiments on Mu spin relaxation $\lambda_{Mu}$ against $O_2$ contents $c(O_2\%)$ in aqueous solution of representative biological molecules. Then, based upon the obtained results, we made predictions for the future development of the muon method. Special emphasis was placed on the Hb aqueous solutions, where by increasing $O_2$ concentration, some fraction of $O_2$ goes to change deoxy-Hb (magnetic) to oxy-Hb (non-magnetic), reducing the Mu spin relaxation rate. Careful measurements were made for $\lambda_{Mu}$ against $(c(O_2\%), c(\text{Hb}))$.

## 2. Experimental arrangement

Experiment was conducted at Port 2 of the RIKEN RAL muon facility in UK, by using 60 MeV/c decay-in-flight polarized positive muons.

Once energetic polarized positive muons are injected and stopped in water, it is known for these $\mu^+$ to take electronic states of diamagnetic $\mu^+$ such as $\mu^+$OH with a fraction of 60%, paramagnetic Mu with a fraction of 20% and a missing fraction of 20% [5, 6, 7]. In the Mu fraction a half becomes ortho state with spin 1, providing a spin rotation signal with 100 times faster precession pattern compared to the diamagnetic $\mu^+$. Spin rotation and its relaxation were detected under 2.1 G transverse magnetic field. All the measurements were conducted at room temperature.

The biological aqueous solutions with controlled $O_2$ concentrations were prepared in a separate flask with magnetically driven stirrer (Fig. 1) and continuously transported into the chamber for muon beam exposure by a circulating flow pump. Oxygen concentration in the gas

Journal of Physics: Conference Series **551** (2014) 012043 doi:10.1088/1742-6596/551/1/012043

space of the flask was monitored by a Vernier Gas monitor and in the liquid phase by a NeoFox oxygen sensor placed inside the muon chamber. The oxygen concentrations of all the data are presented by the reading of the NeoFox monitor. In order to achieve a satisfactorily equilibration between aqueous biological samples and $O_2$ in the gas phase in a short time, Silicone Membrane Modules from Perm Select Co. Ltd. was used. When changing the oxygen concentration ($O_2$%) in the gas phase, equilibration occurred within 10 min for the 0.7 litre solution in the flask.

The concentration of $O_2$ was controlled from 0% to 20.7% by changing the ratio of $N_2$ gas and air. As a reference, He gas flow was used to justify use of the $N_2$ gas for zero $O_2$ concentration.

Biological samples adopted in the present experiment are as follows.

**Albumin** Bovine serum (plasma) albumin which has a molecular weight of around 66,000 and is a single polypeptide chain consisting of about 583 amino acid residues and no carbohydrates purchased from from Sigma-Aldrich Co. Ltd.

**Serum** Donor horse Serum purchased from TCS Biosciences Ltd. It is Sterile filtered Serum screened for Mycoplasma and Adventitous viruses.

**Hemoglobin** Polymerized Hb of bovine origin in a lactated Ringers solution at 13% concentration purchased as OXYGLOBIN from BIOPURE Co. Ltd. It is violet-colored and is mostly deoxy-Hb.

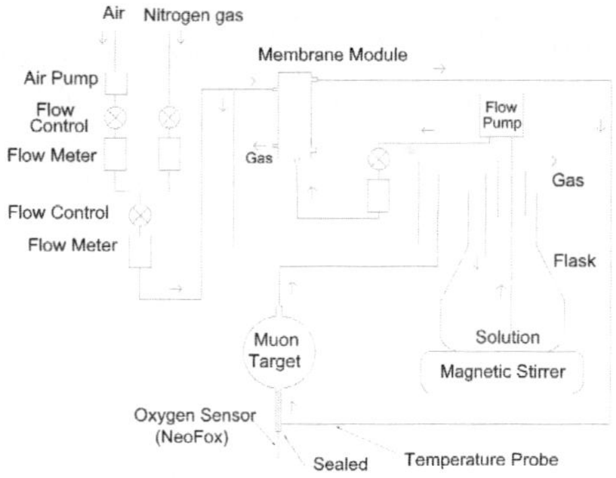


**Figure 1.** Diagram of biological aqueous solution production and supply to the target for the muon spin rotation experiments.

## 3. Experimental results

Typical Mu spin rotation spectrum in pure water at various $O_2$ concentrations are presented in Fig. 2. Superimposed on the spin rotation of the diamagnetic $\mu^+$, a clear Mu spin rotation was observed and it shows a characteristic relaxation rate increase with increasing dissolved $O_2$ concentration. Here the $O_2$ concentration was determined by the NeoFox oxygen monitor that was calibrated for $O_2$ concentration at 0% with $N_2$ gas and at 20.7% with air. Note that in the discussion and figures below the oxygen concentrations are provided as percent $O_2$ (e.g. 10%). These values are the gas compositions that would be in the equilibrium with the actual liquid phase concentrations of free molecular oxygen. Thus, 10% $O_2$ at 23 °C is equivalent to $1.3\times10^{-4}$ mol $L^{-1}$ free molecular oxygen in solution.

The time spectrum is known to be written by the following function,

$$N = N_0 e^{-t/\tau_\mu}\left[1 + A_\mu \cos(\omega_\mu t + \phi_\mu) + A_{Mu} e^{-\lambda_{Mu} t} \cos(\omega_{Mu} t + \phi_{Mu})\right] + B, \tag{1}$$

where $N_0$ is a normalization factor, $B$ is the time-independent background and $\tau_\mu$ is the muon lifetime (2.20 $\mu$s). The terms $A_\mu$ and $A_{Mu}$ are the amplitudes of the spin precession corresponding to the polarization asymmetry for the $\mu^+$ in diamagnetic states and in Mu, respectively. The parameter $\lambda_{Mu}$ is the muonium relaxation rate while the relaxation rate of $\mu^+$ in diamagnetic species is assumed to be negligible, $\omega_\mu$ and $\omega_{Mu}$ are the muon and Mu precession frequencies, $\phi_\mu$ and $\phi_{Mu}$ are the respective initial phases of their precessions. Under transverse field of $H$(G), the spins of $\mu^+$ and Mu take precession with angular velocity of $\omega_\mu$(kHz) $= 2\pi \times 13.553 \times H$(G) and $\omega_{Mu} = 2\pi \times 1390 \times H$(G), respectively. As seen in Fig. 2, the spin precession of the Mu showed faster relaxation with increasing $O_2$ concentration. The data is very consistent with the published data [7].

At first, 7.5 g Tris Buffered Saline (TBS) salt, which is a buffer used in some biochemical techniques to maintain the pH within a relatively narrow range was added to 1 litre of pure water. The Mu spin rotation signal was found to be essentially unchanged. Also, for this TBS solution, Mu was found to show a similar relaxation change with increasing $O_2$ concentration as that for pure water.

Then, 0.4 g albumin was introduced in 1 litre water with 7.5 g TBS buffer and the solution was equilibrated with 10% $O_2$ in the gas phase. Albumin was selected as a typical representative of a biological protein in human tissues. By increasing albumin quantity to 4 g, the Mu signal was found to disappear. In order to observe more clearly, Mu relaxation change with $O_2$ concentration in albumin aqueous solution, systematic measurements were done without TBS. The measurement was extended to 0.5 wt.% serum aqueous solutions. The Mu was found to take a similar relaxation pattern with increasing $O_2$ concentration as previously observed for pure water.

As shown in Fig. 3, before measuring the $O_2$ dependence of the $\lambda_{Mu}$, its dependence on the concentration of each biological molecule was systematically measured with less than 1% $O_2$ in the gas phase. The increasing rate of $\lambda_{Mu}$ with increasing protein concentration was obtained as 1 $\mu s^{-1}/(g\ L^{-1})$ for albumin, 1 $\mu s^{-1}/(g\ L^{-1})$ for serum and 3.1 $\mu s^{-1}/(g\ L^{-1})$ for Hb up to 2.0 g $L^{-1}$. Then the measurements were extended to $O_2$ dependence in 0.05 wt.% Hb aqueous solution (Fig. 4). Here, because of strong $O_2$ absorption by deoxy-Hb to change into oxy-Hb, it was expected that the Mu relaxation pattern would show a reduced response against $O_2$ increase in comparison with pure water and other protein aqueous solutions.

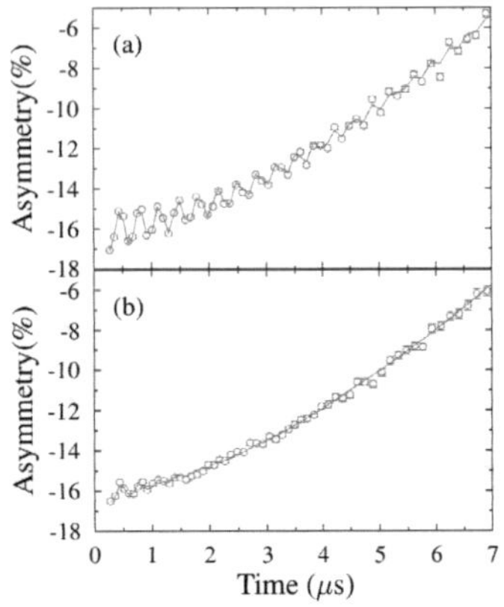


**Figure 2.** Muon spin rotation time spectrum in water with $O_2$ concentration (a) below 1% and (b) 7.5%, at room temperature under 2.1 G transverse field. In most of the measurements on other biological aqueous solutions, almost similar $O_2$ dependence was obtained.

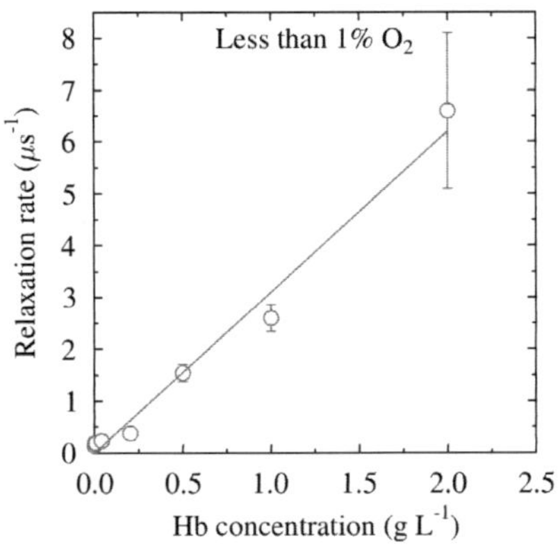


**Figure 3.** Relaxation rate of Mu against Hb concentration in water at less than 1% $O_2$ concentration.

The result of $O_2$ concentration dependence for 0.5 g $L^{-1}$ Hb aqueous solution is shown in Fig. 4. Mu was found to take almost similar relaxation pattern with increasing $O_2$ concentration as that for pure water. Similar results were obtained for the other protein aqueous solutions. The following points were found to be evident.

1. In albumin solution, Mu relaxation takes larger increase against increasing $O_2$ concentration. The result suggests, in the presence of albumin, interaction between Mu and $O_2$ becomes more active in higher concentration.

2. In Hb solution, Mu relaxation increases with increasing $O_2$ concentration is weakened. This weakened response may be due to an influence of decreasing deoxy-Hb concentration with increasing dissolved molecular $O_2$ concentration.

## 4. Discussion

Before clinical application of the muon method for measuring hypoxia, it is important to conduct systematic studies on the response to $O_2$ concentrations in various biological aqueous systems. The response to $O_2$ levels in Hb aqueous solution is quite different than other proteins. Magnetic deoxy-Hb, by absorbing $O_2$, is converted to non-magnetic oxy-Hb so that an increase of $O_2$ may decrease the relaxation of Mu. This situation will be discussed separately in the followings.

### *4.1. Prediction of Mu response to $O_2$ in Hb aqueous solutions*

For a qualitative understandings, let us assume the experimental data of Mu relaxation rate $\lambda_{Mu}$ in aqueous solution of various Hb concentration and $O_2$ concentrations be approximated way as,

$$\lambda_{Mu} = R_{Hb}(Mu) + R_{O_2}(Mu). \tag{2}$$

The $R_{Hb}$(Mu) is relaxation rate of Mu due to deoxy-Hb in solution, which is $k_1(\text{Hb}) \times m_{deoxy-Hb}$.There, $k_1$(Hb) (3.1 $\mu s^{-1}/(g\ L^{-1})$) is the same as experimental data of the best fit of relaxation vs Hb concentration. Here $m_{deoxy-Hb}$ (amount of deoxy-Hb in g $L^{-1}$) is varied by an introduction of $O_2$ in aqueous solution, which is calculated using Hill's equation [10]. It predicts deoxy-Hb fraction (deoxy-Hb/total) $= k^n/(p^n+k^n)$, where $p$ is $O_2$ partial pressure in mmHg, and $n = 2.62$ and $k = 13$ at 25 °C. ($p$ in percentage is $100 \times p/760$ %).

The $R_{O_2}$(Mu) is relaxation rate of Mu due to dissolved free molecular $O_2$ in solution. Dissolved free molecular $O_2$ in solution at $p$ in g $L^{-1}$, $c(O_2) = p \times 8.3 \times 10^{-3}/159$ (the dissolved $O_2$ in water in equilibrium with air is $8.3 \times 10^{-3}$ g $L^{-1}$, which is 159 mmHg) and the oxygen molecules bound to $m_{oxy-Hb}$ in g $L^{-1}$, $(n_{Hb}) = 4\times\ m_{oxy-Hb} \times (w_{O_2}/w_{Hb})$, where $m_{oxy-Hb}$ is amount of oxy-Hb in g $L^{-1}$, and $w_{Hb}$ and $w_{O_2}$ are molecular weight of Hb (=68000 g/mol) and $O_2$ (=32 g/mol), respectively. From the best fit line of experimental data of relaxation rate of Mu vs oxygen concentration (%) in pure water, $R_{O_2}(\text{Mu})\ (\mu s^{-1}) = \alpha_{water} \times c(O_2\%) + \beta_{water}$, where $\alpha_{water} = 0.214(40)\ \mu s^{-1}/(O_2\%)$ and $\beta_{water} = 0.385(363)\ \mu s^{-1}$, which is slightly higher than that in the more purified water [6]. Predicted result for 0.5 g $L^{-1}$ Hb is shown in Fig. 4 (blue line), where a refined experiment is under planning. The total $O_2$ in the solution (g $L^{-1}$) is $c(O_2) + n_{Hb}$.

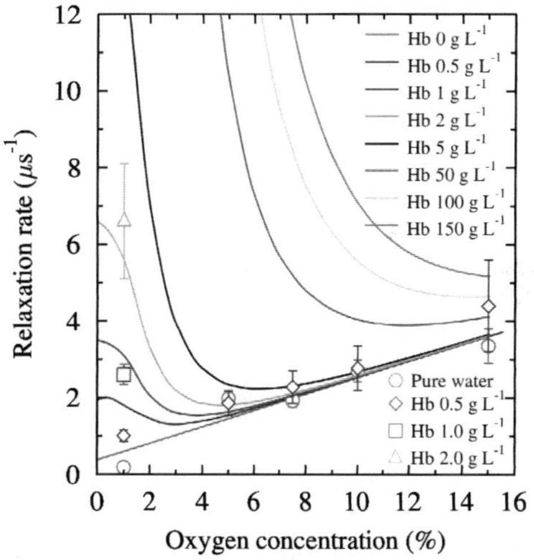


**Figure 4.** The Mu relaxation rates in higher Hb concentration aqueous solution and various $O_2$ concentrations predicted by Eq. (2)(solid lines). As seen in this figure, Mu relaxation rates become observable (below 10 $\mu s^{-1}$) at $O_2$ concentration lower than 10% for the Hb range below 150 g $L^{-1}$. The experimental data (open marks) at less than 1% $O_2$ concentration, zero Hb and 0.5 g $L^{-1}$ Hb are also shown as a reference.

As discussed above, we have estimated the behaviour of Mu at different Hb concentration (upto Hb concentration in human body) as shown in Fig. 5. The observed data at less than 1% $O_2$ concentration, zero Hb and at 0.5 g $L^{-1}$ Hb concentration are also presented in the figure. As shown in Fig. 4, the relaxation rate of Mu increases with increasing Hb at any fixed $O_2$ concentration with slower increasing rate at Hb concentration higher than g $L^{-1}$. It can be predicted that Mu will show undetectable fast relaxation at $O_2$ concentration lower than 5% and at higher Hb concentration around 100 g $L^{-1}$ expected for human body.

### *4.2. Prediction of Mu response to $O_2$ in other biological aqueous solutions*
So far, other than Hb, the Mu relaxation responses to $O_2$ were done for limited cases such as albumin, serum and TBS buffer materials. There should be extended measurements before clinical application.

## 5. Conclusion and future perspectives
Muonium spin rotation signal was directly observed in TBS, albumin, serum and Hb aqueous solution and consistent $O_2$ concentration dependence in spin relaxation was obtained. The result is encouraging to apply the present method to a wide variety of the biological systems including human tissues. Definitely, measurements should be extended to more cases before clinical application. Human blood is composed of cells and plasma, where Hb, albumin and serum are the major components. There, the most important component should be Hb aqueous solution, which was successfully studied in the present study. As indicated in Fig. 4, the most of the Hb aqueous solution, namely, blood in human tissues, deficiency of $O_2$ concentration related to hypoxia can be monitored. Actually, in human tissues, the highest value of $O_2$ concentration is 15% at lung.

The following advantages should be emphasized for the present Mu method; (1) no need of high magnetic field like MRI, (2) no need of injection of radioactive chemicals and (3) increase of spatial resolution of the monitoring spots to 100 $\mu$m at 10 cm depth by employing an advanced muon beam [11, 12].

## Acknowledgments
This work was supported by Grant-in-Aid for Scientific Research on Innovative Areas "Ultra Slow Muon", MEXT KAKENHI Grant Number 23108003 of Japan. We acknowledge Mrs P.M. Odenthal and Ali Haiden for collaboration at the earliest phase of the project.

## References
[1] Mancini D M, Wilson J R, Bolinger L, Li H, Kendrick K, Chance B and Leigh J S 1994 *Circulation* **90** 500
[2] Raleigh J A, Dewhirst M W and Thrall D E 1996 *Semin. Radiat. Oncol.* **6** 37
[3] Kavanagh M C, Tsang V, Chow S, Koch C, Hedley D, Minkin S and Hill R P 1999 *Int. J. Radiat. Oncol. Biol. Phys.* **44** 1137
[4] Tatum J L *et al.* 2006 *Int. J. Radiat. Biol.* **82** 699
[5] Jean Y C, Fleming D G, Ng B W and Walker D C 1979 *Chem. Phys. Lett.* **66** 187
[6] Nagamine K, Nishiyama R, Imazato J, Nakayama H, Yoshida M, Sakai Y, Sato H and Tominaga T 1982 *Chem. Phys. Lett.* **87** 186
[7] Roduner E, Tregenna-Piggott P L W, Dilger H, Ehrensberger K and Senba M 1995 *J. Chem. Soc. Faraday Trans.* **91** 1935
[8] Dean J A 1999 *Lange's Handbook of Chemistry* (New York: McGraw-Hill) 15th Editon pp. 5.6
[9] Pauling L and Coryell C D 1936 *Proc. Nat. Acad. Sci.* **22** 210
[10] Voet D and Voet J 2004 *Biochemistry* (New York: John Wiley and Sons)
[11] Miyadera H *et al.* 2007 *Proceedings of PAC07* 3032
[12] Nagamine K 2014 *JPS Conf. Proc.* **2** 010001

# Interaction of the Mu-cyclohexadienyl radical with metallic (Au, Pt) nanoparticles in mesoporous silica

**J Xiao[1,2], D J Arseneau[2], M D Bridges[3], D Cortie[1,2,4], S P Cottrell[5], M Dehn[6], D G Fleming[1,2], J Kelly[1], R F Kiefl[2,4], W A MacFarlane[1,2,4], M MacLachlan[1] and I McKenzie[2]**

[1] Department of Chemistry, University of British Columbia, Vancouver, BC, V6T 1Z1, Canada
[2] TRIUMF CMMS, 4004 Wesbrook Mall, Vancouver, BC, V6T 2A3, Canada
[3] Department of Chemistry and Biochemistry, California State University, Fullerton, CA 92831, USA
[4] Department of Physics and Astronomy, University of British Columbia, Vancouver, BC, V6T 1Z1, Canada
[5] Rutherford Appleton Laboratory, Swindon, SN2 1SZ, UK
[6] Physik-Department, Technische Universität München, James-Franck-Strasse 1, 85748 Garching bei München, Germany

E-mail: jxiao@triumf.ca;flem@triumf.ca

**Abstract.** $\mu$SR and ALCR techniques have been used to investigate the structure and dynamics of the Mu-cyclohexadienyl radical interacting with Au and Pt metal nanoparticles (MNPs) supported in mesoporous silica (SBA-15). Surprisingly, coherent precession signals are observed and the isotropic hyperfine coupling constants are almost the same in loaded and unloaded samples, implying that the electronic structure of $MuC_6H_6$ is only weakly perturbed by the presence of the MNPs. We propose the observed radicals are shielded from the metallic surfaces by a benzene coating on the MNPs. The $\Delta_1$ resonance is observable in MNP-loaded samples at higher temperatures than in the unloaded SBA-15. This is attributed to stronger binding of $MuC_6H_6$ to the benzene coated MNPs.

## 1. Introduction

In recent years there has been significant interest in the study of metallic nanoparticles (MNPs), which are seen as potential "green" catalysts compared with traditional oxidants in industrial usage. However, it is difficult to monitor and detect chemical reaction products in such a catalytic environment, particularly when free radicals are involved. Muoniated free radicals are relatively easy to study in bulk liquids compared to the protonated versions but the method is typically not useful on surfaces. However, conventional energetic muon beams can be used to probe muonium and its reaction products on surfaces when the specific surface area of the sample is sufficiently large as is the case with MNPs.

The highly reactive MNPs are typically stabilized with protective coatings made of citrates [1] or thiols [2] to prevent their agglomeration. However, for catalysis it is often desireable that the MNPs be uncapped so that guest molecules can bind directly to the MNP surface. In this case the bare MNPs must be supported in inert substrates such as zeolites [3], and silica [4] to prevent agglomeration.

In recent years we have been pursuing the study of muoniated free radicals in faujasitic zeolites [5,6], NaY, HY and USY, which are silica-like substrates that have been used extensively in the petrochemical industry as molecular sieves and heterogeneous catalysts [7]. The USY environment closely resembles that of the present study in mesoporous (SBA-15) silica, both having multi-channel pore structures, although the pore structures in SBA-15 are about ten times larger. This motivated a recent study using mesoporous silica as a reaction substrate [8] and is also one of the main motivations for the present study of $MuC_6H_6$ interacting with MNPs in mesoporous silica.

## 2. Sample Preparation

### *2.1. Bare mesoporous silica synthesis*

SBA-15 was chosen for this experiment because of its more stable channel structure compared to other mesoporous silicas. We used a hydro-thermal method developed by Somorjai *et al.* [9]. Typically, 6.25 g Pluronic P-123 polymer surfactant was dissolved in 127 mL of deionized (DI) water and stirred at 40 °C for about 15 min. Then 0.99 mL of 0.5 M NaF, followed by 9.7 mL of tetramethyl orthosilicate (TMOS) as a silica source was added, resulting in a white slurry. After 3 days aging the sample was washed with ethanol, dried at 100 °C for 24 h and calcined in steps of 5 °C /min up to 500 °C for 5 hours to remove any organic template. The final product was 3.0g of white fluffy powder.

### *2.2. Au and Pt Nanoparticle Synthesis*

Both Au and Pt MNPs were synthesized using a citrate method. Typically, 3.0 g of sodium citrate followed by 6.2 g of polyvinyl pyrrolidone (PVP) was added to 2.0 L of previously boiled DI water. In the case of the Au NPs, 197.0 mg of chloroauric acid ($HAuCl_4{\cdot}3H_2O$) were added. The dark purple colloidal solution was kept boiling for 15 min and then condensed to a 100mL using a rotavapor at 80 C, and then filled into a dialysis bag for 2 days to remove the polymer. Using this method, 6 nm Au NPs (determined from powder XRD) were obtained. For the larger Au NPs (~30 nm), the same procedure was followed with minor changes in the order and amounts. The Pt NPs were obtained by the same method, but starting with the Pt salt (Chloroplatinic acid hexahydrate, $H_2PtCl_6{\cdot}6H_2O$), to give sizes of ~ 7 nm ("small") and 10 nm ("large").

SBA-15 samples loaded with MNPs were synthesized using the same methods as for the bare SBA-15 samples. Colloidal solutions are mixed with the Pluronic P-123 polymer, followed by a waiting period for nanoparticle-polymer stabilization. About 3 g of purple powder was obtained for the Au NPs, with a similar amount grey powder for the Pt-NP loaded mesoporous silica samples. Both MNP loaded samples had large total surface areas, determined from Brunauer-Emmett-Teller (BET) nitrogen absorption analysis, to be in the range ~ 650 $m^2/g$ to 800 $m^2/g$. The average pore size was ~ 10 nm, except for the small (~ 6nm) Au-loaded samples which had pore sizes of ~ 4 nm.

## 3. Results and Discussion

### *3.1. Weak TF-μSR of Muonium in Unloaded SBA-15 Silica*

It is reasonable to expect that muons injected into SBA-15 will form Mu with a high probability and that the Mu will diffuse into the pores on a short time scale, based on previous results on fine silica powders with similar specific surface area (400 $m^2/g$). In that case this behaviour was established by adding paramagnetic $O_2$, which rapidly depolarizes any Mu reaching the free space between grains via electron spin exchange (SE) collisions [10]. The same technique was applied here and the results are shown in Fig. 1. The top panel shows there is a large amplitude Mu precession signal in a sample that had been baked and pumped to remove oxygen and occluded water. The bottom panel shows that the Mu signal depolarizes rapidly in a sample that was not baked or pumped to remove $O_2$. This confirms that nearly all the Mu in SBA-15 reaches the pores. Note the fitted relaxation rate of the Mu signal in the baked sample $\sim 3\ \mu s^{-1}$ is larger than expected from previous studies of Mu interacting with bare silica surfaces [11,12]. More details on the interaction of Mu with silica surfaces are presented in a separate paper at this conference (see M.H. Dehn et al).

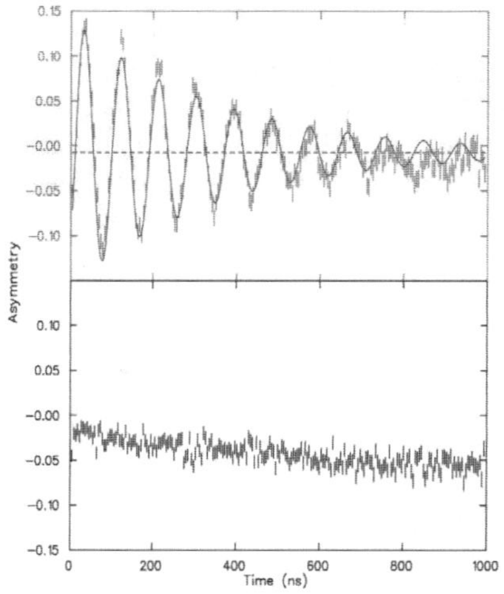


**Figure 1.** Mu precession in a weak TF of 8 G for the SBA-15 powder baked and pumped overnight (top), compared with the same powder but unbaked and unpumped, thus containing both water and oxygen (bottom). Note the Mu precession signal is completely wiped out. This is attributed to rapid spin exchange of Mu in the pores with $O_2$ in the gas phase.

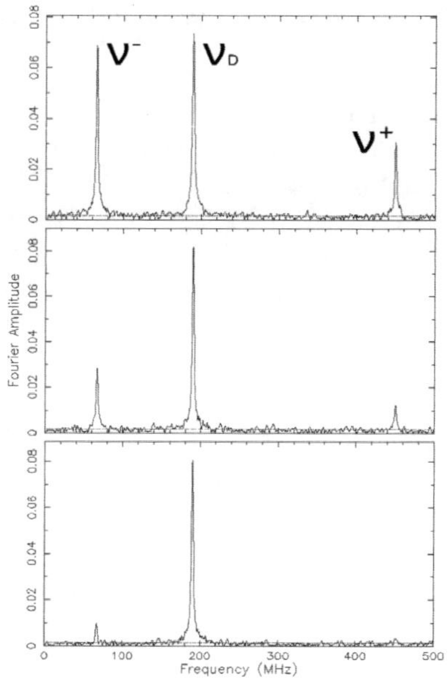


**Figure 2.** Frequency spectra in pure Bz (top) , SBA-15+Bz (middle) and Au loaded SBA-15+Bz (bottom) . In all cases the fourier transform was taken over a time interval from 0.01 to 0.5 microseconds. The relaxation times are all longer than this time so the fourier amplitude is a good measure the initial asymmetry. All the spectra were taken at 300K in a magnetic field of 14 kG. The frequencies labeled $\nu^{\pm}$ are from the $MuC_6H_6$ radical. The large signal labelled $\nu_D$ at 190 MHz is due to muons is a diamagnetic environment.

### *3.2. High Transverse Field $\mu SR$ of the $MuC_6H_6$ Radical*

The $MuC_6H_6$ radicals were formed by adding benzene (Bz) from the vapor phase to the bare mesoporous (SBA-15) silica and also to Au and Pt MNP- loaded samples. We expect that the Mu which reaches the pores will quickly add to the benzene to form the $MuC_6H_6$ radical. This was confirmed by both TF-$\mu$SR and ALC studies. The amount of benzene introduced was about $9 \times 10^{-5}$ moles which is much less than required to cover the silica surface but about 40 times more than required to coat the MNPs with a monolayer. The radical was detected with TF-$\mu$SR in a high magnetic field of 12 kG which is sufficient to decouple the muon-electron hyperfine interaction yielding two well defined muon spin transitions frequencies $\nu^{\pm} = \gamma_\mu B \pm A_\mu/2$ [13] independent of the spin state of the protons. Typical fourier transforms of the TF-$\mu$SR data are shown in Fig. 2. The top shows pure benzene (Bz) for comparison, the middle is the SBA-15+Bz and the bottom is the Au-loaded SBA-15+Bz. The observed signals originate from radicals which form on a time scale of 100 ps or less, assuming that Mu is the precursor to the observed radical. This can be deduced from the fact that in order for the polarization to be transferred coherently from Mu to $MuC_6H_6$ the reaction rate $\lambda_r$ must be faster than the difference in angular frequency between the initial and final state $2\pi(\nu^{\pm}_{\rm Mu} - \nu^{\pm}_{\rm Rad})$ [14]. However, the observed radical signal in SBA-15+Bz is less than seen in pure Bz even though the diamagnetic signal amplitudes are about the same. This suggests that some fraction of the Mu in SBA-15+Bz reacts on a slower time scale and therefore does not contribute to the TF-$\mu$SR signal. Finally we note that the signal amplitude in the Au loaded silica+Bz is less than in the bare SBA-15+Bz. It is interesting to note that the ALC spectra (see below) indicate the radical amplitudes are almost the same in the loaded and unloaded samples, consistent with the reduction seen in Fig. 2 being due to a slightly delayed reaction rate, since this would affect the TF-$\mu$SR amplitude but not the ALC spectra.

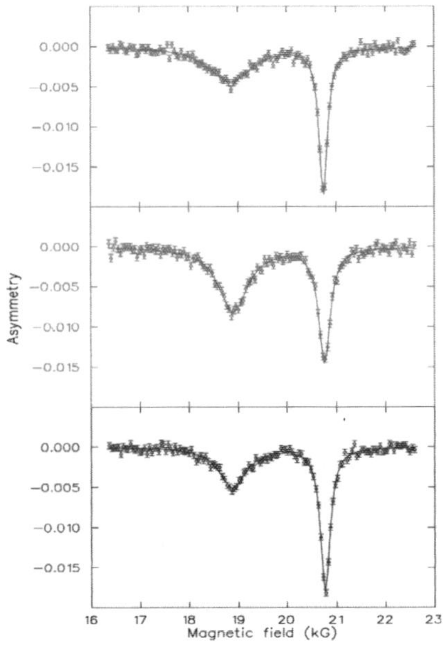


**Figure 3.** ALC resonance at 323 K for the bare silica (top), the Au-loaded (middle) and the Pt-loaded (bottom). The solid lines are Lorentzian fits

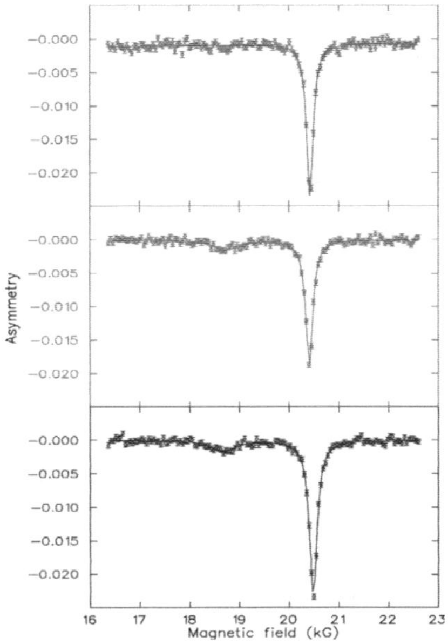


**Figure 4.** As in Fig. 3 but at 418 K (Pt scan at 398 K). Note that only the $\Delta_1$ resonance for the silica has essentially disappeared into the baseline.

*3.3. Avoided Level Crossing Resonance of the $MuC_6H_6$ Radical*

Level crossing resonance or avoided level crossing resonance (ALCR) is a powerful method to investigate electronic structure since it allows one to determine both the muon and all the proton hyperfine constants from the positions of the level-crossing resonances [15,16]. Figs. 3 and 4 shows typical ALC spectra taken on the bare SBA-15 +Bz and the Au and Pt-loaded SBA-15 +Bz. The $\Delta_0$ resonance occurs when the muon transition frequency is matched to

Journal of Physics: Conference Series **551** (2014) 012044 doi:10.1088/1742-6596/551/1/012044

that of the methylene proton next to the muon. Under these conditions there is mixing of the muon-proton-electron hyperfine levels leading to a resonant loss of muon polarization near that particular magnetic field. Note the resonance positions are almost the same in all three samples and shifted only slightly from pure benzene. This confirms the results obtained from TF-$\mu$SR that the electronic structure of $MuC_6H_6$ in the Au and Pt loaded samples is almost the same as for the unloaded SBA-15+Bz. It is clear the observed radicals cannot be in direct contact with a metallic surface since then one would expect a strong exchange interaction between the unpaired electron of the radical and the metallic electrons, which in turn would have dramatic effects on the structure and electron spin dynamics. This expectation is consistent with the fact that there is no EPR signal for radicals absorbed directly onto a metal surface [17]. In this sense our results are very puzzling. The most likely explanation is that the benzene preferentially coats the MNPs and shields the $MuC_6H_6$ radicals from direct contact with the metal surfaces. This is reasonable since the amount of benzene introduced was about 40 times that need to coat the MNP surfaces and there is a strong binding energy for multiple Bz layers on metal surfaces [18,19].

Although there are no significant changes in the electronic structure induced by the MNPs, there are differences in the ALC spectra between the Au and Pt loaded SBA-15+Bz and pure SBA-15+Bz which indicate there is some interaction between the $MuC_6H_6$ and the MNPs. This is most noticeable for the $\Delta_1$ resonances near 19 kG in Figs. 3 and 4. This resonance occurs at a magnetic field where the $\nu^-$ transition frequency goes to zero and the applied field exactly cancels the longitudinal component of the muon hyperfine field. This resonance is especially sensitive to the reorientation dynamics of the radical since it is directly sensitive to the muon hyperfine anisotropy (without any influence from the proton-electron hyperfine interactions.) For example in a gas or liquid, or in the pores of mesoporous silica, the $\Delta_1$ resonance is unobservable due to the rapid reorientation dynamics which severely broadens it. On the other hand if the radical is confined to a surface, such that the reorientation dynamics are impeded, this will give rise to transverse hyperfine fields which result in a decrease in the time averaged muon polarization on resonance.

The observation of the $\Delta_1$ resonance implies that the $MuC_6H_6$ radical is not freely rotating within the SBA-15 nanopores, at least not on a time scale much faster than the inverse of the hyperfine anisotropy [6]. All three samples show similar positions for the $\Delta_1$ (near 19 kG) and $\Delta_0$ (near 21 kG) resonances but the Au and Pt loaded samples exhibit a weaker temperature dependence than the pure silica+Bz sample. In all three samples the $\Delta_1$ resonance broadens and decreases in amplitude as the temperature increases but the effect is more pronounced for the bare silica where it becomes too broad to detect above $\sim$ 350 K. Note also the width of the $\Delta_0$ resonance narrows at high temperature as expected from rapid motion on the surface which averages out any proton and muon hyperfine anisotropy and leads to narrow resonances, similar to that for $MuC_6H_6$ in liquid benzene. It is evident that the binding energy of the $MuC_6H_6$ radical to the Bz coated MNPs is greater than the bare silica surface. This can be seen from the spectra shown in Fig. 4, where the amplitude of the $\Delta_1$ resonance in the unloaded SBA-15 is invisible at 418K, whereas in the Bz loaded samples the $\Delta_1$ resonance persists up to this highest measurement.

## 4. Conclusions

We have used both TF-$\mu$SR and ALCR to study the $MuC_6H_6$ interacting with surfaces in SBA-15 silica loaded with Au and Pt nanoparticles. Surprisingly, coherent precession signals are observed in Au and Pt loaded SBA-15+Bz. Furthermore the electron structure of the radicals, as determined from the muon and proton hypefine interactions, are only weakly perturbed by the presence of the Au and Pt nanoparticles. The most significant effect of the nanoparticles is in the dynamics, which suggest that the $MuC_6H_6$ radical is more strongly bound to surfaces in SBA-15 loaded with MNP than in the unloaded SBA-15. These results suggest that the observed $MuC_6H_6$ radical is protected from the bare metal surface of the MNP by several layers of benzene. Additional studies at different loading of Bz and MNPs are in progress which may help understand these intriguing observations. Nevertheless it is already clear that $\mu$SR and ALC are unique and powerful methods for investigating the interaction of muonium and of muoniated radicals in these catalytic environments.

**Acknowledgments**
We would like to thank Bassam Hitti and Deepak Vyas for their excellent level of technical support.

## References

[1] Zhao P, Li N and Astruc D 2013 *Coordination Chemistry Reviews* **257** 638–65
[2] Balasubramanian R, Kim B, Trpp S L, Wang X, Liebermnan M and Wei A 2002 *Langmuir* **18** 3676–81
[3] Zhang X, Ke X and Zhu H 2012 *Chem. Eur. J.* **18** 8048–56
[4] Kantcheva M and Sayan S 1999 *Catalysis Letters* **60** 27–38
[5] Fleming D G, Arseneau D J, Bridges M D, Chen Y K and Wang Y A 2013 *J. Phys. Chem.* C **117** 16523–39
[6] Fleming D G, Arseneau D J, Shelley M Y, Beck B, Dilger H and Roduner E 2011 *J. Phys. Chem.* C **115** 11177–91
[7] Haw J F 2002 *Phys. Chem. Chem. Phys.* **4** 5431
[8] Scholz J, Walter A and Ressler T 2014 *Journal of Catalysis* **309** 105–14
[9] Song H, Rioux R M, Hoefelmeyer J D, Komor R, Niesz K, Grass M, Yang P and Somorjai G A 2006 *J. Am. Chem. Soc.* **128** 3027–37
[10] Marshall G M, Warren J B, Garner D M, Clark G S, Brewer J H and Fleming D G 1978 *Phys. Lett.* A **65** 351
[11] Kiefl R F, Warren J B, Oram C J, Marshall G M, Brewer J H, Harshman D R and Clawson C W 1982 *Phys. Rev.* B **26** 2432-2441
[12] Harshman D R, Keitel R, Senba M, Ansaldo E J, Brewer J H and Kiefl R F 1984 *Phys. Lett.* A **104** 472–476
[13] Roduner M and Fischer H 1981 *Chem. Phys.* **54** 261
[14] Meier P F 1982 *Phys. Rev.* A **25** 1287
[15] Kiefl R F, Kreitzman S R, Celio M, Keitel R, Luke G M, Brewer J H, Noakes D R, Percival P W, Matsuzaki T and Nishiyama K 1986 *Phys. Rev.* A **34** 681–684.
[16] Heming M, Roduner M, Patterson B D, Odermatt W,Schneider J, Baumeler H, Keller 1986 *Chem. Phys. Lett.* **128** 100.
[17] Zhang Z, Berg A, Levanon H, Fessenden R W, Meisel D 2003 *J. Am. Chem. Soc.* **125** 7959–63
[18] Liu W, Carrasco J, Santra B, Michaelides A, Scheffler M and Tkatchenko A 2012 *Phys. Rev.* B **86** 245405
[19] Wellendorff J, Kelkkanen A, Mortensen J J, Lundqvist B I and Bligaard T 2010 *Top Catal.* **53** 378–83
[20] Schwager M, Dilger H, Roduner E, Reid I D, Percival P W and Baiker A 1994 *Chemical Physics* **189** 697–712

Journal of Physics: Conference Series **551** (2014) 012045 doi:10.1088/1742-6596/551/1/012045

# Muonium states in $Cu_2ZnSnS_4$ solar cell material

**H V Alberto[1], R C Vilão[1], J M Gil[1], J Piroto Duarte[1], R B L Vieira[1], A Weidinger[1], J P Leitão [2], A F da Cunha[2], M G Sousa[2], J P Teixeira[2], P A Fernandes[2], P M P Salomé[3], K Timmo[4], M Loorits[4], A Amato[5], H Luetkens[5], T Prokscha[5], A Suter[5] and Z Salman[5]**

[1] CEMDRX, Department of Physics, University of Coimbra, R. Larga, P-3004-516 Coimbra, Portugal

[2] I3N and Department of Physics, University of Aveiro, Aveiro, Portugal

[3] International Iberian Nanotechnology Laboratory, 4715-330 Braga, Portugal

[4] Department of Materials Science, Tallinn University of Technology, Ehitajatetee 5, Tallinn 19086, Estonia

[5] Laboratory for Muon Spin Spectroscopy, Paul Scherrer Institut, CH-5232 Villigen PSI, Switzerland

E-mail: lena@uc.pt

**Abstract.** We investigated bulk and thin-film samples of the quaternary p-type semiconductor $Cu_2ZnSnS_4$ (CZTS) by $\mu$SR, in order to characterize the existing muonium signals. We find that the majority of the implanted muons form a diamagnetic state broadened by an interaction with the Cu nuclear moments, which we interpret as $Mu^+$ bound to sulphur. A paramagnetic fraction is also present at low temperatures and the ratio between the two muon charge states, $Mu^+$ and $Mu^0$, varies between 20 and 40% prior to the onset of muon diffusion, which occurs at around 150 K. The fraction of $Mu^0$ is found to be sensitive to the defect content of the sample. The paramagnetic fraction has two different contributions and their origin is discussed and related with the muon role as a probe for charge carriers in the material.

## 1. Introduction

$Cu_2ZnSnS_4$ (CZTS) is a p-type semiconductor based on earth-abundant elements and with a band gap energy of 1.5 eV, which is near the optimum value for application in solar cells. The low cost and low environment impact of its component elements make it an advantageous solution for solar cells. Power conversion efficiency of solar cells based on CZTS has reached 8.4% [1] and was observed to surpass 12% when incorporating Se [2], but these values are still significantly below the aimed 15-20% efficiency range needed to be a serious competitor in thin film solar cell technology. Further progress depends on a deeper understanding of the electrical properties of the devices p-n junctions. The non-stoichiometric compositions required for the increase of the solar cell efficiency have a huge influence on the density of defects. The edges of both valence and conduction bands are affected by these changes and, consequently, also the radiative and non-radiative channels [3]. This research effort would greatly benefit from additional information on the local charge carrier dynamics.

Structurally similar I-III-$VI_2$ solar cell materials such as $CuInSe_2$ and $CuInS_2$ were studied by $\mu$SR by our group [4, 5, 6, 7] and it was found that the major fraction of the muons are in a diamagnetic state, most likely $Mu^+$, bound to anion S or Se. At low temperatures, a small portion of the signal (5 to 10%) is missing. Repolarization experiments showed that this missing fraction is due to a paramagnetic center $Mu^0$ with a hyperfine interaction consistent with neutral muonium in a interstitial position [8, 9]. Linewidth analysis suggested that $Mu^+$ is at an anti-bonding position close to S (or Se) [4]. Muonium is a pseudo-isotope of hydrogen and $\mu$SR results should be compared with theoretical calculations for the isolated hydrogen. First-principles total-energy calculation for hydrogen impurities in $CuInSe_2$ [10, 11] concluded that $H^+$ takes up the Cu-Se bond center position, whereas $H^0$ and $H^-$ take up the tetrahedral interstitial site next to In. Hydrogen creates a negative-U center in these compounds, i.e., the neutral state is never stable and in a p-type material it is expected to be in $H^+$ charge state.

Our goal in this work is to characterize and compare the muonium states formed in CZTS both in bulk and in film and discuss the information provided by the muon probe in the material. We aim not only to provide information on the hydrogen impurity in these material, but also to test the sensitivity of the muon to the properties of this new solar cell material that might be used in future studies of solar cell junctions.

## 2. Experimental Details

We have investigated bulk CZTS and two CZTS thin-film samples, named SF and RTP24. Photoluminescence measurements have shown a broader emission, with a much lower relative intensity, for sample SF when compared to the RTP24. This result suggests the presence of a higher number of different types of defects and also a higher total concentration of defects in sample SF.

Transverse field $\mu$SR experiments at a field of $B = 0.01$ T were performed at the Paul-Scherrer Institute, in Switzerland, at GPS and LEM[13, 12] facilities, for, respectively, bulk and thin films.

Figure 1 shows a typical time spectrum for the bulk CZTS material at low temperatures, indicating the presence of a small, fast relaxing component in addition to the main diamagnetic signal. The same general features were also observed in the time spectra for the CZTS thin films.

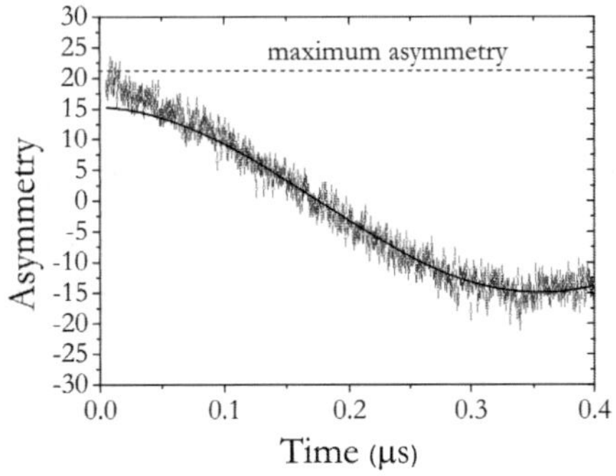


**Figure 1.** Time spectrum of bulk CZTS at T = 5 K, in transverse geometry (B = 0.01 T). The solid line is a one-component fit with a Gaussian-damped cosine function. The deviation at initial times indicates the presence of a fast relaxing component. A missing fraction is also apparent.

All the asymmetry data were analysed with Wimda software [14] considering a two-component fit, where the main component is assumed to have a Gaussian depolarization function and the second component is described with a Lorentzian envelop:

$$A(t) = A_{dia}\, e^{-\frac{1}{2}\sigma^2 t^2} \cos(\omega t + \phi) + A_{fast} e^{-\lambda t} \cos(\omega t + \phi) \qquad (1)$$

where $\sigma$ and $\lambda$ are the Gaussian and Lorentzian depolarization rates, respectively. The frequency $\omega$ and the phase $\phi$ were assumed to be the same in both components.

In dynamic situations where the muon is diffusing, the Gaussian envelope was substituted by a Abragam depolarization function given by

$$P(t) = \exp\left\{-\sigma^2\tau_c^2\left[\exp\left(-\frac{t}{\tau_c}\right) - 1 + \frac{t}{\tau_c}\right]\right\} , \qquad (2)$$

where $\tau_c$ is the correlation time.

In thin films, a sample size correction was used in addition to silver calibration for an accurate determination of muonium fractions.

## 3. Results and Analysis

The dominant feature of the spectrum of bulk CZTS is a diamagnetic signal (Fig.1). At low temperatures, it accounts to about 60 % of the total $\mu$SR signal. The depolarization rate $\sigma$ of this signal is constant below 150 K with an average value $\sigma = 0.12\ \mu s^{-1}$. This value is similar to the value $\sigma = 0.14\ \mu s^{-1}$ found in $CuInS(Se)_2$ chalcopyrites [4] and is consistent with nuclear dipolar broadening caused by Cu nuclear moments when the muon is at the Cu-S bond axis, close to the anion. We therefore assign this component to diamagnetic muons bound to the anion S and call it $Mu_S^+$.

The depolarization rate $\sigma$ of the diamagnetic signal is seen to decrease above 150 K, which is interpreted as motional narrowing due to the onset of muon diffusion. The motional narrowing region was analysed using the Abragam depolarization function, fixing $\sigma$ at the low temperature value. An Arrhenius plot of the hop rate $\nu = 1/\tau_c$ obtained from this analysis is displayed in Fig.2, yielding an attempt frequency $\nu_0 = 1.1 \times 10^7\ s^{-1}$ and an activation energy of 60 meV. Comparing with the $CuInS(Se)_2$ data [4], the onset of muon diffusion occurs at a lower temperature and with a lower activation energy in this family of compounds. The attempt frequency is also smaller. The reason for these differences is unclear. In $Cu_2ZnSnS_4$ the cations are, in average, lighter than in $CuInS(Se)_2$ and it is possible that the lattice is less rigid and the muon potential well at low temperatures is wider and less profound. However, additional studies are needed to identify the parameters affecting muon mobility in CZTS samples.

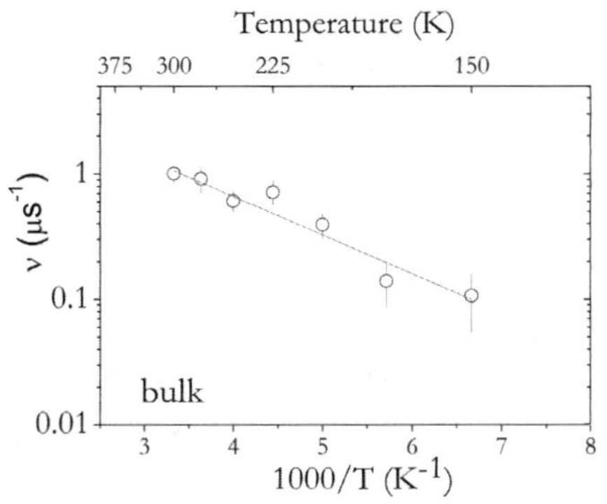


**Figure 2.** Arrhenius plot of the hop rate $\nu$ in the region of motional narrowing for bulk CZTS in transverse field. The straight line is a fitting to an activated hop rate and yields a pre-exponential factor $\nu_0 = 1.1 \times 10^7\ s^{-1}$ and an activation energy of 60 meV.

At early times in the spectrum (Fig.1), a fast relaxing component is seen. The value of the relaxation at low temperatures, $\lambda \sim 3\,\mu s^{-1}$ is about 20 times larger than the nuclear dipolar brodening and clearly indicates an interaction of the muon with an electron to form muonium. This relaxation shows a rapid increase above 200 K (Fig.3).

A missing fraction is also apparent in the spectrum which is attributed to the dephasing caused by a muonium precursor [15, 22]. The temperature dependence of the various fractions is presented in Fig.4 for the bulk sample. At room temperature only the diamagnetic fraction survives.

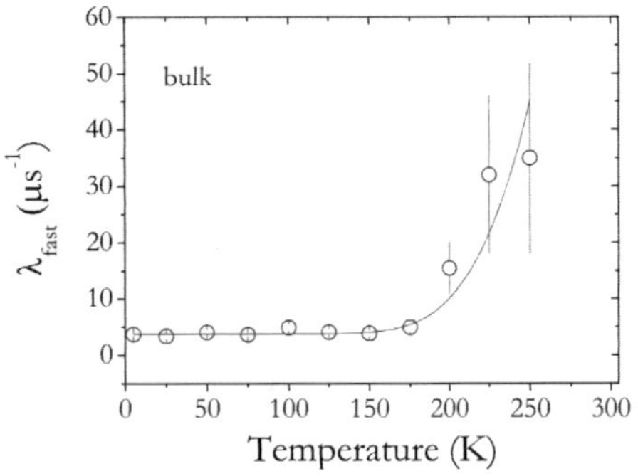


**Figure 3.** Temperature dependence of the relaxation rate of the paramagnetic component, in a transverse field of B = 0.01 T. The line is a fit with an Arrhenius activated function, yielding an activation energy of 0.16(3) eV.

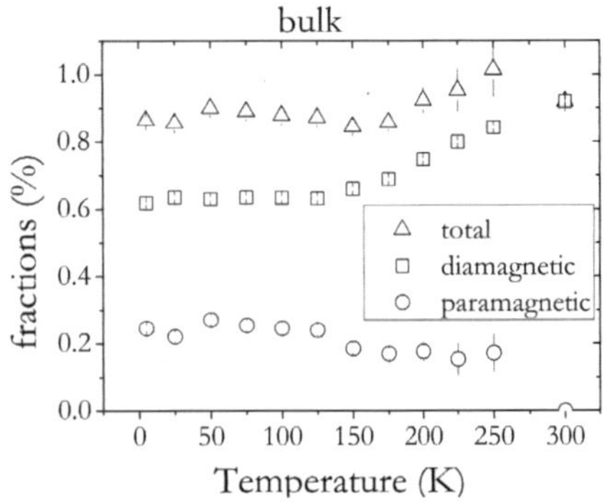


**Figure 4.** Temperature dependence of muonium fractions in bulk CZTS, in transverse geometry (B = 0.01 T).

The results in the CZTS thin films follow the same general trends described above for bulk, namely concerning the values of $\sigma$ and $\lambda$ at low temperatures and their temperature behaviour. The main difference is the relative fraction of the various components below 150 K. A separation of the two non-diamagnetic components is difficult due to fitting uncertainties and therefore we will represent these two paramagnetic fractions together and call it the muonium fraction $f_{Mu}$. However, we have to keep in mind the possibility that they play different roles in the behaviour of $f_{Mu}$.

Figure 5 shows the muonium formation probability $f_{Mu}$ as a function of temperature for bulk and for the two thin films, SF and RTP24. Above 150 K, $f_{Mu}$ follows a similar trend for the three samples: $f_{Mu}$ decreases until it disappears at around room temperature, indicating that muonium becomes unstable. However, below 150 K $f_{Mu}$ is clearly different for the various samples. In both thin films $f_{Mu}$ is lower than the bulk value, being smaller for SF, the film with higher defect concentration. The muonium formation probability is therefore sample dependent and is smaller in samples with higher defect content. Interestingly, the same observation was reported in CdS and Si single crystal samples[15] with very different final muonium states. In a ZnO thin film, no muonium signal was observed[16], in contrast with bulk behaviour [17] and this was also attributed to a higher defect/impurity concentration in the thin film as compared to bulk.

## 4. Discussion and Conclusions

The observation that the formation probability of the muon bound state is sensitive to the defect content of the sample indicates that this quantity is sensitive to interactions occurring during the muonium formation process. In order to interpret the information conveyed by this quantity

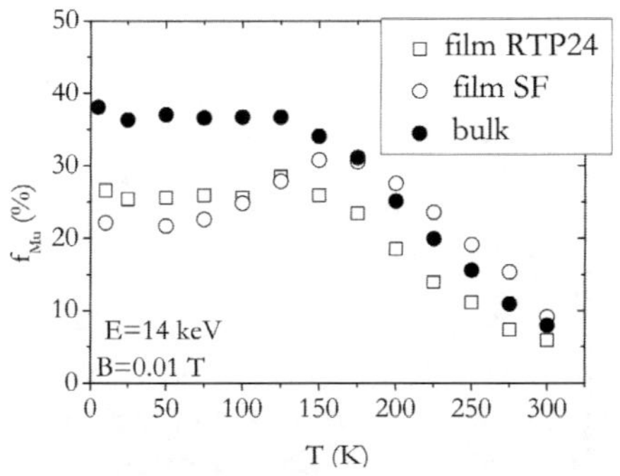


**Figure 5.** Temperature dependence of the Mu fraction in CZTS at B =0.01 T for bulk and for two different thin films.The implantation energy was E=14 keV for the thin films.

it is important to discuss the origin of the two contributions for the paramagnetic fraction: the missing fraction and the fast relaxing component.

As mentioned in the introduction, a missing fraction is observed in similar systems [4] and is known to correspond to a deep muonium state in an atomic-like configuration. We therefore assign the missing fraction to a neutral state thermalizing at an interstitial position, $Mu_I^0$.

The origin of the fast relaxing component is less clear and can be attributed to different mechanisms. One possibility is that it corresponds to unresolved lines [18, 19, 20] due to muonium formed at a donor configuration, bound to the anion S, which ionizes to form $Mu_S^+$:

$$Mu_S^0 \rightarrow Mu_S^+ + e^- \quad . \tag{3}$$

In this hypothesis, the observed relaxation $\lambda \sim 3\,\mu s^{-1}$ at low temperatures (Fig 3) corresponds to the hyperfine interaction of the paramagnetic state. It should be noted that in p-type $CuInSe_2$ a frequency shift of the diamagnetic line was observed in high transverse magnetic field and interpreted as a neutral state with a hyperfine interaction of 3 MHz [7].

An alternative scenario is that the fast relaxing component corresponds to a delayed muonium formation at an acceptor configuration, i.e., in an interstitial position. If the average time $\tau$ for muonium formation is large compared with the period of muonium precession, i.e., if $\omega_{Mu}\tau \gg 1$ there is a coherence loss. In this case, the relaxation rate could be interpret as a muonium formation rate $1/\tau$ [21], which would yield a muonium formation time of the order of $\tau$=0.3 $\mu$s. Two different mechanisms can be envisaged for delayed muonium formation, an electron capture or an electron loss, depending on whether we interpret the diamagnetic component as $Mu^+$ or $Mu^-$, respectively :

$$Mu_I^+ + e^- \rightarrow Mu_I^0 \tag{4}$$

$$Mu_I^- \rightarrow Mu_I^0 + e^- \quad . \tag{5}$$

Muonium is expected to create a negative-U center in this family of compounds [11] and since the sample is p-type, the $Mu^+$ state is predicted to be more stable than $Mu^-$. The diamagnetic component is therefore interpreted as being mainly $Mu_S^+$ but it cannot be ruled out that it includes a contribution of $Mu_I^-$. None of processes 3, 4 or 5 can be excluded based on the existing information. Thus, the fast relaxing component of the $\mu$SR signal may be assigned either to a hyperfine interaction of the muon at the stable anion-bound position, $Mu_S^0$, or to a delayed muonium formation at the metastable interstitial position $Mu_I^0$.

All these observations are consistent with a picture similar to one presented before [15, 22] for muonium formation in semiconductors. In this picture, the muon comes to rest at an interstitial position in the unrelaxed lattice, either as $Mu_I^+$, $Mu_I^0$ or even $Mu_I^-$. The initial formation probabilities of these different charge states are strongly sample- and temperature dependent.

The interstitial bare muon, $Mu_I^+$, will move to the more stable $Mu_S^+$ site or eventually converts to $Mu_I^0$ (process 4). The $Mu_I^-$, if formed, is unstable. In order to convert to $Mu_S^+$ it has to undergo an intermediate step (process 5) and this is likely to be a delayed process. The neutral $Mu_I^0$ is not stable either, but lives long enough to leave a signature in the $\mu$SR signal. The general behaviour of the neutral state may be described as follows: the lattice reacts to the presence of the impurity and starts to relax. The muonium tends to migrate, but along its path its energy level may come close to the conduction band and interact with electrons. The final result of this interaction may be either a drift to the more stable bonded position, as $Mu_S^0$ or $Mu_S^+$, or a return to the metastable interstitial position as $Mu_I^0$. At this intermediate stage, the fate of the muon is likely to be very sensitive to the presence of defects. On one side, the muon interaction with electrons depend on the number of electrons available in the conduction band and on the competition with electron-hole recombination processes. On the other end, defects may affect the muon transition path, favoring the transition to the donor configuration, close to S. The diamagnetic fraction, assigned to muons at the more stable site, is found to be larger in a sample with higher defect content, in agreement with findings in systems like CdS and Si [15], with very different final muonium states. The sensitivity of the muon probe on the local electronic properties of the material depends on the details of the processes occurring during the muonium formation stage. In particular, the dependence of the muonium formation probability on the local charge carrier density and dynamics, opens the possibility to use it as a probe in future studies of CZTS-based solar cell junctions.

**Acknowledgments**
This work was supported from funds from FEDER (Programa Operacional Factores de Competitividade COMPETE) and from FCT - Fundação para a Ciência e Tecnologia - under the project PEst-OE.FIS.UI0036.2014 and through Grants PTDC/CTM-MET/113486/2009, PEST-C/CTM/LA0025/2011 and RECI/FIS-NAN/0183/2012.

## References

[1] Shin B *et al.* 2011 *Prog. Photovolt: Res. Appl.* **21** 72–6
[2] Wang W *et al.* 2014 *Adv. Energy Mater.* **4** 1301465
[3] Leitão J P *et al.* 2011 *Phys. Rev.* B **84** 024120
[4] Gil J M *et al.* F 1999 *Phys. Rev.* B **59** 1912
[5] Gil J M *et al.* 2000 *Physica* B **289-290** 567–9
[6] Vilão R C *et al.* 2003 *Physica* B **326** 181–4
[7] Vilão R C *et al.* 2003 *Physica* B **340–342** 965–8
[8] Vilão R C 2002 *M.Sc. Thesis* Universidade de Coimbra
[9] Ayres de Campos N *et al.* 1998 *ISIS Experimental Report* RB9673
[10] Kiliç Ç and Zunger A 2003 *Phys. Rev.* B **68** 075201
[11] Kiliç Ç and Zunger A 2003 *Appl. Phys. Lett.* **83** 2007
[12] Morenzoni E *et al.* 2000 *Physica* B **289-290** 653
[13] Prokscha T *et al.* 2008 *Nucl. Instr. Meth.* A **595** 317
[14] Pratt F L 2000 *Physica* B **289-290** 710
[15] Alberto H V *et al.* 2012 *Phys. Rev.* B **86** 035203
[16] Alberto H V *et al.* 2009 *Physica* B **404** 870–2
[17] Cox S F J *et al.* 2001 *Phys. Rev. Lett.* **86** 2601
[18] Cox SF J *et al.* 2006 *Journal of Physics: Condensed Matter* **18** 1079
[19] Vilão R C *et al.* 2011 *Phys. Rev.* B **84** 045201
[20] Silva E L *et al.* 2012 *Phys. Rev.* B **85** 165211
[21] Storchak, V Brewer J H and Morris G D 1997 *Phys. Rev.* B **56** 55
[22] Alberto H V *et al.* 2010 *Phys. Rev.* B **81** 245205

13th International Conference on Muon Spin Rotation, Relaxation and Resonance IOP Publishing
Journal of Physics: Conference Series **551** (2014) 012046 doi:10.1088/1742-6596/551/1/012046

# Muonic atom as an acceptor centre in diamond

**T N Mamedov[1], A S Baturin[2], K I Gritsaj[1], A Maisuradze[3,4], V G Ralchenko[5], R Scheuermann[4], K Sedlak[4] and A V Stoykov[4]**

[1] Joint Institute for Nuclear Research, 141980 Dubna, Moscow reg., Russia
[2] Moscow Institute of Physics and Technology, 141700 Dolgoprudny, Moscow reg., Russia
[3] Physik-Institut der Universität Zürich, Winterthurerstrasse 190, CH-8057 Zürich
[4] Paul Scherrer Institut, CH-5232 Villigen PSI, Switzerland
[5] Natural Sciences Center of Institute of General Physics, RAS, 119991 Moscow, Russia

E-mail: tmamedov@jinr.ru

**Abstract.** Polarized negative muons were used to study the behaviour of the boron acceptor centre in diamond produced by the chemical vapour deposition (CVD) method. The temperature dependence of the muon spin relaxation rate and spin precession frequency were measured in the range of 20 – 330 K in a transverse magnetic field of 14 kOe. The muon polarization amplitude $P(t = 0)$ does not depend on the temperature and the muon spin relaxation rate decreases as the temperature increases. For the first time a negative shift of the muon spin precession was observed in diamond. The measurements show that the magnetic susceptibility of the CVD sample is negligible ($\chi = -4.13(2) \cdot 10^{-7}\text{cm}^3/\text{g}$ at 20 K) and it could not be the reason of the negative shift muon spin precession frequency. The negative shift of the muon spin precession frequency is tentatively attributed to an anisotropic hyperfine interaction in the boron acceptor in diamond.

Diamond with its unsurpassed mechanical strength, thermal conductivity, and radiation hardness is a promising semiconductor for particle detectors and electronic components capable of withstanding high heat and radiation loads. Great advances have been made over the last years in the technology of manufacturing synthetic single crystal diamond and diamond films [1, 2].

Boron is the only dopant which forms an acceptor centre (AC) in diamond with an ionization energy of $\approx 370$ meV [3]. The metal-insulator transition occurs at a concentration of $\approx 2 \cdot 10^{20}$ cm$^{-3}$ of boron atoms [4]. The EPR signal of boron impurities in diamond was observed only for uniaxially stressed samples [5], and the electronic state of this acceptor is investigated insufficiently.

The possibility of using negative muons to study the behaviour of acceptor impurities in diamond arises from the fact that capture of a negative muon by a carbon atom results in the formation of a muonic atom $_\mu$B with an electron shell that is analogous to that of the boron atom. The evolution of the polarization of $\mu^-$ in the 1s atomic state depends on the interaction of the muon spin with the electron shell of the muonic atom and on the interactions of this muonic atom as AC with the crystal lattice. Providing that the electron shell of $_\mu$B has a nonzero magnetic moment (paramagnetic $_\mu$B), there is a hyperfine interaction between the muon and the electron shell. The efficiency of the hyperfine interaction depends on the relaxation rate $\nu$ of the magnetic moment of $_\mu$B and on the hyperfine interaction constant $A_{hf}$. According to theoretical calculations [6], relaxation of the muon spin and a paramagnetic shift of its precession frequency are expected for $\nu \gg |A_{hf}|$. Under the assumption of an isotropic hyperfine interaction, the

Journal of Physics: Conference Series **551** (2014) 012046 doi:10.1088/1742-6596/551/1/012046

paramagnetic shift should be of positive sign and inversely proportional to temperature.

The boron atom as AC in diamond may be in the diamagnetic ($B^-$) or paramagnetic ($B^- + h$) state. In the latter case a hole is localized in vicinity of the AC. At equilibrium the boron acceptor in diamond is expected to be paramagnetic below $\approx 300$ K. In earlier experiments [7] we observed no frequency shift of the muon spin precession in diamond at the level of $5 \cdot 10^{-3}$. The goal of the present experiment was to improve the accuracy of the muon spin precession frequency measurement.

## 1. Diamond sample and measurements

The diamond sample was produced on Si substrates by a microwave plasma assisted chemical vapour deposition (CVD) technique in the $CH_4/H_2$ mixtures at the Institute of General Physics of RAS (Moscow, Russia) [9]. The typical growth conditions were as follows: 1.2 or 2.0% methane concentration in the mixture, total gas flow rate 1000 sccm, pressure 90 or 100 Torr, the substrate temperature of 720$^o$C or 820$^o$C. While the grain size on substrate side was about 1 $\mu$m, the crystallites of up to 200 $\mu$m were revealed on the opposite growth side of the films. It was found from the infrared and ultraviolet optical absorption measurements that the main impurities in the sample were hydrogen ($\approx$70 ppm) and substitutional (paramagnetic) nitrogen atoms ($\approx$1.5 ppm) (see [10]). In CVD diamond nitrogen is contained predominantly in the substitutional form. Other impurities were less than 0.1 ppm. The hydrogen atoms (CH bonds) are located mostly on grain boundaries [11]. After chemical etching of the substrate the cylinder-shaped sample D6 with the diameter 25 mm and height 1.3 mm (thickness 0.427 g/cm$^2$) for the $\mu$SR study, and two bricks (samples A and B) with the dimensions 3x1.5x1.5 mm$^3$ (weight 34 mg) for magnetic measurements were cut using a laser. The DC magnetization measurements of the samples A and B were performed with the Quantum Design SQUID magnetometer.

The $\mu$SR experiment was performed with the upgraded ALC instrument [8] located at the $\pi$E3 muon beam line of the proton accelerator of the Paul Scherrer Institute in Switzerland. In the $\pi$E3 beam line the muons were "transversely" polarized – the angle between the muon spin and the muon momentum was $\sim 45^\circ$. The polarization of negative muons in synthetic diamond was studied in the magnetic field of 14 kOe in the temperature range of $20 - 330$ K. One additional measurement was performed in 3.0 kOe at 140 K. The magnetic field at the sample was parallel to the momentum of the incoming muon. The time-differential $\mu$SR spectrum was measured with a CAEN V1190B time-to-digital convertor (TDC), bin width was 0.39 ns and the total number of bins in the spectrum was 26000. About 85% of the incoming muons stopped in the sample and 15% in the copper holder. As a reference, a graphite sample (diameter 26 mm and thickness 0.66 g/cm$^2$) was measured by $\mu$SR under the same conditions as the diamond sample.

The time distribution of electrons from the $\mu^- \rightarrow e^-$ decay was fitted by the function

$$\begin{aligned} N(t) &= N_0(\mathrm{C})[1 + A(t)]e^{-t/\tau(\mathrm{C})} + N_0(\mathrm{Cu}) \cdot e^{-t/\tau(\mathrm{Cu})} + bg \\ A(t) &= a \cdot P(t) = a \cdot P_0 \cdot e^{-R \cdot t} \cdot \cos(\omega t + \varphi) \end{aligned} \quad (1)$$

where $N_0$(C) ($N_0$(Cu)) is proportional to the number of muons stopped in the sample (in the copper holder); $\tau$(C) ( $\tau$(Cu)) is the mean lifetime of the muon in the 1s state of carbon (copper); $a$ is the coefficient of asymmetry of the space distribution of electrons taking into account the actual parameters of the spectrometer; $P_0$ is the muon polarization in the 1s state at $t = 0$; $R$ is the muon spin polarization damping rate; $\omega$ and $\varphi$ are the frequency and the initial phase of the muon spin precession in the magnetic field, and $bg$ is the time-independent background.

The muon polarization $P_0$ in the 1s state of copper is close to zero due to the hyperfine interaction of the muon spin with the spin of the nucleus and due to very fast transitions between the levels $F_+ = I + S_\mu$ and $F_- = I - S_\mu$, where $I$ is the spin of the copper nucleus and $S_\mu$ is the muon spin (see [12, 13, 14]).

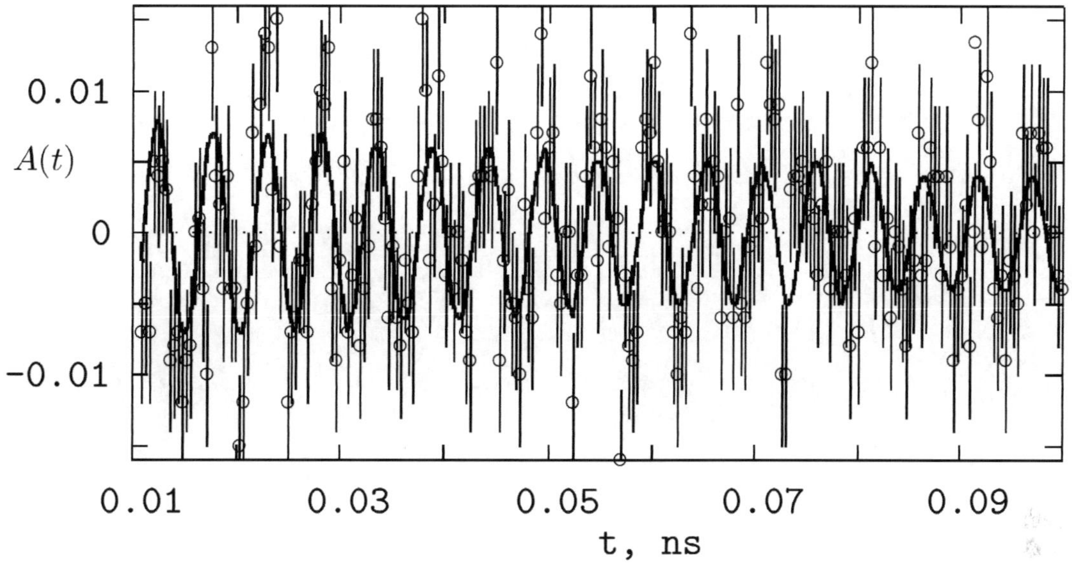


**Figure 1.** The $\mu^-$SR spectrum for the diamond sample D6 at 27 K after background subtraction and correction for the negative muon lifetime. The line corresponds to fitting the data with function (1).

An example of the fragment of $\mu^-$SR spectrum measured for the diamond sample at 27 K in the magnetic field of 14 kOe is presented in Fig 1.

## 2. Results and discussion

The behaviour of the muon spin polarization in the reference graphite sample was studied at temperatures 9.6, 50, 100, 200, and 300 K in a magnetic field of 14 kOe. It was observed that in graphite the muon spin polarization and the frequency of the muon spin precession do not depend on temperature: $[P_0(T) - P_0(300\text{ K})]/P_0(300\text{ K}) \lesssim 0.06 \pm 0.08$ and $\Delta\omega/\omega = [\omega(T) - \omega(300\text{ K})]/\omega(300\text{ K}) \lesssim (6 \pm 4) \cdot 10^{-5}$ (see also [14]). The relaxation of the muon spin in graphite was not observed within the accuracy of the measurements ($R < 0.05$ MHz).

The results of the measurements on the CVD diamond are shown in Fig. 2. The muon polarization damping rate depends on temperature as $R \sim 1/T^{0.8}$ in the range of 20 – 80 K (solid line in Fig. 2) and as $R \sim 1/T^{2.0}$ in the range of 80 – 230 K (dotted line). A strong drop of the damping rate at $\sim$ 250 K and no damping at 300 and 330 K were observed. There is no difference in the damping rate of the muon spin at 140 K in the magnetic field of 3.0 kOe compared to 14 kOe. Generally, the temperature dependence of $R$ in sample D6 is similar to that observed earlier for the CVD film samples in 1.5 and 2.5 kOe magnetic fields [7].

In the sample D6 the muon polarization amplitude $P_0$ does not depend on temperature and it is approximately 16 % larger than in the reference carbon sample.

As seen in Fig. 2, the muon spin precession frequency in the carbon reference sample does not depend on temperature. At room temperature ($T$ = 300 K) the muon spin precession frequency in the D6 sample is close to that in the reference sample: $[\omega(C) - \omega(D6)]/\omega(C) = (1.8 \pm 0.4) \cdot 10^{-4}$. Opposite to the theoretical prediction [6] and to the experimental results in silicon [15], a negative shift of the muon spin precession frequency was observed at temperatures below 250 K. The temperature dependence of the frequency shift can be approximated as

Journal of Physics: Conference Series **551** (2014) 012046 doi:10.1088/1742-6596/551/1/012046

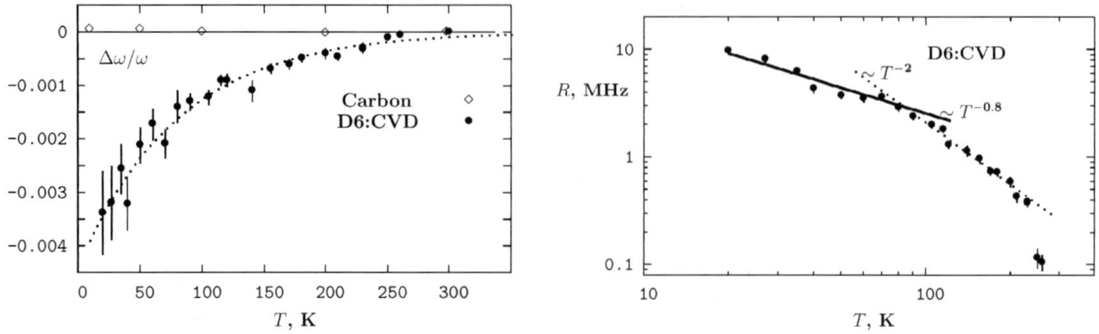


**Figure 2.** Damping rate $R$ and precession frequency shift $\Delta\omega/\omega$ of the negative muon spin measured in the synthetic CVD diamond sample D6 in the magnetic field of 14 kOe.

$[\omega(T) - \omega(300\ \text{K})]/\omega(300\ \text{K}) = \Delta\omega/\omega \sim -1/T$ (dotted line in Fig. 2).

The negative shift of the muon spin precession frequency may be due to three different reasons: a) it may be due to the presence of other paramagnetic defects (excluding the acceptor centres) in diamond. The magnetic field at the muon could differ from the external field because of polarization of these paramagnetic defects. For example, $N^0$ and H1(H2) paramagnetic centres were observed in polycrystalline CVD diamond by EPR [10];
b) let us assume that the muonic atom is formed in the ${}_{\mu}B^0$ state within a time less than $10^{-9}$ s and that the transition rate from this state to the equilibrium one (${}_{\mu}B^{-}$+h) is smaller than the muon decay rate. In the state ${}_{\mu}B^0$ the muonic atom has covalent chemical bonds with three nearest carbon atoms, and the fourth carbon atom has an electron with an unpaired spin (one dangling bond). The polarization of this paramagnetic complex in an external magnetic field could result in the negative shift of the muon spin precession frequency;
c) negative frequency shift can also be due to polarization of the paramagnetic acceptor centre in the magnetic field if the muonic atom is formed in the (${}_{\mu}B^{-}$+h) state in $\lesssim 10^{-8}$ s and the hyperfine interaction in the AC is anisotropic. In the state (${}_{\mu}B^{-}$+h) the muonic atom has covalent chemical bonds with all four nearest carbon atoms and a hole localized in the vicinity of AC.

The frequency shift of the muon spin precession should be proportional to the susceptibility of the paramagnetic defects or to the susceptibility of the paramagnetic acceptor centres if $\nu \gg |A_{hf}|$. Therefore, a simple $1/T$ (Curie type) temperature dependence of the muon spin precession frequency shift is expected.

To elucidate the effect of paramagnetic (magnetic) defects on the muon spin precession frequency we measured the magnetic susceptibility of our sample. The magnetic moments ($M$) of samples A and B were measured in magnetic fields in the range of $0-50$ kOe at the temperatures 20, 100, and 300 K. The temperature dependence $M(T)$ was studied in detail in the magnetic fields of 95.2 Oe and 20.0 kOe for both samples. The results for both samples A and B are similar within the accuracy of the measurements. The results of the measurements for sample A are presented in Fig. 3 and in Fig. 4. There is evident linear dependence of the magnetic moment on the magnetic field in the range from 0 to 50.0 kOe. The susceptibility of the sample was $\chi = -4.13(2)\cdot 10^{-7}$ cm$^3$/g, $-4.29(2)\cdot 10^{-7}$ cm$^3$/g, and $-4.36(2)\cdot 10^{-7}$ cm$^3$/g, at the temperatures 20, 100, and 300 K, respectively. The value of $\chi$ for the synthetic CVD sample is in agreement with that for natural diamond [16, 17]. As seen from Fig. 4, the magnetic moment of the sample in the 95.2 Oe and 20.0 kOe magnetic field practically does not depend on temperature in the range of $20 - 300$ K. Only at temperatures below 20 K the magnetic moment slightly increases

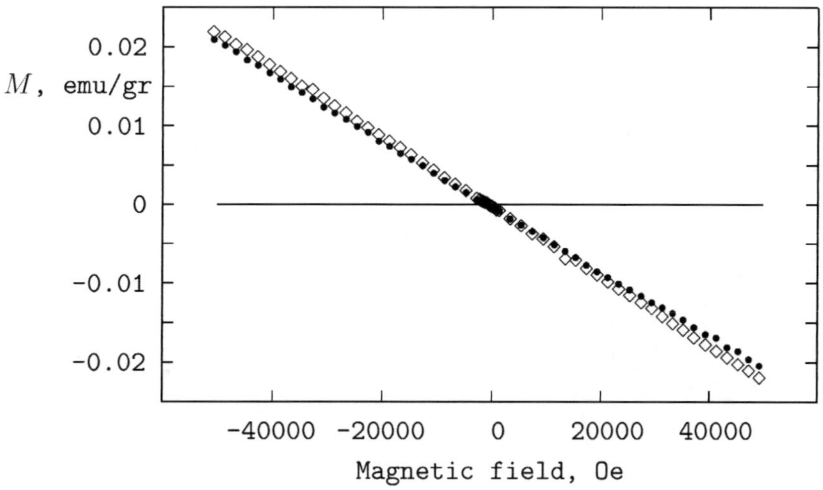


**Figure 3.** Magnetic moment of CVD diamond sample A vs the magnetic field: • at 20 K and ◇ at 300 K.

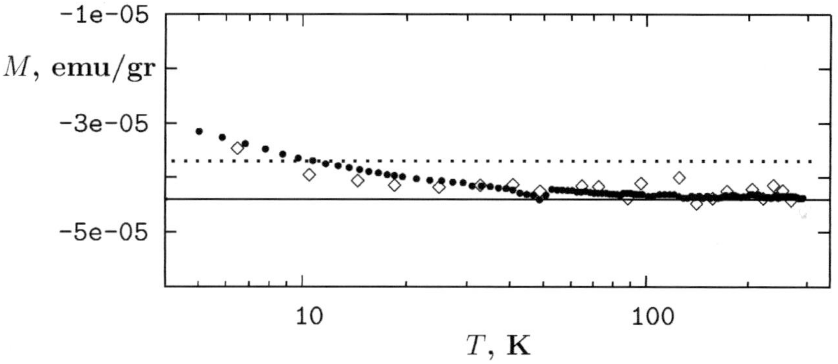


**Figure 4.** Temperature dependence of the magnetic moment of the synthetic CVD diamond: ◇ in the magnetic field of 95.2 Oe and • in 20.0 kOe. For better visualization in one single graph the value of $M$ at the magnetic field of 20.0 kOe was scaled by 1/210

with decreasing temperature. Therefore, the concentration of magnetic impurities (defects) in our CVD sample is negligible and cannot produce the observed frequency shift. Moreover, the muon spin precession frequency shift depends on temperature as $\sim -1/T$ in the range of $20 - 300$ K, where the magnetic moment of the sample is practically constant.

The measured values of $P_0$ at different $T$ are close to the maximum possible muon polarization in the 1s state of a carbon atom. This means that the muonic atom $_\mu$B is formed in one of the chemically bonded $_\mu B^0$ or ($_\mu B^{1-}$+ h) states with a probability close to unity in less than $10^{-9}$ s. It is worth mentioning that the Auger transitions of the muon destroy chemical bonds of the muonic atom with neighbouring carbon atoms. Theoretical calculations [18, 19] suggest that the muonic atom in diamond restores chemical bonds and is formed in the $_\mu B^0$ state within $10^{-10} - 10^{-9}$ s.

At present we do not have an analytical expression for the relaxation of the muon spin if there is an anisotropic hyperfine interaction in the muonic atom. Nevertheless, we may assume that $R$ is inversely proportional to the relaxation rate of the magnetic moment of the acceptor centre in both $_\mu B^0$ and ($_\mu B^{1-}$ +h) states. In diamond relaxation of the magnetic moments of $_\mu B^0$ and ($_\mu B^-$ +h) should be attributed to scattering of phonons by these centres. Direct, Raman, and Orbach processes can contribute to phonon scattering, and the temperature dependence of the probability of these processes is very different [20]. In the present experiment two temperature domains with different temperature dependences of the relaxation rate of the muon spin were observed. Thus, we conclude that there is a contribution of two different processes to scattering of phonons at the acceptor in diamond. The dominance of one process over the other reverses at about 80 K. According to the calculations for the acceptor in silicon, the Raman process overcomes the direct (one phonon) process at $T = 8$ K, and the Orbach process is negligible below 100 K [20].

## 3. Conclusion

A large negative shift of the precession frequency of the negative muon spin in diamond was found for the first time. Simple Curie type $\Delta\omega/\omega \sim -1/T$ temperature dependence of the frequency shift was observed. It is not excluded that the muonic atom as an AC forms in the equilibrium ($_\mu B^-$ +h) state in $10^{-8}$ s and there is an anisotropic hyperfine interaction in AC. The temperature dependence of the relaxation rate of the muon spin evidences that two different processes contribute to phonon scattering by the acceptor centre in the temperature range of $20 - 250$ K.

The magnetic measurements showed that the magnetic susceptibility of the CVD sample was close to that of the purest natural diamond.

The authors are grateful to prof. Yu. M. Belousov for fruitful discussions. We thank the Directorate of the Paul Scherrer Institute for giving us the possibility of carrying out the experiments at PSI.

## References

[1] Tuve C, Angelone M, Bellini V et al 2007 *Nucl. Instr. Meth. Phys. Res.* A **570** 299
[2] `http://www.e6.com` (10.04.2014)
[3] Bezrukov G N, Butuzov V P, Gerasimenko N N et al 1970 *Sov. Phys. Semicond.* **4** 587
[4] Willams A W S, Lightowlers and E C Collins A T 1970 *J. Phys.* C **3** 1727
[5] Ammerlaan C A J 1981 *Inst. Phys. Conf. Series* **59** 81
[6] Gorelkin V N, Mamedov T N and Baturin A S 2000 *Physica* B **289-290** 585
[7] Mamedov T N, Baturin A S, Blank V D et al 2008 *Diamond Relat. Matter.* **17** 1221
[8] Stoykov A, Scheuermann R, Sedlak K, Shiroka T and Zhuk V 2009 *Physica* B **404** 986
[9] Ralchenko V G , Smolin A A, Kononov V I et al 1997 *Diamond Relat. Mater.* **6** 417
[10] Nistor S V, Stefan M, Ralchenko V, Khomich A and Schoemaker D 2000 *J. Appl. Phys.* **87** 8741
[11] Khomich A, Ralchenko V, Nistor L, Vlasov I and Khemilnitsky R 2000 *Phys. Stat. Sol.*(a) **181** 37
[12] Winston R 1963 *Phys. Rev.* **129** 2766
[13] Brewer J H 1984 *Hyperfine Interact.* **19** 873
[14] Mamedov T N, Chaplygin I L, Duginov V N et al 1999 *J.Phys: Condens. Matter* **11** 2849
[15] Mamedov T N, Gorelkin V N and Stoykov A V 2002 *Physics Particle Nuclei* **33** 519
[16] Heremans J, Olk C H and Morelli D T 1994 *Phys. Rev.* B **49** 15122
[17] Yelisseyev A P, Afanasiev V P and Ikorsky V N 2009 *Doklady Earth Sciences* **425(2)** 330
[18] Baturin A S, Gorelkin V N, Rastunkov V S and Soloviev V R 2006 *Physica* B **374-375** 340
[19] Baturin A S, Gorelkin V N, Mamedov T N *et al.* 2007 *JINR communication, Dubna* P14-2007-188
[20] Yafet Y 1965 *J. Phys. Chem. Solids* **26** 647

# Process of an acceptor center ${}_{\mu}B^{-}$ formation in diamond crystals

**Y M Belousov**

Moscow Institute of Physics and Technology, 141700, Dolgoprudny, Institutsky lane 9, Russia

E-mail: theorphys@phystech.edu

**Abstract.** A formation process of an acceptor center ${}_{\mu}B$ induced by negative muon in a diamond crystal is considered in this work. It is shown that the process under consideration could be represented as three consecutive stages: a kinetic stage, connected with a capture of track electrons, a chemical stage, connected with establishing of chemical bounds of the radiation defect with lattice atoms, and a formation phase of the ionized negatively charged acceptor center. Quantitative estimates of characteristic times are obtained for every stage. It is shown that the total time necessary for the acceptor center formation is of the order $10^{-8}$s. So this fact should be taken into account when experimental results are interpreted.

## 1. Introduction

It is well known, that a negative muon is captured on a lattice nuclear forming the so called $\mu$-atom. But just after the muon capture in a lattice arises a positively charged radiation defect. Neutralization process of radiation defects induced by negative muons in crystals with diamond structure (Diamond, Silicon and Germanium) was considered in works [1, 2]. There was shown that this process could be considered as four consistent stages. At the first stage positively charged radiation defect with a charge equal to $Z-1$, where $Z$ is the charge of a host nuclear, captures electrons of the radiation track. At the second stage a neutral defect restores broken chemical bounds with lattice atoms. At the third stage an ionized acceptor center is formed. And at the final stage is formed a neutral (usual) acceptor center induced by the $\mu$-atom. Only the last stage is studied well for the present. In [2] was considered a neutralization kinetics as a result of a capture of electrons from the radiation trace. There was shown that the capture process strongly differs in diamond crystal and in crystals of silicon and germanium. Namely, in a diamond crystal the radiation defect has not a time to be neutralized in a muon life-time $\tau_{\mu} \approx 2.2 \cdot 10^{-6}$s. So, the formation process of a chemically bounded radiation defect differs in crystals under consideration. Chemical bounds are restored in silicon and germanium for the neutral impurity ${}_{\mu}A$ (muonic atom), but in diamond for the positively charged ion ${}_{\mu}B^{+}$. A chemically bounded impurity ${}_{\mu}B^{+}$ is neutralized in a diamond lattice. Estimation of a formation time of a neutral chemically bounded defect caused by a radiation transition was obtained in [1]. Because of this difference in neutralization processes a formation rate of chemically bounded neutral defect ${}_{\mu}A$ in a diamond almost two order higher than in silicon and germanium. In the following a neutral chemically bounded radiation defect ${}_{\mu}A$ forms an acceptor center in a lattice. At the first step of an acceptor center formation one need to consider a capture of an additional electron to fill an empty chemical bound around ${}_{\mu}A$. After the capture of an electron an ionized acceptor center $_{\mu}A^{-}$ is formed. This process is not studied yet, but the following step of a positive hole capture at the ionized acceptor center is well known. A process of an ionized acceptor center $_{\mu}B^{-}$ formation in a diamond crystal is studied in this work.

Experiments with Si and Ge was carried out more than 20 years ago [3]-[7]. Acceptor centers $_{\mu}$Al and $_{\mu}$Ga was observed and hyperfine structure of $_{\mu}$Al was studied in these experiments. Recent experiments with $\mu^{-}$ in synthetic diamond [8, 9] showed unusual behaviour of a muon spin polarization. A negative shift of a precession frequency and slow temperature dependence of a damping rate at temperature $T > 20$K were observed. Full polarization (100%) was observed. Thus, in these experiments no neutral acceptor center $_{\mu}$B was observed. The experimental results show that kinetic processes of a formation of the neutral acceptor center $_{\mu}$B demand more detail theoretical examination.

## 2. Phonon spectrum changing caused by a neutral center $_{\mu}$B

A neutral center $_{\mu}$A enters into a chemical bound with nearest atoms of a lattice and forms a cluster in the first coordination sphere. This cluster possesses a local symmetry $C_{3v}$ and a rather large electric dipole moment [1]. This electric dipole moment gives rise to an interaction necessary to capture a lattice electron and form of an ionized acceptor center. The neutral center $_{\mu}$A has unsaturated chemical bounds because of that a crystal lattice turns out to be deformed. This deformation is a reason to change a phonon spectrum and a local phonon mode appearance. Chemical bounds are saturated after the missing electron capture. The new cluster is an ionized acceptor center and possesses a local crystal symmetry. When the ionized acceptor center is formed an appropriate phonon of the local mode is radiated and crystal deformations disappear. So, the problem is very similar to the problem of a thermalization of molecular ions in molecular crystals and cryocrystals of noble atoms (see e.g. [10, 11, 12]).

An exact solution of the problem taking into account a crystal symmetry is scarcely possible. But we can obtain qualitative results to get an estimation of a formation rate of the ionized acceptor center $_{\mu}B^{-}$ in diamond.

To solve the problem we will carry out some simplifications. They will make our model rather crude but describing qualitatively all necessary effects nevertheless. First of all we will accept an approximation of an elastic continuous media to describe a phonon spectrum changing. In this case a displacement vector $\mathbf{u}$ satisfies the wave equation in an elastic media at the presence of the electric field [13]:

$$\rho \ddot{u}_i = \frac{\partial \sigma_{ik}}{\partial x_k}, \tag{1}$$

where $\rho$ is the density of a media. A tension tensor will be represented as a sum

$$\sigma_{ik} = \sigma_{ik}^{\mathrm{el}} + \sigma_{ik}^{E}. \tag{2}$$

Here $\sigma_{ik}^{\mathrm{el}}$ is determined by constants of the elastic media as usual. The second term in the expression (2) is the tension tensor of the electromagnetic field in a media [14]:

$$\sigma_{ik}^{E} = \frac{1}{8\pi}\left(E_i D_k + E_k D_i\right), \tag{3}$$

where $\mathbf{E}$ and $\mathbf{D}$ are the average (macroscopic) electric field and the induction created by the dipole moment of the cluster $\mathbf{d}$ respectively. We are taking a simplification to determine the electric field $\mathbf{E}$, namely $\varepsilon_{ik} = \varepsilon(\mathbf{r})\delta_{ik}$

$$\mathbf{E} = \frac{1}{\varepsilon}\mathbf{D}. \tag{4}$$

So, for the force acting on the media from the side of the electric field created by the dipole moment we are obtaining

$$F_i = \frac{\partial \sigma_{ik}^{E}}{\partial x_k} = -\frac{1}{4\pi\varepsilon^2}\frac{\partial \varepsilon}{\partial x_k} D_i D_k. \tag{5}$$

Here was taken into account that outside the first coordinative sphere in the lattice div$\mathbf{D} = 0$.

Let's place the coordinate beginning at the center of the nuclear ${}_{\mu}$A and consider that the dielectric susceptibility possesses a spherical symmetry, i.e. $\varepsilon(\mathbf{r}) = 1 + 4\pi\varkappa(r)$, where $\varkappa(r)$ is the dielectric susceptibility. As well we will consider the coordinate dependence of the dielectric susceptibility is stipulated for the specific volume $v$ changing as a result of the deformation. So, we obtain $\varkappa \propto v^{-1}$ because the dielectric susceptibility is proportional to the media density. The changing of the specific volume as the result of the deformation is determined by the trace of the deformation tensor [13]:

$$\delta v = v u_{ll}, \tag{6}$$

where $u_{ll} = \partial u_l/\partial x_l = \mathrm{div}\mathbf{u}$.

In this case the gradient of the dielectric penetration could be represented in a form

$$\frac{\partial \varepsilon}{\partial x_k} = 4\pi\frac{\partial \varkappa}{\partial x_k} \propto \frac{\partial}{\partial x_k}\frac{1}{v} = -\frac{1}{v}\frac{\partial u_{ll}}{\partial x_k} \equiv -\frac{1}{v}\frac{\partial^2 u_l}{\partial x_k \partial x_l} = -\frac{1}{v}\frac{\partial}{\partial x_l}\frac{\partial u_l}{\partial x_k}. \tag{7}$$

Let us rewrite the equation (1) taking into account that in our case rot$\mathbf{u} = 0$

$$\ddot{u}_i - c_{\parallel}^2 \Delta u_i = \frac{\varkappa}{\varepsilon^2 \rho} D_i D_k \Delta u_k, \tag{8}$$

where $c_{\parallel}$ is the longitudinal sound velocity.

The equation (8) is rather complicated yet and we'll make some additional very serious simplifications to get analytical expressions. Let's consider only the radial component of the displacement vector $\mathbf{u}$ and neglect its angular dependence, i.e. $\mathbf{u}(\mathbf{r}) \to u_r(r)$. In this approximation we'll keep in the induction vector only a dependence on the modulus of $r$:

$$D_r = \frac{2(\mathbf{dn})}{r^3} \to \frac{2d}{r^3}.$$

Only in this approximation we can solve the problem analytically. After all these simplifications we have

$$\ddot{u}_r - \left(c_{\parallel}^2 + \frac{4\varkappa d^2}{\varepsilon^2 \rho}\frac{1}{r^6}\right)\Delta_r u_r = 0. \tag{9}$$

It is suitable to search a solution for the radial displacements in a form

$$u_r(r,t) = \frac{\chi(r,t)}{r}. \tag{10}$$

In this case we have the more simple equation for $\chi$:

$$\ddot{\chi} - \left(c_{\parallel}^2 + \frac{4\varkappa d^2}{\varepsilon^2 \rho}\frac{1}{r^6}\right)\chi'' = 0, \tag{11}$$

where $\chi''$ is the second derivative on the radial variable $r$.

For the function $\chi(r,t)$ the one-dimensional Fourier-transformation is valid:

$$\chi(r,t) = \iint_{-\infty}^{\infty} \frac{d\omega dk}{(2\pi)^2} \mathrm{e}^{-i(kr-\omega t)}\chi_{\omega,k}, \quad \chi_{\omega,k} = 2\int_{-\infty}^{\infty} dt\, e^{-i\omega t}\int_0^{\infty} dr\, \mathrm{e}^{ikr}\chi(r,t). \tag{12}$$

For the Fourier-amplitudes (12) we obtain an integral equation taking into account that the expression for the force in the equation (8) is valid only for $r > R_1$, where $R_1$ is the radius of the first coordination sphere:

$$(\omega^2 - k^2 c_{\parallel}^2)\chi_{\omega,k} = \frac{4\varkappa d^2}{\varepsilon^2 \rho}\int_{-\infty}^{\infty}\frac{dk'}{2\pi}k'^2\chi_{\omega',k'}\int_{R_1}^{\infty}\frac{dr}{r^6}e^{i(k-k')r}. \tag{13}$$

Let's make the last serious simplification in the equation (13), changing an integral in the right-hand side by a constant before $\chi_{\omega,k}$. Really, values of the wave-vector $k' \approx k$ give the main contribution to the second integral. This permits us to take out the first integral the amplitude $\chi_{\omega',k'}$ at the point $k$. Values of $k' \sim R_1^{-1}$ give the main contribution to the integral as well. So, for the estimation of the integral in the right-hand side we have $\chi_{\omega,k}/(10\pi R_1^8)$. Finally we obtain a simple dispersion relation:

$$(\omega^2 - k^2 c_{\|}^2 - \Omega^2)\chi_{\omega,k} = 0, \quad \Omega = \frac{d\sqrt{2\varkappa}}{\varepsilon R_1^4 \sqrt{5\pi\rho}}. \tag{14}$$

## 3. Electron capture rate

Electric dipole moment creates a scalar potential and an interaction energy with lattice electrons is $\mathcal{U} = e\varphi = e(\mathbf{dr})/(\varepsilon r^3)$. So, taking into account the displacement $\mathbf{u}$ ($\mathbf{r} \to \mathbf{r} + \mathbf{u}$), we can write an electron-phonon interaction operator

$$\widehat{V}_{\mathrm{e-ph}} \approx \frac{e(\mathbf{d}\hat{\mathbf{u}})}{\varepsilon r^3} = \frac{ed}{\varepsilon r^3}\cos\theta \hat{u}, \tag{15}$$

where the operator of radial displacements is equal to [11]

$$\hat{u} = \frac{1}{r}\sum_n \sqrt{\frac{\hbar}{2\rho\omega(k_n)R_\infty}} e^{-ik_n r}\left(b_n^\dagger + b_n\right). \tag{16}$$

Here $b_n^\dagger$ and $b_n$ are creation and annihilation operators for radial phonons respectively.

An electron capture rate could be calculated by means of well known Fermi "golden rule"

$$dw_{if} = \frac{2\pi}{\hbar}\left|\langle f|\widehat{V}_{\mathrm{e-ph}}|i\rangle\right|^2 \delta(E_f - E_i)d\nu_f \frac{d\mathbf{k}}{(2\pi)^3}, \tag{17}$$

where an initial state $|i\rangle = |\mathbf{k}\rangle|0_{\mathrm{ph}}\rangle$ corresponds to a free (in a valence band) electron with the wave-vector $\mathbf{k}$ and the absence of excited radial phonons, and $\nu_f$ is the density of final states. The final state is determined by the state of the electron captured in the hybridized state of the cluster [1] and an excitation of a radial phonon with the wave-vector $\mathbf{k}_{\mathrm{ph}}$:

$$|f\rangle = (\alpha|2s\rangle + \beta|2p\rangle)|\mathbf{k}_{\mathrm{ph}}\rangle, \quad |\alpha|^2 + |\beta|^2 = 1.$$

Eq.(17) takes into account that all electrons with wave-vectors are in the interval from $\mathbf{k}$ to $\mathbf{k} + d\mathbf{k}$ can participate in the process.

Energies of the initial and the final states are equal to $E_i = E_0 + \hbar^2 k^2/2m^*$, $E_f = E_- + \hbar\omega(k_{\mathrm{ph}})$, $E_0 - E_- = \varepsilon_{ac}$, where $\varepsilon_{ac}$ is the acceptor ionization energy and $m^*$ is an effective mass of the electron in the valence band.

We can carry out a partial integration of the expression (17) and obtain a rather short formula

$$dw_{if} = \frac{2m^*}{3\pi\rho\hbar^2}\left(\frac{ed}{\varepsilon}\right)^2 \left|{}_e\langle f\left|\frac{1}{r^4}\mathrm{e}^{-\imath k_{\mathrm{ph}} r}\right|k_0\mathbf{n}\rangle\right|^2 \frac{k_0}{\omega(k_{\mathrm{ph}})}dk_{\mathrm{ph}}, \tag{18}$$

where $k_0^2 = 2m^*(\Delta - \omega_{\mathrm{ph}})/\hbar$, $\Delta = \varepsilon_{ac}/\hbar$.

For us will be comfortable in the future to introduce dimensionless parameters and variables

$$x = \widetilde{Z}r/(2a_0), \quad \widetilde{k}_{\mathrm{ph}} = 2k_{\mathrm{ph}}a_0/\widetilde{Z}, \quad \widetilde{k}_0 = 2k_0 a_0/\widetilde{Z} \quad \text{and} \quad \omega_0 = \langle c_{\|}\rangle\widetilde{Z}/(2a_0). \tag{19}$$

Here $\widetilde{Z}$ is the effective nuclear charge for electrons at the appropriate atomic energy level and $\langle c_{\parallel}\rangle$ is the average longitudinal sound velocity. For Boron it is equal to $\widetilde{Z} = 2.6$.

In this case we can express a capture cross-section through a dimensionless matrix element:

$$dw_{if} = \frac{2m^*}{3\pi\rho\hbar^2}\left(\frac{ed}{\varepsilon}\right)^2\left(\frac{\widetilde{Z}}{2a_0}\right)^5\left|\mathcal{A}(\widetilde{k}_{\rm ph},\widetilde{k}_0)\right|^2\frac{k_0}{\omega(k_{\rm ph})}dk_{\rm ph}, \tag{20}$$

where

$$\mathcal{A}(\widetilde{k}_{\rm ph},\widetilde{k}_0) =_e \langle f\left|\frac{1}{x^4}e^{-\imath\widetilde{k}_{\rm ph}x}\right|\widetilde{k}_0\mathbf{n}\rangle. \tag{21}$$

We can see, that the matrix element (21) has a peculiarity in the point $x = 0$ and the integrals diverge if the low limit is equal to zero. So we must take into account that expression (15) is valid for $r \gg a_0$ also. Thus we need to determine the low limit $r \gg a_0$. It is reasonable to take it equal to the length of the chemical bound in the lattice. In diamond crystal it is equal to $r_0 \approx 1.54\cdot 10^{-8}$cm and the low limit of the integrals (21) is $x_0 = \widetilde{Z}r_0/2a_0 \approx 3.78$.

Making rather long calculations and taking into account the dispersion law (14) we obtain the final formula for the capture rate

$$w = G\int\limits_{\widetilde{\Omega}}^{\widetilde{\Delta}}\frac{\left|\widetilde{k}_0\mathcal{A}(\widetilde{k}_{\rm ph},\widetilde{k}_0)\right|^2}{\sqrt{(\widetilde{\Delta}-\widetilde{\omega})(\widetilde{\omega}^2-\widetilde{\Omega}^2)}}d\widetilde{\omega}, \quad \text{where} \quad G = \frac{(ed)^2}{6\pi\rho m^*\varepsilon^2}\left(\frac{2m^*}{\hbar\langle c_{\parallel}\rangle}\right)^{3/2}\left(\frac{\widetilde{Z}}{2a_0}\right)^{13/2}, \tag{22}$$

where $\widetilde{\Delta} = \Delta/\omega_0$, $\widetilde{\Omega} = \Omega/\omega_0$, $\widetilde{\omega} = \omega/\omega_0$.

The following calculations could be carried out numerically.

## 4. Conclusion

We shall choose hydrogen-like $2s$ and $2p$ wave functions like in Ref.[1] to calculate matrix elements (21). The value of the rate (22) is proportional to the unknown parameter $d$, but numerical estimates show that $d \lesssim 1$D (Debye, 1D $= 10^{-18}$CGSE). We will regard $m^* = m$ (the free electron mass) and use the following numerical values for diamond parameters [15, 16]: $\rho = 3.51$g/cm$^3$, $\varepsilon = 5.75$, $c_{\parallel} = 1.83\cdot 10^6$cm/s, $\varepsilon_{ac} = 0.37$eV, the lattice constant is $R_1 = 3.57\cdot 10^{-8}$cm. For these parameter values we calculated the value of the rate (22) in two cases: $d = 1.0$D an $d = 0.5$D.

Numerical results are shown in table 1, where $w_s$ and $w_p$ are the capture rates at $2s$ and $2p$-states respectively. In spite of serious approximations we obtained a numerical values which give an expected interval for a formation-time of the negatively charged acceptor center as an intermediate state of a radiation defect in crystals.

Negatively charged cluster $({}_{\mu}BC_4)^-$ is the ionized boron acceptor in a diamond lattice, which captures a hole from the valence band of a crystal because of Coulomb interaction and forms a neutral acceptor. This last stage of a radiation defect formation is well known and strongly depends on temperature of a crystal. Thus we can write a formula to determine the total formation rate of the radiation defect induced by negative muon in diamond crystals as a sum of rates of all stages:

$$W = w_{tr} + w_{cl} + w_0 + w_- + w_h. \tag{23}$$

Definitions of all items are obvious ($w_- = w_s + w_p$).

The numerical results show that the lowest stage is connected with the stage of the negatively charged cluster $({}_{\mu}BC_4)^-$ formation. From the table 1 we see that $0.5\cdot 10^8 s^{-1} < w_- < 2\cdot 10^8 s^{-1}$. So, we can see that a formation-time of the neutral acceptor center ${}_{\mu}$B is rather large and must

**Table 1.** Numerical results

| $d$, D | $\Omega$, $10^{10}\text{s}^{-1}$ | $w_s$, $10^8\text{s}^{-1}$ | $w_p$, $10^7\text{s}^{-1}$ |
|---|---|---|---|
| 1.0 | 1.24 | 2.1 | 1.26 |
| 0.5 | 0.62 | 0.55 | 0.34 |

be taken into account in the interpretation of experimental results [8, 9]. Namely, a computing of experimental results must be carried out on the base of kinetic equations like in chemistry reactions. This kind of the analysis could give us unknown parameters of the total process. But, this is the other problem.

## Acknowledgments
I would like to acknowledge Profs. R. Zaitsev and L. Sukhanov for fruitful discussions.

## References
[1] Belousov Y M 2012 *J. Phys.: Conf. Series* **343** 012013
[2] Antipov S A, Belousov Y M and Soloviev V R 2012 *JETP* **115** 866
[3] Mamedov T N, Duginov V N, Grebinnik V G, Gritsaj K I, Olshevsky V G, Pomjakushin V Y, Zhukov V A, Kirillov B F, Nikolsky B A, Pirogov A V *et al.* 1994 *Hyp. Interactions* **86** 717
[4] Mamedov T N, Chaplygin I L, Duginov V N, Grebinnik V G, Gritsaj K I, Olshevsky V G, Pomjakushin V Y, Stoykov A V, Zhukov V A, Krivosheev V A *et al.* 1997 *Hyp. Interactions* **105** 345
[5] Mamedov T N, Gritsaj K I, Stoykov A V, Andrianov D G, Gorelkin V N, Herlach D, Zimmermann U, Kormann O, Major J and Schefzik M 2000 *Physica* B **289-290** 574
[6] Andreika D, Baturin A S, Herlach D 2006 *Physica* B **374-375** 390
[7] Mamedov T N, Baturin A S and Duginov V N 2012 *JETP Letters* **95** 662
[8] Mamedov T N, Baturin A S, Blank V D, Gritsaj K I, Kuznetsov M S, Nosukhin S A, Ralchenko V G, Scheuermann R, Stoykov A V and Terentiev S A 2008 *Diamond Relat. Mattter* **17** 1221
[9] Mamedov T N, Baturin A S, Gritsaj K I, Maisuradze A, Ralchenko V G, Scheuermann R, Sedlak K and Stoykov A V 2013 `http://arXiv.org/abs/1305.5706`
[10] Belousov Y M and Smilga V P 1994 *JETP* **79** 811
[11] Belousov Y M and Smilga V P 1997 *Hyp. Interactions* **106** 19
[12] Belousov Y M 2007 *JETP* **104**, 215
[13] Landau L D and Lifshitz 1987 *Course of Theoretical Physics, Vol.7: Theory of Elasticity* (Moscow: Nauka)
[14] Landau L D and Lifshitz 1993 *Course of Theoretical Physics, Vol.8: Electrodynamics of Continuous Media* (Oxford: Elsevier)
[15] Madelung O 2004 *Semiconductors - Basic Data. 3rd ed.* (Berlin Heidelberg: Springer)
[16] Adachi S 2005 *Properties of Group-IV, III-V and II-VI Semiconductors. Wiley Series in Materials for Electronic & Optoelectronic Applications* (Chichester: John Wiley & Sons)

# High-field study of muonium states in $HfO_2$ and $ZrO_2$

**R B L Vieira[1] , R C Vilão[1], H V Alberto[1], J M Gil[1], A Weidinger[1], B B Baker[2], P W Mengyan[2] and R L Lichti[2]**

[1] CEMDRX, Department of Physics, University of Coimbra, P-3004-516 Coimbra, PORTUGAL

[2] Department of Physics, Texas Tech University, Lubbock TX 79409-1051, USA

E-mail: `ruivilao@fis.uc.pt`

**Abstract.** We present high-transverse field measurements, as a function of temperature, in monoclinic $ZrO_2$ and $HfO_2$. In monoclinic zirconia and hafnia, a diamagnetic component had been previously reported in low-transverse-fields, but a significant fraction of the total muon polarization was missing in these experiments. We now characterize this missing fraction using the high-field capabilities at TRIUMF: a high relaxation component (above 100 $\mu s^{-1}$ in monoclinic $ZrO_2$, about 10 $\mu s^{-1}$ in $HfO_2$) is observed, which we relate to the formation of compact muonium in these materials. A model for the formation of muonium in these materials is presented.

## 1. Introduction

Oxides with high dielectric permittivity (high-$\kappa$) are currently under intense scrutiny in order to find adequate solutions for the insulator layer in the gates of MOSFET devices. $ZrO_2$ and $HfO_2$ are among the most prominent candidates and $HfO_2$-based transistors have already reached the commercialization stage [1]. Processing of the oxides, whereby hydrogen is often incorporated, is known to have a big impact in its electrical performance. Hydrogen is a strong candidate defect in order to explain the positive fixed charges in the $ZrO_2$ and $HfO_2$ layer [2]. Fundamental studies on the effect of hydrogen in the electrical properties of semiconductors and insulators have known major advances due to the analogy with muonium, a light pseudo-isotope of hydrogen [3].

Previous $\mu$SR measurements on monoclinic $HfO_2$ and $ZrO_2$ [4] revealed the formation of a diamagnetic component with a temperature dependent relaxation, which was attributed to the ionization of a (shallow) neutral muonium state at an oxygen-bonded position. However, most of the muon spin polarization was not observed at these ISIS measurements at room temperature: a missing fraction of about $80-90\%$ was reported, which was correlated to the formation of a deep muonium state through longitudinal-field repolarization measurements.

Lyons *et al.* [2] have calculated the stable configurations for hydrogen in monoclinic $ZrO_2$ and monoclinic $HfO_2$ and found negative-U behaviour (indicating compensating character of the impurity). $H^+$, bonded to a threefold-coordinated oxygen atom, was found to be the stable configuration for lower values of the Fermi level, donating its electron to the oxygen atom and acting as a donor. In the negative charged state, stable for higher values of the Fermi level, hydrogen was found to sit interstitially, being coordinated by three Hf atoms. In these calculations, hydrogen was also found to be possibly incorporated on an oxygen site in the so-called multicenter bond configuration [5].

## 2. Experimental details

Muon spin rotation experiments were performed using the HiTime instrument of the M15 beamline at TRIUMF. High-transverse field experiments (up to 7 T) were performed on high-purity powder samples of monoclinic $HfO_2$ and $ZrO_2$, obtained commercially from Alfa-Aesar. Data analysis was done using the WiMDA program [6]. The data reveal the presence of two components precessing at the diamagnetic Larmor frequency $\omega_L$, with very different relaxations. The data were therefore fitted with two relaxing components, according to

$$A(t) = f_{fast}e^{-\lambda_{fast}t}\cos\left(\omega_L t + \phi_{fast}\right) + f_{slow}e^{-\lambda_{slow}t}\cos\left(\omega_L t + \phi_{slow}\right) \quad (1)$$

We present in Figures 1(a) and 1(c) the first 50 ns of the time spectra obtained at $T = 2$ K, for $B = 2\,$T. In Figures 1 (b) and 1(d) we show the same spectrum after removal of the slowly relaxing component, fitted as described above. Calibration experiments at room temperature were done with a silver sample in order to extract the maximum instrumental muon spin polarization, as well as the phase of the calibration signal.

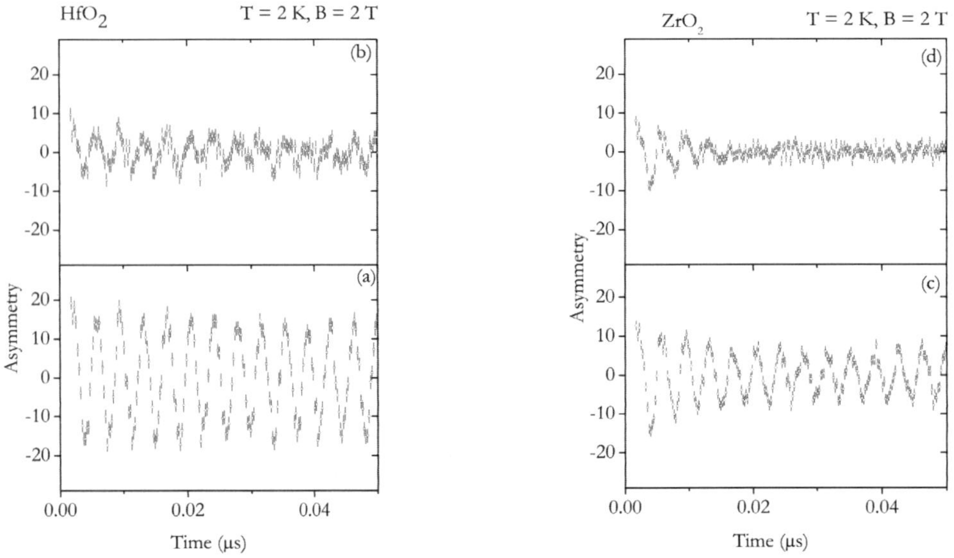


**Figure 1.** Time spectra at 2T (detail of first channels showing the fast relaxation) for $HfO_2$ (left) and $ZrO_2$ (right) at T = 2 K and for transverse magnetic field B = 2 T. The full time spectra are shown in (a) and (c). In (b) and (d) only the fast component is shown, as discussed in the text.

The analysis of the slow component was first performed in the time interval $0.3 - 2.89\,\mu$s and the corresponding parameters were henceforth fixed and used in the two component analysis in the interval $0 - 0.3\,\mu$s. In the final fits, the relaxation $\lambda_{fast}$ of the fast component was fixed to the low-temperature ($T = 2$ K) value. Table 1 summarizes the fitting results for the lowest temperature ($T = 2$ K).

## 3. Results and discussion

We present in Fig. 2 the temperature dependence of the fraction of the muon spin polarization corresponding to each configuration, for $HfO_2$ and for $ZrO_2$, after normalizing to the maximum

**Table 1.** Low-temperature (2 K) fit parameters, as discussed in the text.

| | $\lambda_{slow}$ | $\lambda_{fast}$ | $\phi_{slow}$ | $\phi_{fast}$ | $f_{slow}$ | $f_{fast}$ | missing fraction |
|---|---|---|---|---|---|---|---|
| $HfO_2$ | $0.220(4)\,\mu\,s^{-1}$ | $11.1(7)\,\mu\,s^{-1}$ | 151.8(3) | 152(2) | 54.9(4)% | 20.9(9)% | 24(1)% |
| $ZrO_2$ | $0.272(7)\,\mu\,s^{-1}$ | $101(8)\,\mu\,s^{-1}$ | 152.6(5) | 153(2) | 25.8(3)% | 43(3)% | 31(3)% |

instrumental asymmetry obtained with the silver calibration. The corresponding temperature dependence of the phase $\phi_{slow}$ of the slow component is shown in fig. 3. The dashed line in fig. 3 represents the value $\phi_{Ag} = 140(1)$ deg obtained for the phase of the calibration measurement with silver at room temperature.

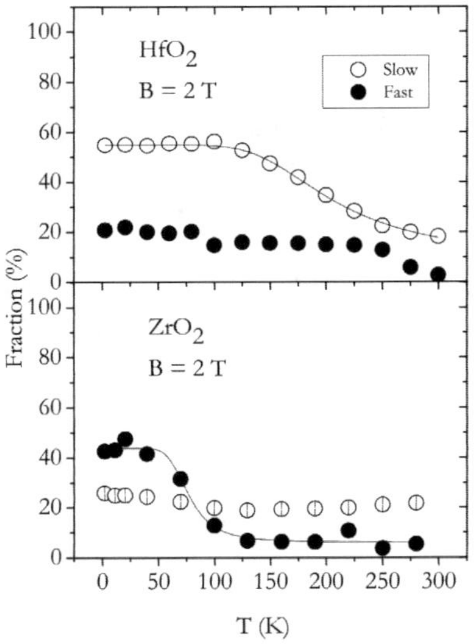


**Figure 2.** Temperature dependence of the fraction $f_{slow}$ and $f_{fast}$, for $ZrO_2$ and $HfO_2$, for a transverse field $B = 2$ T.

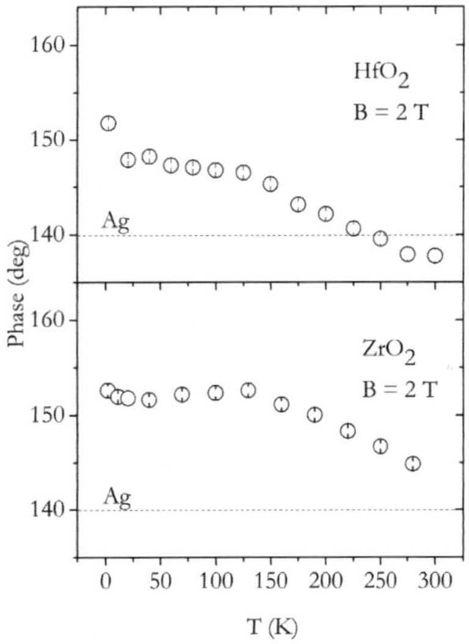


**Figure 3.** Temperature dependence of the phase $\phi_{slow}$ of the slowly relaxing component, for $ZrO_2$ and $HfO_2$, for a transverse field $B = 2$ T. The dashed line indicates the value obtained for the silver calibration measurement.

### *3.1. Slow component*

We assign the slowly relaxing diamagnetic component to muons thermalizing at an oxygen-bonded position, $Mu_{Ob}$, forming a $O-\mu$ bond. This site was already discussed at length by Cox *et al.* [4] and a possible reinterpretation is discussed in an acompanying paper [7]. *Ab-initio* calculations in similar systems [2, 8] also show that this configuration is the more stable for the positively charged $H^+/\mu^+$.

However, the observed phase-shift (Fig. 3) indicates that this state results from the conversion of a previous (unobserved) state. As mentioned already, previous longitudinal-field repolarization measurements in both monoclinic $ZrO_2$ and $HfO_2$ reveal the formation of a deep muonium state at room temperature [4]. The slowly relaxing diamagnetic component is thus likely to arise from delayed formation of a precursor deep muonium state, as observed in other semiconductors [9, 10].

The precursor muonium state is likely to sit at an interstial position. However, at the lowest temperatures, the lattice has unsufficient time to relax around the impurity and the neutral precursor muonium at the interstitial position thus corresponds to a metastable interstitial configuration [9, 10, 11]. From this metastable interstitial configuration, the muon can quickly convert to the oxygen-bonded position, $Mu_{Ob}$.

The alternative conversion of the metastable interstitial configuration to its true minimum at an interstitial site does not appear to dominate at the lowest temperatures. This process leads to the formation of the missing fraction, where the stable interstitial state is strongly dephased due to line broadening in the conversion process. However, the temperature dependence of the slow component and of the missing fraction suggest that temperature increases the conversion rate of the metastable interstitial configuration to its true minimum. This process thus becomes a more competitive alternative to the conversion to the oxygen-bonded configuration. This is also evidenced by the temperature dependence of the phases in Fig. 3, where a strong dephasing of the slow component is observed for the lowest temperatures, which gradually approaches the value obtained for the silver calibration as the temperature increases, suggesting a suppression of the process responsible for the phase-shift.

In the stable interstitial configuration, the barrier for the conversion to the oxygen-bonded site is larger. The final fraction of muons in the slowly relaxing diamagnetic component therefore decreases as temperature increases, with a corresponding increase of the missing fraction.

We propose that this increase with temperature of the competing formation of the true interstitial configuration is mediated by optical phonons in these ionic compounds. When these modes are activated, the formation process of the true interstitial configuration is enhanced. We have thus modelled the decrease of the slow component with a phenomenological Boltzmann-like functional dependence [4, 12, 13]:

$$f_{slow} = \frac{f^0_{slow}}{1 + N\exp\left(-E_a/k_B T\right)} \tag{2}$$

where $E_a$ is the activation energy, $k_B$ the Boltzmann constant and $f^0_{slow}$ is the low temperature value of the slowly relaxing fraction. N is a parameter associated to the effective density of phonon modes.

The lines in fig 2 are fits with this phenomenological Boltzmann-like dependence. We get an activation energy of 79(7) meV for the decrease of the slow fraction in $HfO_2$ above 100 K. We note that the phonon modes in monoclinic $HfO_2$ are calculated in the range $10 - 100$ meV [14]. We also note that the fraction of muons stopping at the slowly relaxing diamagnetic fraction is seen to be relatively temperature independent in $ZrO_2$ (Fig. 2), which is consistent with the much smaller value (around 10 meV) of the phonon mode with maximum density of states in monoclinic zirconia [15].

### *3.2. Fast component*

In Fig. 4 we show the field dependence of the fast relaxation $\lambda_{fast}$ in $ZrO_2$, for $T = 20$ K. $\lambda_{fast}$ is seen to be fairly field independent, suggesting that it corresponds to lifetime broadening associated to delayed muonium formation. We suggest that it corresponds to muons stopping at the interstitial position in a charged state (either $Mu_I^+$ or $Mu_I^-$) and converting to $Mu_I^0$, where the neutral configuration is known to be stable [2, 8]. In the $Mu^+$ case, the electron is captured from the radiolysis products; in the $Mu^-$ case, the second electron is either lost to the conduction band or to the valence band (hole capture).

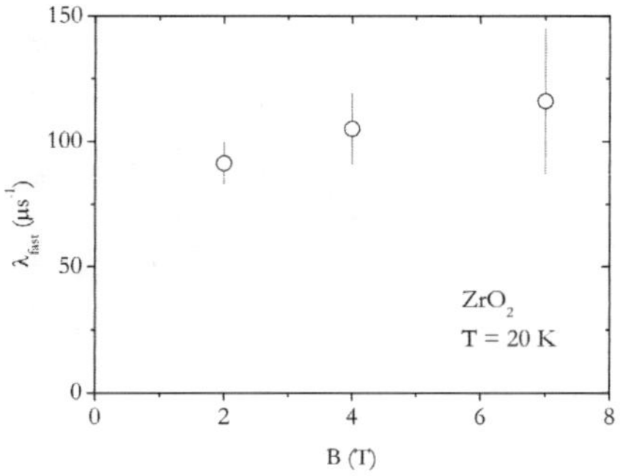


**Figure 4.** Field dependence of the relaxation $\lambda_{fast}$ of the fastly relaxing component in $ZrO_2$, in the transverse geometry, for a temperature $T = 20$ K.

We note that the fast component disappears as the temperature is increased (above 250 K for $HfO_2$; above 50 K for $ZrO_2$). It is not clear from the fits whether this is by line broadening or by a reduction in the fraction $f_{fast}$ of muons stopping in this configuration. In the delayed muonium formation model, it is more likely that the fraction is reduced, so we choose to fix the fast relaxation throughout the temperature dependence.

We note that the relaxation of the fast component is almost one order of magnitude higher for $ZrO_2$ with respect to $HfO_2$. Also, the temperature stabilities of these components are very different: whereas in $HfO_2$ the fast component is observed up to 250 K, in $ZrO_2$ it disappears at about 100 K, with an activation energy of 46(13) meV (again fitting the fast fraction in fig 2 with a Boltzmann-like dependence such as that of eq. 2).

We recall that it is not possible to distinguish $Mu^+$ from $Mu^-$ in $\mu$SR measurements, so it is not possible to establish definitively whether the diamagnetic component associated to the fast fraction is positively or negatively charged. It is however likely that it sits at (or close to) the same interstitial position than $Mu_I^0$. If the fast relaxing diamagnetic component corresponded to $Mu^+$, this would imply delayed electron capture, with an average time of 10 ns (for $ZrO_2$) or 100 ns (for $HfO_2$). This seems to be unlikely high.

It seems therefore plausible that the fastly relaxing diamagnetic fraction corresponds to $Mu^-$. In this case, a small fraction of the incoming muons would capture two electrons in the thermalization process (as suggested for other systems [12, 13, 16]) and would end up in the corresponding stable position at the interstitial site.

## 4. Conclusions

Fig. 5 summarizes our tentative interpretation in a potential-well diagram [4], representing the potential energy for the neutral muonium atom as a function of the configuration coordinate along the path from the interstital site to the oxygen-bonded site. Muons in $HfO_2$ and $ZrO_2$ may stop at the interstitial configuration or at the oxygen-bonded configuration. The formation is however a complex process [9, 10], where a precursor muonium is involved. This precursor is likely at the interstitial position, albeit unrelaxed, and may have an electron or not. The delayed formation of neutral muonium from the charged precursor constitutes the fast component. The precursor stopping as neutral Mu may either relax to the true relaxed interstitial position (missing fraction) or tunnel to the oxygen bonded position. This latter case is progressively hindered as the former case progresses.

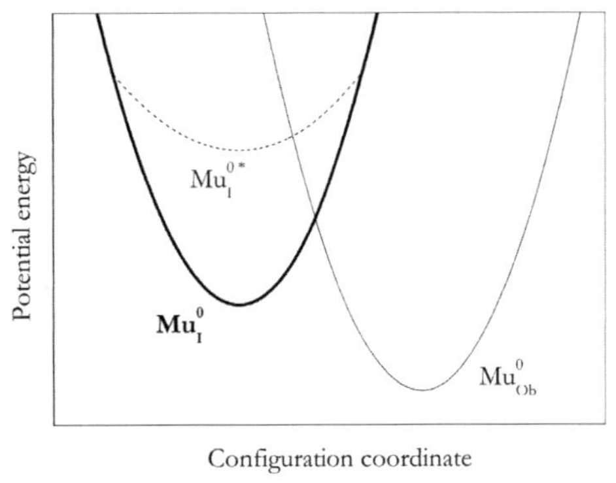


**Figure 5.** Schematic potential energy diagram for the neutral Mu states in $ZrO_2$ and $HfO_2$: a precursor neutral muonium state $Mu_I^{0*}$ is formed in a metastable interstitial position. It either converts from there to the oxygen-bound configuration (corresponding to the slowly relaxing component), or to the stable interstial configuration $Mu_I^0$ (unobserved due to dephasing or line broadening). The potentials of the charged states are not presented here.

**Acknowledgments**

The technical help of the $\mu$SR team at TRIUMF is gratefully acknowledged. This work was supported from funds from FEDER (Programa Operacional Factores de Competitividade COMPETE) and from FCT - Fundação para a Ciência e Tecnologia under projects PEst-C/FIS/UI0036/2014 and PTDC/FIS/102722/2008. RBLV was also supported by PhD grant SFRH/BD/87343/2012 from FCT - Fundação para a Ciência e Tecnologia. Participation of the TTU group is supported by the Welch Foundation (grant D-1321).

## References

[1] Bohr M T, Cox M T, Chau R S, Ghani T and Mistry K 2007 *IEEE Spectrum* **44** 29–35
[2] Lyons J, Janotti A and Van de Walle C 2011 *Microelectronic Eng.* **88** 1452–1456
[3] Cox S F J *et al.* 2013 *Phys. Scripta* **88** 068503
[4] Cox S F J *et al.* 2006 *J. Phys.: Condens. Matter* **18** 1079
[5] Janotti A and de Walle C G V 2007 *Nature Materials* **6** 44–47
[6] Pratt F L 2000 *Physica* B **289/290** 710–714
[7] Vieira R B L *et al.* 2014 *J. Phys.: Conf. Ser.* **551** 012050
[8] Marinopoulos A G 2012 *Phys. Rev.* B **86** 155144
[9] Alberto H V *et al.* 2010 *Phys. Rev.* B **81** 245205
[10] Alberto H V *et al.* 2012 *Phys. Rev.* B **86** 035203
[11] Vilão R C *et al.* 2005 *Phys. Rev.* B **72** 235203
[12] Vilão R C *et al.* 2011 *Phys. Rev.* B **84** 045201
[13] Silva E L *et al.* 2012 *Phys. Rev.* B **85** 165211
[14] Wu R *et al.* 2012 *J. Phys. D: Appl. Phys.* **45** 125304
[15] Fadda G, Zanzotto G and Colombo L 2010 *Phys. Rev.* B **82** 064105
[16] Vilão R C *et al.* 2008 *Phys. Rev.* B **77** 235212

Journal of Physics: Conference Series **551** (2014) 012049 doi:10.1088/1742-6596/551/1/012049

# Simulation of TF-$\mu$SR histograms in germanium in the presence of cyclic charge state transitions of muonium

**Thomas Prokscha**

Laboratory for Muon Spin Spectroscopy, Paul Scherrer Institut, CH-5232 Villigen PSI, Switzerland

E-mail: `thomas.prokscha@psi.ch`

**Abstract.** It has recently been shown that at $T < 270$ K a persistent inversion from $n$- to $p$-type can be induced by illumination in a 200-nm-thick surface layer of lowly-doped commercial germanium wafers [1]. The presence of photo-generated holes is detected by the appearance of a fast relaxing component in TF-$\mu$SR measurements, caused by cyclic muonium charge state transitions $\mathrm{Mu}_\mathrm{T}^- + h^+ \leftrightharpoons \mathrm{Mu}_\mathrm{T}^0$. For a quantitative determination of the photo-induced hole carrier concentration we use a Monte-Carlo simulation to generate muon decay histograms for different hole capture rates (forward reaction), and ionization rates (reverse reaction due to thermal activation) [2]. The hole carrier concentration is determined by comparing simulated and experimental relaxation rates. These results have been used to estimate the photo-induced hole concentration in low-energy $\mu$SR experiments [1].

## 1. Introduction

In semiconductors the presence of free charge carriers can give rise to charge state transitions of muonium. The change of the charged *diamagnetic* state (either $\mathrm{Mu}^+$ or $\mathrm{Mu}^-$) to the neutral *paramagnetic* state $\mathrm{Mu}^0$ causes a loss of muon spin phase coherence due to the hyperfine interaction in $\mathrm{Mu}^0$, which can be the origin of the appearance of a fast relaxing component in either longitudinal (LF) or transverse field (TF) $\mu$SR experiments. Recently, a photo-induced persistent inversion from $n$- to $p$-type of a hundreds of nanometer thick surface layer of commercial germanium wafers has been observed by low-energy TF-$\mu$SR at temperatures $T < 270$ K [1]. Here, the photo-generated holes become noticeable by the occurrence of a fast relaxing component in the diamagnetic precession signal. In view of a potential application of this effect to device technology it is important to know the photo-generated hole carrier concentration $p$. This question has motivated the current study, where we will show that $p$ can be determined from the observed fast relaxation rate $\Lambda_f$ of the diamagnetic signal by means of a Monte-Carlo technique [2].

In germanium, at $T > 180$ K, the thermally activated ionization of $\mathrm{Mu}_\mathrm{T}^0$ (muonium at the tetrahedral interstitial site) to the $\mathrm{Mu}_\mathrm{T}^-$ state leads to an increase of the diamagnetic asymmetry as a function of temperature [3]. In the presence of free holes $h^+$ cyclic charge state transitions $\mathrm{Mu}_\mathrm{T}^- + \mathrm{h}^+ \leftrightharpoons \mathrm{Mu}_\mathrm{T}^0$ cause a fast relaxing component of the diamagnetic $\mu$SR asymmetry, where the fast relaxation rate $\Lambda_f$ depends on $p$. The ionization of $\mathrm{Mu}_\mathrm{T}^0$ can be well described by an Arrhenius rate process with an ionization rate $\Lambda_i = \Lambda_0 \exp(-E_A/k_B T)$, where $\Lambda_0$ is the attempt

Journal of Physics: Conference Series **551** (2014) 012049 doi:10.1088/1742-6596/551/1/012049

frequency, and $E_A$ the activation energy. In germanium $E_A = 170$ meV and $\Lambda_0 = 6.7 \times 10^{13}$/s were determined in previous bulk $\mu$SR studies [4]. The hole capture rate $\Lambda_c$ is related to $p$ by $\Lambda_c = p \cdot v_p \cdot \sigma_c^h$, where $v_p$ is the temperature dependent hole velocity ($\sim 1.9 \times 10^7$ cm/s at room temperature, $v_p \propto \sqrt{T}$), and $\sigma_c^h$ is the hole capture cross section of $\mathrm{Mu}_\mathrm{T}^-$, which is expected to be in the order of "typical" atomic cross sections of $10^{-14} - 10^{-15}$ cm$^2$, and which has a temperature dependence $\sigma_c^h \propto T^{-2}$ [4]. We will show in section 2 that the TF-$\mu$SR $\Lambda_f \propto \Lambda_c$, and therefore $\Lambda_f \propto p$.

In LF-$\mu$SR experiments the longitudinal muon spin $1/T_1$ relaxation in muonium undergoing cyclic charge exchange can be expressed as [4, 5, 6]

$$1/T_1 = 0.5 \cdot \frac{\Lambda_i \Lambda_c}{\Lambda_i + \Lambda_c} \cdot \frac{\omega_0^2}{\Lambda_i^2 + \omega_0^2(1+x^2)}, \qquad (1)$$

where $\omega_0 = 2\pi \cdot 2359.5$ MHz is the hyperfine coupling of $\mathrm{Mu}_\mathrm{T}^0$, and $x = B/B_0$ with $B_0 = \omega_0/(\gamma_\mu - \gamma_e) = 0.0839$ T. Knowing $E_A$ and $\Lambda_0$ allows to calculate $\Lambda_i$ as a function of temperature. Thus, Eq. 1 contains only $\Lambda_c$ as an unknown parameter, which can be determined by fitting Eq. 1 to $1/T_1$ data measured as a function of magnetic field and temperature. Therefore, it appears to be more appropriate to do LF-$\mu$SR $1/T_1$ measurements rather than TF-$\mu$SR to determine $\Lambda_c \propto p$. However, this approach only works if $p$ is constant during the field and temperature scans. This is not the case for the experiments of Ref. [1] showing the photo-induced persistent inversion, where illumination was carried out for some time (minutes to hours) at 220 K, followed by $\mu$SR measurements without light at 220 K and higher temperatures, where holes disappeared by recombination due to the finite lifetime of the photo-induced inversion. Moreover, the temperature and field scans are a time consuming procedure, whereas our aim is to determine $p$ in the end by a single TF-$\mu$SR measurement. In the following section we will describe the method how this can be achieved.

## 2. Simulation procedure and results

We use a Monte-Carlo simulation [2] to calculate the evolution of the muon spin phase in an applied magnetic field in the presence of the cyclic charge state transitions described in the previous section. The ionization of $\mathrm{Mu}_\mathrm{T}^0$ with its effect on the diamagnetic signal is simulated the following way (assuming $p = 0$, i.e. hole capture is neglected):

- at time $t = 0$, the muon is in the $\mathrm{Mu}_\mathrm{T}^0$ state with a probability of 100%. Its decay time $t_d$ is "thrown" according to the exponential muon decay probability $\exp(-t/\tau_\mu)$ ($\tau_\mu$ the muon life time).
- for a given applied transverse magnetic field four precession ("transition") frequencies $\nu_{ij}$ are possible, see the Breit-Rabi diagram in Fig. 1. The probabilities $a_{ij}$ for the precession frequencies $\nu_{ij}$ are field dependent, with $a_{12} = a_{34}$ and $a_{23} = a_{14}$.
- according to the probabilities $a_{ij}$ the simulation "throws" one of the four precession frequencies. Transition frequencies and probabilities for the experimentally applied field of 0.1 T are shown in Tab. 1.
- the time of ionization $t_i$ is "thrown" according to the ionization rate $\Lambda_i$.
- if $t_i > t_d$, the muon precesses with frequency $\nu_{ij}$ until it decays, and the muon spin phase (precession angle) is $\Phi(t_d) = 2\pi\nu_{ij}t_d$. If $t_i < t_d$ the muon precesses with $\nu_{ij}$ until $t_i$, and after ionization the muon precesses now with its Larmor frequency $\nu_\mu$ until it decays. In this case we get $\Phi(t_d) = 2\pi\nu_{ij}t_i + 2\pi\nu_\mu(t_d - t_i)$. The decay positron histogram at time bin $t_d$ is incremented by $1 + A \cdot \cos[\Phi(t_d)]$, where $A$ is the muon decay asymmetry.

For increasing temperature the ionization rate $\Lambda_i$ increases exponentially and once $\Lambda_i \gtrsim \nu_{ij}$ the lifetime of $\mathrm{Mu}_\mathrm{T}^0$ is short enough so that the dephasing of the muon spin in the hyperfine field

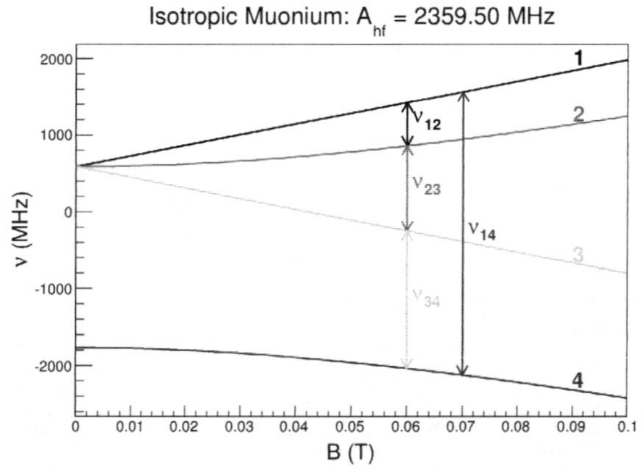


**Figure 1.** Breit-Rabi diagram with possible transitions $\nu_{ij}$ in a transverse magnetic field B of isotropic muonium $\mathrm{Mu}_\mathrm{T}^0$ in Ge with a hyperfine coupling of 2359.5 MHz.

**Table 1.** Transition frequencies and probabilities of isotropic $\mathrm{Mu}_\mathrm{T}^0$ in germanium in an applied transverse field of 0.1 T.

| transition | frequency (MHz) | probability |
|---|---|---|
| $\nu_{12}$ | 736.9 | 0.441 |
| $\nu_{34}$ | −1622.6 | 0.441 |
| $\nu_{23}$ | 2048.8 | 0.059 |
| $\nu_{14}$ | 4408.3 | 0.059 |

of the electron becomes sufficiently small: a coherent precession at the Larmor frequency $\nu_\mu$ occurs, and the diamagnetic asymmetry $\mathrm{A_D}$ of muons precessing at $\nu_\mu$ increases as the lifetime of $\mathrm{Mu}_\mathrm{T}^0$ decreases. An increase of $\mathrm{A_D}$ as a function of temperature is the result, as can be seen in Fig. 2, which shows excellent agreement between simulation and experimental data for the proper choice of $E_A$.

Hole capture of $\mathrm{Mu}_\mathrm{T}^-$ at a rate $\Lambda_c$ leads to a depolarization of the diamagnetic signal, if the lifetime of the generated $\mathrm{Mu}_\mathrm{T}^0$ state is long enough to cause dephasing of the diamagnetic fraction. In the presence of holes the simulation uses $\Lambda_i$ and $\Lambda_c$ as input parameters, and "throws" the times $t_{i,c}$ of successive ionization/capture processes, and follows the muon spin phase in the different charge states with precession frequencies $\nu_{ij}$ and $\nu_\mu$, until the muon decays. For each $\mathrm{Mu}_\mathrm{T}^0$ a different transition $\nu_{ij}$ may occur according to the transition probabilities $a_{ij}$.

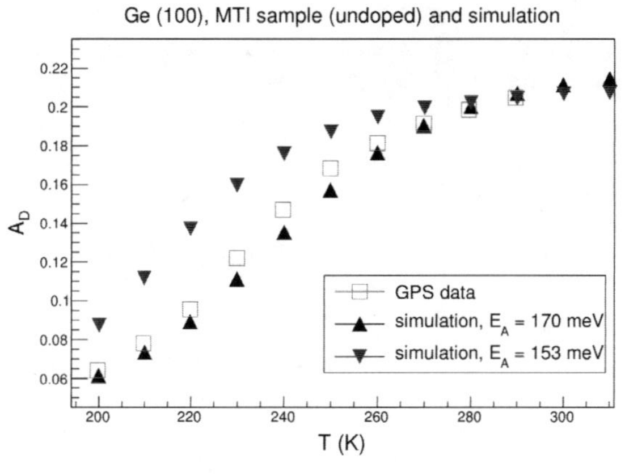


**Figure 2.** Temperature dependence of the TF-$\mu$SR diamagnetic asymmetry $\mathrm{A_D}$ (due to $\mathrm{Mu}_\mathrm{T}^-$ formation at $T > 180$ K), comparison of experimental bulk data (GPS, 0.1 T) of nominally undoped n-Ge (100) with simulation data for two different values of $E_A$: 170 meV (from Ref. [4]), and a lower $E_A$, which is far off the experimental data between 200 K and 260 K. The maximum diamagnetic asymmetry is 0.21.

In order to test the simulation in the presence of hole capture we used a $p$-type Ge wafer with $p \sim 10^{15}$ cm$^{-3}$, where we determined $\Lambda_c(T)$ and $E_A$ for a fixed $\Lambda_0 = 6.7 \times 10^{13}$/s by the temperature and $B$-field dependence of $1/T_1$ in bulk LF-$\mu$SR measurements. The data and a global fit according to Eq. 1 are shown in Fig. 3. The obtained value for $E_A$ is in good

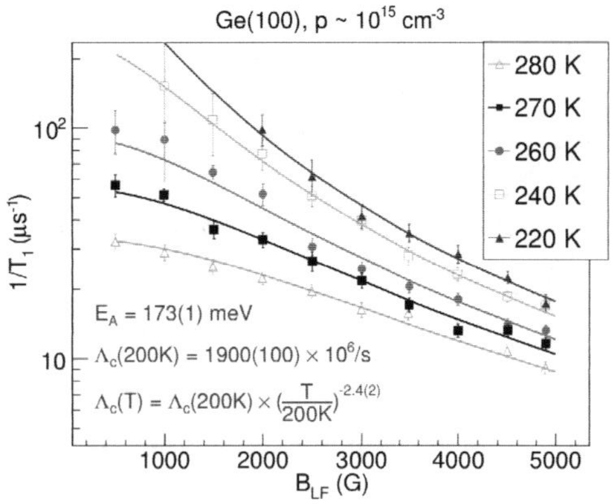


**Figure 3.** Ge (100), $p \sim 10^{15}$ cm$^{-3}$, bulk LF-$\mu$SR $1/T_1$ data (DOLLY). The solid lines are from a global fit of Eq. 1 to the data with the fit parameters in blue. These fit parameters were used in the Monte-Carlo simulation to generate the data shown in Fig. 4.

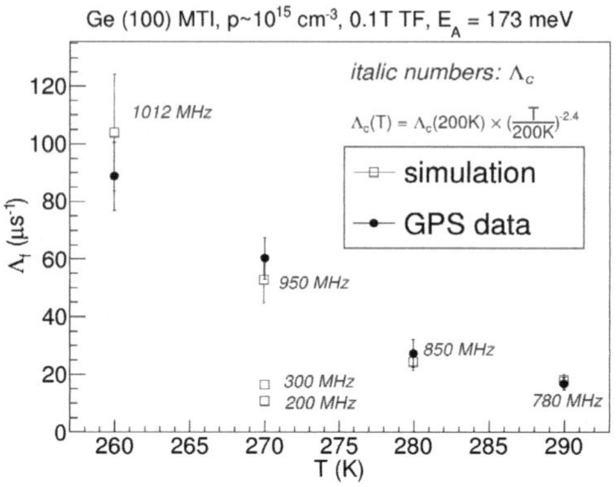


**Figure 4.** Ge (100), $p \sim 10^{15}$ cm$^{-3}$, comparison of bulk TF-$\mu$SR data (GPS) with the simulation. The blue squares are fast relaxation rates $\Lambda_f$ obtained from the Monte-Carlo simulation for the specified $\Lambda_c$, which are determined by the LF-$\mu$SR data in Fig 3.

agreement with the 170 meV from previous bulk studies [4]. In [4] the hole capture cross section has been found to decrease with temperature as $T^{-2}$. Here, we find $\Lambda_c \propto T^{-2.4(2)}$, which means – assuming that $p$ is constant in the temperature range of the experiment – that $v_p \cdot \sigma_c^h \propto T^{-2.4(2)}$. For the usual temperature dependence $v_p \propto \sqrt{T}$ one would expect a weaker temperature dependence $v_p \cdot \sigma_c^h \propto T^{-3/2}$. On the other hand the mobility of holes in germanium scales as $T^{-2.3}$ [7] and this suggests that the temperature dependence of $\Lambda_c$ is governed by the hole carrier mobility. For the Monte-Carlo simulations of the TF-$\mu$SR histograms we use the temperature dependence determined by the LF-$\mu$SR measurements. The TF-$\mu$SR fast relaxation rates as a function of temperature are shown in Fig. 4. Below 260 K the fast relaxation rate is too high ($\Lambda_f > 100\mu$s$^{-1}$) to be reliably fitted. The blue squares are the $\Lambda_f$ from the simulation, where the $\Lambda_c(T)$ are specified for each temperature: there is excellent agreement within errors between simulation and experiment. At 270 K two additional simulated fast relaxation rates for $\Lambda_c = 200$ and $\Lambda_c = 300$ MHz are displayed for comparison. They demonstrate that $\Lambda_f \propto \Lambda_c$, and therefore $\Lambda_f \propto p$ for a given temperature.

We now turn to the low-energy $\mu$SR illumination data of nominally undoped $n$-Ge from two different suppliers (Crystec GmbH and MTI corporation) [1]. The relaxation rates as a function of temperature and illumination are shown in Fig. 5. In the dark, only a slowly relaxing diamagnetic component with relaxation rate $\Lambda_s$ is observable. When turning on the light at 220 K the fast relaxing component with rate $\Lambda_f$ appears, which is revealing of the appearance of holes. Even after turning off the light at 220 K and slowly warming up, the fast component – i.e. the photo-generated holes – persists up to a temperature of $\sim$ 270 K, where the hole carriers quickly disappear, indicated by the fast drop of $\Lambda_f$. This persistent hole accumulation has its origin in the trapping of photo-generated electrons in empty surface acceptor states. The negatively charged surface keeps the holes in the at least 200-nm-thick

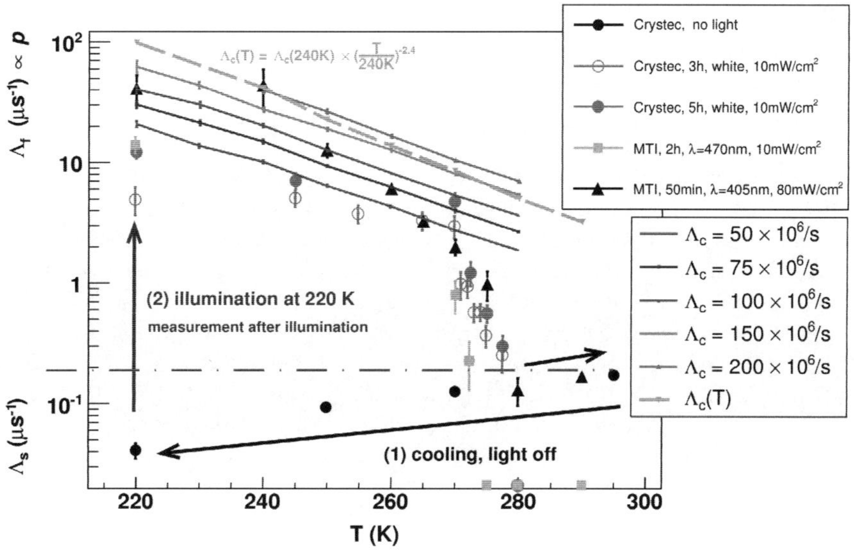


**Figure 5.** TF-$\mu$SR relaxation rates $\Lambda_s$ and $\Lambda_f$ of the slowly and fast relaxing components as a function of temperature $T$ and illumination history for two different commercial wafers (Crystec and MTI), low-energy $\mu$SR data at a mean depth $\langle z \rangle = 85$ nm, $B = 0.1$ T. Without the presence of holes there is only a slowly relaxing component visible: the $\Lambda_s$ data beneath the purple line. Symbols: experimental data from [1]. Solid and dashed lines: simulation results for different hole capture rates.

surface layer. Only after warming to $T > 270$ K, the trapped electrons are released from the surface acceptor states to recombine with the holes. The solid lines illustrate the results of the simulation for fixed (temperature independent) hole capture rates $\Lambda_c$. The decrease of $\Lambda_f(T)$ is caused in this case by the exponentially increasing ionization rate of $\mathrm{Mu}_\mathrm{T}^0$, which means that the muon spends less and less time in the neutral state, thus reducing the phase decoherence, i.e. $\Lambda_f$. The dashed line uses the $T^{-2.4}$ dependence of $\Lambda_c$ determined in the LF-$\mu$SR measurement. Here, $\Lambda_f(T)$ is decreasing faster compared to the fixed $\Lambda_c$ data because of the decreasing hole capture rate.

We now use the 240 K data to quantitatively estimate the photo-induced hole carrier concentration $p_{pi}$. According to Fig. 5 one needs $\Lambda_c(240\mathrm{K})$ to be between $150 \times 10^6$/s and $200 \times 10^6$/s to obtain the experimental value of $\Lambda_f$. Using the $p \sim 10^{15}$ cm$^{-3}$ sample as reference with $\Lambda_c^{p15}(240\mathrm{K}) = 1230 \times 10^6$/s, we obtain $p_{pi} = 200/1230 \times 10^{15}$ cm$^{-3}$ $= 1.6 \times 10^{14}$ cm$^{-3}$, which is in excellent agreement with estimates from resistance measurements [1].

## 3. Discussion and summary

In ultra-pure germanium in Ref. [4] a different charge transition for $\mathrm{Mu}_\mathrm{T}^0$ was suggested, involving a change to the bond-centred (BC) site: $\mathrm{Mu}_\mathrm{T}^0 \leftrightharpoons \mathrm{Mu}_\mathrm{BC}^+ + e^-$. In a recent study [8] on silicon-germanium alloys the formation of $\mathrm{Mu}_\mathrm{T}^-$ has been suggested to result from the process $\mathrm{Mu}_\mathrm{T}^0 \rightarrow \mathrm{Mu}_\mathrm{BC}^0 \rightarrow \mathrm{Mu}_\mathrm{T}^-$. However, these assignments do not naturally explain the observed effects in the presence of holes in the ionization region at $T > 200$ K in TF-$\mu$SR data (in particular the conversion of $\mathrm{Mu}_\mathrm{T}^0$ to $\mathrm{Mu}_\mathrm{BC}^+$ does not explain how $\mathrm{Mu}_\mathrm{BC}^+$ retransforms to $\mathrm{Mu}_\mathrm{T}^0$ in the presence of holes), nor do they represent the most simple model for the charge cycles in the

$p$-type sample LF-$\mu$SR measurements of Fig 3: the simplest charge cycle to interprete these data is the aforementioned $\mathrm{Mu}_\mathrm{T}^- + h^+ \leftrightharpoons \mathrm{Mu}_\mathrm{T}^0$. The temperature dependence of the carrier capture rate of $T^{-2.4(2)}$ also indicates the significance of holes, if one assumes that the carrier capture rate is governed by the carrier mobility: the electron mobility in Ge scales with $T^{-1.7}$, whereas the hole mobility scales with $T^{-2.3}$ [7], in agreement with the present data. In p-type samples with $p$ ranging between $10^{15}$ cm$^{-3}$ and $10^{16}$ cm$^{-3}$ we observe with low-energy muons a depletion of holes at depths of several hundred nanometer (for $p = 10^{15}$ cm$^{-3}$) and tens of nanometer (for $p = 10^{16}$ cm$^{-3}$) [9]. Under illumination – which removes the depletion layer, i.e. holes are moving towards the surface – we observe the appearance of fast relaxing components in LF- and TF-$\mu$SR measurements. This further supports the interpretation that the data can be described by charge-cycles involving only holes. In the end the essential result for this work is that the observed data can be described by simple cyclic transitions between two muonium charge states, where the charge capture rate (formation of the neutral state) is governed by the hole carrier concentration, and the formation of the charged state can be described by an Arrhenius like process. The details of the formation of the muonium states – either with or without site change – are less important for the objective of this work, which is the determination of the hole carrier concentration.

In summary we have shown that – in the presence of cyclic muonium charge state transitions – the extrinsic hole carrier concentration in germanium can be estimated by comparing TF-$\mu$SR relaxation rates with rates obtained from a Monte-Carlo simulation and by comparison with a reference sample. This is useful for low-energy $\mu$SR applications where the charge carrier concentration can be tuned by illumination and/or by electric fields at a semiconductor surface or interface. By means of the Monte-Carlo simulation it is possible to determine the charge carrier concentration by a single TF-$\mu$SR measurement, rather than by performing time consuming LF-$\mu$SR magnetic field and temperature scans.

## Acknowledgments

We would like to acknowledge the support of Hubertus Luetkens and Pabitra Biswas in the Ge measurements on the GPS and DOLLY instruments of S$\mu$S. We are grateful to Kim Chow whose computer code we used to calculate the $\mathrm{Mu}_\mathrm{T}^0$ frequencies and corresponding amplitudes. All data have been analyzed using the program `musrfit` [10]. The Monte-Carlo simulation is based on the `ROOT` toolkit developed at CERN. This work has been performed at the Swiss Muon Source S$\mu$S, Paul Scherrer Institut, Switzerland.

## References

[1] Prokscha T, Chow K H, Stilp E, Suter A, Luetkens H, Morenzoni E, Nieuwenhuys G J, Salman Z and Scheuermann R 2013 *Scientific Reports* **3** 2569
[2] Prokscha T 2012 *Physics Procedia* **30** 50–54
[3] Fan I, Chow K H, Hitti B, Scheuermann R, Mansour A I, MacFarlane W A, Schultz B E, Egilmez M, Jung J, Celebi Y G, Bani-Salameh H N, Carroll B R, Vernon J E and Lichti R L 2008 *Phys. Rev.* B **78** 153203
[4] Lichti R L, Cox S F J, Chow K H, Davis E A, Estle T L, Hitti B, Mytilineou E and Schwab C 1999 *Phys. Rev.* B **60** 1734
[5] Chow K H, Hitti B, Kiefl R F, Dunsiger S R, Lichti R L and Estle T L 1996 *Phys. Rev. Lett.* **76** 3790
[6] Chow K H, Kiefl R F, Schneider J W, Hitti B, Estle T L, Lichti R L, Schwab C, DuVarney R C, Kreitzman S R, MacFarlane W A and Senba M 1993 *Phys. Rev.* B **47** 16004
[7] van Zeghbroeck B 2011 *Principle of Electronic Devices, Chapter 2: Semiconductor Fundamentals* available at `http://ecee.colorado.edu/~bart/book/book/index.html`
[8] Carroll B R, Licht R L, King P J C, Celebi Y G, Yonegga I and Chow K H 2010 *Phys. Rev. B* **82** 205205
[9] Prokscha T *et al.* unpublished data of the PSI LEM group
[10] Suter A and Wojek B 2012 *Physics Procedia* **30** 69–73

Journal of Physics: Conference Series **551** (2014) 012050 doi:10.1088/1742-6596/551/1/012050

# Muon-Spin-Rotation study of yttria-stabilized zirconia ($ZrO_2$:Y): Evidence for muon and electron separate traps

**R B L Vieira[1], R C Vilão[1], P M Gordo[1], A G Marinopoulos[1,2], H V Alberto[1], J Piroto Duarte[3], J M Gil[1], A Weidinger[1] and J S Lord[4]**

[1] CEMDRX, Department of Physics, University of Coimbra, P-3004-516 Coimbra, Portugal
[2] Department of Physics & I3N, University of Aveiro, 3810-193 Aveiro, Portugal
[3] Institute of Shock Physics, Department of Physics, Imperial College London, Exhibition Road, London SW7 2AZ, United Kingdom
[4] ISIS Facility, Rutherford Appleton Laboratory, Chilton, Didcot, Oxon OX11 0QX, United Kingdom

E-mail: vieira.fis@gmail.com

**Abstract.** This paper is part of an extended study of oxide materials with the $\mu$SR technique. As an example, we present here experimental data on yttria-stabilized zirconia ($ZrO_2$ doped with 8% $Y_2O_3$). Three different muon states can be distinguished: i) Deep muonium (less than 17(1)% fraction), seen as a fast-relaxing signal or indirectly via decoupling measurements in high longitudinal fields, ii) $\mu^+$ in a paramagnetic environment 62(6)% fraction), characterized by a very weak but clearly-visible hyperfine interaction, and iii) diamagnetic muon 21(1)% fraction); the diamagnetic signal is broadened only by the interaction with nuclear moments. The state corresponding to $\mu^+$ in a paramagnetic environment and the diamagnetic state are attributed to the same (oxygen-bound) muon configuration, but we assume that they have different electron surroundings (with or without an unpaired electron in the vicinity). The paramagnetic electron is not captured in the Coulomb potential of the positive muon but is self-trapped (polaron formation) at a nearby Zr ion. The distant electron interacts with the muon only via dipolar magnetic fields. This explains the very weak hyperfine interaction felt by the $\mu^+$ state in a paramagnetic environment. A further result of the experiment is that the disappearance of this signal with increasing temperature is not due to ionization of an electron shallowly bound to the muon but is caused by rapid spin fluctuations of the electron, averaging the hyperfine interaction to zero.

## 1. Introduction

Zirconia is used in a multitude of devices such as oxygen sensors [1], thermal barrier coatings [2], medical prostheses [3], gate oxide material [4, 5], solid oxide fuel cells [6], cladding material for nuclear reactors [7]. In this material, protonated forms can occur in the conspicuous presence of hydrogen, and that might play an important role in the use of $ZrO_2$ in the last three applications mentioned. Muonium spectroscopy has now become a standard technique in order to investigate hydrogen configurations through modelling with muonium, a light pseudo isotope of hydrogen [8]. In $\mu$SR experiments on chalcogenides and oxides [9, 10, 11, 12, 13, 14, 15] one often finds

Journal of Physics: Conference Series **551** (2014) 012050 doi:10.1088/1742-6596/551/1/012050

a muonium state with a rather small hyperfine interaction. This signal is sometimes directly identified spectroscopically through the frequencies corresponding to the hyperfine interaction [9, 10, 11]. However, in many cases these lines are not resolvable and only a broad diamagnetic line is visible, with a temperature dependent relaxation larger than that expected from nuclear broadening alone [16, 15, 13, 14, 12]. These signals disappear at low temperatures between about 20 and 50 K. The usual interpretation is that a shallow muonium state is formed in the screened Coulomb potential of the muon. The ionization energy and the contact hyperfine interaction correspond roughly to the calculated values for a hydrogen-like atom with an effective-mass electron and a dielectric medium.

In our experiments in yttria-stabilized zirconia (and other oxide materials) we find severe deviations from this model: The hyperfine interaction is extremely small (only about 50 kHz) and not unique and the paramagnetic signal persists up to rather high temperatures (150 K and higher in some cases). These findings are inconsistent with the shallow-donor concept and require a different interpretation. Recent theoretical calculations [17] show that in these materials the paramagnetic electron is not localized on the oxygen-bound muon but is concentrated on a Zr cation some distance away from the muon. The corresponding single electron energy level lies about 1 eV below the conduction band indicating a rather strong binding energy. This concept of a separate localization of the electron and the muon gives a consistent interpretation of the present experimental data. Yttria-stabilized zirconia has been investigated before and the data have been published [14]. However, the analysis and the interpretation of these earlier measurements were not clear and many questions were left over. We felt that new and more precise data were necessary to solve the open problems. The main result of the present paper is a new interpretation of the paramagnetic signal. We show that it arises from a self-trapped electron in the vicinity of the muon.

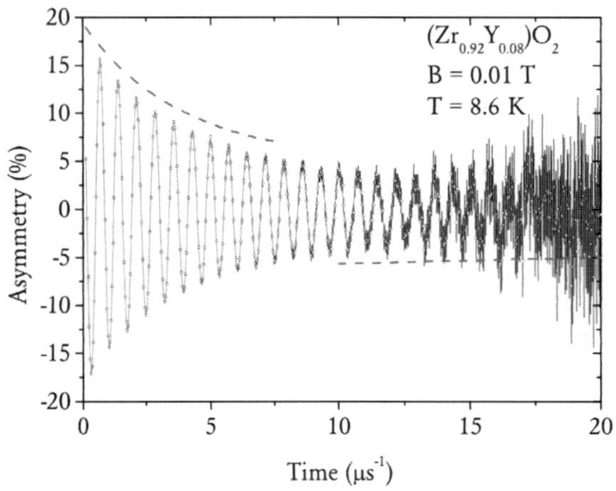


**Figure 1.** $\mu$SR time spectrum of $Zr_{0.92}Y_{0.08}O_2$ measured at ISIS in the transverse geometry, for an external magnetic field $B = 0.01$ T and a temperature $T = 8.6$ K. An oscillation at the Larmor frequency ($\nu$ =1.355 MHz) is observed. The spectrum is decomposed in two components ($f_{para}$ and $f_{dia}$) with slightly different relaxations, as discussed in the text. Dashed lines show the respective trends $f_{para}\exp(-\lambda_{para}\,t)$ (above) and $f_{dia}\exp(-\lambda_{dia}\,t)$ (below). The very-fast relaxing component (deep muonium) is not observed in this experiment.

Journal of Physics: Conference Series **551** (2014) 012050 doi:10.1088/1742-6596/551/1/012050

## 2. Experimental details

The muon-spin rotation experiments took place at the EMU instrument of the ISIS Facility, Rutherford Appleton Laboratory, United Kingdom. A nanocrystalline $(Zr_{0.92}Y_{0.08})O_2$ sample (grain size 17 nm) kindly provided by Innovnano was investigated. In the experiments, 4 MeV positive muons were implanted into the sample, and transverse-field measurements at an external magnetic field of 0.01 T were undertaken in the temperature interval 8.6 K to 600 K. Repolarisation measurements in the longitudinal geometry were also undertaken at room temperature (300 K). A calibration of the maximum instrumental asymmetry was performed with a silver sample. Data analysis was done using the WiMDA program [18]

## 3. Experimental results and analysis

Figure 1 shows the 8.6 K transverse-field time spectrum. In this experiment the presence of a fast-relaxing component is observed indirectly from a missing fraction of the muon spin polarisation. After comparison with the maximum instrumental asymmetry obtained with silver, the missing fraction, corresponding to the fast component, is 17(1)% at this temperature.

The main feature of the spectrum in figure 1 is an oscillation at the Larmor frequency corresponding to the external transverse magnetic field of 0.01 T. An analysis of the spectrum with a single relaxing component (using Lorentian, Gaussian, or Abragam relaxation functions) does not give an acceptable fit (typical $\chi^2$ values are around 2.5). We therefore decomposed the spectrum $P(t)$ into two fractions ( $f_{para}$ and $f_{dia}$) with slightly different relaxations ($\lambda_{para}$and $\lambda_{dia}$):

$$P(t) = f_{para} \exp(-\lambda_{para}\, t) \cos(\omega t + \phi) + f_{dia} \exp(-\lambda_{dia}\, t) \cos(\omega t + \phi) \qquad (1)$$

The angular frequency $\omega$ and the phase $\phi$ were assumed to be the same for the two fractions. The indices *dia* and *para* stand for diamagnetic and for $\mu^+$ in a paramagnetic environment, respectively.

We used exponential decay functions because of mathematical reasons since the relaxations are additive if two mechanisms contribute to the relaxation. This is the case for the fraction of $\mu^+$ in a paramagnetic environment, where electronic and nuclear interactions exist. If at higher temperatures the electronic interaction disappears, the nuclear part remains. In principle, a Gaussian function should be used for the nuclear part of the relaxation if this is the only interaction. But this leads to mathematical problems when two interactions (electronic and nuclear) exist at the same time. Thus choosing a Lorentzian function for the nuclear relaxation has practical reasons. The error introduced by this choice is not too severe since, at low temperatures, it concerns only the correction part of the relaxation but not the main component.

A fit with all parameters free did not give satisfactory results because of too many ambiguities. In particular at higher temperatures where the two components merge together, no clear separation was possible. We therefore used the following strategy: we first fitted the spectrum at 8.6 K with all parameters free, except $\lambda_{dia}$ which was fixed to 0.03 $\mu s^{-1}$. From this fit we obtained a formation probability of the diamagnetic fraction $f_{dia} = 21(1)\%$, a fraction $f_{para} = 62(6)\%$ of $\mu^+$ in a paramagnetic environment, and $\lambda_{para} = 0.29(1)\mu s^{-1}$, all at 8.6 K. We than assumed that the fractions do not change in the low temperature range and fitted the rest of the data with only $\lambda_{para}$ as a free parameter. The results are shown in Fig. 2.

The longitudinal-field data obtained at room temperature were fitted using $P(t) = f_{dia} \exp(-\lambda_{dia}\, t)$ since at that temperature the $\lambda_{para}$ relaxation is practically the same as $\lambda_{dia}$.

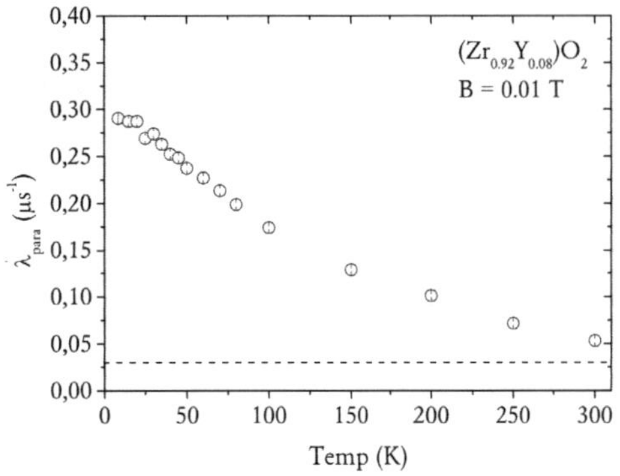


**Figure 2.** Temperature dependence of the relaxation rate $\lambda_{para}$ of the component corresponding to $\mu^+$ in a paramagnetic environment. The value of the corresponding fraction $f_{para}$ is given in the insert of the figure. The transverse magnetic field was 0.01 T. The dashed line at 0.03 $\mu s^{-1}$ corresponds to the estimated dipolar width caused by $^{91}Zr$ and $^{89}Y$ nuclei.

## 4. Discussion

### *4.1. Deep muonium state*

The 17(1)% fast-relaxing component is attributed to the formation of deep interstitial muonium centers at or near the surface of the nano-grains. An alternative interpretation is that muonium formation occurs in the bulk of the grains, despite the higher strain in the nano-crystals compared to samples with large grains. The strength of the hyperfine interaction for this muonium state is estimated to be around 2 GHz (about half the vacuum value) as derived from the decoupling curve in Fig.3. The interaction is possibly anisotropic. The line in figure 3 is a fit with an approximate analytical form for the repolarization of axially symmetrical muonium in a polycrystalline sample [19], yielding $A_{iso} = 2.1(1)$ GHz and $D = 0.129(15)$ GHz.

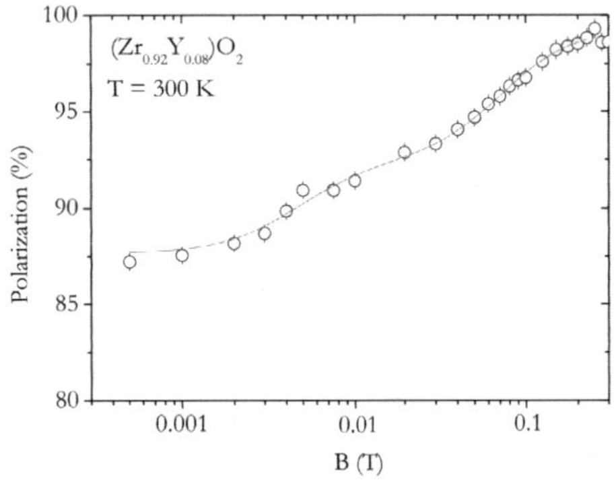


**Figure 3.** Repolarization curve at 300 K. The line is a fit with an approximate analytical form for the repolarization of axially symmetrical muonium in a polycrystalline sample [19], yielding $A_{iso} = 2.1(1)$ GHz and $D = 0.129(15)$ GHz.

### *4.2. $\mu^+$ state in a paramagnetic environment*

The 83(4)% slowly-relaxing signal is attributed to muons bound to oxygen (muoxyl $\mu$-O similar to hydroxyl H-O). As described above, we divided this part into two fractions, 62(6)% corresponding to diamagnetic muons without the influence of electron relaxation, and 21(1)% to a $\mu^+$ state in a paramagnetic environment. The relaxation of the latter component at low temperatures is almost an order of magnitude larger than the expected nuclear relaxation (about 0.03 $\mu s^{-1}$) and must therefore be due to an unpaired electron in the vicinity of the muon. However, the electron cannot be too close either, since this would cause a too strong interaction. Theoretical calculations by Marinopoulos [17] employing hybrid functionals show that the electron is likely to be localized predominantly on a Zr cation 4 Å away from the H nucleus. These calculations demonstrate convincingly this separate localization of hydrogen (analogous for muon) and the electron (Fig. 4): the density distribution of the unpaired electron (in yellow) is not centered on hydrogen (red point) but is localized quite a distance away from it. Thus the contact hyperfine interaction is small (or zero) and the only remaining term is the very small dipolar interaction. The dipolar interaction being angle-dependent, a distribution of hyperfine splittings is expected in a polycrystalline sample. This picture explains the small, non-unique value of the hyperfine interaction observed in the experiment.

The relaxation $\lambda_{para}$ has two contributions, one part $\lambda_e$ due to the nearby electron, the other $\lambda_n$ due to the magnetic field distribution caused by the surrounding nuclear moments: $\lambda_{para} = \lambda_e + \lambda_n$. The decrease with temperature is attributed to line narrowing due to electron spin fluctuations. This allows an estimate of the electron relaxation time $T_1$ [20, 21]. We obtain $T_1$-values in the order of 1 $\mu$s at 100 K. As far as we know, the value of $T_1$ for the corresponding T center observed in EPR measurements has not been measured and only $T_1$ values for the F center (electron trapped in an oxygen vacancy) are known ($T_1 = 7\mu$s at T=100 K) [22].

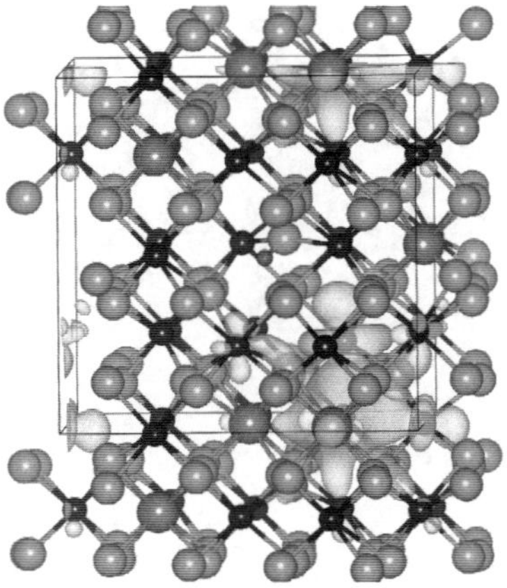

**Figure 4.** Calculated electron-density isosurface (in yellow) for the defect state of the neutral bond-type hydrogen (red point) configuration in yttria-stabilized zirconia. The unit cell, indicated by the cube, is repeated periodically [17].

**Acknowledgments**

The technical help of the $\mu$SR team at ISIS is gratefully acknowledged. We also thank INNOVNANO for providing the sample used in this work. This research project was supported from funds from FEDER (Programa Operacional Factores de Competitividade COMPETE) and from FCT - Fundação para a Ciência e Tecnologia under the project PEst-C/FIS/UI0036/2014, as well as by the European Commission under the $7^{th}$ Framework Programme through the 'Research Infrastructures' action of the 'Capacities' Programme, NMI3-II Grant number 283883, contract n°: CP-CSA_INFRA-2008-1.1.1 Number 226507-NMI3. RBLV was also supported by PhD grant SFRH/BD/87343/2012 from FCT - Fundação para a Ciência e Tecnologia.

## References

[1] Chao C C, Park J S, Tian X, Shim J H, Gür T M and Prinz F B 2013 *ACS Nano* **7** 2186–2191
[2] Chen L B 2006 *Surface Review and Letters* **13** 535–544
[3] Damestani Y, Reynolds C L, Szu J, Hsu M S, Kodera Y, Binder D K, Park B H, Garay J E, Rao M P and Aguilar G 2013 *Nanomedicine: Nanotechnology, Biology and Medicine* **9** 1135–1138
[4] Narayanachari K V L V and Raghavan S 2012 *Journal of Applied Physics* **112** 074910
[5] Lorenz M, Börner H, Semmelhack H C and Hochmuth H 1993 *Fresenius' Journal of Analytical Chemistry* **346** 169–172

[6] Guerra C, Lanzini A, Leone P, Santarelli M and Brandon N P 2014 *Journal of Power Sources* **245** 154–163
[7] Firouzdor V, Brechtl J, Wilson L, Semerau B, Sridharan K and Allen T R 2013 *Journal of Nuclear Materials* **438** 268–277
[8] Cox S F J, Lichti R L, Lord J S, Davis E A, Vilão R C, Gil J M, Veal T D and Celebi Y G 2013 *Physica Scripta* **88** 068503
[9] Gil J M, Alberto H V, Vilão R C, Piroto Duarte J, Mendes P J, Ferreira L P, Ayres de Campos N, Weidinger A, Krauser J, Niedermayer C and Cox S F J 1999 *Phys. Rev. Lett.* **83** 5294–5297
[10] Cox S F J, Davis E A, Cottrell S P, King P J C, Lord J S, Gil J M, Alberto H V, Vilão R C, Piroto Duarte J, Ayres de Campos N, Weidinger A, Lichti R L and Irvine S J C 2001 *Phys. Rev. Lett.* **86** 2601–2604
[11] Gil J M, Alberto H V, Vilão R C, Piroto Duarte J, Ayres de Campos N, Weidinger A, Krauser J, Davis E A, Cottrell S P and Cox S F J 2001 *Phys. Rev.* B **64** 075205
[12] Silva E L, Marinopoulos A G, Vilão R C, Vieira R B L, Alberto H V, Piroto Duarte J and Gil J M 2012 *Phys. Rev.* B **85** 165211
[13] King P D C, Lichti R L, Celebi Y G, Gil J M, Vilão R C, Alberto H V, Piroto Duarte J, Payne D J, Egdell R G, McKenzie I, McConville C F, Cox S F J and Veal T D 2009 *Phys. Rev.* B **80** 081201
[14] Cox S F J, Gavartin J L, Lord J S, Cottrell S P, Gil J M, Alberto H V, Piroto Duarte J, Vilão R C, Ayres de Campos N, Keeble D J, Davis E A, Charlton M and van der Werf D P 2006 *Journal of Physics: Condensed Matter* **18** 1079
[15] Vilão R C, Marinopoulos A G, Vieira R B L, Weidinger A, Alberto H V, Duarte J P, Gil J M, Lord J S and Cox S F J 2011 *Phys. Rev.* B **84** 045201
[16] Davis E A, Cox S F J, Lichti R L and Van de Walle C G 2003 *Applied Physics Letters* **82** 592–594
[17] Marinopoulos A G 2012 *Phys. Rev.* B **86** 155144
[18] Pratt F L 2000 *Physica* B **289-290** 710–714
[19] Pratt F L 1997 *Philosophical Magazine Letters* **75** 371–380
[20] Alberto H V, Weidinger A, Vilão R C, Duarte J P, Gil J M, Lord J S and Cox S F J 2010 *Phys. Rev. B* **81** 245205
[21] Alberto H V, Vilão R C, Duarte J P, Gil J M, Weidinger A, Lord J S and Cox S F J 2012 *Phys. Rev.* B **86** 035203
[22] Costantini J M and Beuneu F 2011 *Journal of Physics: Condensed Matter* **23** 115902

Journal of Physics: Conference Series **551** (2014) 012051 doi:10.1088/1742-6596/551/1/012051

# An Effect of the Supercell Calculation on Muon Positions and Local Deformations of Crystal Structure in $La_2CuO_4$

**B Adiperdana[1,2], E Suprayoga[1,3], N Adam[1,4], S S Mohd-Tajudin[4], A F Rozlan[4], S Sulaiman[4], M I Mohamed-Ibrahim[4], T Kawamata[5], T Adachi[6], I A Dharmawan[2], R E Siregar[2], Y Koike[5] and I Watanabe[1,2,3,4]**

[1]Advanced Meson Science Laboratory, RIKEN Nishina Center, 2-1 Hirosawa, Wako, Saitama 351-0198, Japan,
[2]Physics Department, Universitas Padjadjaran, Jl. Raya Bandung-Sumedang Km21, Jatinangor, Sumedang 45363, Indonesia,
[3]Physics Department, Bandung Institut Teknologi, Jl Ganesha 10, Bandung, Jawa Barat 40132, Indonesia,
[4]Computational Chemistry and Physics Laboratory, School of Distance Education, Universiti Sains Malaysia, 11800 Penang, Malaysia,
[5]Department of Applied Physics, Tohoku University, Sendai 980-8579, Japan,
[6]Department of Engineering and Applied Sciences, Sophia University, Tokyo 102-8554, Japan.

E-mail: nabedon@riken.jp

**Abstract.** Muon positions in $La_2CuO_4$ were examined by using the density functional theory. Potential minimum positions near apical and plane oxygen have been determined as possible initial muon stopping positions. We found that final muon stopping positions were different from those initial positions due to effects of the local deformation of crystal structure which was induced by injected muons. This means that injected muons relax their positions deforming local crystal structures and minimizing the total energy of the system. We also found that the estimation of those final muon positions had to be done in the large scale area as a supercell which contained 27 unit cells in order to achieve realistic situations of the system with the muon as a dilute impurity.

## 1. Introduction

Information of muon sites in materials is of importance to know about local hyperfine interactions in order to discuss detailed electromagnetic properties. The mother material of the Cu-based La high-$T_c$ oxides, $La_2CuO_4$ (LCO), is one of good target material which has been investigated so well by muons. In general, it has been argued that muons prefer positions near to anion due to the attractive force between the positive charge of muon and the negative one of anion. In the case of LCO, muons are expected to stop near apical and/or plane oxygen of the $CuO_6$ octahedra. In order to estimate muon positions, one way is to estimate minimum electrostatic potential by the first principal calculation and then to regard those potential minimum points as muon positions [1]. In past studies on LCO [2, 3, 4, 5, 6, 7, 8, 9], efforts have been paid to estimate muon positions by the minimization of the electrostatic potential and also from nuclear dipole field measurements. In addition to those information, the internal field coming from Cu spins at the muon site which is observed in an antiferromagnetically ordered state of LCO is a good indication to estimate the muon site. This has been measured in the beginning of the history of the $\mu$SR study on LCO to be about 415 G [10]. However, any numerical estimation has not yet succeeded to reproduce the experimental data and estimated muon positions were widely scattered as shown in Figure 1.

Recently, Suter *et al.* has shown an effect of the local deformation of the crystal structure by an injected muon which is a new concept to estimate the final position of the muon. They have indicated a possible relaxation of the muon position to the different final one pushing surrounding local atoms [11]. Following this result, we re-examined muon positions in LCO taking into account more parameters which have not been included in past studies for LCO, like the relaxation of the final muon position and the local deformation of the crystal structure induced by muons. Especially, we examined an effect of the size of the calculation area. This means that we compared results which were obtained from the single unit-cell calculation and supercell calculations which have never been tried in the past due to the limitation of performance and capacity of computing resources.

In this paper, we report the first result of our detailed estimation of final muon positions after taking into account the relaxation of the muon itself and the local deformation of the crystal structure. Especially, we emphasize in this paper the importance of the supercell calculation in order to estimate the final muon position in LCO.

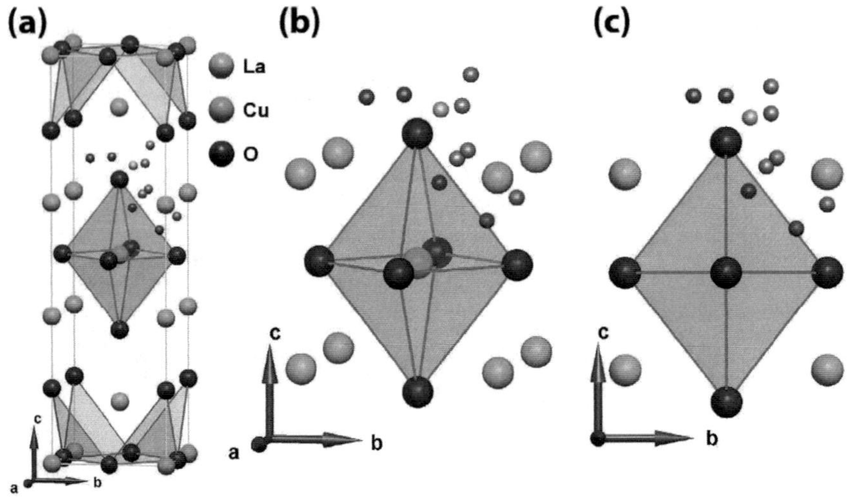


**Figure 1.** Muon positions in $La_2CuO_4$ which were suggested in past studies. (a) View with suggested muon positions in the past. (b) and (c) 3D and 2D views of $CuO_6$ octahedra with muon positions suggested in the past, respectively. Green marks are suggested from Saito *et al.* [2], a pink one is from McMillen *et al.* [3], an orange one is from Hitti *et al.* [4], grey ones are from Torikai *et al.* [5], and red ones are from S. Sulaiman *et al.* [6].

## 2. Calculations

We carried out the potential calculation by using the VASP program [12, 13]. We used the density functional theory (DFT) using generalized gradient approximation (GGA) corrections with Perdew and Wang (PW91) parametrization [14]. The methodology for determining muon sites by DFT calculations in insulators has been described and tested in some ionic fluorides [15, 16]. The projector augmented wave (PAW) method was used for the pseudopotential [17] with cutoff energy 500 eV. The resolution for electrostatic potential grid was $60 \times 60 \times 140$ in a-, b-, and c-axes. The crystal structure was optimized until the total energy and the total force were less than $1 \times 10^{-4}$ eV and $1 \times 10^{-3}$ eV/Å, respectively. At first, the electrostatic potential without the muon was calculated in order to estimate initial positions for injected muons. This first-step estimation showed that there would be more than two minimum potential positions [18]. After putting $H^+$ at those minimum potential positions as a model of the initially trapped muon, we estimated final muon positions taking into account effects of relaxation of the muon itself and the local deformation of the crystal structure. We repeated this operation on the single-unit cell and the supercell to see an effect of the calculation area to the final muon positions. The size of the supercell was $3 \times 3 \times 3$ which contained 27 unit cells. We used a computer cluster system in RIKEN named RIKEN Integrated Cluster of Clusters (RICC) to achieve the stronger computing performance.

## 3. Results and Discussions

Figures 2-(a) and (b) show the results of potential calculations around the $CuO_6$ octahedra of LCO. These calculations were done on the single-unit cell taking into account the orthorhombic crystal structure. Similar minimum potential areas were drawn by isosurfaces. As has been already suggested in past studies, we can see one possible muon position near the apical oxygen. Since the $CuO_6$ octahedra is slightly bent from the $CuO_2$ plane due to the backling of $CuO_6$ octahedra which leads the orthorhombic crystal structure, minimum potential area near the apical oxygen was not symmetric in all direction and tend to stay in one side around the apical oxygen. We found that the typical distance of this area from the apical oxygen was more than 1 Å which was bigger than those predicted in past studies. This suggests a possibility that the muon near the apical oxygen would not form the typical hydrogen bonding with the apical oxygen. At this moment, we have no detail estimation of the electronic state around the muon, so that more calculations are needed to discuss the bounding state of the muon with the apical oxygen.

Other positions were new compared to past studies. One was nearly attached to the surface of the $CuO_6$ octahedra. The distance from the apical oxygen was also more than 1 Å. The other one is on the line between two plane oxygen and almost on the $CuO_2$ plane. This position had the closest distance to the Cu spin. This means that the larger internal field would be expected at this position in the magnetically ordered state compared to other positions.

Those estimated positions have similar minimum potential. Differences in the potential energy were not so big and in the order of a couple of ten kelvin. Thus, all of positions can be candidates to trap injected muons. In terms of the number of muon positions in LCO, it has been believed for a long time to be one because the first $\mu$SR data on LCO has been well analyzed by a single component function [10]. Our current result of the potential calculation seems to be against to this experimental result. But, if the fraction of the muon-stopping ratio on each position is different and one position has large fraction compared to other components, the time spectrum can be analyzed by a single component in case of that the statics of the data is not good enough to realize other components with small amplitude. The zero-point vibration of the muon itself can help and enhance this effect. Since the zero-point vibration energy of the muon is recently estimated to be in the order of a half eV [15], the muon can move around potential minimum positions if the potential barrier between those positions is not so high rather than the zero-point vibration energy. At this moment, more numerical investigations are going on to estimate the difference of the fraction of stopping muons and the zero-point vibration energy.

In order to achieve final muon positions, we put $H^+$ as a model of $\mu^+$ and repeated calculations taking into account effects of the muon-position relaxation and the local deformation of the crystal structure. Since the muon position near the apical oxygen has been argued to be the most possible one in past studies, we concentrated on this position to estimate as the first trial. Figure 3 shows the estimated final state of this position. Figure 3-(a) indicates the result of the single unit-cell calculation and Figure 3-(b) shows the one of the supercell calculation. Light-red marks are initial positions of the muon which are corresponding to the minimum potential position as indicated in Figure 2. Deep-red marks are final positions after the calculation including all effects. Red lines and blue lines show the initial and final positions of the $CuO_2$ plane before and after the calculation, respectively. Arrows in Figure 3 exhibit directions of the local deformation of the $CuO_6$ octahedra induced by the muon. The magnitude of each arrow is corresponding how much distance each atom is shifted from its initial place.

In the case of the single unit-cell calculation, as shown in Figure 3-(a), the injected muon can induce the local modification of the crystal structure. The apical oxygen which was the nearest to the muon was pushed up along the c-axis. The $CuO_2$ plane was forced to be rotated to the same direction with the buckling direction of the $CuO_6$ octahedra. Following this movement, the apical oxygen at the opposite site to the muon rotates along the buckling direction also. It can be said that the injected muon tends to make the buckling angle bigger on the basis of this single unit-cell calculation. A change in the potential state was expected to be induced due to this deformation of the local crystal structure causing the muon position to be relaxed and become stable after the muon found out the final position in order to make the total energy of the system minimized. Accordingly, the muon moved from the initial position to the up-right direction approaching to the apical oxygen.

We have to note that the single unit-cell calculation condition is corresponding to the case of that each unit cell has the muon at the crystallographcally same position. Accordingly, due to this single unit-cell periodicity, the local deformation of the next neighbor unit cell should be treated to be rotated in the opposite direction by the muon leading that the degree of the orthorhombicity of the crystal structure stronger. This operation could change the crystal structural axis by the injected muons. In addition to this, the single unit-cell calculation assumes that each unit cell has one muon, indicating that the injected muon is no longer as a dilute impurity and exists in all unit cells However, these conditions are not realistic. The density

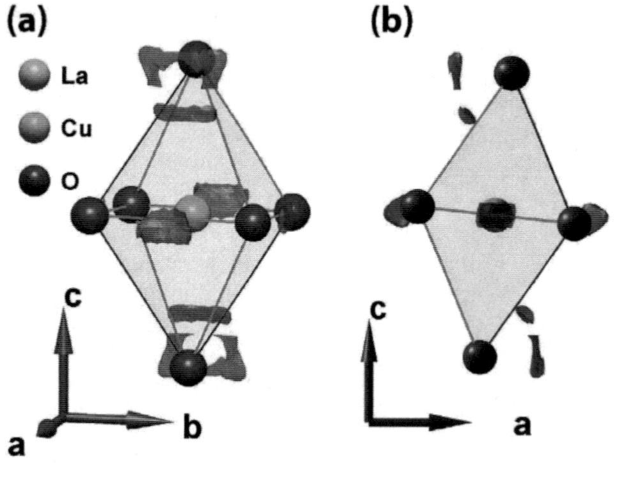


**Figure 2.** Estimated minimum potential positions around the $CuO_6$ octahedra of LCO. (a) A view in the 3 D state and (b) a view along the b axis. The similar minimum potential positions are drawn as isosurfaces with energy scales of about 0.25 eV.

Journal of Physics: Conference Series **551** (2014) 012051 doi:10.1088/1742-6596/551/1/012051

of the muon is very small in materials as very dilute impurities and the crystal structure must not be changed by the injection of muons. Past computational studies adopted this single-unit cell condition and resulted in the scattering of estimated muon positions. Therefore, the single unit-cell calculation could lead to an inappropriate result in terms of the final state of the muon and the local crystal deformation.

Following this result on the single unit-cell calculation, we re-tried the muon-position estimation on the basis of the super-cell calculation. We calculated 27 unit cells which were the limit of the acceptance of RICC at this moment. We put one $H^+$ as the model of the muon in the central unit cell of the 3×3×3 supercell block. We tested the same position with that adopted in the unit-cell calculation which was the nearest to the apical oxygen. No $H^+$ was allocated in neighboring unit cells, although the density of the muon in the supercell was still higher rather than any realistic cases. As shown in Figure 3-(b), the movement of the muon and the local deformation estimated from the supercell calculation were fairly different from those estimated in the single unit cell. The apical oxygen which was nearest to the muon moved to the up-right direction. The rotation angle of the $CuO_2$ plane became bigger rather than the case of the single unit-cell calculation, showing that the injected muon would push the rotation of the $CuO_6$ octahedra more. On the other hand, it was found that almost no local deformation was induced in other neighboring unit cells. Especially, all atoms near the edge of the supercell have almost no local deformation. Those facts mean that the deformation of the crystal structure induced by the injected muon is very limited in one unit cell which contains the muon as an impurity nearby and that this local deformation effect disappears quickly in neighboring unit cells. After the injected muon stopped at the initial position, the muon moved from the initial position to the right and slightly went down approaching to the surface of the $CuO_6$ octahedra. Since the apical oxygen was escaping to the up direction from the initial position, the distance between the apical oxygen and the muon did not change so much. This is naturally understood like that the muon would like to keep the distance to the oxygen not to change the relative potential energy higher.

We are now trying to explain the experimental result of the hyperfine field from Cu spins at the muon site in the magnetically ordered state [10] on the basis of the dipole-field calculation in order to confirm our calculation results. However, we have not yet achieved good agreement. In parallel, we are examining other muon positions which are shown in Figure 2 by using the same method to try to explain experimental results. Since LCO is a good insulator, we believe that the dipole-dipole interaction can explain the experimental data on the basis of our estimated

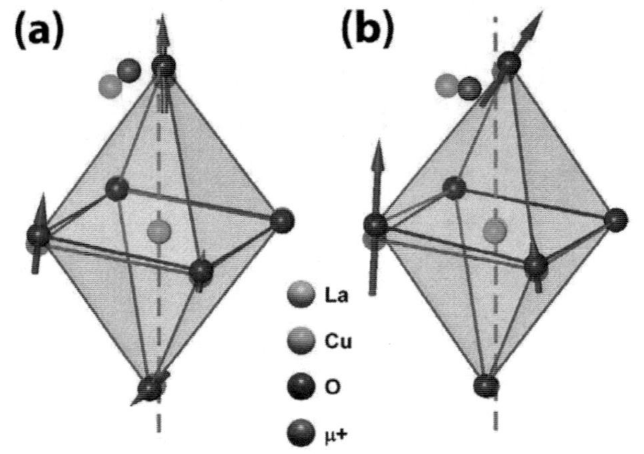


**Figure 3.** Initial and final states of estimated muon positions and local deformation of the $CuO_6$ octahedra. The muon position near the apical oxygen was investigated. Light-red and deep-red marks are the initial and final positions of the muon, respectively. Red and blue lines are positions of $CuO_2$ plane before and after the calculation, respectively. (a) In the case of the unit-cell calculation and (b) in the case of the supercell calculation.

muon positions. If our current trial finally fails, we have to move to think about a different hyperfine interaction which is the Fermi-contact type interaction. In order to think about such a type of the hyperfine interaction, we need to argue the electronic state of the muon which means the charged state of the muon itself and interactions with surrounding electronic orbitals. In parallel, we will try to include the magnetic interaction between ordered spins by using the spin-polarized DFT estimations in order to tune final muon positions which will take more calculation time and not the main matter to discuss in this paper at this moment.

## 4. Conclusion

We have examined possible muon sites in $La_2CuO_4$ (LCO) by using density functional theory calculations. Minimum electrostatic potential positions in the unit cell of LCO were estimated to be as initial positions of injected muons. We found some possible positions which have the similar potential energy. One of them was very similar with that discussed in past computational studies and close to the apical oxygen of the $CuO_6$ octahedra. Taking into account effects of the relaxation of the muon position and the local deformation of the crystal structure induced by the injected muon, final state of the muon was calculated in two models. One was the single unit-cell model and the other is the 3×3×3 supercell one. We found that the single unit-cell calculation would lead to inappropriate results which were different from a realistic situation. It was better to estimate muon positions by adopting the supercell with a single muon as an impurity. We confirmed that the injected muon can deform the local crystal structure. This effect was very limited around the muon and disappeared quickly in neighboring unit cells. One of estimated final muon positions was near the apical oxygen as has been argued in the past. However, this position still cannot explain the experimental result. More estimations on other muon positions are going on.

## Acknowledgments

Some of authors, BA, IAD, RES and IW would like to appreciate Direktorat Jendral Pendidikan Tinggi of Indonesia (DIKTI) for their supports of this study.

## References

[1] Maeter G *et al.* 2009 *Phys. Rev.* B **80** 094524
[2] Saito R *et al.* 1991 *Physica* C **185-189** 1217
[3] McMillen T *et al.* 1991 *Int. J. Mod. Phys.* B **5** 1579
[4] Hitti B 1990 *et al. Hyperfine Interact.* **63** 287
[5] Torikai E *et al.* 1993 *Hyperfine Interact.* **79** 921
[6] Sulaiman S *et al.* 1994 *Hyperfine Interact.* **84** 87
[7] Huang W *et al.* 2012 *Phys. Rev.* B **85** 104527
[8] Adiperdana B *et al.* 2012 *Physics Procedia* **30** 109
[9] Stilp E *et al.* 2013 *Phys. Rev.* B **88** 064419
[10] Uemura Y J *et al.* 1987 *Phys. Rev. Lett.* **59** 1045
[11] Suter H U *et al.* 2003 *Physica* B **326** 329
[12] Kresse G and Futhmüller J 1996 *Phys. Rev.* B **54** 11169
[13] Kresse G and Futhmüller J 1996 *Comput. Mater. Sci.* **6** 15
[14] Perdew J P *et al.* 1992 *Phys. Rev.* B **46** 6671
[15] Bernardini F *et al.* 2013 *Phys. Rev.* B **87** 115148
[16] Möller J S *et al.* 2013 *Phys. Rev.* B **87** 121108(R)
[17] Blöchl P E *Phys. Rev.* 1994 B **50** 17953
[18] Adiperdana B *et al.* 2013 *AIP Conf. Proc.* **1554** 214

# An investigation of muon sites in $YBa_2Cu_3O_6$ by using Density Functional Theory

**S. S. Mohd-Tajudin[1], S. N. A. Ahmad[1], D. F. Hasan-Baseri[1], E. Suprayoga[2,3], N. Adam[1,2], Rozlan A. F.[1], S. Sulaiman[1], M. I. Mohamed-Ibrahim[1], and I. Watanabe[2]**

[1]Computational Chemistry and Physics Laboratory, School of Distance Education, Universiti Sains Malaysia, 11800 Penang, Malaysia.
[2]Advanced Meson Science Laboratory, RIKEN Nishina Center, 2-1 Hirosawa, Wako, Saitama 351-0198, Japan.
[3]Physics Department, Bandung Institut Teknologi, Jl Ganesha 10, Bandung, Jawa Barat 40132, Indonesia.

E-mail: saidah.akin@gmail.com

**Abstract.** The electrostatic potential has been investigated in $YBa_2Cu_3O_6$ by applying the density functional theory in order to estimate possible muon sites. We found five minimum potential positions around the apical oxygen of the $CuO_5$ polyhedral. In addition to those, we also found another minimum potential position near the yttrium atom which is in between the $CuO_2$ layers. Those estimated positions are different from those suggested from previous $\mu$SR studies on the basis of the dipole-field estimation.

## 1. Introduction

The mechanism of high-$T_C$ superconductivity is still beyond a dark shade and more investigations are ongoing by using a lots of techniques. One key direction of the investigation is to understand the magnetic properties of the system. The muon-spin relaxation ($\mu$SR) study has been applied to high-$T_C$ oxides and has achieved a lots of results and suggestions from the beginning of the history of high-$T_C$ oxides [1-4]. One advantage of the $\mu$SR technique is the magnetic phase diagram. For instance, a long-range antiferromagnetically ordered state of Cu spins in $La_2CuO_4$ has been detected by $\mu$SR [1]. The co-existing state between the superconductivity and the magnetically ordered state has been suggested [4] and an exotic spin alignment around the hole concentration of 1/8 per Cu. This so called 1/8 effect has been detected in $La_{2-x}Ba_xCuO_4$ [3]. Such a magnetically ordered state and the 1/8 effect have been confirmed in $YBa_2CuO_{6+\delta}$ by means of $\mu$SR studies [5-7]. Since $\mu$SR data can give us information of hyperfine fields, we can principally discuss magnetic properties of the system. For such discussions, we need to know the muon positions because relative positions of stopped muons to magnetic moments are necessary to clarify hyperfine couplings which are the key information to understanding the magnetic interactions.

We have been trying to clarify the muon sites in Cu-based high-$T_C$ oxides in order to understand the Cu spin states and hyperfine interactions in order to achieve new information on the magnetic properties. One useful way to know the muon positions are to estimate electrostatic interactions between the injected muon and surrounding atoms on the basis of computational techniques. We have carried out computational investigations on $La_2CuO_4$ (LCO) by using the density functional theory (DFT) method taken into account effects of the local crystal deformation and relaxation position of the injected muon itself which have not been included in past studies to estimate muon positions [8]. Another type of high-$T_C$ oxides, $YBa_2Cu_3O_{6+\delta}$ (YBCO), is also a good target to apply our developed method to estimate muon positions. The hole concentration of YBCO can be changed by changing the oxygen content, $\delta$, from 0 to 1 and the system changes from the antiferromagnetic state to the superconducting state [5-7]. The magnetic phase diagram of YBCO is similar to that of LCO. The electronic property is insulating and the antiferromagnetic ordered state appears in the underdoped regime. The antiferromagnetic transition temperature, $T_N$, at $\delta = 0$ is above the room temperature. Clearly, muon-spin precession components have been observed below $T_N$ showing that there are some muon sites in YBCO [5]. The YBCO becomes metallic with increasing $\delta$ and the antiferromagnetic ordered state disappears and the superconducting state appears.

Although DFT investigations on LCO have been carried out by some groups in the past, studies on YBCO to determine muon sites have not yet been done, especially in the underdoped regime near $\delta = 0$. In this paper, we report the first and preliminary trial of the potential calculation by using the DFT method in order to estimate muon positions in YBCO with $\delta = 0$. We found at least six minimum potential positions in the unit cell where the injected muon could be initially trapped. Five of them were around the apical oxygen as argued in previous studies [5,9,10] but the sixth is near the yttrium atom which is in between the $CuO_2$ layers and not yet discuss in the past.

## 2. Results and Discussions

Figure 1 shows the crystal structure of YBCO with $\delta = 0$. The unit cell has the tetragonal structure with the space group P4/mmm. There are two Cu sites in the unit cell. One of the Cu is in the $CuO_2$ plane and the other is outside the $CuO_2$ plane. The Cu atom in the $CuO_2$ plane forms the polyhedral structure with five oxygen atoms and is named as Cu(II). The Cu(II) has a half spin with the charged state of $Cu^{2+}$. The other Cu atom forming the O-Cu-O linear bond is named as Cu(I). The Cu(I) does not have any magnetic moment and has a charged state of $Cu^{1+}$. Cell parameters of this system are $a = b = 3.8715$ Å and $c = 11.738$ Å [11].

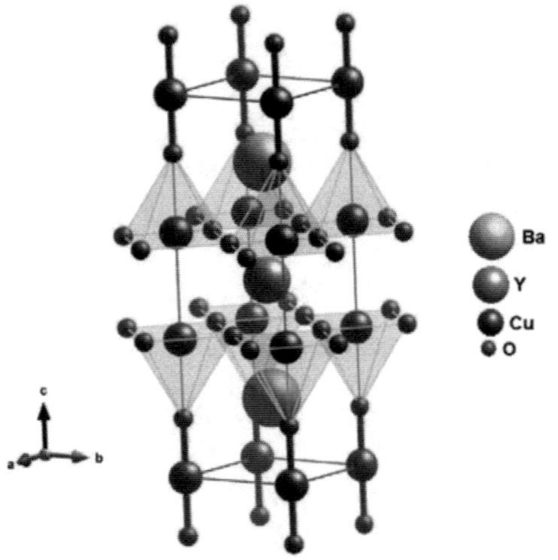


**Figure 1.** Crystal structure of $YBa_2Cu_3O_6$. The Cu(II) atom which is surrounded by five oxygen atoms and has a spin half with the charged state of $Cu^{2+}$. The Cu(II) influences the magnetism of this system and forms the antiferromagnetic structure in the magnetically ordered state, while the Cu(I) is in the non-magnectic state with zero spin and charged state of $Cu^{1+}$.

We have carried out the potential calculations by using the VASP program [12,13]. The total phonon energy of the model system was minimized changing the density of electrons with the threshold energy of $1 \times 10^{-4} eV$. We adopted the plain wave function and the Perdew-Burke-Ernzerhof (PE) function for the electronic correlations. We meshed the unit cell to be $100 \times 100 \times 140$ along each crystal axes and calculated the electrostatic potential without muon in order to estimate the most probable initial positions for the injected muons. We adopted a supercell which contained $2 \times 2 \times 2$ unit cells. All calculations were performed on the RIKEN Integrated Cluster of Clusters (RICC) supercomputer.

Figure 2 indicates the results of the potential calculations. We searched for the local minimum potential throughout the unit cell and found a few possible positions. Fig. 2(a) shows the total view of the unit cell with isosurfaces which show areas of minimum potential positions. Differences in colours exhibit differences in minimum potentials. Fig. 2(b) and 2(c) are expanded views of sections surrounded by black and pink lines in Fig. 2(a). Minimum potentials are marked by M1 to M6, following the order of the depth of the potential energy. Accordingly, the M1 position has the lowest energy and the potential energy increases to M6. Differences of potential energy of those positions were not so large and in the order of 10 Kelvin. It was not possible to find more minimum positions in regions which have the higher potential energy than that of the M6 position.

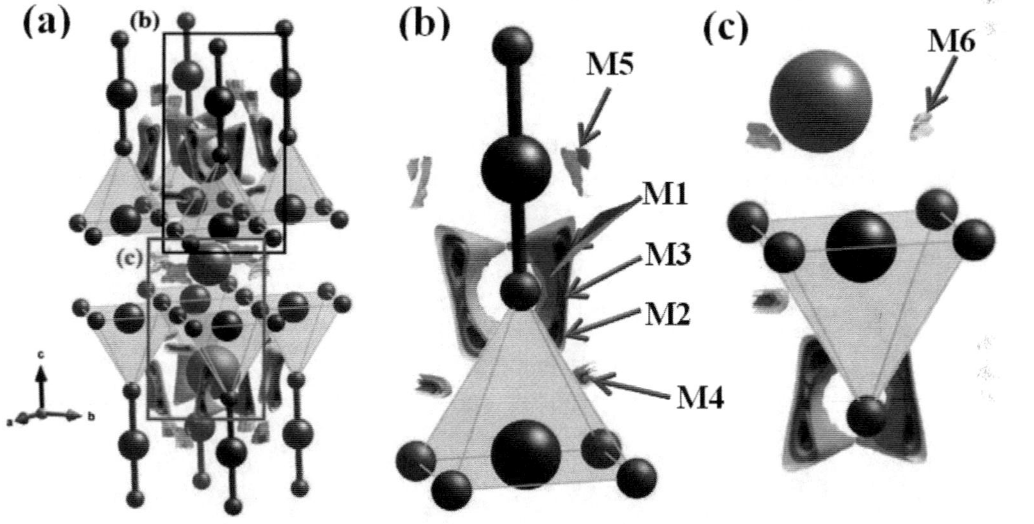


**Figure 2.** Results of the potential calculation by using the DFT method. Minimum potential positions are drawn by isosurfaces with different colours. The M1 position has the lowest potential energy and the M6 one has slightly higher energy compared to that of M1, although the difference of the energy is at most the order of 10 Kelvin.

Positions of M1, M2 and M3 are located around the apical oxygen of the $CuO_5$ polyhedral. The position of M4 is closer to the $CuO_2$ plane with the distance of about 1.10 Å. The fifth minimum potential position M5, is located between the Cu(I) atoms with the distance from Cu(I) of about 1.89Å, The sixth minimum potential M6, is near to the yttrium atom with the distance from the yttrium atom of about 1.95 Å. The angle of Y-M6-O was 94.26°. Detailed coordinates of all the estimated minimum potential positions and calculated potential energies at those positions were listed in Table 1.

Nishida *et al.* has suggested at least two muon positions in YBCO [5]. Plural muon positions in YBCO have also suggested by the other groups [9,10]. One is near the apical oxygen with a smaller internal field from Cu spins of about 300 G and the other is near the planar oxygen with a larger internal field of about 1300 G. Since the positions of M1, M2 and M3 are close to the apical oxygen of the $CuO_5$ polyhedral with the distance to the apical oxygen of about 1 Å, one of those positions can possibly be regarded as the one suggested by Nishida *et al.* [5]. However, the other minimum potential positions from our investigate ions do not match with those near the planar oxygen as suggested by Nishida *et al.* In addition to this, the minimum potential positions of M6 and M5 are new and difficult to think about from the view point of the well-known argument on the muon-binding state with anions. The injected muon prefers to associate with anions and forms like the hydrogen binding state, thus, gaining the binding energy with anions. We tried to find more minimum potential positions in and near the $CuO_2$ plane by changing the calculation resolution. However, we have not yet succeeded to locate any muon positions near the planar oxygen which would correspond to those suggested by Nishida *et al.*. Such a conflict in muon sites between those suggested from the dipole-field calculations and those estimated from the potential calculation was also reported in the case of LCO [14]. This fact means that a simple estimation of the electrostatic potential is not sufficient to estimate muon stopping sites as have been done in previous studies [15]. We can propose to re-investigate the dipolar field interaction between the Cu and muon spins. This is because the Cu spin in high-$T_C$ oxides is strongly hybridized with neighbouring oxygen atoms and could be spatially distributed throughout the $CuO_6$ octahedral. This effect could lead to Fermi-contact type interactions adding more hyperfine fields at the muon sites.

We are still carrying on more detailed potential calculations, including taking into account the local lattice deformations which are caused by the injected muon and the movement of the muon itself after being trapped at the minimum potential positions. The DFT method can include such conditions and shows the final position of the muon which is different from the initial stopping position. We hope to publish those results in a separate paper soon.

**Table 1.** Position of minimum potentials in $YBa_2Cu_3O_6$.

| Minimum potential | Estimated muon coordinate | | | Energy |
|---|---|---|---|---|
| | $x$ | $y$ | $z$ | $(eV)$ |
| M1 | 1.051 | 0.000 | 1.510 | $-4.310 \times 10^{-2}$ |
| M2 | 0.000 | 0.798 | 2.600 | $-4.292 \times 10^{-2}$ |
| M3 | 1.051 | 0.000 | 2.100 | $-4.185 \times 10^{-2}$ |
| M4 | 0.000 | 1.848 | 3.350 | $-4.127 \times 10^{-2}$ |
| M5 | 0.000 | 1.868 | 0.168 | $-4.032 \times 10^{-2}$ |
| M6 | 0.000 | 1.770 | 5.701 | $-3.991 \times 10^{-2}$ |

## 3. Conclusion

We have applied the DFT calculation to the Cu-based high-$T_C$ oxide $YBa_2Cu_3O_{6+\delta}$ in order to clarify possible muon sites. We found at six minimum potential sites and since the difference in potential energy among those sites are small (in the order of 10 Kelvin), all sites can be candidates to trap injected muons. Comparing with the experimentally suggested positions [5], it can be concluded that three positions with the lowest energies, which are located close to the apical oxygen of the $CuO_5$ polyhedral can be regarded as the positions that have been suggested experimentally. However, we still cannot explain the other estimated muon positions in order to match with the experimental data. More detailed calculations including the effects of the local crystal deformations caused by the injected muon and the relaxation position of the muon itself to the final state are now ongoing.

## 4. Acknowledgement

The corresponding author acknowledges financial support from RIKEN and Universiti Sains Malaysia.

## 5. References

[1] Uemura Y J 1987 *Phys. Rev. Lett.* **59** 1045
[2] Uemura Y J, Luke G M, Sternlieb B J, Brewer J H, Carolan J F, Hardy W N, Kadono R, Kempton J R, Kiefl R F, Kreitzman S R, Mulhern P, Riseman T M, Williams D L, Yang B X, Uchida S, Takagi H, Gopalakrishnan J, Sleight A W, Subramanian M A, Chien C L, Cieplak M Z, Xiao G, Lee V Y, Statt B W, Stronach C E, Kossler W J and Yu X H 1989 *Phys. Rev. Lett.* **62** 2317
[3] Watanabe I, Kawano K, Kumagai K, Nishiyama H and Nagamine K 1992 *J. Phys. Soc. Jpn.* **61** 3058
[4] Niedermayer Ch, Bernhard C, Blasius T, Golnik A, Moodenbaugh A and Budnick J I 1988 *Phys. Rev. Lett.* **80** 3843
[5] Nishida N, Miyatake H, Shimada D, Okuma S, Ishikawa M, Takabatake T, Nakazawa Y, Kuno Y, Keitel R, Brewer J H, Riseman T M, Williams D L, Watanabe Y, Yamazaki T, Nishiyama K, Nagamine K, Ansaldo E J and Torikai e 1987 *Jpn. J. Appl. Phys.* **26** L1856
[6] Sanna S, Allodi G, Concas G, Hillier A D and Renzi R De 2004 *Phys. Rev. Lett.* **93** 207001
[7] Akoshima M, Koike Y, Watanabe I and Nagamine K 2000 *Phys. Rev. B* **62** 6761
[8] Adiperdana B, Dharmawan I A, Siregar R E, Watanabe I, Ohishi K, Ishii Y, Suzuki T, Kawamata T, Risdiana, Sheuermann R, Sedlak K, Tomioka Y, Waki T, Tabata Y and Nakamura H 2012 *Physics Procedia* **30** 109-112.
[9] Weber M, Birrer P, Gygax F N, Hitti B, Lippelt E, Maletta H, and Shenck A 1990 *Hperfine Interact.* **63** 207
[10] Pinkpank M, Amato A, Adnreica D, Gygax F N, Ott H R, and Shanck A 1999 *Phyica C* **317-318** 299
[11] Bordet P, Chaillout C, Capponi J J, Chenavas J and Marezio M 1987 *Nature.* **327** 687
[12] Kresse G and Futhmller J 1996 *Phys. Rev. B* **54** 11169
[13] Kresse G and Futhmller J 1996 *Comput. Mat. Sci.* **6** 15
[14] Sulaiman S, Sahoo N, Srinivas S, Hagelberg F, Das T P, Torikai E and Nagamine K 1994 *Hyperfine Interact.* **84** 87
[15] Stilp E, Suter A, Prokscha T, Morenzoni E, Keller H, Wojek B M, Luetkens H, Gozar A, Logvenov G and Bozovic I 2013 *Phys. Rev. B* **88** 064419

Journal of Physics: Conference Series **551** (2014) 012053 doi:10.1088/1742-6596/551/1/012053

# Muon sites in $Ce(Ru,Rh)_2Al_{10}$ investigated by using Density Functional Theory from the view point of electronic potential

**N Adam[1,2], E Suprayoga[1,3], B Adiperdana[1,4], H Guo[1,5], H Tanida[6], S S Mohd-Tajudin[2], R Kobayashi[7,8], M Sera[6], T Nishioka[9], M Matsumura[9], S Sulaiman[2], M I Mohamed-Ibrahim[2] and I Watanabe[1,2,3,4,5]**

[1] Advanced Meson Science Laboratory, RIKEN Nishina Center, 2-1 Hirosawa, Wako, Saitama 351-0198, Japan
[2] Computational Chemistry and Physics Laboratory, School of Distance Education, Universiti Sains Malaysia, 11800 Penang, Malaysia
[3] Physics Department, Bandung Institut Teknologi, Jl Ganesha 10, Bandung, Jawa Barat 40132, Indonesia
[4] Physics Department, Universitas Padjadjaran, Jl. Raya Bandung-Sumedang Km21, Jatinangor, Sumedang 45363, Indonesia
[5] Department of Physics and State Key Laboratory of Silicon Materials, Zhejiang University, Hangzhou 310027, China
[6] Department of Quantum Matter, ADSM, Hiroshima University, Higashi-Hiroshima, Hiroshima 739-8530, Japan
[7] Quantum Condensed Matter Division, Oak Ridge National Laboratory, Oak Ridge, Tennessee 37831, USA
[8] Neutron Science Laboratory, Institute for Solid State Physics, University of Tokyo, Tokai 319-1106, Japan
[9] Graduate School of Integrated Arts and Science, Kochi University, Kochi 780-8520, Japan

E-mail: noraina.adam@riken.jp

**Abstract.** Numerical investigations on muon sites in Ce-based Kondo semiconductors, $Ce(Ru,Rh)_2Al_{10}$ were carried out by using the Density Functional Theory. From the view point of simple electrostatic potential calculations, we found all the previously reported muon sites, suggested by different groups (Kambe S *et al.* 2010 *J. Phys. Soc. Jpn.* **79** 053708 and Khalyavin D D *et al.*, 2010 *Phys. Rev.* B **82** 100405(R)), can be possibly chosen as muon stopping sites. We also investigated the changes in the potential of the Rh-doped case. We discovered that the electronic potential around the nearest Ru atom to the substituted Rh atom is affected and the potential becomes asymmetric around the nearest Ru ion. Although big changes in hyperfine fields at muon sites have been reported (Guo H *et al.* 2013 *Phys. Rev.* B **88** 115206), the muon positions estimated from the potential calculations do not change much.

## 1. Introduction

$CeRu_2Al_{10}$ has the orthorhombic $YbFe_2Al_{10}$–type structure with space group Cmcm in which Ce in $4c$, Ru in $8d$, two Al atoms in $8g$, another two Al atoms in $8f$ while one Al atom in $8e$. A unit cell of $CeRu_2Al_{10}$ consists of 4 Ce, 8 Ru and 40 Al atoms [1]. Figure 1 shows the crystal structure

Journal of Physics: Conference Series **551** (2014) 012053 doi:10.1088/1742-6596/551/1/012053

of $CeRu_2Al_{10}$. Ce atoms are in charge of the magnetism of this system and are surrounded by Al atoms forming a cage-like structure. This Ru-Al cage is suggested to form a two-dimensional plane structure stacking along the *b*-axis [2,3]. Ru atoms are in interstitial sites among those Ce-Al cages. This system has been argued to be a Kondo semiconductor and has drawn attention by its unusual electromagnetic properties [4]. Despite having nearest-neighbour distance, about 5.2 Å between Ce ions, $CeRu_2Al_{10}$ still exhibits an antiferromagnetic transition at $T_0 = 27.3$ K. Such a high magnetic transition temperature cannot be explained by taking into account only the RKKY interaction [5]. Meanwhile, the magnetic susceptibility is largely anisotropic while the direction of magnetic moment is along the *c*-axis in the magnetically ordered phase of $CeRu_2Al_{10}$ [6] and it is expected to flip to the *a*-axis when a slight amount of Ru is substituted by Rh [7]. At this moment, the insight of the magnetic properties is still inadequate to be understood. Therefore, a lot of efforts are being paid to investigate the system including the relative systems which have the same crystal structure but with different components.

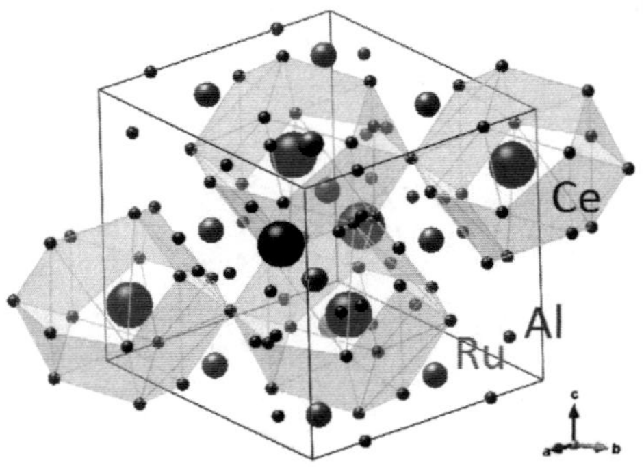


**Figure 1.** Crystal structure of $CeRu_2Al_{10}$. A unit cell which consists of 4 Ce (dark blue), 8 Ru (light blue) and 40 Al (red) atoms is shown. The Ce atom are surrounded by 10 of Al atoms forming a cage structure.

In order to investigate the magnetic properties of $CeRu_2Al_{10}$, a $\mu$SR study has already been conducted by Kambe *et al.* [8]. They found the appearance of the muon-spin precession in the time spectrum below $T_0$. The internal field at the muon site which was calculated from the muon-spin precession frequency was around 30 G. This small internal field could be caused by the cancellation of the internal field at the muon site due to the antiferromagnetic spin alignment. Thus, Kambe *et al.* suggested some possible muon sites which located at symmetric sites between Ce atoms. On the other hand, Khalyavin *et al.* compared numerical estimations and experimental data for nuclear dipole fields and suggested different muon positions from those suggested by Kambe *et al.* [9]. Khalyavin *et al.* also reported two components of internal fields and the unusual temperature dependences of them. The lower-field component found by Khalyavin *et al.* was almost the same with that observed by Kambe *et al.* Another component had the larger internal field in the order of 120 G. In addition, Guo *et al.* has reported detailed $\mu$SR results on Rh-doped systems,$Ce(Ru_{1-x}Rh_x)_2Al_{10}$ [10]. They reported drastic changes in internal fields at muon sites which were related to the spin-flop caused by the Rh-doping. They confirmed that the spin-flop from along the *c*-axis to the *a*-axis was caused by a couple percent of Rh-doping.

Figure 2 summarises muon sites in $CeRu_2Al_{10}$ suggested by Kambe *et al.* [8] and Khalyavin *et al.* [9]. Kambe's suggested positions located at symmetric sites in between Ce atoms around the surface on the unit cell. On the other hand, Khalyavin's suggested sites were not on the surface but at the inside of the unit cell. They were not crystallographically the same with Kambe's ones. Therefore, muon sites in $CeRu_2Al_{10}$ are still in controversial and the hyperfine interactions cannot be deeply discussed. We try to analyse those muon sites from the view point of electrostatic potential calculation by using the Density Functional Theory (DFT) and aim to explain $\mu$SR results on $CeRu_2Al_{10}$. Since the direction of ordered spins changed in the Rh-doped system, we tried to deduce a change in the electrostatic potential as well as the muon sites in the Rh-doped case. In this paper, we report the results of the first trial in estimating muon sites from the potential view point of $CeRu_2Al_{10}$ and the first examination in observing a change in the electrostatic potential of the Rh-doped system.

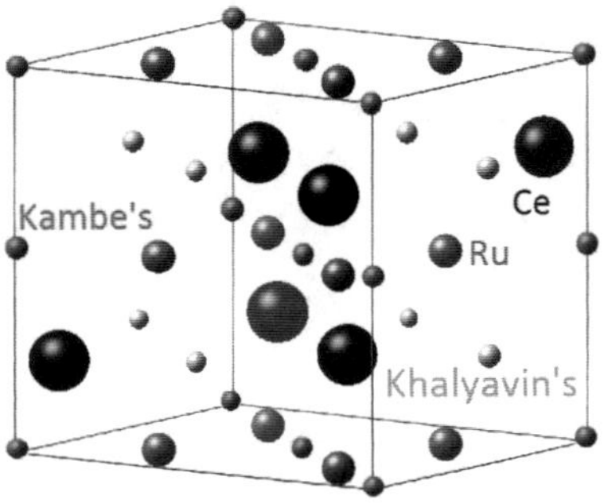


**Figure 2.** Muon sites in $CeRu_2Al_{10}$ suggested by Kambe *et al.* (purple marks [8]) and Khalyavin *et al.*(yellow marks[9]). Kambe *et al.* suggested from the internal field at the muon site while Khalyavin *et al.* suggested from the nuclear-dipole field distribution.

## 2. Calculations

We have carried out the potential calculation by using the VASP program [11,12]. We used DFT by using the LDA electronic correlations since the ground state of the system would be metallic [13,14] even though physical properties of the system are categorized to be the Kondo semiconductor. The resolution of the calculation grid was 108×120×108 in *a*-, *b*-, and *c*-axis. The cutoff energy of the plane wave expansion was set to be 375 eV. The electrostatic potential of one unit cell without muon was calculated in order to estimate the initial positions for injected muons. Other parameters such as the relaxation of the muon position, the local deformation of the crystal structure induced by the injected muon, the spin polarized effect on the total energy of the system and the spin-orbit coupling were not included in order to simplify calculations in this study. It has been pointed out that the zero-point vibration motion of the muon itself is important to determine final state of the muon [15]. However, we did not include this effect as well as other parameters because we would like to show in this report how simple potential estimations can explain experimental results and possible muon sites in $Ce(Ru_{1-x}Rh_x)_2Al_{10}$.

We used a supercomputer system in RIKEN named RIKEN Integrated Cluster of Clusters (RICC) in order to achieve the faster calculation condition. In order to estimate the change in the electrostatic potential of the Rh-doped system, we substituted one Ru with Rh in the cell. This condition was corresponding to the concentration of Rh to be about 12.5 %. In such a high concentration range, the spin alignment in the magnetically ordered state is already along the *c*-axis. Although this Rh concentration value was too high to compare with the realistic sample condition which has been investigated by Guo *et al.* [10], we tried to achieve the basic and the simplest result in order to conceive the idea of next calculations.

## 3. Results and Discussions

While our calculation area covered the whole unit cell with fine meshes, we chose some characteristic areas in order to compare between the non-doped and doped system. This is because those three planes contain all estimated minimums. Figure 3 shows those characteristic planes in the unit cell. Figures 3(a) and 3(b) indicate the case of the non-doped and doped systems, respectively. One Ru atom was substituted by the Rh atom as shown in Figure 3(b) by the orange mark. The plane ① is the Ru-Al plane which is on the boundary of the unit cell and includes muon positions suggested by Kambe *et al.* The plane ② is the Ce-Al plane which includes muon positions suggested by Khalyavin *et al.* The plane ③ is again the Ru-Al plane which is located in the middle of the unit cell and includes muon positions suggested by Kambe *et al.* In the case of the Rh-doped system, we substituted one Ru atom by the Rh atom in this plane. Changes in the potential energy in each plane are also illustrated around atoms. Blue colored areas show the lower potential areas rather than the yellow/light-blue ones.

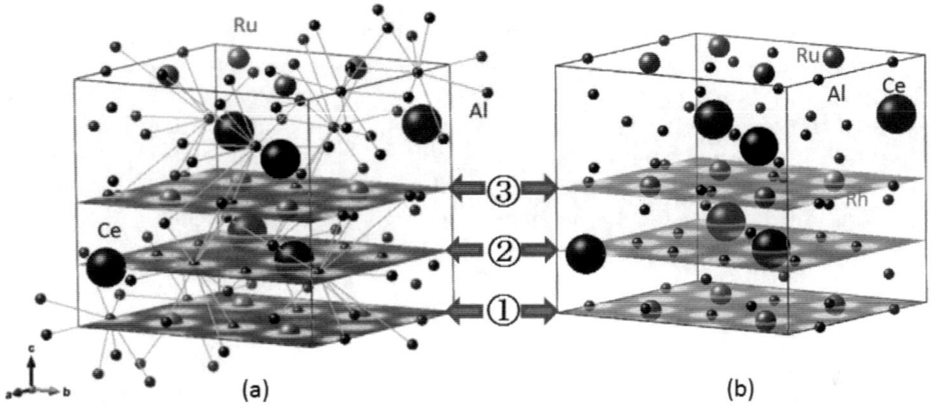


**Figure 3.** Planes contain Ru and Al (①,③), Ce and Al (②), respectively. Dark-blue, light-blue, red marks are corresponding to Ce, Ru and At atoms, respectively. The plane ① and ③ contain muon positions suggested by Kambe *et al.* The plane ② contains muon positions suggested by Khalyavin *et al.* In order to simulate the Rh-doped system, one Ru atom in the plane ③ was substituted by the Rh atom which was corresponding the Rh concentration of 12.5 %.

Figure 4 shows detailed results of the potential calculation on each plane of non-doped system. Muon positions suggested by Kambe *et al.* [8] and Khalyavin *et al.* [9] are drawn by purple and yellow marks, respectively. As indicated in Figure 4, all muon positions are located at the bottoms of potential valleys. From the comparison made between each minimum potential, it

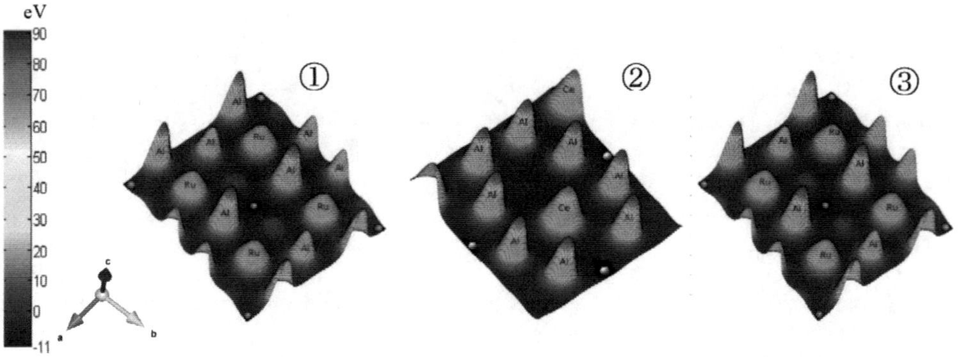


**Figure 4.** Electrostatic potential in each plane as shown in Figure 3. Blue color areas have the higher and lower potential energy, respectively. Purple and yellow marks are muon positions suggested by Kambe *et al.* [8] and Khalyavin *et al.* [9], respectively.

was found that the differences among each minimum potential was less than a couple of Kelvin.

It should be noted that those points shown in Figure 4 could be spatially spread due to the zero-point vibration motion of the muon itself and the initially trapped muon could be hopping around site by site passing through energy barriers between suggested positions [15]. These effects are now being tested and will be published in later papers.

Our simple calculation suggests that electrostatic calculations can explain all positions proposed by Kambe *et al.* [8] and Khalyavin *et al.* [9] as the candidates that trap injected muons. Dipole field calculations from Ce atoms at those muon sites has been carried out and it was found that muon positions suggested by Kambe *et al.* were expected to have the smaller internal field; 20-30 G. [8] Instead, muon positions suggested by Khalyavin *et al.* had the larger internal field with the order of 100-200 G. [9] Since the dipole-field estimation depends on the muon position itself, we are carrying out more detailed calculations which cover more unit cells and include detailed parameters like the local crystal deformation, muon-position relaxation and changing conditions of the Ce magnetic moment as well as the spin alignment.

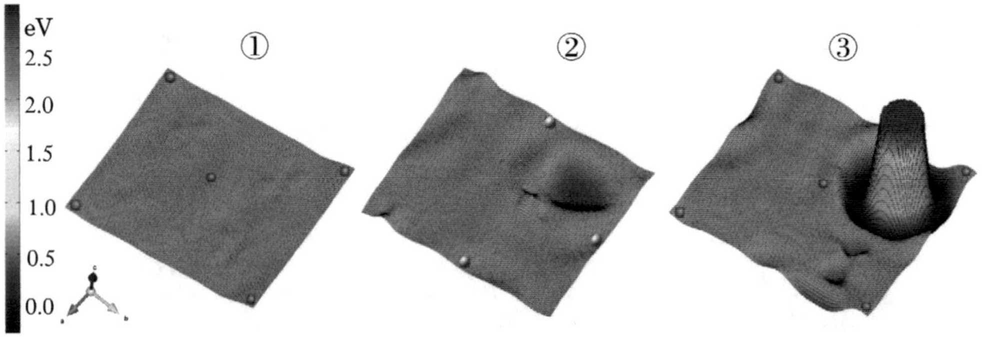


**Figure 5.** Differences in the potential between the non-doped and doped systems in terms of planes ①,② and ③. The location of each plane was as indicated in Figure 3. Purple and yellow marks in each plane are corresponding to muon positions suggested by Kambe *et al.* [8] and Khalyavin *et al.* [9], respectively. The big difference seen in the plane ③ is induced by the substituted Rh. The middle point between two minimums which appear in the plane ③ near the substituted Rh atom is corresponding to the neighboring Ru atom.

In order to investigate the difference in the potential between the non-doped and doped system, we subtracted both potential in each plane (①,② and ③). This operation means that the potential value on each grid of the doped system which was shown as Figure 3(b) was subtracted from that on the same position in the non-doped system which was shown as Figure 3(a). Figure 5 shows the result of differences in potential values between planes of non-doped and doped systems. Both Kambe's and Khalyavin's suggested muon positions are shown together in each plane by purple and yellow marks, respectively. The potential difference of the plane ① did not change so much. This would be due to that this plane was far from the substituted Rh atom. A big change in the potential in the plane ③ appeared around the substituted Rh atom. This large difference in the potential is expected to be caused by the different valence configuration from that of Ru atom. Due to this big change in the potential at the Rh atom, the potential in the plane ② which was just underneath the substituted Rh atom was slightly deformed. However, the electrostatic potential around suggested muon sites seemed not to be affected so much by the Rh substitution. This result leads us to suggest that muon positions would not change so much even in the Rh-doped system since the Rh concentration is much

Journal of Physics: Conference Series **551** (2014) 012053 doi:10.1088/1742-6596/551/1/012053

smaller in realistic systems and that the change in the internal fields at muon positions are caused by the change in the Ce spin alignment as discussed by Guo *et al.* [10]

It was found that two minimums appeared in the plane ③ beside the substituted Rh atom as seen in Fig. 5. Since the neighboring Ru atom and the substituted Rh atom located in the middle of the two minimums, the Rh substitution is expected to change the potential around the neighboring Ru atom to be asymmetric. Such a change may be induced by the modification of electronic outer orbitals of Ru such as $d$ orbitals because the current system has a strong mixing state between conduction and $d$ electrons. If the change in the $d$ orbital expands through the whole crystal by some reason or through a long-range magnetic correlation like the RKKY interaction, a small amount of the substitution of Ru by Rh could lead to a drastic change in the spin direction. In order to give an answer to this hypotheses, more calculations should be done on the charged state of Ru atoms, Rh atoms and on the charge density of the whole system. Those works are on-going projects and beyond the purpose of the current study.

## 4. Conclusion

We have carried out potential calculations by using the DFT method in order to investigate muon positions in $Ce(Ru_{1-x}Rh_x)_2Al_{10}$. We have found that all positions suggested by Kambe *et al.* and Khalyavin *et al.* are located at local minimum positions which could be a good candidate for muon sites. This explained the experimental data of internal fields measured in magnetically ordered phase. Since at least two muon sites have been identified in $CeRu_2Al_{10}$, we proposed that the difference of internal fields at muon sites came from the difference of those suggested muon positions. On the basis of the dipole field calculation, we proposed that the Kambe's suggested positions were corresponding to the sites which had the lower field and the Khalyavin's suggested positions were corresponding to those which had the higher field. We also investigated the changes in the electronic potential in the case of the Rh-doped system. It was found that the Rh doping introduced a change in the electrostatic potential around the Rh site. However, there are no big changes in the potential around the suggested muon sites.

## References

[1] Thiede V M T, Nesterenko S N, Murashova E V, Chernyshev I V, Noel H and Sropegin Y D 2005 *Acta Crystallogr.* E **61** i12
[2] Tanida H, Tanaka D, Sera M, Tanimoto S, Nishioka T, Matsumura M, Ogawa M, Moriyoshi C, Kuroiwa Y, Kim J E, Tsuji N and Takata M, 2011 *Phys. Rev.* B **84** 115128
[3] Sera M, Tanaka D, Tanida H, Moriyoshi C, Ogawa M, Kuroiwa Y, Nishioka T, Matsumura M, Kim J E, Tsuji N and Tanaka M, 2013 *J. Phys. Soc. Jpn.* **82** 024603
[4] Tanida H, Nonaka Y, Tanaka D, Sera M, Kawamura Y, Uwatoko Y, Nishioka T and Matsumura M 2012 *Phys. Rev.* B **85** 205208
[5] Nishioka T, Kawamura Y, Takesaka T, Kobayashi R, Kato H, Matsumura M, Kodama K, Matsubayashi K and Uwatoko Y 2009 *J. Phys. Soc. Jpn.* **78** 123705
[6] Kato H, Kobayashi R, Takesaka T, Nishioka T, Matsumura M, Kaneko K and Metoki N 2011 *J. Phys. Soc. Jpn.* **80** 073701
[7] Kondo A, Kindo K, Kunimori K, Nohara H, Tanida H, Sera M, Kobayashi R, Nishioka T and Matsumura M 2013 *J. Phys. Soc. Jpn.* **82** 054709
[8] Kambe S, Chudo H, Tokunaga Y, Koyama T, Sasaki H, Ito T U, Ninomiya K, Higemoto W, Takesaka T, Nishioka T and Miyake Y 2010 *J. Phys. Soc. Jpn.* **79** 053708
[9] Khalyavin D D, Hillier A D, Advoja D T, Strydom A M, Manuel P, Chapon L C, Peratheepan P, Knight K, Deen P, Ritter C, Muro Y and Takabatake T 2010 *Phys. Rev.* B **82** 100405
[10] Guo H, Tanida H, Kobayashi R, Kawasaki I, Sera M, Nishioka T, Matsumura M, Watanabe I and Xu Z 2013 *Phys. Rev.* B **88** 115206
[11] Kresse G and Futhmüller J 1996 *J Phys. Rev.* B **54** 11169
[12] Kresse G and Futhmüller J 1996 *Comput. Mater. Sci.* **6** 15
[13] Kimura S, Iizuka T, Miyazaki H, Hajiri T, Matsunami M, Mori T, Irizawa A, Muro Y, Kajino J and Takabatake T 2011 *Phys. Rev.* B **84** 165125

[14] Kondo A, Wang J, Kindo K, Takesaka T, Ogane Y, Kawamura Y, Nishioka T, Tanaka D, Tanida H and Sera M 2011 *J. Phys. Soc. Jpn.* **80** 013701
[15] Bernardini F, Bonfa P, Massidda S and De Renzi R 2013 *Phys. Rev.* B **87** 115148

Journal of Physics: Conference Series **551** (2014) 012054 doi:10.1088/1742-6596/551/1/012054

# Search for potential minimum positions in metal-organic hybrids, $(C_2H_5NH_3)_2CuCl_4$ and $(C_6H_5CH_2CH_2NH_3)_2CuCl_4$, by using density functional theory

**E Suprayoga[1,2], A A Nugroho[2], A O Polyakov[3], T T M Palstra[3] and I Watanabe[1]**

[1]Advanced Meson Science Laboratory, RIKEN Nishina Center, 2-1 Wakoshi, Hirosawa, Saitama 351-0198, Japan

[2]Physics Study Program, Faculty of Mathematics and Natural Science, Institut Teknologi Bandung, Jl. Ganesha 10, Bandung 40131, Indonesia

[3]Zernike Institute for Advanced Materials, University of Groningen, Nijenborgh 4, 9747AG Groningen, The Netherlands

E-mail: edi.suprayoga@riken.jp

**Abstract.** The *ab-initio* density functional theory analysis was applied to metal-organic hybrids, $(C_2H_5NH_3)_2CuCl_4$ (EA) and $(C_6H_5CH_2CH_2NH_3)_2CuCl_4$ (PEA), in order to estimate possible muons stopping positions. Six potential minimum positions and eight ones were revealed in PA and PEA, respectively. Those potential minimum positions can be regarded as initial stopping positions of injected muons. All of expected potential minimum points in EA were near and around the apical Cl and the $CuCl_2$ plane of the $CuCl_6$ octahedra. Instead, in the case of PEA, two of eight positions were close to the phenyl ring giving a possibility that there would be muon states which couple surrounding electrons via a radical formation.

## 1. Introduction

Metal-organic hybrids which consist of organic and inorganic components have recently attracted much attention because these materials have lots of possibilities for tailoring functionalities and physical properties such as optical, electrical and magnetic properties by adjusting its organic and/or metal building blocks [1]. The organic component functions not only to bind the metal component but also to control the dimensionality and the connectivity in metal components [2]. One of the interesting series is the magnetic hybrid with perovskite-type transition metal salt with chemical formula such as $(C_2H_5NH_3)_2CuCl_4$ (EA) and $(C_6H_5CH_2CH_2NH_3)_2CuCl_4$ (PEA) [3]. Figure 1 shows the crystal structures of EA and PEA. Both hybrids have the orthorhombic crystal structure with the space group *Pbca*. Lattice constants of EA are $a$ = 7.4789 Å, $b$ = 7.1731 Å, and $c$ = 21.2406 Å [4]. In the case of PEA, those are $a$ = 7.2099 Å, $b$ = 7.2664 Å, and $c$ = 38.238 Å. The Cu atom has a spin half and forms the $CuCl_6$ octahedra with surrounding six Cl atoms. The $CuCl_6$ octahedra forms a 2D network of Cl-Cu-Cl and determines the magnetic property of those systems. The distance between $CuCl_6$ planes is about 10 Å for EA and 20 Å for PEA respectively. The orthorhombic crystal structure is due to the buckling of the $CuCl_6$

octahedra along the c-axis with angles of 7.74° and 6.63° for EA and PEA, respectively. Organic parts of EA and PEA consist of the double layer of ethylanmmonium and phenylethylammonium ions respectively and their charged ends of the ion, which is the nitrogen atom, are oriented to the nearest $CuCl_6$ plane. The EA contains no double bond between C atoms while PEA has aromatic phenyl rings in organic parts.

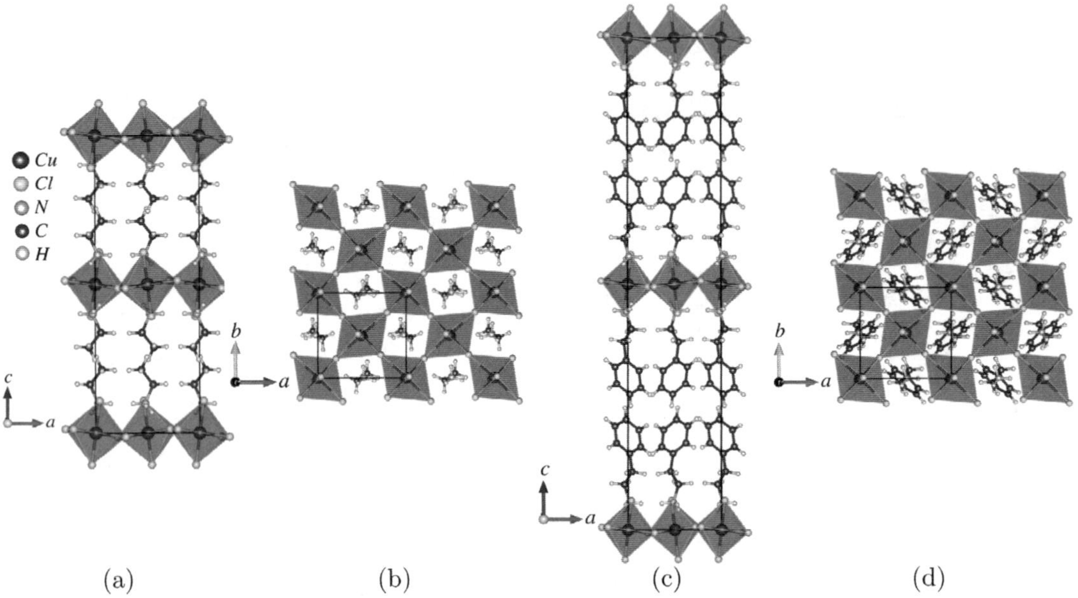


**Figure 1.** Crystal structures of $(C_2H_5NH_3)_2CuCl_4$ (EA) and $(C_6H_5CH_2CH_2NH_3)_2CuCl_4$ (PEA). Views of EA (a) along the b-axis and (b) along the c-axis. Views of PEA (c) along the b-axis and (d) along the c-axis. The distance between $CuCl_6$ octahedra layer is about 10 Å in the case of EA and 20 Å in the case of PEA.

The in-plane magnetic correlation between Cu spins was estimated to be ferromagnetic for EA and PEA while the inter-plane magnetic correlation is expected to be much smaller than that of the in-plane one due to the far away distance between $CuCl_6$ layers. Although the magnetic property of both EA and PEA is expected to be 2D like, the appearance of a 3D magnetic state was confirmed in both systems around 10 K by magnetic susceptibility measurements [5, 6, 7]. Our $\mu$SR measurements on EA and PEA revealed the appearance of the clear muon-spin precession in the zero-field condition in magnetically ordered states. This observation evidenced the formation of a coherent long-range magnetic order of Cu spins. The internal field at the muon site was determined from the muon-spin precession frequency to be around 200 G at 4 K in both systems. In the case of PEA, a fast depolarized component in an earlier time region was observed from room temperature down to 4 K. This fast depolarized component was not decoupled even in fields of a couple of Tesla showing the existence of a possible muon state which strongly coupled with surrounding electrons via a radical formation. On the other hand, such a state was not observed in EA. Those experimental results will be published in a separated paper [8].

The dipole-dipole interaction is too weak to trigger the 3D magnetic ordered state through the far away distance between $CuCl_6$ layers, so that the mechanism of the magnetic interaction between layers is still unknown. In addition to this, the alignment of Cu spins in the magnetically ordered state has not yet been determined for EA and PEA. In order to approach those problems

Journal of Physics: Conference Series **551** (2014) 012054 doi:10.1088/1742-6596/551/1/012054

and to get hints to argue the magnetic properties of EA and PEA, we are now trying to reveal the muon positions by using computational techniques. We can determine the spin alignment from the comparison between experimental results and muon positions estimated from computational techniques on the basis of the assumption that internal fields at the muon site are caused by dipole fields from surrounding magnetic moments. In this paper, we present results of the first trial to apply the density functional theory (DFT) method [9, 10] to EA and PEA in order to reveal minimum energy points which could be possible preferences for injected muons to be initially trapped. We found six potential minimum positions in EA and eight ones in PEA. All of expected potential minimum points in EA were near/around the $CuCl_6$ octahedra. Instead, in the case of PEA, two of eight positions were close to the aromatic phenyl ring giving a possibility that there are muon states which couple surrounding electrons via a radical formation.

## 2. Computational details

The estimation of the potential minimum positions was carried within the DFT framework as implemented in the Vienna Ab-initio Simulation Package (VASP) [11, 12]. We divided the unit cell into 90 × 90 × 265 grids for EA and 90 × 90 × 480 grids for PEA along a-, b-, and c-axes, respectively. The electrostatic potential energy was calculated putting $H^+$ on each grid. The projector augmented wave (PAW) pseudo-potentials was used [13] with the cutoff energy of 500 eV. The generalized gradient approximation (GGA) with the interpolation formula according to Vosko Wilk and Nusair was applied as the exchange correlation potentials [1]. We used the Monkhorst-Pack grid of $k$-points for the Brillouin-zone integrations for the single unit cell [1]. Local lattice relaxation effects which could be driven by injected muons were not included in order to make calculations simpler as the first trial. Potential maps and isosurface plots were generated using Visualization for Electronic and Structural Analysis (VESTA).

## 3. Results and Discussions

Figure 2 shows calculation results of potential minimum positions in EA. The strategy for finding energy minima consisted in a search for a regions enclosed by the lowest-energy isosurface contours. We firstly found three minimums around the apical Cl atom of the $CuCl_6$ octahedra. Figure 2(a) indicates those three positions by using isofurface areas. Figures 2(b) and (c) are expanded pictures of those three positions around the apical Cl atom. We named those positions to be A1, A2 and A3 from the lower energy side. The A1 and A2 are located symmetrically against the apical Cl atom. The difference of the potential energy between those two positions is due to the tilting of the $CuCl_6$ octahedra. With increasing the energy, another minimum appeared near A2 as shown in Fig. 2(c). This position was named as A4 and would be a counter position to A3. Two more positions were achieved in between two Cl atoms on the $CuCl_2$ plane. Since those two other positions were relatively far away from the apical Cl atom, we marked those two positions as B1 and B2. In total, we found six minimum positions in EA. We could not find out any more minimums at the higher energy region. Each position has the similar potential energy of about -6 eV and the difference in the energy among them was about a couple of hundred meV.

Figure 3 exhibits calculation results on PEA. As was done in the case of EA, we searched minimum potential positions from the lowest potential energy. With increasing the energy level, four minimum potential positions were found around the apical Cl atom of the $CuCl_6$ octahedra. Those positions are shown in the unit cell in Fig. 3(a) and in an expanded view of the $CuCl_6$ octahedra, Fig. 3(b), by isosurfaces. In the higher energy region, two more potential minimum positions were found in between two in-plane Cl atoms in the $CuCl_2$ plane. This finding of six potential minimum positions in PEA was the same tendency with that found in EA. Actually, each position has the similar potential energy of around -6 eV and the difference in the energy among them was about a couple of hundred meV as well as the case of EA. Accordingly, those

Journal of Physics: Conference Series **551** (2014) 012054 doi:10.1088/1742-6596/551/1/012054

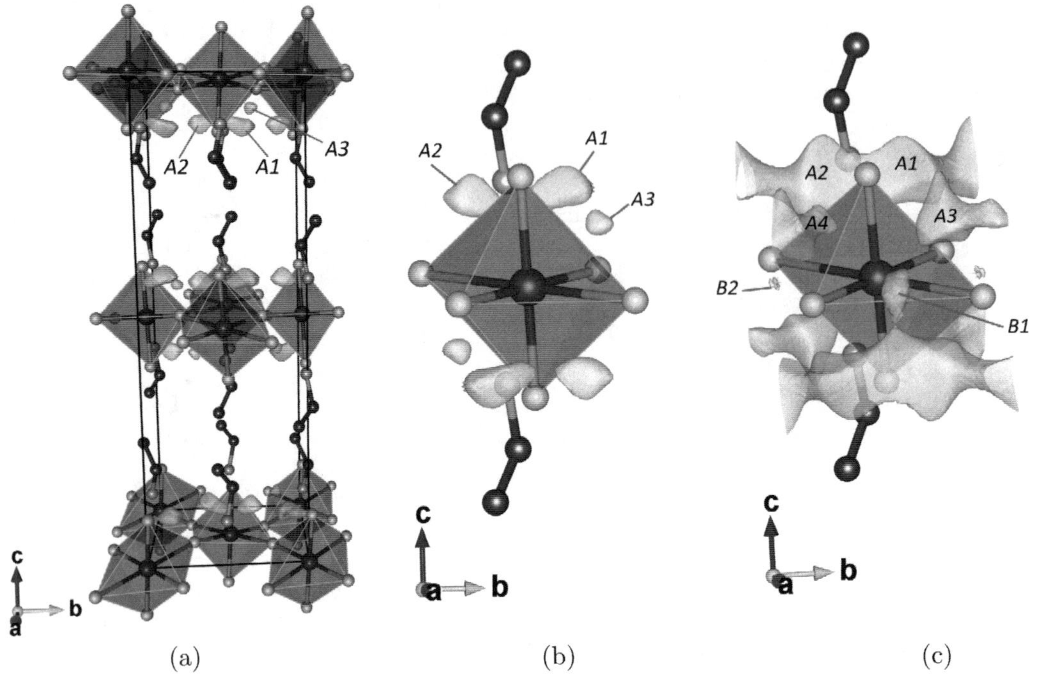


**Figure 2.** (a)Calculation results of the potential energy by using the density functional theory on $(C_2H_5NH_3)_2CuCl_4$ (EA). (b)(c)Expanded picture of the $CuCl_6$ octahedra with estimated minimum potential positions which are drawn by isosurfaces. Isosurfaces in (b) and (c) show the potential energy areas which are 337 and 455 meV higher than from the minimum potential position A1, respectively.

positions were named as A1 to A4 and B1 to B2 following the same definition in the case of EA as indicated in Fig. 3(b) and (c).

In addition to those six potential minimum positions, two more positions appeared in and on the phenyl ring of PEA. Those two positions are shown in Fig. 3(d) named as C1 and C2. The C1 position is almost on the C-H bond of the phenyl ring. The C2 position is within the phenyl ring. Potential energies at C1 and C2 sites were about 0.5 eV higher than others and the difference in the potential energy between the two position was about 0.1 eV.

Comparing results achieved in EA and PEA, it was found that positions from A1 to A3, B1 and B2 are quite similar in those systems. This fact means that the electrostatic potential around the $CuCl_6$ octahedra is similar in both of EA and PEA. It can be supposed that muon positions around $CuCl_6$ octahedra are similar and that surrounding organic parts do not affect muon positions around $CuCl_6$ octahedra. Since differences in the potential energy among estimated positions are not so big and was about a hundred of meV, all positions around $CuCl_6$ octahedra can be preferences for injected muons to be initially trapped. Those positions are expected to correspond to those which show the muon-spin precession in the zero-field condition as observed in our previous $\mu$SR experiment [8]. It should be noted that the final muon positions must be different from those estimated in the current study. We have not yet taken into account local relaxation effects of lattice which are caused by injected muons. We also did not include the effect of the zero-point vibration motion of the muon itself. Those effects relax initial muon positions and lead to the deformation of the crystal structure minimizing the total energy and

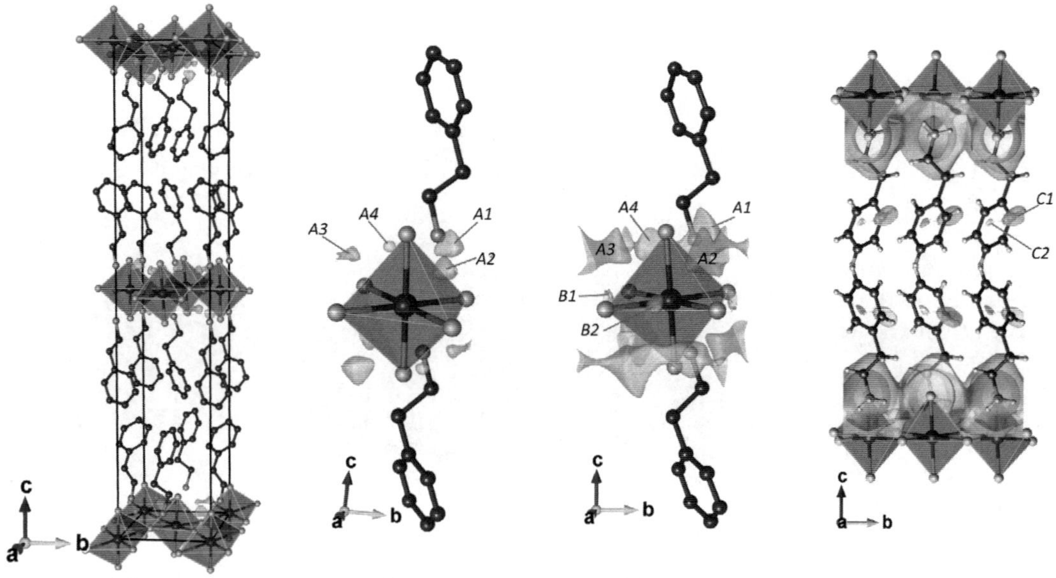


**Figure 3.** (a) Calculation results of the potential energy on $(C_6H_5CH_2CH_2NH_3)_2CuCl_4$ (PEA). (b)(c) Expanded pictures of the $CuCl_6$ octahedra with estimated minimum potential positions which are drawn by isofurfaces. (d) Expanded picture of organic components around the phenyl ring. Two potential minimum positions in and on the phenyl ring are drawn by isosfurfaces as well. Isosurfaces in (b), (c) and (d) show potential energy areas which are 195, 425 and 1,025 meV higher than that of the minimum potential position, respectively.

to the relaxation of muon positions. More calculation efforts are being undertaken to estimate final muon positions in EA and PEA taking into account those local relaxation effects.

On the other hand, C1 and C2 which are estimated in PEA around the phenyl ring can give different muon states from others. As has been well discussed, the injected muon prefers to form a radical state around the phenyl ring on which $\pi$ electrons are delocalized over the ring [14]. Such a radical state can make a large internal field at the muon site via the Fermi contact interaction. Thus, the muon state identified in our $\mu$SR experiment with a large hyperfine field [8], must be the one that forms a radical state around the phenyl ring. However, the estimation of the final muon state around the phenyl ring is not so easy and straightforward. This is because, the phenyl ring has at least about 10 crystallographic orientations which are ortho-, meta- and parapositions and their inequivalences on both sides of the ring. This complicated local structure of the phenyl ring and delocalized $\pi$ electrons over the ring make the estimation difficult. Local relaxation effects are also very important in the more accurate estimation of muon positions around the the phenyl ring in PEA.

## 4. Conclusions

We applied the density functional theory analysis to estimate the electrostatic potential energy in metal-organic hybrids of $(C_2H_5NH_3)_2CuCl_4$ (EA) and $(C_6H_5CH_2CH_2NH_3)_2CuCl_4$ (PEA) in order to estimate local minimum potential positions as preferences for injected muons to be initially trapped. We found six potential minimum positions around the $CuCl_6$ in EA. The similar positions were found in PEA as well. Four of them were around the apical Cl atom and two of them were in between in-plane Cl atoms on the $CuCl_2$ plane. Those muon positions can be candidates to show the muon-spin precession behavior in the zero-field condition in magnetically ordered states in both EA and PEA. Two more minimum potential positions were found in PEA in and on the phenyl ring of the organic part causing possible muon states bound with surrounding electrons via a radical formation having large hyperfine fields as identified in our $\mu$SR study [8]. Differences in the potential energy among those estimated positions were not so large and all estimated positions could be those which injected muon would prefer to be initially trapped.

**Acknowledgments**
Some of the authors, E. Suprayoga and A.A.Nugroho are grateful for the support from the Directorate of Higher Education, Indonesia (DIKTI).

## References

[1] Rao C N R, Cheetham A K and Thirumurugan A 2008 *J. Phys.: Cond. Matter* **20** 15980
[2] Kagan C R, Mitzi D B and Dimitrakopoulos C D 1999 *Science* **286**, 945
[3] Polyakov A O, Arkenbout A H, Baas J, Blake G R, Meetsma A, Caretta A, van Loosdrecht P H M and Palstra T T M 2012 *Chem. Mater.* **24** 133
[4] Kundys B, Lappas A, Viret M, Kapustianyk V, Rudyk V, Semak, Simon C and Bakaimi I 2010 *Phys. Rev.* B **81** 224434
[5] Estes W E, Losee D B and Hatfield W E 1969 *J. Chem. Phys.* **72** 630
[6] De Jongh L J, Botterman A C, de Boer F R and Miedema A R 1969 *J. Appl. Phys.* **40** 1363
[7] Khomskii D I and Kugel K I 1973 *J. Solid State Comm.* **13** 763
[8] Nugroho A A *et al.* in preparation
[9] Hohenberg P and Kohn W 1964 *Phys. Rev.* B **13** 864
[10] Kohn W and Sham L J 1965 *Phys. Rev.* A **140** 1133
[11] Kresse G and Furthmuller J 1996 *Phys. Rev.* **54** 11169
[12] Kresse G and Furthmuller J 1996 *Comput. Mater. Sci.* **6** 115
[13] Blöchl P E 1994 *Phys. Rev.* B **50** 17953
[14] Roduner E and Walton J C 1988 *The Positive Muon As a Probe in Free Radical Chemistry* (Berlin: Springer-Verlag)

# Tuning the spin dynamics of single molecule magnets via dipolar interactions

**A Hofmann and Z Salman**

Laboratory for Muon Spin Spectroscopy, Paul Scherrer Institute, CH-5232 Villigen PSI, Switzerland

E-mail: zaher.salman@psi.ch

**Abstract.** We present calculations of the dipolar field distribution acting on a single molecule magnet due to its neighbours in thin films. The calculations are presented for different packing/configuration scenarios, with different easy axis orientations. The potential for controlling the molecular spin dynamics by tuning the molecule-substrate interaction and its competition with intra-molecular interactions is discussed. We argue that by altering the configuration of the molecular moments, and thus their dipolar interactions, one can enhance or slow down their spin dynamics.

## 1. Introduction

Single molecule magnets (SMMs) [1] have attracted considerable attention in the last decade due to their fascinating magnetic properties [1, 2, 3]. They have also been considered for applications in quantum information processing [4, 5, 6]. As a result, a considerable focus in this field has shifted to the production and engineering of nano-structures of SMMs, e.g. deposition of thin films and monolayers on different substrates [7, 8, 9, 10, 11, 12, 13, 14, 15, 16, 17, 18]. New and novel techniques have proven successful for characterizing the magnetic properties of SMMs in these nano-structures [11, 12, 18, 19]. Depending on the type of studied SMM, some reports show that the bulk magnetic properties may be altered in a thin film or a monolayer due to its intra-molecular or molecule-substrate interactions. However, these studies have mostly focused on the static magnetic and low temperature properties of SMMs. Recent developments in low energy muon spin relaxation (LE-$\mu$SR) [20, 21] provide the possibility for depth-resolved investigations of magnetic properties in thin films [22, 23]. Since the time scale sensitivity of the $\mu$SR technique matches well the spin dynamics of SMMs [24, 25, 26], LE-$\mu$SR provides a perfect tool for studies of spin dynamics in thin film SMMs and the effect of the substrate on them. The depth resolved capability enables one to follow these gradually from the SMM/substrate interface to the surface of the film. However, in order to interpret results of such experiments, it becomes important to take into account effects of the substrate and the resulting SMM packing/configuration of the film.

In this paper, we present simulations of the dipolar field distribution acting on a single SMM, produced by its neighbours. We calculate the variation in the distribution as a function of the packing of SMMs and their proximity to the substrate, and discuss their effect on the molecular spin dynamics. Note that due to the strong correlation between the geometric SMM structure and its magnetic easy axis orientation, a change in the packing of molecules could

Journal of Physics: Conference Series **551** (2014) 012055 doi:10.1088/1742-6596/551/1/012055

drastically alter the dipolar field distribution. As a guiding example, we take thin films of $TbPc_2$, a prototypical SMM, where a competition between molecule-molecule and molecule-substrate interactions results in a clear variation of the SMM packing/configuration as a function of proximity to the SMM/substrate interface [27]. In these films the easy axis orientation of the $TbPc_2$ changes from pointing out of plane near the interface to in plane away from it. Our dipolar field calculations show that a lower fluctuation rate is expected in the region near the substrate, while enhanced dynamics are expected far away from it. This difference is mainly due to the local magnetic field distribution along and perpendicular to the easy axis of the SMM, which in turn influence the probability of transition between spin states due to mixing. We point out here that SMMs may be influenced by other factors as well, which we do not consider here. However, we expect that the ability to control the configuration of SMMs in thin films, e.g. by changing the substrate material, may open a new avenue for controlling and tuning their spin dynamics.

## 2. Simulations

In what follows, we consider as an example the case of a film of $TbPc_2$ deposited on a gold substrate. Near the substrate the molecular easy axis points along the $z$-direction (normal to the substrate) due to interactions with the substrate (Fig. 1, scenario **I**) [27]. Away from the substrate, where intermolecular interactions dominate, the easy axes of the molecules are randomly oriented within the $xy$ plane (Fig. 1, scenario **II**). We calculate the dipolar magnetic field distributions, acting on a single SMM in the layer, along and perpendicular to its easy axis for both scenarios. For these calculations we assume that the SMMs are arranged in a two-dimensional square or triangular lattices, both with a lattice constant $a$ and a magnetic moment per molecule of $\boldsymbol{\mu} = g\mu_B\mathbf{S}$, where $\mathbf{S}$ is the total spin. We also discuss the effect of random disorder in the layer on the field distribution.

Generally, for SMMs with a total ground spin state $S$ and easy axis along $z$, the two spin ground states $|S, S_z\rangle = |S, \pm S\rangle$ are degenerate in zero applied magnetic field. In order to simulate this degeneracy, each molecule is randomly assigned a spin value $S_z = +S$ or $-S$, i.e. we assume a very low temperature limit, $T \to 0$. Since different spin values and orientation lead to different magnetic fields, we average the magnetic fields, $\mathbf{B}$, for different SMM configurations for each scenario.

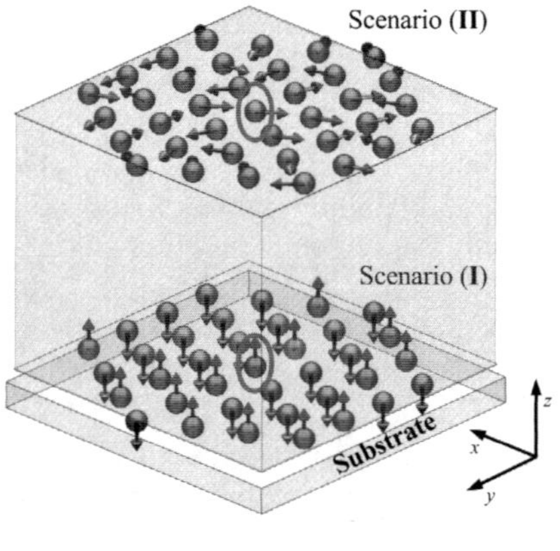


**Figure 1.** Illustration of two different scenarios for SMM packing, these are similar to the case of $TbPc_2$ thin films. The dipolar fields are calculated at the site of the marked SMM.

The dipolar magnetic field of one SMM moment $\boldsymbol{\mu}_i$ produced on another amounts to,

$$\mathbf{B}_i(\mathbf{r}_i) = \frac{\mu_0}{4\pi} \frac{3\mathbf{r}_i(\boldsymbol{\mu}_i \cdot \mathbf{r}_i) - \boldsymbol{\mu}_i r_i^2}{r_i^5}, \tag{1}$$

where $\mathbf{r}_i$ is the spacing vector between them. The resulting field of all contributing moments on a single SMM can be obtained by summing the fields from the individual moments. It is sufficient to take into consideration only a few neighbouring moments within a distance $\rho$ from it [28]. Typically, the calculated field distribution converges for $\rho = 5a$. Note that here, since the average magnetization per unit area is zero in both scenarios, the contribution of moments outside $\rho$ is zero [28]. For all the calculations performed here we use $|\boldsymbol{\mu}| = 1\mu_B$ ($S = 1/2$) with all distances normalized by $a$ in Å, giving the normalized dipolar fields in units of $Å^3$T (to be divided by $a^3$ to obtain the actual field).

## 3. Results

### *3.1. Near the substrate*

We start by considering a layer of molecules which lie directly on the substrate. Assuming their easy axis is normal of the surface, the magnetic moment has only a component along the $z$-direction. Therefore, the magnetic field these molecules produce on an individual molecule in the same layer (highlighted red in Fig. 1) does not have components perpendicular to its easy axis. The results of the calculation described above for a triangular and a square lattice are shown in Fig. 2(a) and (b), respectively. In what follows, we distinguish between field components along

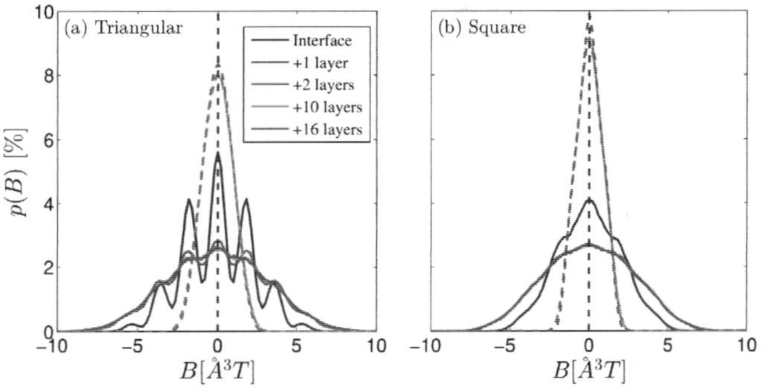


**Figure 2.** The dipolar field distribution calculated for scenario **I**, in the case of a (a) triangular and (b) square lattice. The solid and dashed lines represent $B_{||}$ and $B_{\perp}$, respectively.

the SMM easy axis ($B_{||}$) and perpendicular to it ($B_{\perp}$). Note, for both lattices, the probability for zero field is large. In the triangular lattice, there are also notable peaks at magnetic fields of $B_{||} = 2n\beta$ $Å^3$T (where $\beta = \mu_0\mu_B/4\pi a^3$ and $n = 0, \pm1, \pm2$ and $\pm3$). These are due to the six nearest neighbours and all their possible spin up/down configurations. For example, the zero field is experienced if three spins are assigned $S_z = +S$ and three are assigned $-S$. The probability for $B = 0$ reflects the number of possible configurations of three spins up and three down, i.e. $p(B = 0) = \binom{6}{3}(\sum_{k=0}^{6}\binom{6}{k})^{-1} = 31.25\%$. Magnetic moments further away, will broaden the distribution and a finite probability for intermediate magnetic fields emerge. In contrast, in the square lattice, each SMM has four nearest neighbours, but also four next-nearest neighbours with only slightly bigger distance. Therefore, the maxima are expected at $B_{||} = 2n\beta$ (with $n = 0, \pm1$ and $\pm2$). However, the strong contribution of the next-nearest neighbours produces high probability for intermediate fields as well.

Additional contributions from molecules lying above this first layer also contribute to the magnetic field distribution. Taking these into account produces some broadening of the probability distribution as shown in Fig. 2. Only the first two layers have a visible effect on the calculated field distribution. This is due to the $\frac{1}{r^3}$ dependence of the dipolar field. More importantly, the contribution of additional layers broadens the distributions of both $B_{||}$ and $B_{\perp}$ components, as shown in Fig. 2. Here we assumed that the layers of molecules stack directly on top of each other with spacing $a$ between them. However, different stacking does not change our general conclusions.

*3.2. Far from the substrate*
Next we consider a layer of molecules far away from the substrate, where we assume, as in the case of $TbPc_2$, that their easy axis is randomly oriented in the plane. We calculate the magnetic field distribution on an individual SMM, highlighted red in Fig. 1. Since the easy axes of the molecules are not parallel to one another, the produced dipolar magnetic fields may point in random directions. Therefore, the probability for one molecule to experience magnetic field components perpendicular to its own easy axis is larger compared to scenario **I**. Moreover, the distributions of both $B_{\|}$ and $B_{\perp}$ are identical (Fig. 3). The distributions for both components/lattices are almost Gaussian and featureless since the moments are randomly aligned in the plane. It also remains unchanged when neighbouring layers are also considered.

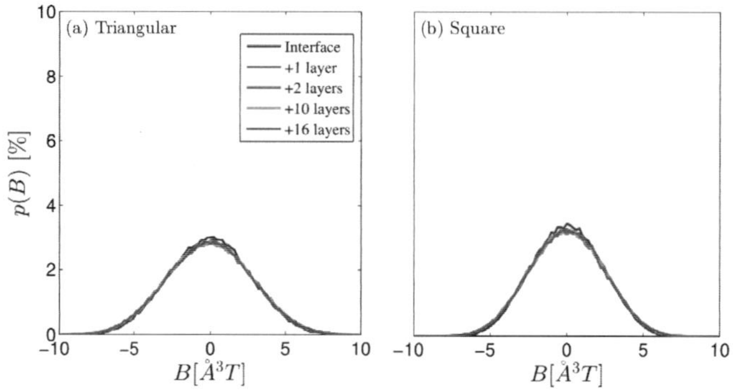


**Figure 3.** The dipolar field distribution calculated for scenario **II**, in the case of a (a) triangular and (b) square lattice. $B_{\perp}$ and $B_{\|}$ components are identical.

*3.3. Random lattice*
The above calculations are based on the assumption of SMMs which are ordered on a regular lattice. In reality, the arrangement can be highly disordered in SMM thin films, especially far from the substrate. To take this into consideration, we introduce as small ($< a$) and random displacement of all SMMs from their lattice sites along the $x$ and $y$ directions,

$$\begin{aligned} x &\mapsto x + R\cos(\phi) \\ y &\mapsto y + R\sin(\phi). \end{aligned} \tag{2}$$

Here, $x$ and $y$ denote the coordinates of a lattice site, $R$ is random (uniform distribution) displacement magnitude and $\phi$ a random angle $\in (0, 2\pi)$ which determines the direction of the displacement. In Fig. 4, we plot as an example the magnetic field distribution, $B_{\|}$, calculated for a triangular lattice in scenario **I** for various values of $R$. In fact, the effect of disorder is almost equivalent to a convolution of the ordered lattice distribution with a function

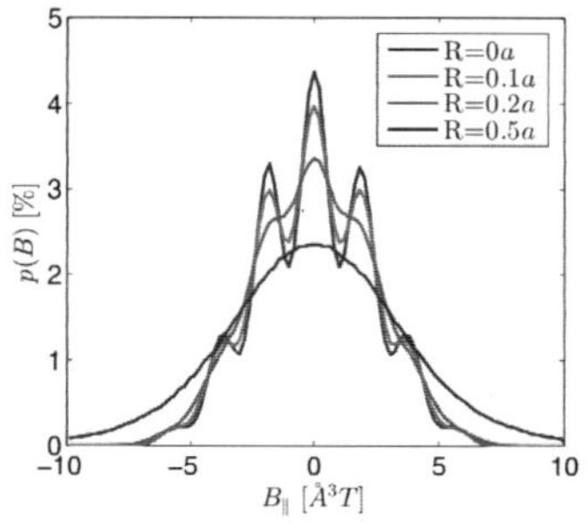


**Figure 4.** The effect of lattice randomization on $B_{\|}$ distribution calculated for scenario **I**.

that mimics the randomization of the lattice,

$$f(B) = \begin{cases} 1 & |B| \leq \beta \left(\frac{R}{a}\right)^3 \\ 0 & \text{otherwise} \end{cases} \tag{3}$$

The features mentioned above, due to the nearest neighbours, gradually disappear for larger $R$. We point out here that other random (non-uniform) displacements will naturally result in a different form of $f(B)$, e.g. a Gaussian distribution leads to a $f(B)$ with a Gaussian form, etc..

## 4. Discussion

In first approximation, the Hamiltonian of a magnetic molecule with easy axis along z is given by

$$\mathcal{H} = -DS_z^2 + g\mu_B \mathbf{B} \cdot \mathbf{S}, \tag{4}$$

where $D$ is the axial anisotropy, $S_z$ the spin operator along the easy axis and $\mathbf{B}$, $\mathbf{S}$ are the field and spin vector operators, respectively. In zero field (ZF), the eigenstates are $|S, M_S\rangle$, where $S$ denotes the total spin quantum number and $M_S$ is the azimuthal spin quantum number $(-S \leq M_S \leq S)$. In the absence of magnetic fields, the ground state is twofold degenerate, $E(M_S = \pm S) = -DM_S^2$. If a field $\mathbf{B} = B\hat{z}$ is applied along the easy axis, then the energy levels of the spin states are $E(M_S) = -DM_S^2 \pm g\mu_B BM_S$, with the same eigenvectors as in the ZF case. However, the ground state degeneracy is removed and quantum tunnelling is prohibited, i.e. no transition between different spin states is possible. In contrast, a field applied perpendicular to the easy axis causes a change (mixing) of eigenstates. For example, if the magnetic field $\mathbf{B} = B\hat{x}$, then the Hamiltonian can be written as

$$\mathcal{H} = -DS_z{}^2 + g\mu_B BS_x = -DS_z{}^2 + \frac{g\mu_B}{2}B(S_+ + S_-) \tag{5}$$

where $S_+$ and $S_-$ are the raising and lowering operators, $S_\pm = S_x \pm iS_y$. Consequently, the eigenstates are a mixture of states with different quantum spin numbers $M_S$. The first order mixing can be calculated using first order perturbation theory. A system, which is initially prepared in the ZF ground spin state, $\Psi_{gs} = \frac{1}{\sqrt{2}}(|S, S\rangle + |S, -S\rangle)$, has a finite probability for a transition to the first excited states, $\Psi_1 = \frac{1}{\sqrt{2}}(|S, S-1\rangle + |S, -S+1\rangle)$,

$$p(\Psi_{gs} \rightarrow \Psi_1) = \frac{\mu_B{}^2 B^2 S}{4D^2(2S-1)^2} \tag{6}$$

which can be understood as an enhancement in the spin dynamics.

From these simple quantum mechanical considerations, we expect that fields along the easy axis reduce tunnelling and slow spin dynamics at low temperatures, while fields perpendicular to it enhance spin dynamics. With this in mind, we compare spin dynamics in the two scenarios **I** and **II**. In scenario **I**, high probability for non-zero $B_{||}$ is observed, while the distribution of $B_\perp$ is narrow (it fits to a Gaussian of width $\sigma \sim 2.509 \mathring{A}^3 T$). On the other hand, in scenario **II** the distribution of both, $B_{||}$ and $B_\perp$ are identical with a Gaussian of width $\sigma \sim 6.799 \mathring{A}^3 T$. Note, the width of $B_{||}$ in scenario **II** is smaller than **I**, while the width of $B_\perp$ is larger. Therefore, the spin dynamics in scenario **I** should be slower than that in **II**. Finally, we note that these general conclusions do not change with the type of lattice or the amount of random disorder. This is due to the fact that the field distribution is primarily affected by the relative orientation of the neighbouring spins rather than their geometric configuration.

## 5. Conclusion

In conclusion, we have modelled the dipolar field distribution experienced by an individual SMM in a film, along and perpendicular to its easy axis. We have also investigated the dependence of the calculated distribution on the arrangement of the SMMs in the films. We find that different molecular configuration could alter significantly the molecular spin fluctuation rate of the SMM, due to the change in distribution of dipolar fields. Assuming a system similar to $TbPc_2$, where the packing changes as a function of proximity to the substrate, we expect the spin dynamics to change as a function of depth. In particular, the dynamics are slower in the layers near the substrate interface, but they are enhanced gradually as the SMMs are further away from it. These results point to a potential tweaking mechanism for tuning SMM spin dynamics by manipulating the packing of SMMs in a thin film, e.g. by the choice of a substrate. We expect that such considerations could provide a new avenue for the design of new tunable SMM nanostructures. Finally, it is important to point out here that in these calculations we consider only dipolar interactions with neighbouring SMMs. In practice, however, other interactions of SMMs with their environment may be present, and may lead to different effects on spin dynamics. In fact, recent LE-$\mu$SR measurements on thin films of $TbPc_2$ have revealed an opposite substrate proximity effect to what is expected from dipolar interactions alone [29]. This is due to the special case of $TbPc_2$, which has an unpaired electron on its organic shell which introduces additional interactions between the SMMs themselves as well as the substrate.

## References

[1] Gatteschi D, Sessoli R and Villain J 2006 *Molecular Nanomagnets* (Oxford: Oxford University Press)
[2] Wernsdorfer W, Aliaga-Alcalde N, Hendrickson D N and Christou G 2002 *Nature* **416** 406–409
[3] Gatteschi D and Sessoli R 2003 *Angew. Chem. Int. Ed.* **42** 268
[4] Leuenberger M N and Loss D 2001 *Nature* **410** 789–793
[5] Tejada J, Chudnovsky E M, del Barco E, Hernandez J M and Spiller T P 2001 *Nanotechnology* **12** 181
[6] Ardavan A, Rival O, Morton J J L, Blundell S J, Tyryshkin A M, Timco G A and Winpenny R E P 2007 *Phys. Rev. Lett.* **98** 057201
[7] Naitabdi A, Bucher J P, Gerbier P, Rabu P and Drillon M 2005 *Adv. Mater.* **17** 1612
[8] Gomez-Segura J, Diez-Perez I, Ishikawa N, Nakano M, Veciana J and Ruiz-Molina D 2006 *Chem. Commun.* 2866
[9] Burgert M, Voss S, Herr S, Fonin M, Groth U and Rudiger U 2007 *J. Am. Chem. Soc.* **129** 14362–14366
[10] Moroni R, Buzio R, Chincarini A, Valbusa U, de Mongeot F B, Bogani L, Caneschi A, Sessoli R, Cavigli L and Gurioli M 2008 *J. Mater. Chem.* **18**(1) 109–115
[11] Margheriti L *et al* 2009 *Small* **5** 1460–1466
[12] Mannini M *at al* 2009 *Nature Mater.* **8** 194–197
[13] Cavallini M, Facchini M, Albonetti C and Biscarini F 2008 *Phys. Chem. Chem. Phys.* **10**(6) 784–793
[14] Cornia A *et al* 2003 *Angew. Chem. Int. Ed.* **42** 1645
[15] Condorelli G G, Motta A, Fragalà I L, Giannazzo F, Raineri V, Caneschi A and Gatteschi D 2004 *Angew. Chem. Int. Ed.* **43** 4081
[16] Coronado E, Forment-Aliaga A, Romero F M, Corradini V, Biagi R, Renzi V D, Gambardella A and del Pennino U 2005 *Inorg. Chem.* **44** 7693–7695
[17] Fleury B, Catala L, Huc V, David C, Zhong W Z, Jegou P, Baraton L, Palacin S, Albouy P A and Mallah T 2005 *Chem. Commun.* 2020
[18] Salman Z *et al* 2007 *Nano Lett.* **7** 1551
[19] Mannini M *et al* 2010 *Nature* **468** 417–421
[20] Morenzoni E, Kottmann F, Maden D, Matthias B, Meyberg M, Prokscha T, Wutzke T and Zimmermann U 1994 *Phys. Rev. Lett.* **72** 2793
[21] Prokscha T, Morenzoni E, Deiters K, Foroughi F, George D, Kobler R, Suter A and Vrankovic V 2008 *Nucl. Instrum. Meth. A* **595** 317–331
[22] Morenzoni E, Glucker H, Prokscha T, Khasanov R, Luetkens H, Birke M, Forgan E M, Niedermayer C and Pleines M 2002 *Nucl. Instrum. Meth. B* **192** 254
[23] Morenzoni E, Prokscha T, Suter A, Luetkens H and Khasanov R 2004 *J. Phys.: Condens. Matter* **16** S4583–S4601
[24] Lascialfari A, Jang Z H, Borsa F, Carretta P and Gatteschi D 1998 *Phys. Rev. Lett.* **81** 3773

[25] Branzoli F, Carretta P, Filibian M, Graf M J, Klyatskaya S, Ruben M, Coneri F and Dhakal P 2010 *Phys. Rev.* B **82** 134401
[26] Salman Z, Giblin S R, Lan Y, Powell A K, Scheuermann R, Tingle R and Sessoli R 2010 *Phys. Rev.* B **82** 174427
[27] Margheriti L *et al* 2010 *Adv. Mater* **22** 5488–5493
[28] Salman Z and Blundell S 2012 *Physics Procedia* **30** 168–173
[29] Hofmann A, Salman Z, Mannini M, Amato A, Malavolti L, Morenzoni E, Prokscha T, Sessoli R and Suter A 2012 *ACS Nano* **6** 8390–8396

# The effect of the demagnetizing field in cylindrical samples in high transverse field $\mu^+$SR experiments

**Franz Lang and Stephen J Blundell**

University of Oxford, Department of Physics, Clarendon Laboratory, Parks Road, OX1 3PU, UK

E-mail: `franz.lang@physics.ox.ac.uk`

**Abstract.** We investigate the spatial variation of the demagnetizing field for uniformly magnetized, cylindrical samples using a recently developed Fourier space approach. We show that the demagnetizing response of the sample leads to a position dependence of the magnetic field, which varies most strongly near the radial boundary of the cylinder. Furthermore, we demonstrate that the demagnetizing field leads to a subsequent broadening of the field distribution, as experienced by the muons implanted in the sample, with a characteristic shape including a low field tail and a sharp high-field cutoff. We present a detailed study of this field broadening as a function of the aspect ratio of the cylinder and find that it is significant and largest for aspect ratios typical for cylindrical samples grown in mirror furnaces. We identify two strategies to minimize this broadening: adding a degrader so that muons implant closer to the surface of the sample and using a circular mask to stop muons from implanting near the radial edge. This could help identify whether an experimentally observed broadening is caused by the demagnetization response or the intrinsic properties of the sample.

## 1. Introduction

The development of new high transverse field (TF) muon spectrometers such as Hifi, HiTime, GPD or HAL-9500 is opening up new possibilities for experimental investigations of samples with exotic magnetic phases. The high magnetic fields involved may magnetize the sample and subsequently induce a demagnetizing field. It is important to understand to what extent the measured variations in the internal magnetic field distribution might be caused by variations of the demagnetizing field rather than by intrinsic effects due to exotic sample physics. Many single crystals grown using mirror furnaces are cylindrical, with experiments carried out on thin cylindrical slices. Therefore, an understanding of the demagnetizing field inside such cylindrical samples is desirable for high transverse-field muon-spin rotation (TF-$\mu^+$SR) studies.

Demagnetizing fields inside magnetized objects are difficult to calculate, but if the object is ellipsoidal in shape then the demagnetizing field is uniform [1]. This simple case includes the sphere, the flat plate, and the infinitely long cylinder as subsets. In other geometries an analytical solution is not so forthcoming. Though a rather unwieldy analytic solution was outlined some time ago [2], a recently developed Fourier space approach [3] has extended the range of analytically solvable geometries in a much more general and elegant fashion. This approach has been used to calculate the demagnetizing tensor of a uniformly magnetized cylinder [4] analytically, but the published solution contains errors (see [5]).

## 2. Magnetic Fields

The total magnetic field acting on a muon implanted into a sample can be written as

$$\boldsymbol{B}_{\text{total}} = \boldsymbol{B}_{\text{ext}} + \boldsymbol{B}_{\text{hf}} + \boldsymbol{B}_{\text{dip}} + \boldsymbol{B}_{\text{L}} + \boldsymbol{B}_{\text{D}},$$

where $\boldsymbol{B}_{\text{ext}}$ respresents the externally applied field, $\boldsymbol{B}_{\text{hf}}$ represents the contact and transferred hyperfine interactions, $\boldsymbol{B}_{\text{dip}}$ and $\boldsymbol{B}_{\text{L}}$ account for the dipolar fields inside and outside the Lorentz sphere respectively, and $\boldsymbol{B}_{\text{D}}$ corresponds to the demagnetizing field. We will restrict our attention to the last two terms, since we are interested in the general behaviour of magnetized samples regardless of the sample's chemical structure, which is required to evaluate $\boldsymbol{B}_{\text{dip}}$. We assume that the sample acquires a uniform magnetization $\boldsymbol{M}$ along the direction of the external field $\boldsymbol{B}_{\text{ext}}$, taken to be along $\hat{\boldsymbol{z}}$, in which case the Lorentz field becomes $\boldsymbol{B}_{\text{L}} = \mu_0 \boldsymbol{M}/3$. We define the shape function $D(\boldsymbol{r})$ which takes the value 1 when $\boldsymbol{r}$ is inside the sample and 0 outside the sample. The demagnetizing tensor is then given by $N_{ij}(\boldsymbol{k}) = D(\boldsymbol{k})k_i k_j/|\boldsymbol{k}|^2$ and hence [3]

$$N_{ij}(\boldsymbol{r}) = \frac{1}{(2\pi)^2}\int d^3\boldsymbol{k}\,\frac{D(\boldsymbol{k})}{k^2}k_i k_j e^{i\boldsymbol{k}\cdot\boldsymbol{r}},$$

with the additional relation $\mathrm{Tr}[N_{ij}(\boldsymbol{r})] = D(\boldsymbol{r})$. Thus, the demagnetizing field becomes $B_{\text{D},i}(\boldsymbol{r}) = -\mu_0 N_{ij}(\boldsymbol{r})M_j$ and we use our analytically computed expression for $N_{ij}(\boldsymbol{r})$ for a cylinder [5]. Note that even if the sample is not ferromagnetic, it will acquire a magnetization $\boldsymbol{M} = \chi\boldsymbol{B}_{\text{ext}}/\mu_0$ in a strong applied field $\boldsymbol{B}_{\text{ext}}$. The purpose of this paper is to provide a guide in assessing whether an observed broadening is due to intrinsic sample physics affecting $\boldsymbol{B}_{\text{dip}}$, or a demagnetization effect due to $\boldsymbol{B}_{\text{D}}$.

## 3. Sample Details

We will assume a cylindrical geometry with a uniform magnetization of magnitude $|\boldsymbol{M}| = M$ along the cylindrical axis, taken to be the z-axis, and the muons being implanted along the same axis. Furthermore, we will assume a typical experimental sample with a radius $R$ of 10mm, a height $h$ of 2mm, and a muon implantation depth of 0.1mm, in line with the stopping range of surface muons of about 110 mg cm$^{-2}$, unless explicitly stated otherwise. Figure 1 shows a schematic of the experimental setup we are simulating.

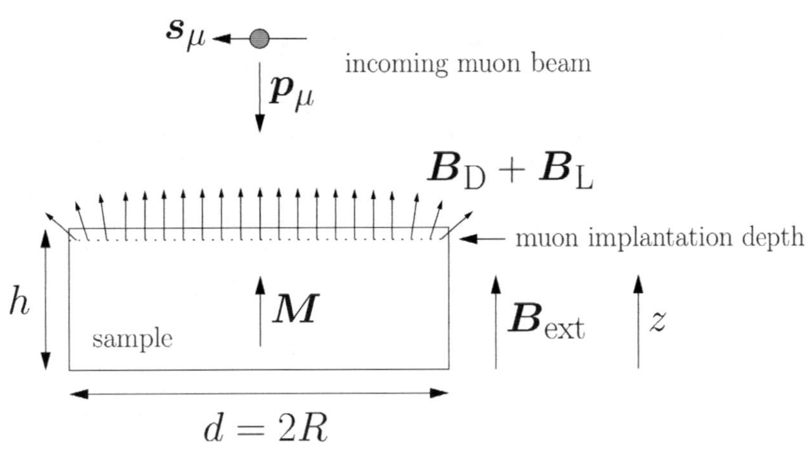


**Figure 1.** Schematic of high-field TF-$\mu^+$SR experiment, showing the demagnetization field variation within a cylindrical sample (shown in cross section) at the muon implantation depth.

Journal of Physics: Conference Series **551** (2014) 012056 doi:10.1088/1742-6596/551/1/012056

## 4. Demagnetizing Field Distribution

We can use the analytic solution for the demagnetization tensor to compute the demagnetizing field at the stopping depth of the muons as a function of the radial position. The resulting fields are plotted in Figure 2 for the demagnetizing field and the sum of Lorentz and demagnetizing field and their respective z-components. The reason for plotting the z-components is that the large external field is applied along $\hat{\boldsymbol{z}}$ and therefore $(\boldsymbol{B}_{\mathrm{D}} + \boldsymbol{B}_{\mathrm{L}}) \cdot \hat{\boldsymbol{z}}$ will add to the total field in first order, whereas the components perpendicular to $\hat{\boldsymbol{z}}$ add in second order. Thus, the former will have the largest effect in a TF-$\mu^{+}$SR experiment.

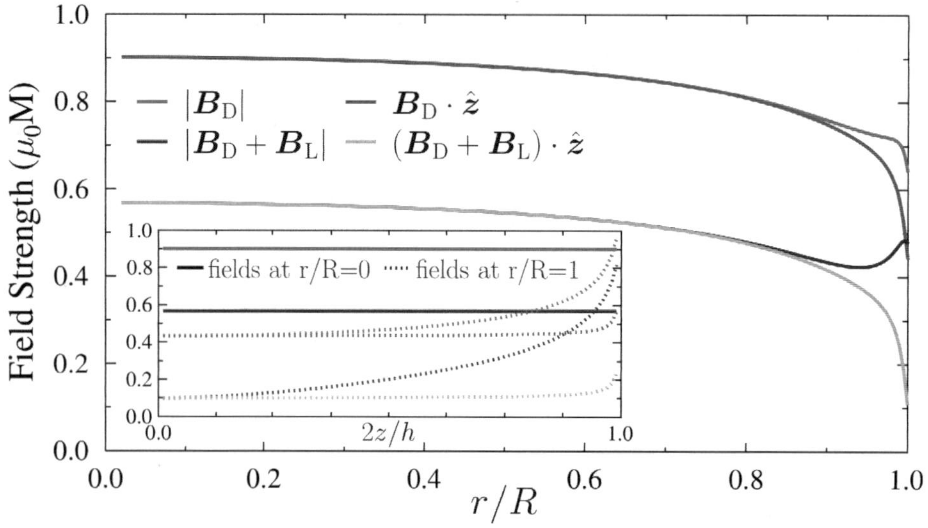


**Figure 2.** Magnetic field strengths and components along the cylinder axis at the muon stopping depth. The inset shows the z dependence of the fields at the centre and edge of the cylinder. $R$ and $h$ are the cylinder radius and height respectively; the cylinder centre is at $z = r = 0$. Note that the Lorentz field $\boldsymbol{B}_{\mathrm{L}} = \mu_0 \boldsymbol{M}/3$ is homogenous and that the demagnetizing field $\boldsymbol{B}_{\mathrm{D}}$ is not.

Figure 2 shows that the demagnetizing field is position dependent and more importantly that its z-component almost vanishes near the radial boundary, while the overall field strength does not. Qualitatively, this is due to a realigning of the field to satisfy the boundary condition at the radial edge, as indicated in Figure 1, but we note that the region near the radial edge actually constitutes a significant portion of the total volume. Thus, even though the field realignment only happens close to the boundary, it can be a significant effect due to the volume weighting.

The impact of the demagnetizing field becomes apparent in the probability distributions, which are plotted in Figure 3 and show a clear and sizeable broadening of the magnetic field. The discrepancy between the two distributions for the field strength and its z-component can be understood in the light of the realigning of the field near the radial boundary and the demagnetizing field being stronger near the flat z-boundary, which is often modelled to contain "magnetic charges" induced by the external field.

The calculations described above assume all muons to stop precisely at a depth of 0.1mm and in a radially uniform way, which is a good approximation if the muon beam covers the entire sample. However, when one repeats the above calculations and includes a Gaussian stopping profile centred around this implantation depth, the outcomes are effectively unchanged. In fact, the field distribution results were largely insensitive to all stopping profiles we tested. Thus, we restrict the subsequent discussion to a simple delta-function stopping profile.

Journal of Physics: Conference Series **551** (2014) 012056 doi:10.1088/1742-6596/551/1/012056

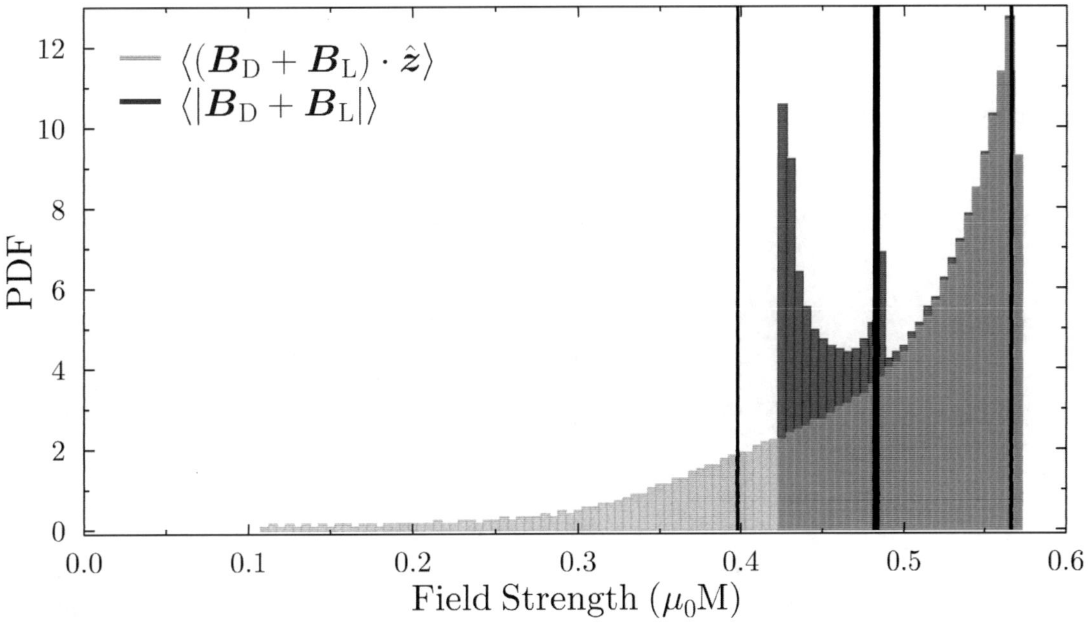


**Figure 3.** Probability density plot of the magnetic fields at the muon stopping depth. The black vertical lines indicate the mean and standard deviation of the z-component's distribution.

## 5. Effect of Aspect Ratio

We now examine the impact of the edge effects on the field broadening as a function of the cylinder's aspect ratio, defined as the ratio of height $h$ to diameter $d = 2R$ (the calculations above assumed $h/d = 0.1$). Computing the field distributions at the muon stopping depth for different values of $h/d$ yields the results shown in Figure 4. Plotted in the figure are the mean and standard deviation of both the field strength and its z-component. Figure 4 shows that the correct results are recovered in the limits of a flat cylinder (disc: $h/d \to 0$, $\boldsymbol{B}_{\rm D} = -\mu_0\boldsymbol{M}$, $\boldsymbol{B}_{\rm D} + \boldsymbol{B}_{\rm L} = -2\mu_0\boldsymbol{M}/3$) or long cylinder (wire: $h/d = \infty$, $\boldsymbol{B}_{\rm D} = 0$, $\boldsymbol{B}_{\rm D} + \boldsymbol{B}_{\rm L} = \mu_0\boldsymbol{M}/3$). Furthermore, it illustrates that the field broadening is strongest when the aspect ratio is of order 1, which applies to most experimental samples, and that it can exceed 20% of the average demagnetizing field in such cases. Therefore, we conclude that the field broadening due to the demagnetizing response and edge effects can in general be significant for the typical samples used for high-field TF-$\mu^+$SR experiments.

## 6. Effect of Muon Stopping Depth

Another possible way of influencing the effect of the demagnetizing field experienced by the muons, besides changing the aspect ratio of the sample, is to reduce the implantation depth of the muons by placing a degrader on top of the sample. This has the effect that more muons stop in regions closer to the flat z-boundary than the radial edge, and Figure 5 shows the two main resulting consequences for the case $h/d = 0.1$. Firstly, the average demagnetizing field strength increases as the stopping distance is reduced, which shows that the field is stronger near the cylinder's flat surfaces where $\nabla \cdot \boldsymbol{M} \neq 0$. Secondly, the broadening of the field distribution decreases as the implantation depth is shortened. Therefore, by adding a degrader it can be possible to reduce the field broadening at the expense of raising the average demagnetizing field strength.

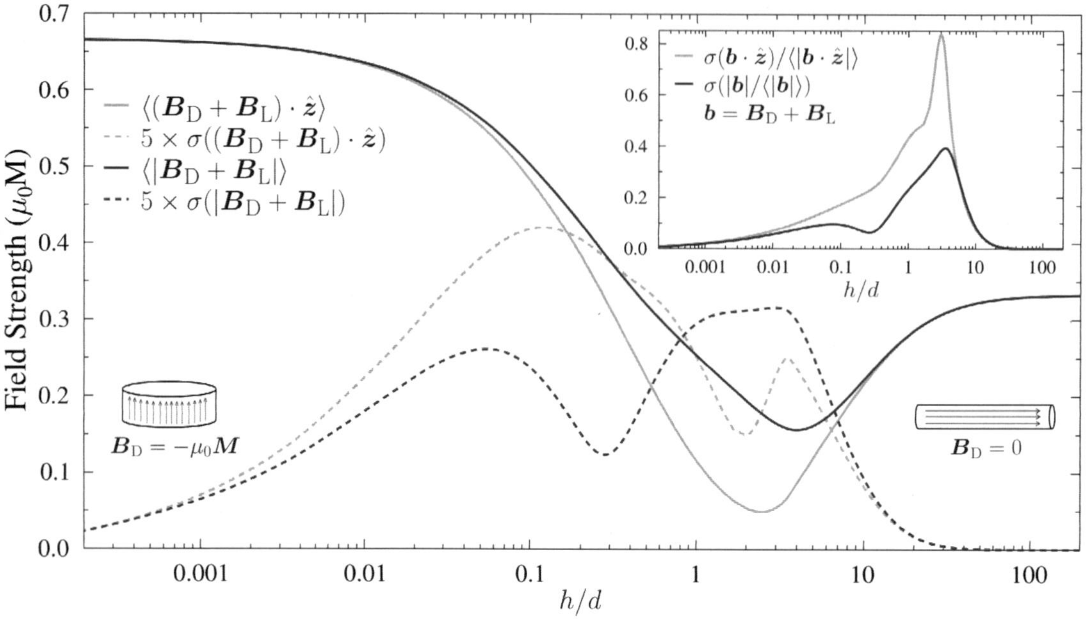


**Figure 4.** Field distribution versus cylinder aspect ratio. The standard deviations are scaled up by a factor of 5. The inset graph shows the ratio of standard deviations and mean values.

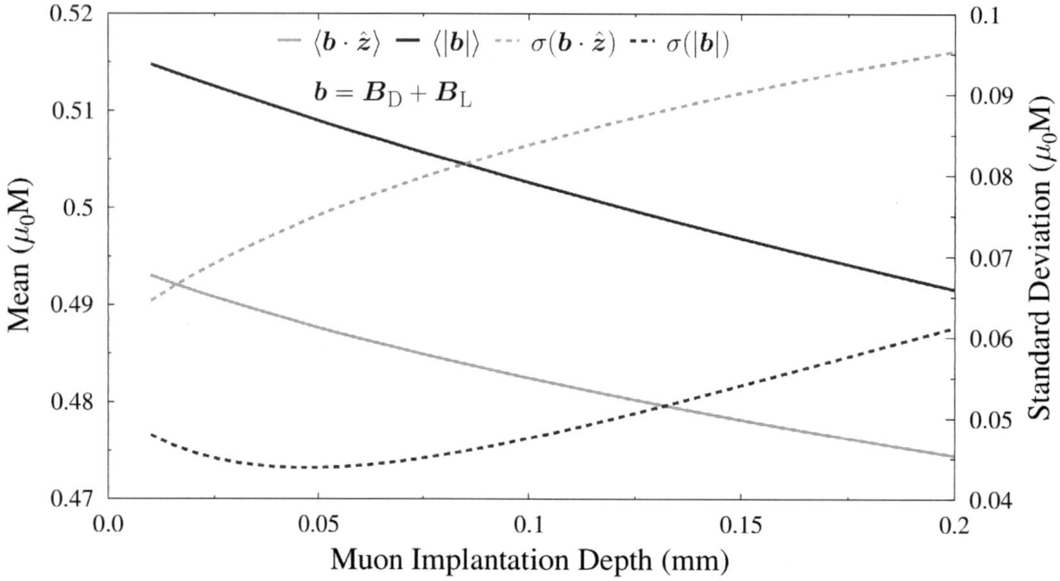


**Figure 5.** Field distribution dependence on the muon implantation depth for the case $h/d = 0.1$. Solid lines and the left hand axis correspond to mean values, dashed lines and the right hand axis to standard deviations.

## 7. Discussion

Our investigations show that the demagnetizing response of typical cylindrical samples can be significant due to edge effects. If we assume the external field $\boldsymbol{B}_{\mathrm{ext}} \parallel \hat{\boldsymbol{z}}$ to be very large then to a good approximation the demagnetizing and Lorentz field components parallel and perpendicular to $\hat{\boldsymbol{z}}$ add to the total field in first and second order respectively. In this case, a possible signature that the observed broadening is caused by the demagnetizing field could be a field distribution resembling that of the z-component of $\boldsymbol{b} = \boldsymbol{B}_{\mathrm{D}} + \boldsymbol{B}_{\mathrm{L}}$ shown in Figure 3, which has a long low-field tail due to the radial edge behaviour and a sharp high-field cutoff (this behaviour is reversed if $\chi < 0$). If an experimentally observed field distribution resembles this characteristic behaviour it could be useful to consider the demagnetization response more closely.

Additionally, we can think of two potential ways of influencing the demagnetization effects experimentally. One possibility is to place a degrader on top of the sample in order to shorten the muon implantation depth and subsequently reduce the demagnetizing field broadening as discussed above. This has the disadvantage that we might then get additional field contributions due to muons stopping in the degrader and the demagnetizing response of the degrader itself. Another possibility to reduce the demagnetizing field broadening is to place a thick circular mask with radius $< R$ in front of the sample, such that the muons that would stop in the regions near the radial boundary are completely stopped in the mask. In Figure 2 this would correspond to only considering regions away from the radial edge, thereby shortening the low-field tail of the $b_z$ distribution shown in Figure 3. However, the presence of the mask will lower the data rate and potentially shift the background signal experienced by muons. An alternative to the mask could be to reduce the muon beamspot radius below $R$, with the sample well centred in the beam.

## 8. Conclusions

We have shown that the demagnetizing field inside a uniformly magnetized cylinder is position dependent, which results in a field broadening caused by edge effects. Furthermore, we demonstrated that this field broadening has a characteristic shape and is non-negligible for sample geometries close to typical experimental ones, with standard deviations in excess of 10% of the mean demagnetizing field. We have identified two strategies to ameliorate this effect. A degrader can be used to shorten the muon implantation depth, at the expense of increasing the mean demagnetizing field experienced by the muons, or a circular mask could be deployed to ensure no muons implant near the radial edge of the sample. These strategies can be used to minimize the demagnetization broadening, or diagnostically, to ascertain how significant the effects are in a particular geometry. It is hoped that our calculations may serve as a useful guide for future high-field TF-$\mu^+$SR experiments and allow an assessment to be made of whether an observed broadening is intrinsic due to exotic sample physics or simply caused by a field-induced artifact.

## Acknowledgments

Thanks go to EPSRC, Merton College and the CR Barber Trust Fund for financial support.

## References

[1] Maxwell J C 1891 *A treatise on electricity and magnetism, Vol. 2* 3rd ed (Oxford: Clarendon Press)
[2] Joseph R I and Schlömann E 1965 *J. Appl. Phys.* **36** 1579
[3] Beleggia M and De Graef M 2003 *J. Mag. Mag. Mater.* **263** L1–L9
[4] Tandon S, Beleggia M, Zhu Y and De Graef M 2004 *J. Mag. Mag. Mater.* **271** 9–26
[5] Lang F and Blundell S J *in preparation*

Journal of Physics: Conference Series **551** (2014) 012057 doi:10.1088/1742-6596/551/1/012057

# Use of the $\mu$SR-technique for studying magnetization processes in nanocrystal films of ferromagnetic metals

**Y M Belousov**

Moscow Institute of Physics and Technology, 141700, Dolgoprudny, Institutsky lane 9, Russia

E-mail: theorphys@phystech.edu

**Abstract.** I show that the $\mu$SR technique allows one to study the occurrence of a magnetic ordering in nanocrystal films made of ferromagnetic metals – so called "scale" phase transition. The relation between the microscopic (local) field with macroscopic characteristics just as the external magnetic field, the average magnetization and the saturation magnetization is determined for a model for which the nanocrystal film consists of crystallographically ordered grains separated by disordered areas. Expressions for the behaviour of the muon spin polarization ensemble in this type of structures are obtained in cases of fast diffusing and nondiffusing muons. It is shown that experiments with "slow" positively charged muons allow one to measure all parameters of this type of structures and obtain important information for the study of phase transition physics.

## 1. Introduction

Magnetic nanocrystalline metals are of interest for the $\mu$SR technique since the use of positive muons and neutron diffraction provides real possibilities to study their bulk properties[1]. Possible applications of the $\mu$SR technique for the study of nanocrystalline ferromagnetic materials were still mentioned few years ago [1]. Nevertheless, no serious experimental researches has been carried out yet. It was emphasized that at least two problems regarding the fundamental physics of magnets could be solved for nanostructured ferromagnets. One of them is the problem of specific "scale" phase transitions, when the existence of a spontaneous magnetization depends both on temperature and size of a crystal. The mechanism of this size-induced phase transition is being actively studied (see, e.g. [2] - [4]). Currently, materials with grain sizes too small to exhibit ferromagnet properties are called superparamagnets. The second problem is related with the structure and magnetic properties of domains in nanostructures. They should be strongly different from the respective characteristics of "ordinary" polycrystalline ferromagnets. Actually, both theory and experiments show that microcrystalline grains ($10^2 - 4\cdot 10^2$ Å) should be single domain. The characteristic thickness of domain walls in a bulk sample[2] is $d \sim 10 - 30$ Å, which is of the same order of magnitude as the thickness of the intercrystalline amorphous interface. Thus, the concept of magnetic domains and domain walls in nanocrystals differs from the one observed for macroscopic polycrystals. In our case we need to consider magnetized regions without domain grains in a nonmagnetic medium. Hence, the macroscopic field $B_{\rm dom}$

[1] Nanocrystalline metals are conventionally accepted to be polycrystals with grain sizes of 10–400Å.

[2] The characteristic thikness of domain walls depends on the ratio between exchange and magnetocrystalline anisotropy energy scale. The presented values are correct for the ferromagnets under consideration.

Journal of Physics: Conference Series **551** (2014) 012057 doi:10.1088/1742-6596/551/1/012057

inside a crystallographically ordered grain and its dependence on the external magnetic field $\mathcal{B}$ should differ significantly from the properties of macroscopic samples.

The possibilities to use the $\mu$SR technique for studying ferromagnetic nanostructured thin films are represented in this report.

## 2. Behaviour of muon spin polarization

Let consider a nanostructured film consisting of crystallographically ordered grains separated by disordered regions. Spontaneous magnetization can arise only in ordered regions. Therefore, the local field sensed by a muon depends on wether the muon stops in a site within a grain or in an intergrain region. Let consider the situation when a spontaneous magnetization does not equal to zero. Thus the spin polarization of the muons ensemble can be written as the sum

$$\mathcal{P}(t) = \mathbf{P}_{\rm cr}(t) + \mathbf{P}_{\rm nc}(t), \tag{1}$$

where $\mathbf{P}_{\rm cr}(t)$ and $\mathbf{P}_{\rm nc}(t)$ are the polarization of the muons fraction stopping in grains and in intergrain regions, respectively.

For non-diffusing muons, the polarization is given by the well-known expression (see e.g. [5])

$$P_i(t) = \mu_{ik}(t)P_k(0) \tag{2}$$

where the tensor $\mu_{ik}(t)$ is

$$\mu_{ik}(t) = n_i n_k + (\delta_{ik} - n_i n_k)\cos\gamma_\mu b_\mu t + e_{ikl} n_l \sin\gamma_\mu b_\mu t\ . \tag{3}$$

Here $\mathbf{b}_\mu$ is the local field sensed by the muon, $\gamma_\mu/(2\pi) = 13.554$ kHz/G is the muon gyromagnetic ratio and $\mathbf{n} = \mathbf{b}_\mu/|\mathbf{b}_\mu|$ is the unit vector along the magnetic field.

The local fields acting on a muon in a grain and an intergrain region differ significantly. We are first interested by the field in a crystallographically ordered grain. In general, by separating the Lorentz sphere around the muon position, we can write [5, 6]:

$$\mathbf{b}_\mu = \mathbf{B} - \frac{8\pi}{3}\mathbf{M} + \mathbf{b}_{\rm dip} + \mathbf{B}_{\rm cont}, \tag{4}$$

where $\mathbf{B}_{\rm cont}$ is the contact field induced by electrons and $\mathbf{b}_{\rm dip}$ is the microscopic field induced by the oriented magnetic dipoles inside the Lorentz sphere. The contact field can always be written as $B_{i\,\rm cont} = K_{ik}B_k$. In cubic crystals, we can set $K_{ik} = \delta_{ik}K$. Therefore, the contact field causes only an isotropic Knight shift[3].

In an intergrain region, the spontaneous magnetization is equal to zero ($\mathbf{M}_{\rm nc} = 0$). Hence, dipole fields $\mathbf{b}_{\rm dip}$ can only be induced by the disordered nuclear magnetic moments. These fields cause an inhomogeneous line broadening, which can be adequately described by a Gaussian in the case of nondiffusing muons and a simple exponent for rapidly diffusing muons (see e.g. [5, 7]). Thus, the behaviour of the muon spin polarization for integrain regions is controlled by the local magnetic field $\mathbf{b}_\mu = \langle\mathbf{B}\rangle + \delta\mathbf{b}$, where $\delta\mathbf{b}$ is the static field inhomogeneity. The characteristic scale of the field inhomogeneity in the intergrain region is determined by the magnetization of grains and the distance between them. Thus, the polarization precession frequency of muons in the intergrain fraction allows one to determine the average magnetic field in the film. In the case of nondiffusing muons a Gaussian damping allows to determine the characteristic scale of the field inhomogeneity, $\sigma \sim \gamma_\mu\langle\delta\mathbf{b}^2\rangle/|\langle\mathbf{B}\rangle|$. In the case of diffusing muons, the depolarization rate depends on the diffusion constant and in the fast diffusing limit becomes negligibly small. It is

[3] For simplicity, we can omit the Knight shift in what follows, although it can make an appreciable contribution in some cases.

known that muons can diffuse rapidly in polycrystal samples (see e.g. [5] Ch.5 and references therein). But an irregular structure of an amorphous intergrain region differs from a structure of polycrystal samples. So, we can assume that a diffusion of a muon in intergrain regions is improbable.

If a crystallographically ordered grain has a nonzero spontaneous magnetization, the microscopic dipolar field is induced by the ordered electron magnetic moments. In this case, inside the Lorentz sphere, this field can always be written as [5, 6]

$$b_{i\,\mathrm{dip}} = -\frac{4\pi}{3}M_i + a_{ik}M_k, \tag{5}$$

where the tensor $a_{ik}$ depends on the type of interstitial sites. Calculations showed (see e.g. [5, 6]) that $a_{ik} = \delta_{ik}4\pi/3$ in an fcc Ni lattice; hence, the microscopic dipolar field is zero, $\mathbf{b}_{\mathrm{dip}}(fcc) = 0$.

In an hcp Co lattice, the dipolar field is also weak but is nonzero and has different values in crystallographically nonequivalent interstitial sites. If we direct the $z$ axis along the hexagonal axis, we have

$$\begin{array}{llll} \delta a^h_{xx} & = \delta a^h_{yy} = \Delta/2, & \delta a^h_{zz} = -\Delta & \text{in octahedral interstitial sites,} \\ \delta a^t_{xx} & = \delta a^t_{yy} = -\Delta, & \delta a^t_{zz} = 2\Delta & \text{in tetrahedral interstitial sites,} \end{array} \tag{6}$$

where $\Delta \approx 0.1$. So, the dipolar field can be written as $\mathbf{b}_{\mathrm{dip}}(hcc) = \delta a_{ik}M_k$. Therefore, we have $b_{\mathrm{dip}} \ll M$.

A more complicated picture is observed in a bcc Fe lattice, where the dipolar field is large and depends not only on the type of interstitial sites but also on the direction of the magnetization vector $\mathbf{M}$. The components of the tensor $a_{ik}(bcc)$ along the principal axes are [5, 6]

$$\begin{array}{llll} a^h_{xx} & = a^h_{yy} = -1.165, & a^h_{zz} = 14.9 & \text{in octahedral interstitial sites,} \\ a^t_{xx} & = a^t_{yy} = \quad 5.707, & a^t_{zz} = 1.152 & \text{in tetrahedral interstitial sites.} \end{array} \tag{7}$$

Hence, the local field acting on a muon in a bcc lattice is given by

$$b_{\mu\,i}(bcc) = B_i - 4\pi M_i + a_{ik}(bcc)M_k. \tag{8}$$

The local field acting on a muon depends on the direction of the magnetization vector in the grain (see Eq. (5)). The polarization of the muons fraction that stop in the crystallographically ordered grains should be defined by averaging over all possible orientations of the principal crystallographic axes.

Let us represent the local field acting on a muon as the sum of the two components parallel and perpendicular to the film plane $\mathbf{b} = \mathbf{b}_{\|} + \mathbf{b}_{\perp}$. Then, the polarization components are defined as

$$\mathcal{P}_{\perp} = \langle \frac{b_{\|}}{b} \mathrm{e}^{i\omega t} \rangle, \quad \mathcal{P}_{\|} = \langle \frac{b_{\perp}}{b} \sin\omega t \rangle, \qquad \omega = \gamma_\mu b = \gamma_\mu \sqrt{\mathbf{b}_{\|}^2 + \mathbf{b}_{\perp}^2} \tag{9}$$

The preexponential factors (the direction cosines) depend on whether the external field is perpendicular or parallel to the film plane.

## 3. Hierarchy of fields

To determine the local field acting on a muon implanted into a target, we need to find the microscopic field in the sample. At first, the hierarchy of fields in films of nanostructured ferromagnetic metals should be refined. As for a multidomain ferromagnet, in addition to the macroscopic fields in *each* grain $\mathbf{M}$, $\mathbf{B}$, $\mathbf{H}$ and the external one with the respect to *entire* sample $\boldsymbol{\mathcal{B}}$, we need to introduce average macroscopic fields in the sample. These fields are the results of an averaging of all mentioned above macroscopic fields and the magnetization over all grains and intergrain regions, $\langle\mathbf{B}\rangle$, $\langle\mathbf{H}\rangle$ $\langle\mathbf{M}\rangle$. The last parameters are determined in macroscopic experiments. The same parameters determine the field acting on a muon in a disordered intergrain region of the sample.

Let us denote the demagnetization factors of the film as $N_{ik}$ and write the relations between the average fields and the external field in the form

$$\mathcal{B}_i = \langle H_i\rangle + 4\pi N_{ik}\langle M_k\rangle, \quad \langle\mathbf{B}\rangle = \langle\mathbf{H}\rangle + 4\pi\langle\mathbf{M}\rangle. \tag{10}$$

Let the $z$ axis of the coordinate system associated with the film be perpendicular to the film plane. Since, by definition, the film thickness $d$ is much smaller than its linear dimensions $L$, we have $N_{zz} \equiv N_\perp \approx 1$ and all other components are negligibly small.

If the external field is perpendicular to the film plane $\boldsymbol{\mathcal{B}}\|z$, we have

$$\mathcal{B}_z \equiv \mathcal{B} = \langle H_z\rangle + 4\pi\langle M_z\rangle, \quad \langle H_{\rm pl}\rangle \approx 0, \tag{11}$$

where $H_{\rm pl}$ are the vector $\mathbf{H}$ components lying in the film plane. From the relation (10) we obtain

$$\langle B_z\rangle = \mathcal{B}, \quad \langle B_{\rm pl}\rangle \approx 4\pi\langle M_{\rm pl}\rangle. \tag{12}$$

If the external field lies in the film plane $\boldsymbol{\mathcal{B}}\perp z$, we have

$$\langle H_z\rangle + 4\pi\langle M_z\rangle = 0, \quad \langle H_z\rangle = -4\pi\langle M_z\rangle, \quad \langle H_{\rm pl}\rangle \approx \mathcal{B} \tag{13}$$

and the average induction is given by

$$\langle B_z\rangle = 4\pi\langle M_z\rangle, \quad \langle B_{\rm pl}\rangle = \mathcal{B} + 4\pi\langle M_{\rm pl}\rangle. \tag{14}$$

In the usual model, a nanocrystal grain in a film is considered as a single-domain particle. Therefore, in contrast to the situation found in an ordinary ferromagnet, each grain has no domain wall and is in both the external (homogeneous) field $\boldsymbol{\mathcal{B}}$ and the field induced by the magnetized grains of the film.

For each grain (with an ellipsoid shape) we can write a relation similar to Eq. (10), $\langle B_i\rangle = H_i + 4\pi n_{ik}M_k$, where $n_{ik}$ are the grain demagnetization factors. Analytical expressions could be obtained for the approximation of oblate ellipsoids, when $n_\| = n_{zz} \approx 1, \quad n_\perp = 1 - n_\| \ll 1$.

If the external field is perpendicular to the film plane, Eqs. (12) give

$$\langle B_z\rangle \approx \mathcal{B}, \qquad H_z = \mathcal{B} - 4\pi M_z \qquad H_{\rm pl} = \langle B_{\rm pl}\rangle = 4\pi\langle M_{\rm pl}\rangle, \quad B_{\rm pl} = 4\pi\big(\langle M_{\rm pl}\rangle + M_{\rm pl}\big). \tag{15}$$

If the external field lies in the film plane, we obtain

$$H_z = -4\pi M_z, \quad H_{\rm pl} \approx \langle B_{\rm pl}\rangle = \mathcal{B} + 4\pi\langle M_{\rm pl}\rangle \qquad B_z = 0, \quad B_{\rm pl} = \mathcal{B} + 4\pi\big(\langle M_{\rm pl}\rangle + M_{\rm pl}\big). \tag{16}$$

Thus, to determine the fields, we should determine the direction of the magnetization vector $\mathbf{M}$ in each grain and $\langle\mathbf{M}\rangle$ in the film. The local field was obtained in [8] in the approximation when the ordered grains are considered as oblate ellipsoids.

## 4. Muon spin polarization description

We can see that the local field acting on a muon (see Eqs. (12) and (16)) could be represented as a sum of two items, one of them is a constant and the other depends on a magnetization orientation in a grain. The averaging of the muon spin polarization overall possible grain orientations leads up to an effective depolarization in accordance with Eq. (9). In the case of a strong external field ($\mathcal{B} \gg M$) the precession frequency of the muon spin polarization could be also written as a sum of two items, $\omega = \omega_0 + \omega(\vartheta)$, where $\vartheta$ is the angle between the grain magnetization and the external field. Only the transverse component of the muon spin polarization does not equal to zero, and it can be approximately written in the form

$$\mathcal{P}_\perp = \langle \frac{b_z}{b} e^{-i\gamma_\mu bt} \rangle \approx e^{-i\omega_0 t} \langle e^{-i\omega(\vartheta)t} \rangle = e^{-i(\omega_0 + \Delta\omega)t} e^{-\sigma^2 t^2}, \tag{17}$$

Here, the frequency shift $\Delta\omega$ and the depolarization rate $\sigma$ are determined by both the grain magnetization value and the muon interstitial position. Analytical expressions for these important parameters were obtained in [8] in cases of rapidly diffusing and nondiffusing muons.

In the case of fast diffusing the muon spin polarization behaviour in fcc and bcc lattices is practically the same. Up to the quadratic terms of $M/\mathcal{B}$, the precession frequency $\omega_0$ and its shift $\Delta\omega$ are given by

$$b_0 = \mathcal{B} - \frac{8\pi}{3} M + \frac{1}{2}\left(\frac{4\pi}{3}\right)^2 \frac{M^2}{\mathcal{B}}, \quad \Delta\omega = \gamma_\mu \frac{1}{63}\frac{2\pi}{3} M \left(\frac{M}{\mathcal{B}}\right)^2. \tag{18}$$

The second moment $\sigma^2_{\rm diff} \propto (M/\mathcal{B})^2$ is rather small and it is unlikely that it can be measured in experiments.

In the case of nondiffusing muons one can obtain more detail information. In a bcc lattice of ferromagnet belongs to the "easy axis" type (e.g. Fe) the precession frequency and its shift are determined by

$$\omega_0^{bcc} = \gamma_\mu(\mathcal{B} - (2\pi + \frac{a_\perp}{2})M), \quad \Delta\omega^{bcc} = -\frac{1}{3}\gamma_\mu d\, M \left[1 - \left(2d - \frac{41}{28}\beta\right)\frac{2M}{15\mathcal{B}}\right], \tag{19}$$

and the second moment

$$\sigma^2_{\rm nd} = \frac{7}{30}\left(\gamma_\mu d\, M\right)^2 . \tag{20}$$

Here, $a_\parallel$ and $a_\perp$ are the components of the tensor $a_{ik}$ defining the dipole field, $d = (a_\parallel - a_\perp)/2$, $\beta$ is the magnetic anisotropy parameter.

The behaviour of the polarization of nondiffusing muons in uniaxial ferromagnets is similar to that presented above for cubic ferromagnets (19)–(20)

$$\omega_0^{hcp} = \gamma_\mu \left[\mathcal{B} - \left(\frac{8\pi}{3} - \frac{\Delta}{2}\right)M\right], \quad \Delta\omega^{hcp} = -\frac{1}{2}\gamma_\mu M\Delta\left[1 - (\beta + \frac{3}{2}\Delta)\frac{4M}{5\mathcal{B}}\right],$$
$$\sigma^2_{\rm nd} = \frac{21}{40}\left(\gamma_\mu M\Delta\right)^2 . \tag{21}$$

## 5. Conclusion

The formulas presented in this report show that the $\mu$SR-technique gives the opportunity to determine the magnetization in a grain by measuring the precession frequency and the depolarization rate and to reveal the existence of muon diffusion in the crystallographically ordered fraction of a sample. In the case of nondiffusing muons in a crystallographically ordered grain the precession frequency measurement allows one to determine the magnetic anisotropy constant, which is of great importance in the physics of the phenomenon under consideration. The measurement of the depolarization rate of the muon spins precessing at the frequency corresponding to the average magnetic field of the film allows determination of the characteristic scale of the magnetic field inhomogeneity in the disordered intergrain region. The ratio between the precession amplitudes of the two fractions of muons allows to determine the ratio of the volumes of the paramagnetic and ferromagnetic phases of the sample. In addition, the local fields induced by nuclear magnetic moments can be taken into account separately using the well known approaches for normal metals [5]. We note that all stable isotopes of Fe and Ni, except $^{57}$Fe (2.21%) and $^{61}$Ni (1.25%), have zero spins (see e.g. [9]) and, hence, experiments with these widespread magnets are preferred.

Recent experiments [10]-[14] showed that low-energy muons (LE-$\mu$SR) give opportunities to study the magnetic properties and inhomogeneity of thin superconducting films. This LE-$\mu$SR technique has a scale resolution up to 10 nm and can be applied successfully to solve the fundamental problem of magnetic scale phase transitions in ferromagnets too. Simultaneous measurements by macroscopic methods (see e.g. [11]) would make it possible to obtain a complete information on the physics of the phase transitions in nanocrystalline ferromagnetic films.

**Acknowledgments**
I would like to acknowledge my colleagues for fruitful discussions.

## References

[1] Belousov Y M and Smilga V P 2000 *Physica* B **289-290** 141
[2] Teplykh A E, Bogdanov S G, Valiev E Z, Pirogov A N, Dorofeev Y A, Ostroushko A A, Udilov A E, Kazantsev V A and Kar'kin A E 2003 *Phys. Solid State* **45** 2328
[3] Ol'khovic L P, Khvorov M M, Borisova N M, Golubenko Z V, Sizova Z I and Shurinova E V 2003 *Phys. Solid State* **45** 675
[4] Frolov G I, Zhigalov V S, Zharkov S M, Pol'skii A I and Kirgizov V V 2003 *Phys. Solid State* **45** 2303.
[5] Smilga V P and Belousov Y M 1994 *The Muon Method in Science* (New York: Nova Science)
[6] Belousov Y M, Gorelkin V N, Mikaelyan A L, Miloserdin V Y and Smilga V P 1979 *Sov. Phys. Usp.* **22** 679
[7] Abraham A 1961 *The principles of nuclear magnetism* (Oxford: Clarendon)
[8] Belousov Y M 2007 *Phys. Solid State* **49** 288
[9] Vonsovskiy S V 1973 *Magnetism of Microparticles* (Moscow: Nauka)
[10] Kozhevnikov E, Suter A, Fritzsche H, Gladilin V, Volodin A, Moorkens T, Trekels M, Cuppens J, Wojek B M, Prokscha T, Morenzoni E, Nieuwenhuys G J, Van Bael M J, Temst K, Van Haesendonck C and Indekeu J O 2013 *Phys. Rev.* B **87** 104508
[11] Charnukha A, Svitcivic A, Prokscha T, Pröpper D, Ocelic N, Suter A, Salman Z, Morenzoni E, Deisenhofer J, Tsurkan V, Loidl A, Keimer B and Boris A V 2012 *Phys. Rev. Lett.* **109** 017003
[12] Hofmann A, Salman Z, Mannini M, Amato A, Malavolti L, Morenzoni E, Prokscha T, Sessoli R and Suter A 2012 *ACS Nano* **6** 8390
[13] Saadaoui H, Salman Z, Prokscha T, Suter A, Huhtinen H, Paturi P and Morenzoni E 2013 *Phys. Rev.* B **88** 180501(R)
[14] Stilp E, Suter A, Prokscha T, Morenzoni E, Keller H, Wojek B M, Luetkens H, Gozar A, Logvenov G and Bozovic I 2013 *Phys. Rev.* B **88** 064419

Journal of Physics: Conference Series **551** (2014) 012058 doi:10.1088/1742-6596/551/1/012058

# Time differential ALC - experiments, simulations and benefits

**J S Lord, F L Pratt and M T F Telling**
ISIS Muon facility, STFC Rutherford Appleton Laboratory, Chilton, OX11 0QX, U.K.
E-mail: james.lord@stfc.ac.uk

**Abstract.** Traditional muon avoided-level-crossing measurements use "integral counting" analysis, which results in a simple 1-D spectrum with resonance lines. At pulsed muon facilities, the time spectrum (TD-ALC) is available "for free" so it should be analysed to extract all relevant information. An ALC line corresponds to resonant transfer of polarisation back and forth between the muon and the other spins in a radical, and this oscillation is often at a convenient frequency of order 1 MHz. Following the oscillation frequency through just one $\Delta m = 0$ line can give the hyperfine constants for both the muon and nucleus and the gyromagnetic ratio for that nucleus.

## 1. Introduction

Muons often form muonated radicals via muonium addition when implanted into molecular solids, liquids, solutions or gases[1]. Measurements can give the hyperfine coupling constants between the unpaired electron and the nuclei in the molecule (including the muon itself which behaves chemically like another hydrogen atom). In addition, dynamics such as electron spin exchange and molecular reorientation will be observed.

A number of variations of the muon technique can be used to measure radicals and muonium. The simplest is repolarisation, where we scan the magnetic field from zero up to typically 0.5 T and measure the average longitudinal asymmetry. This gives a characteristic step at a field proportional to the hyperfine coupling and so gives a quick and simple confirmation of the presence of radicals. However it is unable to resolve distinct radicals with similar hyperfine constants, as may be expected in complex or low symmetry molecules, and is affected by dynamics and anisotropic interactions.

Muon spin rotation spectra taken in a sufficiently high transverse field will give two lines corresponding to the unpaired electron spin being parallel or antiparallel to the applied field, and with spacing equal to the muon's hyperfine constant. Multiple radicals are easily identified. At lower fields, the lines may split if the electron also interacts with other nuclei. However this technique fails if there is a large anisotropy in a powder sample, if the formation of the radicals is slow and the muons dephase, as may happen in a dilute solution, or if there is strong relaxation.

Radio-frequency resonance can be used, with the longitudinal field scanned to bring different transitions into resonance. This is similar to the rotation measurement, where in high fields and frequencies only the muon hyperfine constant is relevant, and low fields show splitting from other nuclei. It is not affected by slow formation, and in some cases can measure the timescale of such dynamics.

Avoided level crossing is where the longitudinal magnetic field is scanned and at certain fields the levels will cross. Usually the levels interact so that they "avoid" each other with a small gap of order 1 MHz. At this point — assuming the two levels concerned have different values of the muon polarisation — the muon spin will oscillate or precess giving a lower average polarisation.

Traditional muon avoided-level-crossing measurements [2, 3] use "integral counting" analysis, which results in a simple 1-D spectrum with resonance lines. It requires only simple instrumentation and can take advantage of high rates on continuous muon sources. The resonance spectra can be fitted to give the central field, linewidth, and in some cases a line shape resulting from dipolar coupling.

At pulsed muon facilities, the time spectrum is available "for free" so it should be analysed to extract all relevant information. A $\Delta m = 1$ ALC line corresponds to the muon spin precessing around an off-axis internal field, while a $\Delta m = 0$ line corresponds to resonant transfer of polarisation back and forth between the muon and another nucleus [5]. These oscillations are often at a convenient frequency of order 1 MHz. The oscillation may still be observable if the radical formation is too slow to obtain a precession spectrum in high transverse field (for example in dilute solutions). The damping rates can be very low in liquids and single crystals, and the oscillation is even seen in powder samples at certain magic fields.

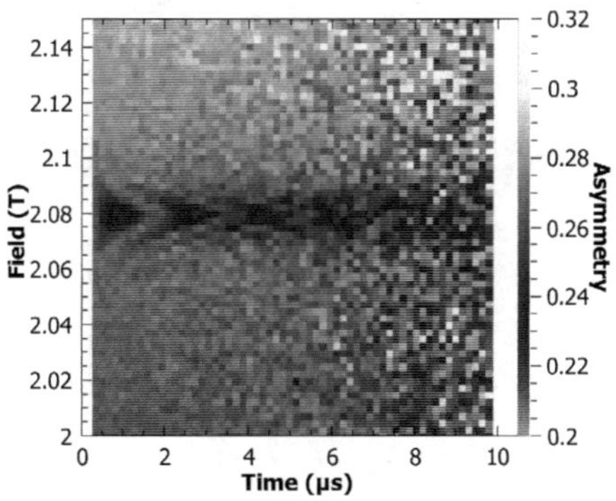


**Figure 1.** Experimental $\Delta m = 0$ level crossing in liquid benzene at 300 K, as a 2-D plot.

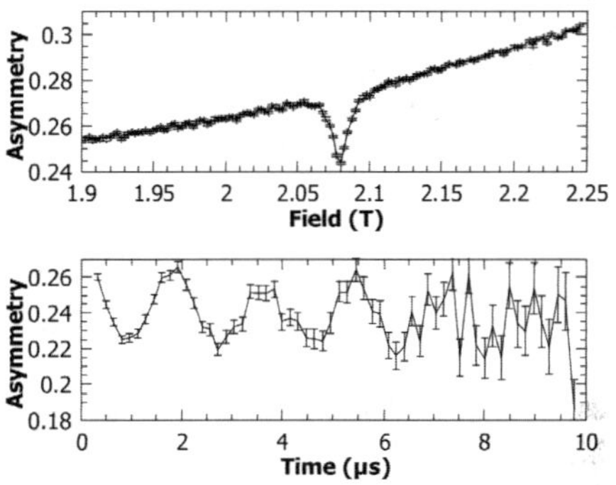


**Figure 2.** Integral asymmetry (upper) and the on-resonance time spectrum (lower) from figure 1.

## 2. Sharp lines

Figure 1 shows one of the $\Delta m = 0$ ALC lines for the cyclohexadienyl radical in liquid benzene neasured on the HiFi instrument at ISIS [4], presented as a 2D intensity plot versus time and magnetic field. Figure 2 shows the integral asymmetry which would be obtained from this data (before background subtraction), and one of the individual time spectra. In addition to the line centre position and width, it is clear that there is an oscillation which has a minimum frequency on-resonance and increases when further away, while the amplitude decreases.

To a rough approximation, for most sharp avoided level crossings (and RF resonances) we have:

$$P(t) = (1 - A(B)) + A(B)\cos(\omega(B)t) \tag{1}$$

with:

$$A(B) = \frac{1}{2}\frac{\omega_0^2}{\omega_0^2 + [\gamma_{\text{eff}}(B - B_0)]^2} \tag{2}$$

Journal of Physics: Conference Series **551** (2014) 012058 doi:10.1088/1742-6596/551/1/012058

$$\omega(B) = \sqrt{\omega_0^2 + \gamma_{\mathrm{eff}}(B - B_0)} \tag{3}$$

For an isotropic $\Delta m = 0$ line, a rough approximation is that the Zeeman splitting for each of the muon and nucleus adds to (or subtracts from) its hyperfine splitting until the two spins have the same energy difference and can exchange polarisation:

$$B_0 = \frac{A_\mu - A_\mathrm{X}}{2(\gamma_\mu - \gamma_\mathrm{X})} \tag{4}$$

$$\gamma_{\mathrm{eff}} = (\gamma_\mu - \gamma_\mathrm{X}) \tag{5}$$

The polarisation is transferred via the electron so the oscillation frequency depends on the coupling of the muon and nucleus to it, versus the field keeping the electron fully aligned:

$$\omega_0 = \frac{A_\mu A_\mathrm{X}}{2\gamma_\mathrm{e} B_0} \tag{6}$$

Therefore, by fitting the whole data set, we should be able to extract values for $A_\mu$, $A_\mathrm{X}$ and $\gamma_\mathrm{X}$. Usually the last of these is known (e.g. X = hydrogen) and fixing it will help improve the accuracy of the fit. If there are several species which might give rise to lines then we may be able to identify them unambiguously.

For accurate fitting, the full Hamiltonian is solved numerically rather than using these approximations (which will serve to give a starting point for the fit). Fit parameters $A_\mu - A_\mathrm{X}$ and $A_\mu + A_\mathrm{X}$ are used, with these being converted internally back to $A_\mu$ and $A_\mathrm{X}$ for the simulation, to avoid instability due to the correlation between them.

## 3. Broadened lines

In an anisotropic environment the behaviour is more complex and the integral line deviates from the simple Lorentzian shape. However, for an axial $\Delta m = 1$ line an oscillation appears near the line centre. The local field seen by an individual muon is $\mathbf{B}_\mu = \mathbf{B} + (\mathbf{A}.\mathbf{S}/\gamma_\mu)$ and in high field the electron spin $\mathbf{S}$ is parallel or antiparallel to $\mathbf{B}$. If we work in coordinates where the hyperfine tensor is diagonal, and an arbitrary field direction $(\theta,\phi)$, and consider only that electron spin direction giving the crossing, $\mathbf{B}_\mu = (B - A_\perp/2\gamma_\mu)\hat{\mathbf{x}}\sin\theta\cos\phi + (B - A_\perp/2\gamma_\mu)\hat{\mathbf{y}}\sin\theta\sin\phi + (B - A_\parallel/2\gamma_\mu)\hat{\mathbf{z}}\cos\theta$ . When $B = (A_\parallel + A_\perp)/4\gamma_{\mathrm{Mu}}$, the magnitude of $\mathbf{B}_\mu$ is independent of $\theta$ and $\phi$ giving $\omega_\mu = \gamma_\mu|\mathbf{B}_\mu| = \frac{1}{4}|A_\parallel - A_\perp|$ . Away from this magic field the frequencies spread out so the oscillation disappears by its damping rate increasing and only a small change in the average frequency.

A non-axial hyperfine tensor does not show a single magic field, but if statistics are sufficient, three fields each give rise to weak oscillations when all field orientations perpendicular to one principal axis have the same frequency.

Figure 3 shows ALC resonances in solid benzene just below the melting point where the $\Delta m = 1$ line and one of the $\Delta m = 0$ lines are both visible. The different characters of the lines are apparent. The $\Delta m = 1$ line (figure 4, lower) has a strongly damped oscillation of about 1.5 MHz indicating anisotropy $|A_\parallel - A_\perp| = 6$ MHz. At this temperature the benzene molecules are rotating about their 6-fold axes so the radical also averages out most of the dipolar interaction by "magic angle spinning" leaving an approximately axial component with the same symmetry. Residual non-axial asymmetry (the molecules are not at high symmetry points in the lattice) and dynamics may contribute to the loss of the longer-lived oscillations sometimes seen.

The $\Delta m = 0$ line at higher field (figure 4, upper) has a frequency 0.6 MHz very similar to that in the liquid, since the isotropic hyperfine couplings are similar. The increase in frequency when off-resonance is now damped out by the anisotropy.

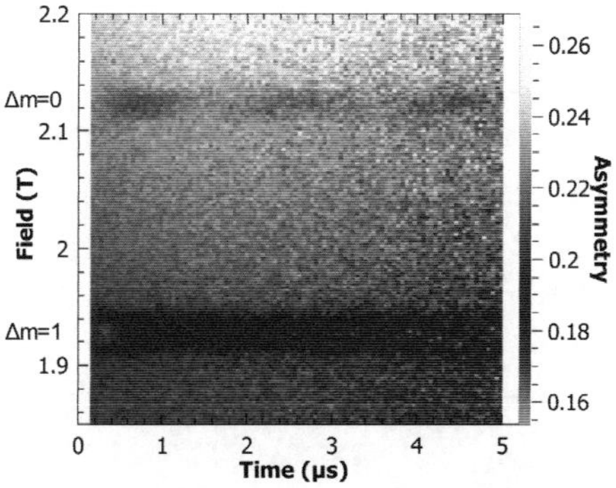


**Figure 3.** Experimental level crossing in solid benzene at 263 K, as a 2-D plot.

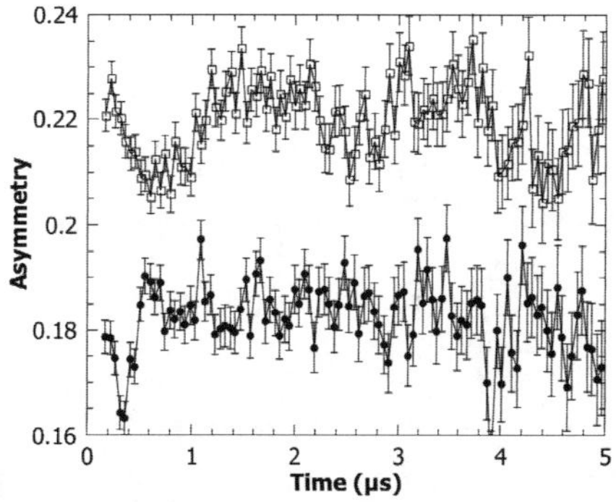


**Figure 4.** On-resonance time spectra from figure 3. Upper: $\Delta m = 0$ at 2.1225 T, lower $\Delta m = 1$ at 1.93 T

Single crystal $\Delta m = 1$ ALCs should show oscillating polarisation in the transverse directions, especially for lower symmetry crystal structures with few molecules per unit cell. Positron spiralling may result in the measured transverse polarisation being rotated compared to that seen with small transverse fields.

## 4. Implementation

The simulation package "Quantum" is now integrated with the data analysis framework "Mantid" so that the results of simulations can be converted back to experimental format, or used in fitting.

A real muon instrument often shows a variable background as a function of field, seen in figure 2, especially at the higher fields where ALC lines are found and with complex or thick walled sample cells. This is because the positrons travel on spiral tracks in the applied field, with radii comparable to the size of the spectrometer and varying with the field magnitude and their remaining momentum after escaping the sample. Fitting only the oscillating part of $P(t)$ allows us to let the constant part of each spectrum be a free parameter in the fit, instead of requiring careful background modelling and subtraction. When following a resonance line as a function of temperature or other parameter there may be no need to measure a large background field range each side of the line. We take advantage of the fact that the polarisation starts at 1.0 at time=0 and only varies away from this as $\cos(\omega t)$.

## 5. Sensitivity

Simulated data can be fitted and the accuracy of the resulting fits estimated. For this example we used the $\Delta m = 0$ line from a radical with $A_\mu = 520$ MHz and $A_\mathrm{p} = 120$ MHz. We generated a series of random data sets with selected spacing and number of field points. Each field has 10 Mevents, full asymmetry of the radical $= 5\%$, and $\alpha = 1.01$.

Table 1 shows the mean error value E reported by the Levenberg-Marquardt fitting routine in Mantid, and the RMS scatter S of the fitted parameter between 50 data sets, for the full time dependent fit and a simple line fit to the integral data.

The 10 kHz accuracy for $A_\mu - A_\mathrm{p}$ comes from the line centre position and requires the magnetic field in each measurement to be known to 0.1 mT, probably requiring a NMR measurement rather than relying on the set current in a superconducting magnet. However $A_\mu + A_\mathrm{p}$ is a measurement

**Table 1.** Fitting accuracy

| Parameter | Time Dependent fit | | Integral Fits (Lorentzian) | | | | | |
|---|---|---|---|---|---|---|---|---|
| | 5 points 50G | | 5 points 50G | | 15 points 50G | | 30 points 25G | |
| | S | E | S | E | S | E | S | E |
| $A_\mu + A_p$ / MHz | 0.54 | 0.89 | – | – | – | – | – | – |
| $A_\mu - A_p$ / MHz | 0.010 | 0.015 | 0.016 | 0.021 | 0.014 | 0.021 | 0.0090 | 0.0144 |
| Amplitude / % | 0.11 | 0.19 | 2.8 | 1.4 | 0.23 | 0.32 | 0.16 | 0.23 |
| Total Mevents | 50 | | 50 | | 150 | | 300 | |

of muon oscillation frequency, determined with much lower systematic errors. The integral fit needs a background region in order to get an accurate amplitude while the time dependent fit can work with the oscillations within the line.

When the nuclear hyperfine coupling is small ($\Delta m = 0$) or the hyperfine tensor is nearly isotropic ($\Delta m = 1$), the oscillation frequency may be too small to see a whole cycle within the time scale of the muon measurement, or fit the frequency. In this case the asymmetry still follows a behaviour $a(t) = a_0 \cos(\omega t) \approx a_0 - a_1 t^2$ and if the data is taken in time differential mode a significant improvement over standard integral analysis is still possible with suitable choice of time window or weighting.

## 6. Dynamics

Observation of the damping of oscillations, and relaxation of the underlying baseline, can allow us to fit the relaxation rates. Different mechanisms may have different relative effects.

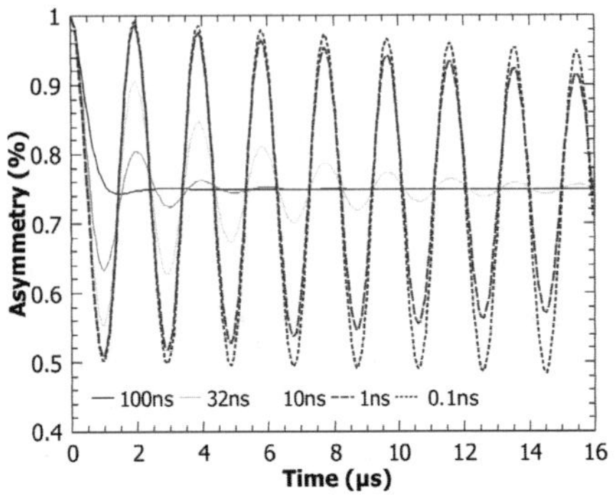


**Figure 5.** Time spectra on-resonance for various conversion rates

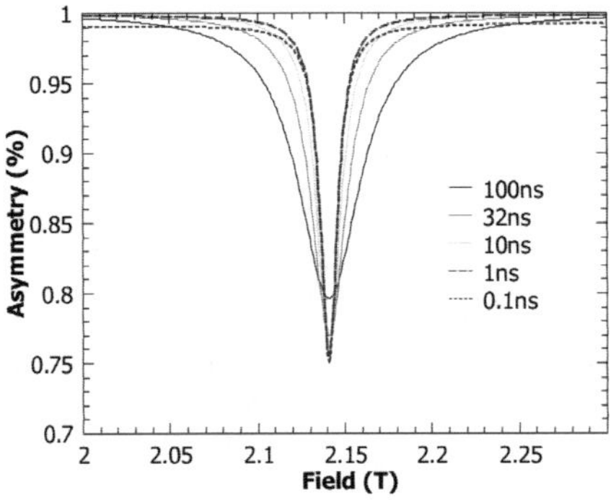


**Figure 6.** Integral ALC resonances for various conversion rates

As an example we modelled the rapid exchange between two states with hyperfine constants differing by ±2.5% from the values earlier. Again the $\Delta m = 0$ line is studied. Figure 5 shows simulated on-resonance time differential data and figure 6 shows integral data for the same conversion rates.

Figure 7 shows the resulting fitted residence time $\tau$ as a function of that used in the simulation, along with the integral asymmetry linewidth. The time dependent fit is able to directly extract much lower times (faster exchange). The residence time can also be extracted from the linewidth but that relies on the width of the static averaged line being known accurately. Each simulation

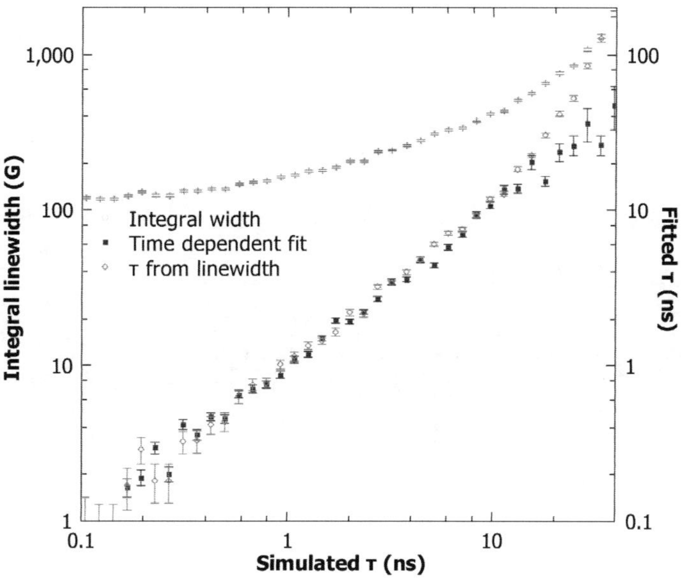


**Figure 7.** Fit results for rapid conversion between two radicals, observed by the $\Delta m = 0$ line.

uses 10 Mevents per run. 5 runs at 50 G spacing are used for the time dependent fit, but 61 runs at 25 G for the integral data since a region of background must be measured in order to get an accurate fit of the linewidth. As can be seen in the individual spectra (figure 5), the background of the integral line varies due to relaxation of the non-oscillating component. The increasing errors at large $\tau$ for either analysis method are because the field scans only include the central motionally narrowed line and not the whole field region including the two individual lines.

## 7. Conclusions

Fitting the time dependent asymmetry is a useful way to extract further information from ALC data, especially when measuring narrow resonances in solution or single crystals. Sharp $\Delta m = 0$ lines start with $P(t = 0) = 1.0$ so can be used to check the instrumental background. However solid powder or polycrystalline samples with large or non-axial anisotropy may not show significant time dependence. Strong dynamics may obscure the oscillations, but the time dependent data may still give useful information.

## References

[1] McKenzie I 2013 *Annu. Rep. Prog. Chem.* C **109** 65–112
[2] Heming M *et al.* 1986 *Chem. Phys. Lett.* **128** 100
[3] Roduner E 1993 *Chem. Soc. Rev.* **22** 337
[4] Lord J S *et al.* 2011 *Rev. Sci. Inst.* **82** 073904
[5] Roduner E 1988 *The Positive Muon as a Probe in Free Radical Chemistry: Potential and Limitations of the μSR Techniques* (Berlin: Springer-Verlag)
[6] Lord J S 2005 *Physica* B **374–375** 423
[7] `http://dx.doi.org/10.5286/SOFTWARE/MANTID`

13th International Conference on Muon Spin Rotation, Relaxation and Resonance IOP Publishing
Journal of Physics: Conference Series **551** (2014) 012059 doi:10.1088/1742-6596/551/1/012059

# The Initial State of Optically Polarized $^8Li^+$ from the $\beta$-NMR in Bismuth

**W A MacFarlane[1], C D P Levy[2], M R Pearson[2], T Buck[3], K H Chow[4], A N Hariwal[5], R F Kiefl[3], F H McGee[3], G D Morris[2] and D Wang[3]**

[1] Chemistry Department, University of British Columbia, Vancouver, BC, V6T 1Z1, Canada
[2] TRIUMF, Vancouver, B.C., V6T 2A3, Canada
[3] Department of Physics, University of British Columbia, Vancouver, BC, V6T 1Z1, Canada
[4] Department of Physics, University of Alberta, Edmonton, AB, T6G 2E1, Canada
[5] Department of Physics and Astrophysics, University of Delhi, Delhi-110007, India

E-mail: wam@chem.ubc.ca

**Abstract.** Unlike the positive muon, $\beta$-NMR probe nuclei must be actively polarized. At the TRIUMF ISAC facility this is accomplished by in-flight collinear optical pumping with resonant circularly polarized laser light. This reliably produces a high degree of polarization, but the detailed state populations in the beam emerging from the optical polarizer are not well known. These populations are significant as they represent the initial state of the ensemble of probe spins implanted in a $\beta$-NMR experiment. Here we use the well-resolved quadrupolar split spectrum of $^8Li^+$ in a high purity single crystal of bismuth to extract the sublevel populations under typical polarizer operating conditions, accounting for the spin relaxation in this semimetal.

## 1. Introduction

Unlike the positive muon, $\beta$-NMR probe nuclei must be actively polarized. At ISAC, this is accomplished by in-flight collinear[1] optical pumping with resonant circularly polarized laser light. Optical polarization is a well-developed technique in the vapour phase[2], particularly for alkali atoms. Noble gases may be spin-exchange polarized using optically polarized alkalis[3] and these have many applications in surfaces physics[4, 5], biology[6] and quantum computation[7]. Optical pumping with polarized light is also used to polarize nuclei in many semiconductors[8, 9, 10, 11]. Optically polarized radioactive nuclei are useful as probes of the solid state analogous to $\mu$SR[12] and also have important applications in nuclear physics[13].

The ISAC optical polarizer reliably produces a high polarization $\sim 70$ % [14], but the detailed state populations in the beam emerging from the polarizer are not well known. These populations are significant as they represent the initial state of the ensemble of probe spins used in a $\beta$-NMR experiment. Moreover, with a spin $I = 2$ nucleus, 4 parameters (rather than 1 for spin 1/2) are required to specify the spin state. Here, we use the well-resolved quadrupolar split NMR spectrum of $^8Li^+$ in a high purity single crystal of bismuth to extract the sublevel populations under typical polarizer operating conditions, accounting for the spin relaxation in this semimetal. The few detailed measurements of the spin states of optically polarized atoms that have been performed serve as points of comparison[15, 16, 17].

## 2. Experiment

A freshly cleaved plate of high purity single crystal bismuth approx 1 $cm^2$ in area was mounted on a sapphire plate in the ultrahigh vacuum high field $\beta$-NMR spectrometer. The superconducting solenoid produced a highly homogeneous magnetic field of 6.55 T along the $^8Li^+$ beam direction (parallel to the trigonal axis of the crystal) providing the static NMR field as well as focusing the ion beam. A continuous beam of $\sim 10^6$ $^8Li^+$ ions per second was implanted at 20 keV into a beamspot $\sim$ 2 mm in diameter centred on the crystal. A CW RF oscillating transverse magnetic field $H_1$ was applied with frequency stepped slowly (250 Hz/s) through a range of frequencies about the Larmor frequency. The beta decay asymmetry was monitored in two scintillation detectors downstream (F) and upstream (B) of the sample. The direction of polarization was alternated by introducing a half-wave plate into the polarizing laser beam. F/B asymmetry spectra were collected in both helicities, and the results are shown in Fig. 1.

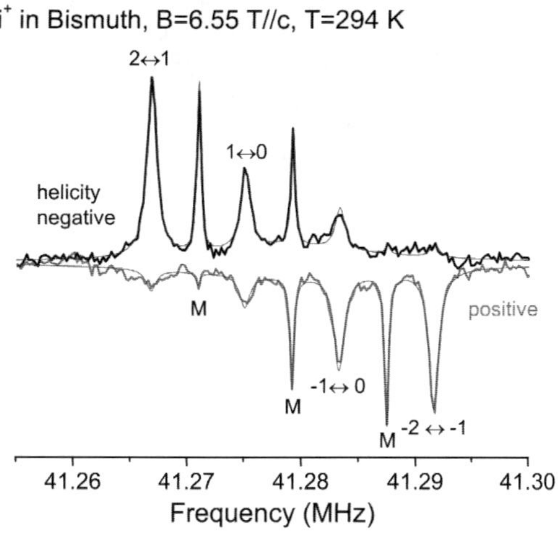


**Figure 1.** The $\beta$-NMR spectrum of $^8Li^+$ in bismuth. Here the two helicities (polarization directions) are shown with their baselines subtracted. The single quantum satellites are labeled with the corresponding $m$ values. The three narrow resonances marked M are multiquantum. The thin blue lines are the fits that yield the satellite amplitudes.

The resonance is split into a highly resolved multiplet consisting of the expected 4 first order (primary) quadrupole satellites[18, 19], interlaced with 3 narrower multiquantum resonances[20]. The multiquantum nature of these resonances is confirmed by their position midway between the single quantum satellites, their narrowness[21], and, most clearly, they are suppressed very strongly as the RF power is reduced (much more so than the single quantum satellites). This is not the first time multiquantum resonances have been observed with $\beta$NMR[22], but it is the first example at TRIUMF, likely because the pattern is much better resolved here than in other materials[18, 19], reflecting the very high quality Bi crystal.

A quadrupole splitting is expected as the $^8Li^+$ site cannot be cubic in the rhombohedral structure of Bi, hence it should experience a finite electric field gradient (EFG). The spectrum was fit with a sum of 7 Lorentzians with the splitting determined by a single parameter $\nu_Q \approx 8$ kHz. In the fits, the widths of the primary satellites are the same. The amplitudes of these satellites, crucial to the analysis in this work, for the two helicities are reported in Table 1.

Finally, in the absence of the RF field, a pulsed beam of $^8Li^+$ was implanted to monitor the small spin-lattice relaxation rate. The detailed results will be published elsewhere[23], but the relaxation is single exponential, with $\lambda = 1/T_1 = 0.088(5)$ $s^{-1}$ at 294 K. Moreover the

Journal of Physics: Conference Series **551** (2014) 012059 doi:10.1088/1742-6596/551/1/012059

**Table 1.** Lorentzian amplitudes of the single quantum quadrupole satellites of $^8Li^+$ in Bi from the fits shown in Fig. 1.

| $m, m-1$ | Negative | Positive |
|---|---|---|
| $2, 1$ | 0.0161(2) | 0.0019(2) |
| $1, 0$ | 0.00715(16) | 0.0029(2) |
| $0, -1$ | 0.0037(2) | 0.0085(2) |
| $-1, -2$ | 0.0011(2) | 0.0123(2) |

temperature dependence of $\lambda(T)$ is linear, implying Korringa relaxation which is magnetic in origin (rather than quadrupolar), simplifying the analysis below.

## 3. Analysis

Most generally, the state of the $^8$Li nuclear spin ($I = 2$) is described by the density matrix $\rho_{m_1 m_2}$, with diagonal elements representing the statistical populations of the 5 magnetic sublevels $m$ and off-diagonal elements representing coherences. In particular, it is the ensemble average spin state for $\sim 10^8$ spins obtained over several minutes of data collection that is relevant to the experiment. For the optically pumped state, we expect no coherences, so we restrict ourselves to spin states described by the vector of populations $p_m = \rho_{mm}$. The spin polarization is then defined as

$$\mathcal{P} = \frac{1}{I} \sum_m m p_m. \tag{1}$$

The polarizer produces an initial polarization $\mathcal{P}(0)$ that we assume is transported (at a typical energy of 20 keV) to the sample and implanted without loss. Once stopped, the $^8Li^+$ spin state evolves, relaxing towards equilibrium $\mathcal{P}_{eq} \approx 0$ by spin–lattice relaxation. With a continuous beam, we measure the time average polarization

$$\bar{\mathcal{P}} = \frac{1}{\tau} \int_0^\infty e^{-s/\tau} \mathcal{P}(s) ds, \tag{2}$$

where $s$ is the time after implantation and $\tau = 1.21$ s is the radioactive lifetime.

Unlike the spin 1/2 $\mu^+$, where specifying the polarization is equivalent to specifying the spin state, for higher spins, there are more degrees of freedom, and many spin states yield the same $\mathcal{P}$. For $^8Li^+$ 4 parameters specify the spin state[1], and specifying $\mathcal{P}$ leaves 3. Also, unlike spin 1/2 particles, $^8Li^+$ has an electric quadrupole moment that couples the spin to the local EFG, which is generally non-zero at sites of lower than cubic symmetry. This coupling splits the $|\Delta m| = 1$ Zeeman transitions, which would otherwise be degenerate at the Larmor frequency $\nu_L$, into a multiplet pattern of quadrupole satellites. The scale of the coupling is given by the quadrupole frequency $\nu_Q = e^2qQ/4h$, where $Q = +31.4$ mb is the nuclear electric quadrupole moment, $e$ the electronic charge and the principal component of the EFG tensor is $eq$. Here $\nu_Q/\nu_L \sim 10^{-4}$ so the quadrupole interaction can be treated in first order, and we expect, for the field along the EFG principal axis, satellite lines at 1/2 and 3/2 $\nu_Q$ on either side of $\nu_L$. In addition, in this experiment, the RF field is strong enough that multiquantum transitions are evident[20, 21, 22]. We thus assign the satellites as shown in Fig. 1. We assume the principle direction of the EFG is the trigonal axis of the crystal, but an orientation dependence is required to confirm this.

[1] The fifth is determined by the normalization $\sum p_m = 1$.

The amplitude of the primary satellites is determined by the magnitude of the RF field and the time average[2] population difference between the two $m$ levels involved,

$$A_{m1,m2} = a_{RF}|\bar{p}_{m_1} - \bar{p}_{m_2}|. \tag{3}$$

In the measurement, the RF amplitude is constant across the spectrum, so the factor $a_{RF}$ is common to all the satellites. Thus, the satellite amplitudes give additional information to help to determine the initial spin state. If $a_{RF}$ was known, then one could simply invert these linear equations using the satellite amplitudes in Table 1, and solve for the time averaged populations $\bar{p}_m$. Instead, we take ratios of the Eqs. (3), reducing the number of constraints from 4 to 3, and treat the polarization $\mathcal{P}$ as a variable over a range near the independently estimated value of 0.7, yielding the inferred average populations as a function of $\mathcal{P}$.

The primary interest is, however, the initial values of the populations $p_m(0)$ that is characteristic of the polarized beam and independent of the sample. To extract these values from the time averages, we adopt a simple model of the relaxation – the master equation, that captures the correct transition matrix elements among the sublevels. The observed Korringa relaxation is magnetic, so the transition matrix elements follow[24]

$$\mathcal{W}_{m \to m-1} = w_{mag}(I+m)(I-m+1), \tag{4}$$

where $w_{mag}$ is the magnetic transition rate. The master equation is a system of first order rate equations for the evolution of the populations[3],

$$\frac{dp_m}{dt} = W_{mn}p_n, \tag{5}$$

where the matrix $W_{mn}$ encodes the transition probabilities of Eq. (4) and the thermal equilibrium state that is the asymptotic limit as $t \to \infty$, $(p_m)_{eq}$, e.g. see Ref. [25]. Because the Zeeman splitting is much less than $kT$ for all experimentally accessible conditions, $(p_m)_{eq} \approx 0.2$, for all $m$. Eq. (5) can be solved by finding the eigenvalues of $W$, which for $I = 2$ are

$$\lambda_i = 10\lambda, 6\lambda, 3\lambda, \lambda, 0, \tag{6}$$

where $\lambda = 1/T_1 = 2w_{mag}$. The solution of Eq. (5) is

$$p_m(t) = E_{mi}c_i e^{-\lambda_i t}, \tag{7}$$

where $E_{mi}$ are the eigenvectors of $W$, and the coefficients $c_i$ are determined by the initial populations. An example of the solutions for a representative initial state and the experimental relaxation rate in Bi is shown in Fig. 2. From these solutions, we compute the time averages and arrive at the relation

$$\bar{p}_m = \frac{E_{mi}E_{ni}p_n(0)}{\lambda_i\tau + 1}. \tag{8}$$

This linear set of equations can then be inverted to yield the initial populations from the time averages deduced from the satellite amplitudes. For the negative helicity values in Table 1, the results of these calculations are shown in Figs. 3 and 4 as a function of the initial polarization $\mathcal{P}(0)$. Note that Eq. (3) does not enforce the physical condition that the populations must lie in the range [0,1].

[2] The time average of the $p_m$ is analogous to Eq. (2).
[3] We use the convention of summation of repeated indices throughout.

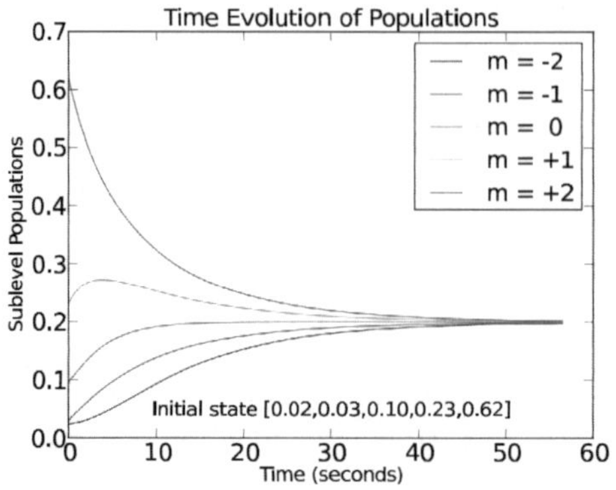


**Figure 2.** The calculated time evolution of the sublevel populations as they approach equilibrium from Eq. (5). Note the time range shown is much longer than the $^8$Li lifetime $\tau$.

## 4. Discussion

First we note that the results do not uniquely determine the spin state, as there is still one degree of freedom, here represented by the initial polarization. The calculated populations are not, however, strong functions of $\mathcal{P}(0)$, and one can use the results to yield well-defined ranges. The results show that the dominant population is in the extremal $m = -2$ sublevel as expected, but that there are substantial populations in both $m = -1$ and 0 states. This will have important implications for modeling quadrupolar lineshapes in other materials, particularly where the satellites are not well-resolved. The result is also important for a detailed treatment of spin relaxation that should account for a realistic initial state. One can also see from Table 1 that the two helicities are not polarized exactly complementarily, likely due to some unintended optical effect of the half-wave plate.

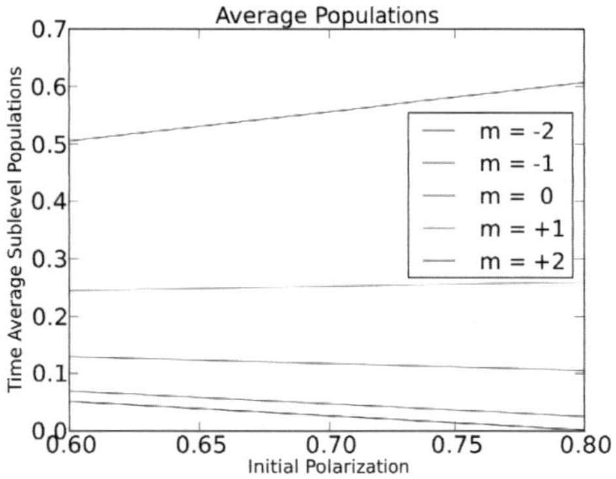


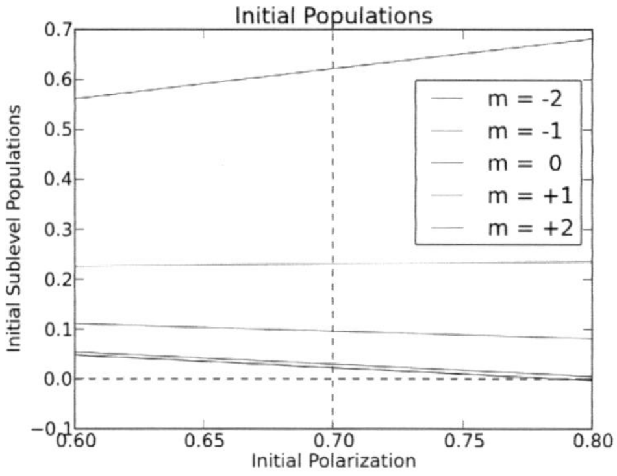


**Figure 3.** Time average sublevel populations deduced from the negative helicity satellite amplitudes in Table 1 using Eq. (3) as a function of the initial polarization $\mathcal{P}(0)$.

**Figure 4.** The corresponding initial state sublevel populations deduced from the average populations using the measured Korringa relaxation rate and Eq. (8)

One can also consider why the polarization is not higher, as the polarizer operates well into the saturation regime, where increasing the laser power does not further increase the polarization.

In contrast, stable alkali beams can be nearly completely polarized[15, 16]. The polarizer is well aligned with the beam axis and with the experimental magnetic field, and one can anticipate a negligible reduction due to misalignment. Imperfect circular polarization (ellipticity) of the laser is a more likely candidate, but even this could only account for a small reduction of polarization (a few %). Coherent effects from the very short pumping time (the beam traverses the polarizer in about 2 $\mu$s) may be responsible. The results presented here can be used as a diagnostic for further optimization of the ISAC polarizer, and this method complements those based on optical or EPR hyperfine spectroscopy[15, 16, 17].

## 5. Summary

We have used the highly resolved quadrupole spectrum of $^8Li^+$ implanted into bismuth to determine the intial state of the $^8Li^+$ spin in the beams used in $\beta$-NMR at TRIUMF. The results show that populations aside from the extremal $m = \pm 2$ are non-negligible and should be taken into account. The results also give a more detailed picture of the effectiveness of the optical polarizer.

## Acknowledgments

We thank D.A. Bonn and B.J. Ramshaw for providing the bismuth crystal and S. Daviel, R. Abasalti and D. Vyas for technical assistance.

## References

[1] Silverans R, Borghs G, Bisschop P and Hove M 1985 *Hyp. Int.* **24** 181–201
[2] Happer W 1972 *Rev. Mod. Phys.* **44** 169–249
[3] Walker T G and Happer W 1997 *Rev. Mod. Phys.* **69** 629–642
[4] Jänsch H J, Gerhard P and Koch M 2004 *Proc. Nat. Acad. Sci. (USA)* **101** 13715–13719
[5] Bifone A, Pietrass T, Kritzenberger J, Pines A and Chmelka B F 1995 *Phys. Rev. Lett.* **74** 3277–3280
[6] Mayo J R and Hayden M E 2002 *Radiology* **222** 8–11
[7] Verhulst A S, Liivak O, Sherwood M H, Vieth H M and Chuang I L 2001 *Appl. Phys. Lett.* **79** 2480–2482
[8] Lampel G 1968 *Phys. Rev. Lett.* **20** 491–493
[9] Barrett S E, Tycko R, Pfeiffer L N and West K W 1994 *Phys. Rev. Lett.* **72** 1368–1371
[10] Tycko R, Barrett S E, Dabbagh G, Pfeiffer L N and West K W 1995 *Science* **268** 1460–1463
[11] Michal C A and Tycko R 1999 *Phys. Rev.* B **60** 8672–8679
[12] Morris G D, MacFarlane W A, Chow K H, Salman Z, Arseneau D J, Daviel S, Hatakeyama A, Kreitzman S R, Levy C D P, Poutissou R, Heffner R H, Elenewski J E, Greene L H and Kiefl R F 2004 *Phys. Rev. Lett.* **93** 157601
[13] Neyens G 2003 *Rep. Prog. Phys.* **66** 633
[14] Hatakeyama A, Hirayama Y, Behr J, Izumi H, Levy C, Melconian D and Shimoda T 2002 *Polarized Sources and Targets, Proceedings of the Ninth International Workshop* ed Derenchuk V P and von Przewoski B (World Scientific) pp 339–343
[15] Schinn G W, Han X L and Gallagher A 1991 *J. Opt. Soc. Am.* B **8** 169–173
[16] Masterson B P, Tanner C, Patrick H and Wieman C E 1993 *Phys. Rev.* A **47** 2139–2145
[17] Tardiff E R 2009 *Towards a Measurement of the Electric Dipole Moment of* $^{223}$*Rn* Ph.D. thesis University of Michigan, Physics Department
[18] MacFarlane W, Morris G, Chow K, Baartman R, Daviel S, Dunsiger S, Hatakeyama A, Kreitzman S, Levy C, Miller R, Nichol K, Poutissou R, Dumont E, Greene L and Kiefl R 2003 *Physica* B **326** 209–212
[19] Parolin T J, Shi J, Salman Z, Chow K H, Dosanjh P, Saadaoui H, Song Q, Hossain M D, Kiefl R F, Levy C D P, Pearson M R and MacFarlane W A 2009 *Phys. Rev.* B **80** 174109
[20] Kaplan J I and Meiboom S 1957 *Phys. Rev.* **106** 499–501
[21] Vega S, Shattuck T W and Pines A 1976 *Phys. Rev. Lett.* **37** 43–46
[22] Dubbers D, Dörr K, Ackermann H, Fujara F, Grupp H, Grupp M, Heitjans P, Körblein A and Stöckmann H J 1977 *Z. Phys.* A **282** 243–248
[23] MacFarlane W *et al.* In preparation
[24] Andrew E R and Tunstall D P 1961 *Proceedings of the Physical Society* **78** 1
[25] McDowell A 1995 *J. Magn. Reson.* A **113** 242–246

# *o'Peer*: open peer review

**J H Brewer**

Professor Emeritus, Department of Physics and Astronomy, University of British Columbia, Vancouver, BC, Canada V6T 1Z1

E-mail: jess@phas.ubc.ca

**Abstract.** I have built a "demonstration" website at http://oPeer.org to illustrate how peer review and publication might be improved relative to the current model, which was designed and implemented in an era when scientific communication was either face-to-face or relied upon human delivery of ink marks on dead trees.

## 1. Introduction

Please excuse my indulgence in a little long-range perspective:

A million years ago, it was difficult for one person to know what another person thought; language was invented to solve that problem. Many millennia ago, it was difficult for one person to know what another person had said; writing was invented to solve that problem. Many centuries ago, it was difficult for one person to know what another person had written; the printing press was invented to solve that problem. Many years ago, it was difficult for one person to discover what other people had written on a particular subject; journals were invented to solve that problem. More recently, the limitations imposed by forest conservation and library shelf space have also been circumvented by the seemingly infinite capacity of electronic storage: arXiv[1] has plenty of room for everything ever written, and the Internet provides convenient access to it all.

Today there is no difficulty in discovering what any number of people have written about any subject; the difficulty is in discovering which of those writings are worth reading. Because one determined individual can only hope to read an infinitesimal fraction of the papers published in his or her area of specialization, there is a desperate need for good judgement in filtering the wheat from the chaff.

Journals employing peer review are currently the best solution to this latest difficulty. However, the ratio of submissions to publications in the most selective journals is so enormous that the selection process demands a great deal of summary judgement: Editors can only afford to request reviews from a small number of referees for a given paper, and many papers are not even reviewed except by the Editors.

The Internet has spawned myriad *open access* journals (OAJ), published at the authors' expense rather than the readers', which address the authors' problem of getting their results published but not the reader's problem of picking what to read. To their credit, some OAJ strive to maintain high standards; but they are like King Canute trying to hold back the tide: if they must turn a profit to continue and the authors are their only source of funds, the temptation to publish indiscriminately will remain as persistent as entropy itself. The ground state of an OAJ is a useless vanity press.

Journal of Physics: Conference Series **551** (2014) 012060 doi:10.1088/1742-6596/551/1/012060

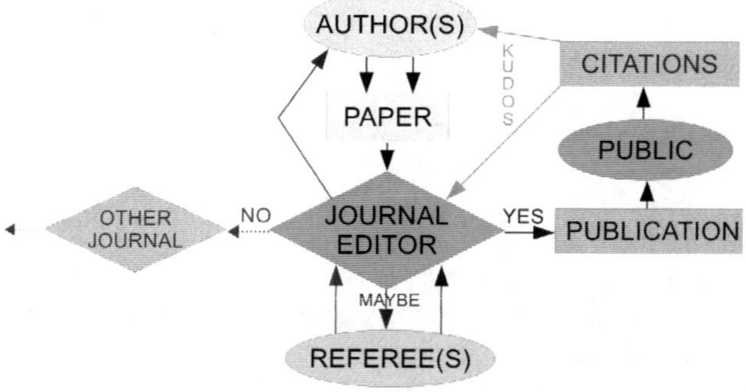


**Figure 1.** Cartoon of conventional publication procedure.

**Figure 2.** Conceptual diagram of open peer review procedure.

Meanwhile, ubiquitous data analysis has facilitated continuous monitoring of all journal publications, combining the number of publications, the prestige of the journals and the numbers of citations by other authors in other papers into a well-known "impact factor" for every journal and every author, which in many cases determines the profits of the journal as well as the career progress of the author. Under these pressures, the peer review process may not be able to maintain its presumed objectivity. Good work may go unpublished and flawed but fashionable work may become recommended reading.

## 2. "Like" is not enough!

Another problem with the current system is that it actually is susceptible to the form of corruption imagined by all "cranks" whose papers are rejected: the "Old Boys' Network" of Editors and their cronies who serve as Referees, excluding all new ideas and protecting their territory. This caricature is patently unfair and usually undeserved, but a sort of chemical potential does exist that drives networks of Editors and Referees in that general direction.

Editors have an extensive database of Referees, either in their heads or on their computers. These databases must be very well protected, as they contain not only sensitive personal information but also intelligence essential to the journal's success. And yet the least interesting information about each Referee is their personal identity. A list of unique serial numbers (such as ORCID[2]) would serve just as well, provided that all potential Referees were registered.

A more democratic model is widely employed by entities as diverse as Slashdot[3], YouTube[4] Facebook[5], ResearchGate[6], Google[7] and essentially all "social networks" that allow posting of material of any kind. This method involves simple indications of preference ("likes") by semi-anonymous viewers. It is, perhaps, useful for measuring appeal to a broad contemporary audience, but it is clearly useless to an intellectually elitist discipline like Physics. We don't care how much non-Physicists "like" a paper; we want to know the opinions of people whose judgement we respect, usually because they have established their own credibility in the same field and are known to exercise their judgement fairly in the interest of better science. That is, we want to know the opinions of the Referees a good Editor would choose from that extensive database of various attributes — which brings us back to peer reviewed journals.

## 3. What Can Be Done?

To paraphrase Winston Churchill, "Democracy is the worst possible form of government, except for all the rest."

Let's try to imagine a system of *Open Peer Review* (herein given the cute name "*o'Peer*" [8]) which encourages universal participation in the process of evaluation and yet weights the opinions of *cognoscenti* more than those of the ignorant. Obviously this entails either stealing the precious information in the journal Editors' databases or generating the equivalent information independently. I prefer the latter. How can this be done?

Allow me to assume that all possible authors and reviewers with any potential interest in *o'Peer* are registered in the *o'Peer* database in a secure and capacious computer facility. The $i^{\rm th}$ registrant has a *Credibility* index $\vec{C}_i$ (actually an array with many components) determined and updated by the algorithms described below.

The database also contains *papers* submitted by the registrants. The location, status and *Quality* $\vec{q}_\mu$ of the $\mu^{\rm th}$ paper are stored in the database and $\vec{q}_\mu$ (also a multi-component array) is updated by similar (in fact, the same) algorithms as $\vec{C}_i$.

### 3.1. The Components of $\vec{C}_i$ and $\vec{q}_\mu$

The choice of characteristics maintained in the *o'Peer* database is subject to revision and will certainly engender intense debate if Open Peer Review ever really catches on. Fortunately it is easy to add extra columns to the database if they are deemed useful. In fact, users can decide for themselves which attributes they consider most important in any given search of the database. My initial guess is

$$\vec{C}_i = \{L_i, W_i, O_i, A_i, E_i, T_i, I_i\} \qquad \text{and} \qquad \vec{q}_\mu = \{\ell_\mu, w_\mu, o_\mu, a_\mu, e_\mu, t_\mu, i_\mu\} \tag{1}$$

where, for registrant $i$, $L_i$ is judgement of appropriate manuscript *length*, $W_i$ is judgement of good *writing*, $O_i$ is judgement of *originality*, $A_i$ is judgement of *creativity* (*art*), $E_i$ is judgement of *execution*, $T_i$ is judgement of *correctness* of the work and $I_i$ is judgement of its *importance*. For paper $\mu$, the analogous lower case components of the $q_\mu$ array have the same meanings as the upper case ones above, minus the phrase "judgement of". For simplicity, in the following I will use $C_i$ and $q_\mu$ to denote arbitrary components of $\vec{C}_i$ and $\vec{q}_\mu$; all components are treated the same way.

### 3.2. The Algorithms

When a new paper is submitted, it becomes open for review by everyone but its own author(s). (To ensure this, it is obviously necessary for *all* the authors of any paper submitted to *o'Peer* to be registered in the database.) When reviewer $j$ evaluates paper $\mu$, s/he may submit a written Review along with the evaluation $Q^j_\mu$ (on a percentage scale from 0 to 100, the latter meaning, "It doesn't get any better than this!"). Said Review will be added to the database as another paper in the reviewer's name (identity can be hidden or revealed at the reviewer's discretion, but the computer always knows who wrote what) and can be in turn evaluated by others. In fact, registrants with no submitted papers of their own are *required* to submit a written Review of paper $\mu$ in order for their evaluation to be counted.

The *Quality* of the paper is then updated as follows:

$$q_\mu = \frac{\sum_j Q^j_\mu C_j}{\sum_j C_j} \tag{2}$$

where the sums $\sum_j Q^j_\mu C_j$ and $\sum_j C_j$ accumulate as various reviewers offer their opinions. Thus $q_\mu$ is the running "Credibility-weighted average" quality of the paper.

Meanwhile, the *Credibility* of the author(s) of the paper is updated in the same way from the same evaluations:

$$C_i = \frac{\sum_j Q^j_\mu C_j}{\sum_j C_j} \tag{3}$$

Journal of Physics: Conference Series **551** (2014) 012060 doi:10.1088/1742-6596/551/1/012060

The only difference is that the latter sums accumulate *for every paper* $\mu$ *of which registrant* $i$ *is an author*. Assuming there are $N_i$ such papers, we can rewrite the above equation as

$$C_i = \frac{1}{N_i} \sum_{\nu} q_{\nu} \tag{4}$$

where the summation runs over only the $N_i$ papers of which registrant $i$ is an author.

Again for simplicity I will pretend that the paper has only a single author. I anticipate some discussion of whether every author of a multi-author paper should receive the same "credit" for its quality, (as is the convention now in most cases), or if the credit should be apportioned fractionally, perhaps with extra credit going to the submitting author. I doubt that this would be a good idea, so for now I would prefer to give every author the same Credibility update. Unfortunately this presents practical difficulties, as discussed below.

## 4. Problems

One obvious problem is that only those authors who are *registered* in the *o'Peer* database can receive bouquets or brickbats for the papers with their names on them. It will therefore be necessary to begin initially with a system that credits *only the submitting author*. Hopefully the other authors will find this inequitable enough that they will register themselves to obtain their share. This still introduces an "update" burden on the system, but it is manageable using Eq. (4) *as long as all the authors of each paper are correctly and uniquely identified*. The best way to ensure this is to eventually require the submitting author to check the database and identify all the authors by their unique index. This same problem exists in all citation databases now, and some promising measures have been taken to ensure reliable identification of authors; one example is ORCID[2], which could provide the necessary uniqueness if all *o'Peer* users were willing to register themselves with ORCID.

Another dilemma is how much credibility to assign to a *new o'Peer* user: if one starts with *zero* credibility, the motivation to review other authors' work is absent until one has one's own papers or reviews evaluated, and even then it grows rather slowly. If all started with 100% credibility, *o'Peer* would be no different from various "like"-collection sites. Since credibility can increase or decrease depending on subsequent work, I plan to start with 50% on all new accounts. (This may be the high point of some careers.)

Clearly there may be a temptation to "play the system" by teaming up with other authors to give superb reviews to one another's work, and/or to systematically "pan" that of third parties. This problem is not unique to *o'Peer*, of course. With computers handling all the evaluations, it should not be difficult to detect such cheating using "metadata" analysis techniques. It would be nice to see a few unprincipled referees get caught and exposed.

What about copyrights? One of the purposes of *journals* is to make money for the publishers and (on *extremely* rare occasions) for the authors. More often the authors pay the journals to publish their work, knowing that this will bring them greater esteem and better jobs with higher salaries; for most scholars, *Intellectual Property* (IP) is something you pay publishers to take from you. The goal of most researchers is to get their work into the public domain. *o'Peer* starts and ends there. It remains perfectly reasonable to write monographs and textbooks which are sold by publishers who share their profits with the authors; such works can be evaluated within *o'Peer*, but they need not be. Similarly with patents and other forms of IP.

## 5. Conclusions and Speculations

Since I was unaware of such a facility, I bought the domain name `oPeer.org` and built a "demo" version "*o'Peer Physics*" which is now open for business[8]. Obviously several developments are required if this is to have any impact: first, enough people have to register to form a critical mass of reviewers; second, enough papers have to be submitted to form a critical mass of material; and third, registrants have to treat it as seriously as they now do their responsibility of refereeing journal papers. While this may seem unlikely, I take heart from the fact that journals do find enough qualified people to perform this essential service without any remuneration other than the satisfaction of keeping their colleagues honest (and having their opinions listened to).

So far there is only *o'Peer Physics.* Since every discipline has a different "culture" and different criteria for "excellence", the database architecture for *o'Peer Chemistry, o'Peer Biology, o'Peer Medicine, o'Peer Literature, o'Peer Art* and *o'Peer Music* should be quite different, as should the procedures for peer evaluation. It may even be that YouTube's "like" option already adequately satisfies the need for democratic evaluation of videos, although one might hope for something better.

Unlike most "open access journals", I have no plans to charge authors *or* readers for use of *o'Peer Physics* — if the cost of maintaining this "hobby site" becomes burdensome I will seek funding from appropriate sources.

The most improbable outcome is that *o'Peer* should "catch on" and still remain under my management, such as it is. It is much more likely that the concept, once demonstrated to work, would be adopted by entities far more capable of refining and exploiting it, such as ResearchGate[6] or Google[7]. That would be fine with me, as long as they don't screw it up!

I dream of a future in which all the products of human endeavor are critically but fairly evaluated because the credibility of reviewers is subject to equally just and public evaluation.

## Acknowledgments

I acknowledge that "proper scientific papers" are supposed to be written in third person passive voice. That's just stupid. Motive, attitude and intent are integral parts of any analysis, and hiding them is the same as suppressing data.

I am well aware that URLs are not considered legitimate references. Get over it. They are *different kinds* of references — ones that are explicitly impermanent and have an implicit expiry date.

I also acknowledge that there is an extensive literature on this and related topics of which I am largely unaware. Therefore I am no scholar of open peer review; I'm just a guy with an idea.

Finally, I admit that I have not even thoroughly read all of those listed below: [9, 10, 11, 12, 13, 14, 15, 16, 17, 18, 19, 20, 21, 22, 23, 24, 25]. So sue me. Said list is offered (like most Reference lists) only to show that such literature does exist and is receiving serious consideration by real experts in the management of scholarly material.

## References

[1] `http://arxiv.org`
[2] `http://orcid.org`
[3] `http://slashdot.org`
[4] `http://www.youtube.com`
[5] `https://www.facebook.com`
[6] `http://www.researchgate.net` (See especially ResearchGate's "*Open Review*" feature at `https://www.researchgate.net/publicliterature.OpenReviewInfo.html`)
[7] `http://google.com` (See especially `http://scholar.google.com` for "Google Scholar".)
[8] `http://oPeer.org`
[9] Hunsinger J, Klastrup L and Allen M 2010 *International Handbook of Internet Research* (Springer Science+Business Media B.V.)
[10] David P A and Uhlir P F 2005 Creating the Global Information Commons for Science `http://www.codata.org/wsis/GICSI-prospectus.html`
[11] Pöschl U 2010 *IFLA Journal* **36** 40–46
[12] David P A, den Besten M and den Besten M 2006 *Proceedings of the Second IEEE International Conference on e-Science and Grid Computing (e-Science'06)*

[13] Pöschl U 2004 *Learned Publishing* **17** 105113
[14] Meir E V 2004 *Nature* **429** 803
[15] Lahiri D 2006 *Nature* **439** 784
[16] de Carvalho L 2006 *Nature* **439** 784
[17] Müller U 2008 *Peer-Review-Verfahren zur Qualitätssicherung von Open-Access-Zeitschriften – Systematische Klassifikation und empirische Untersuchung* Ph.D. thesis Humboldt University, Berlin
[18] Bodenschatz E and Pöschl U 2008 *in Open Access Challenges and Perspectives – A Handbook, European Commission and German Commission for UNESCO)* 54–57
[19] Pöschl U and Koop T 2008 *Information Services & Use* **28** 105107
[20] Pöschl U 2009 *Against the Grain* **21** Art. 11 (Also vailable at: http://docs.lib.purdue.edu/atg/vol21/iss3/11.)
[21] Alberts B, Kirschnerb M, Tilghmanc S and Varmu H 2014 *PNAS Early Ed.*
[22] Gura T 2002 *Nature* **416** 258–260
[23] Harnad S 2004 *in Peer Review: A Critical Inquiry* 235–242 available at http://cogprints.org/1646/.
[24] Harnad S 2010 *D-Lib Magazine* **16** available at http://eprints.ecs.soton.ac.uk/21348/.
[25] Harnad S 2013 *in The Future of the Academic Journal (2nd edition)* Available at http://eprints.soton.ac.uk/353991/.

Journal of Physics: Conference Series **551** (2014) 012061 doi:10.1088/1742-6596/551/1/012061

# Current status of the J-PARC muon facility, MUSE

**Y Miyake[1,2], K Shimomura[1,2], N Kawamura[1,2], P Strasser[1,2], A Koda[1,2], H Fujimori[1,2], Y Ikedo[1,2], S Makimura[1,2], Y Kobayashi [1,2], J Nakamura[1,2], K Kojima[1,2], T Adachi[1,2], R Kadono[1,2], S Takeshita[1,2], K Nishiyama[1,2], W Higemoto [1,3], T Ito[1,3], K Nagamine[1,4], H Ohata[2,5], Y Makida[2,5], M Yoshida[2,5], T Okamura[2,5], R Okada[2,5] and T Ogitsu[2,5]**

[1] Muon Science Laboratory, High Energy Accelerator Research Organization, Tsukuba, Ibaraki 305-0801, Japan

[2] Muon section, Materials and Life Science division, J-PARC Center, 203-1 Shirane Shirakata, Tokai-mura, Naka-gun, Ibaraki 319-1195, Japan

[3] Advanced Science Research Center, Japan Atomic Energy Agency, Tokai, Ibaraki 319-1195, Japan

[4] Advanced Meson Science Laboratory, The Institute of Physical and Chemical Research (RIKEN), Wako, Saitama 351-0198, Japan

[5] J-PARC Cryogenics, KEK, 203-1 Shirane Shirakata, Tokai-mura, Naka-gun, Ibaraki 319-1195, Japan

E-mail: `yasuhiro.miyake@kek.jp`

**Abstract.** The muon science facility (MUSE), along with the neutron, hadron, and neutrino facilities, is one of the experimental areas of the J-PARC project. The MUSE facility is located in the Materials and Life Science Facility (MLF), which is a building integrated to include both neutron and muon science programs. Since the autumn of 2008, users operation is effective and making use of the pulsed muon beam particularly at the D-Line. Unfortunately, MUSE suffered severe damages from the earthquake on March 11, 2011, the so-called "Higashi-Nippon Dai-Shinsai". We managed to have a stable operation of the superconducting solenoid magnet with use of the on-line refrigerator on December, 2012, although we had to overcome a lot of difficulties against components not working properly. But we had to stop again the whole operations on May 2013, because of the radioactive materials leakage accident at the Hadron Hall Experimental Facility. Finally we restarted the users' runs on February 2014.

## 1. J-PARC MUSE

The muon science facility (MUSE), along with the neutron, hadron, and neutrino facilities, is one of the experimental areas of the J-PARC project. The MUSE is located in the Materials and Life Science Facility (MLF), which is a building integrated to include both neutron and muon science programs [1]. The 3 GeV proton beam from the 3 GeV proton synchrotron ring (RCS) is focused on the 1st graphite target with 20 mm thickness. For phase 1, MUSE is designed to extract efficiently either pions or muons from a muon production graphite target to the four muon beam lines, the so-called D-Line, U-Line, S-Line and H-Line, enabling a variety of muon related experiments, at the ten experimental areas (D1, D2, U1A, U1B, S1, S2, S3, S4, H1 and H2) utilizing unique features of the pulsed muon beams. Figure 1 shows a layout of the MUSE facility. The repetition rate of the 3 GeV synchrotron is 25 Hz, with a pulse-separation of 40

Journal of Physics: Conference Series **551** (2014) 012061 doi:10.1088/1742-6596/551/1/012061

ms. Therefore always pulsed muons are delivered to the any experimental area, consisting of two bunches separated by 600 ns with a full width of 100-150 ns.

A superconducting decay/surface beam line, the so-called D-Line, was constructed as the the first muon beam line and has been used as a general use muon beam line since September 2008. It has a modest-acceptance pion injector of about 45 mSr, where either surface muons, or decay muons (5-120 MeV/c, $\mu^+$ or $\mu^-$) can be extracted that are obtained through the in-flight decay of $\pi^+/\pi^-$ confined by a strong longitudinal magnetic field of several tesla from a superconducting decay solenoid magnet. In November 2009, the surface muon extraction rate was significantly increased up to $1.8 \times 10^6$/s with a 120 kW proton beam from the RCS [2]. One of the sales point of the D-Line is its intense $\mu^-$ yield, since the cross section of $\pi^-$ at a 3 GeV proton beam is larger than that at the other meson factories [3]. Therefore, the yields of the decay $\mu^+$ and the decay $\mu^-$ are equivalent for the same momentum. In particular, decay $\mu^-$ as low as 5 MeV/c, which has an implantation depth of 1 $\mu$m in the case of gold, can be extracted with an intensity of one $\mu^-$/pulse.

## 2. Damages by the earthquake Higashi-Nippon Dai-Shinsai

The D-Line as well as the other MUSE facility suffered severe damages from the earthquake on March 11, 2011, the so-called Higashi-Nippon Dai-Shinsai. In particular, damages on the boundaries of the building structure were serious, since severe ground settlements occurred due to the Higashi-Nippon Dai-Shinsai. In the following, we specify the major damages.

- The helium ducts, control cables, power cables, compressed air piping for the superconducting magnet, and support stand for the D-line on-line refrigeration system were damaged, due to ground settlements, about 10 cm to 1.5 m at the boundaries of the MLF building, as is shown in Figure 2. Also a buffer tank for the He gas was tilted by 2-3 degrees. Because of the high-pressure gas regulation, we had to replace all the helium ducts at the boundary. Therefore we fixed all the ducts element as well as a support for the ducts and had a high pressure gas inspection on September, 2011.
- The M1 and M2 air circulation systems for the proton beam tunnels in the vicinity of the muon target were damaged. In particular, water ducts and control cables were found necessary to be removed and re-built, since the wall where they have been fixed was about to fall down located at an expansion joint between MLF and the proton beam transport line from the 3-GeV RCS (3NBT). Finally, the wall and water ducts etc. in the vicinity of the expansion joint were fixed by November, 2011, which enabled us to run a stable operation of the air circulation systems. Figure 3 is showing a picture of the damaged wall which was about to fall down at the expansion joint between 3NBT and MLF, as if water ducts for the M1 and M2 air circulation systems, were supporting the wall.
- The M2 tunnel consists of a target chamber, six quadrupole magnets, four steering magnets, two sets of profile monitor assemblies, 29 sets of pillow-seal assemblies, and seven sets of duct assemblies. When electricity was recovered in a month after the earthquake, it turned out that the M2 line primary beam duct was vented. We suspected something wrong with the pillow-seals caused some leakage in vacuum. By undergoing vacuum leak check for all the pillow-seals, it turned out that the pillow-seals themselves were not damaged, but tightness of a metal gate valve was not appropriate.
- We have been adopting an edge-cooled non-rotating graphite target, a so-called fixed target because of its ease of handling during maintenance. In this target, graphite is indirectly cooled by a copper frame, which surrounds the graphite. In order to reduce stress, a titanium buffer layer of 2 mm is placed between the graphite disk and the copper frame. In order to inspect the soundness of the graphite target, just in case damages were caused by the earthquake, the graphite target assembly was transported to the hot cell. There we

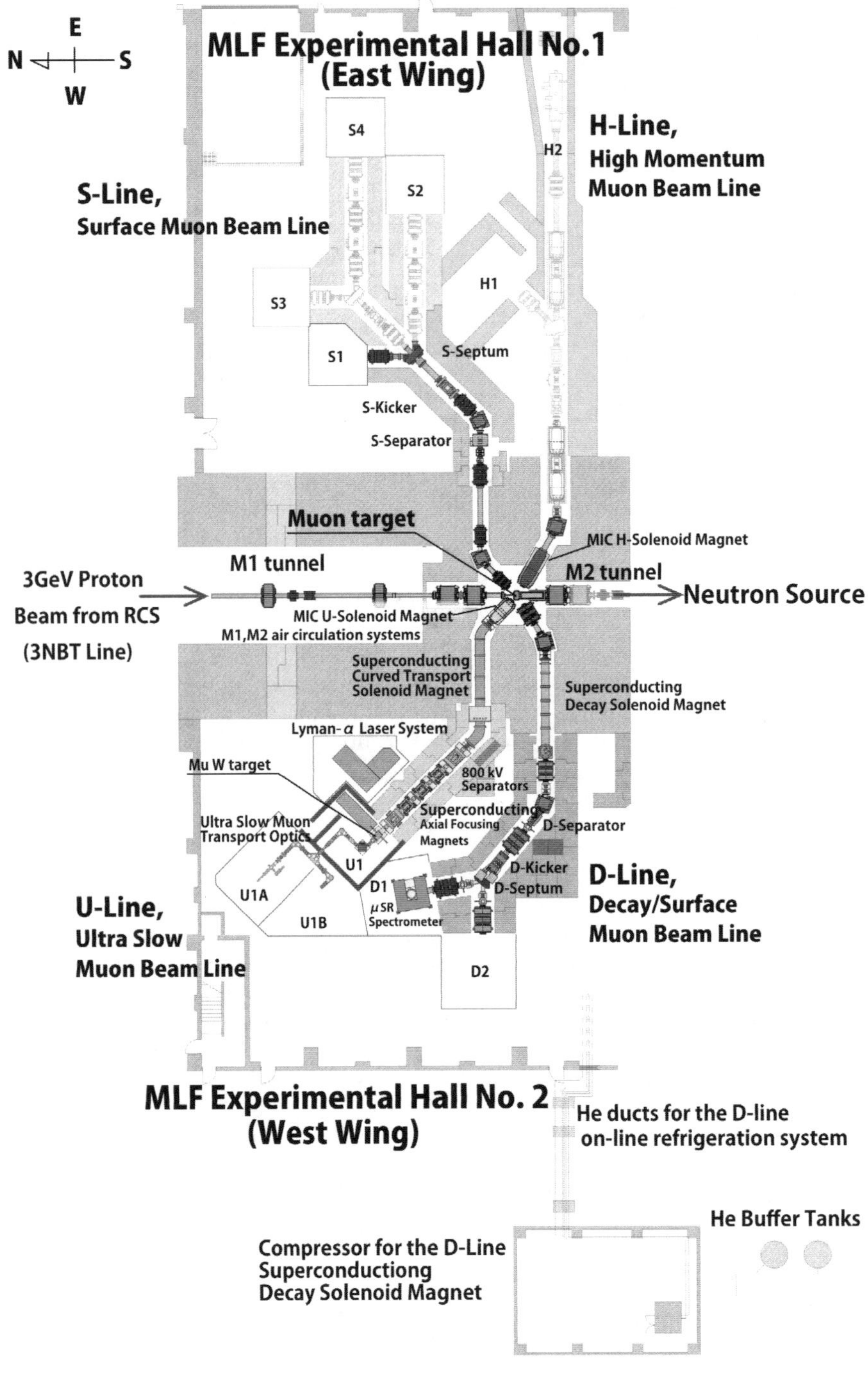


**Figure 1.** A layout of the J-PARC MUSE facility. The D-Line and the U-Line are located in the west wing in the MLF experimental hall No. 2. The S-Line and the H-Line are located in the east wing of the MLF experimental hall No. 1.

investigated not only dimensional changes but also heat conductivity by remote handling techniques. Consequently, it turned out the graphite target was alive and can be operated at least for another year.

- It was surprising that a lot of anchor bolts (M24) were ripped out from the concrete shielding blocks due to slipping off by the strong quake. Therefore several concrete shielding blocks were broken or cracked. All the damaged blocks were temporarily repaired by a concrete maker until November 2011.
- Moreover, the slipping off the concrete blocks by several tens of cm gave us more serious damages to the beam line components. In particular, an air-sealing hatch to prevent activated air in the tunnel from leaking into the user area, was smashed, causing deterioration in the air tightness. It had to be re-fabricated and re-installed because of the radiation safety. Also cable racks in the M2 tunnel struck by a heavy concrete block were found to be pushed away from the wall.

**Figure 2.** Damaged helium ducts for the D-Line on-line refrigerator at the boundary of the MLF building.

## 3. D-Line, recent developments and current status

Although MUSE suffered severe damages from the earthquake on March 11, 2011, consequently, we managed to re-operate the D-line by recovering from almost all the damages by the end of December, 2011. It was our great pleasure to report we could deliver again, surface and decay muons up to 60 MeV/c beam towards D1 and D2 during the commissioning beam time held on January 17, 2012, with an equivalent intensity as before 2011. The regular users runs were restarted from February 2, 2012.

Utilizing a beam-off period, we managed to install a switch yard, a kicker and a septum magnets to allow a single bunched muon beam up to 60 MeV/c towards D1 and D2 area in the D-line. It worked successfully to deliver a single pulse towards the D1 area. We have a plan to install a new $\mu$SR spectrometer at the D2 area in future.

A new $\mu$SR spectrometer was installed at the D1 area. It consists of 1280 scintillators (640 telescopes) with their scintillation light detected by pixeled avalanche photo diodes (MPPCs), covering a total solid angle of 23.4 %, which is 3.3 times larger [4] than the conventional D1 spectrometer, D$\Omega$1. The spectrometer is equipped with a Helmholtz coil of 4 kG with a gap of 135 mm, which can be rotated by 90 degree, allowing us to apply either transverse or longitudinal

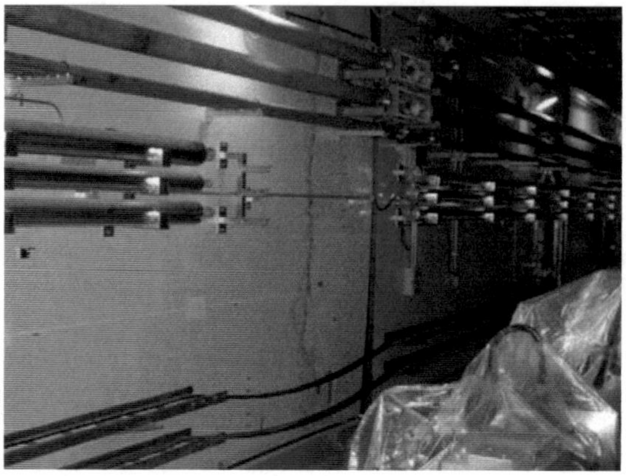

**Figure 3.** The damaged wall was about to fall down at the expansion joint between 3NBT and MLF, as if water ducts for the M1 and M2 air circulation systems, were supporting the wall.

magnetic field. It is also designed to be connected with a fly path chamber with a diameter of 410 mm. During the commissioning, it was demonstrated that 200 M of coincidence positron events per hour were collected for a sample specimen of 15 $\times$ 15 mm$^2$ with a 20 mm collimator. The event rate is 2-3 times higher than that in RIKEN-RAL [3], due to the stronger pulsed muon beam intensity of J-PARC.

## 4. U-Line

In addition to the D-Line, the U-line was constructed as the second muon beam line just beside of the D-Line in the MLF building No. 2. The U-line consists of a large acceptance solenoid made of mineral insulation cables (MIC), a superconducting curved transport solenoid magnet and a superconducting axial focusing magnets system. There, it allows to collect surface muons with a large acceptance solid angle of 400 mSr [5]. Compared to the conventional beam lines such as the D-line, the large acceptance of the front-end solenoid enables to capture more than ten times higher intensity pulsed surface muons of 5 $\times$ 10$^8$/s (proton beam intensity of 1MW operation). The U-line components of the superconducting curved and axial focusing magnets were already fabricated and installed in the end of September of 2012. In the commissioning works held in October, 2012, it turned out that the surface muon intensity achieved at the U-line exceeded the intensities at the D-line by one order of magnitude. Finally, it was succeeded to extract the world's highest intensity pulsed muons, 2,500,000 muons per pulse (6.4 $\times$ 10$^7$/s at 212 kW, corresponding to 3 $\times$ 10$^8$/s at 1MW) to the U1 experimental area. Another important feature of the U-Line adopting all the axial focusing transport system, was that we could extract not only positive muons, but also negative muons with a yield of 1/10 of the surface muons. Simultaneous extraction of positive and negative muons enables us to generate an exotic atom $\mu^+\mu^-$, although those can be easily separated by a pair of the dipole magnets equipped with the superconducting curved transport solenoid [6].

As a next step, it is planned to stop such intense pulsed muons into a hot tungsten target for generating an intense ultra slow muon beams in order to realize gultraslow muon microscopes [7]h that will enable us not only for a variety of surface, and nano-science, but also 3D imaging. Installation of a slow ion optics [8] as well as a $\mu$SR spectrometer dedicated to the ultra slow muons were completed [9]. When the production of intense ultra slow muon is realized, the use of its short-range penetration depth (e.g., 1 nm resolution at a penetration of 1 nm, and 10 nm

Journal of Physics: Conference Series **551** (2014) 012061 doi:10.1088/1742-6596/551/1/012061

at a penetration of 6 nm in Gold) will allow muon science to be expanded towards a variety of new scientific fields [10] such as,

- Surface/boundary magnetism utilizing its spin polarization and unique time-window,
- Sub-Surface chemistry, utilizing a feature of a light isotope of hydrogen,
- Precise fundamental physics such as QED, and
- Ion sources towards $\mu^+\mu^-$ collider experiments in high-energy physics.

## 5. S-Line
We are constructing a new surface muon beam line dedicated to the material sciences, the so-called S-Line. Although we are planning to construct four experimental areas, S1, S2, S3, and S4 areas finally, we are constructing the beam line up to the S1 area now. It consists of large acceptance quadrupole triplets made of MIC as a front-end in the vicinity of the muon target, four sets of normal quadrupole triplet magnets, three sets of bending magnets and an electrical kicker system, associated with a set of 250 kV DC separator [11]. Recently, we managed to complete the installation of a power supply yard with four floors, two users cabins, a dedicated catswalk, and beam line shielding blocks. We are planning to complete the installation of the key components of the S-Line, for the beam commissioning in the coming summer of 2014.

## 6. H-Line
The fourth muon beam line, the so-called H-line, was designed to extract muons or positrons (electrons) up to 120 MeV/c, for fundamental physics such as experiments on the precise Mu hyperfine measurement, muon-electron conversion experiment, or g-2/EDM experiment. Its front-end components such as large acceptance solenoid magnets, HS1 and HS2 made of MIC, and a bending magnet HB1 were installed in the vicinity of the muon target, where radiation is very high in the order of several mSv/h to Sv/h [12].

## 7. Summary
Although the whole J-PARC MUSE operation had to be shut down due to not only the Higashi-Nippon Dai-Shinsai but also the radioactive materials leakage accident at the Hadron Hall Experimental Facility, finally we restarted users run at the D-Line on February, 2014.

We completed the installation of the U-line, which consists of a large acceptance solenoid made of mineral insulation cables (MIC) and a superconducting curved transport solenoid and 6 sets of superconducting axial focusing magnets. There, we managed to extract the world's highest intensity pulsed muons, 2,500,000 muons per pulse ($6.4 \times 10^7$/s) at 212 kW, for the intense ultra slow muon source.

We are planning to complete the installation of the key components of the S-Line, for the beam commissioning in the coming summer of 2014.

## References

[1] Nagamiya S, Nagae, Ooyama Y, Miyake Y, Takano H, Helliwell JR, Peng, JC 1999 *Energy Review* **19** 4

[2] Miyake Y, Nishiyama K, Kawamura N, Strasser P, Makimura S, Koda A, Shimomura K, Fujimori H, Nakahara N, Kadono R, Kato M, Takeshita S, Higemoto W, Ishida K, Matsuzaki T, Matsuda Y and Nagamine K 2009 *Nucl. Instr. and Meth. in Phys. Res.* A **600** 22

[3] Nagamine K, Matsuzaki T, Ishida K, Watanabe I, Nakamura SN, R. Kadono, N. Kawamura, S. Sakamoto, M. Iwasaki, M. Tanase M, Kato M, Kurosawa K, Eaton GH, Jones HJ, Thomas G and Williams WG 1996 *Hyperfine Interact.* **101/102** 521

[4] Kojima K *et al.* 2014 *J. Phys.: Conf. Ser.* this issue

[5] Nakahara K, Miyake Y, Shimomura K, Strasser P, Nishiyama K, Kawamura N, Fujimori H, Makimura S, Koda A, Nagamine K, Ogitsu T, Yamamoto A, Adachi T, Sasaki K, Tanaka K, Kimura N, Makida Y, Ajima Y, Ishida K, Matsuda Y 2007 *AIP conference Proceedings* **981** 312

[6] Miyake Y, Nakahara K, Shimomura K, Strasser P, Kawamura N, Koda A, Makimura S, Fujimori H, Nishiyama K, Matsuda Y, Bakule P, Adachi T and Ogitsu T 2009 *AIP Conference Proceedings* **1104** 47
[7] Torikai E *et al.* unpublished
[8] Strasser P *et al.* 2014 *J. Phys.: Conf. Ser.* this issue
[9] Higemoto W *et al.* unpublished
[10] Bakule P and Morenzoni E 2004 *Contemporary Physics* **45** 203
[11] Koda A *et al.* unpublished
[12] Kawamura N *et al.* 2014 *J. Phys.: Conf. Ser.* this issue

Journal of Physics: Conference Series **551** (2014) 012062 doi:10.1088/1742-6596/551/1/012062

# H line; a beam line for fundamental physics study

**Naritoshi Kawamura[1,2], Akihisa Toyoda[3,4], Masaharu Aoki[5], Koichiro Shimomura[1,2], Tsutomu Mibe[3,4], Yohei Nakatsugawa[2], Masashi Otani[4], Naohito Saito[3,4] and Yasuhiro Miyake[1,2]**

[1] Muon Science Sec., Materials and Life Science Div., J-PARC, Tokai, Ibaraki 319-1106, Japan
[2] Muon Science Labo., IMSS, KEK, Tsukuba, Ibaraki 305-0801, Japan
[3] Hadron Sec., Particle and Nuclear Physics Div., J-PARC, Tokai, Ibaraki 319-1106, Japan
[4] IPNS, KEK, Tsukuba, Ibaraki 305-0801, Japan
[5] Graduate School of Science, Osaka University, Toyonaka, Osaka 560-0043, Japan

E-mail: nari.kawamura@kek.jp

**Abstract.** The muon facility, J-PARC (Muon Science Establishment; MUSE), has been operating since the first beam in 2008. Starting with a 200 kW proton beam, a beam intensity of $3 \times 10^6$ muons/s was reached in 2009 which was the most intense pulsed muon beam in the world. From the 2 cm thick graphite target, four secondary muon beam lines are designed to be extracted. Three beam lines currently exist, the first being operational and the other two undergoing commissioning. The fourth and the last beam line, the H line, is planned to be constructed. This new beam line is designed to have a large acceptance, provides the ability to tune the momentum, and use a kicker magnet and/or a Wien filter. The H line is designed to provide an intense beam of $10^8$ surface muons/s for fundamental physics studies to observe new physics beyond the standard model. Such studies require high statistics and they need to occupy the experimental areas for a relatively long period.

## 1. Introduction

The Japan Proton Accelerator Research Complex (J-PARC) project was proposed jointly by Japan Atomic Energy Agency (JAEA) and High Energy Accelerator Research Organization (KEK) [1]. J-PARC consists of a 400 MeV linac, 3 GeV and 50 GeV synchrotron rings, which provide an intense proton beam to pursue particle and nuclear physics, materials and life science and nuclear technology. A 1 MW proton beam is transported from the 3 GeV synchrotron ring to Materials and Life Science Facility (MLF) which consists of the muon and the spallation neutron facilities, MUSE and JSNS. As shown in Fig. 1 and 2, the proton-beam-line tunnel runs through the center of MLF building from the north to the south. The east and the west wings are experimental halls where neutron and muon beam lines are constructed. In order to avoid the diffusion of radioactive contamination generated around the muon target chamber and along the proton beam transport tunnel, the proton-beam transport tunnel is isolated from the experimental halls dividing the MLF building into east and west wings. This building structure achieved its purpose to confine the contamination and contributed to making the radiation-shield structure simple. However, the severe geometrical constraint which was put on the design of the secondary beam required the number and the directions of muon beam lines to be fixed at the primary stage of the building design.

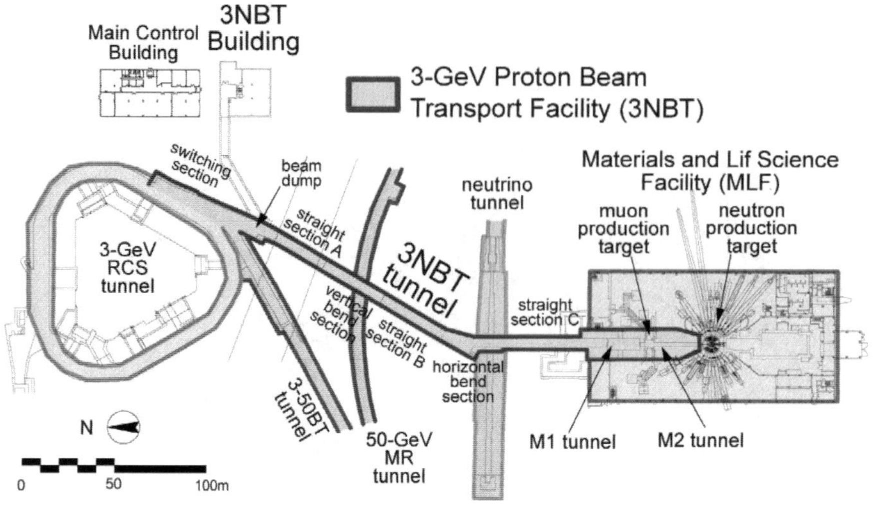


**Figure 1.** A schematic drawing from 3 GeV rapid cycling synchrotron (RCS) to MLF.

The muon-production target is inserted at about 30 m upstream of the neutron target in the proton beam line. From the muon target, we decided to extract four muon beam lines. Each of them provides the intense pulsed muon beam with an individual design concept to be used for a variety of muon science experiments. Two of the secondary beam lines are extracted at the angle of 60 degrees to the proton beam line, and the others are at 135 degrees [2].

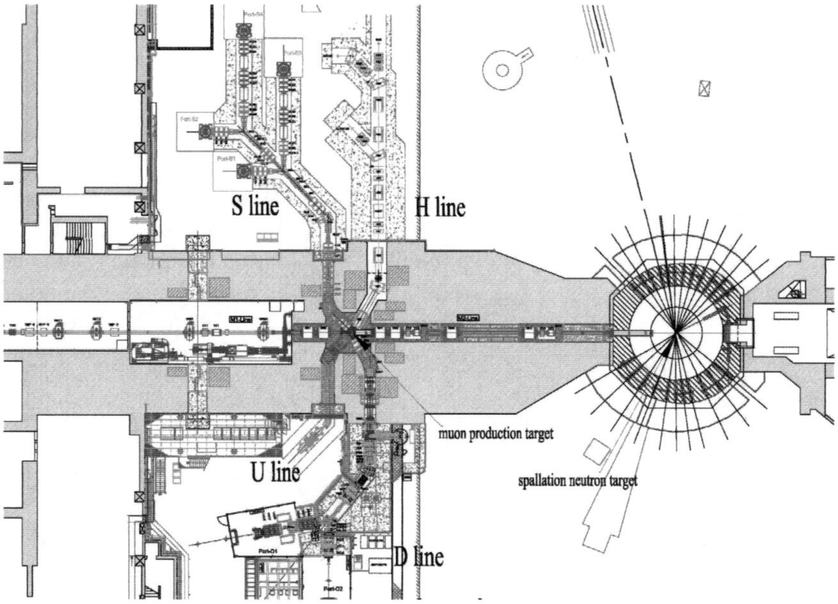


**Figure 2.** The layout plan of MUSE.

The decay muon beam line (D line) was constructed first, and this is the only beam line under operation. The D line extracts both negative and positive decay muons up to a momentum of 120 MeV/c, as well as 30-MeV/c surface and cloud muons [3]. The ability of wide-range momentum-tuning answers a variety of user's programs. The second beam line, the U line, provides the most intense surface-muon beam among the four secondary lines [4] and will be used to generate an ultra-slow muon beam by the laser resonant ionization method [5]. It is currently under commissioning. The remaining two beam lines, the S and the H line, are located in the east experimental hall. The S line, dedicated to $\mu$SR spectroscopy, is designed to distribute the surface muon pulses to several experimental areas simultaneously by using kicker devices. The first beam is planned to be delivered in 2014.

A typical period of an experiment in the field of material science is a few days to obtain enough statistics for $\mu$SR spectrum of sample materials. Therefore, a study requiring much higher statistics [6, 7, 8] does not match the design of the existing S, D and U lines. The fourth beam line, the H line, aims to answer the demands for these fundamental physics studies that need high statistics, *i.e.* the use of a high intensity beam for a long periods, typically more than one year.

## 2. Muon Source

The intense pulsed muon beam is produced by a 3-GeV proton beam injected to the target made of 20 mm thick isotropic graphite [9]. About 5% of proton beam is consumed at the target to produce pions. A surface muon is produced when a positive pion is stopped near the surface inside the production target and a muon, whereas a decay beam is produced when positive or negative pions decay in flight along the beamline.

The surface muon intensity can be estimated from the number of pions stopped near the surface. Calculation was performed utilizing the measured data on pion production cross sections [10] followed by tracking of the pions with GEANT 3.21. By integrating the generator equation for various angles and for pion energies, the total cross section for pion production is about 90 mb, generating about 0.016 per a 3-GeV proton passing through the 2-cm (3.6 g/cm$^2$) graphite target. Thus, $3.2 \times 10^{13}$ pions/s are estimated to be generated from 1 MW (333$\mu$A) proton beam. Taking into account the emitting angle and energy distribution of the pions, $7.4 \times 10^9$ pions/s are simulated to be stopped in 0.1 mm from the target surface and in a diameter of 24 mm which corresponds with the proton beam spot size on the target.

The details of the H line is described in the sections below. Starting from a source distribution of 24 mm in diameter, about 90% of the beam transmission efficiency is able to be achieved for the muons plunging into the H line of which solid angle is about 120 mstr. This transmission efficiency is almost constant for the beam with the momentum range of ±5% from the central momentum so that a beam in the range 27.0 to 29.8 MeV/c, can be transported with the efficiency of 90%. The momentum spread of the surface muon reflects the generated depth from the target surface of the pion stopping position. A muon generated at 0.2 mm in depth has 27.0 MeV/c. Taking into account the effective depth to the H line extracted at 60 degrees $15 \times 10^9$ pions/s contribute to generate the surface muons. The overall beam intensity in the H line is then evaluated to be $15 \times 10^9 \times 120 \times 10^{-3}/4\pi \times 0.9 = 1.2 \times 10^8$ muons/s. Assuming that the efficiency of the final focusing part is 80%, $1 \times 10^8$ muons can be available in the experimental area.

## 3. Beam line optics

The intended experiments in H line require high statistics, *i.e.* high intensity beam. Momentum tune-ability is also needed. Both features will be important for potential experiments in future. The design work on H line is performed along these two concepts. The basic idea of the beam line optics was proposed to use a large-aperture muon-capture solenoid, a wide-gap bending magnet and a pair of two solenoid magnets which have opposite-direction field each other [11]. Conventional matrix calculation is not applicable because the near-axis approximation is not good in a large-aperture solenoid. Thus, G4BEAMLINE [12] is applied to optimize the parameters of the beam line magnets.

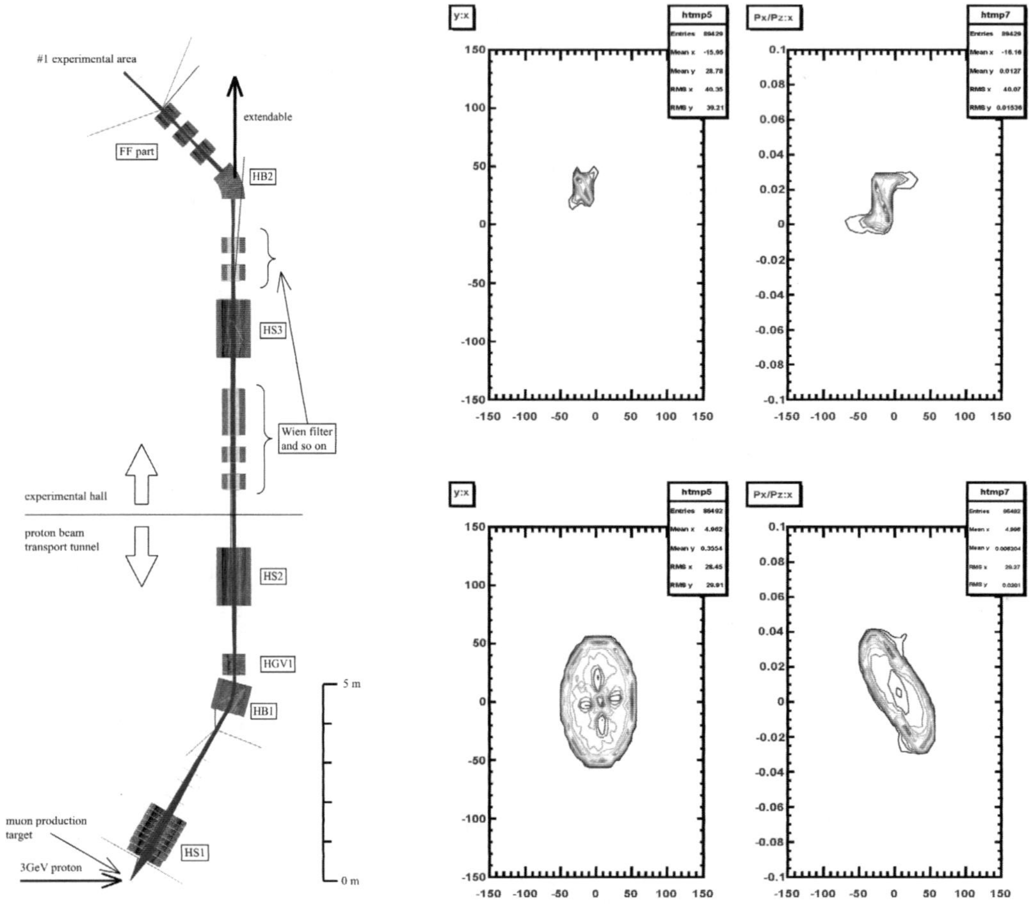


**Figure 3.** A typical result of G4BEAMLINE calculation. The blue lines are beam trajectories (left). The surface muon beam from a point source is transported to the first experimental area through three solenoid magnets and two bending magnets. The wien filter and the kicker magnets are also shown, but they are not in use in this transmission. The beam profile at the exit of HS2 and HS3 are shown (right).

The front-end capture solenoid magnet (HS1) consists of 8 coils, and three kinds of currents are supplied to the first, the second and the remaining coils, respectively. The first and the second coils make the main capture field, and by tuning the rest, the beam profile is minimized at around the first bending magnet (HB1) and the gate valve (HGV1) necessary for beam line maintenance. The second solenoid magnet (HS2) generates a weak field to produce a parallel or slightly-focusing beam. The third solenoid magnet (HS3) makes an opposite-direction field to cancel the beam-profile rotation caused by HS2. This canceling function doesn't depend on

Journal of Physics: Conference Series **551** (2014) 012062 doi:10.1088/1742-6596/551/1/012062

the momentum, and thus beam loss due to momentum dispersion at the second bending magnet (HB2) is reduced. The distance between HS2 and HS3 can be extended to about 5 m, and HB2 which determines the position of the experimental area can be placed where the area doesn't interfere the neighboring S line. In addition, one can install a Wien filter, a kicker magnet and other devices like vacuum components in this 5 m gap. Depending on the intended experiment, installed devices in this gap are selected. For instance, a Wien filter is installed to eliminate electrons/positrons from the beam [6, 7], and the second Wien filter is necessary to obtain full polarization rotation. Kicker magnets are considered to be installed instead of a Wien filter to protect detectors from the prompt burst of muons and electrons right after the proton pulse injection and to detect delayed electrons from $\mu$-e conversion generated in the muon production target [8], although the detector which is tolerant of the burst is under development not to use kicker magnets.

Figure 3 shows a typical transport calculation, and the beam is delivered to the first experimental area which is the nearest to the muon source. Turning off HB2, the beam goes straight and is transported to the extended beam line by adopting another pair of opposite-field solenoid magnets. Namely, H line can provide the intense beam to a few experiments which use different apparatuses and cannot share the experimental area.

In the other high intensity beam line in MUSE, the U line, we adopted only axial focusing magnets to obtain high transmission efficiency [4]. In contrast the H line uses bending magnets, whose non-axial focusing makes beam losses higher. To maintain high transmission efficiency, large aperture magnets and other devices are adopted in the H line. A typical size of the aperture is 600 mm in diameter.

## 4. Construction status

At the primary stage of Muon Facility construction, only the magnets in the D line and the frontend magnets in the S line were installed, and then the frontend magnets in the U line were installed in 2009. In the H line, temporary radiation-shield blocks were placed. J-PARC has been operated since 2008, and thus the activation around the muon production target and the shield in the H line becomes serious year by year. According to the evaluation by a Monte-Carlo code, the dose rate beside the target chamber was estimated to be close to 1 Sv/h, and the summer shutdown in 2012 would be the actual time limit to install the frontend magnets in the H line. Thus, installation of the frontend devices, *i.e.* the muon-capture solenoid, HS1, the first bending magnet, HB1, and the vacuum components took precedence in 2012.

The magnets and other devices in the downstream are under designing. These devices are placed in the experimental hall, and thus they have many design options regarding material selection, geometrical size and position and so on. In order to provide the satisfactory muon beam to all the planed experiments in the H line, designing work is performed.

The designing work of the radiation shield is also in progress according to the same manner in the other beam lines [13]. Along the beam line a few meter thick concrete shield will be necessary to enclose the streaming neutrons and other radiation sources. Because the H line adopts large aperture devices, the effect of the streaming neutrons are more serious than the other beam lines. The evaluation of the streaming neutron is important not only for the radiation safety but also to check its effect on the detectors and other devices in the experimental area.

## 5. Summary

The H line is designed to be dedicated to fundamental physics studies to observe a new physics beyond the standard model, and they require high intensity beam and need to occupy an experimental area for a long period of more than a year.

The beam optics calculation predicts high transmission efficiency, and more than 80% of the captured muons can be transported. The detailed design work of the magnets and the other beam-line components is in progress, and the result will be fed back to the optics calculation. In total, $1 \times 10^8$ surface muons/s us calculated to be available.

The muon facility has been under operation since 2008. The temporary radiation shields had been placed in H line, and they have been activated. The radiation dose from the residual activities was expected to reach the order of 1 Sv/h in 2012. Thus HS1, HB1 and the other front-end devices were needed to be installed as early as possible, and their installation work was successfully completed. The devices in the downstream are under designing at present.

## References

[1] Joint project team of JAERI and KEK 1999 KEK Report 99-4; JAERI-Tech 99-056; JHF-99-3
[2] Miyake Y *et al.* 2005 *Nucl. Phys. B* **149** 393
[3] Shimomura K *et al.* 2009 *Nucl. Instrum. Meth. A* **600** 192
[4] Nakahara K *et al.* 2010 *AIP Conf. Proc.* **1222** 420
[5] Bakule P *et al.*, 2003 *Spectrochimica Acta Part B* **58** 1019
[6] Shimomura K *et al.* 2011 *AIP Conf. Proc.* **1382** 245
[7] Saito N *et al.* 2012 *AIP Conf. Proc.* 2012 **1467** 45
[8] Aoki M *et al. IFMF2010 Proceedings*
[9] Makimura S *et al.* 2009 *Nucl. Instrum. Meth. A* **600** 146
[10] HARP Collaboration 2008 *Eur. Phys. J.* **C53** 177
[11] Doornbos J *private communication*
[12] `http://www.muonsinc.com/`
[13] Kawamura N *et al.* 2009 *Nucl. Instrum. Meth. A* **600** 114
[14] Kawamura N *et al.* 2014 *Nucl. Instrum. Meth. A* **741** 33

Journal of Physics: Conference Series **551** (2014) 012063 doi:10.1088/1742-6596/551/1/012063

# New $\mu$SR spectrometer at J-PARC MUSE based on *Kalliope* detectors

**K M Kojima[1,2,7], T Murakami[3,7], Y Takahashi[1,7], H Lee[1,7], S Y Suzuki[4,7], A Koda[1,2,7], I Yamauchi[1], M Miyazaki[1], M Hiraishi[1], H Okabe[1], S Takeshita[1], R Kadono[1,2,7], T Ito[5], W Higemoto[5], S Kanda[3,7,6], Y Fukao[3,7], N Saito[3], M Saito[3,7], M Ikeno[3,7], T Uchida[3,7] and M M Tanaka[3,7]**

[1] Institute of Materials Structure Science, High Energy Accelerator Research Organization (KEK-IMSS), and J-PARC Material and Life Science Experimental Facility (J-PARC MLF), Tsukuba and Tokai Ibaraki, 319-1195, Japan

[2] Department of Materials Structure Science, The Graduate University for Advanced Studies, Tsukuba, Ibaraki, 305-0801, Japan

[3] Institute of Particle and Nuclear Study, High Energy Accelerator Research Organization (KEK-IPNS), Tsukuba, Ibaraki, Japan

[4] Computing Research Center, High Energy Accelerator Research Organization (KEK-CRC), Tsukuba, Ibaraki 305-0801, Japan

[5] Advanced Science Research Center, Japan Atomic Energy Agency (JAEA), Tokai, Ibaraki, 319-1195, Japan

[6] University of Tokyo, Graduate School of Science, Hongo, 113-8654, Japan

[7] Open source consortium of instrumentation, Open-it, KEK, Tsukuba, Ibaraki, 305-0801, Japan

E-mail: kenji.kojima@kek.jp

**Abstract.** We developed a new positron detector system called Kalliope, which is based on multi-pixel avalanch photo-diode (m-APD), application specific integrated circuit (ASIC), field programmable gated array (FPGA) and ethernet-based SiTCP data transfer technology. We have manufactured a general-purpose spectrometer for muon spin relaxation ($\mu$SR) measurements, employing 40 Kalliope units (1280 channels of scintillators) installed in a 0.4 T longitudinal-field magnet. The spectrometer has been placed at D1 experimental area of J-PARC Muon Science Establishment (MUSE). Since February of 2014, the spectrometer has been used for the user programs of MUSE after a short commissioning period of one week. The data accumulation rate of the new spectrometer is 180 million positron events per hour (after taking the coincidence of two scintillators of telescopes) from a 20×20 mm sample for double-pulsed incoming muons.

## 1. Introduction

At pulsed muon facilities, muons come in a pulse of a time width of several 10s of nano-seconds. Time differential muon spin relaxation ($\mu$SR) measures the elapsed time between each muon's arrival and the decay positron in the mu-e decay $\mu^{\pm} \rightarrow e^{\pm} + \nu$. The requirement for the positron/electron detectors for $\mu$SR is the minimal pile-up and dead time to handle the huge event rate right after the arrival of muons. The decay muon channel (D-line) of MUSE at

J-PARC delivers 180,000 positive muons per pulse at D1 experimental area for the 300 kW operation regularly achieved nowadays [1]. These muons decay into positrons with an average life-time of 2.2 micro-seconds, resulting to the instantaneous count-rate of $180,000/2.2\ \mu s \approx 100$ giga counts/second (Gcps) for the entire solid angle.

The most reliable method of counting high-rate events is to reduce the size of the individual counters and tile them to cover the desired solid angle. The general purpose $\mu$SR spectrometer ($D\Omega$-1) previously installed at D1 experimental area had 256 scintillators (128 telescope pairs), covering approximately 8% of the solid angle around the sample [2]. With the full beam, the instantaneous count-rate for each counter yields $100\times10^9\times0.08/128 = 60$ Mcps. This is still a too high-rate, because each counter experiences approximately 60 Mcp$\times$2.2 $\mu$s$\approx$120 counts/pulse, which corresponds to $2.2\ \mu s/120 \approx 20$ ns for the average event separation. This event separation is shorter than the typical recovery time of photo-multiplier-tubes (PMTs: $\sim$20 ns) or avalanch photo diodes (APDs: $\sim$40 ns), and we reduce the muon beam intensity to $\approx$1/10-1/20 by collimators and slits, so that the high event-rate may be handled by the existing detectors. In order to enjoy the high intensity of pulsed muons at MUSE, the segmentation of the positron counters has to be higher. Recent development of multi-pixel avalanch photo diodes (m-APD), Application Specific Integrated Circuit (ASIC) and Field Programmable Gated Arrays (FPGA) has made such spectrometer affordable. Employing these new technologies, we developed a new positron/electron detector system called Kalliope (KEK Advanced Linear and Logic-board Integrated Optical detectors for Positrons and Electrons), manufactured a 1280 channel $\mu$SR spectrometer and installed to D1 experimental area.

## 2. Kalliope detector system

In Fig.1, we show one unit of Kalliope detector. It is composed of three blocks: the detector block, the analogue board and the digital board. Current design of the detector block is a plastic scintillator of $10\times10\times12$ mm cube which has a wave-length shifting (WLS) fiber of diameter 1 mm running through. At the end of the WLS fiber, a m-APD (Multi-Pixel Photon Counter (MPPC) by Hamamatsu Photonics Co. Ltd.) is equipped to detect the scintillating light. The structure is shown in Fig.2. Signals from the two cubic scintillators aligned in series are taken coincidence, in order to identify positron/electron signals from the thermal noises. The m-APD has major advantages of lower cost ($\sim$ 10 dollars/channel) and higher tolerance to the magnetic fields ($\sim$Tesla) over usual PMTs.

What distinguishes Kalliope detector system most from other APD-based positron counters is that the electronics (the discriminator, TDC and memory) are built in as the front-end circuit. The analogue board is equipped with two of 16-channel preamplifier and comparator circuits in ASIC (gold chip in Fig.1). The diagram of the VOLUME2012 ASIC [3], which is currently employed, is shown in Fig.3. It has two voltage amplifiers, each with $\times10$ gain in series, equipped with the selection switch (SW) which may bypass the second amplifier. Each amplifier is equipped with a 4-bit digital to analog converter (DAC4N in Fig.3) which controls the offset level of the amplifier to adjust the pulse shape and the response speed. After the second amplifier, a comparator is installed which outputs a digital pulse if the analog input exceeds the preset threshold level determined by a pair of DACs (ThDAC: denoted as DAC4-1 and -2 in Fig.3) and the external voltage level (vref) adjusted by a variable resister (blue box in Fig.1).

The bias voltage applied to the m-APD, which determines the detector gain, may be finely controlled by another DAC installed at the input line of the first amplifier. This BiasDAC supplies a DC current going through the termination register outside of ASIC to give a DC voltage offset from the ground level. The termination impedance is composed of a 10 k$\Omega$ register in parallel with 0.1 $\mu$F +100 $\Omega$ in series, so that the fast signal sees 100$\Omega$, whereas DC current sees 10 k$\Omega$ as the termination.

Journal of Physics: Conference Series **551** (2014) 012063 doi:10.1088/1742-6596/551/1/012063

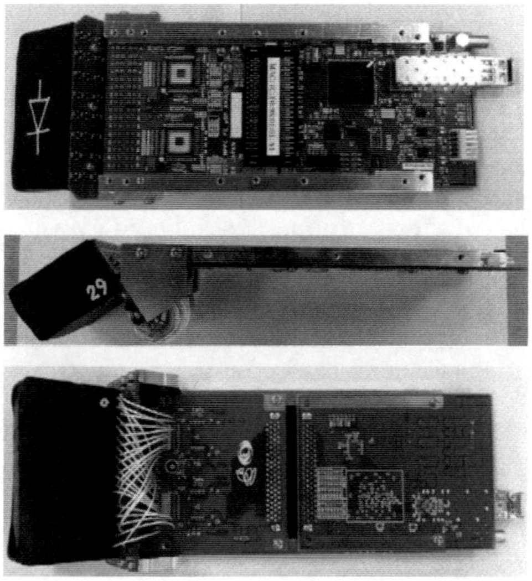

**Figure 1.** One unit of Kalliope detector.

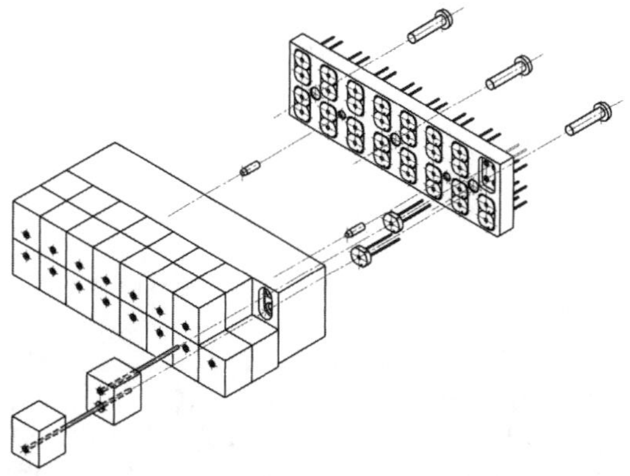

**Figure 2.** Structure of the detector block which appears as the black one in Fig.1. Plastic scintillators (cubes) and wave-length shifting fiber (rods) are painted by white reflective paint for light shield.

Each channel of the ASIC has a 20 bit digital control, which consists of one BiasDAC(4bit), two AmpDACs(4bit×2), one ThDAC(4bit) and four control bits. The last control bits are On/Off switchs of BiasDAC, 2nd stage amplifier, analog output and digital output. The digital control must be supplied in pulse series before the usage of ASIC.

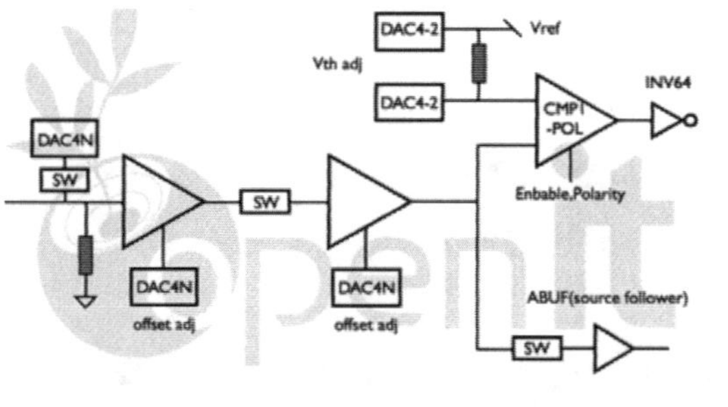


**Figure 3.** Diagram of VOLUME2012 ASIC used in analog-board of Kalliope.

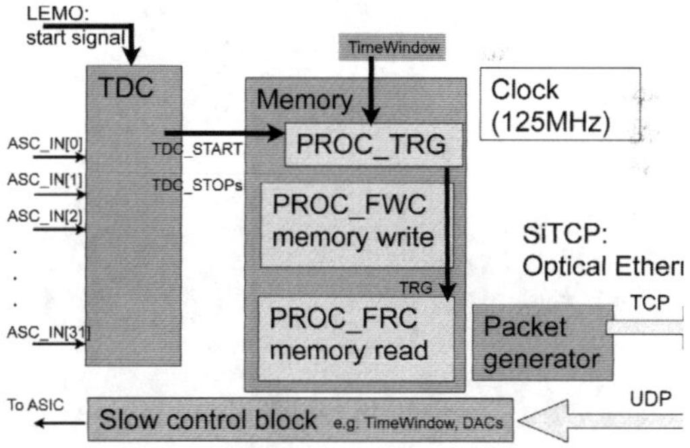


**Figure 4.** Diagram of digital signal processing in FPGA of digital-board of Kalliope system.

The digital outputs as well as the serial control of the ASIC is connected to the digital board of Kalliope system. The main component of the digital board is the FPGA which is the largest black chip in Fig.1. It is a programable digital chip with a pre-compiled firmware downloaded via a special cable prior to the usage. The current logic diagram running on FPGA is schematically shown in Fig.4. The time-to-digital converter (TDC) counts up the four of 250 MHz clocks which are phase shifted by 90°, and whenever the positron/electron pulse comes to the ASC_IN, which is 32 channel in parallel, the timing is recorded into the memory. The timing of the LEMO input (start) signal is independently recorded and the time difference between the start and the multiple stops are calculated with the timing resolution of 1ns. There is a preset time window of measurement (default 64 $\mu$s) which may be changed from the data acquisition (DAQ) PC, if necessary. When the time reaches after the start signal to this time-window, the stored hit events are read from the memory, formed into a series of ethernet packets and sent to DAQ-PC using SiTCP protocol [4]. The ethernet cables from each Kalliope units are combined in a commercially available ethernet hub and data are transferred to the DAQ-PC. Since the electronics are built in, Kalliope unit has only three wires connected: one optical ethernet fiber, a NIM trigger cable and a power-supply cable. Compared to the traditional PMT-based system, the reduction in the number of the cables is a great advantage of Kalliope.

## 3. New spectrometer at D1 experimental area

Employing 40 units of Kalliope detectors installed to a 0.4 T magnet, we have manufactured a 1280 channel (640 telescope pairs) spectrometer. The spectrometer is installed at D1 experimental area of Material and Life science experimental Facility (MLF) as shown in Fig.5. Detailed arrangements of Kalliope units inside the magnet is shown in Fig. 6.

**Figure 5.** Full view of the new D1 spectrometer at J-PARC.

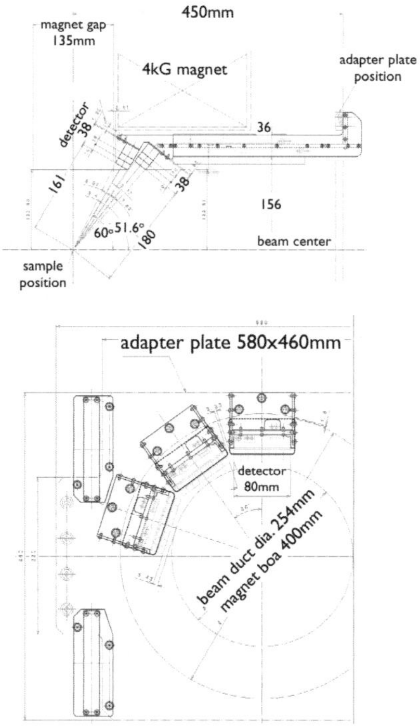


**Figure 6.** Diagram of the side and the tail view of the new D1 spectrometer.

The ethernet hub, the NIM start-pulse distributor as well as the power supply for Kalliope detectors are placed underneath the magnet, so that cables are confined within the spectrometer. Connections to the spectrometer are only the two cooling water lines, power cables for the 0.4 T magnet, one NIM trigger source line, one optical ethernet fiber from DAQ-PC and three 100 V power supply cables. The minimized number of the connection lines makes this spectrometer portable.

We start a commissioning of the new spectrometer on February 18th 2014. The procedure of the commissioning is (1) determination of the approprqiate bias voltage (HV) for m-APD, threshold DAC (ThDAC) and amplifier DAC (AmpDAC) of ASIC for the best performance in (2) characterization of the $\mu$SR spectra such as the full asymmetry and the spectrum distortion. The m-APD are provided with a list of minimum operation voltages which distributes within 0.2 V in our spectrometer (66.6-66.8 V). In order to perform (1), we have developed a computer program to scan the DAC values and take counts for a preset number of accelerator pulses. In Fig.7, we show an example of the efficiency curve, the signal count for the threshold DAC (ThDAC) scan. We choose the appropriate ThDAC as a few steps higher than the peak position which corresponds to the noise level.

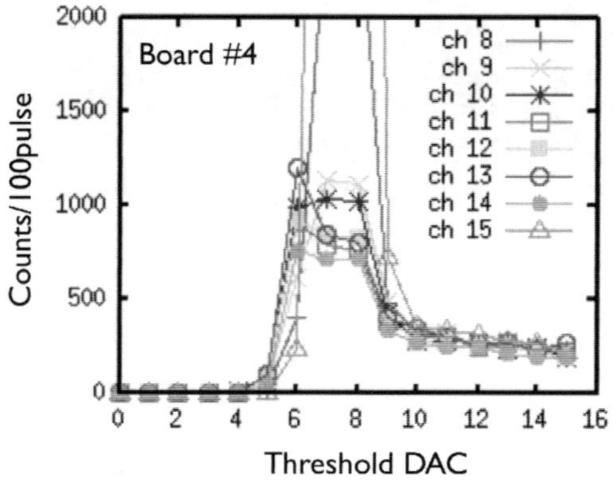


**Figure 7.** An example of efficiency curve, which is a count of each counters for a preset number of accelerator pulses. The condition of measurement is HV=67.92V (over-voltage of +1.1-1.3V), AmpDAC=3B. BiasDAC is not used.

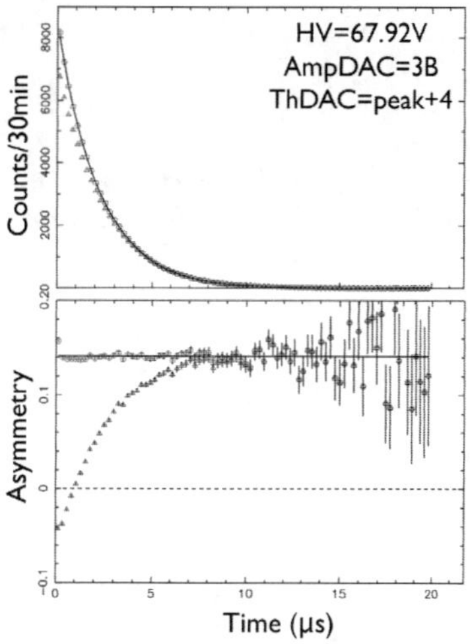


**Figure 8.** $\mu$e-decay spectrum (top panel) and asymmetry (bottom panel) of up-stream counter. Symbols are for before (triangle) and after (circle) the pile-up correction.

After determination of ThDAC for a set of a selective combinations of AmpDACs, we take the time spectrum of $\mu$e decay (Fig.8) and tried to find the appropriate combination of DACs and HV to give the minimum distortion. Since we receive 660k positron single hits integrated for 120 accelerator pulses, each counter receives ~4 positron events (660k/120/1280=4.29) per pulse in the average. This high rate of decay positron signals result to a pile-up loss of counts in early times. We employ a theory of pile-up correction [5], and found that the simple non-extended correction scheme (eq.(2) of Ref. [5]) is practically enough to restore the $\mu$e-decay with the detector deadtime $\tau$ as an adjustable parameter. The characteristic detector deadtime is estimated to be $\tau$ ~100 ns for single counters. We phenomenologically uses the same pile-up correction scheme for coincidenced counts which results to a longer deadtime of $\tau$ ~300 ns; this is because the registered coincidence counts is reduced to ~1/3 of the single counts, whereas the countloss occurs in accordance to the single hit rate. The deadtime could be improved eventually in electronics, because the recovery tail of the m-APD signal is typically ~40 ns.

In table 1, we compare specification of existing spectrometers in pulsed muon facilities. The new spectrometer has a similar solid angle and longitudinal-field capability with Argus at RIKEN-RAL, but the count rate is more than 4 times higher, thanks to the high muon intensity of J-PARC MUSE.

Journal of Physics: Conference Series **551** (2014) 012063 doi:10.1088/1742-6596/551/1/012063

**Table 1.** Specifications of general-purpose spectrometers in pulsed-muon facilities.

| Spec. | $D\Omega$-1 (J-PARC) | New D1 (J-PARC) | Argus (RIKEN-RAL) |
|---|---|---|---|
| maximum longitudinal field (T) | 0.15 | 0.4 | 0.4 |
| solid angle (%) | 8 | 23 | 25 |
| counter channels (pair) | 256 (128pair) | 1280 (640pair) | 192 |
| data acquisition rate (Mhits/hour)* | 40 | 180 | ~40 |

* Coincidenced counts of pair counters (if applicable) for 20x20 mm sample, double pulse.

In Fig.9, we show forward-backward asymmetry of a standard sample (silver), after adjusting HV, DACs and deadtime parameters in the new D1 spectrometer. The Full asymmetry ~0.16 is significantly smaller than that of other general-purpose $\mu$SR spectrometers (0.2-0.25), because the positron counters are placed in high-angle (51-60°: see Fig.6) relative to the beam axis, in order to maximizing the detector solid angle while accommodating various kinds of inserts. Since the muon beam intensity is high enough at MUSE, we plan to re-position the positron counters to gain a higher asymmetry.

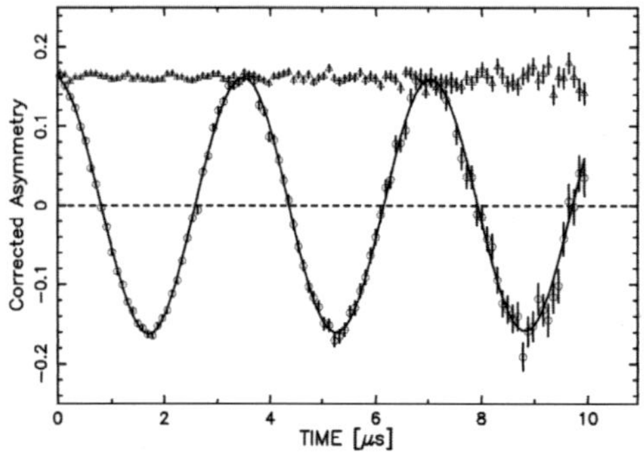


**Figure 9.** Forward-backward asymmetry of silver 20×20 mm sample in transverse-field of 2 mT and longitudinal-field of 10 mT. The data were taken in 10 minutes with kicker sliced single pulse operation.

## 4. Summary

We have developed a new positron detector called Kalliope, built and installed a general purpose $\mu$SR spectrometer at D1 experimental area of J-PARC MUSE. The spectrometer has been employed for user programs of MUSE since February of 2014. The data acquisition rate of the new spectrometer is 4 times higher than that of $D\Omega$-1 spectrometer of J-PARC and/or Argus spectrometer of RIKEN-RAL.

## References

[1] Kadono R and Miyake Y 2012 *Rep. Prog. Phys.* **75** 026302
[2] Tanaka H K M *et al.* 2005 *Nucl. Inst. Meth. A* **554** 201
[3] Tanaka M *et al.* 2014 *preprint*
[4] Uchida T 2008 *IEEE Trans. on Nucl. Sci.* **55** 1631
[5] Ida T and Iwata Y 2005 *J. Appl. Cryst.* **38** 426

Journal of Physics: Conference Series **551** (2014) 012064 doi:10.1088/1742-6596/551/1/012064

# Cosmic-ray muon spin rotation in Fe and industrial application

**K Nagamine[1,2,3*], T Fujimaki[4], T Hashimoto[4,5], M Tsukamoto[4], S Kubota[4], T Hirai[4,6], A Manabe[4], Y Tomisawa[4,7], A D Pant[4] and E Torikai[4]**

[1] Atomic Physics Laboratory, RIKEN, 2-1 Hirosawa, Wako, Saitama 351-0198, Japan
[2] Department of Physics and Astronomy, University of California, Riverside, USA
[3] Muon Science Laboratory, IMSS. KEK, Oho, Tsukuba, Ibaraki, Japan
[4] Interdisciplinary Graduate School of Medicine and Engineering, University of Yamanashi, 4-3-11 Takeda, Kofu, 400-8511, Japan

E-mail: `k.nagamine@riken.jp`

**Abstract.** Spin polarized positive muons contained in the cosmic-rays were stopped in the Fe plates providing a characteristic spin rotation signal of decay positrons. This signal along with the decay lifetime of the negative muons can be used as a non-invasive radiographic measurement method for a characterization of the inner structure of the aged architectures. Principle, results of test experiments and future prospects are described.

## 1. Introduction

After a history of more than 500 years of the modern life, most of the large-scale architectures such as buildings, bridges, highway roads, dams, etc. are coming to the end of their lives. There, usual happenings inside these architectures are corrosions of the iron rods in the reinforced concretes. However, it is not easy to examine the status of these iron rods non-invasively, namely, without breaking the architectures.

In order to inspect the inner structure of the large scale architectures, a use of cosmic-ray muon is quite effective. In fact, by measuring the positional distribution of the intensity attenuation of the penetrating cosmic-rays, the inner-structure of blast furnaces have been studied [1], based upon the experiences of radiography of the volcanic mountains [2, 3]. Recently, in order to explore the inner status of the damaged Fukushima nuclear reactors, use of this cosmic-ray muon transmission radiography was proposed [4, 5] along with the multiple scattering method [6].

Here, we propose a different and new use of the cosmic-ray muons to explore near-surface (20 cm thick from the surface) structure of the architecture by detecting the $\mu$SR signal of decay positrons and the lifetime of decay electrons from the cosmic-ray muons stopped inside the objectives.

[5] Present address: Yazaki Cooperation
[6] Present address: Yamato Scientific Co., Ltd.
[7] Present address: Kowa Pharmaceutical Co., Ltd.

## 2. Principle

### 2.1. *Nature of cosmic-ray muons*

Cosmic-ray muons at the sea level of the earth are known to have the following properties; a characteristic Zenith-angle dependent energy spectrum with common natures of high energy (GeV to TeV) and low intensity ($1/(cm)^2/min$ in vertical direction) [7], a mixed charge state ($\mu^+$ (0.6) and $\mu^-$ (0.4)) and a partial spin polarization (-0.33 for $\mu^+$ [8]). Because of the high energy nature, a stopping distribution of the cosmic-ray muon in concrete extends from cm to more than 100 m in depth.

### 2.2. *The present method: reflection radiography of stopping cosmic-ray muons*

Thus, by detecting a signal of the decay positrons and decay electrons whose energies are up to 50 MeV from various selected position of the stopping muons inside the objective, one can obtain a radiographic imaging information such as a spatial distribution of microscopic magnetism from the $\mu^+$SR signal and an element distribution from the amplitude distributions of components with different $\mu^-$ free-decay lifetime [9]. As for a classification of the radiography method, we can consider this signal as a reflection signal which is a signal from the objective in the direction opposite to the direction of the injected probing radiation.

The signal N($t$) of time dependent intensity change of decay positrons and decay electrons with reference to the time of stopping of the incoming cosmic-ray muons detected for an area of each stopping position of the cosmic-ray muons can be written as follows. Here, the stopping position can be obtained by the direction of the incoming cosmic-ray muons determined by a telescope of more than two arrays of the segmented counters [3].

$$N(t) = N_0 \left[ \alpha\ \mathrm{e}^{-t/\tau(0)} \{1 + \sum_i (A_i \mathrm{e}^{-\lambda(i)t} \cos(\omega_i t + \phi_i)\} + (1-\alpha) \sum_j P_j\ \mathrm{e}^{-t/\tau(j)} + C \right], \quad (1)$$

where $\alpha$: a fraction of positive muon in the cosmic-ray, $\tau(0)$: $\mu^+$ lifetime, $A_i$: decay positron asymmetry parameter in the ith material including $\mu^+$ polarization, $\omega_i$: $\mu^+$ spin precession due to internal field of the $i$th material, $\lambda(i)$: relaxation rate of the $\mu^+$ spin precession due to the $i$th internal field, $P_j$: fraction of the $\mu^-$ stopping in the $j$th material and $\tau(j)$: $\mu^-$ lifetime in the $j$th material, respectively.

## 3. Test experiments

As the first step, time-spectrum of the decay positrons and electrons from the cosmic-ray muons stopping in Fe plate of regular 99% purity Fe plate (2 cm thick) and in concrete plate of commercial use (6 cm thick) were measured separately as shown in Fig. 1, where each counter is a plastic scintillator with a size of 10 cm width 100 cm long and 3 cm thick. The results are presented in Fig. 2 and 3. There, the data for the FWD (BWD) direction where decay electrons/positrons were detected along (opposite to) the direction of incoming vertical cosmic-ray muons were presented.

In the present set-up, the following weak points existed. Because of a relatively slow fall off time of the output signals of the plastic scintillator, the BWD event which needs anti-coincidence of the incoming muon, effective dead-time existed up to 150 ns after muon stop. Because of this weak point, the $\mu^+$SR data was taken only by the FWD data. Also, stopping events in D1~D6 counters are included in the stopped muon events in the data for either Fe or concrete.

In Fe, $\mu^-$ takes strong nuclear capture resulting free-decay lifetime of 200 ns [9], while polarized $\mu^+$ is expected to feel the internal field of ferromagnetic metal causing spin precession of 50 MHz [10]. On the other hand, in concrete whose major elements are Si (~30%) and O (~50%), $\mu^-$ takes weak nuclear capture resulting free-decay lifetime of 760 ns for Si and 1830 ns for O [9], while polarized $\mu^+$ is expected to feel no internal field causing no spin precession.

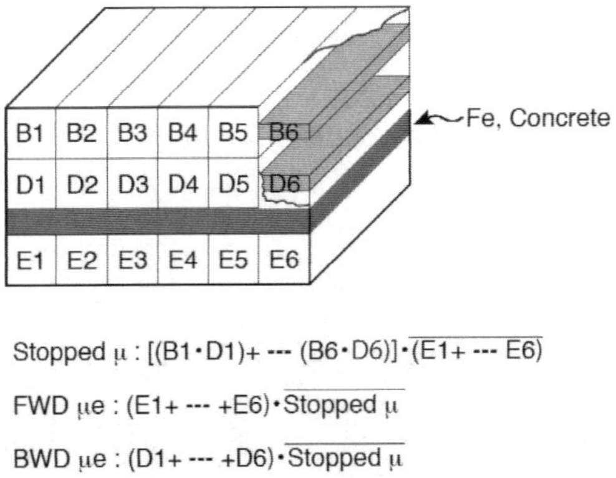


**Figure 1.** Counter geometry and logics of decay electron/positron detection from stopped muons in the test experiment. Inside each counter box, a plastic scintillator with the size of 3 cm thick, 10 cm wide and 100 cm long is located.

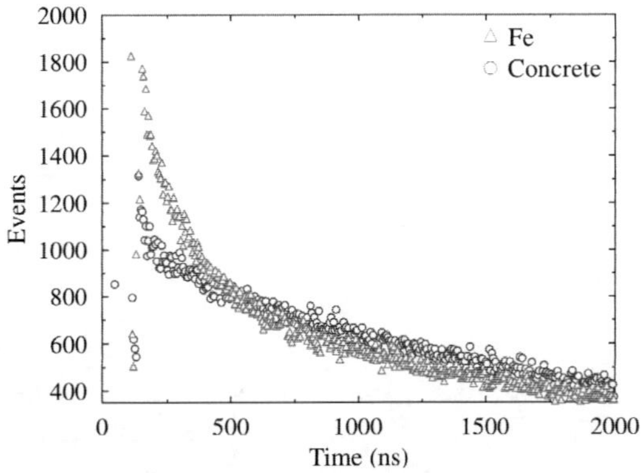


**Figure 2.** The time spectrum in BWD geometry from stopped cosmic-ray muons in concrete and Fe observed in the test experiment. The 200 ns lifetime component from negative muons stopped in Fe is clearly seen.

Fitting analysis for Eq. (1) were made by assuming $\tau(0)$=2.2 $\mu$s and $\tau(Fe)$ =0.2 $\mu$s [9], yielding the values listed in Table 1. Thus, $\mu$SR in Fe with a precession frequency of around 50 MHz [10] was obtained as expected. The obtained values of $A_{Fe}$ and $B_{Fe}$ ($\mu^+$ internal magnetic field in Fe converted from $\omega_{Fe}$ ) obtained from $\omega_{Fe}$ are also consistent as expected; 1/3 (reduction by asymmetry average) $\times$ 0.6 (charge fraction)$\times$ 0.33 ($\mu^+$ polarization) = 0.066 and 3595 Oe obtained in the accelerator muon experiment [10], respectively. On the other hand, the measured relaxation rate of the $\mu^+$ in usual iron plates looks faster than that of the $\mu^+$ in pure Fe at room temperature observed in accelerator muon experiment (0.80 $\mu s^{-1}$) [10], suggesting the effect of corrosion in usual iron plates.

At the same time, as seen in Fig. 2, earlier part of the time spectrum corresponding to the $\mu^-$ free-decay from Fe with the lifetime of 200 ns was correctly detected. The consistent data was also obtained for concrete.

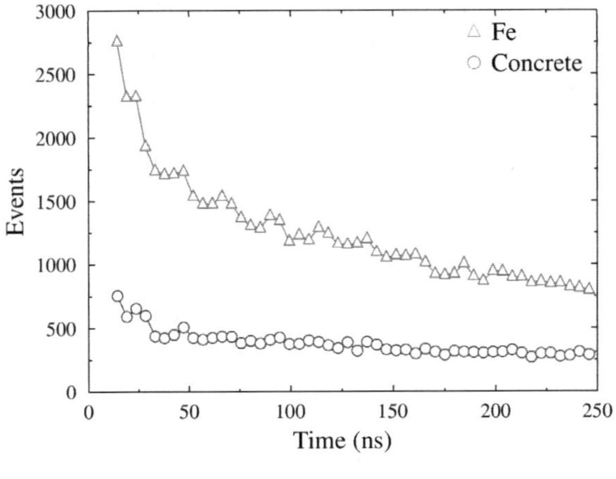


**Figure 3.** Observed ZF spin rotation time spectrum from cosmic-ray muons stopped in Fe (46 days of measurement time) and concrete (20 days). Data in the FWD geometry are shown. Existence of 50 MHz positive muon spin rotation signal in Fe is obvious.

**Table 1.** Summary of the cosmic-ray $\mu^+$ spin rotation in Fe.

| $A_{Fe}$(%) | $B_{Fe}$(Oe) | $\lambda(i)(\mu s^{-1})$ |
|---|---|---|
| 5.7(2.2) | 3390(130) | 25(13) |

## 4. Some considerations

### *4.1. Next steps toward the goal of the present method*

The goal of the present new reflection radiography method by detection of the decay $e^+/e^-$ from stopped cosmic-ray muons can be summarized as follows,

1. Element analysis by the $\mu^-$ lifetime method; mapping of the profile of the Fe rods without any destruction.

2. Material analysis by $\mu^+$SR method; by measuring spin relaxation rate of the $\mu^+$, one can learn the corrosion status in comparison with the systematic reference data to be taken at the accelerator facilities.

3. Monitoring 20 cm depth in concrete with a spatial resolution of close to $(1\,\text{cm})^2$

Following success of the present test experiment, we are considering to take the following next steps. (1) Imaging of the reinforced concrete to detect the Fe rod status; by placing two telescope counters for muon tracking in front of the decay $e^+$ and $e^-$ detector (see Fig. 4), where all the signals of the decay $e^+/e^-$ will be obtained with reference to each muon stopping position; (2) systematic detection of corrosion status of the Fe rod through a change of the $\mu^+$SR data, combined with the planned studies at the accelerator facilities of the continuous muons as described later. (3) design optimization of the imaging device for the re-enforced architectures will be conducted. Then direct monitoring experiment of the Fe rods status will be started for the big architectures like buildings, highways, dams, etc.

### *4.2. Ambiguities in stopping position determination*

In the present method, stopping position of the cosmic-ray muons is considered to be determined by the trajectory determination just in front of the objective. A lateral and longitudinal spread ($D_\perp$, $D_\parallel$) of the parallel pencil beam after the range $R_0$ of stopping can be estimated following the phenomenological formula [11]; $D_\perp = 7\times10^{-2}\times R_0^{0.92}$, $D_\parallel = 2.6\times10^{-2}\times R_0^{0.94}$. Thus, ambiguity in determining the stopping position at 20 cm depth corresponds to the range in concrete of 50 MeV decay $e^+/e^-$ is around 1 cm.

### *4.3. Expected role of the $\mu^+$SR data of the imbedded Fe*

The most important feature of the $\mu^+$SR in Fe is the fact that one can probe a microscopic molecular level characteristic from outside. By stopping cosmic-ray muons in reinforced concrete, this probing can be done up to the depth of 20 cm from where the decay $e^+/e^-$ can be coming out to the E-counter placed as shown in Fig. 4.

On the other hand, corrosion of the steel in reinforced concrete (RC) is the central problem causing a serious damage in big architectures. The alkalinity of the concrete usually protects the imbedded Fe from corrosion. However, during a long time use in 100 years or longer, various electrochemical reactions proceed between iron and dissolved ions in the vicinity of the iron to create corrosion [12]. This corrosion causes a damage of the reinforced concrete in the form of expansion, cracking and destruction of the covered concrete. It is expected that magnitude/dynamics of the microscopic internal magnetic fields at the interstitial sites of Fe under corrosion is significantly different from those of pure iron crystal.

Journal of Physics: Conference Series **551** (2014) 012064 doi:10.1088/1742-6596/551/1/012064

In order to establish the reference data, the $\mu^+$SR experiment will be conducted at accelerator facility of the continuous muon beam for the following Fe materials; Fe rod extracted from fresh RC, that from expanded RC and that from destroyed RC. Although the relevant experimental data do not exist at all, considering from theoretical understanding of the $\mu^+$ hyperfine fields in Fe [13], a significant difference might be expected for these different stages of RC.

### *4.4. Feasibility considerations*

In order to confirm the feasibility, let us consider the necessary measurement time length (years) for the present method to monitor the inside of the actual building with reference to the test experiment mentioned above.

First, the number of the $\mu^+$SR events and the decay $\mu^-$ to $e^-$ events from the cosmic-ray muons stopped in the sample can be written as follows,

$$N_{\mu e,\mu SR} = N_{stopped\,\mu} \times \Delta\Omega_{\mu e,\mu SR} \times \delta t = [I_{stopped\,CR\mu}(\theta_z, \Delta\Omega, \Delta R) \times \delta A] \times \Delta\Omega_{\mu e,\mu SR} \times \delta t \quad (2)$$

where $\theta_z, \Delta\Omega$ and $\Delta R$ are cosmic-ray Zenith-angle, solid angle and stopping range, respectively.

In the test experiment, the parameters in Eq. (2) can be written as follows,

$[I_{stopped\,CR\mu}](0°, 5/6 \times 2\pi, 5\,\text{cm Fe})$, $\delta A = 60$ cm $\times$ 100 cm, $\Delta\Omega_{\mu e,\mu SR} = (1/2)2\pi$, $\delta t = 9$ days

In the case of an inspection of the re-enforced concrete building by employing the detection system as seen in Fig. 4, it is intended to identify the stopping position of the cosmic-ray muons in the unit of 3 cm$\times$3 cm by employing tracking detector with the resolution accordingly. There, the parameters of Eq. (2) become as follows,

$I_{stopped\,CR\mu}(80°, 0.45\times 2\pi, 20\,cm\ ReEnf.Concrete) \cong 0.27\times I_{stopped\,CR\mu}(0°, 5/6\times 2\pi, 5\,cm\ Fe)$, $\delta A = 3$ cm $\times$ 3 cm, $\Delta\Omega_{\mu e,\mu SR} = (1/2)2\pi$,

$\delta t = 9$ days $\times (1/0.27) \times (60 \times 100/(3 \times 3)) = 61$ years for the same $N_{\mu e,\mu SR}$ of obtained in the test experiment.

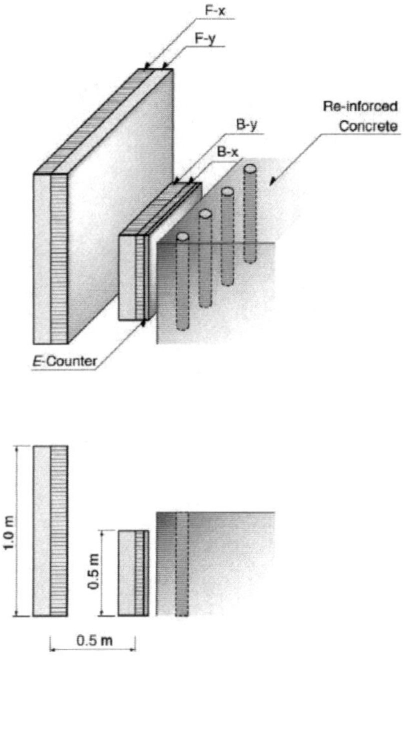


**Figure 4.** Proposed set-up for the reflection radiography using decay electron/positron from the stopped cosmic-ray muon. Muon beam tracking hodoscope is placed in front of decay $e^+/e^-$ to determine muon stopping position.

Journal of Physics: Conference Series **551** (2014) 012064 doi:10.1088/1742-6596/551/1/012064

There are several ideas to increase the event rates and decreasesubstantially the measurement time from 61 years. The followings are the present suggestions.

(1) Increase the sensitive area of the whole detection system by a factor of 4.

(2) Data rearrangement by summing up similar Fe region to inspect average chemical status of the Fe rods ($\times 10$).

Thus, required measurement time $\delta t$ becomes 1.5 year. Obviously, once the application becomes to be proved as a feasible method, industrial application will be started, promoting a production of cheap and larger-scale detection systems. This development scenario should be started within a near-future.

## 5. Conclusion

As presented here, based upon the results of test experiments, feasibility considerations and combined use of well-established transmission radiography instrumentation, the cosmic-ray muons are ready to be used to explore inner structure of the reinforced concrete architectures. A large-scale industrial application will soon be opened.

## References

[1] Nagamine K *et al.* 2005 *Proc. Jpn. Acad. Sci.* B **81** 257–260
[2] Nagamine K *et al.* 1995 *Nuclear Instruments and Methods* A **356** 585
[3] Nagamine K 2003 *Introductory Muon Science* (Cambridge: Cambridge University Press) pp 1–208
[4] Nagamine K 2011 *Urgent Proposal to KEK Director General*
[5] Fujii H *et al.* 2013 *Prog. Theor. Exp. Phys.* 073C01
[6] Morris C *et al.* 2012 *AIP Advances* **2** 042128
[7] Adair R K and Kasha H 1976 *Muon Physics* vol 1 ed V W Hughes and C S Wu (New York: Academic Press) p 323
[8] Turner R *et al.* 1971 *Phys. Rev.* D **4** 17
[9] Measday D F *et al.* 2001 *Phys. Reports* **354** 243
[10] Nishida N *et al.* 1977 *Solid State Comm.* **22** 235
[11] Fowler P H *et al.* 1961 *Nature* **189** 524
[12] Verbeck G J *et al.* 1975 *American Concrete Institute Special Publication* **49** 21
[13] Kanamori J *et al.* 1981 *Hyperfine Interactions* **8** 573

Journal of Physics: Conference Series **551** (2014) 012065 doi:10.1088/1742-6596/551/1/012065

# Design and construction of the ultra-slow muon beamline at J-PARC/MUSE

**P Strasser[1,2], Y Ikedo[1,2], S Makimura[1,2], J Nakamura[1,2], K Nishiyama[1,2], K Shimomura[1,2], H Fujimori[1,2], T Adachi[1,2], A Koda[1,2], N Kawamura[1,2], Y Kobayashi[1,2], W Higemoto[3,2], TU Ito[3,2], T Nagatomo[4], E Torikai[5], R Kadono[1,2] and Y Miyake[1,2]**

[1] Muon Science Laboratory, Institute of Materials Structure Science, High Energy Accelerator Research Organization (KEK), 1-1 Oho, Tsukuba, Ibaraki 305-0801, Japan
[2] Muon Science Section, Materials and Life Science Division, J-PARC Center, 2-4 Shirane Shirakata, Tokai-mura, Naka-gun, Ibaraki 319-1195, Japan
[3] Advanced Science Research Center, Japan Atomic Energy Agency (JAEA), 2-4 Shirane Shirakata, Tokai-mura, Naka-gun, Ibaraki 319-1195, Japan
[4] Ion Source Team, Accelerator Group, RIKEN Nishina Center, RIKEN, Hirosawa-2-1, Wako, Saitama 351-0198, Japan
[5] Interdisciplinary Graduate School of Medicine and Engineering, University of Yamanashi, Takeda 4-3-11, Kofu, Yamanashi 400-8511, Japan

E-mail: `patrick.strasser@kek.jp`

**Abstract.** At the J-PARC Muon Science Facility (MUSE), a new Ultra-Slow Muon beamline is being constructed to extend the $\mu$SR technique from bulk material to thin films, thus empowering a wide variety of surface and nano-science studies, and also a novel 3D imaging with the "ultra-slow muon microscope". Ultra-slow muons will be produced by the re-acceleration of thermal muons regenerated by the laser resonant ionization of muonium atoms evaporated from a hot tungsten foil, a method that originated from the Meson Science Laboratory at KEK. The design parameters, construction status and initial beam commissioning are reported.

## 1. Introduction

The Muon Science Facility (MUSE) [1] is now under construction at J-PARC (Japan Proton Accelerator Research Complex) in the Materials and Life Science Facility (MLF) building. A new Ultra-Slow Muon beamline is being developed with the goal to expand the scope of the $\mu$SR technique from bulk material to thin films, multi-layers, surfaces and extremely small samples, enabling not only for a wide variety of surface and nano-science studies, but also a novel 3D imaging with the "ultra-slow muon microscope" [2].

Ultra-slow muons have been successfully produced at PSI using the technique of moderating a continuous surface muon beam in simple van de Waals bound solids to produce ultra-slow muons with epithermal energies down to 15 eV without any loss of the spin polarization [3]. This new beamline however is based on an alternative concept that was delevopped at the ultra-slow muon beamlines built at KEK in 1990 [4] and RIKEN-RAL in 1999 [5]. Intense ultra-slow muons will be produced by the re-acceleration of thermal muons regenerated by the laser resonant ionization of muonium atoms evaporated from the surface of a hot tungsten foil.

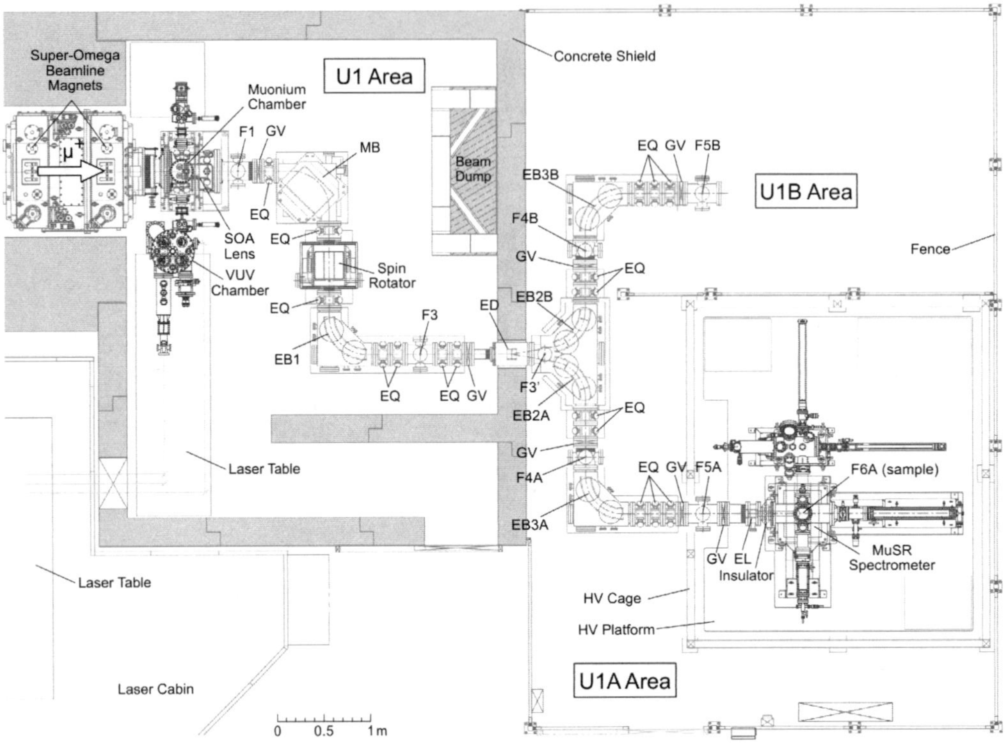


**Figure 1.** Layout of the new ultra-slow muon beamline at J-PARC/MUSE.

To efficiently ionize muonium atoms, a resonant ionization scheme via the "1s-2p-unbound" transition by a pulsed nanosecond laser was adopted. A new state of the art all solid-state laser system produces the VUV light of 122.088 nm (Lyman-alpha) needed to induce the 1s-2p transitions [6]. After laser ionization, the free muons have a mean thermal kinetic energy of only 0.2 eV with 50% spin polarization. They are extracted, re-accelerated up to 30 keV and focussed by an electrostatic lens. This re-accelerated beam has better energy resolution, time resolution and space distribution compared to the initial surface muon beam. The beam is then transported with a series of electric quadrupoles and electric bends to two experimental areas U1A (Ultra-Slow Muon $\mu$SR) and U1B (Re-acceleration, Microbeam $\mu$SR), respectively. The layout of this new ultra-slow muon beamline is shown in Fig. 1.

The now completed Super-Omega muon beamline (U-Line) in the MLF experimental hall No. 2 will be used as the primary muon source [7]. Muons up to 45 MeV/c can be extracted with a large acceptance solid angle of 400 mSr upstream from the muon target with an angle of 45 degrees by a normal-conducting capture solenoid. Captured muons are then transported to the experimental hall by a superconducting curved transport solenoid and an axial focusing solenoid to the experimental area U1 (see Fig. 1 of Ref. [8] for a layout of the Super-Omega muon beamline). The muon beam has a double-pulsed structure of 100 ns wide, separated by 600 ns with a repetition rate of 25 Hz. We achieved the world most intense pulsed muon source at the U-Line with a time-averaged surface muon intensity of 6.4 $\times$ $10^7$/s at 212 kW proton beam, 20 times more intense than at the D-Line, which is already in operation since 2009, and corresponding in the future to 3.0 $\times$ $10^8$/s with a beam power of 1MW.

The expected ultra-slow muon intensity is obtained by scaling the production rate from the R&D at RIKEN-RAL [5] where 20 $\mu^+$/s were produced with an initial surface muon beam of $1.2\times10^6$/s. Comparing this intensity to the one that will be available at the U-Line, we gain a factor of 250. The new laser system was designed to achieve roughly 100 times the power of the previous laser [6]. And a factor of two is gained because the laser and the muon beam are synchronized to 25 Hz, whereas at RIKEN-RAL the laser was synchronized to every second muon pulse (50 Hz). Consequently, we can expect a maximum ultra-slow muon production rate of $1.0\times10^6$ $\mu^+$/s, and about 50% at the sample position due to muon decay in-flight.

An overview of the different aspects of this new ultra-slow muon beamline and the present status of the beam tuning and commissioning are presented.

## 2. Thermal muonium production chamber

The thermal muonium production chamber shown in Fig. 2 contains the hot tungsten target (W target) that will be used to produce muonium atoms. Thermal muons will be regenerated by the laser resonant ionization of the muonium atoms evaporated from its surface. Surface muons enter the chamber through a 50-$\mu$m thick stainless steel window placed at the end of a capillary device (conical shape) that enhances slightly the muon stopping distribution on the target. The muons are first moderated by a 50-$\mu$m thick W foil (moderator) and then stopped in a 50-$\mu$m thick W foil (target) that is heated up by a pulsed DC current. The moderator foil is also used to shield the stainless steel window from the heat of the target. The W foil dimensions (70-mm wide by 45-mm high) are determined by the beam parameters of the Super-Omega muon beamline at the focussing point to maximize the muon stopping distribution, while the total foil thickness is optimized to stop muons near the rear surface of the hot W target to maximize the muonium emission in vacuum. The W target was designed to be heated up to ~2300 K in an ultra-high vacuum of $10^{-7}$ Pa, and have $O_2$ flow control for surface treatment of carbon removal. A pulsed interrupt heater power supply provides a DC current of up 1000 A (15 V max.) to heat the W target. It is turned off for 1 ms around the muon pulse arrival time to eliminate induced magnetic field from the target, and is also electrically insulated to apply 30 kV to the target for beam extraction. The entrance and exit flanges as well as the wall of the the muonium chamber are water cooled to remove the radiation heat from the target. A vacuum of $8.7\times10^{-8}$ Pa with the W target at room temperature and a degassing rate of $3.2\times10^{-7}$ Pa m$^3$/s were achieved.

A highly pure tungsten foil is essential because a clean tungsten surface is required to obtain intense muonium emission. The foil dimensions required for mounting on the target holder are 70 mm by 54 mm and 50-$\mu$m thick. Unfortunately, the companies, which had supplied highly pure (6 N, 99.9999%) tungsten foil in the past, discontinued their manufacturing. It used to be manufactured by a hot rolling technique of highly pure tungsten bulk material. Therefore the development of manufacturing methods was started instead of fabricating a new and expensive hot rolling apparatus. The feasibility was determined by utilizing 0.2-mm thick plates made of ordinarily pure tungsten material (3 N, 99.95%). The thickness was reduced up to 50 $\mu$m by two methods, etching and mechanical polishing, respectively. As a result, uniform 50-$\mu$m thick foils could be successfully obtained by the mechanical polishing method. Then mechanical polishing was introduced to highly pure tungsten bulk. The development of the manufacturing method of highly pure tungsten foil is described in details in Ref. [9].

The extraction and focusing optics for the ionized muons consists a SOA immersion lens [10] similar to that of the previous ultra-slow muon beamlines [4, 5]. All electrical electrodes can withstand a high-voltage of up to 30 kV. The ionized muons are initially extracted by a low electrical field gradient of typically 200 V between the W target held at 30 kV and the mesh electrode S1 held at 29.8 kV (mesh transmission efficiency of 90%). The gap between the target and S1 is 14 mm. It should be noted that a 1-mm thick laser ionization volume in front of the target will give to the muon beam an additional energy spread of 14 eV and a corresponding

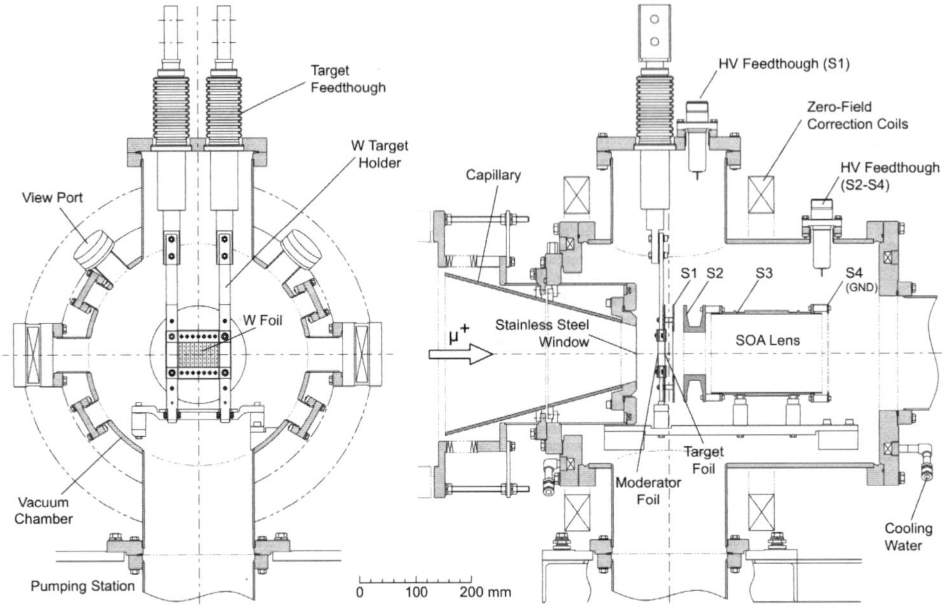


**Figure 2.** Schematic view of the thermal muonium production chamber with the hot W target.

time spread of 1.9 ns. Lower field gradient will reduce the energy spread but increase the time spread, and vice versa. A possible workaround is using a short HV pulse between the target and S1 to accelerate muons before they reach S1 so that all muons have the same initial extraction energy when passing the mesh. A typical 10-ns long pulse of 2 kV would give almost no energy spread while limiting the time spread to 1.0 ns. The required accuracy of the pulse should be $< 10^{-3}$. The mechanical alignment of the target, laser and S1 is also important.

## 3. Stray field from Super-Omega muon beamline

The W target for the production of ultra-slow muons is located 700-mm downstream from the last solenoid of the Super-Omega muon beamline. The residual magnetic field at the W target position has a critical effect to the muon beam properties, such as extracting and focusing the beam. In addition, the magnetic field normal to the beam direction reduces seriously the muon spin polarization. To minimize these effects, a magnetic shield was installed to enclose almost entirely the muonium production chamber. The magnetic shield is made of 20-mm thick iron plates, three plates normal to the beam axis, and two plates on the sides. The first plate (1.2-m wide and 2.2-m high) made of ordinary iron is placed right downstream of the last magnet of the axial focusing solenoid. The other plates (1.0-m wide and 2.4-m high) made of extra-low carbon steel are located 250 mm away from the first plate, 150 mm further downstream of the second one, and 500 mm from the beam axis on both sides, respectively. Each plate normal to the beam has a hole for the beamline duct with a diameter of 500, 400, and 300 mm, respectively.

The code OPERA-3D (TOSCA) was used for the magnetic field analysis. Figure 3 shows a top view of the magnetic shield around the W target illustrating the last two solenoid magnets of the Super-Omega muon beamline and the two zero-field correction coils. The muon beam is coming from left, and the W target is located at the center position (Z=0) in between the two zero-field correction coils. Initially, the calculated stray field at the W target position is 240 G, but can be reduced down to ~40 G with the magnetic shield. This residual stray field

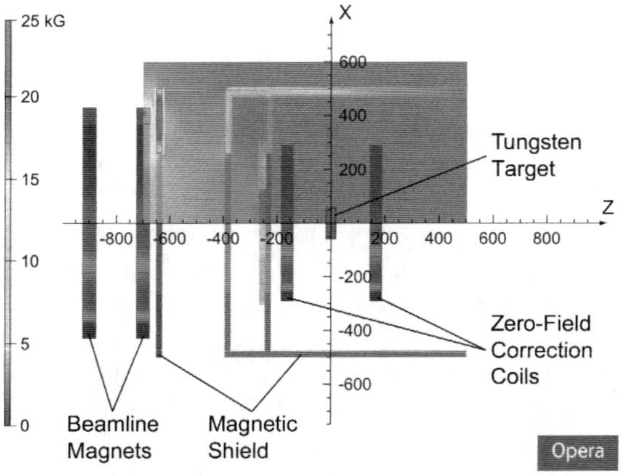


**Figure 3.** Opera-3D model of the magnetic shield showing the last beamline solenoid magnets and the zero-field correction coils.

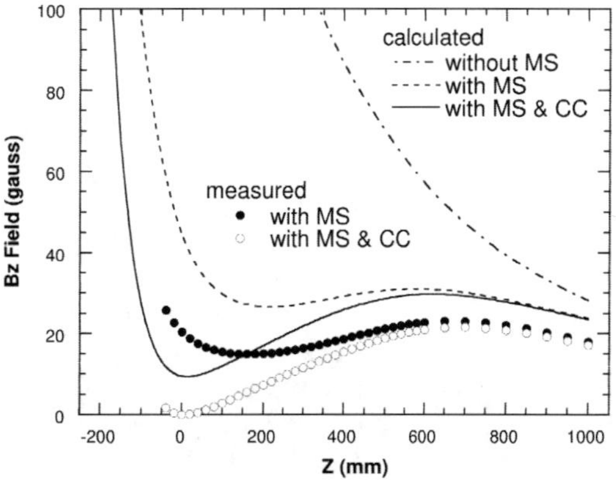


**Figure 4.** Magnetic field distribution results without and with magnetic shield (MS) and zero-field correction coils (CC), respectively.

originates mainly from the large hole in the magnetic shield for the beamline duct. The field can be further canceled by using three orthogonal zero-field correction coils placed on the W target chamber. Measurements were performed after the installation of the W target chamber and the magnetic shield using a three-dimensional Hall probe mounted on a XYZ positioning stage. The measured and calculated stray fields along the beam axis are shown in Fig. 4. Although an offset exists between measured and calculated fields, the data are in good agreement. The offset originates probably from the magnetization effect of the magnetic shield. The residual stray field was suppressed by the zero-field correction coils to less than 1.5 G over the W target region. Then the field gradually increases with increasing Z, showing a peak structure of 20 G, and then decreases again. The major field component is along the beam axis, in the same direction as the muon spin, and will therefore not affect the spin depolarization of the muons. The normal component is sufficiently small ($< 0.2$ G) to affect any changes to the muon spin.

## 4. Beam Transport Line

The complete layout of the new ultra-slow muon beamline is shown in Fig. 1. After laser ionization, the thermal muons are extracted, re-accelerated and focussed to an intermediate focusing point, F1, located 600 mm from the W target. From there, the beam is transported with a series of electric quadrupoles (EQ) and electric bends (EB) to two experimental areas U1A and U1B, respectively. A magnetic bend (MB) placed just after F1 carries out the mass separation of the beam to eliminate positrons and also select $^2H^+$ and $^7Li^+$ ions for off-line beam tuning. The C-shaped magnetic bend allows the direct monitoring of the hot W target through a viewport located at the end of the magnet vacuum chamber. A 90 degrees magnetic bend was preferred so that the muon spin rotates perpendicular to muon beam direction, allowing $\mu$SR measurements with transverse field similarly to RIKEN-RAL and PSI low energy muon beamlines. A dedicated spin rotator will soon be installed in between MB and EB1 to rotate back the muon spin for longitudinal field measurements.

The general shape of the beamline was decided based on the space available in the MLF experimental Hall No. 2 and to satisfy the requirement for two experimental areas for "Ultra-Slow Muon $\mu$SR" and "Reacceleration/Microbeam $\mu$SR", respectively, with sufficient distance in between for experimental instruments like a spectrometer, a high-voltage platform, etc. The ultra-slow muon beam is sent to the desired experimental area by an electric deflector (ED) that deflects the beam by 10 degrees and directs it to one of the entrance of the double electric bend (EB2). At first, the operation of the ED will be static directing the full beam (25 Hz) either to U1A or U1B. In the future, a pulsed operation is thought to allow every other pulse to be sent to both areas with a repetition rate of 12.5 Hz. Five beam monitor and slit assemblies are located along the beamline (F1–F5) for monitoring and tuning purposes. Both horizontal slit mechanisms (left and right) are remotely activated from the same side, while the vertical slits (top and bottom) are both mounted from below. This compact arrangement leaves space for a beam monitor transfer mechanism to be placed at the top of the vacuum chamber. Presently, either a standard MCP detector (micro-channel plate, $\phi$40 mm, single anode) or a two-dimensional MCP detector (with delay-line anode for high resolution 2D-imaging) can be mounted. The detector is lowered behind the slits that can be used to evaluate the beam profile with a single-anode MCP. The total beamline length becomes around 8.5 meters from the hot W target to F5. The ultra-slow muon beam energy of 30 keV was decided to minimize the beam loss due to muon decay in-flight. At 30 keV, the time-of-flight is approximately 1.3 $\mu$s, and about 45% of the ultra-slow muons will decay before reaching the sample position.

At present, only the construction of U1A area is almost completed. A new $\mu$SR spectrometer has been constructed and installed on a high-voltage platform ($\pm$ 30 kV) allowing the muon beam to be decelerated down to a few tens of eV on the sample, or further accelerated up to 60 keV. Because of the residual magnetic field from the $\mu$SR spectrometer along the beam axis, an Einzel lens (EL) was preferred to mirror the beam image from F5A to the sample position (F6A). This Einzel lens is also used to adjust the focussing position when changing the high-voltage at the sample. This new ultra-slow muon $\mu$SR apparatus is described in details in Ref. [11].

The optimization of the initial beam extraction and acceleration with the SOA lens was performed using a Monte Carlo simulation with calculated electric field by the code RELAX3D. A source of 4 cm in diameter can be focused at F1 with a FWHM of 1 mm at 30 keV. A distance of 600 mm to the intermediate focusing point F1 was decided to minimize the beam aperture to 30 mrad for easier beam transport, while increasing its size to about 2 mm (FWHM). Then from F1 the beam was optimized using the transport codes WINGIOS2 and GICOSY. A 30-keV beam with a source radius of 2 mm and an aperture of 30 mrad can be fully transported with very little beam aberration.

## 5. Present status and future plans

The full Geant4 beamline transport simulation of the ultra-slow muon beam is in progress using the PSI simulation package musrSim [12]. Magnetic and electric field distributions of each individual component were calculated using the code OPERA-3D (TOSCA), and incorporated into the simulation code. Preliminary simulation results [13] are consistent with the initial beam calculation. In parallel, off-line beam tuning and commissioning is proceeding using a $^7Li^+$ beam. Lithium atoms originate from impurities in the W foil that are evaporated and surface ionized when the target is heated up. Results were already reported in Ref. [13]. Li beam has now been successfully transported to the last focal plane (F5A) located just before the $\mu$SR spectrometer. Optimization of the beam transportation and properties is ongoing. As soon as the laser system will be operational, beam tuning will continue with ionized $^2H^+$, followed by the first ultra-slow muon beam expected in Autumn of 2014.

## Acknowledgments

This work was supported by a Grant-in-Aid for Scientific Research in Innovative Area "Ultra Slow Muon" (No. 23108002) of the Ministry of Education, Culture, Sports, Science, and Technology, Japan.

Journal of Physics: Conference Series **551** (2014) 012065 doi:10.1088/1742-6596/551/1/012065

## References

[1] Miyake Y *et al. J. Phys.: Conf. Ser.* this issue
[2] Miyake Y *et al.* 2014 *JPS Conf. Proc.* **2** 010101; Torikai E *et al* unpublished
[3] Prokscha T *et al.* 2001 *Appl. Surf. Sci.* **172** 235
[4] Nagamine K *et al.* 1995 *Phys. Rev. Lett.* **74** 4811
[5] Bakule P *et al.* 2008 *Nucl. Instr. and Meth.* B **266** 335
[6] Oishi Y *et al.* 2014 *JPS Conf. Proc.* **2** 010105; Nakamura J *et al. J. Phys.: Conf. Ser.* this issue
[7] Y. Ikedo *et al.* 2014 *JPS Conf. Proc.* **2** 010103, 2013 *Nucl. Instr. Meth.* B **317** 365, 2012 *Phys. Procedia* **30** 34
[8] Strasser P *et al.* 2013 *Nucl. Instr. Meth.* B **317** 361
[9] Makimura S *et al.* 2014 *JPS Conf. Proc.* **2** 010104
[10] Canter K F *et al.* 1986 Modified Soa immersion lens positron gun *Positron Studies of Solids, Surfaces and Atoms* ed A P Mills Jr *et al* (Singapore: World Scientific) pp 199-206
[11] Higemoto W *et al.* unpublished
[12] Sedlak K. *et al.* 2009 Physica B **404** 970 (`http://www.psi.ch/lmu/geant4-simulations`)
[13] Nagatomo T *et al.* 2014 *JPS Conf. Proc.* **2** 010102

# Optimal crossed overlap of coherent vacuum ultraviolet radiation and thermal muonium emission for $\mu$SR with the Ultra Slow Muon

**J Nakamura[1,2], Y Oishi[3], N Saito[4], K Miyazaki[4], K Okamura[4], W Higemoto[2,5], Y Ikedo[1,2], K M Kojima[1,2] , P Strasser[1,2], T Nagatomo[3], S Makimura[1,2], Y Miyake[1,2], N Kawamura[1,2], K Yokoyama[6], D Tomono[7], K Shimomura[1,2], S Wada[4], A Koda[1,2], Y Kobayashi[1,2], H. Fujimori[1,2], R Kadono[1,2] and K Nishiyama[1,2]**

[1] Muon Science Laboratory, Institute of Materials Structure Science, High Energy Accelerator Research Organization (KEK), Tsukuba, Ibaraki 305-0801, Japan
[2] Muon Science Section, Materials and Life Science Division, J-PARC Center, Tokai, Ibaraki 319-1195, Japan
[3] RIKEN Nishina Center, RIKEN, Wako, Saitama 351-0198 Japan
[4] RIKEN Center for Advanced Photonics, Wako, Saitama 351-0198, Japan
[5] Advanced Science Research Center, Japan Atomic Energy Agency (JAEA), Tokai, Ibaraki 319-1195, Japan
[6] School of Physics and Astronomy, Queen Mary University of London, London E1 4NS, UK
[7] Graduate School of Science, Kyoto University, Kyoto 606-8502 Japan

E-mail: jumpei.nakamura@kek.jp

**Abstract.** For $\mu$SR with ultra slow muon, we are constructing U line in materials and life science facility (MLF), J-PARC at present. Generation of ultra slow muon requires thermal muonium generation and laser resonant ionization process with vacuum ultraviolet radiation (1S→2P) and 355-nm radiation (2P→unbound). For laser resonant ionization, the coherent radiations and the thermal muonium emission must be coincident in time and space. The radiations can be steered in a chamber for reasonable overlap in space, and they can be easily overlapped in time because they are generated from one laser source. The trigger signal of the accelerator is useful for stable overlap in time.

## 1. Introduction

For $\mu$SR experiment, a surface muon (slow muon, we called in this article) with the 4-MeV kinetic energy is usually used. $\mu$SR experiment with brand-new muon, ultra slow muon (USM) willl be carried out at J-PARC. The USM is defined as a low energy muon generated by laser resonant ionization of thermal muonium (Mu, $\mu^+e^-$). Figure 1 shows a schematic diagram of generating USM and the energy level diagram of Mu. A 2100-K heated tungsten (hot W) foil plays the roles of not only degrader, but also generation target for Mu. In the W foil, muons are stopped and emitted to ultra high vacuum as thermal energy Mu (thermal Mu). The thermal Mu is ionized by coherent radiations to generate the USM. The USM beam has been developed during the quarter century at Muon Science Laboratory / KEK in Japan [2] and the RIKEN-

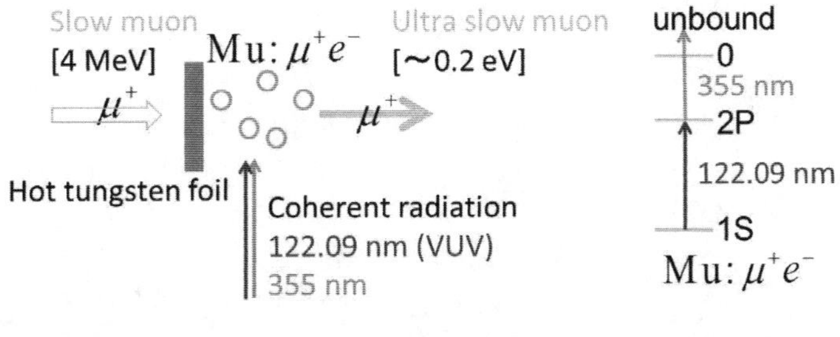


**Figure 1.** Schematic diagrams of generating ultra slow muon (left), and energy level diagram of muonium (right). The USM is a low energy muon generated by laser ionization of thermal Mu. Slow muon is injected into a hot tungsten foil, and mounium with thermal energy can escape to ultra high vacuum. Muonium is excited by 122.09-nm radiation, and ionized by 355-nm radiation. Wavelength of the second radiation is required shorter than 364 nm for taking electron away.

RAL Muon Facility / ISIS in UK [3]. However, the achieved beam intensity was not enough to conduct a $\mu$SR experiment in a practical time period.

The USM beam intensity is required to be more than 1000 /s for $\mu$SR spectroscopy in the condensed and solid-state physics. To maximize the intensity, several new ideas are proposed. The USM beam intensity is determined by two factors. One is generation of the USM, and the other is extraction and transportation efficiency of the USM [4]. The former is determined by next three factors.

(1) Generation of thermal Mu [4] with the most intense pulsed slow muon beam [5]

(2) High-intensity coherent radiation from solid-state source [6]

(3) Crossed overlap of coherent radiation and thermal Mu emission.

In this article, we report this crossed overlap. The thermal Mu emitted in vacuum is shot by the coherent radiation to ionize and dissociate muon. The coherent radiation is introduced from a perpendicular direction to the muon beam.

## 2. Configuration

At our U line / MUSE [7], thermal Mu generation and laser resonant ionization process are required for generation of the USM. Coherent vacuum ultraviolet (VUV, 122.09-nm) radiation and 355-nm radiation are used for resonant ionization of the thermal Mu. All-solid-state sources were developed at RIKEN [6]. They are generated at the laser cabin and introduced to the U line [8] by a VUV steering chamber [9]. The U line is planned to be operated with more than $10^4$ times higher USM intensity than that at the RIKEN-RAL Muon facility [3, 7]. $10^4$ means $10^2$ times higher slow muon intensity and $10^2$ times higher coherent radiation intensity. All factors are important, because they act as the multiplication rather than the addition. Figure 2 shows top view of the area of USM generation (U1 shielding area), where the USM will be generated and transported to the experimental area.

## 3. Space overlap

Figure 3 shows schematic diagram of space overlap at the RIKEN-RAL Muon facility and J-PARC, MUSE. At Port 3 / RIKEN-RAL, the 355-nm radiation was introduced at 30° from two VUV radiations. The 355-nm radiation profile is 3 mm ($W$) × 25 mm ($H$). Each of the VUV radiation profiles is 1-2 mm ($W$) × 5-8 mm ($H$) in front of W foil, which size is 40 mm ($W$) × 35 mm ($H$). A volume of the overlap region of coherent radiations and thermal Mu emission

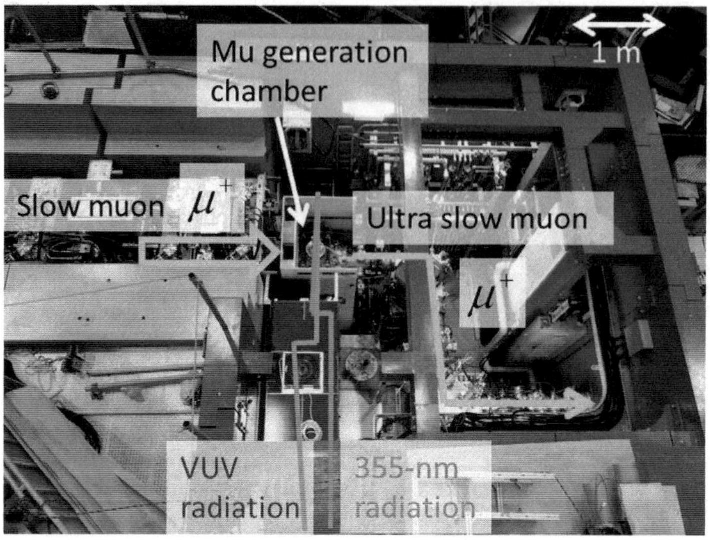


**Figure 2.** Top view of U1 area at U line. The USM is generated at Mu generation chamber. The area is a place where coherent radiations and thermal Mu emission are crossed overlapped.

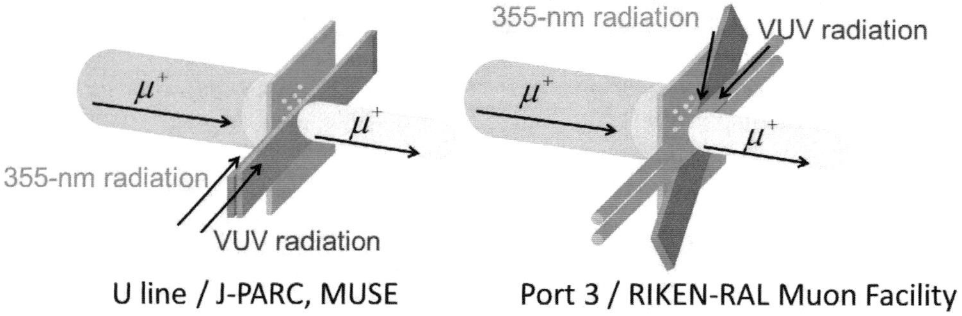


**Figure 3.** Schematic diagrams of space crossed overlap at J-PARC (left), and at RIKEN-RAL (right). The most significant difference between these diagrams is an angle between coherent radiations. A small angle is realized at J-PARC, by the benefit of a VUV steering chamber.

for interaction was ideally estimated $\sim 1\times10^3$ mm$^3$ [3]. However, quite high intensity (80 mJ / pulse) of 355-nm radiation was required for ionization[3], which was not consistent with previous results [2]. The overlap status at the RIKEN-RAL was not estimated so good.

At the U line / J-PARC, the 355-nm radiation is introduced similarly to the VUV radiation. Angular difference between these radiations is less than 0.5°. The effective crossed overlap in space will be realized at the U line by using the VUV steering chamber. The VUV steering chamber was newly designed for transportation of coherent radiation. The chamber contains four mirrors and two mirror control units (Fig. 4). VUV radiations and 355-nm radiations are steered respectively. VUV radiation is generated in krypton-argon gas mixture from fundamental radiations: 212.556-nm and 820.648-nm radiation. This process is two-photon resonance four-

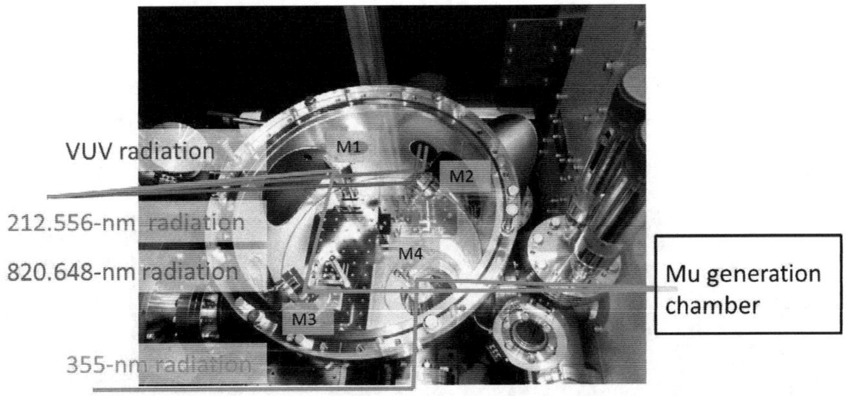


**Figure 4.** VUV steering chamber with four mirrors (M1-M4). VUV radiation is generated by frequency mixing of 212.556-nm and 820.648-nm radiations in Kr-gas cell. The wedge plate of lithium fluoride is not only a window between the cell and the VUV steering chamber, but also a disperser of radiations. By M1 mirror, VUV radiation is separated from bright fundamental radiations, which have different incident angles to the chamber. 212.556-nm and 820.648-nm radiations are kicked out of the chamber by M2. VUV radiation is steered by M3, and injected into the Mu generation chamber. 355-nm radiation is steered by M4, and injected, too. M1 and M3 are high reflectivity (83%) mirrors for 122.09-nm radiation.

wave mixing with phase matching gas mixture [13]. Usually, Piezo actuators with cables are applied for mirror-control in vacuum. To reduce costs and the outgassing from the cables, the brand-new mirror control units were installed. Two mirrors, M3 and M4, can be steered by these control units. In this method, the conventional mirror mounts, which can be aligned manually and can stay in vacuum, are controlled from the outside of the vacuum chamber. The adjuster screws on the mirror mounts in vacuum are rotated outside of the chamber through magnetic coupling mechanisms. Solid lubricants are applied for drive in vacuum. Then, ultra high vacuum can be achieved ($\sim 10^{-6}$ Pa) in the VUV steering chamber. This chamber is connected to Mu generation chamber ($\sim 10^{-8}$ Pa) with differentially pumping [10]. The profile of VUV radiation and 355-nm radiation can be controlled by setting a cylindrical lens and other optical components in the VUV steering chamber. The vertical hight of Mu emission area is estimated to be the same as the height of the W foil (70 mm ($W$) $\times$ 40 mm ($H$)) [4]. A gap between the W foil and a first electrode for extraction is 14 mm [4]. Then, a vertically long profile of high-intensity coherent radiation ($\sim$2 mm$\times$40 mm) is effective for ionizing Mu. The volume of the overlap region of coherent radiations and thermal Mu emission will be expected $\sim 5\times 10^3$ mm$^3$ which is 5 times bigger than that in the RIKEN-RAL.

An absolute intensity measurements of VUV radiation will be performed by a gas cell with nitrogen monoxide [9]. The gas cell is located after the Mu generation chamber which contains the W foil. The profile of VUV radiation will be estimated by using the cell and slits for coherent radiation. A fluorescent plate with sodium salicylate is also used to align of VUV radiation and to confirm the profile easily.

## 4. Time overlap

To operate an accelerator and to conduct an experiment, various synchronizations are necessary in general. At J-PARC, some trigger signals are prepared for the synchronizations [11]. At D line / MLF [12] a rapid cycling synchrotron (RCS) kicker trigger, which is one of the 25-Hz

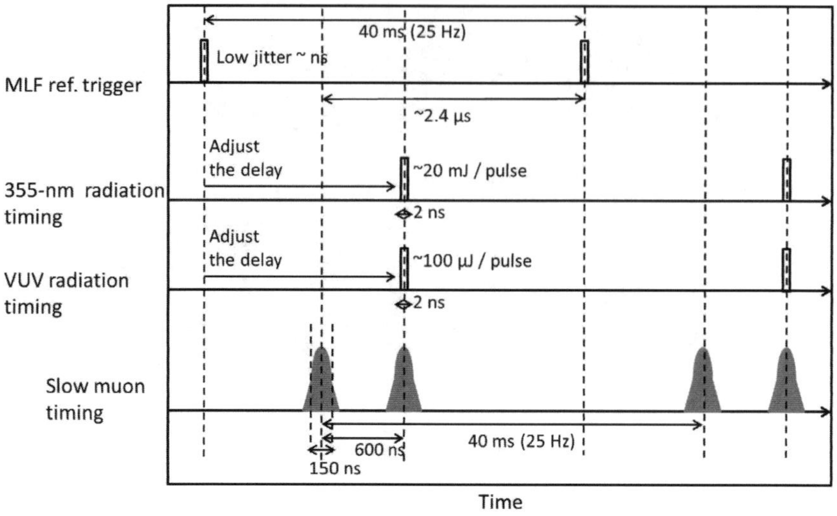


**Figure 5.** Schematic diagram of time overlap. MLF reference trigger is a low jitter and high stable scheduled timing. Timing of coherent radiations is depend on the signal. 355-nm radiation can be easily synchronized with VUV radiation, because they are generated from one laser source. Timing of thermal Mu emission is depend on slow muon timing with some delay. The delay, which is derived from diffusion and evaporation processes, is necessary to examine.

synchronized timings, is used for $\mu$SR. The RCS kicker trigger is generated by the rf system at RCS, and sometimes missing similarly to the real proton beam. The missing is not good for the coherent radiation source based on fiber laser technology.

At the U line, synchronizing between the coherent radiation and the thermal Mu emission is important for laser resonant ionization (Fig. 5). The slow muon beam has time structure similarity to the proton beam [12]. A pulse repetition rate is 25 Hz, *i. e.* the pulse repetition interval is 40 ms, the double pulse interval is 600 ns, and the pulse width is 150 ns. Time structure of thermal Mu interacting with radiations depends on this slow muon beam, muon diffusion in the W-foil target, evaporation from surface and Mu diffusion in vacuum. Therefore, it is necessary to examine the details. For stable timing of coherent radiation, 2 trigger signals are used. A scheduled timing for operating J-PARC accelerator is a 12-MHz master clock [11]. A scheduled timing generated at MLF via 50-Hz trigger clock from a master clock is a 25-Hz MLF reference trigger. This 25 Hz trigger has an about 2.4-$\mu$s delay after a secondary-beam injection. The master clock and the MLF reference trigger were used for operating coherent radiation sources at the U line. Low jitter ($\sim$ns) and high stability of these trigger signals are suitable for these sources. Furthermore, a control unit with an internal clock is made. Receiving some alarm signal, this unit can trigger slow shutdown system to protect the coherent radiation sources.

Compared with the RIKEN-RAL, effective overlap in time will be also realized. At the RIKEN-RAL, three coherent radiations (212.556-nm, 820.648-nm and 355-nm) are generated from four sources (Nd:YAG laser). At the U line, these coherent radiations are generated from one 1.06278-$\mu$m laser source with wavelength conversion by nonlinear optical crystals, LBO and CLBO [6]. Then, 212.556-nm radiation can be easily synchronized with 820.648-nm radiation for generation of coherent VUV radiation, and 355-nm radiation can be also easily synchronized with VUV radiation for resonant ionization of thermal Mu. In particular, an optical path length for timing adjustment can be reduced. The accuracy of the optical path length is required at least 1 cm, because the pulse widths and lengths of radiations are 2.0 ns, 60 cm, respectively.

## 5. Summary

The coherent radiations and the thermal Mu emission will be overlapped in time and space for laser resonant ionization. The VUV steering chamber is taken advantage of overlap in space. The scheduled timing at J-PARC is useful for stable overlap in time. Furthermore, the all-solid-state coherent radiation source takes advantage of overlap in time.

The slow muon source and coherent radiation source are improved respectively in the present work. We will be actually controlling the overlap in the autumn of 2014 when operation of J-PARC is restarted. Absolute intensities of radiations and ionization rate of the thermal Mu with changing overlap status will be monitored.

## Acknowledgments

This work was supported by a Grant-in-Aid for Scientific Research on Innovative Areas "Ultra Slow Muon" (No.23108002) of the Ministry of Education, Culture, Sports, Science, and Technology, Japan.

## References

[1] Makimura S *et al.* 2009 *Nucl. Instrum. Methods Phys. Res.* A **600** 146–149
[2] Miyake Y *et al.* 1995 *Nucl. Instrum. Methods Phys. Res.* B **95** 265–275
Nagamine K *et al.* 1995 *Phys. Rev. Lett.* **74** 4811
[3] Bakule P *et al.* 2003 *Spectrochimica Acta Part B* **58** 1019–1030
Bakule P *et al.* 2008 *Nucl. Instrum. Methods Phys. Res.* B **266** 335–346
Matsuda Y *et al.* 2003 *Physica* B **326** 217–221
[4] Strasser P *et al.* 2014 *J. Phys.: Conf. Series* this issue
Nagatomo T *et al.* 2014 *JPS Conf. Proc.* **2** 010102
Nakamura J *et al.* 2014 *J. Phys.: Conf. Ser.* **502** 012042
[5] Ikedo Y *et al.* 2014 *JPS Conf. Proc.* **2** 010103
Ikedo Y *et al.* 2013 *Nucl. Instrum. Methods Phys. Res.* B **317** 365
Strasser P *et al.*, 2013 *Nucl. Instrum. Methods Phys. Res.* B **317** 361
[6] Saito N *et al.* 2012 *Reports on the Topical meeting of the Laser Soc. of Japan* **427** 49
Oishi Y *et al.* 2014 *JPS Conf. Proc.* **2** 010105
Okamura K *et al.* 2014 *JPS Conf. Proc.* **2** 010106
Miyazaki K *et al.* 2014 *JPS Conf. Proc.* **2** 010107
[7] Miyake Y *et al.* 2014 *JPS Conf. Proc.* **2** 010101
Miyake Y *et al.* 2013 *Hyperfine Interact.* **216** 79-83
Miyake Y *et al.* 2009 *Nucl. Instrum. Methods Phys. Res.* A **600** 22–24
Kadono R and Miyake Y 2012 *Rep. Prog. Phys.* **75** 026302
[8] Nakamura J *et al.*, 2012 KEK Progress Report 2012-2 KEK-MSL REPORT 2011 17
[9] Nakamura J *et al.* 2014 *JPS Conf. Proc.* **2** 010108
[10] Nakamura J *et al.* 2013 KEK Progress Report 2013-3 KEK-MSL REPORT 2012 27
[11] Tamura F *et al.* 2013 *Nucl. Instrum. Methods Phys. Res.* A **647** 25
Tamura F *et al.* 2003 *Proc. of ICALEPCS2003* 247–249
[12] Fujimori H *et al.* 2011 KEK Progress Report 2011-3 KEK-MSL REPORT 2010 10
Strasser P *et al.* 2012 *Physics Procedia* **30** 65–68
[13] Louchev O A *et al.* 2011 *Phys. Rev.* A **84** 033842

# Developments at the ISIS muon source and the concomitant benefit to the user community

**A D Hillier, D J Adams, P J Baker, A Bekasovs, F C Coomer, S P Cottrell, S D Higgins, S J S Jago, K G Jones, J S Lord, A Markvardsen, P G Parker, J N T Peck, F L Pratt, M T F Telling and R E Williamson**

ISIS Facility, STFC Rutherford Appleton Laboratory, Chilton, Oxfordshire, OX11 0QX, UK

E-mail: `adrian.hillier@stfc.ac.uk`

**Abstract.** For the last 27 years, muon experiments at ISIS have been making a significant contribution to a number of scientific fields. However, as a community of researchers, we are always aiming to improve and extend the instruments' capabilities. In this paper we will review the current significant developments at the ISIS muon facility, namely the primary beamline upgrade, proton pulse compression and the MANTID muon analysis package.

## 1. Introduction

The ISIS Muon Facility has been operating a science program for the past 27 years which has resulted in a significant contribution to a number of scientific fields. Over this time, the facility has been developed, upgraded and enhanced from one beamline to three and a new facility, namely RIKEN-RAL. Recently the ISIS muon facility has undergone significant instrument upgrades: the new instrument HiFi [1] and the new EMu spectrometer [2] and now the installation of an high-power laser on HiFi [3]. Along with instrument upgrades, smaller developments with data acquisition electronics have enabled data rates in excess of 120 Mev/h. In this paper we describe the latest developments and the benefits to the user community.

## 2. Primary Beamline Upgrade

### *2.1. Current Beamline*

The ISIS muon beamline was constructed in the mid-1980s and first muons were delivered on 23rd March 1987 [4, 5] into the original MuSR area. Later, with EU funding, the beamline was split into three independent instruments: DEVa (subsequently replaced by HiFi), EMu and MuSR (see Fig. 1a) [6]. The beamline was constructed using components from previous accelerators and experiments of various types which might not have been optimal. Next to the muon target there are two quadrupoles which focus the beam into two bending magnets with a quadrupole triplet and momentum slits between them. After this there is a separator and two more quadrupole doublets. Finally, this leads into the electrostatic kicker which 'kicks' the first of two closely spaced pulses from the accelerator into the two side instruments (EMU and HiFi via a septum and three more quadrupoles) and the second goes through into MuSR.

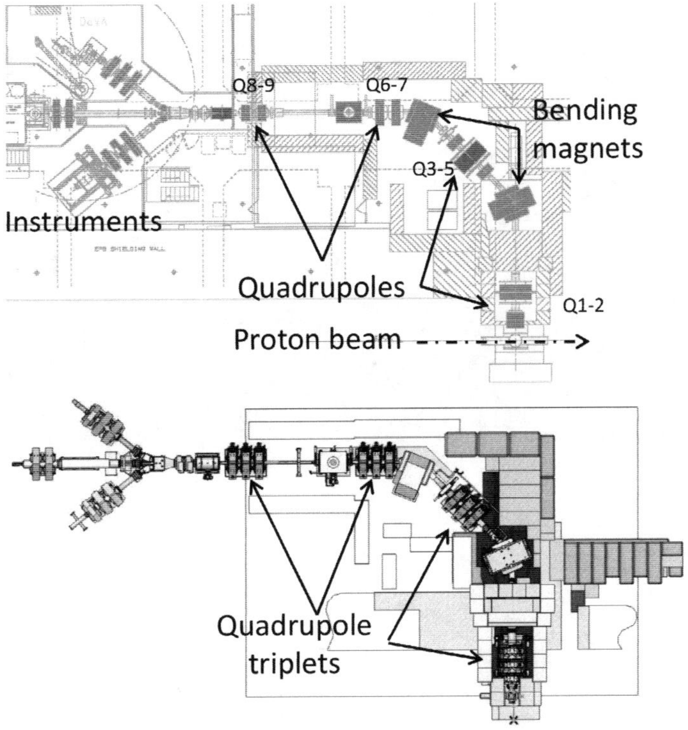


**Figure 1.** The upper figure (a) is the current ISIS beamline after the EU funded upgrade where the three instruments were installed. The lower figure (b) is the layout for the new primary beamline.

### *2.2. Possible Beamline Configurations*

The primary beamline has remained largely untouched and is starting to show signs of ageing. Given the age and location of these components, we started a project to upgrade the primary beamline with purpose-designed components, with a planned phased installation between 2014 and 2016. This new beamline has to fit into the existing space and deliver muons to the present instrument locations. The current design was simulated using TRANSPORT [7] and TURTLE [8]. These results (see Fig. 2a) reproduce those reported by Eaton *et al* [6] and the observed performance of the current suite of instruments. Therefore, we explored the following options: 1) Keeping same optics but with standard types of quadrupoles and 2) upgrade to use quadrupole triplets instead of doublets.

Now considering option 1) the standard quadrupoles proposed are the M9 type as on MuSR and larger, radiation-hard, quadupoles used for Q1 and Q2 (the first two quadrupoles of the primary beamline), and also for Q4 where the beam envelope is larger. The opportunity has been taken to have Q1 and Q2 centred nearer the muon target in order to increase the acceptance angle, since they will be shorter than the current ones. This setup gives a slightly lower flux for the same spot sizes. The second option is to use quadrupole triplets everywhere, which gives a more circular beam envelope. Here we could replace Q1 and Q2 with a triplet of radiation-hard quadrupoles (see Fig. 2b). Similarly the Q6-Q7 and Q8-Q9 groups become triplets of M9s. The beam still has to be focused through the kicker and septum magnet apertures. Triplets at the end of each beamline give some opportunity to vary spot height versus width while keeping a focus

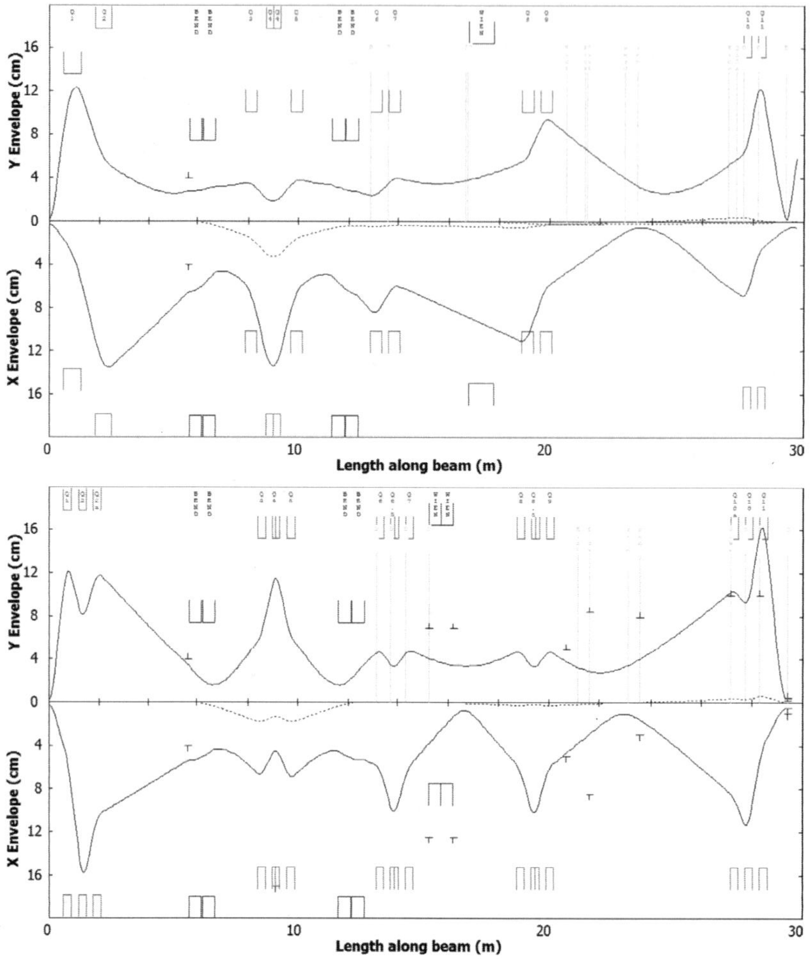


**Figure 2.** The upper graph (a) gives the TRANSPORT envelope for the current ISIS beamline and the lower graph (b) gives the TRANSPORT envelope for the new beamline. The red blocks represent quadrupoles and blue blocks bending magnets. Their positions represent the aperture. The black lines through the figure are the beam envelope for vertical (upper half) and horizontal (lower half).

at the sample, and adjusting the X collimation slits, in front of the final focusing components, will allow both the spot height and width to be tailored to suit an experiment.

*2.3. Instrument Performance*

The triplet configuration gives an increase in acceptance to the sample position. Therefore, we simulated the instrumental performance by considering the spot size and rate using TURTLE. The simulations used a standard flux of muons from the target and gives a relative performance for each design, not absolute figures (which depend on target thickness and ISIS beam current). The plot (see Fig. 3) shows spot size (rms radius) and rate at the MuSR instrument, with slit setting as an implicit parameter. The minimum spot size achievable is similar for both designs.

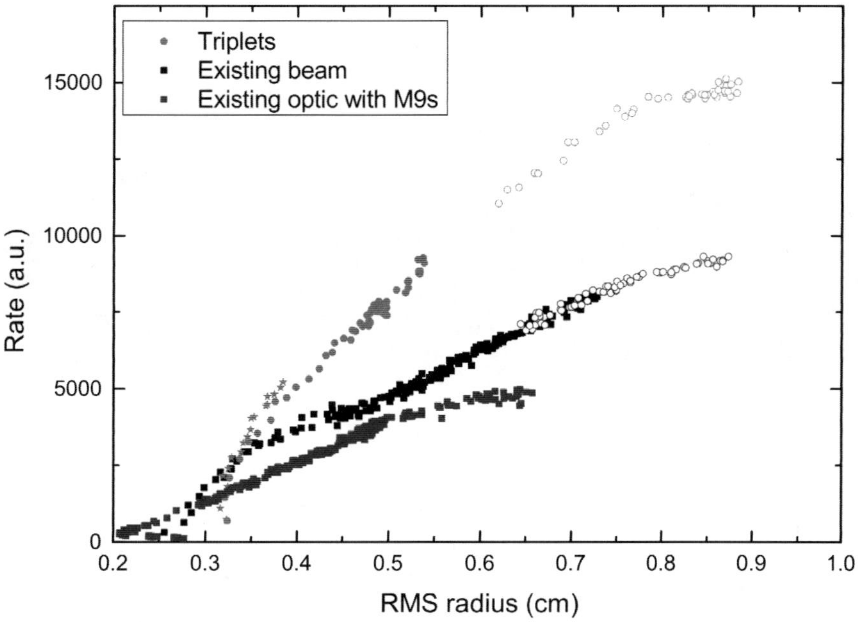


**Figure 3.** This graph shows the spot size and rate dependence for the various different beamline setups (current (black), M9s (blue), triplet (red)).

A rate increase around a factor of 2 should be obtained by going to a triplet-based beamline. Therefore, the triplet based design has been chosen for installation (see Fig. 1b).

## 3. Proton Pulse Compression

The ISIS muon facility is a pulsed source and operates with a repetition rate of 40 frames per second, with each frame containing two proton pulses with a width of ≈60 ns (FWHM) each, separated by 330 ns. The frequency response of the muon instruments at ISIS is determined, primarily, by the width of the proton pulse, but also by the pion decay process. One method of improving the frequency limit, currently around 10 MHz, is to shorten the extracted proton pulse length. An experimental trial to compress the proton pulse was carried out at ISIS and its effect on the muon instruments frequency response measured.

Pulse compression, shortening the length of an accelerated particle beam, has become a common technique in accelerator physics. Several methods can be used, but many involve variation of the synchrotron RF cavity parameters [9] to induce a rotation in the longitudinal phase space. The technique used in this experiment involved an adiabatic reduction of the cavity voltages over 2.75 ms to increase the bunch length whilst also reducing the spread of particle momenta. These voltages were then abruptly increased over 0.25 ms to induce rotation in phase space where the reduction in particle momenta rotated to deliver a shorter bunch length. The bunch was then extracted from the synchrotron at its shortest length and transported toward the targets. As seen in figure 4 the bunch length was visibly reduced by an average of 20.75 %

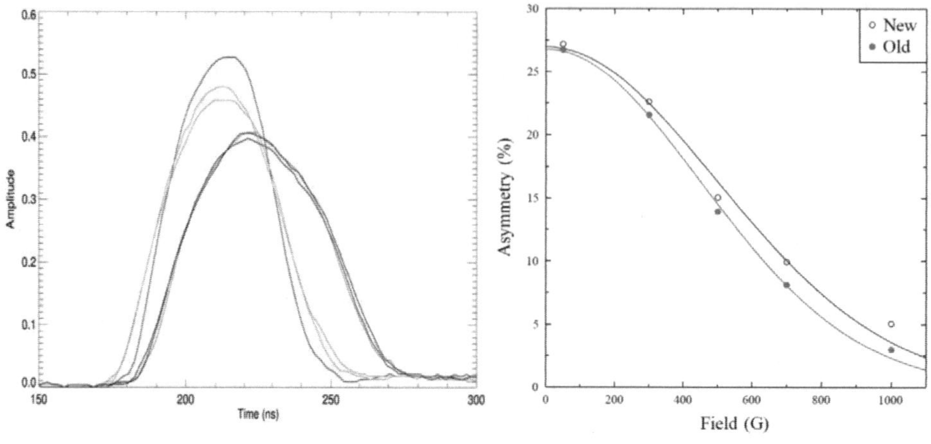


**Figure 4.** The left graph gives pulse charge density in time and the right graph gives the frequency response of the MuSR instrument. The frequency response was determined by measuring the asymmetry of the transverse field precession of a Ag sample at room temperature.

(FWHM) from 58.8 ns to 45.8 ns. The frequency response of the MuSR spectrometer was measured with and without pulse compression (see Fig. 4). From these results it is clear that the measurable frequency response for MuSR was increased by 20-25 % with the shorter pulse. Future studies are planned to reduce the bunch length further including making full use of the ISIS second harmonic RF system.

## 4. MANTID Analysis Software

The MANTID project provides a framework that supports high-performance computing and visualisation of scientific data[10, 11]. MANTID has been created to manipulate and analyse neutron scattering and muon spectroscopy data, but could easily be applied to many other techniques. This framework is open source and supported on multiple target platforms (Windows, Linux and Mac). The development of codes running under the MANTID framework for the analysis of muon data continues. An interface focused on time domain analysis is now included as part of the MANTID distribution, is available on all the ISIS instruments and was successfully used on the recent muon training school. Currently, in the time domain analysis, the spectra can be reduced, fitted (with any model the user cares to fit) and the results from the fits easily obtained, along with the relevant logging information. Using this, models can be fitted to the results. Recently, Fourier/maximum entropy has been added for analysis of both transverse and longitudinal spectra. Developments are planned to augment the capabilities of this interface over the coming months, such as simultaneous fitting of multiple groups and runs with shared parameters. Improvements to simplify data analysis of Radio Frequency and Avoided Level Crossing experiments are to be included as part of the work of the Muon JRA, with existing algorithms being developed into a single interface guiding users through the process of reduction and analysis of these datasets. Moreover, developments with simulation codes are leading to greater exploitation of data. It seems clear that establishing links to codes modelling both sample phenomena and instrument characteristics should bring an additional degree of sophistication to the analysis and interpretation of muon data. This development will be fed into MANTID.

## 5. Conclusion

In conclusion, the primary beam upgrade, the proton pulse compression and the software development will clearly lead to an improvement to the muon facility at ISIS, enabling the next 27 years of muon production. For our user community, these developments will give improved data rates, increased frequency response and the software tools to exploit these enhancements.

## References

[1] Lord J S *et al.* 2011 *Rev. Sci. Instrum.* **82** 073904
[2] Giblin S R *et al.* 2014 *Nuclear Instrum. Methods* A **751** 70
[3] Yokoyama K *et al.* 2013 *Physica Scripta* **88** 068511
[4] Carne A *et al.* 1984 *Hyperfine Interactions* **17-19** 945
[5] King P J C *et al.* 2013 *Phys. Scr.* **88** 068502
[6] Eaton G H *et al.* 1988 *Nucl. Instrum. Methods* A **269** 483
[7] Rohrer U *PSI Graphic Transport Framework* based on a CERN-SLAC-FERMILAB version by Brown K L *et al.*
[8] Rohrer U *PSI Graphic Turtle Framework* based on a CERN-SLAC-FERMILAB version by Brown K L *et al.*
[9] Garoby R *RF gymnastics in synchrotrons, CERN Accelerator School* 2010 Denmark pp 431–446
[10] `http://dx.doi.org/10.5286/SOFTWARE/MANTID`
[11] Arnold O *et al.* 2014 *Nucl. Instrum. Methods* A **764** 156

# AUTHOR INDEX

# AUTHOR INDEX

# AUTHOR INDEX

# AUTHOR INDEX

**Institute of Physics**
Dirac House, Temple Back
Bristol BS1 6BE UK

CURRAN ASSOCIATES INC
proceedings
.com